Digital Systems Design and Practice
Using Verilog HDL and FPGAs
Second Edition

—

Digital Systems Design and Practice
Using Verilog HDL and FPGAs
Second Edition

Ming-Bo Lin

Department of Electronic and Computer Engineering
National Taiwan University
   of Science and Technology
Taipei, Taiwan

The author of this book has done his best efforts in preparing this production. These efforts include the development, research, and testing of the programs included in this book, as well as the documentation of the book. The author makes no warranty of any kind with regard to these programs. The author shall not be liable in any event in connection with the use of these programs.

Second edition © 2015 Ming-Bo Lin
All rights reserved. No part of this book may be reproduced in any format or by any means without the written permission from the author.

ISBN: 1514313308
ISBN-13: 9781514313305
This book was typeset in 12/10.5-pt Times Roman.

Cover designer: Frank
Cover photographer: Alice

First printing, July 2015
Fourth printing, August 2019

To Alfred, Fanny, Alice, and Frank
and in memory of my parents

Contents

Preface		xvi
About the Author		xxi

1 Introduction — 1
  1.1 Introduction — 1
    1.1.1 Digital System Implementations — 1
    1.1.2 Popularity of Verilog HDL — 3
    1.1.3 Simple Examples of Verilog HDL — 3
    1.1.4 HDL-Based Approach — 5
  1.2 Introduction to Verilog HDL — 7
    1.2.1 Module Concept — 7
    1.2.2 Lexical Conventions — 7
    1.2.3 Value Set — 9
    1.2.4 Constants — 9
    1.2.5 Data Types — 10
    1.2.6 Primitives — 11
    1.2.7 Attributes — 11
  1.3 Module Modeling Styles — 12
    1.3.1 Modules — 12
    1.3.2 Structural Modeling — 14
    1.3.3 Dataflow Modeling — 16
    1.3.4 Behavioral Modeling — 17
    1.3.5 Mixed-Style Modeling — 18
  1.4 Simulation — 19
    1.4.1 Basic Simulation Constructs — 19
    1.4.2 Related Compiler Directive and System Tasks — 20
    1.4.3 A Tutorial Example — 22
    1.4.4 Concluding Remarks — 25
  1.5 Summary — 26
  References — 26
  Problems — 27

2 Structural Modeling — 29
  2.1 Gate-Level Structural Modeling — 29
    2.1.1 Gate Primitives — 29
    2.1.2 Tristate Buffers — 36
    2.1.3 Wired Logic — 39
  2.2 Gate Delays — 41
    2.2.1 Delay Models — 41
    2.2.2 Delay Specifications — 44
  2.3 Hazards — 45

		2.3.1	Static Hazards	46
		2.3.2	Dynamic Hazards	47
	2.4	Switch-Level Structural Modeling		49
		2.4.1	MOS Switches	49
		2.4.2	CMOS Switches	54
		2.4.3	Bidirectional Switches	55
		2.4.4	Delay Specifications	56
		2.4.5	Signal Strength	57
		2.4.6	The **trireg** Net	59
	2.5	Summary		62
	References			63
	Problems			63
3	**Dataflow Modeling**			**71**
	3.1	Dataflow Modeling		71
		3.1.1	Continuous Assignments	71
		3.1.2	Expressions	73
		3.1.3	Delays	74
	3.2	Operands		75
		3.2.1	Constants	76
		3.2.2	Variable Types	79
		3.2.3	Bit-Selects and Part-Selects	81
		3.2.4	Array and Memory Elements	83
	3.3	Operators		84
		3.3.1	Bit-wise Operators	84
		3.3.2	Arithmetic Operators	87
		3.3.3	Concatenation and Replication Operators	89
		3.3.4	Reduction Operators	91
		3.3.5	Logical Operators	93
		3.3.6	Conditional Operator	94
		3.3.7	Relational Operators	95
		3.3.8	Equality operators	96
		3.3.9	Shift Operators	98
	3.4	Case Study — A Two-Digit BCD Adder		100
		3.4.1	Principles of BCD Addition	100
		3.4.2	Modeling Examples of BCD Adders	101
	3.5	Summary		103
	References			103
	Problems			104
4	**Behavioral Modeling**			**107**
	4.1	Procedural Constructs		107
		4.1.1	**initial** Blocks	107
		4.1.2	**always** Blocks	109
		4.1.3	Concluding Remarks	110
	4.2	Procedural Assignments		111
		4.2.1	Procedural Assignments	111
		4.2.2	Blocking Assignments	112
		4.2.3	Nonblocking Assignments	113
		4.2.4	Blocking versus Nonblocking Assignments	114
	4.3	Timing Control		119
		4.3.1	Delay Timing Control	119

Contents

		4.3.2	Event Timing Control	124
	4.4	Selection Statements		130
		4.4.1	**if-else** Statement	130
		4.4.2	**case** Statement	132
		4.4.3	**casex** and **casez** Statements	136
	4.5	Iteration (Loop) Statements		137
		4.5.1	The **while** Loop Statement	138
		4.5.2	The **for** Loop Statement	139
		4.5.3	The **repeat** Loop Statement	141
		4.5.4	The **forever** Loop Statement	142
	4.6	Summary		143
	References			144
	Problems			144
5	**Tasks, Functions, and UDPs**			**151**
	5.1	Tasks		151
		5.1.1	Task Definition and Calls	151
		5.1.2	Types of Tasks	155
	5.2	Functions		157
		5.2.1	Function Definition and Calls	157
		5.2.2	Types of Functions	160
		5.2.3	Constant Functions	160
		5.2.4	Sharing Tasks and Functions	162
	5.3	System Tasks and Functions		163
		5.3.1	Simulation-Related System Tasks and Functions	164
		5.3.2	File I/O System Tasks and Functions	167
		5.3.3	String Formatting System Tasks and Functions	173
		5.3.4	Conversion System Functions	174
		5.3.5	Probabilistic Distribution System Functions	176
		5.3.6	Stochastic Analysis System Tasks and Functions	177
		5.3.7	Math System Functions	178
		5.3.8	Command Line Arguments	178
	5.4	User-Defined Primitives		181
		5.4.1	UDP Basics	181
		5.4.2	Combinational UDPs	182
		5.4.3	Sequential UDPs	185
	5.5	Summary		189
	References			189
	Problems			190
6	**Hierarchically Structural Modeling**			**193**
	6.1	Modules		193
		6.1.1	Module Definition	193
		6.1.2	Parameters	196
		6.1.3	Module Instantiation	199
		6.1.4	Overriding Parameter Values	201
		6.1.5	Hierarchical Path Names	204
	6.2	Generate Regions		205
		6.2.1	Generate-Loop Statement	205
		6.2.2	Generate-If Statement	207
		6.2.3	Generate-Case Statement	213
	6.3	Configurations		215

		6.3.1	Library	215
		6.3.2	Basic Configuration Elements	216
	6.4	Summary	222
	References	223
	Problems	223
7	**Advanced Modeling Techniques**	**227**
	7.1	Sequential and Parallel Blocks	227
		7.1.1	Sequential Blocks	227
		7.1.2	Parallel Blocks	229
		7.1.3	Special Features of Blocks	231
		7.1.4	The disable Statement	233
	7.2	Procedural Continuous Assignments	234
		7.2.1	**assign** and **deassign** Statements	234
		7.2.2	**force** and **release** Statements	235
	7.3	Delay Models and Timing Checks	237
		7.3.1	Delay models	237
		7.3.2	Specify Blocks	240
		7.3.3	Timing Checks	250
	7.4	Compiler Directives	260
		7.4.1	**'define** and **'undef** Compiler Directives	260
		7.4.2	**'include** Compiler Directive	261
		7.4.3	**'ifdef, 'else, 'elsif, 'endif,** and **'ifndef** Compiler Directives	261
		7.4.4	**'timescale** Compiler Directive	263
		7.4.5	Miscellaneous Compiler Directives	264
	7.5	Summary	265
	References	266
	Problems	266
8	**Combinational Logic Modules**	**273**
	8.1	Decoders	273
		8.1.1	Decoders	273
		8.1.2	Expansion of Decoders	277
	8.2	Encoders	278
		8.2.1	Encoders	278
		8.2.2	Priority Encoders	280
	8.3	Multiplexers	283
		8.3.1	Multiplexers	284
		8.3.2	Expansion of Multiplexers	287
	8.4	Demultiplexers	289
		8.4.1	Demultiplexers	289
		8.4.2	Expansion of Demultiplexers	293
	8.5	Magnitude Comparators	294
		8.5.1	Magnitude Comparators	295
		8.5.2	Cascadable Magnitude Comparators	295
	8.6	Case Study — Seven-Segment LED Displays	298
		8.6.1	Seven-Segment LED Displays	298
		8.6.2	Multiplexed Seven-Segment LED Display Module	301
	8.7	Summary	305
	References	305
	Problems	305

Contents

9 Sequential Logic Modules — 309
- 9.1 Flip-Flops — 309
  - 9.1.1 Flip-Flops — 310
  - 9.1.2 Basic Timing and Metastable State — 315
  - 9.1.3 Synchronizers — 321
  - 9.1.4 A Switch-Debounce Circuit — 326
- 9.2 Memory Elements — 328
  - 9.2.1 Registers — 328
  - 9.2.2 Register Files — 330
  - 9.2.3 Synchronous RAM — 331
  - 9.2.4 Asynchronous RAMs — 332
- 9.3 Shift Registers — 338
  - 9.3.1 Shift Registers — 338
  - 9.3.2 Universal Shift Registers — 340
- 9.4 Counters — 343
  - 9.4.1 Ripple Counters — 343
  - 9.4.2 Synchronous Counters — 346
- 9.5 Sequence Generators — 349
  - 9.5.1 PR-Sequence Generators — 349
  - 9.5.2 CRC Code Generators/Detectors — 354
  - 9.5.3 Ring Counters — 357
  - 9.5.4 Johnson Counters — 359
- 9.6 Timing Generators — 360
  - 9.6.1 Multiphase Clock Generators — 360
  - 9.6.2 Digital Monostable Circuits — 362
- 9.7 Summary — 363
- References — 364
- Problems — 365

10 Implementation Options of Digital Systems — 371
- 10.1 Implementation Options of Digital Systems — 371
  - 10.1.1 Hierarchical System Design — 372
  - 10.1.2 Implementation Options of Digital Systems — 373
  - 10.1.3 ASIC Approaches — 377
  - 10.1.4 Field-Programmable Devices — 382
- 10.2 PLD Structure and Modeling — 386
  - 10.2.1 Read-Only Memory — 387
  - 10.2.2 Programmable Logic Arrays — 389
  - 10.2.3 Programmable Array Logic — 391
  - 10.2.4 PLA Modeling — 392
- 10.3 CPLDs and FPGAs — 400
  - 10.3.1 CPLDs — 400
  - 10.3.2 FPGAs — 403
  - 10.3.3 Advanced Structures of FPGAs — 406
- 10.4 Practical Issues — 409
  - 10.4.1 I/O Standards — 409
  - 10.4.2 Voltage Tolerance — 412
- 10.5 Case Study — A Keypad Scanner and Decoder — 414
  - 10.5.1 A Keypad Scanner — 414
  - 10.5.2 A Keypad Scanner and LED Display System — 417
  - 10.5.3 FPGA Implementation — 421
- 10.6 Summary — 424

	References	425
	Problems	425

11 System Design Methodologies — 429
- 11.1 Finite-State Machines — 429
  - 11.1.1 Types of Sequential Circuits — 429
  - 11.1.2 FSM Modeling Styles — 432
  - 11.1.3 Implicit versus Explicit Finite-State Machines — 437
- 11.2 RTL Design — 439
  - 11.2.1 ASM Charts — 440
  - 11.2.2 ASM Modeling Styles — 445
  - 11.2.3 Datapath-and-Controller Approach — 453
- 11.3 RTL Implementation Options — 467
  - 11.3.1 Single-Cycle Structure — 468
  - 11.3.2 Multicycle Structure — 470
  - 11.3.3 Pipeline Structure — 471
  - 11.3.4 FSM versus Iterative Logic — 475
- 11.4 Case Study — Liquid-Crystal Displays — 481
  - 11.4.1 Principles of LCDs — 481
  - 11.4.2 Commercial Dot-Matrix LCD Modules — 483
  - 11.4.3 Datapath Design — 488
  - 11.4.4 Controller Design — 491
- 11.5 Summary — 498
- References — 499
- Problems — 499

12 Synthesis — 509
- 12.1 Synthesis Flow of ASICs and FPGA-Based Systems — 509
  - 12.1.1 General Synthesis Flow — 510
  - 12.1.2 Timing-Driven Placement — 513
- 12.2 RTL Synthesis Flow — 516
  - 12.2.1 Design Environment — 516
  - 12.2.2 Design Constraints — 517
  - 12.2.3 Architectures of Logic Synthesizers — 521
  - 12.2.4 Optimization of Logic Synthesis — 523
- 12.3 Technology-Independent Synthesis — 525
  - 12.3.1 Two-Level Logic Synthesis — 525
  - 12.3.2 Multilevel Logic Synthesis — 528
- 12.4 Technology-Dependent Synthesis — 534
  - 12.4.1 Network Covering — 534
  - 12.4.2 A Two-Step Approach — 538
  - 12.4.3 The FlowMap Method — 540
  - 12.4.4 Shannon's Expansion Approach — 541
- 12.5 Language Structure Synthesis — 543
  - 12.5.1 General Considerations of Language Synthesis — 543
  - 12.5.2 Synthesis of Selection Statements — 544
  - 12.5.3 Delay Values — 546
  - 12.5.4 Synthesis of Positive and Negative Signals — 547
  - 12.5.5 Synthesis of Loop Statements — 549
  - 12.5.6 Memory and Register Files — 550
- 12.6 Coding Guidelines — 551
  - 12.6.1 Guidelines for Clock Signals — 552

Contents xiii

			12.6.2	Guidelines for Reset Signals	553

- 12.6.2 Guidelines for Reset Signals — 553
- 12.6.3 Partitioning for Synthesis — 555
- 12.6.4 Synthesis for Power Optimization — 557
- 12.7 Summary — 561
- References — 562
- Problems — 563

13 Verification — 569
- 13.1 Functional Verification — 569
  - 13.1.1 Models of Design Units — 570
  - 13.1.2 Simulation-Based Verification — 571
  - 13.1.3 Formal Verification — 573
- 13.2 Simulation — 574
  - 13.2.1 Types of Simulation and Simulators — 574
  - 13.2.2 Architecture of HDL Simulators — 576
  - 13.2.3 Event-Driven Simulation — 578
  - 13.2.4 Cycle-Based Simulation — 580
- 13.3 Test Bench Design — 582
  - 13.3.1 Design of Test Benches — 582
  - 13.3.2 Clock Signal Generation — 586
  - 13.3.3 Reset Signal Generation — 590
  - 13.3.4 Verification Coverage — 592
- 13.4 Dynamic Timing Analysis — 594
  - 13.4.1 Basic Concepts of Timing Analysis — 594
  - 13.4.2 Standard Delay Format Files — 595
  - 13.4.3 Delay Back-Annotation — 596
  - 13.4.4 Details of Standard Delay Format Files — 597
- 13.5 Static Timing Analysis — 599
  - 13.5.1 Fundamentals of Static Timing Analysis — 599
  - 13.5.2 Timing Specifications — 601
  - 13.5.3 Timing Exceptions — 601
- 13.6 Value Change Dump (VCD) Files — 604
  - 13.6.1 Four-State VCD Files — 604
  - 13.6.2 VCD File Format — 606
  - 13.6.3 Extended VCD Files — 606
- 13.7 Case Study — FPGA Design and Verification Flow — 609
  - 13.7.1 ISE Design Flow — 609
  - 13.7.2 Dynamic Timing Simulation — 615
  - 13.7.3 An RTL-Based Synthesis and Verification Flow — 619
- 13.8 Summary — 620
- References — 621
- Problems — 622

14 Arithmetic Modules — 625
- 14.1 Addition and Subtraction — 625
  - 14.1.1 Carry-Lookahead Adders — 625
  - 14.1.2 Parallel-Prefix Adders — 631
- 14.2 Multiplication — 635
  - 14.2.1 Unsigned Multiplication — 636
  - 14.2.2 Signed Multiplication — 642
- 14.3 Division — 646
  - 14.3.1 Restoring Division Algorithm — 646

	14.3.2 Non-Restoring Division Algorithm	647
	14.3.3 Non-Restoring Array Dividers	648
14.4	Arithmetic-and-Logic Units	651
	14.4.1 Shifts	651
	14.4.2 ALUs	654
14.5	Digital-Signal Processing Modules	657
	14.5.1 Finite-Impulse Response Filters	658
	14.5.2 Infinite-Impulse Response Filters	659
14.6	Summary	661
	References	662
	Problems	663

15 Design Examples — 669

15.1	Bus	669
	15.1.1 Bus Structure	670
	15.1.2 Bus Arbitration	673
15.2	Data Transfer	676
	15.2.1 Synchronous Data Transfer	677
	15.2.2 Asynchronous Data Transfer	679
	15.2.3 Multiple-Clock Domains	683
15.3	General-Purpose Input and Output	696
	15.3.1 Basic Principles	697
	15.3.2 A Design Example	699
15.4	Timers	701
	15.4.1 Basic Timer Operations	701
	15.4.2 Advanced Timer Operations	707
15.5	Universal Asynchronous Receiver and Transmitter	712
	15.5.1 UART	712
	15.5.2 Transmitter	714
	15.5.3 Receiver	716
	15.5.4 Baud-Rate Generator	719
	15.5.5 UART Top-Level Module	721
15.6	A Simple CPU Design	723
	15.6.1 Fundamentals of CPU	723
	15.6.2 Datapath Design	728
	15.6.3 Controller Design	730
15.7	Summary	735
	References	736
	Problems	736

16 Design for Testability — 743

16.1	Fault Models	743
	16.1.1 Fault Models	743
	16.1.2 Fault Detection	746
	16.1.3 Test Vectors	747
16.2	Automatic Test Pattern Generation	749
	16.2.1 Path Sensitization	749
	16.2.2 A Simplified D-Algorithm	751
16.3	Testable Circuit Design	754
	16.3.1 Ad hoc Approach	754
	16.3.2 Scan-Path Method	756
	16.3.3 BIST	757

	16.3.4 Boundary-Scan Standard—IEEE 1149.1 Standard	761
16.4	System-Level Testing	773
	16.4.1 SRAM BIST and March Test	773
	16.4.2 Core-Based Testing	775
	16.4.3 SoC Testing	776
16.5	Summary	777
References		778
Problems		778
Index		781

Preface

WITH the advance of semiconductor and communication technologies, the use of system-on-chip (SoC) has become an essential technique to decrease product costs. To design and implement an SoC-level product, it proves necessary to totally or partly rely on the hardware description language (HDL) synthesis flow and field programmable gata array (FPGA) devices or cell libraries. As a consequence, it has become an important attainment for electrical engineers to develop a good understanding of the key issues of HDL design flows based on FPGA devices or cell libraries. To achieve this, this book addresses the need for teaching such a topic based on Verilog HDL and FPGAs.

This book, *Digital System Designs and Practices: Using Verilog HDL and FPGAs*, aims to be used as a text for students and as a reference book for professionals or a self-study book for readers. For classroom use, each chapter includes many worked examples and review questions for helping readers test their understanding of the contents. In addition, throughout the book, a lot of worked examples are provided for helping readers realize the basic features of Verilog HDL and grasp the essentials of digital system designs as well.

The contents of this book largely stem from the course *FPGA System Designs and Practices*, offered at our campus over the past decade. This course is an undergraduate elective and the first-year graduate course. This book is so structured that it can be used as a variety of courses, including *Hardware Description Language*, *FPGA System Designs and Practices*, *Digital System Designs*, *Advanced Digital System Designs*, and others.

New to This Edition

Even though the pedagogical approach of the previous edition has been retained, almost all paragraphs are substantially modified or rewritten to update the materials and several critical changes are made in this edition. Among these, the most important one is to rewrite all Verilog HDL examples in compliance with the most recent IEEE standard, IEEE Std. 1364-2001/2005. These updates and changes not only reflect a better presentation of the book but also include more practical worked examples and case studies.

- **More examples:** Almost all paragraphs of the book are rewritten to give a better presentation and many worked examples are also added in place to make the material much more understandable.
- **More end-of-chapter problems:** A lot of additional end-of-chapter problems are included to each chapter to provide readers more opportunities to practice what they have learned from the chapter. Many such problems may be assigned as student projects, in particular, the problems of Chapters 11, 14, and 15.
- **More case studies:** Many case studies are included in this edition to explore the problems that the reader often encounters in practical systems. Through the case studies of the book,

the reader may grasp the essentials of practical digital systems to be designed using Verilog HDL and FPGAs.
- **Highlighting the Verilog HDL syntax:** The Verilog HDL syntax introduced in the book is highlighted to make it distinguishable from the text. This is proved much helpful for readers to refer the desired syntax as they need.
- **Printing all keywords in boldface:** All Verilog HDL keywords are printed in boldface throughout the book to emphasize the language structures and improve the readability of Verilog HDL modules.

Contents of This Book

The contents of this book can be roughly divided into four parts. The first part includes Chapters 1 to 7 and introduces the basic features and capabilities of Verilog HDL. This part can also be used as a reference for Verilog HDL. The second part covers Chapters 8 to 10 and contains basic combinational and sequential modules. In addition, a broad variety of options for implementing a digital system are discussed in detail. The third part consists of Chapters 11 to 13 and examines the three closely related topics in developing a digital system product: design, synthesis, and verification. The last part focuses on the design examples at both the RTL and system level, and the techniques for testing and testable design. This part comprises Chapters 14 to 16.

Chapter 1 introduces the features and capabilities of Verilog HDL along with a tutorial example to illustrate how to use it to model a design at various abstraction levels and in various modeling styles. In addition, we also demonstrate the use of Verilog HDL to verify a design after finishing the description of a design.

Chapter 2 deals with how to model a design in structural style. In this style, a module is described as a set of interconnected components, including modules, user-defined primitives (UDPs), gate primitives, and switch primitives. In this chapter, we introduce the structural modeling at both the gate and switch levels. The UDPs and modules are dealt with in Chapters 5 and 6, respectively.

Chapter 3 describes the essentials of the dataflow modeling style. In this modeling style, the most basic statement is the continuous assignment, which in turn consists of operators and operands. The continuous assignment continuously drives a value onto a net and is usually used to model combinational circuitry.

Chapter 4 is concerned with the behavioral modeling style, facilitating the user with the capability of modeling a design in a way like that of most high-level programming languages. In this modeling style, the most common statements include procedural assignments, selection statements, and iterative (loop) statements. In addition, timing control methods are dealt with in detail.

Chapter 5 describes three additional behavioral ways provided by Verilog HDL that are widely used to model designs. These include tasks, functions, and user-defined primitives (UDPs). Tasks and functions provide the ability to reuse the same piece of code from many places in a design and UDPs provide a means to model a design with a truth table. In addition, the predefined system tasks and functions are introduced. These system tasks and functions are useful in abstractly modeling a design in behavioral style or writing test benches for designs.

Chapter 6 discusses three closely related issues of hierarchically structural modeling. These include instantiation, generate regions, and configurations. Instantiation is a mechanism through which a hierarchical structure can be constructed by embedding modules into other modules. The generate regions can conditionally create declarations and instantiate modules into a design. By using configurations, we may specify a new set of target libraries in a way such that we may change the mapping of a design without having to modify the source description.

Preface xix

Chapter 7 deals with the additional features of Verilog HDL. These features include block constructs, procedural continuous assignments, specify blocks, timing checks, and compiler directives.

Chapters 8 and 9 examine some basic combinational and sequential modules that are often used as basic building blocks to construct a complex design. In particular, these modules are the basic building blocks of datapaths when the datapath and controller approach is employed to design complex systems.

Chapter 8 concerns the most commonly used combinational logic modules, covering encoders and decoders, multiplexers and demultiplexers, and magnitude comparators. Besides, a multiplexed seven-segment light-emitting diode (LED) display system combining the use of decoders and multiplexer is discussed in detail.

Chapter 9 involves several basic sequential modules widely used in digital systems. These include flip-flops, synchronizers, a switch-debouncing circuit, registers, data registers, register files, shift registers, counters (binary, BCD, Johnson), CRC generators and detectors, and sequence generators, as well as timing generators.

Chapter 10 describes various design and implementation options of digital systems, including application-specific integrated circuits (ASICs) and field-programmable devices. ASICs are devices that must be fabricated in IC foundries and can be designed with one of the following methods: full-custom, cell-based, and gate-array-based. Field-programmable devices are the ones that can be personalized in laboratories and include programmable logic devices (PLDs), complex PLDs (CPLDs), and field-programmable gate arrays (FPGAs). In addition, the issues of interfacing two logic modules or devices with different logic levels and power-supply voltages are treated in detail.

The next three chapters consider three closely related issues in developing a digital system: design, synthesis, and verification. Chapter 11 introduces two useful techniques by which a system can be designed. These techniques include the finite-state machine (FSM) and register-transfer level (RTL) design approaches. The former may be described by using a state diagram or an algorithmic-state machine (ASM) chart; the latter may be described by an ASM chart or by using the datapath and controller (DP+CU) paradigm described by an FSM with datapath (FSMD) diagram or an ASM with datapath (ASMD) chart. For a simple system, a three-step paradigm introduced in the chapter may be used to derive the datapath and the controller of a design from its ASM chart. For a complex system, the datapath and the controller are often obtained from specifications in a state-of-the-art manner. This approach is exemplified by a commercial dot-matrix liquid-crystal display (LCD) module capable of displaying 4-digit data at a time. In addition, we also emphasize in this chapter the concept that a hardware algorithm can be usually realized by using either a multicycle or a single-cycle structure. The choice of a multicycle or a single-cycle structure for the hardware algorithm is based on the trade-off among area (hardware cost), performance (operating frequency or propagation delay), and power consumption.

Chapter 12 is concerned with the principles of logic synthesis and the general architecture of synthesis tools. The function of logic synthesis is to transform an RTL representation into a gate-level netlist. In order to make effective use of a synthesis tool, we need to provide the synthesis tool the design environment and design constraints along with RTL codes and technology libraries of the underlying digital system. Moreover, some guidelines about how to write a good Verilog HDL code acceptable by most logic synthesis tools and to achieve the best compile times and synthesis results are given. These guidelines include clock signals, reset signals, how to partition a design, and power optimization.

Verification is a necessary process to ensure that a design can meet its specifications in both functionality and timing. Chapter 13 deals with this issue in depth and gives a comprehensive example based-on the FPGA design flow to illustrate how to enter, synthesize, implement, and configure the underlying FPGA device of a design. Along the design flow, static timing analysis is also given and explained. In addition, design verification through dynamic timing simulation, incorporating the delays of logic elements and interconnect, is introduced.

The next two chapters involve with more complex modules. Chapter 14 exposes many frequently used arithmetic modules, including addition, multiplication, division, arithmetic-and-logic units (ALUs), shift, and two digital-signal processing (DSP) filters. Along the introduction of these arithmetic operations and their algorithms, we also reemphasize the concept that a hardware algorithm can often be realized by using either a multicycle or a single-cycle structure.

Chapter 15 describes the design of a small microcontroller (μC) system, which is the most complex design example in the book. This system includes a 16-bit central processing unit (CPU), a general-purpose input and output (GPIO) module, timers, and a universal asynchronous receiver and transmitter (UART) module connected by a system bus, which is in turn composed of an address bus, a data bus, and a control bus.

The final chapter is concerned with the topic of testability and testable design. Testing is the only way to ensure that a system or a circuit may function properly. The goal of testing is to find any existing faults in a system or a circuit. In this chapter, we examine fault models, test vector generations, and testable circuit design or design for testability. In addition, the system-level testing, including SRAM, a core-based system, and system-on-chip (SoC), is also briefly dealt with.

Supplements

The instructor's supplements, including a solution manual and lecture notes in PowerPoint and pdf (generated by Latex beamer class) files, are available for teachers who adopt this book.

Student Projects

Many end-of-chapter problems may be assigned as student projects, in particular, the problems of Chapters 11, 14, and 15. Of course, many other chapters may also contain problems that may be used for the same purpose.

Acknowledgments

Most material of this book has been taken from the course ET5009 offered at our campus over the past decade. Thanks are due to the students of the course, who suffered through many of the experimental class offerings based on the draft of this book. Valuable comments from the participants of the course have helped in evolving the contents of this book and are greatly appreciated. Special thanks are due to my mentor, Ben Chen, a cofounder of Chuan Hwa Book Co., who brought me into this colorful digital world about forty years ago. I also gratefully appreciate to Frank for his exquisite design of the excellent book's cover and to Alice for her wonderful photos used in the cover. Finally but most sincerely, I would like to thank my wife, Fanny, and my children, Alice and Frank, for their patience in enduring my absence from them during the writing of this book.

M. B. Lin
Taipei, Taiwan

About the Author

Ming-Bo Lin is a Professor of Department of Electronic and Computer Engineering at National Taiwan University of Science and Technology. Professor Lin received his B.Sc. degree in electronic engineering from the National Taiwan Institute of Technology, Taipei, his M.Sc. degree in electrical engineering from the National Taiwan University, Taipei, and his Ph.D. degree in electrical engineering from the University of Maryland, College Park. He has been teaching the courses related to the field of Computer Engineering and Microelectronics for over thirty years. He was an adjunct Professor at National Taiwan University. His research interests include VLSI system designs, mixed-signal integrated circuit designs, parallel architectures and algorithms, and embedded computer systems. He has published numerous journal and conference papers in these areas. In addition, he has directed the designs of over fifty ASICs and has been consulted in industry extensively in the fields of microelectronics, ASIC, SoC, and embedded system designs. He is a senior member of IEEE. He received the Distinguished Teaching Award in 2007 from National Taiwan University of Science and Technology. He chaired the Workshop on Computer Architectures, Embedded Systems, and VLSI/EDA in National Computer Symposium (NCS) 2009. During the past three decades, Professor Lin has translated two books and authored over forty books (include revisions), especially including the following **textbooks** in English:

1. **Ming-Bo Lin**, *Digital System Designs and Practices: Using Verilog HDL and FPGAs*, John Wiley & Sons, 2008. (ISBN: 978-0470823231)
2. **Ming-Bo Lin**, *Introduction to VLSI Systems: A Logic, Circuit, and System Perspective*, CRC Press, 2012. (ISBN: 978-1439868591)
3. **Ming-Bo Lin**, *Digital System Designs and Practices: Using Verilog HDL and FPGAs*, 2nd ed., CreateSpace Independent Publishing Platform, 2015. (ISBN: 978-1514313305)
4. **Ming-Bo Lin**, *An Introduction to Verilog HDL,* CreateSpace Independent Publishing Platform, 2016. (ISBN: 978-1523320974)
5. **Ming-Bo Lin**, *Principles and Applications of Microcomputers: 8051 Microcontroller Software, Hardware, and Interfacing*, CreateSpace Independent Publishing Platform, 2016. (ISBN: 978-1537158372)
6. **Ming-Bo Lin**, *Principles and Applications of Microcomputers: 8051 Microcontroller Software, Hardware, and Interfacing,* Vol. I: *8051 Assembly-Language Programming*, CreateSpace Independent Publishing Platform, 2016. // (ISBN: 978-1537158402)
7. **Ming-Bo Lin**, *Principles and Applications of Microcomputers: 8051 Microcontroller Software, Hardware, and Interfacing,* Vol. II: *8051 Microcontroller Hardware and Interfacing*, CreateSpace Independent Publishing Platform, 2016. (ISBN: 978-1537158426)
8. **Ming-Bo Lin**, *Digital Logic Design: With An Introduction to Verilog HDL*, CreateSpace Independent Publishing Platform, 2016. (ISBN: 978-1537158365)
9. **Ming-Bo Lin**, *FPGA-Based Systems Design and Practice—Part I: RTL Design and Prototyping in Verilog HDL*, CreateSpace Independent Publishing Platform, 2018. (ISBN: 978-1721530199)

10. **Ming-Bo Lin**, *FPGA-Based Systems Design and Practice—Part II: System Design, Synthesis, and Verification*, CreateSpace Independent Publishing Platform, 2018. (ISBN: 978-1721530106)
11. **Ming-Bo Lin**, *A Tutorial on FPGA-Based System Design Using Verilog HDL: Intel/Altera Quartus Version—Part I: An Entry-Level Tutorial*, CreateSpace Independent Publishing Platform, 2018. (ISBN: 978-1721530380)
12. **Ming-Bo Lin**, *A Tutorial on FPGA-Based System Design Using Verilog HDL: Intel/Altera Quartus Version—Part II: ASM Charts and RTL Design*, CreateSpace Independent Publishing Platform, 2018. (ISBN: 978-1721530571)
13. **Ming-Bo Lin**, *A Tutorial on FPGA-Based System Design Using Verilog HDL: Intel/Altera Quartus Version—Part III: A Clock/Timer and a Simple Computer*, CreateSpace Independent Publishing Platform, 2018. (ISBN: 978-1721530496)
14. **Ming-Bo Lin**, *A Tutorial on FPGA-Based System Design Using Verilog HDL: Xilinx ISE Version—Part I: An Entry-Level Tutorial*, CreateSpace Independent Publishing Platform, 2018. (ISBN: 978-1721530441)
15. **Ming-Bo Lin**, *A Tutorial on FPGA-Based System Design Using Verilog HDL: Xilinx ISE Version—Part II: ASM Charts and RTL Design*, CreateSpace Independent Publishing Platform, 2018. (ISBN: 978-1721530809)
16. **Ming-Bo Lin**, *A Tutorial on FPGA-Based System Design Using Verilog HDL: Xilinx ISE Version—Part III: A Clock/Timer and a Simple Computer*, CreateSpace Independent Publishing Platform, 2018. (ISBN: 978-1721530830)
17. **Ming-Bo Lin**, *An Introduction to Cortex-M0-Based Embedded Systems — Cortex-M0 Assembly Language Programming*, CreateSpace Independent Publishing Platform, 2019. (ISBN: 978-1721530885)
18. **Ming-Bo Lin**, *An Introduction to Cortex-M3-Based Embedded Systems — Cortex-M3 Assembly Language Programming*, CreateSpace Independent Publishing Platform, 2019. (ISBN: 978-1721530946)
19. **Ming-Bo Lin**, *An Introduction to Cortex-M4-Based Embedded Systems — TM4C123 Microcontroller Principles and Applications*, CreateSpace Independent Publishing Platform, 2019. (ISBN: 978-1721530984)

and the following popular textbooks (in Traditional Chinese):

1. **Ming-Bo Lin**, *Digital System Design: Principles, Practices, and Applications*, 5th ed., Taipei, Taiwan: Chuan Hwa Book Ltd., 2017. (ISBN: 978-986-4635955)
2. **Ming-Bo Lin** and Shu-Tyng Lin, *8051 Microcomputer Principles and Applications*, Taipei, Taiwan: Chuan Hwa Book Ltd., 2012. (ISBN: 9789572183755)
3. **Ming-Bo Lin**, *Microprocessor Principles and Applications: x86/x64 Family Software, Hardware, Interfacing, and Systems*, 6th ed., Taipei, Taiwan: Chuan Hwa Book Ltd., 2018. (ISBN: 978-986-4637713)
4. **Ming-Bo Lin** and Shu-Tyng Lin, *Basic Principles and Applications of Microprocessors: MCS-51 Embedded Microcomputer System, Software, and Hardware*, 3rd ed., Taipei, Taiwan: Chuan Hwa Book Ltd., 2013. (ISBN: 9789572191750)
5. **Ming-Bo Lin**, *Digital Logic Design—With an Introduction to Verilog HDL*, 6th ed., Taipei, Taiwan: Chuan Hwa Book Ltd., 2017. (ISBN: 978-986-4635948)
6. **Ming-Bo Lin**, *An Entry-Level Tutorial on FPGA-Based System Design Using Verilog HDL: Intel/Altera Quartus Version*, Taipei, Taiwan: Chuan Hwa Book Ltd., 2013. (ISBN: 978-986-4638901)
7. **Ming-Bo Lin**, *Digital Logic Principles,* Taiwan: Chuan Hwa Book Ltd., 2018. (ISBN: 978-986-4638895)

1
Introduction

USING hardware description languages (HDLs) to describe a digital system has been popularized as an essential procedure in modern electronic engineering and hence it has become the fundamental attainment of digital system designers. With the use of HDLs, a design can be described at many levels of abstractions, ranging from the algorithm level to the gate level, even to the switch level, and the functional verification of the design can be done early in the design cycle.

At present, the two most commonly used HDLs in industry are Verilog HDL (or Verilog for short) and very high-speed integrated circuit (VHSIC) hardware description language (VHDL). Describing a design with HDLs is analogous to computer programming in the sense that the design is also described in text. However, HDL programming is fundamentally different from computer programming in that HDLs also need to capture the peculiar features of hardware: timing and concurrency. Because of the inherent simplicity of the language and the intimate relationship to the hardware features, Verilog HDL has been popular in industry over the past three decades. As a consequence, in this book we will focus our attention on the study of this HDL.

Just like C language, Verilog HDL is very closely related to the behavior of hardware modules and hence is easy to learn. To introduce the reader the beautiful and useful features of Verilog HDL, we address in this introductory chapter the basic concepts of Verilog HDL programs (usually called modules), such as how to describe a design in various modeling styles, including structural, dataflow, behavioral, and hybrid. After this, we give a tutorial example in the final section to demonstrate how to verify the description of a design.

1.1 Introduction

In this section, we first point out the differences between both the traditional and one modern approach for digital system implementations. Then, we introduce the features of Verilog HDL. Finally, we briefly describe the HDL-based approach in order to expose the reader early to the field of the so-called *synthesis flow*, including the use of *programmable logic devices* (PLDs), *complex programmable logic devices* (CPLDs) and *field-programmable gate arrays* (FPGAs), and *gate arrays* (GAs) and *standard cells*.

1.1.1 Digital System Implementations

Traditionally, the design of a digital system can be proceeded in a top-bottom way as a design hierarchy consisting of the system level, register-transfer level (RTL), gate level, and circuit level. Details of this design hierarchy will be addressed in Section 10.1.1.

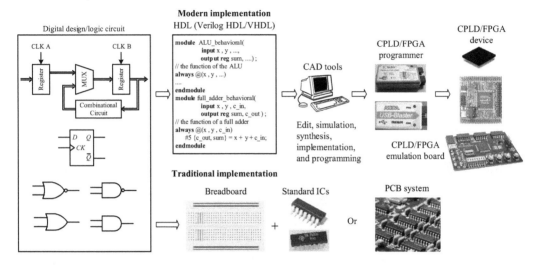

Figure 1.1: The differences between traditional and modern digital system implementations with CPLDs/FPGAs.

After the design of a digital system is done, it can be implemented with either a breadboard plus a number of discrete components–standard integrated circuits (ICs)–or a *printed circuit board* (PCB) system, as illustrated in Figure 1.1. The resulting system is then tested by applying stimuli generated from a signal generator to it and observing its responses with a logic analyzer. Here, the word "design" has a different meaning from the word "implementation," to be defined later in this chapter. However, for the moment it is sufficient to keep in mind that a design may generally have several different implementations.

With the advance of information technology, modern digital systems have functionality whose complexity is much more than ever before. To provide a simple way to implement such systems, a broad variety of field-programmable devices (such as CPLDs and FPGAs) with each may containing up to several million, even more, gates, are available nowadays. It is reasonable to expect that much more complex functionality will be incorporated into a single CPLD/FPGA along the reduction of VLSI feature sizes. With such devices, it proves very difficult to implement a design without the aid of *computer-aided design* (CAD) tools (i.e., computer programs). Nevertheless, to make the CAD tools understand the design (i.e., the logic circuit), a language that can be understood by computers is needed to describe the design. For this purpose, HDLs are proposed to capture the features of hardware—*timing* and *concurrency*—and equip a bridge between human beings and CAD tools.

As illustrated in Figure 1.1, one modern approach for implementing a digital system is based on a CPLD or an FPGA; such an approach is referred to as the *synthesis flow*. In this approach, the CAD tool receives an HDL description of a digital design and generates a programming file. This programming file can be used by a CPLD/an FPGA programmer to program a proper CPLD/FPGA suitable for the design or downloaded to a CPLD or an FPGA emulation board to verify the design in a realistic world. Generally speaking, CPLDs and FPGAs are used in products whereas CPLD and FPGA emulation boards are used to prototype a product during the development cycle of the product. Other modern approaches for implementing digital systems will be described in detail later in Chapter 10.

Before the programming file for configuring a CPLD or an FPGA is generated, a number of tasks are often needed to carry out in order to ensure that the design can meet its specifications. The most common tasks include simulation in both behavioral and gate levels, logic synthesis to convert an RTL description of a design into its equivalent gate-level netlist, con-

Section 1.1 *Introduction* 3

sisting of gates, wires, and so on, and implementation which maps the resulting gate-level netlist onto the structure of an underlying CPLD/FPGA. Thus, the CAD tools at least perform the following tasks: edit, simulation, synthesis, implementation, and the generation of programming information. The details of these tasks will be described in detail later in two dedicated chapters—Chapters 12 and 13. In the rest of this chapter, we will give a brief tutorial on the synthesis and design flow based on the use of CPLDs/FPGAs.

■ Review Questions

1-1 What are the differences between traditional and modern implementations of digital systems?
1-2 Why is HDL needed in modern digital system design?
1-3 What tasks do CAD tools usually perform?

1.1.2 Popularity of Verilog HDL

Verilog HDL originated in 1983 at Gateway Design Automation,[1] as a hardware modeling language associated with their simulator products. Verilog HDL has gradually gained popularity in industry since then. In order to increase its popularity, Verilog HDL was placed in the public domain in 1990. In 1995, it was standardized by IEEE as IEEE Std. 1364-1995, which has been promoted by Open Verilog International (OVI) since 1992. An updated version of the language was standardized by IEEE in 2001 as IEEE Std. 1364-2001. The new version covers many new features, including the port-list declaration style, configurations, and generate regions. The revision proposed in 2005, referred to as IEEE Std. 1364-2005, does not include new features but instead corrects and clarifies features ambiguously described in the 1995 and 2001 editions.

The main features of Verilog HDL are as follows:

- Verilog HDL is a general-purpose, an easy-to-learn, and an easy-to-use HDL.
- Verilog HDL can model a design in different design levels of abstractions in the design hierarchy.
- Verilog HDL allows different modeling styles to be mixed in the same module.
- Verilog HDL allows modeling a design entirely at the switch level by only using its built-in switch-level primitives.
- Almost all popular logic synthesis tools support Verilog HDL.
- All fabrication vendors provide Verilog HDL libraries for post-logic synthesis simulation.

In addition, Verilog HDL provides a powerful *programming language interface* (PLI), allowing users to develop their own computer-aided design (CAD) tools, such as a delay calculator.

1.1.3 Simple Examples of Verilog HDL

A simple Verilog HDL module is shown in Example 1.1, which displays a text line on the screen of the system that you used.

■ Example 1.1: A simple example of displaying a text line.

This example demonstrates a simplest module in Verilog HDL that contains merely one **$display** system task to print a message on the screen.

---
[1] Gateway Design Automation has been acquired by Cadence Design Systems.

```
module simple_example_1;
// a simple example to display a text line
initial begin
   $display("Welcome to Verilog HDL World!\n");
end
endmodule
```

The above example reveals many syntactic features of a Verilog HDL module: First, the Verilog HDL module begins with the keyword **module** followed by a module name, an **initial** block (or construct), and ends with the keyword **endmodule**. Second, the **$display** system task with the function similar to the **printf** library function in C language can be used to display something on basic I/O devices. Third, the texts after // are ignored by Verilog HDL compilers and simulators. They are only used for illustrating purposes; that is, they are comments.

The following example is another simple Verilog HDL module, which displays the sum of two integers on the screen.

■ Example 1.2: Displaying the sum of two numbers.

Like in other high-level programming languages, a variable in Verilog HDL must be declared before it can be referred to. Hence, we declare variables a and b as the **integer** type. To print out their values, we have to declare their output formats. For simplicity, both variables a and b and their sum are assumed to be two decimal digits. Therefore, an output format %2d is used to print out each of these three values.

```
module simple_example_2;
// a simple example to display the sum of two numbers
integer a, b;
initial begin
   a = 5; b = 8;
   $display("The sum of %2d and %2d is: %2d\n",a,b,a+b);
end
endmodule
```

Roughly speaking, a Verilog HDL module is much like a high-level language program except that it is used to describe a hardware module rather than a software function. The major differences between a hardware module and a software program are the *timing* and *concurrency* features inherently associated with the hardware module. Hence, Verilog HDL as well as other practical HDLs must provide some relevant mechanisms to describe these features. The following example illustrates how to describe the timing feature of a hardware module. As for the concurrency feature, we defer its details to Chapter 4.

■ Example 1.3: A simple Verilog HDL module.

In this example, an adder module is designed to add two n-bit numbers and yield an $(n+1)$-bit result. In order to precisely model the operations of the adder hardware, we need to include the timing characteristics of the adder. Hence, a delay #5 is placed after the keyword **assign** and before the expression of the **assign** continuous statement to indicate the propagation delay of the adder hardware. The **assign** continuous statement adds up the two input numbers together with the carry-in and yields a carry-out as well as a sum in 5 time units.

```
module nbit_adder
       #(parameter N = 4)(   // set default width
         input   [N-1:0] x, y,
```

```
       input  c_in,
       output [N-1:0] sum,
       output c_out );
// specify the n-bit adder using an assign statement
assign #5 {c_out, sum} = x + y + c_in;
endmodule
```

In summary, we can see from the above three examples that a more complete Verilog HDL module usually begins with the keyword **module** followed by a module name and a port list (if any), declarations of each item in the port list (if any), a number of **assign** continuous statements and/or other statements, and ends with the keyword **endmodule**.

1.1.4 HDL-Based Approach

Because the HDL-based approach may provide the capability of design reuse and easily handle a design with a large number of gate counts, it has become an indispensable method in most modern digital system designs. Consequently, to expose the reader to this approach early, we concisely describe in this section the general concepts and related issues of the HDL-based approach.

The general HDL-based approach can be illustrated in terms of digital implementation options shown in Figure 1.2. The design flow of this approach can be subdivided into the target-independent and target-dependent phases. The target-independent phase begins with a design specification and ends with the functional verification of the design. Following the design specification, different design entries, including a variety of available options, such as schematics, ABEL programs, and Verilog HDL, SystemVerilog, or VHDL modules, are then represented as a unified data structure for use in the next stage—functional verification. Functional verification ensures that the result of the design entry step is correct and meets the design specification in addition to performing some basic checks, such as syntax error in HDL. After this, the work of the target-independent phase is complete.

In the target-dependent phase, a target device or library, such as a programmable logic device (PLD), a complex PLD (CPLD), a field-programmable gate array (FPGA), or a cell library, is chosen. Most work of this phase is carried out by CAD tools automatically. The simplest devices are PLDs, which are usually used to implement designs requiring only from a few gates to several hundred gates. For such devices, the target-dependent phase only consists of two steps: device selection and programming.

For the CPLD/FPGA synthesis flow, because of the complex structure of devices, the target-dependent phase is further divided into three major steps: *synthesis*, *implementation*, and *programming*, as shown in Figure 1.2. The synthesis step is composed of selecting device, synthesis and optimization, and post-synthesis verification. The implementation step consists of the place-and-route and the timing analysis substeps. The function of the synthesis step is to optimize the switching functions[2] designed by the designer, based-on the selected device, and to map the abstract logic elements to actual logic blocks provided by the selected device. After the completion of the synthesis and optimization substep, it is usually necessary to verify the synthesized result through post-synthesis simulation or formal verification.

The implementation step starts with the place and route substep, which places the logic blocks into realistic logic elements and then sets up the related interconnect.[3] Because of the inherent delays associated with logic elements and interconnect, it needs to perform a timing analysis to verify whether the timing constraints meet the desired specification. The widely used timing analysis may be either dynamic or static. *Dynamic timing analysis* (DTA)

---
[2]In this book, we use "switching function" instead of "Boolean function" to mean explicitly that each input variable of the switching function can only have two values: 0 and 1.

[3]The interconnect means wires that link together transistors, circuits, cells, modules, and systems.

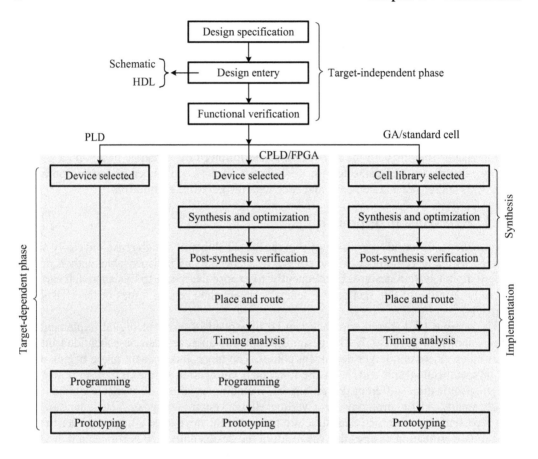

Figure 1.2: An HDL-based approach illustrated with digital system implementation options.

is performed by simulation whereas *static timing analysis* (STA) is by analyzing the timing paths of the design without doing any actual simulation.

Once the implementation step is finished, the next step is to generate the programming file in order to program the selected device. Of course, after a device is programmed, it is often tested in real world to see if it indeed works as expected.

The synthesis flow of both the cell-based (standard cell) and gate-array-based design is almost the same as that of the CPLD/FPGA synthesis flow, with only one exception that it does not need the programming step. In fact, these implementations must be "programmed" in an IC (integrated circuit) foundry; namely, they must be programmed by photomasks.

A final comment on the above HDL-based approach is that the target-independent phase together with synthesis is often referred as the *front-end design* and the remaining part as the *back-end design* in industry. We will return to this topic later in Chapter 12.

■ Review Questions

1-4 Why is HDL so popular in modern digital system design?
1-5 How would you distinguish a hardware module from a software program?
1-6 What can a Verilog HDL module be?
1-7 How could we introduce a comment into a Verilog HDL module?

Section 1.2 *Introduction to Verilog HDL*

1.2 Introduction to Verilog HDL

In this section, we introduce the basic syntax and some most basic components of Verilog HDL.

1.2.1 Module Concept

The basic unit of a digital system is a *module*, which consists of a *body* (also called an *internal* or a *core circuit*) and an *interface* (called *ports*). The body performs the required function of the module while the interface carries out the needed communication between the body and the outside world. In addition, the power supply and ground [4] also need to be provided by the interface. A simple example of a hardware module is shown in Figure 1.3. The module embodies four 2-input OR gates along with their power supply(V_{CC}) and ground (GND). To be consistent with the concept of a hardware module, the basic building block in Verilog HDL is also called a *module*, which describes a specific hardware module.

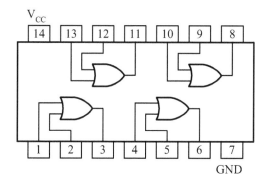

Figure 1.3: A simple example of a hardware module.

Like a hardware module, a Verilog HDL module can be an element or a collection of lower-level design blocks (modules). The basic structure of a Verilog HDL module is shown in Figure 1.4, which includes many different declarations, statements, tasks, and functions, as well as others. Each of these will be described in appropriate sections later in the book.

1.2.2 Lexical Conventions

Like any other high-level programming languages, Verilog HDL uses almost the same lexical conventions except that it needs to describe the unique features of hardware modules, including timing and concurrency.

1.2.2.1 Identifiers An identifier is a sequence of alphanumeric characters, _ (underline), and $, with the restriction that the first character must be a letter, _, or $. The length of an identifier is limited by implementations but it should be at least 1024 characters. If an identifier exceeds the implementation-specified length limit, an error will be reported. The identifiers beginning with $ are reserved for system tasks and functions as well as timing checks, which will be described in Sections 5.3 and 7.3.3, respectively. The following examples are some valid identifiers.

```
counter
```

---

[4] In a Verilog HDL module, the power supply and ground are usually not expressed explicitly; only the signals are revealed.

> **module** [#(parameters)] module name
>
> Parameters (if any)
>
> Port List, Port Declarations (if any)
>
> Declarations of **wire**, **reg**, and other variables
>
> Instantiation of lower level modules or primitives
>
> Dataflow statements (**assign**)
>
> **always** and **initial** blocks. (all behavioral statements go into these blocks).
>
> Tasks and functions.
>
> **endmodule**

Figure 1.4: The basic structure of a Veriolg HDL module.

```
four_bit_adder
a12
_4b_adder
```

Note that Verilog HDL is a *case-sensitive language* just like C. Hence, `adder` and `Adder` are two different identifiers.

For a multiword identifier, three methods are often used to compress a multiple word into a single lexical unit. These are the *underline method*, *Pascal-case method*, and *camel-case method*.

- The *underline method*: In this method, the underline character is utilized to connect words, yielding names such as `four_bit_adder`.
- The *Pascal-case method*: In this method, uppercase letters are used to connect words and the first letter in each word in an identifier is capitalized, such as `FourBitAdder`.
- The *camel-case method*: In this method, uppercase letters are employed to connect words and the first letter in each word except the first letter in an identifier is in upper case, such as `fourBitAdder`.

In the rest of this book, we will use the underline method.

An escaped identifier begins with a \ (backslash) character and ends with a *white space*, which includes a blank space (\b), a tab (\t), a new line (\n), and the eof (end of file).

1.2.2.2 Keywords Like most high-level programming languages, such as C, in Verilog HDL there is a list of reserved identifiers called *keywords*, which may only be used in some contexts. Some examples of keywords are listed as follows:

and	assign	begin	case	else
end	endcase	endmodule	for	if
initial	inout	input	integer	module
nand	negedge	nor	not	or
output	parameter	posedge	real	realtime
reg	repeat	signed	unsigned	xnor
xor				

A keyword may also be escaped and treated as not having the same meaning as its original one. For example, \always will have a different meaning to the keyword **always**. However, to make a module clear and readable, it should avoid using keywords with a different case or escaped keywords as identifiers.

Section 1.2 Introduction to Verilog HDL

1.2.2.3 Comments Verilog HDL has two ways to introduce comments: a *one-line comment* and a *block comment*. A one-line comment starts with two slashes // and ends with a new line. A block comment begins with /* and ends with */. Block comments may not be nested and the one-line comment token // has no special meaning in a block comment.

■ Coding Style

1. *It is good practice to:*
   - *use lowercase letters for all signal names, variable names, and port names;*
   - *use uppercase letters for names of constants and user-defined types;*
   - *use meaningful names for signals, ports, functions, and parameters;*
   - *use the same name for both the module and file.*

1.2.3 Value Set

It is well known that a logic circuit usually outputs a logic value 0 or 1. Nevertheless, in some realistic hardware circuits, a high-impedance state may exist. To model such a high-impedance value of a node or net, a new value, z, is used in Verilog HDL. In addition, some circuit nodes, such as the outputs of D flip-flops, may have unknown logic values as their power supplies are just on and before they sample their input signals. For this case, another logic value, x (unknown), is used to specify an unknown logic value of a net or node. Based on these observations, the resulting system is called a *four-valued logic* system.

- 0: a logic 0, the false condition
- 1: a logic 1, the true condition
- x: an unknown logic value
- z: a high impedance

1.2.4 Constants

There are three types of constants provided in Verilog HDL, including *integers*, *real numbers*, and *strings*. Integer constants can be specified in decimal, hexadecimal, octal, or binary format. An integer value in this form represents a signed number. A negative number is represented in 2's-complement form. For example,

```
-123        // is decimal -123
12345       // is decimal 12345
4'b1011     // a 4-bit binary number
8'habc      // an 8-bit hexadecimal number
2012        // unsized number -- a 32-bit decimal number
```

Real numbers may be specified in either decimal notation or scientific notation. When a real number is represented in decimal notation, there must be at least one digit on each side of the decimal point. For instance,

```
3.4         // legal
 .2         // illegal
294.872     // legal
1.44E9      // the exponent symbol can be e or E
```

A string is a sequence of characters enclosed by double quotes ("") and cannot be split into multiple lines. Each character is represented as an 8-bit ASCII code. Hence, a string can be considered as an unsigned integer constant represented by a sequence of 8-bit ASCII codes. To represent a string, we need to declare a string variable of the **reg** type with a width equal to eight times the number of characters in the string. For example, to store the string "Verilog HDL" requires a **reg** variable of 8*11, or 88 bits.

1.2.5 Data Types

Verilog HDL has two classes of data types: *nets* and *variables*. A net means any *hardware connection point* and a variable represents a *data storage element*. Note that the "net" is not a keyword but indeed represents a class of data types listed in Table 1.1. A variable holds a value from one assignment to the next. In this section, we only deal with the **wire** net and the **reg** variable type. The details of nets and various variable types will be addressed in Chapter 2 and Section 3.2.2, respectively.

Table 1.1: Data types of Veriolg HDL.

Net		Variable
wire	supply0	reg
tri	supply1	integer
wand	tri0	real
wor	tri1	time
triand	trireg	realtime
trior	uwire	

1.2.5.1 The wire Net Type A **wire** net represents a physical connection between structural entities, such as gates and modules. It does not store a value. To use a **wire** net, the following syntax can be used.

> **wire** [**signed**] [[msb:lsb]] net_id{, net_id};

where net_id denotes a net identifier and [msb:lsb] is optional and indicates the range of the net. The item inside square brackets ([]) is optional and within curly brackets ({}) can repeat zero or more times. A net is defaulted to a single bit known as a *scalar* if no range is specified and is referred to as a *vector* if a range of more than one bit is specified. Some examples of **wire** net declarations are given as follows:

```
wire a, b, c_in;       // 1-bit wires
wire [7:0] data_a;     // an 8-bit wire, the msb is bit 7
wire [0:7] data_b;     // an 8-bit wire, the msb is bit 0
wire signed [7:0] d;   // an 8-bit signed wire
```

1.2.5.2 The reg Variable Type A **reg** variable holds a value between assignments; it can be used to model connection nodes and hardware registers, such as edge-sensitive (i.e., flip-flops) and level-sensitive (i.e., latches) storage elements. A form of **reg** declarations is as follows

> **reg** [**signed**] [[msb:lsb]] variable_id{, variable_id};

where variable_id denotes a **reg** identifier. Like the net, a **reg** variable is a scalar if no range, [msb:lsb], is specified, and is a vector if a range of more than one bit, indicated by [msb:lsb], is specified. Some examples of **reg** declarations are given below.

```
reg a, b, c_in;       // 1-bit reg variables
reg [7:0] data_a;     // an 8-bit reg, the msb is bit 7
reg [0:7] data_b;     // an 8-bit reg, the msb is bit 0
reg signed [7:0] d;   // an 8-bit signed reg
```

Section 1.2 *Introduction to Verilog HDL*

1.2.6 Primitives

Verilog HDL has two types of primitives: *built-in primitives*, including 12 gate primitives and 16 switch primitives, and *user-defined primitives* (UDPs), which can be used to define a new combinational or sequential logic function. The gate primitives support gate-level structural modeling while switch primitives support switch-level structural modeling. User-defined primitives (UDPs) provide a convenient way for users to define their own primitive functions. Two major classes of user-defined primitives are combinational UDPs and sequential UDPs. To instantiate a gate primitive, the following syntax can be used.

```
gate_primitive [instance_name] (out, in{, in})
               {, [instance_name] (out, in{, in}};
```

where `gate_primitive` can be any of {**and**, **nand**, **or**, **nor**, **xor**, **xnor**} and `instance_name` is optional. The first port of an instantiated gate primitive is always an output and the other ports are inputs. Primitives can be instantiated only within modules.

1.2.7 Attributes

An attribute is a mechanism provided by Verilog HDL for annotating information about objects, statements, and groups of statements in the source description to be used by various related tools. Almost all statements can be attributed in Verilog HDL. There are no standardized attributes. An attribute can be attached as a prefix to a declaration, a module item, a statement, or a port connection and as a suffix to an operator or a function name in an expression. Attributes have the following syntax:

```
(*attr_spec {, attr_spec}*)
```

where `attr_spec` can be either `attr_name` or `attr_name = const_expr`. If an attribute is not assigned to a value, its value defaults to 1. Some examples are given below.

```
// example 1: attach full_case attribute only
(* full_case=1, parallel_case = 0 *)
case (selection)
   <rest_of_case_statement>
endcase

// example 2: attach an attribute to a module
(* dont_touch *) module array_multiplier(x, y, product);
   <the body of array_multiplier>
endmodule

// example 3: attach an attribute to a reg variable
(* fsm_state *)   reg [1:0] state1;
(* fsm_state=1 *) reg [1:0] state2, state3;

// example 4: attach an attribute to a function call
x = add(* mode = "cla" *)(y, z);

// example 5: attach an attribute to an operator
x = y + (* mode = "cla" *) z;
```

■ Review Questions

1-8 How many possible values may a net or **reg** variable have?
1-9 How many data types are included in net types? What are these?
1-10 How many data types are included in variable types? What are these?
1-11 What kinds of primitives are provided in Verilog HDL?

1.3 Module Modeling Styles

To understand the design methodologies of modern digital systems, it is important to distinguish among *design*, *model*, *synthesis*, and *implementation* or *realization*.

- Design is a series of transformations from one representation of a system to another until a representation that can be fabricated exists; that is, it can be created, fashioned, executed, or constructed according to a plan.
- Model is a system of postulates, data, and inferences presented as a mathematical description of an entity or the state of affair. Modeling is a process that converts a specification document into an HDL description.
- Synthesis is a process for converting an HDL description into a structural representation. Synthesis can be subdivided into *logic synthesis* and *high-level synthesis*. Logic synthesis is a process that converts an RTL description into a gate-level netlist while high-level synthesis is a process that converts a high-level description (i.e., specification) into RTL results.
- Implementation (or realization) is the process of transforming design abstraction into physical hardware components such as FPGAs or cell-based ICs (integrated circuits).

Generally speaking, *a design can often have many different implementations*. For instance, a 4-bit adder may be implemented by using a PLD, a CPLD, an FPGA, or other devices.

1.3.1 Modules

As mentioned, a Verilog HDL module, regardless of at which level in the design hierarchy that the module is designed, consists of two major parts: the *body* (i.e., core circuit) and the *interface* (ports). In this section, we briefly deal with the styles used to describe the body of a module as well as the types of ports, and how to declare ports in a module.

1.3.1.1 Modeling the Body of a Module For each module in Verilog HDL, the body may be described as one of the following styles:

1. *Structural style*: A design is described in this style as a set of interconnected components. The components can be modules, UDPs, gate primitives, and/or switch primitives.

    (a). *Hierarchically structural level*: A design is said to be modeled at the hierarchically structural level when it comprises a set of interconnected modules along with some necessitated gate primitives.

    (b). *Gate level*: A design is said to be modeled at the gate level when it only comprises a set of interconnected gate primitives.

    (c). *Switch level*: A design is said to be modeled at the switch level when it only consists of a set of interconnected switch primitives.

    The structural style is a mechanism that can be used to construct a hierarchical design for a large digital system. The details will be described in Chapter 6.

Section 1.3 *Module Modeling Styles*

2. *Dataflow style*: In dataflow style, the module is described by specifying the data dependence between registers and how the data are processed.

    (a). A module is specified as a set of **assign** continuous assignments.

3. *Behavioral or algorithmic style*: In this style, the module is described in terms of the desired design algorithm without concerning the hardware implementation details.

    (a). Designs can be described in any high-level programming language.

4. *Mixed style*: The module is described in terms of the combination of above three modeling styles.

    (a). The mixed style is most commonly used in modeling large designs.

In industry, the term *register-transfer-level* (RTL) *code* is used to mean that an RTL module (i.e., bulit with RTL components) is modeled in behavioral, dataflow, structural, or mixed style on condition that the resulting description must be acceptable by logic synthesis tools. Also, it is instructive to note that a module can be designed at any level of the design hierarchy (see Section 10.1.1) and the above modeling styles do not concern the design level of the underlying module. Hence, these modeling styles can be applied equally well to describe modules at any level, except for the circuit level that can only be described in structural style at the switch level, of the design hierarchy of digital systems.

1.3.1.2 Port Declaration The interface signals (not including supply and ground) of a Verilog HDL module can be categorized into one of the following three types:

- *Input ports*: The keyword **input** declares a group of signals as input ports.
- *Output ports*: The keyword **output** declares a group of signals as output ports.
- *Bidirectional ports*: The keyword **inout** declares a group of signals as bidirectional ports; that is, they can be used as input or output ports but not at the same time.

The simplest way to describe the complete interface of a module is to subdivide it into three parts: a port list, port declarations, and the data type declaration of each port. An output port, except that it is a **wire** net type, must be declared a data type associated with it. However, an input port is often left with its data type undeclared. For example,

```
// the port-list style
module adder(x, y, c_in, sum, c_out);
input [3:0] x, y;      // 4-bit wire nets
input c_in;            // 1-bit wire net
output [3:0] sum;
output c_out;
reg [3:0] sum;         // 4-bit reg variable
reg c_out;             // 1-bit reg variable
```

This style of port declarations is known as the *port-list style*. The declaration of a port and its associated data type may be combined into a single line. Based on this idea, the above interface portion of the module can be rewritten as follows:

```
// the port-list style
module adder(x, y, c_in, sum, c_out);
input [3:0] x, y;          // 4-bit wire nets
input c_in;                // 1-bit wire net
output reg [3:0] sum;      // 4-bit reg variable
output reg c_out;          // 1-bit reg variable
```

Of course, the port list, port declarations and their associated data types may also be put together into a single list. This style is often called the *port-list declaration style*. Hence, the above module interface can be rewritten as

```
// the port-list declaration style
module adder(input  [3:0] x, y,      // 4-bit wire nets
             input  c_in,            // 1-bit wire net
             output reg [3:0] sum,   // 4-bit reg variable
             output reg c_out        // 1-bit reg variable
); // sometimes called the ANSI style
```

This style is the same as that of the ANSI-C language so that we often refer to it as the *ANSI style*. ANSI is the acronym of American National Standards Institute. Note that all of above interface styles are valid in Verilog HDL. In this book, we prefer to use the port-list declaration style.

1.3.1.3 Port Connection Rules The port connection (also called *port association*) rules of Verilog HDL modules are consistent with those of realistic hardware modules. That is, Verilog HDL allows ports to remain unconnected and to be of different sizes. In addition, unconnected inputs are driven to the "z" state; unconnected outputs are not used. Connecting ports to external signals can be accomplished by either of the following two methods:

- *Named association*: Ports to be connected to external signals are specified by listing their names. The port order is not important.
- *Positional (or ordered) association*: Ports are connected to external signals by an ordered list of the ports. The signals to be connected must have the same order as the ports in the port list, leaving the unconnected port blank.

However, these two methods cannot be mixed in the same instantiation of a module. Moreover, Verilog HDL primitives (including both built-in and user-defined primitives) can be only connected by positional association.

The operation to "call" (much like the *macro expansion* in assembly language) a built-in primitive, a user-defined primitive, or another module, is called *instantiation* and each copy of the called primitive or module is called an *instance*. An illustration of this is shown in Figure 1.5, which reveals how to instantiate gate primitives and user-defined modules as well as how to connect their ports through nets and input/output ports. Note that built-in primitives can be only connected by using positional association; they cannot be connected through named association. In addition, as mentioned before, the instance names of these primitives are optional.

The `full_adder` module depicted in Figure 1.5 indicates how to instantiate an already defined module and how to connect their ports through nets. The ports of user-defined modules can be connected by using either named association or positional association. In addition, the instance names of these instances are necessary.

■ Coding Style

1. *A module cannot be declared within another module.*
2. *A module may instantiate other modules.*
3. *A module instance must have a module identifier (instance name) except for built-in primitives, gate and switch primitives, and user-defined primitives (UDPs).*
4. *It should use named association at the top-level (also called top) module to avoid confusion that may arise from synthesis tools.*

1.3.2 Structural Modeling

As mentioned previously, the structural modeling of a design is by connecting required instances of built-in primitives, user-defined primitives, or other (user-defined) modules through

Section 1.3 *Module Modeling Styles*

```
module half_adder (input x, y, output s, c);
// -- half adder body-- //
// instantiate primitive gates
   xor xor1 (s, x, y);      ← Can only be connected by using positional association
   and and1 (c, x, y);
endmodule                   — Instance name is optional.


module full_adder (input x, y, cin, output s, cout);
wire s1,c1,c2; // outputs of both half adders
// -- full adder body-- //         ← Connecting by using positional association
// instantiate the half adder
   half_adder ha_1 (x, y, s1, c1);
   half_adder ha_2 (.x(cin), .y(s1), .s(s), .c(c2));   ← Connecting by using named association
   or (cout, c1, c2);               — Instance name is necessary.
endmodule
```

Figure 1.5: Port connection rules.

ports (i.e., nets). Many instances demonstrate the structural modeling are given in the following example.

■ Example 1.4: Structural modeling at the gate level.

The half_adder module instantiates two gate primitives, and the full_adder module instantiates two half_adder modules and one gate primitive. Finally, the four_bit_adder module is in turn constructed by four full_adder instances.

```verilog
// a 4-bit adder described hierarchically at the gate level
// a 4-bit adder at the gate level
module four_bit_adder (
       input   [3:0] x, y,
       input   c_in,
       output  [3:0] sum,
       output  c_out);
wire   c1, c2, c3; // intermediate carries
// -- four_bit adder body -- //
// instantiate four full adders
   full_adder fa_1 (x[0], y[0], c_in,   c1, sum[0]);
   full_adder fa_2 (x[1], y[1], c1,     c2, sum[1]);
   full_adder fa_3 (x[2], y[2], c2,     c3, sum[2]);
   full_adder fa_4 (x[3], y[3], c3, c_out, sum[3]);
endmodule

// a full adder at the gate level
module full_adder(input x, y, c_in, output sum, c_out);
wire   s1, c1, c2;  // outputs of both half adders
// -- the full adder body-- //
// instantiate the half adder
   half_adder ha_1 (   x,   y, c1, s1);
   half_adder ha_2 (c_in, s1, c2, sum);
   or (c_out, c1, c2);
```

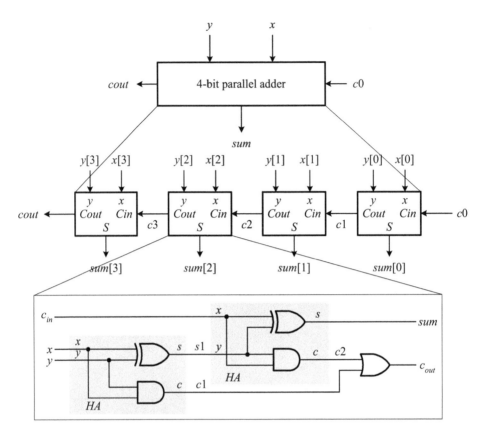

Figure 1.6: A hierarchical 4-bit adder.

```
endmodule
// a half adder at the gate level
module half_adder(input x, y, output c, s);
// instantiate gate primitives
   xor (s,x,y);
   and (c,x,y);
endmodule
```

In fact, the structural style is the way to model a complex digital system in a hierarchical manner. This is exemplified by the 4-bit adder depicted in Figure 1.6. The 4-bit adder is composed of four full-adders and then each full-adder is in turn built by using basic logic gates. Even though this example is quite simple, it manifests many important features in designing a large digital system.

1.3.3 Dataflow Modeling

The essential construct used to model a design in dataflow style is the **assign** continuous assignment. In an **assign** continuous assignment, a value is assigned onto a net. It must be a net because **assign** continuous assignments are used to model the behavior of combinational circuits. An **assign** continuous assignment starts with the keyword **assign** and has the syntax

```
assign [delay] net_lvalue = expression{, net_lvalue = expression};
```

Section 1.3 *Module Modeling Styles*

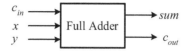

Figure 1.7: The block diagram of a full adder.

Whenever the value of an operand in `expression` changes, `expression` is evaluated and the result is assigned to `net_lvalue` after the specified `delay`. The `delay` specifies the amount of time between the change of an operand in the expression and the assignment to `net_lvalue`. If no `delay` is specified, the default delay is 0 time units. All **assign** continuous assignments in a module execute concurrently regardless of the order they appear.

The following example illustrates how an **assign** continuous assignment is used to describe the 1-bit adder (i.e., full adder) depicted in Figure 1.7.

■ Example 1.5: A full adder modeled in dataflow style.

In this example, we assume that the full adder shown in Figure 1.7 requires 5 time units to complete its operations. The delay will be ignored by synthesis tools when the module is synthesized because the delay will be replaced by the realistic propagation delays of gates used to realize the adder.

```
module full_adder_dataflow(
      input   x, y, c_in,
      output  sum, c_out);
// specify the function of a full adder
assign #5 {c_out, sum} = x + y + c_in;
endmodule
```

1.3.4 Behavioral Modeling

The behavioral style uses two procedural blocks (or constructs): **initial** and **always** blocks (or called constructs). The **initial** block can be executed only once and hence is usually used in test benches to set up initial values of variable types whereas the **always** block, as the name implies, is executed repeatedly, as depicted in Figure 1.8. The **always** block is used to model combinational or sequential logic. Each **always** block corresponds to a piece of logic. All other statements used to model a logic circuit in behavioral style must be placed within **initial** and/or **always** blocks.

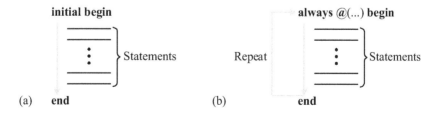

Figure 1.8: The behavior of (a) **initial** blocks and (b) **always** blocks.

The above mentioned **assign** continuous statement used in dataflow style can also be applied equally well to model a logic circuit in behavioral style with the following two exceptions: First, the keyword **assign** must be removed. Second, the `l_value` used in an expression

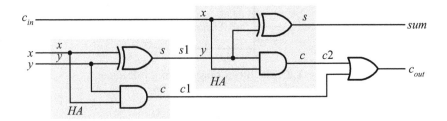

Figure 1.9: A full-adder constructed with basic logic gates.

within an **initial** or **always** block must be a variable type, such as **reg**, which retains its value until it is assigned a new value. All **initial** and **always** blocks begin their execution at simulation time 0 concurrently and are independent of their order appearing in the module.

The following example illustrates how a procedural construct is used to describe the 1-bit adder (i.e., full adder) depicted in Figure 1.7.

■ **Example 1.6: A full adder modeled in behavioral style.**

Basically, the expression used to describe the operations of a full adder in behavioral style is the same as the one in dataflow style except that it is needed to put inside an **always** block. In addition, the sensitivity list @(x, y, c_in) is used to sensitize the changes of input signals. The resulting module is as follows:

```
module full_adder_behavioral(
       input  x, y, c_in,
       output reg sum, c_out); // should be reg type
// specify the function of a full adder
always @(x, y, c_in) // always@(x or y or c_in)
   #5 {c_out, sum} = x + y + c_in;
endmodule
```

1.3.5 Mixed-Style Modeling

Recall that the mixed-style modeling is usually used to construct a hierarchical design in a large system. Nonetheless, we are still allowed to model a simple design in mixed style. As an illustration of this, consider the full-adder depicted in Figure 1.9, which is constructed with two half adders and an OR gate. The demonstration of how to model this full adder in mixed style is given in the following example.

■ **Example 1.7: A full adder modeled in mixed style.**

The first half adder is modeled in structural style, the second half adder is in dataflow style, and the OR gate is in behavioral style.

```
module full_adder_mixed_style(
       input  x, y, c_in,
       output s,
       output reg c_out);
wire s1, c1, c2;
// model HA 1 in structural style
   xor xor_ha1(s1, x, y);
```

Section 1.4 Simulation

```
    and and_ha1(c1, x, y);
// model HA 2 in dataflow style
assign  s = c_in ^ s1;
assign  c2 = c_in & s1;
// model the output OR gate in behavioral style
always @(c1, c2)    // or use always @(*)
    c_out = c1 | c2;
endmodule
```

■ Review Questions

1-12 What are the differences among design, model, synthesis, and implementation?
1-13 Describe the features of the structural style.
1-14 What is the basic statement used in the dataflow style?
1-15 What are the basic constructs used in the behavioral style?
1-16 Could we write a module by mixing the use of various modeling styles?

1.4 Simulation

For a design to be useful, it must be verified to ensure that it can correctly operate according to the design specifications. Verilog HDL not only provides capabilities to model a design but also equips facilities to verify the design, namely, to generate and control stimuli, to monitor and store responses, and to check the results. In this section, we use a 4-bit adder as an example to illustrate how to verify a design entirely through the mechanism provided by Verilog HDL.

1.4.1 Basic Simulation Constructs

Two basic simulation structures in Verilog HDL are shown in Figure 1.10. The first structure is to take the *unit under test* (UUT) as an instantiated module in the stimulus module, which also serves as the top-level module. This is often used in simple or small projects because it is intuitively simple to write. The second structure considers both the stimulus block and UUT as two separate instantiated modules at the top-level module. This is suitable for large projects. In this book, we will use the first structure when writing a test bench for simplicity.

In general, a *test bench* comprises a variety of basic parts: an instance of the UUT, stimulus generation and control, response monitoring and storing, and result checking. Except for the instance of the UUT, the remaining parts are usually modeled in behavioral style. A test bench usually does not have input and output ports. Its general format is as follows:

```
`timescale time_unit/time_precision
module test_module_name;
// local declaration
reg  identifiers;     // if needed
wire identifiers;     // if needed
integer identifiers;  // if needed
    // instantiate the module under test
    // use initial and always to generate stimuli
    // monitor and output the responses of the module under test
endmodule
```

In the rest of this section, we deal with these features of a typical test bench by using an example. The details of how to write test benches along with the other viewpoints and features of verification will be considered later in a dedicated chapter.

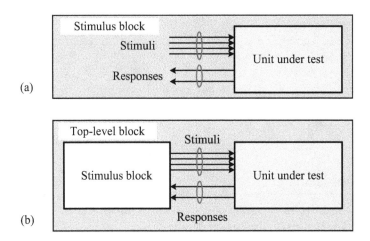

Figure 1.10: Two basic constructs for the simulation setups: (a) stimulus block as the top-level module and (b) stimulus block as a separate module.

1.4.2 Related Compiler Directive and System Tasks

In this subsection, we introduce the most basic compiler directive, **`timescale**, display system tasks, **$display** and **$monitor**, simulation time system functions, **$time** and **$realtime**, and simulation control system tasks, **$finish** and **$stop**.

1.4.2.1 `timescale Compiler Directive When carrying out simulation, we need to specify the physical unit of measure, or time scale, of a numerical time delay value and the resolution of the time scale, i.e., the minimum step size of the scale during simulation. In Verilog HDL, these are done by using the **`timescale** compiler directive.

`timescale time_unit/time_precision

where " ` " is the *back quote*, `time_unit` specifies the units of measure for simulation times and delays, and `time_precision` specifies how simulation times and delays are rounded during simulation. Only integers 1, 10, and 100 may be used to specify both `time_unit` and `time_precision`; the valid units are s, ms, $\mu$s, ns, ps, and fs. Notice that `time_precision` must not exceed `time_unit`. Hence, the **`timescale 1 ns /10 ns** statement is illegal.

■ **Example 1.8: The use of the `timescale compiler directive.**

To illustrate the use of the **`timescale** compiler directive, consider the following statement:

`timescale 10 ns/1 ns

and the outputs from a simulator:

```
#3.55 a = b + 1; // corresponds to 36 ns
#3.54 b = c + 1; // corresponds to 35 ns
```

The `timescale 10 ns / 1 ns statement specifies that the time unit is 10 ns and the time resolution is 1 ns. As a result, the time values in the module are multiples of 10 ns, rounded to the nearest 1 ns because the `time_precision` is 1 ns. Therefore, the delay 3.55 and 3.54 are scaled and rounded to 36 ns and 35 ns, respectively.

■ Example 1.9: Another use of the 'timescale compiler directive.

As another illustration of the use of the **'timescale** compiler directive, consider the following statement:

```
'timescale 1 ns /10 ps
```

and the outputs from a simulator:

```
#3.55 a = b + 1; // corresponds to 3.55 ns
#3.54 b = c + 1; // corresponds to 3.54 ns
```

Here, all time values are multiples of 1 ns because the time_unit is 1 ns. Delays are rounded to real numbers with two decimal places because the time_precision is 10 ps.

It should use the same time unit in both RTL and the gate-level simulations; otherwise, the incorrect results might be obtained. As a rule of thumb, for FPGA designs, it is suggested to use 1 ns as the first attempt for the time unit in the test bench because the propagation delays of gates for current FPGA/CPLD devices are about in the order of nanoseconds. In addition, the effect of the **'timescale** compiler directive lasts for all modules that follow this compiler directive until another **'timescale** compiler directive specifies otherwise.

1.4.2.2 Display System Tasks During simulation, we need to display information about the design for debugging or other useful purposes. To achieve this, Verilog HDL provides two widely used system tasks, **$display** and **$monitor**, for displaying information on the standard output. The **$display** system task displays information only when it is called but the **$monitor** system task continuously monitors and displays the values of any variables or expressions. They have the same form

```
task_name[(arguments)];
```

where task_name is either the **$display** or **$monitor** system task and arguments are used to specify strings to be displayed literally. To print information with a specified format, an escape sequence is used, much like that of C language. For example,

```
$display($time,"ns %d %d %h %h",x,y,c_in,{c_out,sum});
$monitor($realtime,"ns %d %d %h %h",x,y,c_in,{c_out,sum});
```

where %d and %h are used to specify variable values to be displayed in decimal and hexadecimal, respectively.

Only one **$monitor** display list can be activated at any time; however, a new **$monitor** system task with a new display list can be issued any number of times during simulation. In addition to the **$monitor** system task, **$monitoron** and **$monitoroff** system tasks are two related tasks widely used to enable and disable the monitoring operations, respectively.

1.4.2.3 Simulation Time System Functions There are two system functions that provide accesses to the current simulation time: **$time** and **$realtime**. The **$time** system function returns a 64-bit integer of time and the **$realtime** system function returns a real number of time. All of the returned values from these two system functions are scaled to the time unit of the module that calls them.

■ Example 1.10: The use of the $time system function.

As an illustration of the use of the **$time** system function, consider the following module in which the **$time** system function is used to display the simulation time.

```
`timescale 10 ns/ 1 ns
module time_usage;
reg a;

initial begin
    #3.55 a = 0;
    #3.55 a = 1;
    $monitor($time, "a = ", a);
end
endmodule
```

The simulation times are 36 ns and 72 ns. They are scaled to 4 time units and 7 time units, respectively, because the time_unit of the module is 10 ns and the time_precision is 1 ns. Consequently, the simulation results are

```
#0 a = x
#4 a = 0
#7 a = 1
```

To obtain a more precise result, the **$realtime** system function may be used to replace the first argument, **$time**, of the **$monitor** system task in the above module. The simulation results would then be as follows:

```
#0    a = x
#3.6  a = 0
#7.2  a = 1
```

1.4.2.4 Simulation Control System Tasks There are two simulation control system tasks: **$finish** and **$stop**. The **$stop** system task suspends the simulation whereas the **$finish** system task terminates the simulation.

1.4.3 A Tutorial Example

In this subsection, we use a simple synthesizable example to demonstrate how to verify a design through a test bench.

■ **Example 1.11: A 4-bit adder module.**

In this example, four full-adder instances are instantiated to construct a 4-bit adder. The full-adder module is described in the preceding section.

```
// a 4-bit adder at the gate level
module four_bit_adder(
        input  [3:0] x, y,
        input  c_in,
        output [3:0] sum,
        output c_out);
wire    c1, c2, c3; // intermediate carries

// -- four_bit adder body-- //
// instantiate the full adder
    full_adder fa_1 (x[0], y[0], c_in,  c1, sum[0]);
    full_adder fa_2 (x[1], y[1], c1,    c2, sum[1]);
    full_adder fa_3 (x[2], y[2], c2,    c3, sum[2]);
    full_adder fa_4 (x[3], y[3], c3, c_out, sum[3]);
```

Section 1.4 Simulation

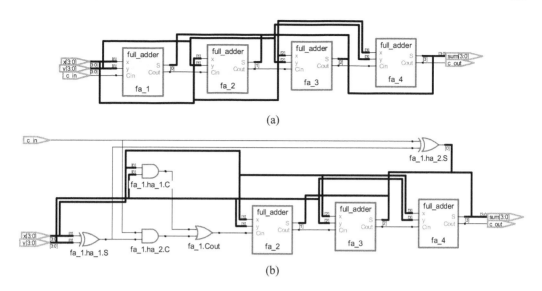

Figure 1.11: (a) The synthesized result of the 4-bit adder module and (b) the result after dissolving the first full_adder module.

endmodule

It is found much useful to verify and/or debug a design with the aid of the synthesized results of the design from synthesis tools because we may discern the differences between the results from our "mind" and those from the synthesis tools. In addition, we might also be able to control the synthesized structure from the synthesis tools by properly structuring the source description of the design. The synthesized results of the preceding example from a synthesis tool are shown in Figure 1.11. Figure 1.11(a) shows the original synthesized result and Figure 1.11(b) is the result after dissolving the first full-adder module.

■ Example 1.12: A test bench for the 4-bit adder module.

In general, a test bench consists of three major parts, as depicted in Figure 1.10, an instance of the UUT, the stimulus generation block, and the response monitoring and checking block. The result checking may be done by comparing the text outputs with the expected results or viewing the waveform generated by a waveform viewer through reinterpreting the text outputs.

```
`timescale 1 ns/100 ps
module four_bit_adder_tb;
//internal signals declarations
reg [3:0] x;
reg [3:0] y;
reg c_in;
wire [3:0] sum;
wire c_out;
// Unit Under Test instance and port map
   four_bit_adder UUT (.x(x), .y(y), .c_in(c_in),
                      .sum(sum), .c_out(c_out));
reg [7:0] i;
initial begin // use in post-map and post-par simulation
// $sdf_annotate("four_bit_adder_map.sdf",four_bit_adder);
```

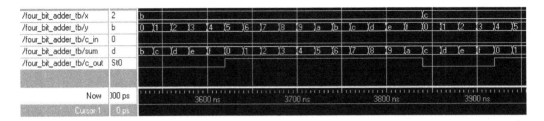

Figure 1.12: The waveforms of the simulation results.

```
// $sdf_annotate("four_bit_adder_timesim.sdf", four_bit_adder);
end
initial      // stimulus generation block
   for (i = 0; i <= 255; i = i + 1) begin
      x[3:0] = i[7:4]; y[3:0] = i[3:0]; c_in =1'b0;
      #20 ;
   end
initial #6000 $finish;
initial      // response monitoring block
   $monitor($realtime,"ns %h %h %h %h", x, y, c_in, {c_out, sum});
endmodule
```

The simulation results are as follows:

0ns 0 0 0 00	# 280ns 0 e 0 0e
# 20ns 0 1 0 01	# 300ns 0 f 0 0f
# 40ns 0 2 0 02	# 320ns 1 0 0 01
# 60ns 0 3 0 03	# 340ns 1 1 0 02
# 80ns 0 4 0 04	# 360ns 1 2 0 03
# 100ns 0 5 0 05	# 380ns 1 3 0 04
# 120ns 0 6 0 06	# 400ns 1 4 0 05
# 140ns 0 7 0 07	# 420ns 1 5 0 06
# 160ns 0 8 0 08	# 440ns 1 6 0 07
# 180ns 0 9 0 09	# 460ns 1 7 0 08
# 200ns 0 a 0 0a	# 480ns 1 8 0 09
# 220ns 0 b 0 0b	# 500ns 1 9 0 0a
# 240ns 0 c 0 0c	# 520ns 1 a 0 0b
# 260ns 0 d 0 0d	# 540ns 1 b 0 0c

The above results can also be viewed through a waveform viewer as shown in Figure 1.12.

It is worth noting that in this tutorial example, we are only concerned with the functional verification of the UUT. In practical applications, it is often desirable to consider the timing verification of the UUT too because the correctness of a hardware module must be assured in both functionality and timing. We will deal with the verification issues later in a dedicated chapter.

■ Review Questions

1-17 What are the major components of a test bench?

1-18 Why is the **'timescale** compiler directive required during simulation?

1-19 How would you monitor the responses of a design during simulation?

1-20 Describe the differences between **$display** and **$monitor** system tasks.

Section 1.4 *Simulation*

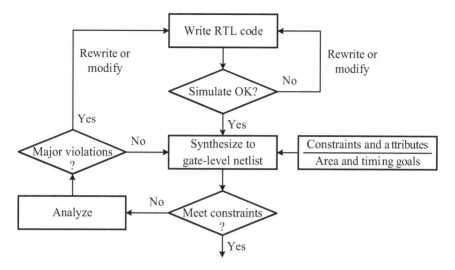

Figure 1.13: A practical point of view of describing and verifying a digital design.

1-21 Describe the functions of the **$time** and **$realtime** system functions.
1-22 Describe the functions of the **$stop** and **$finish** system tasks.

1.4.4 Concluding Remarks

Once we have learned how to describe a digital design and verify the design with Verilog HDL, we give in this section a practical viewpoint of describing and verifying a digital design, as illustrated in Figure 1.13. To get more insight into this, in what follows we first consider the quality of source codes and then give some big-picture guidelines for writing a Verilog HDL module.

1.4.4.1 Quality of Source Codes The most important thing in writing a Verilog HDL module is to guarantee the *quality of source codes*. To achieve this, we should always bear in mind the following key points: First, we could not solely rely on the synthesis tool to fix a poorly coded design. Second, we should always remember that coding style always determines synthesis results. Two functionally equivalent codes with different coding styles may result in different synthesized results. Third, we should try our best to understand the hardware structure being modeled to give the synthesis tool the best possible starting point.

1.4.4.2 Big-Picture Guidelines Generally speaking, the big-picture guidelines in writing a synthesizable RTL code are that we should always *think hardware, think RTL,* and *think synchronous.*

- *Think hardware*: As describing a hardware module in Verilog HDL, we need to identify the hardware topology implied by the code to describe the hardware module. In addition, we should realize that describing a hardware module is not exactly the same as writing a software program since there exist no explicit delays and no file I/O operations.
- *Think RTL*: As writing a synthesizable RTL code, we should be aware of the register architecture and the circuit topology of the design as well as the functionality between registers in the design to be described. Moreover, it should be noted that synthesis tools merely optimize the logic between registers; they do not optimize the register placement.
- *Think synchronous*: We should try to use synchronous designs to their maximum extent since synchronous designs run smoothly through various stages of synthesis, simulation,

verification, and place and route. As asynchronous logic is indeed needed, it is better to isolate the asynchronous logic from synchronous one and to construct it into a separately compiled block because asynchronous logic may need to be instantiated manually and verified by extensive simulation.

The reader is referred to in Section 12.6 for more details about coding guidelines.

1.5 Summary

Using HDLs to describe a digital system has become an essential process in modern electronic engineering. The major feature inherent in HDLs is that they have the capability of modeling a digital system at a variety of levels of abstraction, ranging from the algorithmic level to the gate level, and even to the switch level. Describing a design with HDLs is analogous to computer programming in the sense that they both describe their designs in text.

The basic program structure in Verilog HDL is called a module. A Verilog HDL module is much like a high-level language program except to describe a hardware module rather than a software function. The major differences between a hardware module and a software program are the timing and concurrency features inherently associated with the hardware module.

A Verilog HDL module consists of two major parts: a body and an interface. The body of a module can be modeled in structural style, dataflow style, behavioral or algorithmic style, or a mix of them. In structural style, a design is described as a set of interconnected components. The components can be modules, UDPs, gate primitives, and/or switch primitives. The structural style is also a mechanism used to construct a hierarchical design for a large digital system. In dataflow style, the module is described by specifying the data flow (i.e., data dependence) between registers and how the data are processed. A module is specified as a sequence of **assign** continuous assignments. In behavioral or algorithmic style, the design is described in terms of the desired algorithm without concerning the details of its hardware implementation. A design can be described in any high-level programming language. In mixed style, the design is described in terms of the combination of above three styles. The mixed style is most commonly used in specifying a large design. In industry, the term *register-transfer level* (RTL) *code* is often used to mean that an RTL module (i.e., constructed with RTL components) is modeled in behavioral, dataflow, structural, or mixed style on condition that the resulting description must be acceptable by logic synthesis tools.

For a design to be useful, it must be verified to ensure that both the functionality and timing of the design are correct. Verilog HDL not only provides capabilities to model a design, but also equips facilities to generate and control stimuli, store responses, and check the results. The basic component of simulation-based verification is a test bench. Generally, a test bench comprises three basic parts: an instance of the UUT, the stimulus generation block, and the response monitoring and checking block. Both the stimulus generation and the response monitoring and checking blocks are usually modeled in behavioral style. The result checking may be done by comparing the text outputs with the expected results or viewing the waveform generated by a waveform viewer through reinterpreting the text outputs.

References

1. IEEE 1364-2001 Standard, *IEEE Standard Verilog Hardware Description Language*, 2001.
2. IEEE 1364-2005 Standard, *IEEE Standard for Verilog Hardware Description Language*, 2006.

3. IEEE 1800-2005 Standard, *IEEE Standard for System Verilog—Unified Hardware Design, Specification, and Verification Language*, 2005.

4. IEEE 1800-2009 Standard, *IEEE Standard for System Verilog—Unified Hardware Design, Specification, and Verification Language*, 2009. (Revision of IEEE Std 1800–2005)

5. M. B. Lin, *Digital System Design: Principles, Practices, and Applications*, 4th ed., Chuan Hwa Book Ltd. (Taipei, Taiwan), 2010.

6. S. Palnitkar, *Verilog HDL: A Guide to Digital Design and Synthesis*, 2nd ed., SunSoft Press, 2003.

Problems

1-1 **(Problems 1-1 to 1-5 are a problem set.)** A half subtractor (also subtracter) is a device that accepts two inputs, x and y, and produces two outputs, b and d. The full subtractor is a device that accepts three inputs, x, y, and b_{in}, and produces two outputs, b_{out} and d, according to the truth table shown in Table 1.2.

Table 1.2: The truth table of a full subtractor.

x	y	b_{in}	b_{out}	d
0	0	0	0	0
0	0	1	1	1
0	1	0	1	1
0	1	1	1	0
1	0	0	0	1
1	0	1	0	0
1	1	0	0	0
1	1	1	1	1

(a) Derive the minimal expressions of both b_{out} and d of the full subtractor.

(b) Draw the logic diagram of switching expressions, b_{out} and d, in terms of two half subtractors and one two-input OR gate.

1-2 (Structural style)

(a) Write a Verilog HDL module to describe the full subtractor obtained from Problem 1-1 in structural style using the instances of gate primitives.

(b) Write a test bench to verify if the module behaves as a full subtractor.

1-3 (Dataflow style)

(a) Write a Verilog HDL module to describe the full subtractor obtained from Problem 1-1 in dataflow style using bit-wise operators.

(b) Write a test bench to verify if the module behaves as a full subtractor.

1-4 (Behavioral style)

(a) Write a Verilog HDL module to describe the full subtractor obtained from Problem 1-1 in behavioral style.

(b) Write a test bench to verify if the module behaves as a full subtractor.

1-5 (Mixed-style modeling)

(a) Model the first half subtractor obtained from Problem 1-1 in structural style and the second one in dataflow style.

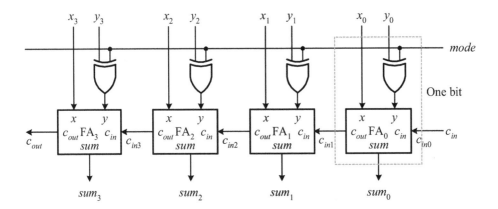

Figure 1.14: A 4-bit 2's-complement adder.

  (b) Model the switching expression b_{out} of the full subtractor obtained from Problem 1-1 in behavioral style.
  (c) Write a test bench to verify if the module behaves as a full subtractor.

1-6 **(Problems 1-6 to 1-9 constitute a problem set.)** (Structural style) Referring to Figure 1.14, consider only 1-bit 2's-complement adder.
  (a) Write a Verilog HDL module to describe the 1-bit 2's-complement adder in structural style using the instances of gate primitives.
  (b) Write a test bench to verify if the module behaves as a 1-bit 2's-complement adder.

1-7 (Dataflow style)
  (a) Write a Verilog HDL module to describe the 1-bit 2's-complement adder shown in Figure 1.14 in dataflow style.
  (b) Write a test bench to verify if the module behaves as a 1-bit 2's-complement adder.

1-8 (Behavioral style)
  (a) Write a Verilog HDL module to describe the 1-bit 2's-complement adder shown in Figure 1.14 in behavioral style.
  (b) Write a test bench to verify if the module behaves as a 1-bit 2's-complement adder.

1-9 (Mixed-style modeling)
  (a) Model the first bit of the 4-bit 2's-complement adder shown in Figure 1.14 in structural style, the second bit in dataflow style, and the other two bits in behavioral style.
  (b) Write a test bench to verify if the resulting module behaves as a 4-bit 2's-complement adder.

2

Structural Modeling

IN structural style, a module is described as a set of interconnected components. The components can be modules, user-defined primitives (UDPs), gate primitives, and switch primitives.

1. *Modules*: A module can instantiate any number of other modules. Whether a module is synthesizable depends on the contents of the module. A gate-level module is usually synthesizable.
2. *Gate primitives*: In Verilog HDL, there are 12 gate primitives. These primitives may be used in any module at any time whenever they are appropriate. Gate primitives are synthesizable.
3. *Switch primitives*: In Verilog HDL, there are 16 switch primitives. These primitives may be used in any module at any time whenever they are appropriate. Switch primitives are usually used to model a new logic gate at the switch level and are generally not synthesizable.
4. *UDPs*: A UDP is like a module but it cannot instantiate any other UDPs or modules. UDPs can only be instantiated in modules. In addition, UDPs are generally not supported by synthesizers.

In this chapter, we introduce the structural modeling at both gate and switch levels. The UDPs and modules will be separately dealt with in Chapters 5 and 6.

2.1 Gate-Level Structural Modeling

Remember that a design can be described at the gate level only using gate primitives. In this section, we first address gate primitives provided in Verilog HDL in detail and then give some more insight into the use of gate primitives for modeling a realistic logic circuit in structural style.

2.1.1 Gate Primitives

In Verilog HDL, there are 12 predefined gate primitives. These primitives can be classified into two groups: *and/or* and *buf/not*. The and/or group includes **and**, **nand**, **or**, **nor**, **xor**, and **xnor**. The buf/not group contains **buf**, **not**, **bufif0**, **bufif1**, **notif0**, and **notif1**. Two gates from the buf/not group, **buf** and **not**, are also described in this subsection because they are often combined with the gates from the and/or group to form useful logic circuits.

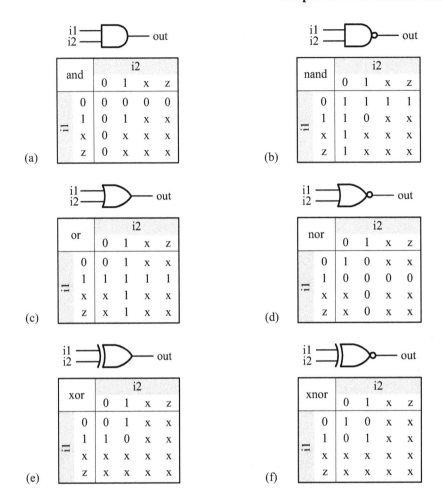

Figure 2.1: The gates of the and/or group: (a) **and** gate; (b) **nand** gate; (c) **or** gate; (d) **nor** gate; (e) **xor** gate; (f) **xnor** gate.

2.1.1.1 The and/or Group All gates of the and/or group have one scalar output and multiple scalar inputs and are used to realize basic logic operations, such as AND, OR, or XOR. In addition, their complementary parts are provided. The symbols and truth tables of these six gates are depicted in Figure 2.1.

To instantiate a gate from the and/or group, the following syntax is used.

```
gate_type [instance_name] (out, in{, in})
        {, [instance_name] (out, in{, in})};
```

where `gate_type` can be any of the set {**and**, **nand**, **or**, **nor**, **xor**, **xnor**}. The first port is always the output and the other ports are inputs. The item inside square brackets ([]) is optional and within curly brackets ({}) can repeat zero or more times. Thus, `instance_name` is optional and {, in} can be repeated zero or more times. Multiple instances of the same gate can be instantiated in one statement as follows:

```
gate_type [instance_name1] (out, in{, in}),
        [instance_name2] (out, in{, in}),
        ...
        [instance_namen] (out, in{, in});
```

Section 2.1 *Gate-Level Structural Modeling*

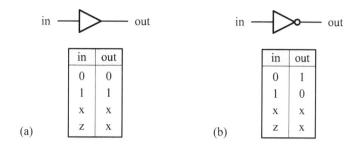

Figure 2.2: The (a) buffer (**buf**) and (b) **not** gates.

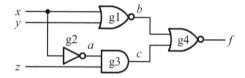

Figure 2.3: A simple application of basic gates.

2.1.1.2 The buf and not Gates Both **buf** and **not** gates have one scalar input and one or multiple scalar outputs. They are used to realize the NOT operation or to buffer the output of an AND/OR gate. The symbols and truth tables of these gates are depicted in Figure 2.2.

To instantiate a **buf** or **not** gate, the following syntax is used.

```
buf_not [instance_name] (out{, out}, in)
    {, [instance_name] (out{, out}, in)};
```

where instance_name is optional. The last port is always the input and the other ports are outputs. Multiple instances of the same gate can be instantiated in one statement as follows:

```
buf_not [instance_name1] (out{, out}, in),
        [instance_name2] (out{, out}, in),
        ...
        [instance_namen] (out{, out}, in);
```

The following example is a simple application of basic gates to model the logic circuit shown in Figure 2.3. In this example, four basic gates are instantiated and connected according to the logic circuit depicted in the figure exactly. This is an example to demonstrate how a logic circuit can be described in text.

■ **Example 2.1: A simple application of basic gates.**

In order to describe the logic circuit shown in Figure 2.3, we need to declare three internal nets: a, b, and c, which are used to connect the output port of one gate to an input port of another. By carefully labeling the port list of each instance of gate primitives, the resulting module is obtained as follows. It is worth noting that instance names are often used to make a Verilog HDL module more readable.

```
module basic_gates (input x, y, z, output f);
wire a, b, c; // internal nets
// model the logic circuit with basic gates
    nor g1 (b, x, y);
    not g2 (a, x);
```

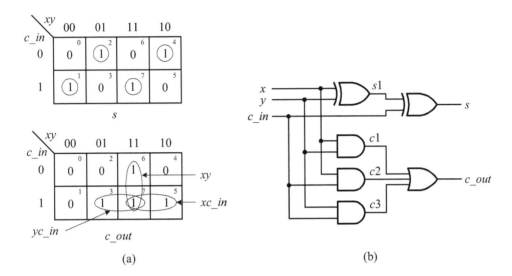

Figure 2.4: A 1-bit (full) adder: (a) Karnaugh maps for s and c_out and (b) logic diagram.

```
    and g3 (c, a, z);
    nor g4 (f, b, c);
endmodule
```

Gate-level structural modeling is usually used to model a logic circuit with gate primitives. Figure 2.4 shows a full adder constructed from basic gates along with the Karnaugh maps used to reduce both sum (s) and carry-out (c_out) functions. The full adder can be modeled in a straightforward manner using basic gates as in the following example.

■ Example 2.2: An example of 1-bit full adders.

From the Karnaugh maps shown in Figure 2.4(a), we obtain the logic diagram shown in Figure 2.4(b). Hence, by instantiating basic gate primitives and connecting them together according to the logic circuit, we have the following module.

```
module full_adder_structural(
       input  x, y, c_in,
       output s, c_out);
wire s1, c1, c2, c3;
   // model the 1-bit full adder in structural style
   xor xor_s1(s1, x, y);      // compute sum
   xor xor_s2(s, s1, c_in);
   and and_c1(c1, x, y);      // compute carry-out
   and and_c2(c2, x, c_in);
   and and_c3(c3, y, c_in);
    or or_cout(c_out, c1, c2, c3);
endmodule
```

A 4-to-1 multiplexer is a device used to route the data from one of its four inputs to its output node. The logic symbol and function table of a 4-to-1 multiplexer are shown in Figures 2.5(a) and (b), respectively. From the function table, we can obtain the following switch-

Section 2.1 Gate-Level Structural Modeling

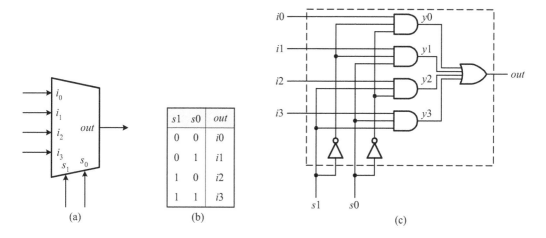

Figure 2.5: A 4-to-1 multiplexer: (a) logic symbol; (b) function table; (c) logic diagram.

ing expression:

$$out = i0 \cdot \overline{s1} \cdot \overline{s0} + i1 \cdot \overline{s1} \cdot s0 + i2 \cdot s1 \cdot \overline{s0} + i3 \cdot s1 \cdot s0$$

The resulting logic circuit is depicted in Figure 2.5(c). To illustrate how to model this 4-to-1 multiplexer structurally using basic gates, let us consider the following example.

■ Example 2.3: A 4-to-1 multiplexer.

To describe the 4-to-1 multiplexer depicted in Figure 2.5(c), two **not** gates and four **and** gates are required to uniquely select one out of the four inputs. In addition, an **or** gate is needed to join together the four possible chosen routes. By properly labeling each port in the port lists of gate instances, the resulting module immediately follows.

```
module mux_4to1_structural (
      input   i0, i1, i2, i3, s1, s0,
      output  out);
wire s1n, s0n;    // internal wires
wire y0, y1, y2, y3;
  // gate instantiation
  not (s1n, s1);  // create s1n and s0n signals
  not (s0n, s0);
  and (y0, i0, s1n, s0n);  // 3-input and gates
  and (y1, i1, s1n, s0);
  and (y2, i2, s1, s0n);
  and (y3, i3, s1, s0);
   or (out, y0, y1, y2, y3); // 4-input or gate
endmodule
```

The final example of this subsection is a 9-bit parity generator, which can be described by the following switching expressions:

$$ep = x[0] \oplus x[1] \oplus x[2] \oplus x[3] \oplus x[4] \oplus x[5] \oplus x[6] \oplus x[7] \oplus x[8]$$
$$op = x[0] \oplus x[1] \oplus x[2] \oplus x[3] \oplus x[4] \oplus x[5] \oplus x[6] \oplus x[7] \odot x[8]$$

These two expressions can be realized using XOR gates according to the sequence of the inputs $x[0], x[1], \cdots, x[8]$, i.e., $(\cdots((x[0] \oplus x[1]) \oplus x[2]) \cdots x[8])$. However, this kind of linear

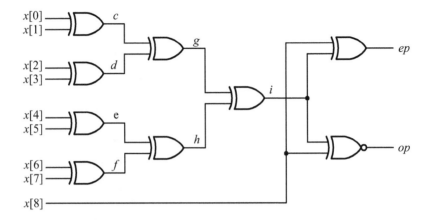

Figure 2.6: A 9-bit parity generator.

structure takes $(n-1)t_{g,xor}$, where $t_{g,xor}$ is the propagation delay of one XOR gate. A much faster realization at no extra cost is to rearrange the XOR gates as a binary tree, as given in the following two expressions:

$$ep = \{[(x[0] \oplus x[1]) \oplus (x[2] \oplus x[3])] \oplus [(x[4] \oplus x[5]) \oplus (x[6] \oplus x[7])]\} \oplus x[8]$$
$$op = \{[(x[0] \oplus x[1]) \oplus (x[2] \oplus x[3])] \oplus [(x[4] \oplus x[5]) \oplus (x[6] \oplus x[7])]\} \odot x[8]$$

With the binary-tree structure, the total propagation delay is reduced from $(n-1)t_{g,xor}$ to $\lceil \log_2 n \rceil t_{g,xor}$. Thus, the total propagation delay of the 9-bit parity generator is only $4t_{g,xor}$ when the binary-tree structure is used and is $8t_{g,xor}$ when the linear structure is used.

The logic circuit corresponding to both expressions ep and op is depicted in Figure 2.6. This 9-bit parity generator can be modeled in structural style using basic gates as described in the following example.

■ Example 2.4: A 9-bit parity generator.

To model the logic circuit shown in Figure 2.6, eight **xor** gates and one **xnor** gate are required. For the purpose of easy reading, we describe the logic circuit in a layered manner, from the primary inputs toward the primary output, and label each layer clearly in the instance name. The resulting module is given as follows:

```
module parity_gen_9b_structural(
       input  [8:0] x,
       output ep, op);
wire c, d, e, f, g, h, i;
   xor xor_11(c, x[0], x[1]);   // first level
   xor xor_12(d, x[2], x[3]);
   xor xor_13(e, x[4], x[5]);
   xor xor_14(f, x[6], x[7]);
   xor xor_21(g, c, d);         // second level
   xor xor_22(h, e, f);
   xor xor_31(i, g, h);         // third level
   xor xor_ep(ep, i, x[8]);     // fourth level
   xnor xnor_op(op, i, x[8]);
endmodule
```

Section 2.1 *Gate-Level Structural Modeling*

Although gate primitives can be used to describe any desired logic function, it is often a tedious task in describing a complicated logic function, such as the above 9-bit parity generator. To make such work easier and more readable, the dataflow style or behavioral style may be used. In these two modeling styles, an abundance of operators of Verilog HDL may be used to replace the tedious work caused by instantiating and connecting gate primitives. See Chapter 3 for more details.

2.1.1.3 Array Instantiation It is possible to instantiate an array of gates from the and/or group as well as **buf** and **not** gates by using the following syntax

```
gate_type instance_name[msb:lsb] (out, in{, in})
    {, instance_name[msb:lsb] (out, in{, in})};
buf_not  instance_name[msb:lsb] (out{, out}, in)
    {, instance_name[msb:lsb] (out{, out}, in)};
```

where `instance_name` is necessary and `[msb:lsb]` denotes the range of the gate array. The most significant bit (`msb`) is always the leftmost bit and the least significant bit (`lsb`) is always the rightmost bit regardless of which sequence, `[high:low]` or `[low:high]`, is used to specify the range. The array instantiation of a gate primitive might be a synthesizer dependent! So it is better to check this feature from the synthesizer that you are using before using array instantiation.

■ Example 2.5: An example of array instantiation.

The following statements demonstrate how to instantiate an array of **nand** gates. Here, we also show that the effect of array instantiation is actually a concise representation of multiple instantiations using the same vector arguments.

```
wire [3:0] out, in1, in2;
  // an array instantiation of nand gate
  nand n_gate[3:0] (out, in1, in2);
  // this is equivalent to the following statements
  nand n_gate0 (out[0], in1[0], in2[0]);
  nand n_gate1 (out[1], in1[1], in2[1]);
  nand n_gate2 (out[2], in1[2], in2[2]);
  nand n_gate3 (out[3], in1[3], in2[3]);
```

2.1.1.4 Concepts of Controlled Gates In designing digital systems, it often proves quite useful to consider the AND/OR/XOR gates and their complementary parts as controlled gates. As shown in Figure 2.7, these basic gates can be classified into three groups: AND/NAND, OR/NOR, and XOR/XNOR. In what follows, we deal with each of these in more detail.

AND/NAND controlled gates. The AND/NAND controlled gates are depicted in Figure 2.7(a). From their truth tables shown in Figures 2.1(a) and (b), we know that for these two gates, as one of its inputs is fixed to 0, the output is a constant 0 if the gate is AND and 1 if the gate is NAND. To pass the input data through the gates to their outputs, all inputs except the one of interest must be set to 1. The data passing to the output will be in true form if the gate is AND and in complemented form if the gate is NAND.

OR/NOR controlled gates. The OR/NOR controlled gates are shown in Figure 2.7(b). From their truth tables shown in Figures 2.1(c) and (d), we know that for these two gates, as one of its inputs is fixed to 1, the output is a constant 1 if the gate is OR and 0 if the gate is NOR. To pass the input data through the gates to their outputs, all inputs except the one of interest must be set to 0. The data passing to the output will be in true form if the gate is OR and in complemented form if the gate is NOR.

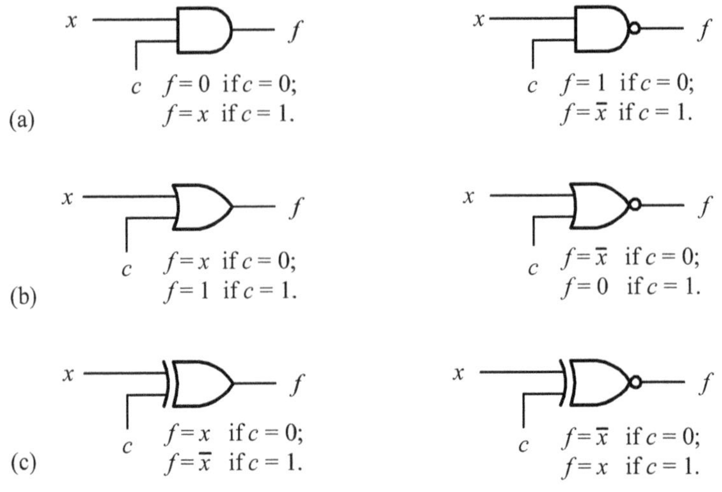

(a)
$f = 0$ if $c = 0$;
$f = x$ if $c = 1$.

$f = 1$ if $c = 0$;
$f = \bar{x}$ if $c = 1$.

(b)
$f = x$ if $c = 0$;
$f = 1$ if $c = 1$.

$f = \bar{x}$ if $c = 0$;
$f = 0$ if $c = 1$.

(c)
$f = x$ if $c = 0$;
$f = \bar{x}$ if $c = 1$.

$f = \bar{x}$ if $c = 0$;
$f = x$ if $c = 1$.

Figure 2.7: Regarding (a) AND/NAND, (b) OR/NOR, and (c) XOR/XNOR gates as controlled gates.

XOR/XNOR controlled gates. The XOR/XNOR controlled gates are shown in Figure 2.7(c). From their truth tables shown in Figures 2.1(e) and (f), we know that for these two gates, the output is in true or complemented form of its input, depending on the status of the other input and the gate type. For a two-input XOR gate, the output is the true form of one input while the other input is set to 0 and is the complemented form, otherwise. For the case of the XNOR gate, the polarity of output is reversed.

■ **Review Questions**

2-1 Describe the meaning of the structural style.
2-2 Define the gate-level structural modeling.
2-3 What types of gates does the and/or group contain?
2-4 What types of gates does the buf/not group contain?
2-5 What is the meaning of array instantiation?

2.1.2 Tristate Buffers

When the output of a logic circuit can also be stayed at an extra state, a *high impedance*, in addition to its two normal states, 0 and 1, the logic circuit is said to have a *tristate* (three-state) output. Each tristate (noninverting) buffer/inverter (i.e., inverting buffer) has three ports: a scalar output, a scalar input, and an enable input. The output is a logic 0 or a logic 1 if the enable input is activated and a high-impedance value if the enable input is disabled.

Tristate buffers and inverters are usually used as controlled buffers/inverters to control the data flow or buffer the data on the data path. The symbols and truth tables of these tristate buffers/inverters are summarized in Figures 2.8(a) to (d), respectively. The symbol L represents 0 or z, and symbol H denotes 1 or z. It is instructive to note that in FPGAs, tristate buffers/inverters are only available in the input/output (I/O) blocks; thereby, they are not allowed to use in the core logic blocks. See Section 10.3.2 for more details.

Since the tristate function may be associated with a buffer or an inverter and the enable input can be activated at a logic 0 or 1, four types of tristate buffers can be obtained. These tristate buffers work as follows:

Section 2.1 Gate-Level Structural Modeling

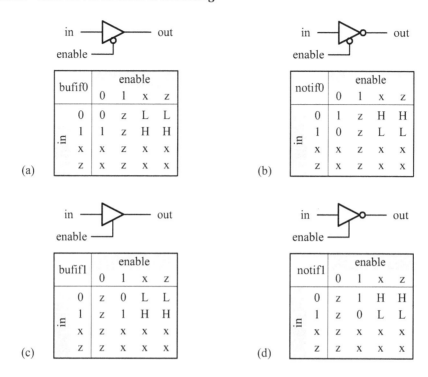

Figure 2.8: Tristate buffers and inverters: (a) **bufif0**; (b) **notif0**; (c) **bufif1**; (d) **notif1**.

- The *active-low buffer* (**bufif0**) acts as a normal buffer if its enable input is at a logic 0 and maintains its output at a high impedance if its enable input is at a logic 1. The other combinations of both the input and the enable input are shown in Figure 2.8(a).
- The *active-low inverter* (**notif0**) acts as a normal inverter if its enable input is at a logic 0 and maintains its output at a high impedance if its enable input is at a logic 1. The other combinations of both the input and the enable input are shown in Figure 2.8(b).
- The *active-high buffer* (**bufif1**) acts as a normal buffer if its enable input is at a logic 1 and maintains its output at a high impedance if its enable input is at a logic 0. The other combinations of both the input and the enable input are shown in Figure 2.8(c).
- The *active-high inverter* (**notif1**) acts as a normal inverter if its enable input is at a logic 1 and maintains its output at a high impedance if its enable input is at a logic 0. The other combinations of both the input and the enable input are shown in Figure 2.8(d).

It is worth noting that for the active-low buffer/inverter the keyword is ended with "**if0**" (i.e., if 0) and for the active-high buffer/inverter the keyword is ended with "**if1**" (i.e., if 1).

To instantiate a buffer from the tristate buf/not group, the following syntax is used.

```
tri_buf [instance_name] (out, in, enable)
    {, [instance_name] (out, in, enable)};
```

where `tri_buf` may be any from the set {**bufif0, bufif1, notif0, notif1**} and `instance_name` is optional. The first port is always the output port, the second port is the input port, and the last port is the enable input. To instantiate multiple instances of the same tristate buffer in one statement may use the following syntax

```
tri_buf [instance_name1] (out, in, enable),
        [instance_name2] (out, in, enable),
        ...
```

Table 2.1: The truth table of **wire** and **tri** nets.

wire/tri	0	1	x	z
0	0	x	x	0
1	x	1	x	1
x	x	x	x	x
z	0	1	x	z

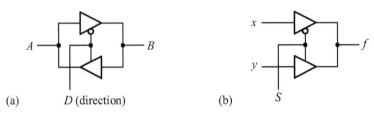

(a)　　　　D (direction)　　　(b)　　S

Figure 2.9: Two simple applications of tristate buffers: (a) a bidirectional buffer and (b) a 2-to-1 multiplexer.

```
[instance_namen] (out, in, enable);
```

Like gate primitives, an array of tristate buffers may also be instantiated by using the following syntax

```
tri_buf instance_name[msb:lsb] (out, in, enable)
    {, instance_name[msb:lsb] (out, in, enable)};
```

where `instance_name` is necessary in this case and `[msb:lsb]` denotes the range of the array.

2.1.2.1 Net Types The nets represent physical connections between structural entities, such as gates and modules. Note that the "net" is not a reserved word but indeed represents a class of data types, as listed in Table 1.1. A net does not store a value (except the **trireg** net). It can be referenced anywhere in a module and must be driven by a primitive (including gates, switches, and UDPs), an **assign** continuous assignment, a **force/release** statement, or a module port. If no driver is connected to a net, the net value is a high impedance (z) unless the net is a **trireg** net, which holds the previously driven value.

A form of net declarations is as follows

```
net_type [signed] [[msb:lsb]] net_id{, net_id};
```

where `net_type` is any type except the **trireg** net in the column of nets in Table 1.1. A net is defaulted to be unsigned but may be declared to represent a signed value by using the keyword **signed** following `net_type`. The range `[msb:lsb]` is an optional part and used to specify the range of the net. Both `msb` and `lsb` are constant expressions. A net is defaulted to a single bit known as a *scalar* if no range is specified and is called a *vector* if a range of more than one bit is specified. The vector is useful in describing a bundle of signals as a unit.

2.1.2.2 wire and tri Nets Both **wire** and **tri** nets connect elements and have the same syntax and function except that a **wire** net is driven by a single driver, such as a gate or an **assign** continuous assignment, but the **tri** net may be driven by multiple drivers. The effective value of a **wire** or **tri** net is determined by the truth table shown in Table 2.1.

As illustrated in Figure 2.9, each simple application of tristate buffers is resulted by properly combining a **bufif0** gate with a **bufif1** gate. Figure 2.9(a) shows a 1-bit bidirectional buffer and Figure 2.9(b) is a 2-to-1 multiplexer. The following example demonstrates how to model the 2-to-1 multiplexer depicted in Figure 2.9(b).

Section 2.1 Gate-Level Structural Modeling

■ Example 2.6: A 2-to-1 multiplexer based on tristate buffers.

As shown in Figure 2.9(b), the output f is needed to declare as a **tri** net because it is driven by two tristate buffers. By combining two tristate buffers with opposite polarities of enable inputs, **bufif0** and **bufif1**, the 2-to-1 multiplexer is obtained. The tristate buffer **bufif0** is enabled if s is 0 whereas the tristate buffer **bufif1** is enabled if s is 1.

```
// a data selector --- i.e., a 2-to-1 multiplexer
module mux_2to1_tristate(
      input x, y, s,
      output tri f);   // f needs to be tristate

   // the data selector body
   bufif0 b1 (f, x, s);   // enabled if s = 0
   bufif1 b2 (f, y, s);   // enabled if s = 1
endmodule
```

The similar approach can be applied to construct the bidirectional buffer shown in Figure 2.9(a). Because of its intuitive simplicity, it is left for the reader as an exercise.

■ Review Questions

2-6 Explain the operation of active-low inverters.
2-7 Explain the operation of active-high buffers.
2-8 How would you describe a bundle of signals in Verilog HDL?
2-9 Define the following two terms: scalar and vector.
2-10 Explain the meaning of the **tri** net type.
2-11 What is the difference between **wire** and **tri** net types?

2.1.3 Wired Logic

In some applications, a logic circuit with an open-collector (for TTL) or open-drain (for CMOS) output stage is preferred because fewer logic gates or less area is required. As an illustration, consider the wired NAND-AND logic circuit depicted in Figure 2.10. The logic function is as follows

$$f(w,x,y,z) = \overline{w \cdot x} \cdot \overline{y \cdot z}$$

which requires two NAND gates and one AND gate. However, by using wired logic, it only needs two wired NAND gates and hence greatly reduces the hardware cost.

To model wired logic circuits, some extra net types are needed. These net types include **triand/wand**, **trior/wor**, and **tri0/tri1**. In what follows, we describe these net types in more detail.

2.1.3.1 triand/wand and trior/wor Nets In Verilog HDL, four net types, including **triand**, **wand**, **trior**, and **wor**, can be used to model a variety of wired logic circuits. The truth tables of these net types are shown in Table 2.2. The syntax and functionality of both **triand** and **wand** net types are identical and the same situation applies for the **trior** and **wor** net types.

Because a wired net may be driven by multiple drivers, a confliction resolution rule is required and set as follows. The value of a **triand** net or a **wand** net is 0 when any of drivers is 0, and the value of a **trior** net or a **wor** net is 1 when any of the drivers is 1. The other situations are resolved according to Table 2.2.

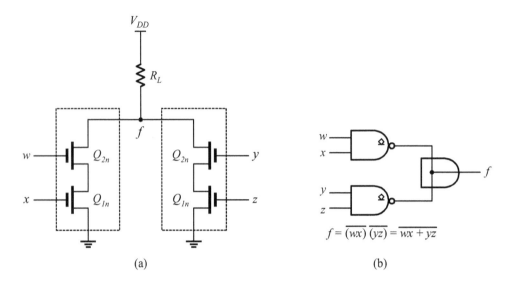

Figure 2.10: A wired NAND-AND logic circuit: (a) logic circuit and (b) logic symbol.

Table 2.2: The truth tables of **triand/wand** and **trior/wor** nets.

triand/wand	0	1	x	z	trior/wor	0	1	x	z
0	0	0	0	0	0	0	1	x	0
1	0	1	x	1	1	1	1	1	1
x	0	x	x	x	x	x	1	x	x
z	0	1	x	z	z	0	1	x	z

With the aid of the **wand** net type, Figure 2.10(b) can be modeled as in the following example.

■ Example 2.7: A wired nand-and logic circuit.

To model the wired NAND-AND logic circuit, the output port is needed to declare as a **wand** net so that it may be driven by two **nand** gates. The resulting module is as follows:

```
module open_drain (
      input  w, x, y, z,
      output wand f ); // internal wired-AND net

// the wired-AND logic gate
   nand n1 (f, w, x);
   nand n2 (f, y, z);
endmodule
```

2.1.3.2 tri0 and tri1 Nets The **tri0** and **tri1** net types are also used to model nets with more than one driver. The **tri0** and **tri1** net types model nets with resistive pulldown and pullup devices on them. In other words, a **tri0** net is equivalent to a **wire** net with a continuous 0 value of pull strength driving it; a **tri1** net is equivalent to a **wire** net with a continuous 1 value of pull strength driving it. The salient features of these nets are as follows. The value of a **tri0** net is 0 when no driver is driving it and the value of a **tri1** net is 1 when no driver

Section 2.2 *Gate Delays*

Table 2.3: The truth tables of **tri0** and **tri1** nets.

tri0	0	1	x	z	tri1	0	1	x	z
0	0	x	x	0	0	0	x	x	0
1	x	1	x	1	1	x	1	x	1
x	x	x	x	x	x	x	x	x	x
z	0	1	x	0	z	0	1	x	1

is driving it. The effective values of the **tri0** and **tri1** nets are determined by the truth tables shown in Table 2.3, respectively, when more than one driver drives the net.

Sometimes, the **uwire** net type is used to model a net which allows only a single driver. Any connection of a **uwire** net to more than one driver or to a bidirectional terminal of a bidirectional switch will raise an error. Nonetheless, when using this net type, one should consult his synthesis tool first because not every synthesis tool supports this net type.

■ Review Questions

2-12 Explain the meaning of the **wand** net type.
2-13 Explain the meaning of the **wor** net type.
2-14 Explain the meaning of a wired-and logic circuit.
2-15 Explain the meaning of a wired-or logic circuit.
2-16 Explain the meanings of **tri0** and **tri1** net types.

2.2 Gate Delays

Because of the existence of resistance (R) and capacitance (C) in electronic circuits, all logic circuits and wires have definite *propagation delays*. In addition, the time of flight of a signal passing through a wire (i.e., a net), referred to as the *transport delay* of the wire, also exists. To model these two types of delays, *inertial* and *transport* delay models are widely used in HDLs. In this section, we first describe both delay models briefly and then introduce the delay specifications of gate primitives and nets.

2.2.1 Delay Models

In this section, we describe the essential features of the inertial delay and transport delay models in detail and investigate the differences between them.

2.2.1.1 Inertial Delay Model Because of the inherent resistances (R) and capacitances (C) existing in logic circuits, all logic circuits have a certain amount of inertia (analogous to inertia in physics); namely, it takes a finite amount of time and a certain amount of energy for the output of a logic circuit to respond to a change on the input signal. This implies that the signal events that do not persist long enough are filtered out and not propagated to the output of the circuit. In other words, the circuit inertia has the effect of suppressing the input signal pulses whose durations are not longer than the propagation delay of the underlying logic circuit. Such an inertial model is often referred to as the *inertial delay model* and is used to model the propagation delay of logic circuitry in HDLs, including Verilog HDL, SystemVerilog, and VHDL.

As an illustration of the inertial delay model, consider the **not** gate shown in Figure 2.11(a). The input signal pulse occurred between time units 2 and 3 is filtered out because it is shorter than the propagation delay of the **not** gate, which is 3 time units, as illustrated in Figure 2.11(b).

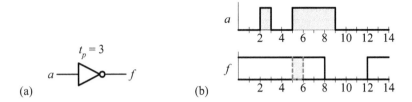

Figure 2.11: An illustration of the effects of the inertial delay model: (a) not gate and (b) timing diagram.

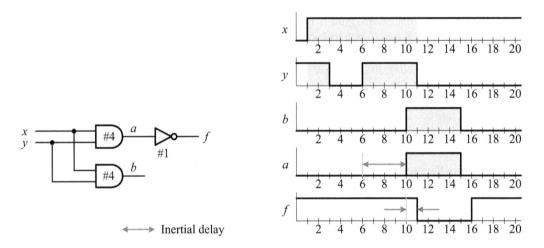

Figure 2.12: An example of the effects of the inertial delay model.

The other pulse occurred between time units 5 and 9 is present at the output of the **not** gate after the propagation delay of 3 time units.

■ Example 2.8: An illustration of inertial delays.

Another more complicated example is shown in Figure 2.12, where two inputs x and y are applied to the inputs of both AND gates. The output f is obtained by passing the output a through an inverter. Assume that the propagation delays of both AND gates are 4 time units and the propagation delay of the inverter is 1 time unit.

From the definition of the inertial delay model, a signal appearing at the input of a gate with a pulse width less than the specified propagation delay will not appear at the output of the gate. Hence, although a pulse with a width of 2 time units (from time units 1 to 3) is resulted from the application of an AND operation to both inputs x and y, it will not appear at the output node a of the AND gate. On the other hand, the pulse with a width of 5 time units (from time units 6 to 11) will propagate to the output node f after 5 time units.

2.2.1.2 Transport Delay Model The inherent feature of a wire is that any signal event appearing at its input will be propagated to the output. This is also the essential feature of the *transport delay model*. The transport delay model is usually used to model net (i.e., wire) delays, that is, the *time of flight* of a signal passing through a wire. The default delay of a net (wire) is zero. For example, as shown in Figure 2.13, both input pulses are present at the output after a transport delay of 3 time units regardless of their pulse widths.

Section 2.2 Gate Delays

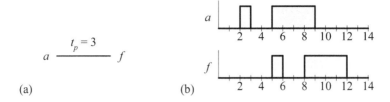

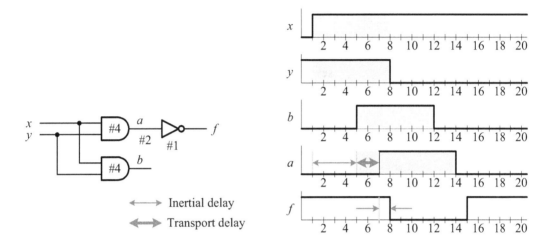

Figure 2.13: An illustration of the effects of the transport delay model: (a) a wire and (b) timing diagram.

Figure 2.14: An example of the effects of both inertial and transport delay models.

A more complicated example showing the effects of both inertial and transport delay models is as follows.

■ **Example 2.9: Effects of both inertial and transport delays.**

As shown in Figure 2.14, the logic circuit of this example is the same as that of Figure 2.12 except that the wire a has a transport delay of 2 time units. According to the transport delay model, any signal events appearing at the input of the wire a will be propagated to its output after the specified delay. Hence, the pulse of width 7 time units (from time units 1 to 8) resulted from the application of an AND operation to both inputs x and y will appear at the output node a after 6 time units, starting from time unit 7, where 6 time units are a combination of the inertial delay (4 time units) of the AND gate and the transport delay of the wire a (2 time units). This signal will appear at the output node f one time unit later due to the propagation delay of the inverter.

■ **Review Questions**

2-17 What are the two delay models used in HDL?
2-18 Define the inertial delay model. When would you use it?
2-19 Define the transport delay model. When would you use it?

2.2.2 Delay Specifications

Because *RC* delays also exist in wires in addition to transport delays and dominate the overall delay of wires, the inertial delay model is defaulted in Verilog HDL to capture both the propagation delay of logic circuits and the net delays. The propagation delay of a gate and the net delay may be specified by a *delay specification*, consisting of zero, one, two, or three *delay specifiers*.[1] In this section, we deal with these two issues in order.

2.2.2.1 Gate-Delay Specifications The propagation delay of a signal from any input to the output of a gate can be specified by a delay specification. The syntax of gate instantiation with the delay specification is as follows

```
gate_type [delay2] [instance_name[range]] (...);
buf_not   [delay2] [instance_name[range]] (...);
tri_buf   [delay3] [instance_name[range]] (...);
```

where gate_type can be any of the set {**and**, **nand**, **or**, **nor**, **xor**, **xnor**}, buf_not means **buf** or **not**, and tri_buf denotes any of the set {**bufif0**, **bufif1**, **notif0**, **notif1**}.

The propagation delay of a gate primitive may be specified in one of the following ways.

1. Specify no delay

    ```
    gate_type [instance_name[range]] (...);
    buf_not   [instance_name[range]] (...);
    tri_buf   [instance_name[range]] (...);
    ```

2. Specify the propagation delay only

    ```
    gate_type #(prop_delay) [instance_name[range]] (...);
    buf_not   #(prop_delay) [instance_name[range]] (...);
    tri_buf   #(prop_delay) [instance_name[range]] (...);
    ```

3. Specify both rise and fall times

    ```
    gate_type #(t_rise,t_fall) [instance_name[range]] (...);
    buf_not   #(t_rise,t_fall) [instance_name[range]] (...);
    tri_buf   #(t_rise,t_fall) [instance_name[range]] (...);
    ```

4. Specify the rise, fall, and turn-off times (only for tristate buffers)

    ```
    tri_buf #(t_rise, t_fall, t_off) [instance_name[range]] (...);
    ```

where t_rise refers to the transition to the 1 value, t_fall refers to the transition to the 0 value, and t_off refers to the transition to a high-impedance value. Note that t_off can only be applied to tristate buffers. It does not make any sense for other gates.

Each delay value within a delay specification may be specified by a *delay specifier* with one of the following formats

```
typical_value
minimum:typical:maximum (min:typ:max)
```

where the minimum, typical, and maximum values[2] for each delay are specified as constant expressions separated by colons (:).

Some examples for specifying gate delays are as follows:

---

[1] We use delay, delay2, and delay3 to mean that up to one, two, and three delays can be specified in the delay specification, respectively.

[2] These values can be invoked on the Verilog (vsim) command line with the runtime option **+mindelays**, **+typdelays**, or **+maxdelays**.

```
// only specify one delay
and #(5) and1 (b, x, y);   // also and #5 and1 (b, x, y);

// only specify one delay using min:typ:max
not #(10:12:15) not1 (a, x);

// specify two delays using min:typ:max
and #(10:12:15, 12:15:20) and2 (c, a, z);

// specify three delays using min:typ:max
bufif0 #(10:12:15, 12:15:20, 12:13:16) buf2 (f, x, c);
```

2.2.2.2 Net-Delay Specifications The signal passes from the input of a net to the output node of the net may be specified by *net delays*. To specify the delay value of specific nets, the following syntax may be used.

```
net_type [signed] [[msb:lsb]] [delay3] net_id{, net_id};
```

Like the delay of a tristate buffer, up to three delays may be applied to a net. The default delay of a net is zero when no delay specification is given.

Some examples for specifying net delays are as follows:

```
// only specify one delay
wire #5 x;   // also wire #(5) x;
wire signed [3:0] #(6) x, y;

// specify two delays using min:typ:max
wire #(10:12:15, 12:15:20) x;
wire signed [3:0] #(10:12:15, 12:15:20) x, y;

// specify three delays using min:typ:max
wire #(10:12:15, 12:15:20, 12:13:16) x;
wire signed [3:0] #(10:12:15, 12:15:20, 12:13:16) x, y;
```

■ Review Questions

2-20 Describe the ingredients of a delay specifier.
2-21 What does the one-delay specification specify?
2-22 What does the two-delay specification specify?
2-23 What does the three-delay specification specify?
2-24 How would you specify a net delay value?

2.3 Hazards

The output signal of a combinational circuit is generally a combination of many signals propagated from different paths. Because of different propagation delays of these paths, the output signal before it is stable must experience an amount of time during which fluctuations occur. This duration is called the *transient time* of the output signal and may result in many undesired short pulses called *glitches*. A *hazard* (or called *timing hazard*) is said to occur when fluctuations appear during the transient time. Hazards can be divided into *static hazards* and *dynamic hazards*, as shown in Figure 2.15. Static hazards are further subdivided into *static-0 hazards* and *static-1 hazards*. In this section, we deal with these two types of hazards in more detail.

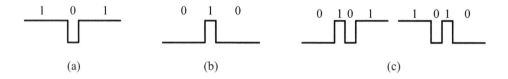

Figure 2.15: Static and dynamic hazards: (a) a static-1 hazard; (b) a static-0 hazard; (c) dynamic hazards.

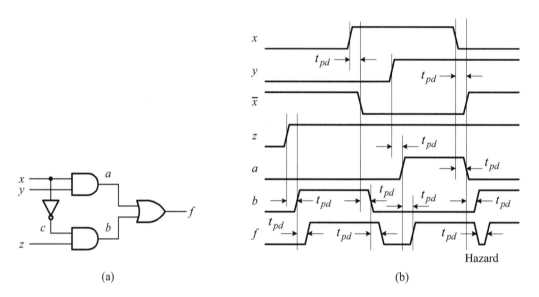

Figure 2.16: An example of static hazards: (a) logic circuit and (b) timing diagram.

2.3.1 Static Hazards

A static hazard represents the situation where a circuit output may momentarily go to 0 (or 1) when it should remain a constant 1 (or 0) during the transient time. In other words, a static hazard is a situation that the output produces a "0" glitch when its stable value is 1 and a "1" glitch when its stable value is 0. The former is called the *static*-1 *hazard* while the latter is denoted by the *static*-0 *hazard*, as shown in Figures 2.15(a) and (b), respectively.

In order to investigate the occurrence of static hazards, consider the logic circuit shown in Figure 2.16(a). There are four paths that can reach the output node f of the logic circuit from inputs, x, y, and z. For simplicity, we assume that all gates have the same propagation delay t_{pd}. As illustrated in Figure 2.16(b), the output node f will momentarily go to 0 for an amount of time t_{pd} when both inputs y and z are 1, and x changes from 1 to 0. As a result, it yields a static-1 hazard.

■ **Example 2.10: An example of static hazards.**

This example simply models the logic circuit shown in Figure 2.16(a) in structural style. All gates are assumed to have the same propagation delay of 5 time units. By appropriately applying stimuli to the module, we can see that the static-1 hazard occurs exactly in the same way as that described previously.

```
// a static hazard example
module hazard_static(
```

Section 2.3 Hazards

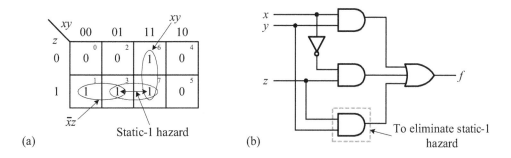

Figure 2.17: A hazard-free logic diagram: (a) Karnaugh map and (b) hazard-free logic circuit.

```
        input  x, y, z,
        output f );
// internal declaration
wire a, b, c;    // internal nets
    // the logic circuit body
    and   #5   a1  (b,x,y);
    not   #5   n1  (a,x);
    and   #5   a2  (c,a,z);
    or    #5   o2  (f,b,c);
endmodule
```

To see the reason why glitches may occur in a combinational circuit, consider the Karnaugh map shown in Figure 2.17(a) of the logic circuit depicted in Figure 2.16(a). As mentioned before, the "0" glitch occurs when variables y and z are 1 and x changes from 1 to 0. This situation corresponds to the case that the output function f switches from the product term xy to another $\bar{x}z$. From the Karnaugh map, we can see that the output yields no glitches when an input combination switches from a minterm to another within the same product term and may produce a glitch when an input combination switches from a minterm to another in a different product term. Based on this observation, the glitch can be removed by adding a redundant product term to cover the gap between two prime implicants. For instance, by adding the product term yz to the logic circuit, the output function f will remain at 1 during the switching time between the two prime implicants, xy and $\bar{x}z$.

From the hardware point of view, each product term is realized by an AND gate and each gate has a definite amount of propagation delay. The glitch is caused by the switching period between two AND gates and the propagation delays of the gates. The outputs of two AND gates may be at 1 or 0 at the same time for a definite amount of time, thereby leaving a signal gap between them. To remedy this, an extra AND gate may be added to realize the redundant product term, i.e., yz, so as to sustain the signal level during the switching period between the two AND gates so that the output function f may still stay at its signal level that it should be. The resulting logic diagram is shown in Figure 2.17(b).

2.3.2 Dynamic Hazards

A dynamic hazard is a situation where the output of a combinational circuit changes from 0 to 1 and then to 0 (or 1 to 0 and then to 1). In other words, the output changes three or more times, as shown in Figure 2.15(c). Because three or more signal changes are required to have a dynamic hazard, a signal must arrive at the output node at three different times at least. That is, there must exist at least three paths from the signal input to the output of the underlying combinational circuit.

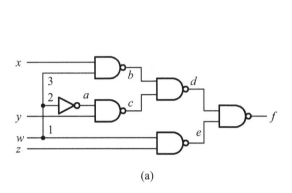
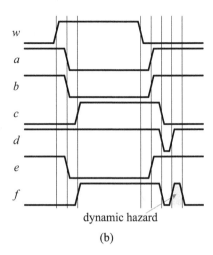

(a) (b)

Figure 2.18: An example of dynamic hazards: (a) logic circuit and (b) timing diagram ($x = y = z = 1$).

An example of a logic circuit with a dynamic hazard is shown in Figure 2.18 along with its associated timing diagram. As illustrated in Figure 2.18(a), the input signal w goes to the output node f through three different paths with different numbers of gates. Consequently, the input signal w may reach the output node f at three different times and hence may cause a dynamic hazard. This is indeed the case. As we can see from the timing diagram depicted in Figure 2.18(b), the dynamic hazard occurs when variables x, y, and z, are equal to 1, and w changes from 1 to 0.

■ Example 2.11: An example of dynamic hazards.

This example simply models the logic circuit shown in Figure 2.18(a) in structural style. For simplicity, all gates are assumed to have the same propagation delay of 5 time units. By appropriately applying stimuli to the module, it can be seen that the dynamic hazard occurs exactly in the same way as that described previously.

```verilog
// a dynamic hazard example
module hazard_dynamic(
        input   w, x, y, z,
        output  f );
// internal declaration
wire   a, b, c, d, e;    // internal nets
    // logic circuit body
    nand  #5   nand1 (b, x, w);
     not  #5   n1    (a, w);
    nand  #5   nand2 (c, a, y);
    nand  #5   nand3 (d, b, c);
    nand  #5   nand4 (e, w, z);
    nand  #5   nand5 (f, d, e);
endmodule
```

A test bench used to drive the preceding module is described in the following example.

Example 2.12: A test bench for the hazard_dynamic module.

From the analysis of Figure 2.18(b), we know that in order to observe the dynamic hazard, we have to set the signals x, y, and z to a logic 1 and then change the input signal w from 1 to 0. Based on this, the resulting test bench module is as follows:

```
`timescale 1ns / 1ns
module hazard_dynamic_tb;
reg  w, x, y, z; // internal signals declarations
wire f;
// Unit Under Test port map
   hazard_dynamic UUT (.w(w),.x(x),.y(y),.z(z),.f(f));
initial begin
      w = 1'b0; x = 1'b0; y = 1'b0; z = 1'b0;
   #5  x = 1'b1; y = 1'b1; z = 1'b1;
   #30 w = 1'b1;
   #20 w = 1'b0;
   #190 $finish;   // terminate the simulation
end
initial
   $monitor($realtime,"ns %h %h %h %h %h ",w,x,y,z,f);
endmodule
```

Review Questions

2-25 Define static hazards.
2-26 Define static-0 and static-1 hazards.
2-27 In what situation may a static hazard occur?
2-28 Define dynamic hazards.
2-29 In what situation may a dynamic hazard occur?

2.4 Switch-Level Structural Modeling

As a logic circuit is modeled at the switch level,[3] each of its MOS transistors is regarded as an ideal or a nonideal switch. An ideal switch does not cause any degradation when a signal passes through it, whereas a nonideal (resistive) switch will cause some degradation when a signal passes through it. With the notation of Verilog HDL, all nonideal switches are prefixed with a letter "r" while all ideal switches do not.

2.4.1 MOS Switches

There are two types of MOS switches: **nmos/rnmos** and **pmos/rpmos**. These switches are used to model unidirectional switches through which data may be allowed to pass from input to output or be blocked by appropriately setting the control input(s). The **nmos** and **pmos** switches pass signals from their inputs to their outputs without degradation, whereas the **rnmos** and **rpmos** switches reduce the strength of signals passing through them.

Each **nmos/rnmos** or **pmos/rpmos** switch has three ports: an input, an output, and an enable input. These switches work as follows:

---
[3]This section may be omitted without loss of continuity.

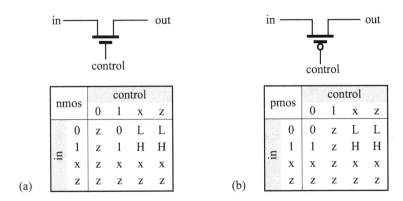

(a)   (b)

Figure 2.19: Both (a) **nmos/rnmos** and (b) **pmos/rpmos** switches.

- The **nmos/rnmos** switch turns on if its enable input is at a logic 1 and maintains its output at a high impedance if its enable input is at a logic 0. The other combinations of both the input and the enable input are shown in Figure 2.19(a).
- The **pmos/rpmos** switch turns on if its enable input is at a logic 0 and maintains its output at a high impedance if its enable input is at a logic 1. The other combinations of both the input and the enable input are shown in Figure 2.19(b).

Figure 2.19 shows both **nmos** and **pmos** switches along with their truth tables. Since some combinations of input values and control values may cause these switches to output either of two values without a preference for either, the symbols L and H are used to denote the results that have a value 0 or z and a value 1 or z, respectively.

To instantiate a MOS switch element, the following syntax may be used.

```
mos_sw [delay3] [instance_name[range]] (out, in, control)
        {, [instance_name[range]] (out, in, control)};
```

where `mos_sw` can be any of the set {**nmos, rnmos, pmos, rpmos**}. `instance_name` is optional but necessary when the range is specified. The first port is always the output, the second is the input port, and the last is the control signal.

2.4.1.1 supply0 and supply1 Nets In Verilog HDL, the **supply0** and **supply1** nets are used to model the *ground* and *power-supply* nets, respectively. The logic value of the **supply0** net is 0 and the logic value of the **supply1** net is 1. These two nets can be declared using the following syntax:

supply0|**supply1** [[msb:lsb]] net_id{, net_id};

A simple application of MOS switches for describing the CMOS inverter shown in Figure 2.20 is given in the following example.

■ **Example 2.13: A CMOS inverter.**

As depicted in Figure 2.20, the inverter consists of an **nmos** switch and a **pmos** switch with the same input and output nets. To properly model this circuit, we need another two nets, **supply1** and **supply0**, to provide the required vdd and gnd signals. By properly combining these four components, the resulting module is obtained.

module not_cmos(**input** x, **output** f);
// internal declaration
supply1 vdd;

Section 2.4 Switch-Level Structural Modeling

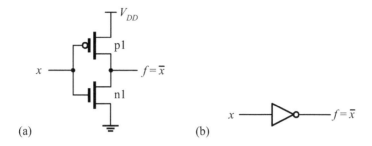

Figure 2.20: A CMOS inverter: (a) logic circuit and (b) logic symbol.

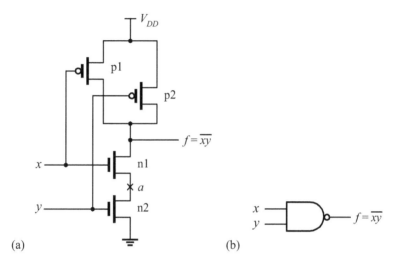

Figure 2.21: A two-input CMOS NAND gate: (a) logic circuit and (b) logic symbol.

```
supply0 gnd;
   // the NOT gate body
   pmos p1 (f, vdd, x);   // source is connected to vdd
   nmos n1 (f, gnd, x);   // source is connected to ground
endmodule
```

A more complex application of MOS switches for modeling the two-input NAND gate depicted in Figure 2.21 is given in the following example.

■ Example 2.14: A two-input CMOS NAND gate.

To model the NAND gate circuit shown in Figure 2.21, we need another two nets, **supply1** and **supply0**, to provide the required vdd and gnd signals. The two **nmos** switches are connected in series and the two **pmos** switches are connected in parallel. One end of the series-connected **nmos** switches is connected to the gnd signal and the other end is connected to the parallel-connected **pmos** switches, which is also the output node. The other end of the parallel-connected **pmos** switches is connected to the vdd signal. The control inputs of both types of the **pmos** and **nmos** switches are connected together in pairs to form inputs x and y, respectively. The resulting module is as follows:

module nand_cmos (**input** x, y, **output** f);

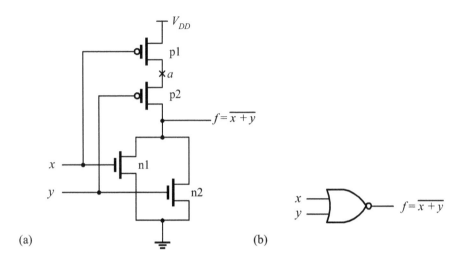

Figure 2.22: A two-input CMOS NOR gate: (a) logic circuit and (b) logic symbol.

```
// internal declaration
supply1 vdd;
supply0 gnd;
wire    a;              // terminal between two nMOS
   // the NAND gate body
   pmos p1 (f, vdd, x); // source is connected to vdd
   pmos p2 (f, vdd, y); // parallel connection
   nmos n1 (f, a, x);   // serial connection
   nmos n2 (a, gnd, y); // source is connected to ground
endmodule
```

The following example applies MOS switches to model the two-input CMOS NOR gate shown in Figure 2.22.

■ Example 2.15: A two-input CMOS NOR gate.

Like in the case of the two-input NAND circuit, we need two nets, **supply1** and **supply0**, to provide the required vdd and gnd signals for modeling the NOR gate circuit shown in Figure 2.22. The two **nmos** switches are connected in parallel and the two **pmos** switches are connected in series. Then one end of the series-connected **pmos** switches is connected to the vdd signal and the other to the **nmos** switches, which is also the output node. The other end of the parallel-connected **nmos** switches is connected to the gnd signal. The control inputs of both types of the **pmos** and **nmos** switches are connected together in pairs to form inputs x and y, respectively. The resulting module is as follows:

```
module nor_cmos(input x, y, output f);
// internal declaration
supply1 vdd;
supply0 gnd;
wire    a;              // terminal between two pMOS
   // the NOR gate body
   pmos p1 (a, vdd, x); // source is connected to vdd
   pmos p2 (f, a, y);   // serial connection
```

Section 2.4 Switch-Level Structural Modeling

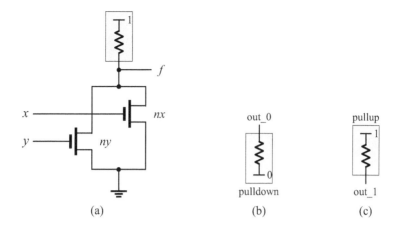

Figure 2.23: A two-input pseudo-nMOS NOR gate: (a) logic circuit; (b) pulldown ; (c) pullup.

```
   nmos n1 (f, gnd, x);  // parallel connection
   nmos n2 (f, gnd, y);  // source is connected to ground
endmodule
```

2.4.1.2 pullup and pulldown Nets The **pullup** and **pulldown** nets separately place logic values 1 and 0 on the nets connected in its terminal list. No delay specification can be applied to these nets. The **pullup** and **pulldown** nets have the following form:

pullup | **pulldown** [(strength)] [instance_name[range]] (out_id)
{, [instance_name[range]] (out_id)};

For example, the following statement declares two **pullup** instances

pullup (**strong1**) p1 (neta), p2 (netb);

where the p1 instance drives neta and the p2 instance drives netb. Both nets have the signal strength of **strong1**. The definition of signal strength will be described in Section 2.4.5.

Figure 2.23(a) shows a two-input pseudo-nMOS NOR gate which consists of two **nmos** switches and one **pullup** net. A Verilog HDL module for describing such a gate is demonstrated in the following example.

■ Example 2.16: A two-input pseudo-nMOS NOR gate.

To model the pseudo-nMOS NOR gate shown in Figure 2.23(a), we need two nets, **pullup** and **supply0**, to provide the required pull-up and gnd signals. The two **nmos** switches are connected in parallel. One end of the resulting circuit of **nmos** switches is connected to the **pullup** signal, which is also the output node, and the other end connected to the gnd signal. The control inputs of **nmos** switches are inputs x and y, respectively.

```
module pseudo_nmos_nor(input x, y, output f);
supply0  gnd;
   // the body of the pseudo-nMOS nor gate
   nmos    nx (f, gnd, x);  // parallel connection
   nmos    ny (f, gnd, y);  // source is connected to ground
   pullup  a (f);           // pull up output f
endmodule
```

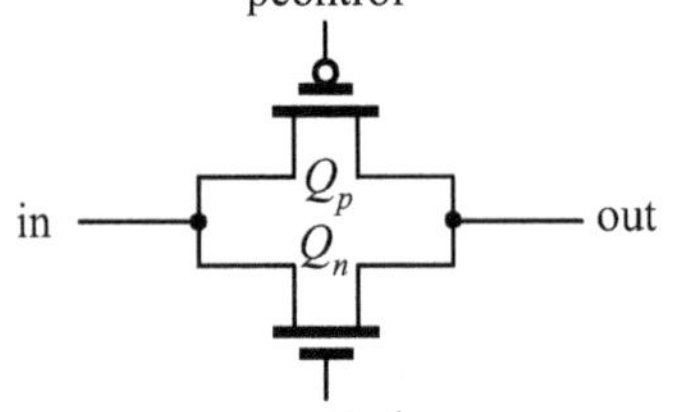

control		data			
n	p	0	1	x	z
0	0	0	1	x	z
0	1	z	z	z	z
0	x	L	H	x	z
0	z	L	H	x	z
1	0	0	1	x	z
1	1	0	1	x	z
1	x	0	1	x	z
1	z	0	1	x	z
x	0	0	1	x	z
x	1	L	H	x	z
x	x	L	H	x	z
x	z	L	H	x	z
z	0	0	1	x	z
z	1	L	H	x	z
z	x	L	H	x	z
z	z	L	H	x	z

Figure 2.24: The (a) logic symbol and (b) truth table of the **cmos/rcmos** switch.

2.4.2 CMOS Switches

The **cmos/rcmos** switch has one data input, one data output, and two control inputs. Figure 2.24 shows the **cmos** switch and its truth table. The **cmos** switch passes signals without degradation whereas the **rcmos** switch reduces the strength of signals passing through it. The **cmos/rcmos** switch operates as follows:

- The **cmos/rcmos** switch turns on if its ncontrol input is at a logic 1 or the pcontrol input is at a logic 0, and maintains its output at a high impedance if its ncontrol input is at a logic 0 while the pcontrol input is at a logic 1. The other combinations of both the input and the enable input are shown in Figure 2.24.

Like the case of the **nmos** or **pmos** switch, some combinations of input values and control values may cause this switch to output either of two values, without a preference for either. Hence, both L and H symbols are also present in the truth table. As before, the symbols L and H are used to denote the value 0 or z and the value 1 or z, respectively.

To instantiate a CMOS switch element, the following syntax may be used.

```
cmos_sw [delay3][instance_name[range]](out,in,ncontrol,pcontrol)
        {, [instance_name[range]](out,in,ncontrol,pcontrol)};
```

where cmos_sw can be either **cmos** or **rcmos** and instance_name is optional but necessary when range is specified. The first port is the output, the second is the input, and the other two ports are ncontrol and pcontrol, which are connected to the control inputs of nMOS and pMOS transistors, respectively.

The **cmos** switch is virtually a combination of a **pmos** switch and an **nmos** switch. The **rcmos** switch is a combination of an **rpmos** switch and an **rnmos** switch. Hence,

    cmos (out, in, ncontrol, pcontrol);

is equivalent to

Section 2.4 Switch-Level Structural Modeling

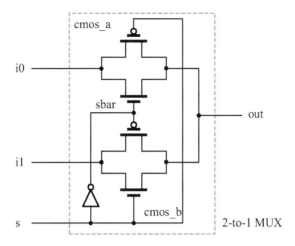

Figure 2.25: A 2-to-1 multiplexer constructed by **cmos** switches.

```
nmos (out, in, ncontrol);
pmos (out, in, pcontrol);
```

A simple application of **cmos** switches for modeling the 2-to-1 multiplexer depicted in Figure 2.25 is given in the following example.

■ Example 2.17: A 2-to-1 multiplexer based on CMOS switches.

To model the 2-to-1 multiplexer depicted in Figure 2.25, we need a **not** gate and two **cmos** switches. These three elements are then connected in the way as illustrated in the following module. The cmos_a switch turns on when the selection input s is 0; the cmos_b switch turns on when the selection input s is 1. Therefore, the circuit is a 2-to-1 multiplexer.

```
module mux_2to1_cmos (
      input i0, i1, s,
      output out);
// an internal wire
wire sbar; //complement of s
// the body of the 2-to-1 multiplexer
not (sbar, s);
// instantiate cmos switches
cmos cmos_a (out, i0, sbar, s);
cmos cmos_b (out, i1, s, sbar);
endmodule
```

2.4.3 Bidirectional Switches

There are six bidirectional switches: **tran, tranif0, tranif1, rtran, rtranif0,** and **rtranif1**, as shown in Figure 2.26. The **tran, tranif0,** and **tranif1** switches pass signals without degradation whereas the **rtran, rtranif0,** and **rtranif1** switches reduce the strength of signals passing through them. Both **tran** and **rtran** switches cannot be turned off but the other four switches can be turned on and off by properly setting the values of their control inputs. Hence, they may be used to model bidirectional switches through which data may be allowed to flow in both directions.

These bidirectional switches operate as follows:

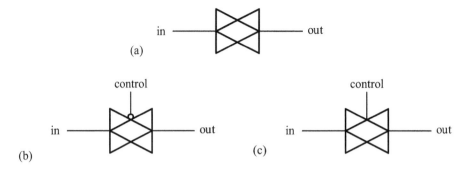

Figure 2.26: The CMOS bidirectional switches: (a) **tran/rtran**; (b) **tranif0/rtranif0**; (c) **tranif1/rtranif1**.

- The *always-on bidirectional switch* (**tran/rtran**) always turns on as a normal bidirectional switch, as shown in Figure 2.26(a). It cannot be turned off in any way.
- The *active-low bidirectional switch* (**tranif0/rtranif0**) turns on if its enable input is at a logic 0 and maintains its output at a high impedance if its enable input is at a logic 1, as shown in Figure 2.26(b).
- The *active-high bidirectional switch* (**tranif1/rtranif1**) turns on if its enable input is at a logic 1 and maintains its output at a high impedance if its enable input is at a logic 0, as shown in Figure 2.26(c).

As in the case of tristate buffers, for active-low bidirectional switches, their keywords are ended with "**if0**," including **tranif0** and **rtranif0**, and for active-high bidirectional switches, their keywords are ended with "**if1**," including **tranif1** and **rtranif1**.

To instantiate a bidirectional switch, the following syntax may be used.

```
bd_sw    [instance_name[range]] (in/out, out/in)
     {, [instance_name[range]] (in/out, out/in)};

bd_sw_en [delay2][instance_name[range]](in/out,out/in,control)
     {, [instance_name[range]](in/out,out/in, control)};
```

where bd_sw can be either **tran** or **rtran**, and bd_sw_en can be any of the set {**tranif0, rtranif0, tranif1, rtranif1**}. instance_name is optional but necessary when range is specified. For **tran** and **rtran** switches, there are only two bidirectional data ports. For **tranif1, tranif0, rtranif1,** and **rtranif0** switches, the first two are bidirectional ports that propagate signals to and from the switches, and the third port is a control input.

■ Review Questions

2-30 What are ideal switches and nonideal switches?
2-31 What are the four MOS switches?
2-32 What are the two CMOS switches?
2-33 Describe the meanings of **pullup** and **pulldown** nets.
2-34 Describe the meanings of **supply0** and **supply1** nets.

2.4.4 Delay Specifications

Like gate primitives, all switch primitives except **tran** and **rtran** may also specify delays.

Section 2.4 Switch-Level Structural Modeling

2.4.4.1 MOS and CMOS Switches The delay of ideal and resistive MOS and CMOS switches can be specified in one of the following ways:

1. Specify no delay

   ```
   mos_sw  [instance_name[range]] (...);
   cmos_sw [instance_name[range]] (...);
   ```

2. Specify the propagation delay only

   ```
   mos_sw  #(prop_delay) [instance_name[range]] (...);
   cmos_sw #(prop_delay) [instance_name[range]] (...);
   ```

3. Specify both rise and fall times

   ```
   mos_sw  #(t_rise,t_fall) [instance_name[range]] (...);
   cmos_sw #(t_rise,t_fall) [instance_name[range]] (...);
   ```

4. Specify the rise, fall, and turn-off times

   ```
   mos_sw  #(t_rise,t_fall,t_off) [instance_name[range]] (...);
   cmos_sw #(t_rise,t_fall,t_off) [instance_name[range]] (...);
   ```

where `t_rise` refers to the transition to the 1 value, `t_fall` refers to the transition to the 0 value, and `t_off` refers to the transition to a high-impedance value.

2.4.4.2 Bidirectional Switches The bidirectional switches **tran** and **rtran** cannot specify delays because they are always on. The delay of ideal and resistive bidirectional switches, **tranif1**, **tranif0**, **rtranif1**, and **rtranif0**, can be specified in one of the following three ways:

1. Specify no delay

   ```
   bd_sw_en  [instance_name[range]] (...);
   ```

2. Specify one turn-on and turn-off delay

   ```
   bd_sw_en #(t_on_off) [instance_name[range]] (...);
   ```

3. Specify separately turn-on and turn-off delays

   ```
   bd_sw_en #(t_on,t_off) [instance_name[range]] (...);
   ```

When two delays are specified, the first one is the turn-on delay and the second is the turn-off delay. If only one delay is specified, it specifies both turn-on and turn-off delays. If no delay is specified, then both turn-on and turn-off delays are defaulted to zeros.

2.4.5 Signal Strength

In Verilog HDL, in addition to the four basic values, 0, 1, x, and z, a net can also have signal strength associated with it. Signal strength represents the ability of the source device to supply energy to drive the signal. There are two types of signal strength that can be specified to a scalar net: *drive strength* and *charge strength*. Signals with drive strength are propagated from gate outputs and **assign** continuous assignment to outputs. Drive strength has generally four different levels: **supply**, **strong**, **pull**, and **weak**. Signals with charge strength are originated in the **trireg** net type. There are three types of charge strength: **large**, **medium**, and **small**. The signal strength levels defined in Verilog HDL are shown in Table 2.4.

The signal strength of a net is specified by a *strength specification*. A strength specification has two components, (strength1, strength0) or (strength0, strength1), where strength0 can be

Table 2.4: Strength levels for scalar net signal values.

Strength	Strength0	Strength1	Type	Degree
supply	supply0	supply1	driving	strongest
strong	strong0	strong1	driving	
pull	pull0	pull1	driving	
large	large0	large1	storage	
weak	weak0	weak1	driving	
medium	medium0	medium1	storage	
small	small0	small1	storage	
highz	highz0	highz1	high Z	weakest

one of **supply0**, **strong0**, **pull0**, **weak0**, and **highz0**, and strength1 can be one of **supply1**, **strong1**, **pull1**, **weak1**, and **highz1**. The combinations (**highz0**, **highz1**) and (**highz1**, **highz0**) are not allowed. The default strength specification is (**strong0**, **strong1**).

The drive strength can be specified by any of the following ways: a net in a net declaration assignment, the output port of a gate primitive instance, and in an **assign** continuous assignment. A scalar net declaration assignment with drive strength has the following syntax

```
net_type [strength] [signed]
         [delay3] net_id = expression{, net_id = expression};
```

where net_type is any type except the **trireg** net in the column of nets in Table 1.1. It should be noted that the **trireg** net may also have charge strength. More details can be referred to in Section 2.4.6.

An output port of a gate primitive with drive strength has the following syntax:

```
gate_type [strength] [delay2] [instancs_name[range]] (...);
buf_not   [strength] [delay2] [instancs_name[range]] (...);
tri_buf   [strength] [delay3] [instancs_name[range]] (...);
```

Note that every gate primitive may have drive strength associated with it. Nonetheless, all switch primitives except **pullup** and **pulldown** may not have drive strength. The use of **pullup** and **pulldown** can be referred to in Section 2.4.1.

An **assign** continuous assignment may also have drive strength associated with it. It has the following syntax

```
assign [strength] [delay3] net_lvalue = expression
                  {, net_lvalue = expression};
```

The signal strength may be printed by using the %v format specification in the **$display**, **$strobe**, and **$monitor** system tasks.

The signal strength can be weakened or attenuated by the resistance of wires, thereby giving rise to signals of different strength levels. The reduction rules of signal strength when signals passing through resistive switches are shown in Figure 2.27.

Signal contention. When multiple drivers drive a net at the same time, a contention occurs on the net. There are many rules applicable to resolve such a contention. In what follows, we only consider the two most widely used cases:

1. *Combined signals with the same value and unequal strength levels*: If two signals with the same known value but different strength levels drive the same net, the stronger signal dominates.
2. *Combined signals with an opposite value and equal strength level*: If two signals with the opposite known values but equal strength level drive the same net, the result is an unknown value, x.

The details of the other rules can be referred to LRM [2, 3].

Section 2.4 Switch-Level Structural Modeling

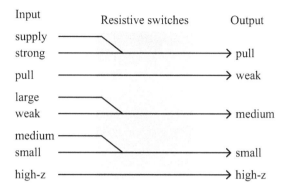

Figure 2.27: The reduction rules of signal strength with resistive switches.

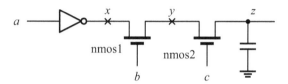

Figure 2.28: The relationship between a **trireg** net and its driver.

2.4.6 The trireg Net

The **trireg** net stores a value and is used to model charge storage nodes. A **trireg** net may have charge strength and drive strength associated with it. To designate these, the following syntax may be used.

    trireg [charge-strength] [**signed**] [delay3] net_id{, net_id};
    trireg [drive-strength] [**signed**]
        [delay3] net_id= expr{, net_id= expr};

A **trireg** net may be in one of two states:

1. *Driven state*: At least one driver drives a value of 1, 0, or x on the net. The value retains on the net with the strength of the driver. The strength can be **supply**, **strong**, **pull**, or **weak**.
2. *Capacitive state*: When all drivers to a **trireg** net are at their high-impedances (z), the net retains its last driven value. The strength can be **small**, **medium** (default), or **large**.

The following example employs the circuit depicted in Figure 2.28 to demonstrate the use of a **trireg** net in modeling a pass-element logic circuit.

■ Example 2.18: The effects of the trireg net.

Referring to Figure 2.28, at simulation time 0, control signals a, b, and c are set to 1 so that the output, x, of the inverter is 0, causing net y to change its value to 0. The **trireg** net, z, then enters the driven state and discharges to **strong0**. At simulation time 10, control signal b is cleared to 0, causing net y to change to a high-impedance state. Net z enters the capacitive state and stores its last driven value 0 with **medium** strength.

```
module triregexample;
reg   a, b, c;
wire  x, y;
```

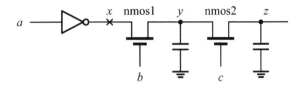

Figure 2.29: The charge-sharing problem.

```
trireg (medium) z;        // declared as medium strength

    not  not1  (x, a);
    nmos nmos1 (y, x, b);
    nmos nmos2 (z, y, c); // nmos that drives the trireg

initial begin
    $monitor("%0d a=%v b=%v c=%v x=%v y=%v z=%v ",
             $time, a , b, c, x, y, z);
    a = 1;    // simulation time 0
    b = 1;
    c = 1;

    // toggle the control input b
    #10 b = 0; // simulation time 10
    #30 b = 1; // simulation time 40
    #10 b = 0; // simulation time 50
    #100 $finish;
end
endmodule
```

The simulation results are as follows:

```
#  0  a=St1 b=St1 c=St1 x=St0 y=St0 z=St0
# 10  a=St1 b=St0 c=St1 x=St0 y=HiZ z=Me0
# 40  a=St1 b=St1 c=St1 x=St0 y=St0 z=St0
# 50  a=St1 b=St0 c=St1 x=St0 y=HiZ z=Me0
```

2.4.6.1 Charge-Sharing Problems The charge-sharing problem is often encountered in practical circuits. Hence, we use a simple circuit, as shown in Figure 2.29, to explore this problem and give an example to illustrate how to model it.

■ Example 2.19: The charge-sharing problem.

Referring to Figure 2.29, at simulation time 0, control signal a is 0, and control signals b and c are 1. Hence, nodes x, y, and z are driven to **strong1**. At simulation time 10, control signal b changes to 0. Net y enters the capacitive state and stores its last driven value 1 with **large** strength. Net z is still in the driven state and is driven to value 1 with **large** strength. At simulation time 20, control signal c changes to 0, causing net z to enter the capacitive state and store a value 1 of **small** strength. At simulation time 30, control signal c changes to 1 again, connecting together the two **trireg** nets, y and z. These two nets now share the same charge. At simulation time 40, control signal c changes to 0 one more time, causing net z to enter the capacitive state and store a value 1 of **small** strength.

Section 2.4 *Switch-Level Structural Modeling*

```
module triregChargeSharing;
reg   a, b, c;
wire  x;
trireg (large) y; // declared as large strength
trireg (small) z; // declared as small strength

  not   not1  (x, a);
  nmos  nmos1 (y, x, b);
  nmos  nmos2 (z, y, c); // nmos that drives the trireg
initial begin
  $monitor("%0d a=%v b=%v c=%v x=%v y=%v z=%v ",
           $time, a , b, c, x, y, z);
  a = 0; b = 1; c = 1;   // simulation time 0

  // toggle the control input c
  #10 b = 0;   // simulation time 10
  #10 c = 0;   // simulation time 20
  #10 c = 1;   // simulation time 30
  #10 c = 0;   // simulation time 40
  #100 $finish;
end
endmodule
```

The simulation results are listed as follows:

```
#  0  a=St0 b=St1 c=St1 x=St1 y=St1 z=St1
# 10  a=St0 b=St0 c=St1 x=St1 y=La1 z=La1
# 20  a=St0 b=St0 c=St0 x=St1 y=La1 z=Sm1
# 30  a=St0 b=St0 c=St1 x=St1 y=La1 z=La1
# 40  a=St0 b=St0 c=St0 x=St1 y=La1 z=Sm1
```

2.4.6.2 trireg Net Charge Decay Like all nets, a **trireg** net declaration may have up to three delays with the following form

`#(t_rise, t_fall, t_decay)`

The t_decay parameter specifies the charge-decay time of the **trireg** net, which is the time duration between when its drivers turn off and the point that its stored charge can no longer be determined.

The charge-decay process begins when the drivers turn off, and the **trireg** net starts to hold charge. It ends whenever either of the following two conditions is satisfied: the charge-decay time elapses and the net makes a transition to x, or the drivers turn on and send a 1, 0, or x to the net.

The following is an example for illustrating the effect of charge decay.

■ Example 2.20: The effect of charge decay.

This example declares a **trireg** net, cap1, with **medium** charge strength. The delay specifications of both rise and fall times are 0 and the charge-decay time is 10 time units. After simulation time 25, the **trireg** net, cap1, has decayed to an unknown value x.

```
module capacitor_decay;
reg x, a;
```

```
// trireg declared as a charge-decay time of 10 time units
trireg (medium) #(0, 0, 10) cap1;
    nmos nmos1 (cap1, x, a); // nmos that drives the trireg
initial begin
    $monitor("%0d x=%v a=%v cap1=%v", $time, x, a, cap1);
    x = 1;          // simulation time 0

    // toggle the control input a
    a = 1;          // turn on the nMOS switch
    #05 a = 0;      // simulation time 05
    #05 a = 1;      // simulation time 10
    #05 a = 0;      // simulation time 15
    #25 a = 1;      // simulation time 40
    #25 $finish;
end
endmodule
```

The simulation results are as follows:

```
# 0   x=St1  a=St1  cap1=St1
# 5   x=St1  a=St0  cap1=Me1
# 10  x=St1  a=St1  cap1=St1
# 15  x=St1  a=St0  cap1=Me1
```

■ Review Questions

2-35 Why the delay specification cannot be applied to the bidirectional switches, **tran/rtran**?

2-36 What are the two types of signal strength that can be specified to a scalar net?

2-37 Describe the function of the **trireg** net.

2-38 Describe the meaning of the driven state and capacitive state.

2-39 How would you model the effects of charge decay in a capacitor?

2.5 Summary

In structural style, a module is described as a set of interconnected components. The components can be modules, UDPs, gate primitives, and/or switch primitives.

Gate-level structural modeling describes a design only using gate primitives in structural style. In Verilog HDL, there are 12 predefined gate primitives. These gate primitives can be cast into and/or and buf/not groups. The and/or group includes **and, nand, or, nor, xor,** and **xnor**, and the buf/not group contains **buf, not, bufif0, bufif1, notif0,** and **notif1**.

Because of the inherent resistances (R) and capacitances (C) associated with electronic circuits, all logic circuits and wires have definite propagation delays. In addition, for every wire between two logic circuits there always exists a transport delay caused by the time of flight of the signal passing through it. With the inertial delay model, any signal event that does not persist longer than the RC propagation delay of a logic circuit is filtered out and not propagated to the output of the logic circuit. With the transport delay model, any signal event appearing at the input of a wire will be propagated to the output of the wire. Since the RC propagation delay dominates the overall delay of a wire, the inertial delay model is also used in Verilog HDL to capture the effects of net delays.

Hazards can be categorized into static hazards and dynamic hazards. A static hazard represents the situation where a circuit output may momentarily go to 0 (or 1) when it should remain a constant 1 (or 0) during the transient period. Static hazards can be further subdivided into static-0 hazards and static-1 hazards. A dynamic hazard is a situation where the output of a combinational circuit changes from 0 to 1 and then back to 0 (or 1 to 0 and then back to 1); namely, the output changes three or more times. Because three or more signal changes are required to have a dynamic hazard, there must exist at least three different paths for a signal to pass from the input to the output of the underlying combinational circuit.

When a logic circuit is modeled at the switch level in structural style, each MOS transistor in the logic circuit is regarded as an ideal or a nonideal switch. An ideal switch does not cause any degradation when a signal passes through it, whereas a nonideal (resistive) switch will cause an amount of degradation when a signal passes through it. In Verilog HDL, there are 16 built-in switch primitives. These switch primitives can be grouped into MOS switches, CMOS switches, bidirectional switches, **pullup** and **pulldown**, and **supply0** and **supply1**.

The propagation delay of a signal from any input to the output of a gate can be specified by a delay specification. Propagation delays of gate primitives can be specified in a no-delay, one-delay, two-delay, or three-delay specification. The one-delay specification specifies the propagation delay only, the two-delay specification specifies both rise and fall times, and the three-delay specification specifies the rise, fall, and turn-off times. Like gate primitives, all switch primitives, except **tran** and **rtran**, may also specify delays.

References

1. J. Bhasker, *A Verilog HDL Primer*, 3rd ed., Star Galaxy Publishing, 2005.
2. IEEE 1364-2001 Standard, *IEEE Standard Verilog Hardware Description Language*, 2001.
3. IEEE 1364-2005 Standard, *IEEE Standard for Verilog Hardware Description Language*, 2006.
4. M. B. Lin, *Digital System Design: Principles, Practices, and Applications*, 4th ed., Chuan Hwa Book Ltd. (Taipei, Taiwan), 2010.
5. S. Palnitkar, *Verilog HDL: A Guide to Digital Design and Synthesis*, 2nd ed., SunSoft Press, 2003.

Problems

2-1 Model the following switching expression at the gate level in structural style:

$$f(x,y,z) = \overline{\overline{(x+y)} + \bar{x}z}$$

Write a test bench to verify whether the module behaves correctly.

2-2 Model the following switching expression at the gate level in structural style:

$$f(w,x,y,z) = \overline{\overline{(wx + \bar{y}z)}xy}$$

Write a test bench to verify whether the module behaves correctly.

2-3 Model the logic circuit shown in Figure 2.30 at the gate level in structural style. Write a test bench to verify whether the module behaves correctly.

2-4 Model the logic circuit shown in Figure 2.31 at the gate level in structural style. Write a test bench to verify whether the module behaves correctly.

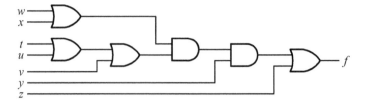

Figure 2.30: The logic diagram for Problem 2-3.

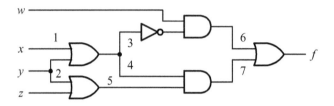

Figure 2.31: The logic diagram for Problem 2-4.

2-5 Recall that a half adder is a circuit that adds its two inputs without concerning the carry-in from its previous stage. In this problem, we first design a 2-bit half adder and then use this circuit to construct a 2-bit full adder. Finally, we in turn use this 2-bit full adder to build a 4-bit adder.

(a) Supposing that the inputs are $x_1 x_0$ and $y_1 y_0$ and the outputs are c and $s_1 s_0$, derive the truth table of a 2-bit half adder and derive the switching expressions of the outputs: c, s_1, and s_0.

(b) Model the 2-bit half adder circuit at the gate level in structural style. Write a test bench to verify whether the module behaves correctly.

(c) Construct a 2-bit full adder by using two 2-bit half adders and an OR gate. Assume that the inputs are $x_1 x_0$, $y_1 y_0$, and c_{in}, and the outputs are c_{out} and $s_1 s_0$. Model this 2-bit full adder in structural style. Write a test bench to verify whether the module behaves correctly.

(d) Construct a 4-bit adder by cascading two 2-bit full adders. Model this 4-bit adder in structural style. Write a test bench to verify whether the module behaves correctly.

2-6 Recall that a half subtracter is a circuit that calculates the difference between its two inputs without concerning the borrow-in from its previous stage. In this problem, we first design a 2-bit half subtracter and then use this circuit to construct a 2-bit full subtracter. Finally, we in turn use this 2-bit full subtracters to build a 4-bit subtracter.

(a) Suppose that the inputs are $x_1 x_0$ and $y_1 y_0$ and the outputs are b and $d_1 d_0$. Derive the truth table of a 2-bit half subtracter and derive the switching expressions of the outputs: b, d_1, and d_0.

(b) Model the 2-bit half subtracter circuit at the gate level in structural style. Write a test bench to verify whether the module behaves correctly.

(c) Construct a 2-bit full subtracter by using two 2-bit half subtracters and an OR gate. Assume that the inputs are $x_1 x_0$, $y_1 y_0$, and b_{in}, and the outputs are b_{out} and $d_1 d_0$. Model this 2-bit full subtracter in structural style. Write a test bench to verify whether the module behaves correctly.

Table 2.5: The relationship between the BCD code and excess-3 code.

BCD code				Excess-3 code				BCD code				Excess-3 code			
d_3	d_2	d_1	d_0	e_3	e_2	e_1	e_0	d_3	d_2	d_1	d_0	e_3	e_2	e_1	e_0
0	0	0	0	0	0	1	1	0	1	0	1	1	0	0	0
0	0	0	1	0	1	0	0	0	1	1	0	1	0	0	1
0	0	1	0	0	1	0	1	0	1	1	1	1	0	1	0
0	0	1	1	0	1	1	0	1	0	0	0	1	0	1	1
0	1	0	0	0	1	1	1	1	0	0	1	1	1	0	0

(d) Construct a 4-bit subtracter by cascading two 2-bit full subtracters. Model this 4-bit subtracter in structural style. Write a test bench to verify whether the module behaves correctly.

2-7 Referring to the relationship between the BCD code and excess-3 code listed in Table 2.5, design a module to convert the input BCD code, $d_3d_2d_1d_0$, into an equivalent excess-3 output code, $e_3e_2e_1e_0$.
   (a) Using Karnaugh maps, derive the output switching expressions, e_3, e_2, e_1, and e_0.
   (b) Describe the module in structural style and write a test bench to verify the functionality of the module.

2-8 Referring to the relationship between the BCD code and excess-3 code listed in Table 2.5, design a module to convert the input excess-3 code, $e_3e_2e_1e_0$, into an equivalent BCD output code, $d_3d_2d_1d_0$.
   (a) Using Karnaugh maps, derive the output switching expressions, d_3, d_2, d_1, and d_0.
   (b) Describe the module in structural style and write a test bench to verify the functionality of the module.

2-9 Referring to the relationship between the 4-bit binary code and Gray code listed in Table 2.6, design a module to convert the input binary code, $b_3b_2b_1b_0$, into an equivalent Gray output code, $g_3g_2g_1g_0$.
   (a) Using Karnaugh maps, derive the output switching expressions, g_3, g_2, g_1, and g_0.
   (b) Describe the module in structural style and write a test bench to verify the functionality of the module.

Table 2.6: The relationship between the 4-bit binary code and Gray code.

Binary code				Gray code				Binary code				Gray code			
b_3	b_2	b_1	b_0	g_3	g_2	g_1	g_0	b_3	b_2	b_1	b_0	g_3	g_2	g_1	g_0
0	0	0	0	0	0	0	0	1	0	0	0	1	1	0	0
0	0	0	1	0	0	0	1	1	0	0	1	1	1	0	1
0	0	1	0	0	0	1	1	1	0	1	0	1	1	1	1
0	0	1	1	0	0	1	0	1	0	1	1	1	1	1	0
0	1	0	0	0	1	1	0	1	1	0	0	1	0	1	0
0	1	0	1	0	1	1	1	1	1	0	1	1	0	1	1
0	1	1	0	0	1	0	1	1	1	1	0	1	0	0	1
0	1	1	1	0	1	0	0	1	1	1	1	1	0	0	0

Table 2.7: The relationship between the 4-bit binary code and BCD code.

Binary code				BCD code								Binary code				BCD code							
b_3	b_2	b_1	b_0	d_3	d_2	d_1	d_0	d_3	d_2	d_1	d_0	b_3	b_2	b_1	b_0	d_3	d_2	d_1	d_0	d_3	d_2	d_1	d_0
0	0	0	0	0	0	0	0	0	0	0	0	1	0	0	0	0	0	0	0	1	0	0	0
0	0	0	1	0	0	0	0	0	0	0	1	1	0	0	1	0	0	0	0	1	0	0	1
0	0	1	0	0	0	0	0	0	0	1	0	1	0	1	0	0	0	0	1	0	0	0	0
0	0	1	1	0	0	0	0	0	0	1	1	1	0	1	1	0	0	0	1	0	0	0	1
0	1	0	0	0	0	0	0	0	1	0	0	1	1	0	0	0	0	0	1	0	0	1	0
0	1	0	1	0	0	0	0	0	1	0	1	1	1	0	1	0	0	0	1	0	0	1	1
0	1	1	0	0	0	0	0	0	1	1	0	1	1	1	0	0	0	0	1	0	1	0	0
0	1	1	1	0	0	0	0	0	1	1	1	1	1	1	1	0	0	0	1	0	1	0	1

Figure 2.32: The seven-segment LED display patterns.

2-10 Referring to the relationship between the 4-bit binary code and Gray code listed in Table 2.6, design a module to convert the input Gray code, $g_3g_2g_1g_0$, into an equivalent binary output code, $b_3b_2b_1b_0$.

 (a) Using Karnaugh maps, derive the output switching expressions, b_3, b_2, b_1, and b_0.
 (b) Describe the module in structural style and write a test bench to verify the functionality of the module.

2-11 Referring to the relationship between the 4-bit binary code and BCD code listed in Table 2.7, design a module to convert the input binary code, $b_3b_2b_1b_0$, into an equivalent BCD output code, $d_4d_3d_2d_1d_0$, where the d_4 bit is the LSB (i.e., d_0) of the second BCD digit.

 (a) Using Karnaugh maps, derive the output switching expressions, d_4, d_3, d_2, d_1, and d_0.
 (b) Describe the module in structural style and write a test bench to verify the functionality of the module.

2-12 Referring to the relationship between the 4-bit binary code and BCD code listed in Table 2.7, design a module to convert the input BCD code $000d_0\ d_3d_2d_1d_0$ into an equivalent binary output code, $b_3b_2b_1b_0$.

 (a) Using Karnaugh maps, derive the output switching expressions, b_3, b_2, b_1, and b_0.
 (b) Describe the module in structural style and write a test bench to verify the functionality of the module.

2-13 Referring to the display patterns shown in Figure 2.32, design a BCD-to-seven-segment decoder. Suppose that to turn on an LED, we have to apply a logic 0 to the corresponding LED. The inputs are $d_3d_2d_1d_0$ with values from 0 to 9 and the outputs, a, b, c, d, e, f, and g, are active-low. The outputs for the rest of input combinations are treated as don't care.

 (a) Derive a truth table for the outputs and then use Karnaugh maps to simplify the output switching expressions, a, b, c, d, e, f, and g.

Problems

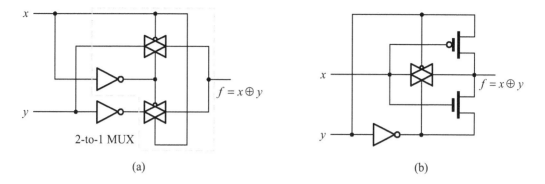

Figure 2.33: The logic diagrams for Problems 2-17 and 2-18.

(b) Describe the module in structural style and write a test bench to verify the functionality of the module.

2-14 Referring to the display patterns shown in Figure 2.32, design a BCD-to-seven-segment decoder. Suppose that to turn on an LED, we have to apply a logic 1 to the corresponding LED. The inputs are $d_3d_2d_1d_0$ with values from 0 to 9 and the outputs, a, b, c, d, e, f, and g, are active-high. The outputs for the rest of the input combinations are treated as don't care.

(a) Derive a truth table for the outputs and then use Karnaugh maps to simplify the output switching expressions, a, b, c, d, e, f, and g.

(b) Describe the module in structural style and write a test bench to verify the functionality of the module.

2-15 Referring to the display patterns shown in Figure 2.32, design a binary-to-seven-segment decoder. Suppose that to turn on an LED, we have to apply a logic 0 to the corresponding LED. The inputs are $b_3b_2b_1b_0$ and the outputs, a, b, c, d, e, f, and g, are active-low.

(a) Derive a truth table for the outputs and then use Karnaugh maps to simplify the output switching expressions, a, b, c, d, e, f, and g.

(b) Describe the module in structural style and write a test bench to verify the functionality of the module.

2-16 Referring to the display patterns shown in Figure 2.32, design a binary-to-seven-segment decoder. Suppose that to turn on an LED, we have to apply a logic 1 to the corresponding LED. The inputs are $b_3b_2b_1b_0$ and the outputs, a, b, c, d, e, f, and g, are active-high.

(a) Derive a truth table for the outputs and then use Karnaugh maps to simplify the output switching expressions, a, b, c, d, e, f, and g.

(b) Describe the module in structural style and write a test bench to verify the functionality of the module.

2-17 Considering the logic diagram of a CMOS XOR gate shown in Figure 2.33(a), answer each of the following problems.

(a) Model the logic circuit using gate and switch primitives in structural style.

(b) Write a test bench to verify whether the module behaves correctly.

2-18 Considering the logic diagram of a CMOS XOR gate shown in Figure 2.33(b), answer each of the following problems.

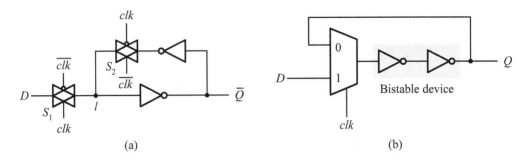

Figure 2.34: The logic diagrams of two CMOS latches for Problems 2-19 and 2-20.

  (a) Model the logic circuit using gate and switch primitives in structural style.
  (b) Write a test bench to verify whether the module behaves correctly.

2-19 Considering the logic diagram of a TG-based latch shown in Figure 2.34(a), answer each of the following problems.

  (a) Model the logic circuit using gate and switch primitives in structural style.
  (b) Write a test bench to verify whether the module behaves correctly.

2-20 Considering the logic diagram of a multiplexer-based latch shown in Figure 2.34(b), answer each of the following problems.

  (a) Model the logic circuit using gate and switch primitives in structural style.
  (b) Write a test bench to verify whether the module behaves correctly.

2-21 Considering the logic diagram of a positive-edge-triggered D flip-flop shown in Figure 2.35, answer each of the following problems.

  (a) Model the logic circuit using gate and switch primitives in structural style.
  (b) Write a test bench to verify whether the module behaves correctly.

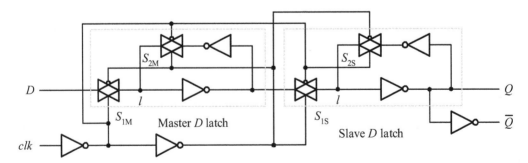

Figure 2.35: The logic diagram of a positive-edge-triggered D flip-flop for Problem 2-21.

2-22 A complex CMOS logic gate implements the following switching expression:
$$f(w,x,y,z) = \overline{(x+y)z+w}$$

  (a) Draw the logic circuit.
  (b) Model the logic circuit at the switch level.
  (c) Write a test bench to verify whether the module behaves correctly.

Problems

2-23 A complex CMOS logic gate implements the following switching expression:

$$f(w,x,y,z) = \overline{xy + y(z+w)}$$

(a) Draw the logic circuit.
(b) Model the logic circuit at the switch level.
(c) Write a test bench to verify whether the module behaves correctly.

2-24 A complex CMOS logic gate implements the following switching expression:

$$f(w,x,y,z) = \overline{wx + yz}$$

(a) Draw the logic circuit.
(b) Model the logic circuit at the switch level.
(c) Write a test bench to verify whether the module behaves correctly.

2-25 Considering the logic diagram shown in Figure 2.36, answer each of the following questions.

(a) Model the logic circuit using gate and switch primitives in structural style.
(b) Write a test bench to verify whether the module behaves correctly.

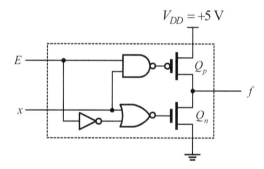

Figure 2.36: The logic diagram for Problem 2-25.

2-26 A pseudo-nMOS logic circuit implements the following switching expression:

$$f(x,y,z) = \overline{x + yz}$$

(a) Draw the logic circuit.
(b) Model the logic circuit at the switch level.
(c) Write a test bench to verify whether the module behaves correctly.

2-27 A pseudo-nMOS logic circuit implements the following switching expression:

$$f(w,x,y,z) = \overline{w(x+y) + z}$$

(a) Draw the logic circuit.
(b) Model the logic circuit at the switch level.
(c) Write a test bench to verify whether the module behaves correctly.

2-28 Use ideal MOS switches to model a switch with both rise and fall times of 0 and a turn-off time of 2 time units.

2-29 Consider the logic diagram shown in Figure 2.36 again. Assume that the rise and fall times of the output node, f, are 5 and 6 time units, respectively. The charge-decay time of the output node, f, is 12 time units.

(a) Model the logic circuit using gate and switch primitives in structural style.

(b) Write a test bench to verify whether the module behaves correctly.

2-30 Consider the logic diagram shown in Figure 2.37. Assume that the rise and fall times of the soft node, x, are 1 and 2 time units, respectively. The charge-decay time of the soft node, x, is 6 time units.

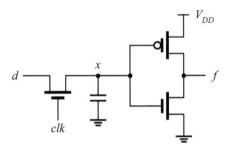

Figure 2.37: The logic diagram for Problem 2-30.

(a) Model the logic circuit using gate and switch primitives in structural style.

(b) Assume that the sampling clock signal, clk, has a period of 20 time units. Write a test bench to test the module and observe the value of output node, f. Explain what you observed.

(c) Assume that the sampling clock signal, clk, has a period of 10 time units. Write a test bench to test the module and observe the value of output node, f. Explain what you observed.

2-31 Consider the TG-based dynamic D flip-flop shown in Figure 2.38. Assume that the rise and fall times of the soft nodes, x and y, are 1 and 2 time units, respectively. The charge-decay time of both soft nodes, x and y, is 6 time units.

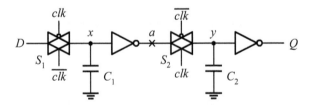

Figure 2.38: The TG-based dynamic D flip-flop for Problem 2-31.

(a) Model the logic circuit using gate and switch primitives in structural style.

(b) Assume that the sampling clock signal, clk, has a period of 20 time units. Write a test bench to test the module and observe the values of output, Q. Explain what you observed.

(c) Assume that the sampling clock signal, clk, has a period of 10 time units. Write a test bench to test the module and observe the values of output, Q. Explain what you observed.

3

Dataflow Modeling

THE rationale behind the design of a digital system in dataflow style is on the fact that any digital system can be constructed by interconnecting registers and putting a combinational circuit in between them to perform the required logic functions. Modeling a design in dataflow style has the following features:

1. Dataflow modeling provides an efficient way to model a combinational circuit.
2. Dataflow modeling provides a more powerful way than structural modeling to implement a complex combinational logic design.
3. Automated tools are used to create a gate-level circuit from the description of a design in dataflow style. This process is often called *logic synthesis*.

In industry, RTL (register-transfer level) modeling means that an RTL module (i.e., constructed with RTL components) is modeled in behavioral, dataflow, structural, or mixed style on condition that the resulting description must be acceptable by logic synthesis tools.

The most basic statement of dataflow modeling is the **assign** continuous assignment. It continuously drives a value onto a net; thereby, it is suitable for modeling a combinational circuit.

3.1 Dataflow Modeling

The **assign** continuous assignment is the essential element of dataflow modeling. Hence, we introduce in this section its basic syntax and give some insight into it. In addition, a variant of the **assign** continuous assignment, referred to as a *net declaration assignment*, is described.

3.1.1 Continuous Assignments

The **assign** continuous assignment is the most basic statement of dataflow modeling. It continuously drives a value onto a net. The continuous assignment begins with the keyword **assign** and has the following syntax

```
assign [delay3] net_lvalue = expression
              {, net_lvalue = expression};
```

where net_lvalue is a scalar or vector net, or their concatenation. The delay3 inside square brackets ([]) is optional. The assignment surrounded by curly brackets ({ }) means that it can repeat zero or more times. The operands used in the expression can be variables, nets, or function calls. Variables or nets can be scalar or vectors. Any logic function can be realized with **assign** continuous assignments. The **assign** continuous assignment can only update the value of a net type, such as **wire**, **tri**, **triand**, and so on.

During simulation, an **assign** continuous assignment is always active and hence all **assign** continuous assignments are executed concurrently. That is, whenever the value of a right-hand-side operand in the expression changes, the expression is evaluated and the resulted value is assigned to net_lvalue. Consequently, it is suitable for modeling the behavior of combinational logic. For example, the following **assign** continuous assignment

```
assign {c_out, sum[3:0]} = a[3:0] + b[3:0] + c_in;
```

describes a 4-bit adder with the carry-in. The braces ({ }) on the left-hand side of the expression is a concatenation operator, which will be described in detail later.

It is worth noting that net_lvalue cannot be on the right-hand side of an **assign** continuous assignment; otherwise, a warning message, like "Found combinational loop during mapping," will be obtained from synthesis tools.

3.1.1.1 Net Declaration Assignments Usually, we declare a net and then use an **assign** continuous assignment to assign a value onto it. In Verilog HDL, there is a shortcut for this. A continuous assignment without the keyword **assign** can be placed on a net when it is declared. For convenience, we call it a *net declaration assignment*, which has the syntax as follows:

```
net_type [delay3] net_id = expression{, net_id = expression};
```

As an illustration, consider the following example

```
wire out = in1 & in2;    // net declaration assignment
```

which is equivalent to

```
wire out;                // normal continuous assignment
assign out = in1 & in2;
```

In other words, a net declaration assignment is virtually a combination of a net declaration and an **assign** continuous assignment using the net as its net_lvalue. Like regular **assign** continuous assignments, net declaration assignments are always active. Moreover, there can only be one net declaration assignment per net because a net can only be declared once.

3.1.1.2 Implicit Net Declarations Implicit net declarations are also a feature of Verilog HDL. An implicit net declaration will be inferred for a signal name when it is used as the net_lvalue of an **assign** continuous assignment. For example,

```
wire in1, in2;
assign out = in1 & in2;
```

Note that the left-hand-side net, out, is not declared as a **wire**, but an implicit **wire** net type declaration for out is done automatically by the simulator. It is good practice to declare the net type explicitly and to use regular **assign** continuous assignments to avoid any unintentional errors. However, implicit net declarations are often used with the input and output ports of modules (see Section 6.1.1).

3.1.1.3 Assignments and Truncation In an assignment (**assign** continuous assignment, net declaration assignments, and procedural assignments), the bit widths of both left-hand and right-hand sides need not be the same. The right-hand side is truncated by keeping the lower significant bits to match the width of the left-hand-side net (or variable) when it has more bits. The right-hand side is zero-extended in the higher significant bits before assigning to the left-hand-side net (or variable) when it has fewer bits. It should be noted that truncating the sign bit of a signed expression may change the sign of the result.

Section 3.1 *Dataflow Modeling*

Table 3.1: The summary of operators in Verilog HDL.

Arithmetic	Bitwise	Reduction	Relational
+: add	~: negation (not)	&: and	>: greater than
-: subtract	&: and	|: or	<: less than
*: multiply	|: or	~&: nand	>=: greater than or equal to
/: divide	^: xor	~|: nor	<=: less than or equal to
%: modulus	^~, ~^: xnor	^: xor	
**: power (exponent)		^~, ~^: xnor	
Logical equality	**Logical**	**Shift**	**Miscellaneous**
==: equality	&&: logical and	<<: logical left shift	{ , }: concatenation
!=: inequality	||: logical or	>>: logical right shift	{const{expr}}: replication
Case equality	!: logical negation (not)	<<<: arithmetic left shift	? : : conditional
===: equality		>>>: arithmetic right shift	
!==: inequality			

3.1.2 Expressions

The essence of dataflow modeling is the use of expressions instead of gate primitives, or instances of modules or UDPs used in the structural style to perform the desired logic functions. An expression is a construct that combines operators with operands to produce a result. In general, an expression has the following format

```
expression = operators + operands
```

where `operands` can be any of allowed data types and function calls, and `operators` act on the operands to yield desired results.

The allowed operands in an expression can be any of the following ones: constants, nets, variables (**reg, integer, time, real, realtime**), bit selects, part selects, array elements, and function calls. These operands will be described in more detail in the next section.

Like C language, Verilog HDL has a rich set of operators, including arithmetic, bitwise, reduction, relational, logical equality, case equality, logical, shift, and miscellaneous operators. The operators of Verilog HDL are summarized in Table 3.1 in terms of groups consisting of similar operators. These operators will be discussed in groups in a dedicated section later in this chapter.

The *precedence of operators* in Verilog HDL is listed in Table 3.2. Operators in the same row have the same precedence. Rows are arranged in the order of decreasing precedence for operators. For example, *, /, and % operators have the same precedence and are higher than the binary + and - operators. The left-to-right *associativity* applies to all operators except the conditional operator, which associates from right to left. Associativity refers to the order in which the operators having the same precedence are evaluated. Thus, in the following example, y is added to x and then z is subtracted from the result of $x + y$.

```
x + y - z
```

When operators differ in precedence, the operators with higher precedence associate first. In the following example, y is divided by z (division has higher precedence than addition) and then the result is added to x.

```
x + y / z
```

Parentheses can be used to change the operator precedence. For instance, $(x+y)/z$ is not the same as $x+y/z$. It is good practice to use parentheses whenever there may be ambiguous.

Table 3.2: The precedence of operators in Verilog HDL.

Operators	Symbols	Operation
Unary (plus, minus)	+ −	Highest
Unary (logical negation, bit-wise negation)	! ~	
Reduction (and, nand, or, nor, xor, xnor)	& ~& | ~| ^ ^~(~^)	
Power (exponent)	**	
Multiply, divide, modulus	* / %	
Add (binary plus), subtract (binary minus)	+ −	
Logical shift (left, right)	<< >>	
Arithmetic shift (left, right)	<<< >>>	
Relational (less than, less than or equal to)	< <=	
Relational (greater than, greater than or equal to)	> >=	
Logical equality (equality, inequality)	== !=	
Case equality (equality, inequality)	=== !==	
Bit-wise (and, xor, xnor, or)	& ^ ^~(~^) |	
Logical (and, or)	&& ||	
Conditional	?:	
Concatenation and replication	{ } {{}}	Lowest

3.1.3 Delays

The time between when an operand on the right-hand side of an expression changes and when the new value of the right-hand side is assigned to the left-hand side is controlled by a delay value. The delay value is specified right after the keyword **assign**. The inertial delay model is used as the default model for both gate circuits and nets. The delay value of an **assign** continuous assignment can be specified by using the following syntax.

```
assign     #delay3 net_lvalue = expression
       {, #delay3 net_lvalue = expression};
```

where `#delay3` is specified exactly in the same way as that of gate primitives. That is, it may be a zero-delay, one-delay, two-delay, or three-delay specification using a delay specifier in `typical_value` or `min:typ:max` form. For example,

```
wire in1, in2, out;
assign #10 out = in1 & in2;       //a delay of 10 time units
assign #(5:6:7) out = in1 & in2;//a delay of (5:6:7) time units

assign #15 out = in1 & in2,       //a delay of 15 time units
       #(5:6:8) out = in1 & in2;//a delay of (5:6:8) time units
```

3.1.3.1 Net Declarations A net can be declared associated with a delay value. Like gate primitives, the inertial delay model is also used by default in this case. Net declaration delays can also be used in the structural modeling.

```
// net delays
wire #10 out;
assign out = in1 & in2;

// regular assignment delay
wire out;
assign #10 out = in1 & in2;
```

Section 3.2 *Operands* 75

The net, out, in the above two examples receives its value 10 time units later after any of the inputs, in1 and in2, changes. Any pulse with a width less than 10 time units will be filtered out and not be propagated to the output, out, in accordance with the inertial delay model.

3.1.3.2 Net Declaration Assignments A net declaration assignment is used to specify both the net type and the assignment on the net. Since the inertial delay model is used for both nets and gate circuits, the following two examples yield the same result.

```
// net declaration continuous assignment with delay
wire #10 out = in1 & in2;
```

```
// regular continuous assignment with delay
wire out;
assign #10 out = in1 & in2;
```

The following example demonstrates the situation when both the net and gate have delays associated with them. The net, out, has a delay of 5 time units while the gate has a delay of 10 time units. As a consequence, the net, out, will receive its value 15 time units later after any of the inputs, in1 and in2, changes its value.

```
// regular assignment delay
wire #5 out;
assign #10 out = in1 & in2;
```

Like **assign** continuous assignments, delay3 can be applied to a specified net; that is, the net can be associated with a propagation delay, both rise and fall times, or both rise and fall times as well as a turn-off time. Refer to in Section 2.2.2 for details.

3.1.3.3 Modeling Transport Delay To model the delay of a net with the transport delay model, a nonblocking assignment with an intra-assignment delay may be used. As an illustration, consider the following example.

```
// transport delay model
reg out;
always @(in1 or in2)
    out <= #5 in1 & in2;
```

Any change in inputs, in1 and in2, gets scheduled on the output, out, 5 time units in the future. More details about the nonblocking assignment may be referred to in Section 4.3.1.

■ Review Questions

3-1 Explain the meaning of dataflow modeling.
3-2 What is the net declaration assignment?
3-3 What is the basic statement in the dataflow style?
3-4 Describe how to assign delays to **assign** continuous assignments.
3-5 Explain the meaning of the implicit net declaration.
3-6 Which delay model can be applied to net declaration delays in Verilog HDL?

3.2 Operands

The operands in an expression can be any of constants, parameters, nets, variables (**reg, integer, time, real, realtime**), bit-selects, part-selects, array elements, and function calls. The various net types have been discussed in the previous chapters and hence we do not repeat them here again. A parameter is like a constant and is declared using a **parameter** or **localparam** declaration, referring to Section 6.1.2 for details. A function call may also be used as an operand in an expression. It can be either a user-defined function or a system function (see Sections 5.2 and 5.3 for details). In this section, we address constants, variable types, bit-selects and part-selects, array and memory elements.

3.2.1 Constants

There are three types of constants in Verilog HDL. These are *integers*, *real numbers*, and *strings*. In this section, we deal with each of these in detail.

3.2.1.1 Integer constants Integer constants can be specified in decimal, hexadecimal, octal, or binary format. There are two forms to express integer constants: the *simple decimal form* and the *base format notation*.

Simple decimal form. In the simple decimal form, a number is specified as a sequence of digits 0 through 9, optionally beginning with a plus (+) or a minus (-) unary operator. For example,

```
-123            // is decimal -123
12345           // is decimal 12345
```

An integer value in this form represents a signed number. A negative number is internally represented in 2's-complement form.

Base format notation. When the base format notation is used, a number is composed of up to three parts: an optional size constant, a single quote followed by a base format character, and the digits representing the value of the number. The complete syntax of the base format notation is as follows

```
[size]'[s/S][base_format]base_value
```

where " ' " is a single quote. `size` specifies the size of the constant in the number of bits. It is specified as a non-zero unsigned decimal number. For example, the size specification for two hexadecimal digits is 8 because one hexadecimal digit contains 4 bits.

The base format, `base_format`, consists of a case-insensitive character specifying the base for the number, optionally preceded by an "s" or "S" qualifier to indicate a signed (in 2's-complement form) quantity and/or a single quote character ('). The allowed bases are decimal (d or D), hexadecimal (h or H), octal (o or O), and binary (b or B).

The base value, `base_value`, consists of digits (0 to 9 and a to f) that are legal for the specified base format. `base_value` should immediately follow `base_format`, optionally preceded by white space. The hexadecimal digits a (A) to f (F) are case insensitive. To represent a signed (2's-complement) number, an "s" or "S" qualifier must be prefixed with `base_format`. For example,

```
4'b1001            // a 4-bit binary number
16'habcd           // a 16-bit hexadecimal number
    2009           // unsized number -- a 32-bit decimal
   'habc           // unsized number -- a 32-bit hexadecimal
4'sb1001           // a 4-bit signed number -- -7
-4'sb1001          // a 4-bit signed number -- -(-7) = 7
```

In order to improve readability, Verilog HDL allows us to use the underscore character (_) anywhere in a number except as the first character. The underscore character is ignored when evaluating the value of the number.

```
16'b1101_1001_1010_0000// a 16-bit number in binary form
 8'b1001_0001         // an 8-bit number in binary form
```

Section 3.2 *Operands*

An x represents an unknown value and a z denotes a high-impedance value. An x (z) represents 4 x (z) bits in the hexadecimal base, 3 x (z) bits in the octal base, and 1 x (z) bit in the binary base. In Verilog HDL, the question mark (?) may also be used to improve readability in the case where the high-impedance value is a 'don't-care' condition. Hence, when used in a number, the question-mark character is an alternative for the z character.

```
16'hxxbc                  // = 16'bxxxx_xxxx_1011_1100
16'hzzbc                  // = 16'bzzzz_zzzz_1011_1100
16'b01??_1001_11?0_??00   // a binary 16-bit number
8'b01??_11??              // = 8'b01zz_11zz
```

If the size specified for a constant is larger than that of base_value, base_value will be padded to the left with zeros when base_value is an unsigned quantity or the sign bit when base_value is a signed quantity. If the leftmost bit in base_value is an x or a z, then the left bits of base_value will be padded with x's or z's.

```
8'bx001          // = 8'bxxxx_x001
8'bzz011         // = 8'bzzzz_z011
16'hx8           // = 16'bxxxx_xxxx_xxxx_1000
16'b1101_1001    // = 16'b0000_0000_1101_1001
16'sb1101_1001   // = 16'b1111_1111_1101_1001
```

In the last one, the left bits are padded with the sign bit ("1") because base_value is a 2's-complement number, declared by the s qualifier.

If the size specified for a constant is smaller than that of base_value, the leftmost bits of base_value are truncated in order to fit the specified size. For example,

```
8'b1110_1101_1001    // = 8'b1101_1001
10'sb1001_1001_0001  // = 10'b01_1001_0001
```

For an unsized constant, the size of base_value is at least 32 bits by default. Hence, base_value will be padded to the left with something like that when the size is specified.

```
// 32'b0000_0000_0000_0000_0000_1110_1101_1001
'b1110_1101_1001        // an unsized unsigned number
// 32'b1111_1111_1111_1111_1111_1001_1001_0001
'sb1001_1001_0001       // an unsigned signed number
```

3.2.1.2 Real constants Real numbers can be specified either in decimal notation or in scientific notation. Real numbers expressed with a decimal point must have at least one digit on each side of the decimal point.

```
1.5              // legal
.3               // illegal ---
1294.872         // legal
1.44E9           // the exponent symbol can be e or E
1.50e-7
15E12
26.176_45_e-12   // underscores are ignored
```

The conversion of real numbers to integers is implicitly defined by the language. Real numbers are converted to integers by rounding the real number to the nearest integer. Implicit conversion takes place when a real number is assigned to an integer. For example,

```
 1.5      // yields 2 when converted
 0.3      // yields 0 when converted
 23.445   // yields 23 when converted
-245.56   // yields -246 when converted
```

3.2.1.3 String constants A string is a sequence of characters enclosed by double quotes ("") and all characters should be on the same line. One character is represented as an 8-bit ASCII code. Hence, a string is considered as an unsigned integer constant represented by a sequence of 8-bit ASCII codes.

A string variable is a **reg** variable type with a width equal to eight times the number of characters in the string, such as the illustration in the following example.

■ **Example 3.1: An example of string manipulation.**

In this example, assume that a string str1 is employed to store the string "Welcome to the Digital World!". Hence, it requires a **reg** variable of 29×8, or a width of 232 bits. Another string str2 is used to store the string "Hello!". It needs a **reg** variable of 6*8 or a width of 48 bits. After the execution of the program, the output is

```
Welcome to the Digital World!
Hello! is stored as: 48656c6c6f21
```

module string_test;
// internal signal declarations:
reg [29*8:1] str1;
reg [6*8:1] str2;
initial begin
    str1 = "Welcome to the Digital World!";
    $**display**("%s\n", str1);
    str2 = "Hello!";
    $**display**("%s is stored as: %h\n", str2, str2);
end
endmodule

The backslash (\) character can be used to escape certain special characters.

```
\n        \\ new line character
\t        \\ tab character
\\        \\ \ character
\"        \\ " character
\ddd      \\ a character specified in 3 octal digits.
```

3.2.1.4 Declaration of Constants There are three ways in Verilog HDL that may be used to declare a constant in the source description of a design. These are the **'define** compiler directive, and module parameters, **parameter** and **localparam**.

The **'define** compiler directive is used to create a macro for text substitution. For instance,

'define BUS_WIDTH 8

Any place in the source description with the occurrence of 'BUS_WIDTH will be replaced with the number 8. The **'define** compiler directive is usually placed at the head of a file or in a separated file and can be used in both the inside and outside of a module definition.

The module **parameter** is the most common approach used to define parameters that can be overridden by a **defparam** statement or a module instance parameter value assignment. For instance,

parameter BUS_WIDTH = 8;

The issues of parameter overrides will be discussed in Section 6.1.4 in more detail.

Another module parameter **localparam** is identical to the module **parameter** except that it cannot be modified with a **defparam** statement or by a module instance parameter value assignment. Hence, the **localparam** is usually employed to define constants only locally used in the module. For example,

localparam BUS_WIDTH = 8;

■ **Review Questions**

3-7 What is the meaning of a parameter?
3-8 Can the expression of an **assign** continuous assignment contain a function call?
3-9 What are the three constants provided by Verilog HDL?
3-10 Explain the meaning of the s qualifier in the base format notation of integer constants.
3-11 How many bits are defaulted at least when a constant is unsized?

3.2.2 Variable Types

A variable represents one data storage element and stores a value from one assignment to the next. The variable types include five different kinds: **reg**, **integer**, **time**, **real**, and **realtime**. All **reg**, **time**, and **integer** variables are initialized with an unknown value, x, whereas both **real** and **realtime** variables are initialized with a 0.0.

3.2.2.1 The reg Variable A **reg** variable holds a value between assignments. It may be used to model hardware registers, including edge-sensitive (i.e., flip-flops) and level-sensitive (i.e., latches) storage elements. However, a **reg** variable does not actually correspond to a hardware storage element when synthesized because it can also be used to denote a combinational logic net. A **reg** variable may be declared by the following syntax

```
reg [signed] [[msb:lsb]]
    variable_id[ = const_expr]{, variable_id[ = const_expr]};
```

where the initialization part, const_expr, is optional. A **reg** variable defaults to be unsigned but may be made to be signed by using the keyword **signed** following the keyword **reg**. The range [msb:lsb] is an optional part and used to specify the number of bits of the **reg** variable. The use and meaning of msb and lsb are the same as in the case of the net type and hence we omit them here. If no range is specified, a 1-bit **reg** variable is defaulted.

```
reg a, b, c_in;          // 1-bit reg variables
reg [7:0] data_a;        // an 8-bit reg, the msb is bit 7
reg [0:7] data_b;        // an 8-bit reg, the msb is bit 0
reg signed [7:0] d;      // an 8-bit signed reg
```

Vector versus scalar. It is more convenient to describe *a bundle of signals* as a basic unit when describing a hardware module. This bundle of signals is referred to as a *vector* (multiple-bit width) in Verilog HDL. All nets and the **reg** variable have the default type of 1-bit vector, called a *scalar*, but can be declared as vectors as needed.

Comparison between reg variables and wire nets. A **reg** variable is one of the variable types while a **wire** net is one of the net types. Upon initialization, a **reg** variable has an unknown value, x, while a **wire** net has a high impedance, z. A **reg** variable can only be assigned values in an **initial** or **always** block while a **wire** net can merely be assigned values in an **assign** continuous assignment, a net declaration assignment, a **force/release** statement, or through the **output** or **inout** port of a module instance. Furthermore, a **wire** net can be assigned a strength value but a **reg** variable cannot.

3.2.2.2 The integer Variable

An **integer** variable contains integer values and can be used as a general-purpose variable for modeling the high-level behavior of a design. The syntax for declaring **integer** variables is as follows

```
integer variable_id[ = const_expr]{, variable_id[ = const_expr]};
```

where the initialization part, const_expr, is optional. The **integer** variable uses the same assignment rules as the **reg** variable. An **integer** variable is treated as a signed **reg** variable with the least significant bit (LSB) being bit 0 and has at least 32 bits, regardless of implementations. Arithmetic operations performed on **integer** variables yield 2's-complement results.

```
integer i,j;          // declare two integer variables
integer data[7:0];    // an array of integer
```

3.2.2.3 The time Variable

A **time** variable is used for storing and manipulating simulation time quantities. It is typically used in conjunction with the **$time** system task. The syntax for declaring **time** variables is as follows

```
time variable_id[ = const_expr]{, variable_id[ = const_expr]};
```

where the initialization part, const_expr, is optional. A **time** variable holds only unsigned values and is at least 64 bits, with the LSB being bit 0.

```
time events;          // hold one time value
time current_time;    // hold one time value
```

3.2.2.4 The real and realtime Variables

Verilog HDL supports **real** and **realtime** variable data types in addition to the **integer** and **time** variables. The **real** and **realtime** variables are identical and can be used interchangeably. The syntax for declaring **real** or **realtime** variables is as follows:

```
real     real_id[ = const_expr]{, real_id[ = const_expr]};
realtime real_id[ = const_expr]{, real_id[ = const_expr]};
```

Both **real** and **realtime** variables cannot use the range declaration and their initial values are defaulted to zero (0.0).

```
real     events;           // declare a real variable
realtime current_time;     // hold current time as real
```

In summary, a value in a **reg** variable is interpreted as an unsigned number by default unless the keyword **signed** is used. A value in an **integer** variable is interpreted as a signed 2's-complement number whereas a value in a **time** variable is regarded as an unsigned number. Values in both **real** and **realtime** variables are regarded as signed floating-point numbers.

■ Review Questions

3-12 Which of variable types can be used to model a hardware storage element?
3-13 What are the differences between **reg** and **integer** variables?
3-14 What are the features of the **time** variable?
3-15 Describe the features of **real** and **realtime** variables.
3-16 What are the differences between the **wire** net and the **reg** variable?
3-17 What is the major difference between **integer** and **time** variables?

3.2.3 Bit-Selects and Part-Selects

A bit-select extracts a particular bit from a vector. Recall that all net types and the **reg** variable type can be declared as vectors. Although **integer** and **time** variables cannot be declared as vectors, they can also be accessed by bit-selects or part-selects. However, the bit-select or part-select of **real** and **realtime** variables is not allowed. The bit-select has the following form

```
vector_name [bit_select_expr]
```

where vector_name can be any vector of nets as well as **reg**, **integer**, and **time** variable types. bit_select_expr may be an expression. If it is evaluated to an x or a z, or out of bounds, then the bit-select value is an x.

In a part-select, a sequence of consecutive bits of a vector is selected. The part-select can be done by either a *constant part-select* or an *indexed part-select*. A constant part-select has the following form:

```
vector_name [msb_const_expr:lsb_const_expr]
```

Both msb_const_expr and lsb_const_expr expressions must be a constant expression. Some examples are given as follows:

```
reg  [15:0] data_bus;  // declarations
wire [7:0] a;
integer response_time;

data_bus[3:0]        // reg variable part-select
a [4:3]              // net part-select
response_time[5:0]   // integer variable part-select
```

An indexed part-select has the following form

```
vector_name [<starting_bit>+:const_width]
vector_name [<starting_bit>-:const_width]
```

where starting_bit can vary at run time and const_width has to be constant. The range of bits selected is the index specified by starting_bit plus or minus the numbers specified by const_width. "+:" indicates that the part-select increases from starting_bit while "-:" indicates that the part-select decreases from starting_bit. For example,

```
data_bus[8+:8]       // select data_bus[15:8]
data_bus[15-:4]      // select data_bus[15:12]
```

Like the bit-select, if either the range index is out of bounds or evaluates to an x or a z, the part-select value is an x.

As an illustration of a simple application of the indexed part-select, consider the problem of converting a number in big-endian form into its equivalent in little-endian form and vice versa. In digital systems, the byte order for representing a multiple-byte word can be cast into two standard forms: the *big-endian form* and the *little-endian form*. In the big-endian form, the least significant byte is on the highest memory address while the most significant byte is on the lowest memory address. In the little-endian form, the byte order is reversed; that is, the least significant byte is on the lowest memory address whereas the most significant byte is on the highest memory address.

To explore how to convert a number in big-endian form into its equivalent in little-endian form and vice versa by using the indexed part-select, consider the following example. For simplicity, the underlying word is assumed to be 32 bits.

■ Example 3.2: Conversion between big endian and little endian.

By properly setting the starting bits and ranges, the following module converts a number in big-endian form into its equivalent in little-endian form and vice versa. It is of interest to note that this module is virtually an arrangement of nets. It does not cost any hardware. Try to synthesize it and see the synthesized results. Moreover, you are encouraged to write a test bench to verify the functionality of the module.

```
module swap_bytes (input   [31:0] in,
                   output  [31:0] out);
// using indexed part-select
assign out [31 -: 8] = in [0  +: 8],
       out [23 -: 8] = in [8  +: 8],
       out [15 -: 8] = in [16 +: 8],
       out [7  -: 8] = in [24 +: 8];
endmodule
```

Note that a net or a **reg** variable is defaulted to be unsigned and may be declared as a signed data type with the keyword **signed**. However, both bit-select and part-select results are unsigned regardless of whether the operands are signed or unsigned.

3.2.3.1 Vectored and Scalared Usually, a vector net can be accessed by both bit-selects and part-selects. However, when a vector net does not want to be accessed by a bit-select or a part-select, the keyword **scalared** may be specified in the net declaration. The syntax is as follows:

```
net_type [vectored|scalared] [signed]
         range [delay3] net_id{, net_id};
net_type [vectored|scalared] [signed]
         range [delay3] net_id=expr{, net_id=expr};
```

When no keyword **scalared** is specified, the default is **vectored**, which is the same as when the keyword **vectored** is specified, and both bit-selects and part-selects are permitted to access the vector net. For example,

```
wire scalared [63:0] bus64;  // treated bus64 as a unit
tri  vectored [31:0] data;   // allow partial accesses
```

where bus64 is regarded as a single unit and not allowed to access partially by way of bit-selects and part-selects. data is a 32-bit vector and allows to be accessed partially.

■ Review Questions

3-18 Explain the meanings of a bit-select and a part-select.
3-19 How would you prohibit a vector to be accessed by a bit-select or a part-select?
3-20 What is a constant part-select?
3-21 What is an indexed part-select?
3-22 Could we apply a bit-select or a part-select to an **integer** or a **time** variable?
3-23 Could we apply a bit-select or a part-select to a **real** or **realtime** variable?

3.2.4 Array and Memory Elements

Arrays can be used to group elements into multi-dimensional objects. Although only nets and the **reg** variable can be declared as vectors, all net and variable types are allowed to be declared as multi-dimensional arrays. A multi-dimensional array is declared by specifying the address ranges after the declared identifier, called *dimensions*, one for each. In addition, an array element can be a scalar or a vector if the element is a net or a **reg** variable. The vector size (defined by range) specifies the number of bits in each element and the dimensions (i.e., address ranges) specify the number of elements in each dimension of the array. The syntax is as follows

```
net_type [signed] [range] net_id [msb:lsb] {[msb:lsb]};
reg      [signed] [range] variable_id [msb:lsb] {[msb:lsb]};
integer|time    variable_id [msb:lsb] {[msb:lsb]};
real|realtime   real_id [msb:lsb] {[msb:lsb]};
```

where msb and lsb are constant-valued expressions indicating the range of indices of a dimension. If no dimensions are specified, each net or variable only stores a value. For example,

```
wire a[3:0];                // a 4-element scalar wire array
reg  d[7:0];                // an 8-element scalar reg array
wire [7:0] x[3:0];          // a 4-element 8-bit wire array
reg  [31:0] y[15:0];        // a 16-element 32-bit reg array
reg  [7:0] datab [3:0][3:0]; // 2-D array of 8-bit vector
wire sum [7:0][3:0];        // 2-D array of scalar wire
integer states [3:0];       // a 4-element integer array
time    current[5:0];       // a 6-element time array
```

To access an element from an array, we use the following format

```
array_name [addr_expr] {[addr_expr]}
```

where addr_expr can be any expression. As with the bit-select or part-select, if addr_expr is out of bounds, or any bit in addr_expr is an x or a z, then the reference value is an x. Only an element of an array can be assigned a value in a single assignment; an entire or a partial array dimension cannot be assigned to another by a single assignment. However, a bit-select or a part-select of an element of an array may be accessed and assigned. To assign a value to an element of an array, it needs to specify an index for each dimension. The index can be an expression. For instance,

```
states[3] = 33559;          // assign decimal number
current[t_index] = $time;   // current simulation time

datab[1] = 0;               // illegal -- it needs two indices
datab[1][3:1] = 0;          // illegal -- write to partial array
datab[1][0] = 3;            // assigns 3 to the datab[1][0]
```

3.2.4.1 Memory Memory is a basic module in any digital system; in Verilog HDL, it is simply declared as a one-dimensional **reg** array. Memory can be used to model a read-only memory (ROM) module, a random access memory (RAM) module, and a register file. Reference to memory may be made to a whole word or a portion of a word.

To access a memory word, we use the following format

```
mem_name[addr_expr]
```

where addr_expr can be any expression. For example,

```
reg [7:0] mem [7:0];    // 1-D array of 8-bit vector

mem = 0;                // illegal -- write to entire array
mem[1] = 0;             // assigns 0 to the 2nd element of mem
```

Memory indirection is allowed and can be specified in a single expression, such as the following example.

```
mem_name[mem_name[23]]  // use memory indirections
```

meaning that mem_name[23] addresses the 23th word of the memory mem_name. The value at the word is then used as the address to access mem_name.

3.2.4.2 Bit-Selects and Part-Selects Both bit-select and part-select are allowed to access a memory element or an array element of arrays. To do this, the desired element is first selected by using a normal array access. Then, a bit-select or a part-select is applied to the selected element in the same way as in the case of vectors.

```
mem[4][3]               // the 3rd bit of the 5th element
mem[5][7:4]             // the higher nibble of the 6th element

sum[5][0]               // the [5][0]th element
datab[3][1][1:0]        // bits 1 and 0 of the [3][1]th element
```

Note that a memory declaration of n 1-bit **reg** variables is different from an n-bit **reg** vector.

```
reg [n:1] rega;         // an n-bit register is not the same
reg mema [n:1];         // as a memory of n 1-bit registers
```

■ Review Questions

3-24 Which data types can be declared as an array?

3-25 How would you declare a multi-dimensional array in Verilog HDL?

3-26 How would you distinguish the vector size from the address range in an array declaration?

3-27 What is a memory declaration from the viewpoint of Verilog HDL?

3-28 Can an array be assigned to another in a single statement?

3-29 How would you access a bit in a memory word?

3.3 Operators

As shown in Table 3.1, Verilog HDL has a rich set of operators, including bit-wise operators, arithmetic operators, concatenation and replication operators, reduction operators, logical operators, a conditional operator, relational operators, equality operators, and shift operators. In this section, we describe these operators in detail and give examples to demonstrate how to use them to model realistic logic circuits.

3.3.1 Bit-wise Operators

Bit-wise operators perform a bit-by-bit operation on two operands and yield a vector result. In bit-wise operations, a z is treated as an unknown x. The set of bit-wise operators includes five operators: ~(negation), & (and), | (or), ^ (xor), and ^~(xnor) as shown in Table 3.3. The functions of these operators are listed as follows:

Section 3.3 *Operators*

Table 3.3: The bit-wise operators.

Symbol	Operation
~	Bit-wise negation
&	Bit-wise and
|	Bit-wise or
^	Bit-wise xor
^~, ~^	Bit-wise xnor

- ~(negation): If the input bit is 1 the result is 0, or else if the input bit is 0, the result is 1; otherwise, the result is an x.
- & (and): If any bit is 0, the result is 0, or else if both bits are 1, the result is 1; otherwise, the result is an x.
- | (or): If any bit is 1, the result is 1, or else if both bits are 0, the result is 0; otherwise, the result is an x.
- ^ (xor): If one bit is 1 and the other is 0, the result is 1, or else if both bits are 0 or 1, the result is 0; otherwise, the result is an x.
- ^~(xnor): If one bit is 1 and the other is 0, the result is 0, or else if both bits are 0 or 1, the result is 1; otherwise, the result is an x.

Except for the ~(negation), all the other bit-wise operators need their operands to have the same length. If two operands do not have the same length and either operand is an unsigned operand, the shorter operand is *zero-extended* to match the length of the longer one. If both operands are signed, the shorter operand is *sign-extended* to match the length of the longer one before the operation takes place.

■ **Example 3.3: An illustration of unequal-length operands.**

In this example, the first value of w is 1111_1101, which is resulted from the OR operation of the x and the sign-extended z. The second value of w is 0001_1110, which is obtained from the OR operation of the x and the zero-extended y because y is an unsigned number.

```
'timescale 1 ns/100 ps
module bit_wise_test_tb;
reg signed [7:0] w, x;
reg [3:0] y;
reg signed [3:0] z;
initial begin
   x = 8'sb0001_1100;  // signed number
   z = 4'sb1001;       // signed number
   y = 4'b1010;        // unsigned number
   w = x | z;          // both are signed
   #10;
   w = x | y;          // y is unsigned
end
initial #50 $finish;
initial                 // response monitoring block
   $monitor($time,"ns %b %b %b %b", w, x, y, z);
endmodule
```

What are the differences between 4'sb1001 and 4'b1001? It is of interest to replace the constants assigned to both **reg** variables x and z in the above module with 8'b0001_1100 and

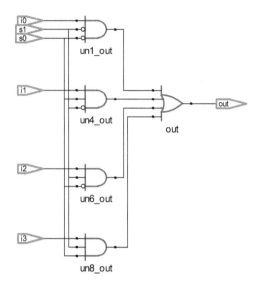

Figure 3.1: The synthesized result from the dataflow model of a 4-to-1 multiplexer.

4'b1001, respectively, and observe the simulation results. Explain the reasons for what you obtained.

■ Example 3.4: A 4-to-1 multiplexer modeled in dataflow style.

As an illustration of using bit-wise operators to model a 4-to-1 multiplexer, consider the switching expression derived from Figure 2.5. The expression can be described by an **assign** continuous assignment directly with & (and), | (or), and ~(negation) bit-wise operators. The resulting module is as follows:

```
module mux_4to1_dataflow(
       input i0, i1, i2, i3,  // inputs
       input s1, s0,          // source selection inputs
       output out);           // output

// using bit-wise operators: and, or, not
assign out = (~s1 & ~s0 & i0) |
             (~s1 &  s0 & i1) |
             ( s1 & ~s0 & i2) |
             ( s1 &  s0 & i3) ;
endmodule
```

The synthesized result of the above `mux_4to1_dataflow` module is shown in Figure 3.1. Actually, this is exactly the same as that obtained from being modeled in structural style.

■ Review Questions

3-30 Describe the basic operations of the set of bit-wise operators.
3-31 What operations are taken when two operands are unequal in length?
3-32 Describe the operation of & (and) using the four-valued set.
3-33 Describe the operation of ^ (xor) using the four-valued set.
3-34 Describe the operation of ~(negation) using the four-valued set.

Section 3.3 *Operators*

Table 3.4: The arithmetic operators.

Symbol	Operation
+	Add
−	Subtract
*	Multiply
/	Divide
**	Exponent (power)
%	Modulus

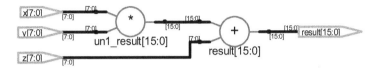

Figure 3.2: A multiply and accumulate unit.

3.3.2 Arithmetic Operators

The set of arithmetic operators contains six members {+, −, *, /, %, **}, as shown in Table 3.4. The operators + and − perform addition and subtraction of two numbers, respectively. In Verilog HDL, negative numbers are represented in 2's-complement form. The operators + and − can also be used as unary operators to denote signed numbers. The operators * and / compute the multiplication and division of two numbers, respectively. The integer division operator (/) truncates any fractional part toward zero while the modulus operator (%) produces the remainder from the division of two numbers. The exponent operator computes the power of a number. The result of the exponent operator ** is **real** if either operand is **real**, **integer**, or **signed**.

■ Example 3.5: A multiply and accumulate unit.

In this example, we directly use arithmetic operators, + and *, to construct a multiply and accumulate unit. The multiply operator (*) may be synthesized by using a hardware macro, basic gates, or cells organized by an internal algorithm associated with the synthesis tool, depending on the synthesis tool and the target technology library used. The synthesized result is shown in Figure 3.2. It is suggested to synthesize this module with different operand sizes and see what happens.

```
// an example to illustrate arithmetic operators
module multiplier_accumulator(
       input  [3:0] x, y, z,
       output [7:0] result);
//
assign result = x * y + z ;
endmodule
```

■ Example 3.6: An unsigned divider.

In this example, we directly use the divide arithmetic operator to construct an unsigned divider. The divide operator / may be synthesized by using a hardware macro, basic gates, or cells organized by an internal algorithm associated with the synthesis tool, depending on the synthesis tool and the target technology library used. The reader is encouraged to synthesize this module with different operand sizes and see what happens.

```verilog
// an example to illustrate the divide operator
module divide_operator(
        input   [7:0] x, y,
        output  [7:0] result);
//
assign result = x / y;
endmodule
```

The following two statements show the use of the exponent (power) operator (**):

parameter ADDR_SIZE = 4;
localparam ROM_SIZE = 2 ** ADDR_SIZE − 1;

The ROM capacity is related to its address width by way of an exponent operation. Through the use of both **parameter** and **localparam** parameters, we can specify a ROM only with one parameter, namely, the number of bits of its address. However, it should be noted that the exponent (power) operator is usually not supported by synthesis tools except that it can be evaluated to an integer power of 2. The details of both **parameter** and **localparam** parameters will be dealt with in Section 6.1.2.

A **reg** variable is treated as an unsigned value unless it is declared to be signed explicitly with the keyword **signed**. An **integer** variable is treated to be signed by default. In addition, for an arithmetic operation, if any bit of an operand is an x or a z, then the entire result value will be an x. The size of the result of an arithmetic expression is determined by the size of the longest operand and the left-hand-side target as well if it is an assignment. This rule is also applied equally well to all intermediate results of an expression.

■ Example 3.7: The result size of an expression.

In the following module, both sizes of the result and intermediate results of the first **assign** continuous assignment are determined by inputs a and b, and the **wire** net c, which are all 4 bits. The size of the result of the second **assign** continuous assignment is 8 bits because the largest operand is 8 bits. The synthesized result is shown in Figure 3.3. It is instructive to synthesize this module by changing the widths of the input and output ports and to explore the results.

```verilog
// an example to illustrate arithmetic operators
module arithmetic_operators(
        input   [3:0] a, b,
        input   [6:0] e,
        output  [3:0] c,
        output  [7:0] d);

//
assign c = a + b;
assign d = a + b + e;
endmodule
```

In general, an unsigned value is stored in a net or a **reg** variable, or an integer in base format without a signed qualifier (s). A signed value is stored in a signed net, a signed **reg** variable, or an **integer** variable, or an integer in decimal form, or an integer in base format with a signed qualifier (s).

In an expression combining signed with unsigned operands, all operands are converted to unsigned ones before any operation takes place. The conversion of a number between signed

Section 3.3 Operators

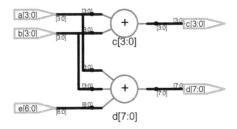

Figure 3.3: The synthesized result of the `arithmetic_operator` module.

and unsigned formats can also be explicitly accomplished by **$signed** and **$unsigned** system functions, which are dealt with in Section 5.3.4.

■ Example 3.8: Mixing signed with unsigned operands.

The first expression is carried out in 2's-complement numbers because all operands are signed but the second expression is performed in unsigned numbers because the input e is an unsigned number.

```
// an example to illustrate arithmetic operators
module arithmetic_signed(
        input   signed [5:0] a, b,
        input   [6:0] e,
        output  signed [3:0] c,
        output  [7:0] d);
// test signed arithmetic operations
assign c = a + b;
assign d = a + b + e;
endmodule
```

The reader is encouraged to synthesize the above module and check the result. It is also suggested to write a test bench to verify the above module.

■ Review Questions

3-35 Describe the operation of the % (modulus) arithmetic operator.
3-36 Describe the operation of the ** (exponent) arithmetic operator.
3-37 Which data types are unsigned by default?
3-38 What will be the result type (signed or unsigned) when both signed and unsigned operands are mixed in an expression?
3-39 How would you determine the size of the result of an arithmetic expression?

3.3.3 Concatenation and Replication Operators

Concatenation is an operation that combines many shorter expressions or bits to form a longer one. A concatenation operator is expressed by braces { and }, with commas separating the expressions within it, as shown in Table 3.5. The operands of a concatenation operator must be sized. Operands can be scalar nets or **reg** variables, vector nets or **reg** vector variables, bit-selects, part-selects, or sized constants. For example,

```
y = {a, b[0], c[1]};
```

Table 3.5: The concatenation and replication operators.

Symbol	Operation
{ , }	Concatenation
{constant_expr{expr}}	Replication

concatenates a, b[0], and c[1] into a group in that order and assigns to y. Hence, y contains three bits if all a, b[0], and c[1] are one bit. Another example is as follows:

y = {a, b[0], 4'b1111};

which concatenates a, b[0], and 4'b1111 into a group and assigns to y. Hence, y contains six bits if both a and b[0] are one bit.

A replication operator specifies how many times to replicate the object inside the braces. It is expressed as {const_expr{expr}}, as shown in Table 3.5. The first expression, const_expr, must be a non-zero, non-x, and non-z constant expression while the second expression, expr, follows the rules for concatenation. For instance, {4{a}} replicates the object "a" 4 times. Another example,

y = {a, {4{b[0]}}, c[1]};

The right-hand side has six bits in total if all a, b[0] and c[1] are a single-bit net or a single-bit **reg** variable.

The following example shows how the concatenation operator is used to concatenate a scalar and a vector net operands.

■ **Example 3.9: A 4-bit adder using the concatenation operator.**

In this example, a 4-bit adder is described in dataflow style. The left-hand side of the **assign** continuous assignment is a concatenation of a scaler, c_out, and a vector, sum. The result is 5 bits.

```
module four_bit_adder(
        input   [3:0] x, y, // declare as a 4-bit vector
        input   c_in,
        output  [3:0] sum,  // declare as a 4-bit vector
        output  c_out);
// specify the function of a 4-bit adder.
assign {c_out, sum} = x + y + c_in;
endmodule
```

The following example illustrates how to use the replication operator to describe a 4-bit 2's-complement adder.

■ **Example 3.10: A 4-bit 2's-complement adder.**

In this example, the carry-in, c_in, is duplicated into four copies by using a replication operator and then combines with an input operand, y, through an **xor** operator to form a true/1's-complement circuit. The resulting circuit is combined with the other input operand, x, and the carry-in, c_in, to construct the desired 4-bit 2's-complement adder.

```
module twos_adder(
        input   [3:0] x, y, // declare as a 4-bit vector
        input   c_in,
```

Section 3.3 Operators

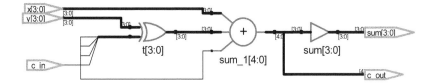

Figure 3.4: The synthesized result of the 4-bit 2's-complement adder.

Table 3.6: The reduction operators.

Symbol	Operation
&	Reduction and
~&	Reduction nand
|	Reduction or
~|	Reduction nor
^	Reduction xor
^~, ~^	Reduction xnor

```
       output [3:0] sum,   // declare as a 4-bit vector
       output c_out);
wire [3:0] t; // outputs of xor gates
// specify the function of a 2's-complement adder
assign t = y ^ {4{c_in}}; // what's wrong with y^c_in?
assign {c_out, sum} = x + t + c_in;
endmodule
```

The synthesized result of the above twos_adder module is shown in Figure 3.4. It is interesting to note what would happen when the replication operator is removed, namely, only using the carry-in, c_in. Try to synthesize it and explain the reason about the synthesized result.

3.3.4 Reduction Operators

The set of unary reduction operators carries out bit-wise operations on a single vector operand and yields a 1-bit result. Reduction operators only perform on one vector operand and work in a bit-by-bit fashion from right to left. The set of reduction operators is shown in Table 3.6. It includes the following six basic operations: {and, nand, or, nor, xor, xnor}. Reduction nand, reduction nor, and reduction xnor are computed by inverting the results of reduction and, reduction or, and reduction xor, respectively.

A simple application of reduction operator, ^ (xor), to model the 9-bit parity generator described in the previous chapter is demonstrated in the following example. It is interesting to compare this with the one modeled in structural style described in Section 2.1.1.

■ Example 3.11: A 9-bit parity generator.

This example uses the reduction operator ^ (xor) to reduce the vector operand, x, to a single-bit result. This results in the even-parity output (ep). The odd-parity output (op) is obtained by complementing the even-parity output.

```
module parity_gen_9b_reduction(
       input [8:0] x,
       output ep, op);
```

92 Chapter 3 Dataflow Modeling

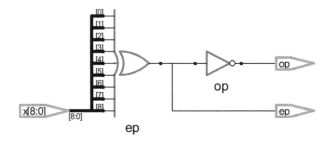

Figure 3.5: A 9-bit parity generator using the reduction operator.

```
// dataflow modeling using reduction operator
assign ep = ^x;  // even-parity generator
assign op = ~ep; // odd-parity generator
endmodule
```

The synthesized result is shown in Figure 3.5. The reader is encouraged to examine the details of the synthesized results and to explore how the synthesis tool realizes the reduction operator ^ (xor). Does it synthesize the reduction operator as a linear structure or a tree structure?

From the preceding example, we can see the power of reduction operators. By using these operations, we may write a very compact and readable module. Another simple but useful application is to detect whether all bits in a byte are zeros or ones. An illustration of this is given in the following example. Here, we assume that both results are desired.

■ **Example 3.12: An all-bits-zero/one detector.**

The outputs, zero and one, are assigned to 1 if all bits of the input vector, x, are zeros and ones, respectively. These two detectors are easily implemented by the reduction operators: ~| (nor), and & (and). The resulting module is as follows:

```
module all_bits_01_detector_reduction(
      input  [7:0] x,
      output zero, one );
// dataflow modeling
assign zero = ~|x;  // all-bit zero detector
assign one = &x;    // all-bit one detector
endmodule
```

The synthesized result is shown in Figure 3.6. It is instructive to see the details of the synthesized results and to explore how the synthesis tool realizes the reduction operators, | and &. Does it synthesize the reduction operator as a linear structure or a tree structure?

■ **Review Questions**

3-40 Describe the operation of the concatenation operator {}.
3-41 Describe the operation of the replication operator {constant_expr{expr}}.
3-42 What are the differences between bit-wise operators and reduction operators?
3-43 Describe the operation of the reduction operator & (and).
3-44 Describe the operation of the reduction operator ^ (xor).

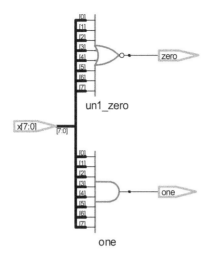

Figure 3.6: An all-bits-zero/one detector.

Table 3.7: The logical operators.

Symbol	Operation
!	Logical negation
&&	Logical and
||	Logical or

3.3.5 Logical Operators

There are three logical operators: ! (negation), && (and), and || (or), as shown in Table 3.7. Each of these operates on two logic values, false (0) and true (1), and produces a 1 bit value 0, 1, or an unknown x. Any operand bit with a value of x or z is treated as an unknown x and is regarded as a false condition by simulators. For vector operands, a zero vector is treated as a logic 0 and a non-zero vector is regarded as a logic 1.

As an illustration, consider the following example. If the **reg** variable c holds the integer value 123 and the **reg** variable d holds the value 0, then the **reg** variable a is a logic 1 and the **reg** variable b is a logic 0.

```
reg a, b;
reg [7:0] c, d;

a = c || d;  // a is set to 1
b = c && d;  // b is set to 0
```

For a naive reader, it is often to misuse the & (bit-wise and) operator to replace the && (logical and) operator, in particular, in the condition expression of the **if-else** statement. As this is the case, the results might be different if the operands in the condition expression are vectors rather than scalars. To illustrate this, let c be 4'b0101 and d be 4'b1010, then (c & d) is 4'b0000 and thus is false (0) but (c && d) is true (1) since both c and d are non-zero vectors. Nonetheless, the results will be the same if all operands in the condition expression are scalars. Despite this, it should use the && operator rather than the & operator in the condition expression of the **if-else** statement.

A comment on the negation operator (!) is that we often use it in a construct like the **if** (!reset)... statement to represent an equivalent one: **if** (reset == 0). Besides, the **if** (!x)... statement is not equivalent to the **if** (~x)... statement if x is not a single bit.

3.3.6 Conditional Operator

The conditional operator (? :) selects an expression based on the value of a conditional expression. It has the following form

```
condition_expr ? true_expr: false_expr
```

where `condition_expr` is evaluated first. If the result is true, `true_expr` is executed; if the result is false, `false_expr` is executed; if the result is an unknown x (ambiguous) or a z, the result is a bit-wise operation of both `true_expr` and `false_expr` expressions with the following rule: each bit position gets the value of the bits if they are identical and an unknown x if they are different. The conditional operator (? :) can be nested.

As we can assure that `condition_expr` is true or false, which is often the case in practice, the following **assign** continuous assignment

```
assign net_lvalue = condition_expr ? true_expr : false_expr;
```

can be rewritten as an **if-else** statement as follows:

```
always @(*)
   if   (condition_expr) variable_lvalue = true_expr;
   else  variable_lvalue = false_expr;
```

The details of the **if-else** statement will be introduced in Chapter 4 when we deal with the details of behavioral modeling.

A simple example of using a conditional operator (? :) to describe a 2-to-1 multiplexer is as follows:

```
assign out = selection ? input1 : input0;
```

The `out` output receives `input1` if `selection` is 1 and `input0` otherwise.

A more complicated application of using a nested conditional operator to model a 4-to-1 multiplexer is illustrated in the following example.

■ **Example 3.13: An example of 4-to-1 multiplexers.**

In this example, we use a nested conditional operator in which both true and false expressions contain their own conditional operators. The reader is encouraged to check the correctness of this module by using a synthesizer and observing the synthesized result. Although a construct like this is quite concise, it is not easy to be understood by a naive reader. In the next chapter, we will introduce a more readable way to describe such a multiplexer.

```
module mux_4to1_conditional(
       input   i0, i1, i2, i3, // inputs
       input   s1, s0,         // source selection inputs
       output  out);           // output
// using conditional operator (?:)
assign out = s1 ? ( s0 ? i3 : i2) : (s0 ? i1 : i0) ;
endmodule
```

Table 3.8: The relational operators.

Symbol	Operation
>	Greater than
<	Less than
>=	Greater than or equal to
<=	Less than or equal to

3.3.7 Relational Operators

The set of relational operators includes four operators, > (greater than), < (less than), >= (greater than or equal to), and <= (less than or equal to), as shown in Table 3.8. Relational operators return a logic 1 if the expression (i.e., operand1 relational operator operand2) is true and a logic 0 if the expression is false. The expression results in a 1-bit unknown x if either operand contains an x or a z bit.

Fundamentally, a relational operator needs its two operands to have equal length in bit. As two operands are not of equal length in bit, the following remediation is carried out. If any operand is unsigned, the shorter operand is zero-extended to match the size of the longer one and the operation is performed between two unsigned values. If both operands are signed, the shorter operand is sign-extended and the operation is carried out between two signed values. In addition, as either operand is a real value then the other operand is converted to an equivalent real value and the operation is performed between two real values.

All relational operators have the same precedence and their precedence is lower than that of arithmetic operators. As an illustration of the use of relational operators, consider a simple example that models a circuit which converts an excess-3 code to an equivalent *binary-coded decimal* (BCD) code.

Like the BCD code, the excess-3 code is also a decimal code, used to represent decimal digits. In the excess-3 code, each codeword is formed by adding 4'b0011 to the BCD codeword representing the same digit. For example, the excess-3 codewords denoting digits 0 and 5 are 4'b0011 and 4'b1000, respectively.

■ **Example 3.14: An excess-3-to-BCD converter.**

This example describes a code converter to convert an excess-3 code to a BCD code. The code converter works as follows. It first checks to see whether the input codeword is a valid excess-3 code or not. If yes, it converts the input codeword into its equivalent BCD codeword by subtracting 4'b0011 from the input codeword and sets the valid signal true to indicate this situation; otherwise, it directly passes the input codeword to the output and sets the valid signal false. The resulting module is as follows:

```
module excess3_to_BCD(
        input  [3:0] x,    // excess-3 input
        output [3:0] y,    // BCD output
        output valid);     // indicate a valid excess-3 input

// the body of the excess-3-to-BCD converter
assign valid = ((x >= 3) && (x <= 12)) ? 1'b1 : 1'b0;
assign y = (valid) ? (x - 4'b0011) : x;
endmodule
```

The synthesized result is depicted in Figure 3.7. It is apparent that two limit comparators and an adder (indeed, a subtracter) are inferred. The two limit comparators are used to check

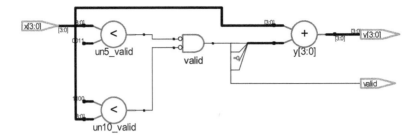

Figure 3.7: The synthesized result of the excess-3-to-BCD converter.

Table 3.9: The equality operators.

Symbol	Operation
==	Logical equality
!=	Logical inequality
===	Case equality
!==	Case inequality

whether the input is a valid excess-3 codeword and to generate the valid signal. The adder performs the desired code conversion by adding 4'1101 (the 2's complement of 3) to the input codeword to yield the output BCD codeword if the valid signal is true and outputs the input codeword directly, otherwise.

■ Review Questions

3-45 Describe the operations of logical operators?
3-46 Describe the operations of the conditional operator (? :).
3-47 Describe the operations of relational operators?
3-48 What is the advantage of using the conditional operator (? :) over an **if-else** statement?

3.3.8 Equality operators

The set of equality operators includes four operators, == (logical equality), != (logical inequality), === (case equality), !== (case inequality), as shown in Table 3.9. Equality operators return a logic 1 if the expression (i.e., operand1 equality_operator operand2) is true and return a logic 0 if the expression is false. The logical equality operators (== and !=) yield an unknown x if either operand contains an x or a z bit. The case equality operators (=== and !==) yield a logic 1 if their two operands match exactly, including values of x and z, and yield a logic 0, otherwise.

The equality operators compare their two operands in a bit-by-bit fashion. Before the comparison is performed, the shorter operand is zero-extended if both operands are not of equal length and either operand is unsigned, and sign-extended if both operands are not of equal length and are signed.

■ Example 3.15: Equality of two numbers.

A simple application of the logical equality operator is to detect whether or not two input numbers, A and B, are equal. If yes, the output `equality` is set to 1 and otherwise it is cleared to 0. The resulting module is quite simple and shown below.

Section 3.3 *Operators*

```
module two_numbers_equality(
    input [3:0] A, B, // inputs
    output equality); // output
// the use of the logical equality operator
assign equality = (A == B);
endmodule
```

Even though we may rely on the precedence of operators and write a correct expression, such as the following example

a < s − 1 && b != c && d != f;

it is good practice to use parentheses to indicate very clearly the intended precedence for improving readability, as in the following rewriting example

(a < s − 1) && (b != c) && (d != f);

The following example rewrites the nested conditional operator used in the 4-to-1 multiplexer described in Section 3.3.6 in another more readable form. It is instructive to compare both constructs and see the differences between them. In fact, both constructs produce the same gate-level circuit exactly. Check them.

■ Example 3.16: Another example of 4-to-1 multiplexers.

In this example, we explicitly use the case equality operator (===) to determine which input will be routed to the output. The selection signal, sel, is compared with a constant, i, where i runs from 0 to 3, in order. While the comparison is exactly matched, the input is assigned to the output, out. The synthesized result of this 4-to-1 multiplexer module is the same as that obtained in Section 3.3.1, as depicted in Figure 3.8. It is instructive to note that if we change the case equality operator (===) into a logical equality (==), what will happen?

```
module mux_4to1_equality(
    input i0, i1, i2, i3, // inputs
    input [1:0] sel,      // source selection inputs
    output out);          // output

// using case equality and conditional operators
assign out = (sel === 0) ? i0 :
             (sel === 1) ? i1 :
             (sel === 2) ? i2 :
             (sel === 3) ? i3 : 4'bz;
endmodule
```

Could we just declare an input vector of 4 bits and then use the source selection inputs as the bit-select to pick up the desired input and route to the output? That is,

    input [3:0] i;
    ...
    assign out = i[sel];

Explain it. What would happen if the i input is declared as an array instead of a vector?

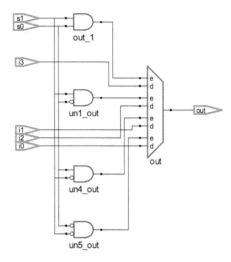

Figure 3.8: A 4-to-1 multiplexer modeled by using the conditional operator.

Table 3.10: The shift operators.

Symbol	Operation
<<	Logical left shift
>>	Logical right shift
<<<	Arithmetic left shift
>>>	Arithmetic right shift

3.3.9 Shift Operators

There are two types of shift operators, including logical shift operators, << (logical left shift) and >> (logical right shift), and arithmetic shift operators, <<< (arithmetic left shift) and >>> (arithmetic right shift), as shown in Table 3.10. Each shift operator shifts the left operand a number of bit positions specified by the right operand. The vacant bits are filled with zeros for logical and arithmetic left shifts, and filled with the sign bit (i.e., the most significant bit, MSB) for the arithmetic right shift.

■ **Example 3.17: An example of the logical left-shift operator.**

In this example, both start and result are declared as **wire** nets and a logical left-shift operator is applied to start. The result variable is assigned the binary value 0100, which is the binary value 0001 shifted left two bit positions and zero-filled.

```
// a logical left-shift example
module logical_shift;
wire [3:0] start, result;
//
assign start = 4'b0001;
assign result = (start << 2);
endmodule
```

Example 3.18: Arithmetic right-shift operator.

In this example, both `result` and `start` are declared as signed **wire** nets and an arithmetic right-shift operator is applied to `start`. The `result` variable is assigned the binary value 1110, which is the binary value 1000 shifted right two bit positions and sign-extended.

```
// an arithmetic right-shift example
module arithmetic_shift;
wire signed [3:0] start, result;
//
assign start = 4'b1000;
assign result = (start >>> 2);
endmodule
```

The following example compares the differences between logical and arithmetic right shifts. The reader is encouraged to write a test bench, simulate the program, see what happens, and indicate their differences.

Example 3.19: Logical and arithmetic shifts.

This example explores the differences between logical and arithmetic right shifts. The logical right shift fills the vacant bits with 0s and the arithmetic right shift fills the vacant bits with the sign bit.

```
// an example showing logical and arithmetic shifts
module arithmetic_logical_shift(
       input  signed [3:0] x,
       output        [3:0] y,
       output signed [3:0] z);
//
assign y = x >> 1;   // logical right shift
assign z = x >>> 1;  // arithmetic right shift
endmodule
```

Note that the nets, x and z, in the above module must be declared with the keyword **signed**. It is instructive to replace the signed net, x or z, with an unsigned one (i.e., remove the keyword **signed**) and see what happens.

Review Questions

3-49 Describe the operations of the set of equality operators?
3-50 What is the distinction between the logical equality (==) and the case equality (===)?
3-51 What is the distinction between the logical inequality (!=) and the case inequality (!==)?
3-52 Could we replace the case equality with the logical equality in Example 3-15?
3-53 What is the difference between the arithmetic right shift (>>>) and the logical right shift (>>)?

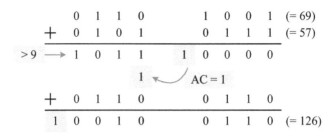

Figure 3.9: An illustration of two-digit BCD addition.

3.4 Case Study — A Two-Digit BCD Adder

In designing a hardware module, one should think it always in hardware. In this case study, we reemphasize this concept by way of designing and modeling a decimal adder. The decimal adder, also known as a *binary-coded decimal* (BCD) *adder*, has found its widespread use in most microprocessors. Generally speaking, a decimal adder has an 8-bit word width, which contains two BCD digits. As a consequence, the problem that we are concerned with is the design and modeling of such a two-digit BCD adder.

3.4.1 Principles of BCD Addition

A two-digit BCD adder is indeed composed of two single-digit BCD adders. Each single-digit BCD adder employs a 4-bit binary adder to add up two BCD input numbers along with a carry-in, and then proceeds the following correction step to yield a single-digit BCD output and a possible carry-out.

- If the result from the binary adder is greater than 9, it is needed to adjust into a valid BCD digit by adding 6, and the generated carry is the desired carry-out.

■ **Example 3.20: A numerical example of BCD addition.**

As an illustration of BCD addition, assume that we want to find the sum of two numbers, 69 and 57. According to the above description, a 4-bit binary adder is used for each BCD digit to find the sum of its two inputs, as shown in Figure 3.9. This results in a carry-out from the least significant digit (LSD) and a value greater than 9 from the most significant digit (MSD). By applying the correction step to both digits, the final correct sum, 126, is obtained.

Of course, one may write a Verilog HDL module simply based on the above description of the decimal adder. However, one should always remember that HDL is only a *hardware description language*. While using it to describe or model a module, one should always have something about the module in his mind. Consequently, a better way to model the two-digit BCD adder is first to design and draw its logic block diagram.

In order to obtain a hardware block diagram of the two-digit BCD adder under consideration, we refine the correction step for each single-digit BCD adder as the following two steps:

- If the carry is 1, it is the carry-out and the sum is added by 6.
- If the carry is 0 and the sum is greater than 9, the sum is added by 6 and the generated carry is the carry-out.

Based on this refinement, each single-digit BCD adder indeed performs the following three steps:

Section 3.4 *Case Study — A Two-Digit BCD Adder* 101

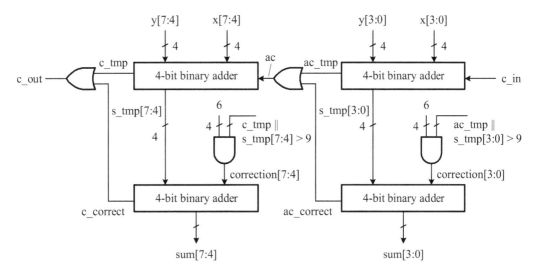

Figure 3.10: A two-digit BCD adder based on binary adders.

1. Perform the addition of the two input numbers along with a carry-in using a 4-bit binary adder.
2. Determine whether the result is needed to add with 6 according to the correction steps.
3. Carry out the required correction and manage the carry-out properly. The carry-out is the carry generated from both correction steps.

The resulting logic block diagram is shown in Figure 3.10.

3.4.2 Modeling Examples of BCD Adders

As illustrations of modeling the aforementioned BCD adder, consider the following two examples that describe the logic block diagram depicted in Figure 3.10 in dataflow style.

■ Example 3.21: A two-digit BCD adder.

The module shown as follows directly describes the logic circuit in Figure 3.10. It is of interest to note that some necessary internal nets are required to avoid the undesired feedback loops of the combinational circuit. You are encouraged to write your own module in order to get more insight into it.

```
// a two-digit BCD adder modeled in dataflow style
// version 1
module BCD_adder_dataflow (,
        input [7:0] x, y,
        input c_in,
        output [7:0] sum,
        output c_out);
wire    [7:0] s_tmp, correction; // temporary sum and correct
wire    ac, ac_tmp, c_tmp, ac_correct, c_correct;

// the least significant digit
assign {ac_tmp, s_tmp[3:0]} = x[3:0] + y[3:0] + c_in,
        correction[3:0]=(ac_tmp||s_tmp[3:0] > 9)?4'b0110:4'b0000,
```

```
        {ac_correct, sum[3:0]} = s_tmp[3:0] + correction[3:0],
        ac = ac_correct | ac_tmp;

// the most significant digit
assign {c_tmp, s_tmp[7:4]} = x[7:4] + y[7:4] + ac,
       correction[7:4]=(c_tmp||s_tmp[7:4] > 9)?4'b0110:4'b0000,
       {c_correct, sum[7:4]} = s_tmp[7:4] + correction[7:4],
       c_out = c_correct | c_tmp;
endmodule
```

The above module may also be described in a little more different way. According to the correction steps, a carry is generated either by the 4-bit binary adder due to the sum being greater than 15 or by the addition of 6 to the sum due to its value being greater than 9, which also always yields a carry. In other words, whenever a carry is generated in either case, the sum is needed to add with 6. Based on this, the following module immediately follows.

■ Example 3.22: A two-digit BCD adder.

According to the above description, the carry-out for each digit is generated either by the 4-bit binary adder or by adding 6 to the sum whenever its value is greater than 9. Therefore, we first perform the binary addition, then compute the output carries, ac and c_out, for each digit, and finally perform the corrections of sum[3:0] and sum[7:4]. The resulting module is as follows:

```
// a two-digit BCD adder modeled in dataflow style
// version 2
module BCD_adder_dataflow (
        input [7:0] x, y,
        input c_in,
        output [7:0] sum,
        output c_out);
wire    [7:0] s_tmp; // temporary sum and correct
wire    ac, ac_tmp, c_tmp;

// the least significant digit
assign  {ac_tmp, s_tmp[3:0]} = x[3:0] + y[3:0] + c_in,
        ac = (ac_tmp || s_tmp[3:0] > 9) ? 1'b1 : 1'b0,
        sum[3:0] = s_tmp[3:0] + (4'b0110 & {4{ac}});

// the most significant digit
assign  {c_tmp, s_tmp[7:4]} = x[7:4] + y[7:4] + ac,
        c_out = (c_tmp || s_tmp[7:4] > 9) ? 1'b1 : 1'b0,
        sum[7:4] = s_tmp[7:4] + (4'b0110 & {4{c_out}});
endmodule
```

In the next chapter, we will rewrite the BCD adder module (version 1) in behavioral style. Remember that a hardware module can always be described in a variety of styles, including the structural, dataflow, behavioral, or their combination. In practice, we often prefer to use the behavioral style because it is a higher level and much easier to model a design directly from its associated algorithm than the other styles.

■ Coding Style

1. While modeling a hardware module, we need to think it always in hardware rather than in software.
2. Before modeling a hardware module, we should already have in our mind a workable block diagram of the hardware module at the RTL.
3. In designing and modeling a hardware module, we always need to take into account the size of each component being used in the hardware module.

■ Review Questions

3-54 Explain how to perform a single-digit BCD addition with binary addition.

3-55 Give why removing common sub-expressions does not simply mean the reduction of hardware resources.

3-56 How would you design a single-digit BCD adder with 4-bit binary adders?

3.5 Summary

The rationale behind the design of a digital system in dataflow style is on the fact that any digital system can be constructed by interconnecting registers and putting a combinational circuit in between them to perform the desired logic functions. The **assign** continuous assignment is the most basic statement of dataflow modeling. It is used to continuously drive a value onto a net. The **assign** continuous assignment is always active.

A continuous assignment begins with the keyword **assign** followed by an expression. An expression is a construct that combines operators with operands to produce a result. It in turn consists of sub-expressions, operators, and operands. The operands in an expression can be any of the following ones: constants, parameters, nets (such as **wire**, **tri**), variables (**reg**, **integer**, **time**, **real**, and **realtime**), bit-selects, part-selects, array elements, and function calls. The set of operators includes the following subsets of operators: arithmetic, bit-wise, reduction, relational, equality (logical and case), logical, shift, and miscellaneous (concatenation, replication, and conditional).

Most operators can be accepted by synthesis tools. Only a few of them cannot or only a limited form can be accepted. The exponent and modulus operators might be two such examples—check them. Through using the dataflow style, a combinational circuit can be modeled by using **assign** continuous assignments rather than structural connections between instances of modules, and/or gate primitives.

This chapter is concluded with a case study—two-digit BCD adders—to reveal the importance of hardware mind in describing a hardware module. Through the illustration of this BCD adder, the concept of hardware mind is explored and emphasized. After all, while designing and modeling a hardware module, one should think it always in hardware mind.

References

1. J. Bhasker, *A Verilog HDL Primer*, 3rd ed., Star Galaxy Publishing, 2005.
2. IEEE 1364-2001 Standard, *IEEE Standard Verilog Hardware Description Language*, 2001.
3. IEEE 1364-2005 Standard, *IEEE Standard for Verilog Hardware Description Language*, 2006.

4. M. B. Lin, *Digital System Design: Principles, Practices, and Applications*, 4th ed., Chuan Hwa Book Ltd. (Taipei, Taiwan), 2010.

5. S. Palnitkar, *Verilog HDL: A Guide to Digital Design and Synthesis*, 2nd ed., SunSoft Press, 2003.

Problems

3-1 Simplify the following switching expression and use bit-wise operators to model it.

$f(w,x,y,z) = \Sigma(5,6,7,9,10,11,13,14,15)$

3-2 Simplify the following switching expression and use bit-wise operators to model it.

$f(w,x,y,z) = \Sigma(0,4,5,7,8,9,13,15)$

3-3 Using bit-wise operators, model the logic circuit shown in Figure 3.11.

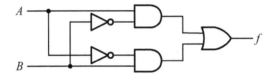

Figure 3.11: The logic diagram for Problem 3-3.

3-4 Using bit-wise operators, model the logic circuit shown in Figure 3.12.

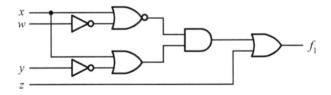

Figure 3.12: The logic diagram for Problem 3-4.

3-5 Using the multiply operator (*), write a module to compute $x * y$, where x and y are two signed numbers. Synthesize the module and check the synthesized result. Write a test bench to verify whether the module behaves correctly.

3-6 Using the divide operator (/), write a module to compute x/y, where x and y are two signed numbers. Synthesize the module and check the synthesized result. Write a test bench to verify whether the module behaves correctly.

3-7 Using the modulus (%) operator, write a module to compute $x\%y$, where x and y are two unsigned numbers. Synthesize the module and check the synthesized result. Write a test bench to verify whether the module behaves correctly.

3-8 By assuming that both numbers are signed, redo Problem 3-7.

3-9 Suppose that an arithmetic-and-logic unit (ALU) has the function shown in Table 3.11. The output is a function of six mode-selection inputs, $m5$ to $m0$.

Table 3.11: The function table for Problem 3-9.

m5	m4	m3	m2	m1	m0	ALU output
0	0	0	1	0	0	A .and B
0	0	1	0	0	0	A .and .not B
0	0	1	0	0	1	A .xor B
0	1	1	0	0	1	A plus .not B plus carry
0	1	0	1	1	0	A plus B plus carry
1	1	0	1	1	0	.not A plus B plus carry
0	0	0	0	0	0	A
0	0	0	0	0	1	A .or B
0	0	0	1	0	1	B
0	0	1	0	1	0	.not B
0	0	1	1	0	0	zero

(a) Derive the 1-bit slice switching expression of the output as the function of the mode selection inputs $m5$ to $m0$.

(b) Write a module in dataflow style to perform the output function. Write a test bench to verify whether the module behaves correctly.

(c) By instantiating the above module, construct a 4-bit ALU. Write a test bench to verify whether the module behaves correctly.

3-10 Write a module to convert a BCD code into an excess-3 code. Also, write a test bench to verify the functionality of the module.

3-11 Using the conditional operator, write a module to shift the input, data, right logically by the number of bit positions specified by another input, shift, ranging from 0 to 3.

3-12 Using the conditional operator, write a module to shift the input, data, right arithmetically by the number of bit positions specified by another input, shift, ranging from 0 to 3.

3-13 Using the conditional operator, write a module to describe a 2's-complement adder. Write a test bench to verify whether the module behaves correctly.

3-14 Use the conditional and case equality operators, respectively, as the selector to model the functions shown in Table 3.12. Write a test bench to verify whether the module behaves correctly.

Table 3.12: Table for Problem 3-14.

Function	Operation
0	$z = x + y$
1	$z = x - y$
2	$z = x * y$

3-15 A 4-to-2 priority encoder is a device with four inputs: in[0] to in[3] and a two-bit output code: out[1] and out[0]. In addition, a signal valid is used to indicate that an input signal is valid. The two-bit output code {out[1], out[0]} is i if input in[i] is 1 and no other inputs in[j] are 1, where $j < i$. The signal valid is 0 if all inputs are 0 and is 1, otherwise. Describe the module in dataflow style and write a test bench to verify whether the module behaves correctly.

3-16 Considering the following two-digit BCD adder module, answer each of the following questions.

```
// a two-digit BCD adder modeled in dataflow style
// version 3
module BCD_adder_dataflow (
        input  [7:0] x, y,
        input  c_in,
        output [7:0] sum,
        output c_out );
    wire   [7:0] s_tmp; // temporary sum and correct
    wire   ac, ac_correct, ac_tmp, c_correct, c_tmp;
    // the least significant digit
       assign {ac_tmp, s_tmp[3:0]} = x[3:0] + y[3:0] + c_in,
              {ac_correct, sum[3:0]} = (ac_tmp || s_tmp[3:0] > 9) ?
                  s_tmp[3:0] + 4'b0110 : {ac_tmp, s_tmp[3:0]},
              ac = ac_tmp | ac_correct;
    // the most significant digit
       assign {c_tmp, s_tmp[7:4]} = x[7:4] + y[7:4] + ac,
              {c_correct, sum[7:4]} = (c_tmp || s_tmp[7:4] > 9) ?
                  s_tmp[7:4] + 4'b0110 : {c_tmp, s_tmp[7:4]},
              c_out = c_tmp | c_correct;
endmodule
```

(a) Synthesize the above module. How many 4-input LUTs are required?

(b) In comparison with the two-digit BCD adder version 1 (Example 3-20), explain the reason why so many 4-input LUTs are required for the synthesized result.

3-17 Assume that we want to design a two-digit binary-coded decimal (BCD) subtracter. A two-digit BCD subtracter is indeed cascaded together two single-digit BCD subtractors, in which each single-digit BCD subtracter is in turn carried out with a 4-bit binary subtracter to find the difference of the two BCD-input operands and then proceeds the following correction steps:
- If the borrow (i.e., the complement of the carry) of the resulting digit is 1, the borrow is the borrow-out and the resulting digit is subtracted by 6.
- If the resulting digit is greater than 9, the resulting digit is subtracted by 6 and the generated borrow is ignored.

(a) Write a module in dataflow style to describe this two-digit BCD subtracter.

(b) Write a test bench to verify whether the BCD subtracter behaves correctly.

3-18 Assume that a two-digit BCD adder/subtracter is desired. The BCD adder/subtracter functions as a BCD adder if the mode selection input, mode, is 0 and functions as a BCD subtracter, otherwise.

(a) Write a module in dataflow style to describe this two-digit BCD adder/subtracter.

(b) Write a test bench to verify whether the BCD adder/subtracter behaves correctly.

4

Behavioral Modeling

WE have discussed the structural style, at both gate and switch levels, and the dataflow style in the previous two chapters. In this chapter, we address the behavioral style. To make use of the full power of Verilog HDL, the description of a design usually combines the use of these three modeling styles.

Like software programming languages, such as C/C++ and Java, Verilog HDL also provides assignments, selection statements, and iteration (loop) statements as basic constructs for behavioral modeling. Assignment statements cover the **assign** continuous assignment, procedural assignments, and procedural continuous assignments. The procedural assignments include blocking assignments (=) and nonblocking assignments (<=). The procedural continuous assignments have two pairs of statements: **assign-deassign** and **force-release**.

Selection statements are used to make a selection according to a given condition. They contain two statements: **if-else** and **case**. The **if-else** and **case** statements can be nested. The **case** statement has two variants: **casex** and **casez**. Iteration (loop) constructs are used to execute a set of statements repeatedly. Four types of iteration (loop) statements are **repeat**, **for**, **while**, and **forever**. All these four iteration (loop) statements are discussed in detail in this chapter.

Timing control provides a way to specify the simulation time at which procedural assignments will execute. In Verilog HDL, if there are no timing control statements, the simulation time will not advance. Hence, it is necessary to include some types of timing control statements within Verilog HDL modules. There are two timing control methods provided in Verilog HDL: delay timing control and event timing control.

4.1 Procedural Constructs

An HDL must have the capability of capturing and describing the hardware features: *concurrency* and *timing*. In behavioral style, the capability of concurrency is equipped with **initial** and **always** blocks and the timing feature is provided by the timing control mechanism.

4.1.1 initial Blocks

An **initial** block (also **initial** construct) consists of all statements inside an **initial** construct. These statements are enclosed by a pair of keywords: **begin** and **end** or **fork** and **join**. An **initial** block starts at simulation time 0 and executes exactly once during simulation. It is used in test benches to contain statements to initialize signals, monitor waveforms, and so forth. The **initial** block has the following syntax

```
initial statement
```

where `statement` may be a single procedural statement (or statement for short) or a complex statement consisting of a set of procedural statements enclosed by a **begin-end** (sequential) or **fork-join** (parallel) block. The procedural statements that can be used within an **initial** block as `statement` are as follows:

```
blocking_assignment;
nonblocking_assignment;
procedural_timing_control_statement
wait_statement
event_trigger
conditional_statement (if-else)
case_statement (case)
loop_statement
task_enable
system_task_enable
seq_block (begin-end)
par_block (fork-join)
disable_statement
procedural_continuous_assignment;
```

The `procedural_timing_control_statement` has the following syntax

```
timing_control statement_or_null
```

where `timing_control` may be either delay timing control or event timing control, to be discussed in Section 4.3, and `statement_or_null` may be a procedural statement or a null statement (;).

As an example, consider the following two **initial** blocks. The first **initial** block is a single statement containing only one procedural assignment and the second statement is a complex statement consisting of six procedural assignments enclosed by a **begin-end** block.

```
reg  x, y, z;
   initial x = 1'b0;       // single statement
   initial begin           // complex statement
        x = 1'b0; y = 1'b1; z = 1'b0;
   #10  x = 1'b1; y = 1'b1; z = 1'b1;
   end
```

If a delay specification, #delay, appears before a procedural assignment, it indicates that the expression will be executed #delay time units after the current simulation time. Hence, the register variable x in the preceding example will be set to 1 after 10 time units.

As multiple **initial** blocks exist, all **initial** blocks start to execute concurrently at simulation time 0 and finish their execution independently. For example, both of the following **initial** blocks start to execute at simulation time 0. The first one ends after 10 time units whereas the second one terminates after 30 time units.

```
reg  x, y, z;
   initial #10 x = 1'b0; // finish at simulation time 10
   initial begin
      #10 y = 1'b1;      // finish at simulation time 10
      #20 z = 1'b1;      // finish at simulation time 30
   end
```

4.1.1.1 Combining Variable Declaration with Initialization Although the initial values of variables can be set by using an **initial** block, it is possible to combine a variable declaration with initialization into one statement. For example,

Section 4.1 Procedural Constructs

```
reg clk;        // regular declaration
   initial clk = 0;
// can be declared as a single statement
reg clk = 0;    // can be used only at the module level
```

The same way can be applied to the port list of a module, which is called the port-list style. For example,

```
module adder(x, y, c , sum, c_out);
input   [3:0] x, y;
input   c_in;
output reg [3:0] sum = 0; // initialize sum
output reg c_out = 0;     // initialize c_out
```

Of course, this can be applied to the port-list declaration style (also called the ANSI style) as well. For instance,

```
module adder(input [3:0] x, y,
             input c_in,
             output reg [3:0] sum = 0,
             output reg c_out = 0
);  // the ANSI style
```

Note that a variable declared with initialization is usually not accepted by synthesis tools. Try to synthesize it and see what happens. Moreover, **initial** blocks are not accepted by synthesis tools in general; they are usually used in test benches only.

4.1.2 always Blocks

An **always** block (also called **always** construct) consists of all behavioral statements within an **always** construct. These statements are enclosed by a pair of keywords: **begin** and **end**. The **always** block starts at simulation time 0 and repeatedly executes the statements within it in a loop fashion during simulation. The **always** block has the following syntax

    always statement

where statement is the same as that of the **initial** block.

An **always** block is used to model a block of activities that are continuously repeated in a digital circuit. For example, the following program segment

```
reg clock;                  // a clock generator
initial clock = 1'b0;       // initial clock = 0
always #5 clock = ~clock;   // period = 10
```

models a clock generator, which produces a symmetric clock signal with a period of 10 time units. Note that we need to set the initial value of clock to either 0 or 1; otherwise, the generated clock signal will be always an unknown value, x. It is worth noting that the above clock generator can only be used in test benches to generate the required clock signal during simulation. It cannot be synthesized into a realistic hardware circuit.

A simple use of the **always** block can be illustrated by the following example which describes a 4-to-1 multiplexer modeled in behavioral style.

■ Example 4.1: A 4-to-1 multiplexer.

In this example, the entire body of the module is merely an **always** block. A procedural assignment is used within the **always** block to implement the operations of the 4-to-1 multiplexer, as described in Section 2.1.1. The procedural assignment is the same as the **assign** continuous assignment introduced in Chapter 3 except that the keyword **assign** is no longer needed here.

```
module mux_4to1_behavioral(
      input i0, i1, i2, i3,  // inputs
      input s1, s0,          // source selection inputs
      output reg out);       // output
// using basic bit-wise operators, and, or, and not
always @(i0 or i1 or i2 or i3 or s1 or s0)
   out = (~s1 & ~s0 & i0)|
         (~s1 &  s0 & i1)|
         ( s1 & ~s0 & i2)|
         ( s1 &  s0 & i3);
endmodule
```

The following example illustrates another use of the **always** block in which a 4-bit adder is described.

■ **Example 4.2: A 4-bit adder.**

In this example, a 4-bit adder is modeled in behavioral style. Because of its intuitive simplicity, we will not further explain it here.

```
module four_bit_adder(
      input   [3:0] x, y,    // declare as a 4-bit vector
      input   c_in,
      output reg [3:0] sum,  // declare as a 4-bit vector
      output reg c_out);

// specify the function of a 4-bit adder
always @(x or y or c_in)
   {c_out, sum} = x + y + c_in;
endmodule
```

4.1.3 Concluding Remarks

An **initial** or **always** construct is usually called an **initial** or **always** block. As the name implies, **initial** blocks are used to initialize variables and set values into variables or nets; **always** blocks are used to model the repeated operations required in the hardware modules. All other behavioral statements must be placed within an **initial** or **always** block.

In a module, an arbitrary number of **initial** and **always** blocks are allowed to use. All **initial** and **always** blocks execute concurrently with respect to each other; namely, their relative order in a module is not important. Each of **initial** and **always** blocks denotes an individual activity flow. Each activity flow starts at simulation time 0. Notice that each **always** block corresponds to a piece of logic. In addition, **initial** and **always** blocks cannot be nested. Each **initial** or **always** construct must form its own block.

■ **Review Questions**

4-1 Explain the operations of an **initial** block.
4-2 Explain the operations of an **always** block.
4-3 Explain why Verilog HDL uses **begin-end** to replace { and } used in C language.
4-4 Distinguish between the port-list style and the port-list declaration style.
4-5 Explain and correct the following clock generator:

Section 4.2 *Procedural Assignments* 111

```
always begin
    initial clock = 1'b0;
    #5 clock = ~clock;
end
```

4-6 Explain and correct the following clock generator:

```
always begin
    #5 clock = ~clock;
end
```

4.2 Procedural Assignments

Recall that an **assign** continuous assignment is used to continuously assign values onto a net in a way similar to that a logic gate drives a net. In contrast, a procedural assignment puts values in a variable, which holds the value until the next procedural assignment updates the variable. Procedural assignments must appear within procedures, including **initial**, **always**, **task**, and **function** constructs. Note that tasks and functions must be invoked within **initial** or **always** blocks but functions can also be used as operands in **assign** continuous assignments.

4.2.1 Procedural Assignments

Procedural assignments are placed inside **initial** or **always** blocks. They update the values of variable types, including **reg**, **integer**, **time**, **real**, and **realtime**, or array (memory) elements declared with variable types. Two types of procedural assignments in Verilog HDL are *blocking* assignment, using the = assignment operator, and *nonblocking* assignment, using the <= assignment operator . The general syntax of procedural assignments is as follows

```
variable_lvalue  = [timing_control] expression
variable_lvalue <= [timing_control] expression

[timing_control] variable_lvalue  = expression
[timing_control] variable_lvalue <= expression
```

where `variable_lvalue` can be any of variable types {**reg**, **integer**, **time**, **real**, **realtime**}, or an array (memory) element, a bit-select, a part-select, a concatenation of any of the above; `timing_control` can be either delay timing control or event timing control.

4.2.1.1 Procedural versus assign Continuous Assignments There is an essential difference between **assign** continuous assignments and procedural assignments. **assign** continuous assignments drive nets, which are evaluated and updated whenever any operand changes its value. In contrast, procedural assignments update the values of variables under the control of the procedural flow constructs that surround them. As a consequence, the left-hand sides of **assign** continuous assignments are nets but the left-hand sides of procedural assignments are variables.

In summary, procedural assignments have several distinct features in comparison with **assign** continuous assignments:

1. They do not use the keyword **assign**.
2. They can only update variable types: **reg**, **integer**, **time**, **real**, **realtime**, array (memory) elements, or any element derived from variable types.
3. They can only be used within **initial** and **always** blocks. Notice that since tasks and functions must be invoked within **initial** and **always** blocks, procedural assignments can also appear within these constructs.

Besides, it is worth noting that `variable_lvalue` may be on the right-hand side of a procedural nonblocking (<=) assignment. However, it may not be on the right-hand side of a procedural blocking (=) assignment because like the **assign** continuous assignment, the blocking assignment is also used to describe combinational logic and hence as `variable_lvalue` is put on the right-hand side it would cause a combinational loop.

4.2.2 Blocking Assignments

Blocking assignments use the "=" assignment operator and are executed in the order they are specified. In other words, a blocking assignment is executed before the execution of the statements that follow it within a *sequential* block (statements grouped with a pair of keywords: **begin** and **end**). Nevertheless, a blocking assignment will not block the execution of statements following it within a *parallel* block (statements grouped with a pair of keywords: **fork** and **join**). For now, we only consider the cases of sequential blocks. The parallel block will be discussed in Section 7.1.2.

An illustration of the use and operations of the blocking assignment is given in the following example, which describes a 4-bit 2's-complement adder.

■ **Example 4.3: A 4-bit 2's-complement adder.**

In this example, a 2's-complement adder is modeled by using the principle that x minus y is equal to x plus the 2's complement of y. The 2's complement of y can be easily obtained by complementing y and then adding it with 1. As a consequence, a two-step procedure immediately follows. In the first step, we compute the complement of y if the mode control input, `c_in`, is 1 and the true value of y if `c_in` is 0. In the second step, we add the complement of y and 1 or the true value of y and 0 and x. The complete operations are given in the following module.

```
module twos_adder_behavioral(
      input    [3:0] x, y,
      input    c_in,
      output reg  [3:0] sum,
      output reg  c_out);
// the body of module
reg    [3:0] t;      // outputs of xor gates
// specify the function of a 2's-complement adder
always @(x, y, c_in) begin
   t = y ^ {4{c_in}}; // what is wrong with: t = y ^ c_in?
   {c_out, sum} = x + t + c_in;
end
endmodule
```

Note that a **reg** variable does not necessarily correspond to a register element when the circuit is synthesized. Whether it is synthesized into a register totally depends on the **always** block with which it associates. A register will be inferred if the sensitivity list of the **always** block contains the keyword **posedge** or **negedge** and a combinational circuit is inferred otherwise. Based on this, the above example is synthesized into a combinational circuit as expected.

Another illustration of using the blocking assignment showing how to model a two-digit BCD adder in behavioral style is given in the following example. As expected, this example is also synthesized into a combinational circuit.

Section 4.2 *Procedural Assignments*

■ Example 4.4: A two-digit BCD adder.

This example is indeed the behavioral version of the two-digit BCD adder described in Section 3.4. Here, we only replace the **assign** continuous assignments with an **always** block. The detailed description about the two-digit BCD adder can be referred to that section again.

```
// a two-digit BCD adder modeled in behavioral style
module BCD_adder_behavior(
        input [7:0] x, y,
        input c_in,
        output reg [7:0] sum,
        output reg c_out);
reg [7:0] s_tmp, correction; // temporary sum and correct
reg ac, ac_tmp, c_tmp,       // temporary carries
    ac_correct, c_correct;

always @(*) begin
  // the least significant digit
  {ac_tmp, s_tmp[3:0]} = x[3:0] + y[3:0] + c_in;
  correction[3:0]=(ac_tmp || s_tmp[3:0] > 9) ? 4'b0110 : 4'b0000;
  {ac_correct, sum[3:0]} = s_tmp[3:0] + correction[3:0];
  ac = ac_correct | ac_tmp;
  // the most significant digit
  {c_tmp, s_tmp[7:4]} = x[7:4] + y[7:4] + ac;
  correction[7:4]=(c_tmp || s_tmp[7:4] > 9) ? 4'b0110 : 4'b0000;
  {c_correct, sum[7:4]} = s_tmp[7:4] + correction[7:4];
  c_out = c_correct | c_tmp;
end
endmodule
```

Note that the "=" assignment operator used by the blocking assignment is also used by the procedural continuous assignment (see Section 7.2 for details) and the **assign** continuous assignment.

4.2.3 Nonblocking Assignments

Nonblocking assignments use the <= assignment operator and are executed without blocking the other statements in a sequential block. Nonblocking assignments provide a method to model several concurrent data transfers taking place after a common event. In other words, nonblocking assignments are used whenever several variable assignments within the same time step are needed to perform, regardless of the order or the dependence on each other.

An illustration of the use and operations of the nonblocking assignment is given in the following example, which describes a 4-bit right-shift register. The conceptual operation of a 4-bit right-shift register is depicted in Figure 4.1.

■ Example 4.5: A 4-bit right-shift register without reset.

In this example, we use a nonblocking assignment along with a concatenation operator to perform the right shift of a 4-bit right-shift register, as shown in Figure 4.1. As stated, because of using the keyword **posedge** in the sensitivity list of the **always** block, the **reg** variable, qout, is synthesized into a 4-bit register. The resulting module indeed exactly models a 4-bit right-shift register.

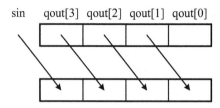

Figure 4.1: The conceptual operation of a 4-bit right-shift register.

```
// an example of a right-shift register without reset
module shift_reg_4b(
      input   clk, sin,
      output reg [3:0] qout);
// the body of 4-bit shift register
always @(posedge clk)
   qout <= {sin, qout[3:1]};   // right shift
endmodule
```

What would happen when a blocking assignment is used in place of the nonblocking assignment in the above module? The answer is that nothing will happen because only one statement within the **always** block. Nevertheless, we prefer to use nonblocking assignments when describing a sequential circuit. The reasons is to be explained later in this section.

Nonblocking assignments are widely used to model pipelined and mutually exclusive data transfers. It is strongly recommended to use nonblocking assignments rather than blocking assignments whenever concurrent data transfers take place after a common event. Nonblocking assignments are executed last in the time step during simulation in which they are scheduled in the mixed use with blocking assignments.

■ Review Questions

4-7 Is "{c_out, sum} = x + y ^ {4{c_in}} + c_in;" correct? If not, correct it.
4-8 What are the differences between procedural and continuous assignments?
4-9 Could we use the "<=" assignment operator in continuous assignments?
4-10 What are the differences between blocking and nonblocking assignments?
4-11 What would happen if we replace the blocking assignments with nonblocking ones in Example 4.3?

4.2.4 Blocking versus Nonblocking Assignments

To get more insight into the differences between blocking and nonblocking assignments, in the remainder of this section, we begin to address the three-step procedure of nonblocking assignments and discuss the race problem. Then, we are concerned with a number of examples to further reveal the differences between blocking and nonblocking assignments.

4.2.4.1 Three-Step Procedure To understand how a nonblocking assignment works, we can dissect the detailed operations of nonblocking assignments carried out by simulators. Generally speaking, a simulator performs the following three-step procedure to execute nonblocking assignments during simulation.

1. *Read*: Read the values of all right-hand-side nets and/or variables of nonblocking assignments.

Section 4.2 Procedural Assignments

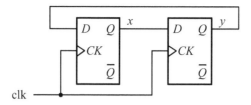

Figure 4.2: The operations of swapping the contents of two registers.

2. *Evaluation and schedule*: Evaluate the right-hand-side expressions and store the results in temporary variables that are scheduled to assign to the left-hand-side variables later.
3. *Assignment*: Assign the values stored in the temporary variables to the left-hand-side variables.

Based on this procedure, it is easy to explain the differences between the result yielded by a blocking assignment and that generated by a nonblocking assignment.

4.2.4.2 Race Problems In C language or other high-level languages, as we want to exchange the contents of two variables, a and b, we need another variable, say, $temp$, and perform the following three operations:

```
temp = a;
a = b;
b = temp;
```

However, as shown in Figure 4.2, we are able to swap the contents of two registers directly without the help of the third register. To carry out this, we might use the following two **always** blocks, with each containing a blocking assignment at the first attempt.

```
// using blocking assignment statements
always @(posedge clk) // has race condition
    x = y;
always @(posedge clk)
    y = x;
```

Nevertheless, this will result in a *race* problem, meaning that the final result depends on the order that the two **always** blocks are executed. If the first **always** block is executed first and then the second, the final result will be the case that both registers, x and y, have the same content as the original y. If both statements are executed in reverse order, the result is that both registers, x and y, have the same content as the original x. To avoid such a race problem, we may replace both blocking assignments with nonblocking assignments within the two **always** blocks, as shown below.

```
// using nonblocking assignment statements
always @(posedge clk) // has no race condition
    x <= y;
always @(posedge clk)
    y <= x;
```

Here, at the positive edge of each clock signal, the contents of registers, x and y, are read and then written into individual temporary variables. Finally, the contents of the temporary variables are separately assigned to registers, x and y, at the next positive edge of the clock signal. As a consequence, the contents of two registers are properly swapped. It is worth noting that the above nonblocking assignments can be emulated by using blocking assignments as follows:

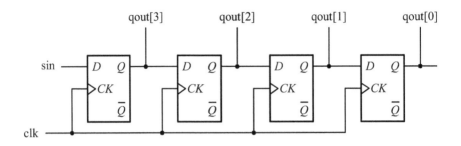

Figure 4.3: The basic structure of a 4-bit right-shift register.

```
always @(posedge clk) begin
   temp_x = x;  // read operation
   temp_y = y;

   y = temp_x;  // write operation
   x = temp_y;
end
```

4.2.4.3 Blocking versus Nonblocking Assignments The three-step procedure proves much useful in exploring the differences between blocking and nonblocking assignments. To get more insight into this, consider the 4-bit right-shift register depicted in Figure 4.3.

■ **Example 4.6: An incorrect 4-bit right-shift register module.**

From the logic diagram shown Figure 4.3, it is straightforward to describe the logic circuit by using four blocking assignments, with each describing one flip-flop. Unfortunately, the resulting module cannot work properly. To see this, consider the synthesized result shown in Figure 4.4(a). At first glance, we may wonder about the result obtained. Nevertheless, after we examine Figure 4.4(a) and the module written more carefully, we can realize that the synthesizer indeed works well even though it produces an unexpected result. The reason is that blocking assignments are executed in the order that they appear, one after the other, and after they are optimized by the synthesis tool, the final result is exactly the same as the one shown in Figure 4.4(a).

```
// a 4-bit right-shift register module --- an incorrect version
module shift_reg_blocking(
       input   clk,
       input   sin,           // serial data input
       output reg [3:0] qout);
// the body of a 4-bit right-shift register
always @(posedge clk) begin  // using blocking assignments
   qout[3] = sin;
   qout[2] = qout[3];
   qout[1] = qout[2];
   qout[0] = qout[1];
end
endmodule
```

To solve the above problem, let us return to Figure 4.3 and reexamine the operations of the 4-bit right-shift register more carefully. From this figure, we realize that the essential op-

Section 4.2 Procedural Assignments

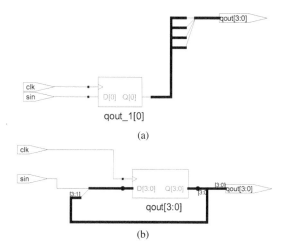

Figure 4.4: A 4-bit right-shift register generated by using (a) blocking and (b) nonblocking assignments.

eration of the 4-bit right-shift register is that at each positive edge of the clock signal, each D flip-flop is assigned a new value, i.e., the output of its previous stage. Thus, the value to be assigned to a D flip-flop at each positive edge of the clock signal is like the one taken from a temporary variable, which stores the output of its previous stage in the previous clock cycle. This is exactly the operation that nonblocking assignments perform. Therefore, nonblocking assignments rather than blocking assignments should be used, as in the following module.

■ Example 4.7: A 4-bit right-shift register module.

In this example, we replace the four blocking assignments with nonblocking assignments. The synthesized result is correct and shown in Figure 4.4(b). Of course, an even better approach is to use a single nonblocking assignment: qout <= {sin, qout[3:1]}. The reader is encouraged to explain the operations of this module using the three-step procedure of nonblocking assignments.

```
// a 4-bit right-shift register module --- a correct version
module shift_reg_nonblocking(
      input   clk,
      input   sin,            // serial data input
      output reg [3:0] qout);
// the body of a 4-bit right-shift register
always @(posedge clk) begin  // using nonblocking assignments
   qout[3] <= sin;            // it is even better to use
   qout[2] <= qout[3];        // qout <= {sin, qout[3:1]};
   qout[1] <= qout[2];
   qout[0] <= qout[1];
end
endmodule
```

From the above two examples, we can realize that nonblocking assignments in Verilog HDL are indeed designed particularly for use in modeling sequential circuits. Therefore, their operations exactly match those of sequential circuits.

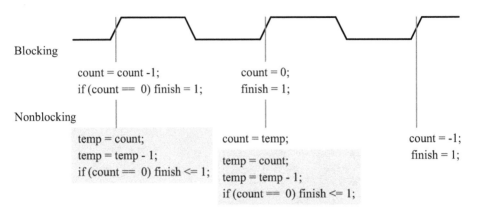

Figure 4.5: The differences between blocking and nonblocking assignments.

4.2.4.4 More Insight In what follows, we explore another important feature that distinguishes nonblocking assignments from blocking assignments. To get more insight into the differences between these two types of assignments, consider the following **always** block using blocking assignments. Assume that the value of count is 1 and finish is 0 before entering the **always** block.

```
always @(posedge clk) begin: block_a
    count = count - 1;
    if (count == 0) finish = 1;
end
```

The result is finish = 1 when count is equal to 0, which is different from that of the gate level. Why? See Problems 4-36 and 4-37. The reason is that according to the inherent features of blocking assignments, the two statements within the **always** block are executed one after the other, just as the order specified, as shown in Figure 4.5. As a consequence, the above result is obtained. However, this is not the result obtained from the gate-level circuit because real-world gates always execute their operations concurrently.

Now, we replace the blocking assignments with nonblocking assignments in the above **always** block. The resulting **always** block is as follows:

```
always @(posedge clk) begin: block_b
    count <= count - 1;
    if (count == 0) finish <= 1;
end
```

The result is finish = 0 when the count value equals 0, which is exactly the same as that obtained from the gate-level circuit. In fact, in this case, the finish is 1 when the count equals -1. The reason is as follows. Based on the features of nonblocking assignments, the two statements within the **always** block are executed as follows. As the block, block_b, is entered, the count is 1 and at the next positive edge of the clock signal, clk, the count is assigned to 0. At that time, the condition is true, and the finish will be assigned to 1 and the count to -1 at the following positive edge of the clock signal, clk. Therefore, the above result is obtained naturally. Figure 4.5 shows the detailed operations of block_b and block_a.

■ **Coding Style**

1. *It should not mix blocking with nonblocking assignments in the same **always** block.*

2. *In an* **always** *block, it is good practice to use nonblocking assignments (<=) when the block describes a piece of sequential logic. Otherwise, the result of RTL behavior may be inconsistent with that of the gate level.*

3. *In an* **always** *block, it is good practice to use blocking assignments (=) when the block describes a piece of combinational logic.*

■ Review Questions

4-12 Which of procedural assignments is more appropriate in modeling a pipelined data transfer?

4-13 Describe the three-step procedure that simulators will perform when they encounter a nonblocking statement.

4-14 Write a correct program segment to swap the contents of two **reg** variables.

4.3 Timing Control

Timing control provides a way to specify the simulation time at which procedural statements will execute. In Verilog HDL, if there are no timing control statements, the simulation time will not advance. Hence, it is necessary to include some kinds of timing control statements in a module. For this purpose, two timing control methods are provided in Verilog HDL:

1. Delay timing control
2. Event timing control (including **repeat** (expression) event timing control)

According to the place of timing control in a procedural statement, timing control can be divided into two types: *regular timing control* and *intra-assignment timing control*. The former can be applied to any procedural statement while the latter can only be applied to procedural assignments. These two methods are investigated in detail in this section.

4.3.1 Delay Timing Control

Delay timing control in a procedural statement specifies the time duration between encountering and executing the procedural statement. It can be specified by a one-delay specifier with either of the following two forms:

```
#delay_value
#(min:typ:max_expr)
```

Because delay timing control is most often applied to procedural assignments, in what follows we will exemplify it with procedural assignments.

4.3.1.1 Regular Delay Control Regular delay control defers the execution of the entire procedural statement by a specified number of time units. It is specified by prefixing a non-zero delay to a procedural statement. Regular delay control has the following forms

```
#delay null_statement
#delay statement
```

where `#delay` can be a constant or an expression. Note that if the delay expression, `#delay`, evaluates to an unknown, x, or a high impedance, z, it is treated as the zero delay. If the delay expression evaluates to a negative value, the 2's-complement unsigned integer is used as the delay. Notice that procedural assignments are one type of procedural statement.

If no procedural statement (i.e., a null statement) is specified, regular delay control only causes a wait for the specified delay before the next procedural statement is executed. Thus,

```
#25;            // wait 25 time units
x = a + 6;      // execute immediately
```

is equivalent to

```
#25 x = a + 6;  // wait 25 time units and execute
```

Because regular delay control defers the execution of the entire statement by an amount of the specified delay, the result is the same regardless of whether a blocking assignment or a nonblocking assignment is used.

■ Example 4.8: An example of regular delay control.

As an illustration of the regular delay control associated with blocking and nonblocking assignments, consider the following module. Since the effects of regular delay control are identical for both blocking and nonblocking assignments, the use of a blocking or nonblocking assignment does not make the results different. That is, both assignments defer the execution of the entire statement by the specified delay, thereby leading to the sequential execution of all assignments. The execution time of each assignment is indicated in its associated comment.

```
// an illustration of regular delay control
module regular_delay_control;
reg   x, y, z;
integer count;
// the same results are obtained regardless of
// whether "<=" or "=" operator is used
initial begin
        count = 0;              // execute at time 0
    #25 y <= 1'b1;              // execute at time 25
    #15 count <= count + 1;     // execute at time 40
    #10 x = ~y;                 // execute at time 50
    #5  z <= 1'b0;              // execute at time 55
end
endmodule
```

Even though there is no difference between blocking and nonblocking assignments under regular delay control, the combination of blocking and nonblocking assignments in an **always** block is usually not allowed by synthesis tools. If you are doing this, you will get a warning message from the synthesizer. However, it is fine in behavioral verification. Verify this on your system.

4.3.1.2 Intra-Assignment Delay Control The intra-assignment delay control of a procedural assignment defers the assignment to the left-hand-side variable of the procedural assignment by a specified number of time units but the right-hand side expression is evaluated at the current simulation time. The procedural assignment with intra-assignment delay control has the following forms

```
variable_lvalue =  #delay expression
variable_lvalue <= #delay expression
```

where #delay can be a constant or an expression. Depending on whether the blocking or nonblocking assignment is used, the resulting effects are different. To illustrate this, consider the following two examples, one using blocking assignments and the other using nonblocking assignments.

Section 4.3 *Timing Control* 121

■ **Example 4.9: The use of blocking assignments.**

Because of the sequential features of blocking assignments, the **reg** variables, x, y, and z, are assigned new values at simulation time 5, 8, and 14, respectively.

```
// an example of blocking assignments
module blocking_intra_assign_delay;
reg   x, y, z;
// blocking assignments
initial begin
   x = #5 1'b0;  // x is assigned 0 at time 5
   y = #3 1'b1;  // y is assigned 1 at time 8
   z = #6 1'b0;  // z is assigned 0 at time 14
end
endmodule
```

Nevertheless, the results will be different when nonblocking assignments are used, as illustrated in the following example.

■ **Example 4.10: The use of nonblocking assignments.**

Because of the inherent nonblocking features of nonblocking assignments, the **reg** variables, x, y, and z, are assigned new values at simulation time 5, 3, and 6, respectively. Once again, the three-step procedure proves useful to explain the above results successfully.

```
// an example of nonblocking assignments
module nonblocking_intra_assign_delay;
reg   x, y, z;

// nonblocking assignments
initial begin
   x <= #5 1'b0;  // x is assigned 0 at time 5
   y <= #3 1'b1;  // y is assigned 1 at time 3
   z <= #6 1'b0;  // z is assigned 0 at time 6
end
endmodule
```

A combination of blocking and nonblocking assignments is illustrated in the following example. The three-step procedure coupled with the inherent nonblocking feature of nonblocking assignments can be used to explain the operations of the module easily.

■ **Example 4.11: Mixing blocking with nonblocking assignments.**

When the first procedural assignment is encountered, the variable a is scheduled to assign a value of 0 at simulation time 5, and then the simulator executes the blocking assignment to assign a value 1 to b at simulation time 3. The third statement is encountered at simulation time 3 and the variable c is scheduled to assign a value 0 at 6 time units later, i.e., at simulation time 9. At the same time step, the last blocking assignment is executed so as to assign a value 1 to d at simulation time 10.

```
// a combination of blocking and nonblocking assignments
module blocking_nonblocking_mixed;
reg   a, b, c, d;
```

```verilog
initial begin
   a <= #5 1'b0;  // a is assigned 0 at time 5
   b  = #3 1'b1;  // b is assigned 1 at time 3
   c <= #6 1'b0;  // c is assigned 0 at time 9
   d  = #7 1'b1;  // d is assigned 1 at time 10
end
endmodule
```

What would happen when we interchange the assignment operators of nonblocking and blocking (i.e., <= ↔ =) in the preceding example? Try this and explain the results.

4.3.1.3 Zero-Delay Control The delay expression can have a zero value, namely, #0. Such a case is called an *explicit zero delay*. An explicit zero delay with a null statement will cause a wait until all procedural assignments to be executed at the current simulation time finish their executions. The simulation time does not advance at all.

■ Example 4.12: An example of explicit zero-delay control.

In this module, the procedural statement, "x = 4'b1101;," after the explicit zero delay will be executed after the procedural statement "#10 x = 4'b1011;" at simulation time 22. As a consequence, the value of the **reg** variable, x, is 4'b1101.

```verilog
`timescale 1 ns/100 ps
module explicit_zero_delay;
reg [3:0] x, y, z;
initial begin
         x = 4'b0101;
   #12   y = 4'b1000; // execute at time 12
   #10   x = 4'b1011; // execute at time 22
   #00;  x = 4'b1101; // execute at time 22
end
initial begin
   y = 4'b0111; // execute at time 0
   z = 4'b1111; // execute at time 0
   x = 4'b1110; // execute at time 0
end
initial #100 $finish;
initial $monitor ($realtime,"ns %h %h %h", x, y, z);
endmodule
```

Remember that all **always** and **initial** blocks are executed concurrently in a nondeterministic order. To ensure an **always** or **initial** block to be executed last, namely, after all other **always** and/or **initial** blocks in that simulation time are executed, a method known as explicit *zero-delay control* may be used. To use explicit zero-delay control, statements within an **always** or **initial** block are prefixed with an explicit zero-delay value, #0. The following example gives an illustration of this.

■ Example 4.13: Zero-delay control.

In this module, the first two **initial** blocks are ensured to execute after the other two **initial** blocks because they contain explicit zero-delay control. As a result, both **reg** variables, x and y, are guaranteed to have values of 4'b1001 and 4'b1010, respectively.

Section 4.3 *Timing Control*

```verilog
`timescale 1 ns/100 ps
module zero_delay_control;
reg [3:0] x, y;

initial #0 x = 4'b1001;
initial #0 y = 4'b1010;

initial x = 4'b0001;
initial y = 4'b0100;

// define the execution time and monitor the signals of interest
initial #100 $finish;
initial $monitor ($realtime,"ns %h %h", x, y);
endmodule
```

When an **always** or **initial** block contains many statements, the block containing one or more explicit zero-delay control statements will be scheduled to execute last, after all other blocks have been executed at that simulation time. However, if multiple explicit zero-delay control statements appear in different blocks, these blocks will be still executed in a nondeterministic order. Hence, the race problem still exists. The following example further explores the applications of explicit zero-delay control statements.

∎ Example 4.14: Zero-delay control.

In the following module, the first **initial** block is executed after the second one because an explicit zero-delay control statement appears in the block. As a result, both **reg** variables, x and y, have the values of 4'b1001 and 4'b1010, respectively. It is of interest to note that the execution order will be nondeterministic if both **initial** blocks contain one or more explicit zero-delay control statements. The reader is encouraged to verify this.

```verilog
`timescale 1 ns/100 ps
module zero_delay_control;
reg [3:0] x, y;
initial begin
   #0 x = 4'b1001;
      y = 4'b1010;
end
initial begin
   x = 4'b0001;
   y = 4'b0100;
end
// define the execution time and monitor the signals of interest
initial #100 $finish;
initial $monitor ($realtime,"ns %h %h", x, y);
endmodule
```

Event though explicit zero-delay control may be used to control the execution order of **always** and/or **initial** blocks, it is not a recommended practical technique.

■ Review Questions

4-15 Explain why regular delay control has the same effect on blocking and nonblocking assignments.

4-16 Explain why intra-assignment delay control causes the different results on blocking and nonblocking assignments.

4-17 What is the meaning of explicit zero-delay control?

4.3.2 Event Timing Control

A procedural statement can be executed in synchronism with an event. In other words, a procedural statement with event control will defer its execution until the occurrence of the specified event. An event is the value change on a net or variable or the occurrence of a declared event, called a *named event*, to be addressed later in this section. There are two kinds of event timing control:

1. *Edge-triggered* event control
2. *Level-sensitive* event control

Like delay timing control, event timing control can also be cast into regular event timing control and intra-assignment event timing control according to the place of the timing control in a procedural statement. Thus, event control has the following general syntax

```
event_control null_statement
event_control statement

variable_lvalue  = event_control expression
variable_lvalue <= event_control expression
```

4.3.2.1 Edge-Triggered Event Control In edge-triggered event control, a procedural statement defers its execution until a specified transition of a given signal occurs. The symbol @ is used to specify edge-triggered event control and has the following form

```
@event null_statement
@event statement
```

where statement is executed whenever the specified event occurs. For example,

```
@(enable) #5;    // controlled by enable
@(enable) a = b;
```

mean that the procedural statements, #5; and a = b, are controlled by the enable event. These procedural statements are executed once the value of enable changes, such as from 1 to 0 or from 0 to 1. The first statement is only to delay 5 time units while the second statement is to assign b to a.

In many applications, we need to specify the values to be changed (i.e., the event) at the positive edge or negative edge of a clock signal. In this case, the keyword **posedge** is used for positive (rising) transitions and the keyword **negedge** for negative (falling) transitions. The behavior of **posedge** and **negedge** events is described as follows:

1. A **posedge** event means the transition from 0 to x, z, or 1, and from x or z to 1.
2. A **negedge** event means the transition from 1 to x, z, or 0, and from x or z to 0.

For example, the following **always** block

Section 4.3 *Timing Control* 125

```
always @(posedge clock) begin
    reg1 <= #25 in_1;  // intra-assignment timing control
    reg2 <= @(negedge clock) in_2 ^ in_3; // event control
end
```

means that the value of in_1 is assigned to reg1 after 25 time units whenever the positive edge of the clock signal occurs; the evaluated result of in_2^in_3 is assigned to reg2 at the oncoming negative edge of the clock signal after the positive edge of each clock signal occurs.

4.3.2.2 Named Event Control In Verilog HDL, a new data type called the **event** can be declared in addition to nets and variables. This provides users a capability to declare an event and then to trigger and recognize it.

An identifier declared with an **event** data type is called a *named event*. A named event does not hold any data and has no time duration. A named event is triggered explicitly by using the symbol -> and the triggering of the named event is recognized by using the symbol @. A named event has the same usage as event control. A named event should be declared explicitly before it is used. To declare a named event, the following syntax can be used.

```
event identifier1, identifier2, ..., identifiern;
```

where **event** is a keyword and identifiier1 to identifiiern are identifiers to be used as named events. The event trigger uses the following form

```
->name_event_id;
```

which means the name_event_id is triggered. The following three-step scenario demonstrates the use of a named event: declaration, triggering, and recognition.

```
// step 1: declare a received_data event
event  received_data;
// step 2: trigger the received_data event
always @(posedge clock)
    if (last_byte) -> received_data;
// step 3: recognize the received_data event
always @(received_data) begin
    // put any required operations here
end
```

An illustration of the use of a named event based on the above three-step scenario is given in following example.

■ Example 4.15: The three-step scenario of using a named event.

In this example, a named event, ready, is declared. To illustrate the three-step scenario of using a named event, a counter, count, that counts up its value by 1 every 2 time units is used to trigger this event, ready, whenever its values are the multiples of 5. Once the ready event is triggered, the current count value is displayed with the **$display** system task. Note that what would happen if the **$display** system task is replaced by the **$monitor** system task? Try it and see what happens.

```
`timescale 1 ns/100 ps
// the use of a named event
module named_event;
reg [7:0] count;
// step 1: declare an event ready
```

```verilog
event ready;
initial begin
   count = 0;
   #400 $finish;
end
// step 2: trigger the event on condition
always begin
   #2 count <= count + 1;
   if (count % 5 == 0) -> ready;
end
// step 3: recognize the event and do something
always @(ready) // wait for the ready event
   $display($realtime,"The count is %d", count);
endmodule
```

The combination of a named event and event control provides a powerful and efficient means to synchronize two or more concurrent processes, such as handshaking control, within the same module, or between two modules with different clock domains. An illustration of the combination of named events and event control within the same module is given in the following example. The case that crosses different modules is left as an exercise for the reader.

■ Example 4.16: A simple handshaking example.

In this example, a sender repeatedly sends the data that it generated randomly to a receiver. After the sender produces data, it triggers the ready event to tell the receiver that data are ready. Once the receiver gets the data, it triggers the ack event to notify the sender that it has accepted the data and is ready for the next one. To initiate the handshaking operations, we trigger the ack event at the startup during simulation.

```verilog
'timescale 1 ns/100 ps
// an example illustrates the use of a named event
module handshaking_event;
reg [7:0] data, data_output;
event ready, ack;
integer seed = 1;
initial begin
   #5 ->ack;   // initiate the operation
   #50 $finish;
end
// the source device part starts from here
always @(ack) begin: sender
   // once the event ack is triggered
   data = $random(seed) % 13;
   $display ($realtime,"The source data is: %d", data);
   seed = seed + 7;
   #5 ->ready;
end
// the destination device part starts from here
always @(ready) begin: receiver
   // once the event ready is triggered
   data_output = data;
   $display ($realtime,"The output data is: %d",data_output);
```

```
    #5 ->ack;
end
endmodule
```

What would happen if all blocking assignments within the two **always** blocks are replaced with nonblocking ones? (*Answer:* The data sent by the sender will start from the second event.)

4.3.2.3 Event or Control In many applications, any of multiple signals or events can trigger the execution of a procedural statement or a block of procedural statements. The signals or events represented as logical **or** (||) are also called a *sensitivity list* or an *event list*. The keyword **or** or a comma (,) is used to specify multiple triggers in a sensitivity list. For example, the following **always** block

```
always @(a or b or c_in) // use the keyword or
    {c_out, sum} = a + b + c_in;
```

is equivalent to

```
always @(a, b, c_in)     // use comma (,)
    {c_out, sum} = a + b + c_in;
```

Another example of event or control which describes a D flip-flop with an asynchronous reset is as follows:

```
always @(posedge clock or negedge reset_n)
    if (!reset_n) q <= 1'b0; // asynchronous reset
    else          q <= d;
```

The output of the D flip-flop is 0 if the reset_n is set to 0 and a sample of its input otherwise.

4.3.2.4 Implicit Event List When the number of input variables for a combinational circuit is very large, users often forget to add some of the nets or variables in the sensitivity list. To avoid such a problem, the wildcard @* or @(*) may be used as a shorthand notation to implicitly add all nets and variables which are read by any statement within the associated **always** block.

The following example uses the wildcard character "*" to mean changes on any signal (a, b, or c_in) on the right-hand side of the procedural assignment within the **always** block. The use of @* and @(*) means the same thing.

```
always @(*)       // use *
    {c_out, sum} = a + b + c_in;
```

Another example of using the wildcard @(*) is shown below. Here, the @(*) means @(a, b, c, d, x, z).

```
always @(*) begin // @(a or b or c or d or x or z),
    x = a & b;    // @(a, b, c, d, x, z), or @*
    z = c | d;
    y = x ^ z;
end
```

4.3.2.5 Level-Sensitive Event Control In level-sensitive event control, the execution of a procedural statement is delayed until a specified condition becomes true. The level-sensitive event control uses the keyword **wait** and has the following forms

```
wait (condition) null_statement
wait (condition) statement
```

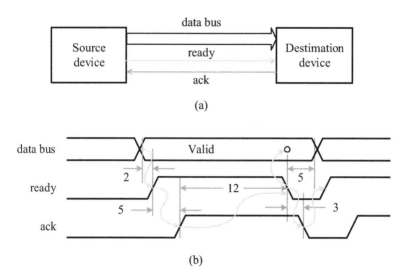

Figure 4.6: An example of handshaking operations between two asynchronous devices: (a) block diagram and (b) timing diagram.

where statement is executed if condition is true and is deferred to be executed until condition becomes true. statement is executed without any additional delay if condition is already true when it is encountered. If the null statement is specified, the level-sensitive event control only causes a wait for the specified condition to become true before the next procedural statement is executed.

As an illustration of the use of level-sensitive event control, consider the following **always** block.

always
    wait (count_enable) count = count − 1;

If the condition, count_enable, is 0, the **wait** statement delays the execution of the procedural statement count = count − 1 until count_enable becomes 1. Generally speaking, the statement controlled by a **wait** statement can also contain a delay specification. For example,

always
    wait (!count_enable) #10 count = count − 1;

If the condition, count_enable, is 1, the **wait** statement delays the execution of the statement count = count − 1 until count_enable becomes 0, and then defers 10 time units and executes the statement count = count − 1. If count_enable is already 0 when the **always** block is entered, then the assignment count = count − 1 is executed after a delay of 10 time units.

Level-sensitive event control can be used to model the handshaking operations between two asynchronous devices abstractly. To illustrate this, consider the block diagram shown in Figure 4.6(a), where the source device wants to transfer data to the destination device asynchronously. The source device first places its data on the data bus, and then activates the ready signal to notify the destination device of this situation. Once the destination device receives the ready signal, it activates the ack signal to acknowledge this case. Next, the source device deactivates its ready signal and the destination device uses this event to latch the data on the data bus into its internal register. Finally, the destination device deactivates the ack signal to signal the completion of one transaction. The detailed timing is shown in Figure 4.6(b).

Section 4.3 *Timing Control*

An illustration of modeling the handshaking operations between two devices in accordance with the timing diagram shown in Figure 4.6(b) by using **wait** statements is given in the following example.

■ Example 4.17: A handshaking module using wait statements.

Two **always** blocks corresponding to the source and destination devices, respectively, are used in the module. It is straightforward to see that both source and destination blocks exactly model the handshaking operations between the two devices in Figure 4.6(a) in accordance with the timing diagram shown in Figure 4.6(b).

```
'timescale 1ns/100 ps
// an example illustrates the use of wait
module handshaking_wait;
reg [7:0] data, data_output;
reg ready = 0, ack = 0;
integer seed = 5;
initial begin
   #400 $finish;
end
// the source device part starts from here
always begin: source
   #5 data <= $random(seed) % 13;
   $display ("The source data is : %d", data);
   #2 ready <= 1;
   wait (ack);    // wait until ack is active
   #12 ready <= 0;
   seed = seed + 3;
end
// the destination device part starts from here
always begin: destination
   wait (ready); // wait until ready is active
   #5 ack <= 1;
   @(negedge ready) data_output <= data;
   $display ("The output data is : %d", data_output);
   #3 ack <= 0;
end
endmodule
```

Of course, the above handshaking module is only used to model the data transfers between two asynchronous devices in behavioral style. It cannot be synthesized into a real-world hardware circuit by synthesis tools. Try to design a handshaking scheme in hardware and then model it.

■ Review Questions

4-18 Define the sensitivity list and the event list.
4-19 What are the three ways that can be used to represent the event or control?
4-20 What are the meanings of the keywords **posedge** and **negedge**?
4-21 What does a named event mean?
4-22 Describe the three-step scenario in using a named event.

4.4 Selection Statements

Selection statements are used to make a selection according to a given condition. Two types of selection statements provided by Verilog HDL are **if-else** and **case** statements. The **if-else** statement can also be nested so as to provide both two-way and multiway selections. The **case** statement supports the multiway selection and has two variants: **casex** and **casez**. The **case** statement and its variants are usually used to model decoders/encoders, multiplexers/demultiplexers, and the like. In this section, we address these statements in detail.

4.4.1 if-else Statement

The selection statement (i.e., the **if-else** statement) is used to make a decision in accordance with a given condition. It has the following the syntax

```
if (condition) true_statement_or_null

if (condition) true_statement_or_null
else false_statement_or_null

if (condition1) true_statement1_or_null
{else if (condition2) true_statement2_or_null}
[else false_statement_or_null]
```

where condition can be a constant or an expression and must always be in parentheses. Both true_statement_or_null and false_statement_or_null can be any procedural statement or a null statement (i.e., ;). If condition is true (namely, has a nonzero known value), true_statement_or_null is executed. If condition is false (namely, has a 0, an x, or a z), true_statement_or_null is not executed. If there is an **else** statement and condition is false, the false_statement_or_null associated with the **else** statement is executed. Note that the **else** statement is always associated with the closest previous **if** statement that lacks an **else** statement except when a **begin-end** block is used to force the proper association.

Basically, an **if-else** statement is used to perform a two-way selection according to a given condition. Nonetheless, since the **if-else** statement can be nested, it can perform a multiway selection as well.

An illustration of using an **if-else** statement to model a 4-to-1 multiplexer is demonstrated in the following example. The reader is encouraged to explain the operations of this module. In addition, it is instructive to synthesize it and compare the result with that modeled in dataflow style.

■ Example 4.18: A 4-to-1 multiplexer using selection statements.

A 4-to-1 multiplexer is modeled using an **if-else** statement. Both true and false statements contain another **if-else** statement to further select the associated inputs. Because of its straightforward simplicity, we do not explain its operations furthermore here.

```
module mux_4to1_ifelse (
      input   i0, i1, i2, i3,   // inputs
      input   s1, s0,           // source selection inputs
      output reg out);          // output

// using an if-else statement
always @(*) // triggered for all signals used
   if (s1) begin
```

Section 4.4 *Selection Statements*

```
      if (s0)  out = i3;  else out = i2;  end
   else begin
      if (s0)  out = i1;  else out = i0;  end
endmodule
```

Another illustration of the use of the **if-else** statement is given in the following example. Here, a 4-bit synchronous binary counter with asynchronous reset is modeled.

■ Example 4.19: A 4-bit synchronous binary counter.

In this example, a 4-bit synchronous binary counter with asynchronous reset is modeled. The counter is cleared if the positive-edge `clear` signal is activated and counts up by 1 at every negative edge of the `clock` signal otherwise. The modulus of the counter is 16 because `qout` is declared to be 4 bits. In addition, the clear operation is done asynchronously because the `clear` signal is put inside the sensitivity list of the **always** block to serve as a trigger source. For many practical applications, it often requires that the reset (i.e., clear) operation be carried out synchronously. Refer to Section 9.1.1 for more details.

```
// a modulo-16 synchronous counter with asynchronous reset
module counter(
        input  clock, clear,       // inputs
        output reg [3:0] qout );   // define 4-bit flip-flops

// the body of 4-bit binary counter
always @(negedge clock or posedge clear)
   if (clear) qout <= 4'h0;
   else       qout <= (qout + 1); // qout <= (qout + 1) % 16;
endmodule
```

Another illustration of the use of the **if-else** statement is given in the following example to model the two-digit BCD adder explored in Section 3.4.

■ Example 4.20: A two-digit BCD adder.

This example is another version of the two-digit BCD adder described in Section 4.2.2 being modeled in behavioral style. Here, we not only replace the conditional operator ?: with an **if-else** statement but also describe the module in a straightforward way just in accordance with the correction steps described in Section 3.4.

```
// a two-digit BCD adder modeled in behavioral style
module BCD_adder_behavior_if (
        input  [7:0] x, y,
        input  c_in,
        output reg [7:0] sum,
        output reg c_out);
reg     ac; // temporary carries
// the body of the two-digit BCD adder
always @(*) begin
   // the least significant digit
   {ac, sum[3:0]} = x[3:0] + y[3:0] + c_in;
   if (ac) sum[3:0] = sum[3:0] + 4'b0110;
   else if (sum[3:0] > 9) {ac,sum[3:0]} = sum[3:0] + 4'b0110;
   // the most significant digit
```

```verilog
    {c_out, sum[7:4]} = x[7:4] + y[7:4] + ac;
    if (c_out) sum[7:4] = sum[7:4] + 4'b0110;
    else if (sum[7:4] > 9) {c_out,sum[7:4]} = sum[7:4] + 4'b0110;
end
endmodule
```

A better result in terms of area (e.g., the number of LUTs) would often be obtained if we rethink the design in a different way. As an illustration, consider the following example.

■ Example 4.21: A two-digit BCD adder in behavioral style.

In this example, we combine the operations required for both correction steps into a single conditional statement to add the result by 6 conditionally and combine the carries generated in both steps into the carry-out. Because only the **if** part without the **else** part of the **if-else** statement is used, ac_tmp and c_tmp are needed to initialize in order to avoid the latches inferred by the synthesizers. More details about the latch inference can be referred to in Section 12.5.2.

```verilog
// a two-digit BCD adder modeled in behavioral style
module BCD_adder_behavior_if(
        input [7:0] x, y,
        input c_in,
        output reg [7:0] sum,
        output reg c_out);
reg     ac, ac_tmp, c_tmp; // temporary carries
// the body of the two-digit BCD adder
always @(*) begin
   // the least significant digit
   ac_tmp = 1'b0;
   {ac, sum[3:0]} = x[3:0] + y[3:0] + c_in;
   if (ac || sum[3:0] > 9){ac_tmp,sum[3:0]} = sum[3:0] + 4'b0110;
   ac = ac | ac_tmp;
   // the most significant digit
   c_tmp = 1'b0;
   {c_out, sum[7:4]} = x[7:4] + y[7:4] + ac;
   if (c_out || sum[7:4] > 9){c_tmp,sum[7:4]} = sum[7:4] + 4'b0110;
   c_out = c_out | c_tmp;
end
endmodule
```

It is instructive to synthesize the above two modules with a specific FPGA device and compare their results in terms of the number of LUTs needed. It is also interesting to compare the results with those obtained in Section 3.4. In addition, it is worth recalling that the variable_lvalue may be on the right-hand side of a procedural assignment, such as ac = ac | ac_tmp and c_out = c_out | c_tmp, without causing any trouble when it is synthesized. However, this is not the case for net_lvalue in an **assign** continuous assignment. Refer to Section 3.1.1 for more details.

4.4.2 case Statement

The **case** statement is used to perform a multiway selection according to a given input condition and is equivalent to a nested **if-else** statement. It is often used to describe the truth table or a function table of a logic circuit in behavioral style. The **case** statement has the syntax

Section 4.4 Selection Statements

```
case (case_expr)
   case_item1_expr{,case_item1_expr}: statement1_or_null
   case_item2_expr{,case_item2_expr}: statement2_or_null
      ...
   case_itemn_expr{,case_itemn_expr}: statementn_or_null
                       [default: statement_or_null]
endcase
```

where `statement_or_null` can be any procedural statement or a null statement (i.e., ;). The **default** statement is optional. Only one **default** statement may be placed inside one **case** statement. In addition, if there is no default statement and all comparisons fail, then none of the procedural statements associated with `case_item_expr` are executed. Furthermore, multiple `case_item_expr` expressions can be specified in one branch.

The execution of the **case** statement is as follows: First, `case_expr` is evaluated. Then, the evaluated result of `case_expr` is compared with `case_item_expr` expressions one by one in the order given. The `statement_or_null` associated with the `case_item_expr` that first matches the `case_expr` is executed. As no `case_item_expr` matches the `case_expr`, the **default** statement is executed if it exists.

The **case** statement acts like a multiplexer. Therefore, it is often used to model a multiplexer, such as the following 4-to-1 multiplexer. You may compare this with the one that is modeled in dataflow style described in the previous chapter.

■ Example 4.22: A 4-to-1 multiplexer using a case statement.

In this example, a **case** statement is used to model a 4-to-1 multiplexer. We can see from the module that it is quite straightforward when using a **case** statement to describe an n-to-1 multiplexer.

```verilog
// a 4-to-1 multiplexer using a case statement
module mux_4to1_case (
      input i0, i1, i2, i3,  // inputs
      input [1:0] s,          // source selection input
      output reg  y);         // output
// the body of the module
always @(i0 or i1 or i2 or i3 or s) // or use always @(*)
   case (s)
      2'b00: y = i0;
      2'b01: y = i1;
      2'b10: y = i2;
      2'b11: y = i3;
   endcase
endmodule
```

The **default** statement is necessary when the set of desired `case_item_expr` expressions cannot cover the whole space spanned by the `case_expr`. In such an incompletely specified **case** statement, the lack of the **default** statement will cause the synthesis tool to infer a latch. The following module reveals how the **default** statement is used to completely specify the **case** statement so as to avoid the unwanted latch inferred by synthesis tools.

■ Example 4.23: A 3-to-1 multiplexer using a case statement.

In this example, a 3-to-1 multiplexer is described. Because only three case_item_expr expressions are needed in this example, we add the **default** statement to complete the **case** statement. Thus, the **default** statement is executed when none of the case_item_expr expressions matches the case_expr (i.e., s). Nevertheless, care must be taken in this case. When assigning the output y with a value, we must take into account the actual requirement, such as to leave the output an unknown x, a z, or a definite known value 0 or 1.

```verilog
// a 3-to-1 multiplexer using a case statement with
// the default statement
module mux_3to1_case_default(
        input i0, i1, i2, // inputs
        input [1:0] s,    // source selection input
        output reg  y);   // output
// the body of the module
always @(i0 or i1 or i2 or s) // or use always @(*)
   case (s)
      2'b00: y = i0;
      2'b01: y = i1;
      2'b10: y = i2;
      default: y = 1'b0; // to completely specify the case
   endcase
endmodule
```

An illustration of the use of multiple case_item_expr expressions in one branch of a **case** statement is given in the following example to describe a 3-to-1 multiplexer.

■ Example 4.24: Multiple case-items in one case branch.

In this example, a 3-to-1 multiplexer is described. Although only three branches are used, all of four combinations of the case_expr are specified. Consequently, no latch will be inferred by the synthesizer even that we do not include the **default** statement in the **case** statement.

```verilog
// a 3-to-1 multiplexer using multiple case items
module mux_3to1_case (
        input i0, i1, i2, // inputs
        input [1:0] s,    // source selection input
        output reg  y);   // output

// the body of the module
always @(i0 or i1 or i2 or s) // or use always @(*)
   case (s)
      2'b00: y = i0;
      2'b01: y = i1;
      2'b10, 2'b11: y = i2;   // multiple case items
   endcase
endmodule
```

The **case** statement compares the logic value in the case_expr expression with that in the case_item_expr expression in a bit-by-bit fashion. The zero-extended approach is used when both case_expr and case_item_expr have different number of bits.

Section 4.4 Selection Statements

■ **Example 4.25: A 5-to-1 multiplexer using a case statement.**

As an illustration of the effect of the zero-extended approach, consider a 5-to-1 multiplexer. Assume that in this multiplexer, there are five inputs, named i0 to i4. The i4 input is routed to the y output whenever the LSB of the source selection, s, is 1. The other four inputs are routed to the y output according to the combinations of the next two higher bits of the source selection, s. If we describe this multiplexer as the module shown below, then according to the zero-extended approach used in Verilog HDL, the 5th case_item_expr that includes only 1 bit is extended to 001. This results in that the i4 input is selected only when the source selection, s, is 3'b001 rather than what we have expected. The reader may check this from Figure 4.7.

```
// a 5-to-1 multiplexer using a case statement
module mux_5to1_case(
       input i0, i1, i2, i3, i4, // inputs
       input [2:0] s,  // source selection signal
       output reg  y); // output

// the body of the 5-to-1 multiplexer
always @(i0 or i1 or i2 or i3 or i4 or s) // or always @(*)
   case (s)
      3'b000: y = i0;
      3'b010: y = i1;
      3'b100: y = i2;
      3'b110: y = i3;
      1'b1  : y = i4;
      default: y = 1'b0; // include all other possible cases
   endcase
endmodule
```

Note that in the above module, the i4 input is selected only when the source selection, s, is 3'b001. If what we really want is that whenever the LSB is 1, the i4 input is routed to the y output, then the most straightforward way to solve this is by using multiple case_item_expr expressions as we have done in the mux_3to1_case module. Because of its intuitive simplicity, it is left as an exercise for the reader.

It is instructive to note that the **case** statement may also be nested. An illustration of this is explored in the following example, which is intended to rewrite a 4-to-1 multiplexer with a nested **case** statement.

■ **Example 4.26: A 4-to-1 multiplexer.**

In this example, a nested **case** statement is used to model a 4-to-1 multiplexer. The outer **case** statement uses the source selection, s1, to select the groups, i0 and i1, as well as i2 and i3, while the inner **case** statement uses the source selection, s0, to further distinguish the elements in each group.

```
module mux_4to1_nested_case(
       input i0, i1, i2, i3, // inputs
       input s1, s0,         // source selection inputs
       output reg out);      // output
// using a nested case statement
always @(*)
   case (s1)
```

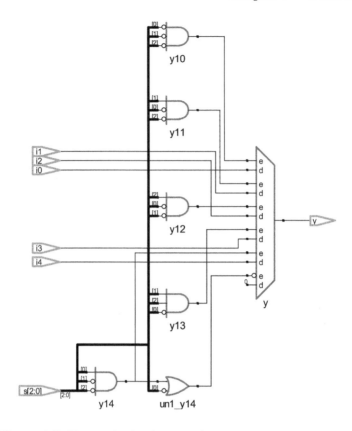

Figure 4.7: The synthesized result of the mux_5x1_case module.

```
        1'b0: case (s0)
                1'b0: out = i0;
                1'b1: out = i1;
              endcase
        1'b1: case (s0)
                1'b0: out = i2;
                1'b1: out = i3;
              endcase
    endcase
endmodule
```

4.4.3 casex and casez Statements

Both **casex** and **casez** statements are used to perform a multiway selection like that of the **case** statement except that the **casez** statement treats all z values as don't care while the **casex** treats all x and z values as don't care. These two statements only compare non-x or -z positions in both case_expr and case_item_expr expressions.

An illustration of using the features of the **casex** statement to model a circuit that counts trailing zeros in a nibble (i.e., a half byte) is given in the following example.

Example 4.27: The number of trailing zeros in a nibble.

In this example, a circuit that counts the number of trailing zeros in a nibble is described. Because the basic feature of the **casex** statement is that it only compares non-x or -z positions in both `case_expr` and `case_item_expr` expressions, by properly arranging each `case_item_expr` and its associated expression in the order as shown in the module, we may obtain the required operations.

```verilog
// count the number of trailing zeros in a nibble
module trailing_zero_4b (
      input   [3:0] data,
      output reg [2:0] out);
// the body of the module
always @(data)
   casex (data)  // treat both x and z as don't care
      4'bxxx1: out = 0;
      4'bxx10: out = 1;
      4'bx100: out = 2;
      4'b1000: out = 3;
      4'b0000: out = 4;
      default: out = 3'b111; // all other possible cases
   endcase
endmodule
```

Similarly, using the same technique as the above example, the circuit that counts the number of trailing zeros in a byte, even more bits, can be readily described. Nevertheless, although this approach is conceptually simple, it is not good practice when the operand width is large. In the next section, we illustrate how to carry out this in a more efficient fashion with the use of loop statements. More examples about the use of the powerful features of the **casex** statement to model priority encoders and many others are explored in Chapter 8.

Review Questions

4-23 Explain the basic syntax of the **if-else** statement.
4-24 Can a block of statements be used as a `case_item` expression?
4-25 Explain the role of the **default** statement in a **case** statement.
4-26 What are the differences between **case** and **casex** statements?
4-27 What are the differences between **casex** and **casez** statements?
4-28 How would you count the number of leading zeros in a byte?

4.5 Iteration (Loop) Statements

Like any general-purpose high-level programming language, Verilog HDL also provides a set of powerful and efficient iteration (loop) statements, including **while**, **for**, **repeat**, and **forever**. These iteration (loop) statements provide a means to control the execution of a procedural statement or a block of procedural statements zero, one, or more times. Like the other procedural statements, iteration (loop) statements may only appear inside an **initial** or **always** block and may contain timing control expressions.

4.5.1 The while Loop Statement

The **while** loop executes a procedural statement until a given conditional expression becomes false. The **while** loop statement uses the keyword **while** and has the following form:

```
while (condition_expr) statement
```

If condition_expr starts out being false, statement will not be executed at all. If condition_expr is an x or a z, it is treated as a 0 (i.e., a false condition).

For example, a simple condition is used in the following **while** loop

```
while (count < 12) count <= count + 1;
```

The procedural assignment count <= count + 1 is repeatedly carried out until the condition (count < 12) is no longer valid.

Another example with a more complicated condition is shown in the following **while** loop.

```
while (count <= 50 && flag) // a complex condition
begin
    // put statements wanted to be carried out here
end
```

Here a complex condition with logical and (&&) of two simple conditions, count <= 50 and flag == 1, is used to control the loop. The procedural statements inside the block enclosed with a pair of keywords **begin** and **end** are executed only if both conditions are true, and are not executed, otherwise.

An illustration of the use of a **while** loop to compute the number of zeros in a byte is explored in the following example. In this example, we illustrate how an algorithm can be directly modeled in behavioral style. Of course, like the dataflow or structural style, it is still synthesized into a combinational circuit.

■ Example 4.28: The number of zeros in a byte.

In this example, we want to count the number of zeros in a byte. For this purpose, a **while** loop is used to control the number of iterations and an **if** statement is used to check and add up the number of nonzero bits. The **while** loop terminates once it reaches the maximum number of iterations. The bit value is checked in sequence from bit 0 to bit 7, which is controlled by an **integer** variable, i. Note that the **if** statement may be replaced with out = out + ~data[i]. Why? Give the reasons for this.

```verilog
// a module to count the number of zeros in a byte
module zero_count_while(
       input  [7:0] data,
       output reg [3:0] out);
// the body of the module
integer i;      // loop counter
always @(data) begin
   out = 0; i = 0;
   while (i <= 7) begin  // a simple condition
      // may use out = out + ~data[i] instead
      if (data[i] == 0) out = out + 1;
      i = i + 1; end
end
endmodule
```

Section 4.5 *Iteration (Loop) Statements*

Another illustration of the use of a **while** loop is to compute the number of trailing zeros in a byte. This is another illustration of how an algorithm can be described in behavioral style. Because this algorithm is combinational logic in nature, it is indeed synthesized into a combinational circuit.

■ **Example 4.29: The number of trailing zeros in a byte.**

In this example, the number of trailing zeros in a byte is to be counted. To achieve this, a **while** loop is used to schedule the related operations: checking the bit value in sequence from bit 0 to bit 7 and counting the number of zero bits. The **while** loop terminates whenever it finds a nonzero bit or reaches the maximum number of iterations. It is worth comparing this example with the one using the **casex** statement described in the previous section and discerning the differences between them.

```
// count the number of trailing zeros in a byte
module trailing_zero_while(
      input   [7:0] data,
      output reg [3:0] out);
integer i;           // loop counter
always @(data) begin
   out = 0; i = 0;
   while (data[i] == 0 && i <= 7) begin  // a complex condition
      out = out + 1;
      i = i + 1;
   end
end
endmodule
```

Note also the differences between this example and the previous one. Especially, what are their requirements and how do we treat them? A popular problem closely related to the trailing-zero problem is to compute the leading zeros in a byte. Since it is intuitively simple, we would like to leave it for the reader as an exercise.

4.5.2 The for Loop Statement

The **for** loop repeatedly executes a procedural statement a fixed number of times or until a given condition becomes false. It is usually used to perform a *counting loop* although it is much powerful than this. A counting loop is a loop repeated with a fixed number of times, or a known number of times. The behavior of the **for** loop is much like the **for** statement in C language. A **for** loop statement contains three parts: an initial condition, the termination condition checking, and the control variable updating. The **for** loop uses the keyword **for** and has the following syntax

```
for (init_expr; condition_expr; update_expr)
    statement
```

where `init_expr` sets the initial condition, `condition_expr` checks the termination condition, and `update_expr` updates the control variable. As long as `condition_expr` is true, `statement` will be executed.

The **for** loop is equivalent to the **while** loop with the following construct:

```
init_expr
while (condition_expr) begin
   statement
```

Two very simple applications of the **for** loop statement are as follows:

```
for (i = 0; i < 32; i = i + 1)
   state[i] = 0;      // initialize to zeros
for (i = 1; i < 32; i = i + 2) begin
   state[i] = 0;
   count = count + 1;
end
```

The following example is an illustration of using a **for** loop to compute the number of zeros in a byte, which has been described previously using a **while** loop. It is instructive to compare and explore the differences between the two loop statements used in both modules.

■ Example 4.30: The number of zeros in a byte.

In this example, the **while** loop of the module described in the previous subsection is rewritten into an equivalent **for** loop. The resulting module is given as follows:

```
// a module counts the number of zeros in a byte
module zero_count_for (
      input   [7:0] data,
      output reg [3:0] out);
// the body of the module
integer i;           // loop counter
always @(data) begin
   out = 0;
   for (i = 0; i <= 7; i = i + 1) // a simple condition
      // may use out = out + ~data[i] instead
      if (data[i] == 0) out = out + 1;
end
endmodule
```

Another illustration of the use of a **for** loop is to compute the number of trailing zeros in a byte, which has been described previously using a **while** loop. It is instructive to compare both modules and explore the differences between the two loop statements.

■ Example 4.31: The number of trailing zeros in a byte.

This example is to simply convert the **while** loop of the module described in the previous subsection into an equivalent **for** loop. The resulting module is as follows:

```
// a module counts the number of trailing zeros in a byte
module trailing_zero_for (
      input   [7:0] data,
      output reg [3:0] out);
integer i;           // loop counter
always @(data) begin
   out = 0;
   // using a complex condition
   for (i = 0; data[i] == 0 && i <= 7; i = i + 1)
      out = out + 1;
```

```
end
endmodule
```

Note the differences between this example and the previous one. In particular, what are their requirements and how do we treat them?

4.5.3 The repeat Loop Statement

The **repeat** loop executes a procedural statement a fixed number of times. It uses the keyword **repeat** and has the following syntax

```
repeat (counter_expr) statement
```

where `counter_expr` may be a constant, a variable, or a signal value. The `counter_expr` expression is evaluated only once before starting the execution of the procedural statement. In addition, if the `counter_expr` expression is an x or a z, then it is treated as a 0 (i.e., a false condition).

The following program segment repeatedly executes the **begin-end** block 42 times.

```
i = 0;
repeat (42) begin
   state[i] = 0;     // initialize to zeros
   i = i + 1;        // next item
end
```

Sometimes, we may use an expression or an identifier as the `counter_expr` in the **repeat** statement. An illustration is given in the following program segment.

```
repeat (cycles) begin
   @(posedge clock) buffer[i] <= data;
   i <= i + 1;       // next item
end
```

where the **begin-end** block will be repeatedly executed `cycles` times. The `cycles` must be a constant or evaluated to a number before entering the loop.

The **repeat** statement shown below means to wait for the positive edge of the clock signal and then increases the `count` variable in a number of `loop_count` times, one for each positive edge of the clock signal.

```
repeat (loop_count)
   @(posedge clock) count = count + 1;
```

The following example illustrates a simple application of the **repeat** loop to generate a reset signal in a test bench during the verification of a design.

■ Example 4.32: A simple application of repeat loops.

The repeat loop is often employed to generate a reset signal in verifying a design. In the following module, we first use an **always** block to generate a 50% duty-cycle clock signal with a period set by the **parameter** statement. Then, a **repeat** loop statement is utilized to generate an active-high reset signal with a period of two clock cycles starting at the negative edge of the clock signal, `clk`. Note the use of the **wait** statement in the module. Try to explain its function.

```
`timescale 10 ns / 100 ps
module repeat_loop;
parameter clk_period = 20;
```

```verilog
reg clk;
reg reset;
// generate a clock signal with a 50% duty cycle
always begin
   #(clk_period/2) clk <= 1'b0;
   #(clk_period/2) clk <= 1'b1;
end
// activate a reset signal for two clock cycles
initial begin
   reset = 1'b0;
   wait (clk !== 1'bx);
   // activate the reset signal for two clock cycles
   // starting at falling edge of the clk
   repeat (3) @(negedge clk) reset <= 1'b1;
   reset <= 1'b0;
end
initial #200 $finish;
initial $monitor($realtime,"ns %h %h", clk, reset);
endmodule
```

4.5.3.1 Repeat Event Control In addition to being used as loop control, the **repeat** statement can also be used as intra-assignment event control, called *repeat event control*. The procedural assignment with **repeat** event control has the following syntax:

```
variable_lvalue  = repeat (expression)
                   @(event_expression) expression
variable_lvalue <= repeat (expression)
                   @(event_expression) expression
```

This form of control is usually used to specify a delay that is based on the number of occurrences of one or more events. The following procedural statement means to compute count + 1 first, then waits for the positive edge of the clock signal for loop_count times, and finally assigns the computed result to count.

```
count = repeat (loop_count) @(posedge clock) count + 1;
```

This **repeat** event control is equivalent to the following program segment:

```
temp = count + 1;
@(posedge clock); // repeat loop_count times
@(posedge clock);
   ...
@(posedge clock);
count = temp;
```

4.5.4 The forever Loop Statement

The **forever** loop continuously performs a procedural statement until the **$finish** system task is encountered or a **disable** statement within the procedural statement is executed to break it out. The **forever** loop uses the keyword **forever** and has the following syntax:

```
forever statement
```

It is equivalent to a **while** loop with an always true condition_expr, such as **while (1)**. A **forever** loop can be exited by the use of a **disable** statement. In addition, some form of timing control must be used inside `statement` to prevent the **forever** loop from looping in zero delay infinitely.

As an example of the use of the **forever** loop, consider the following simple **forever** loop, which is used to generate a clock signal with a period of 10 time units and a duty cycle of 50%.

```
initial begin
   clock <= 0;
   forever #5 clock <= ~clock;
end
```

Another example of the use of the **forever** loop is as follows

```
reg clock, x, y;

initial
   forever @(posedge clock) x <= y;
```

which means that at every positive edge of the `clock` signal the value of the y variable is assigned to the x variable.

■ **Review Questions**

4-29 What are the four types of loop statements provided in Verilog HDL?

4-30 Describe the operations of the **while** loop statement.

4-31 Describe the operations of the **for** loop statement.

4-32 Describe how to replace a **for** loop with a **while** loop.

4-33 Describe the operations of the **repeat** loop statement.

4-34 What is **repeat** event control?

4-35 Describe the two basic ways used to exit a **forever** loop.

4.6 Summary

In this chapter, we have discussed two basic procedural constructs: **initial** and **always**. An **initial** or an **always** construct is usually called an **initial** or an **always** block. All other behavioral statements must be placed within **initial** and/or **always** blocks. The **initial** blocks are used to initialize variables and set values into variables or nets, while **always** blocks are used to model the continuous operations required in hardware modules. Each **always** block corresponds to a piece of logic.

Procedural assignments are used within procedures, including **initial**, **always**, **task**, and **function**. Procedural assignments put values in variables. A variable holds the value assigned by a procedural assignment until the next procedural assignment updates the variable again. There are two procedural assignments: blocking and nonblocking. Blocking assignments use the = assignment operator and are executed in the order specified. Nonblocking assignments use the <= assignment operator and are executed without blocking the other procedural statements in a sequential block. Nonblocking assignments provide a method to model several concurrent data transfers that take place after a common event.

Timing control provides a way to specify the simulation time at which procedural statements are to be executed. In Verilog HDL, there are two timing control methods: delay timing control and event timing control. In accordance with the place of timing control in a procedural statement, timing control can be divided into two types: regular timing control and intra-assignment timing control. Regular timing control defers the execution of the entire procedural statement by a specified time units (i.e., regular delay control) or until the occurrence of the specified event (i.e., regular event control). Intra-assignment timing control defers the assignment to the left-hand-side variable by a specified time units (i.e., intra-assignment delay control) or until the occurrence of the specified event (i.e., intra-assignment event control) but the right-hand-side expression is evaluated at the current simulation time.

Event timing control contains two types: edge-triggered event control and level-sensitive event control. In edge-triggered event control, a procedural statement defers its execution until a specified transition of a given signal occurs. In level-sensitive event control, the execution of a procedural statement is delayed until a condition becomes true.

Verilog HDL also provides selection statements and iteration (loop) statements. Selection statements are used to make a selection according to a given condition. They contain two statements: **if-else** and **case**. The **case** statement also has two variants: **casex** and **casez**.

Iteration (loop) constructs include the following four types of statements: **repeat**, **for**, **while**, and **forever**. These loop statements provide a means of controlling the execution of a procedural statement or a block of procedural statements zero, one, or more times. The **while** loop executes a procedural statement until a given condition becomes false. The **for** loop repeatedly executes a procedural statement a fixed number of times or until a given condition becomes false. The **repeat** loop executes a procedural statement a fixed number of times. The **forever** loop continuously executes a procedural statement infinitely.

References

1. J. Bhasker, *A Verilog HDL Primer*, 3rd ed., Star Galaxy Publishing, 2005.
2. IEEE 1364-2001 Standard, *IEEE Standard Verilog Hardware Description Language*, 2001.
3. IEEE 1364-2005 Standard, *IEEE Standard for Verilog Hardware Description Language*, 2006.
4. S. Palnitkar, *Verilog HDL: A Guide to Digital Design and Synthesis*, 2nd ed., SunSoft Press, 2003.

Problems

4-1 Consider the six basic logic gates: AND, NAND, OR, NOR, XOR, and XNOR.
  (a) Model these basic logic gates in behavioral style.
  (b) Write a Verilog HDL module for each of the gates in behavioral style.

4-2 The following module is an example of a combination of blocking and nonblocking assignments.

```
module blocking_nonblocking_mixed;
reg   a, b, c, d, e;
// blocking assignments
initial begin
    a  = #3 1'b0;
    b <= #4 1'b1;
    c <= #6 1'b0;
    d  = #7 1'b1;
    e  = #8 1'b0;
end
endmodule
```

Problems

(a) At what simulation time does each procedural statement execute?
(b) Interchange the blocking and nonblocking assignment operators and redo part (a).
(c) Replace all nonblocking assignments with blocking assignments and redo part (a).
(d) Replace all blocking assignments with nonblocking assignments and redo part (a).

4-3 Without using intra-assignment delay control, rewrite each of the following two procedural statements:

```
a = #25 b;
c = #12 d;
```

4-4 Without using intra-assignment event control, rewrite each of the following two procedural statements:

```
a = @(posedge clk) b;
c = @(negedge clk) d;
```

4-5 Without using intra-assignment event control, rewrite each of the following two **repeat** event control procedural statements:

```
a = repeat(3) @(posedge clk) b;
c = repeat(4) @(negedge clk) d;
```

4-6 Consider the following simple Verilog HDL module:

```
module xyz(input[1:0] din, output reg [3:0] dout);

always @(din)
    dout = {{3{1'b0}}, 1'b1} << din;
endmodule
```

(a) What does the above module function as?
(b) Rewrite the above module with a **case** statement.

4-7 Suppose that conditions, $c1$ and $c2$, are a single-bit value of 0 or 1. Using an **if-else** statement, rewrite each of the following two **assign** continuous assignments.
(a) A simple statement:

```
assign y = c1 ? expr1 : expr2;
```

(b) A complex statement:

```
assign y = c1 ? (c2 ? expr1 : expr2):
               (c2 ? expr3 : expr4);
```

4-8 Suppose that conditions, $c1$ and $c2$, are a single-bit value of 0 or 1. Using a **case** statement, rewrite each of the following two **assign** continuous statements.
(a) A simple statement:

```
assign y = c1 ? expr1 : expr2;
```

(b) A complex statement:

```
assign y = c1 ? (c2 ? expr3 : expr4):
               (c2 ? expr1 : expr2);
```

4-9 Suppose that conditions, $c1$ and $c2$, are a single-bit value of 0 or 1. Using a **case** statement, rewrite each of the following procedural statements.

  (a) A simple procedural statement:

  ```
  if (c1) statement1; else statement2;
  ```

  (b) A complex procedural statement:

  ```
  if (c1) statement1;
  else if (c2) statement2;
      else statement3;
  ```

4-10 What are the differences between the following **always** blocks?

  ```
  always @(x or in_x)
     x = x + 7 * in_x;

  always @(posedge clk )
     y <= y + 7 * in_y;
  ```

4-11 A majority circuit is a device that outputs 1 whenever more than half of its inputs are 1.

  (a) Describe the behavior of an 8-input majority circuit.
  (b) Model the majority circuit in behavioral style.
  (c) If the inputs are checked only when a `ready` signal is 1, redo part(b).

4-12 A minority circuit is a device that outputs 1 whenever less than half of its inputs are 1.

  (a) Describe the behavior of an 8-input minority circuit.
  (b) Model the minority circuit in behavioral style.
  (c) If the inputs are checked only when a `ready` signal is 1, redo part(b).

4-13 Assume that a square-root circuit has an 8-bit input, x, and a 4-bit output, y. The output value is the square root of its input value, i.e., $y = \sqrt{x}$. Describe the behavior of this circuit and model it in behavioral style on condition that your module must be synthesizable.

4-14 Assume that a square circuit has a 4-bit input, x, and an 8-bit output, y. The output value is the square of its input value, i.e., $y = x^2$. Describe the behavior of this circuit and model it in behavioral style on condition that your module must be synthesizable.

4-15 Assume that a binary-to-BCD (binary coded decimal) code converter is needed. Describe the behavior of this circuit and model it in behavioral style. Write a test bench to verify it.

4-16 Assume that a BCD-to-binary code converter is wanted. Describe the behavior of this circuit and model it in behavioral style. Write a test bench to verify it.

4-17 Assume that an excess-3 code checker is desired. The output of the circuit is 1 if its input is a valid excess-3 code and is 0 otherwise. Describe the behavior of this circuit and model it in behavioral style. Write a test bench to verify it.

4-18 The relationship between the BCD code and the 2-out-of-5 code is shown in Table 4.1. Design a BCD code to 2-out-of-5 code converter and model it in behavioral style. Write a test bench to verify it.

Problems 147

Table 4.1: The relationship between the BCD code and 2-out-of-5 code.

BCD	2-out-of-5 code	BCD	2-out-of-5 code
0	11000	5	01010
1	00011	6	01100
2	00101	7	10001
3	00110	8	10010
4	01001	9	10100

4-19 Design a 2-out-of-5 code to BCD code converter and model it in behavioral style. Write a test bench to verify it.

4-20 Assume that a clock generator is required in a test bench. The clock output, clk, is initialized to 0 and toggles its value every 25 time units. Describe this clock generator without using any **always** block.

4-21 Assume that a clock generator with a burst output is required in a test bench. As the enable input is set to high, the clock generator outputs 50 pulses with a period of 20 time units and a duty cycle of 50%. The output pulses should start from a logic 0. Model this clock generator.

4-22 Assume that a clock generator is required in a test bench. The clock output, clk, is initialized to a logic 0 and toggles its value every 10 time units. Describe this clock generator using a **while** loop.

4-23 Assume that a clock generator is required in a test bench. The clock output, clk, is initialized to a logic 0 and has a period of 20 time units and a duty cycle of 25%. Describe this clock generator using a **for** loop.

4-24 Assume that a clock generator is required in a test bench. The clock output, clk, is initialized to a logic 0 and has a period of 100 time units and a duty cycle of 40%. Describe this clock generator using a **forever** loop.

4-25 Assume that a reset signal is required in a test bench. The reset signal has to last for 3 clock cycles and starts from the negative edge of the clock signal. Design and model the reset signal generator.

4-26 Assume that we want to design a module that counts the number of ones in a byte. Describe the module in behavioral style and write a test bench to verify its functionality.

4-27 Assume that we want to design a module that counts the number of trailing ones in a byte. Describe the module in behavioral style and write a test bench to verify its functionality.

4-28 Rewrite the handshaking example (Example 4.16) in Section 4.3.2 using two modules; namely, place each **always** block into a separate module.

4-29 Professor Bob wrote a test bench to verify the functionality of a 4-bit 2's-complement adder exhaustively. Unfortunately, he found something is wrong in the test bench. The test bench is as follows:

```
// problem 4-28
'timescale 1 ns / 100 ps
module twos_adder_behavioral_tb;
```

```verilog
    reg [3:0] x, y;
    reg c_in;
    wire [3:0] sum;
    wire c_out;
    // Unit Under Test port map
        twos_adder_behavioral UUT (
            .x(x), .y(y), .c_in(c_in), .sum(sum), .c_out(c_out));
    reg [7:0] i;

    initial begin
       for (i = 0; i <= 255; i = i + 1) begin
          x[3:0] = i[7:4]; y[3:0] = i[3:0]; c_in = 1'b0;
          #20; end
       for (i = 0; i <= 255; i = i + 1) begin
          x[3:0] = i[7:4]; y[3:0] = i[3:0]; c_in = 1'b1;
          #20; end
    end
    initial
       #12000 $finish;
    initial
       $monitor($realtime,"ns %h %h %h %h", x, y, c_in, {c_out, sum});
    endmodule
```

(a) What is the bug that Professor Bob found? Give your reasons.

(b) Remove the bug and give your reasons.

4-30 In general, the data transfer between two asynchronous devices using handshaking control can be initiated by the source or the destination device. Figure 4.6 shows the source-initiated handshaking control. Give an example of destination-initiated handshaking control and draw its timing diagram.

4-31 Assume that we want to design a two-digit BCD subtracter, which is indeed composed of two single-digit BCD subtractors cascaded together. The details of a single-digit BCD subtracter can be referred to Problem 3-17 of Chapter 3.

(a) Write a module to describe this two-digit BCD subtracter in behavioral style.

(b) Write a test bench to verify whether the BCD subtracter behaves correctly.

4-32 Assume that a two-digit BCD adder/subtracter is desired. The BCD adder/subtracter functions as a BCD adder if the mode selection input mode is 0 and as a BCD subtracter otherwise.

(a) Write a module to describe this two-digit BCD adder/subtracter in behavioral style.

(b) Write a test bench to verify whether the BCD adder/subtracter behaves correctly.

4-33 Using a **for** loop, rewrite the conversion between the big-endian and little-endian representations given in Example 3.2. Write a test bench to verify it.

4-34 Design a converter that converts an incoming BCD code to an excess-3 code. Write a Verilog HDL module to describe it in behavioral style and write a test bench to verify it.

4-35 Design a converter that converts an incoming excess-3 code to a BCD code. Write a Verilog HDL module to describe it in behavioral style and write a test bench to verify it.

4-36 Professor Bob wants to study the features of both blocking and nonblocking procedural assignments. He writes a 4-bit binary down counter module as follows:

```
module counter_blocking(
        input   clock, clear,
        output reg [3:0] qout,
        output reg finish);
// the body of the 4-bit counter
always @(posedge clock or posedge clear)
   if (clear) qout = 4'h0;
   else       qout = qout - 1;
always @(posedge clock or posedge clear)
   if (clear) finish = 0;
   else if (qout == 13) finish = 1;
endmodule
```

(a) What is the value of qout as finish is 1? What is the value of qout as finish is 1 if all blocking assignments are replaced with nonblocking ones?

(b) Perform behavioral simulations with an available simulator and compare the results with those you obtained in (a).

(c) Synthesize the module separately with blocking and nonblocking assignments and then perform post-synthesis (i.e., gate-level netlist) simulations with an available simulator and compare the results with those you obtained in (b).

4-37 Replace the two **always** blocks in the module of Problem 4-36 with the following **always** block and redo Problem 4-36.

```
always @(posedge clock or posedge clear)
   if (clear) begin
      qout = 4'h0;
      finish = 0; end
   else begin
      qout = qout - 1;
      if (qout == 13) finish = 1; end
```

5

Tasks, Functions, and UDPs

THERE are three additional components that can help model a design in behavioral style. These include tasks, functions, and user-defined primitives (UDPs). Tasks and functions provide the ability to reuse the same piece of code from many places in a design. In addition, they provide a means of breaking up a large design into smaller ones so as to make the source descriptions easier to be read and debugged.

In addition to user-defined tasks and functions, there is a rich set of predefined tasks and functions called system tasks and system functions, whose identifiers are prefixed with a symbol "$." Although these tasks and functions cannot be accepted by synthesis tools, they are useful when modeling a design in behavioral style or writing test benches for designs abstractly.

UDPs provide a means to model a design with a truth table. The truth table is one of the three most widely used approaches to describing a digital logic circuit. The other two approaches are schematics and Boolean expressions. These three approaches correspond to the UDP, the structural style, and the dataflow style in Verilog HDL, respectively.

5.1 Tasks

A procedure (including a task and a function) in high-level programming languages provides the ability to reuse a common piece of code from several different places in a design. This common piece of code is encapsulated as a task/function by using the task/function definition in Verilog HDL. Generally speaking, a task is used when the procedure has to contain timing control, has at least one output argument, or has no input arguments. The features of tasks are listed in Table 5.1.

5.1.1 Task Definition and Calls

A task can contain timing control, and it can invoke tasks and functions. A tasks is declared with a pair of keywords **task** and **endtask** and may be declared by either of the following two forms

```
// port-list style
task [automatic] task_id;
   {task_port_declaration};   // port declarations
   {block_item_declaration}   // local variables or constants
   statement_or_null
endtask
```

Table 5.1: Comparison of task and function features.

Item	Tasks	Functions
Arguments	May have zero, one, or more **input**, **output**, and/or **inout** ports.	At least one **input** port, and cannot have **output** and **inout** ports.
Return value	May have multiple values via **output** and **inout** ports.	Only a single value via the function name.
Timing control	Yes	No
Execution	In non-zero simulation time.	In zero simulation time.
Invoke function/tasks	Functions and tasks.	Functions only.

```
// port-list declaration style
task [automatic] task_id ([task_port_list]);
   {block_item_declaration}  // local variables or constants
   statement_or_null
endtask
```

where the keyword **automatic** declares an automatic task with all declarations allocated dynamically for each concurrent task entry. The port (i.e., argument) declarations specify the following arguments: **input**, **output**, and/or **inout** ports. The task_port_declaration has the following syntax

```
port_type [reg] [signed] [range] port_id{, port_id}
port_type port_data_type port_id{, port_id}
```

where port_type can be any of {**input, output, inout**} and port_data_type can be any of {**integer, real, realtime, time**}. range specifies the vector size in the form of [msb:lsb].

The format of task_port_list is as follows

```
task_port_declaration{, task_port_declaration}
```

Since a task must be invoked within an **initial** or **always** block, not all data types and constant declarations are allowed to use within a task definition. Hence, the data types and constant declarations that may be included in block_item_declaration are variable data types (including **reg, integer, real, realtime,** and **time**), **event, localparam,** and **parameter**. statement_or_null can be any procedural statement or a null statement (i.e., ;) (see Section 4.1.1).

A task definition must be placed within a module definition and can have zero, one, or more arguments. Values are passed to and from a task through arguments. It is worth noting that only procedural statements can be used within a task. Because **initial** and **always** blocks cannot be nested and a task must be called within an **initial** or **always** block, the **initial** or **always** block cannot be used in task definitions.

5.1.1.1 Task Calls A task is called (or enabled) by a task enabling statement, which is a procedural statement that can only appear inside an **initial** or **always** block. The task enable (also call or invoke) statement has the following syntax

```
task_id [(argument{, argument})];
```

The task enable statement specifies the input arguments, passing values to the called task, and the output arguments, receiving the results from the called task. The argument list is an ordered list that must match the order of the argument list in the task definition of the called task. Note that all arguments are passed by value rather than by reference.

Section 5.1 Tasks

■ Example 5.1: A task for counting the number of 0s in a byte.

As an example to demonstrate the task declaration in port-list declaration style, consider the module, described in Section 4.5.2, that counts the number of zeros in a byte. According to the operations described in that module, we may simply rewrite the **for** loop in the module by a task definition. This results in the following module.

```
// a module counts the number of zeros in a byte
module zero_count_task(
       input [7:0] data,
       output reg [3:0] out);
always @(data)         // task calling
   count_0s_in_byte(data, out);

// the task declaration starts from here
task count_0s_in_byte(
     input [7:0] data,
     output reg [3:0] count);
integer i;
begin    // the task body
   count = 0;
   for (i = 0; i <= 7; i = i + 1)
      if (data[i] == 0) count = count + 1;
end
endtask
endmodule
```

In the above module, the following statement

```
if (data[i] == 0) count = count + 1;
```

can be replaced by

```
count = count + ~data[i];
```

Why? The reason is as follows. Because the **reg** variable count increases its value by 1 whenever the current testing bit data[i] is zero, it is equivalent to adding the complement of the testing bit to count. This idea is further demonstrated by the following module, which also illustrates the use of the port-list style of a task.

■ Example 5.2: A task for counting the number of 0s in a byte.

In this example, we rewrite the task definition in the preceding example by changing the task declaration from the port-list declaration style to the port-list style. In addition, in the task definition, the **if** statement in the task definition in the preceding example is replaced with the statement described above.

```
// a module counts the number of zeros in a byte
module zero_count_task2 (
       input  [7:0] data,
       output reg [3:0] out);
always @(data)         // task calling
   count_0s_in_byte(data, out);

// the task declaration starts from here
```

```verilog
task count_0s_in_byte;
input    [7:0] data;
output reg [3:0] count;
integer i;
begin      // the task body
   count = 0;
   for (i = 0; i <= 7; i = i + 1)
      count = count + ~data[i];
end
endtask
endmodule
```

A task may have timing control associated its procedural statements. However, an assignment to an output argument is not passed to the calling argument until the task exits. The actual simulation time required for a task depends on the timing control used within it. To see this, consider the following example.

■ Example 5.3: An example of a task having timing control.

In this example, regular delay timing control is used within the task. Because 10 time units are required for the task to complete its operations, the first **$display** system task is executed at simulation time 3 and the second **$display** system task at simulation time 19.

```verilog
// the use of a static task
module static_task;
reg [3:0] result;
initial begin
   #3 check_counter(1'b1, result);
   $display($realtime,"The value of count is %d.", result);
   #6 check_counter(1'b0, result);
   $display($realtime,"The value of count is %d.", result);
end

// the task definition starts from here
task check_counter(input reset, output reg [3:0] count);
// the body of the reentrant task
begin
   if (reset) count = 0;
   else begin
      # 2 count = count + 1;
      # 3 count = count + 1;
      # 5 count = count + 1;
   end
end
endtask
endmodule
```

The simulation results of the above module are as follows:

```
#   3 The value of count is 0
#  19 The value of count is 3
```

Section 5.1 *Tasks*

■ **Review Questions**

5-1 When is a task usually used to describe a piece of code?
5-2 Can timing control be used within a task?
5-3 Can **always** or **initial** blocks be used within the definition of a task?
5-4 Can a task contain no argument?
5-5 Can a task call another task?

5.1.2 Types of Tasks

Tasks can be cast into two types: (static) tasks and automatic (also called reentrant, dynamic) tasks. A (static) task is declared without the keyword **automatic** while an automatic task is declared with the keyword **automatic**.

Static tasks have the feature that all declared items are statically allocated. The items in a static task can be shared across all concurrent uses of the task. Automatic tasks have the feature that all declared items are dynamically allocated for each recursive call. These dynamically allocated items cannot be accessed by their hierarchical path names. Nevertheless, automatic tasks, like static tasks, can be called through the use of their hierarchical path names.

A task can be called more than once concurrently. All variables of a static task are static in that there is a single variable corresponding to each declared local variable in a module instance, regardless of the number of concurrent calls of the task. All variables of an automatic task are duplicated on each concurrent call in order to store the state specific to that call. However, static tasks in different instances of a module have separate storage from each other. Variables declared within a static task retain their values between calls whereas variables declared within an automatic task are initialized to the default value x whenever execution enters their scope and are deallocated at the end of the task call.

■ **Example 5.4: An example of using tasks.**

In this example, the check_counter task needs 7 time units to accomplish its operations at each call. In the test bench, the task is called at simulation time 5, while the reset signal is deactivated at 10 time units later after the task returns, which is at simulation time 12. Therefore, the reset signal is assigned 0 at simulation time 22. At simulation time 32, the task is called once again. Hence, the simulation results are as follows:

```
#  5 At the beginning of task, count = x
#  5 After reset, count = 0
# 12 At the end of task, count = 2
# 32 At the beginning of task, count = 2
# 39 At the end of task, count = 4

// the use of a non-automatic task
module non_automatic_task;
reg reset;

// enable the task two times
initial begin
      reset = 1'b1;      // reset count
   #5  check_counter;
   #10 reset = 1'b0;
   #10 check_counter;
end
```

```
// the task definition starts from here
task check_counter;
reg [3:0] count;

// the body of the task
begin
   $display ($realtime,"At the beginning of task, count = %d", count);
   if (reset) begin
      count = 0;
      $display ($realtime,"After reset, count = %d",count);
   end
   # 2 count = count + 1;
   # 5 count = count + 1;
   $display ($realtime,"At the end of task, count = %d", count);
end
endtask
endmodule
```

From the simulation results, we can see that the value of the count variable is not destroyed after the task returned. This is also a basic feature of static tasks: variables declared within a static task retain their values between calls. Hence, the second call to the check_counter task uses the value from the previous call to the check_counter task. However, if we use an automatic task, the situation is different, as illustrated in the following example.

■ Example 5.5: An example of using automatic tasks.

In this example, the check_counter task is declared as an automatic task by using the keyword **automatic**. It still needs 7 time units to complete its operations at each call. The rest of the timing information is exactly the same as the preceding example. However, for each task call, a separate **reg** variable count is allocated and is initialized to the default value x. Hence, the simulation results are as follows:

```
# 5 At the beginning of task, count = x
# 5 After reset, count = 0
# 12 At the end of task, count = 2
# 32 At the beginning of task, count = x
# 39 At the end of task, count = x

// the use of an automatic task
module automatic_task;
reg reset;

initial begin
      reset = 1'b1;     // reset count
   #5   check_counter;
   #10  reset = 1'b0;
   #10  check_counter;
end

// the task definition starts from here
task automatic check_counter;
reg [3:0] count;
```

```
// the body of the task
begin
   $display($realtime,"At the beginning of task, count = %d",count);
   if (reset) begin
      count = 0;
      $display ($realtime,"After reset, count = %d",count);
   end
   # 2 count = count + 1;
   # 5 count = count + 1;
   $display ($realtime,"At the end of task, count = %d", count);
end
endtask
endmodule
```

■ Review Questions

5-6 What are the differences between static and automatic tasks?
5-7 Describe the features of static tasks.
5-8 Describe the features of automatic tasks.
5-9 Could automatic tasks be synthesized by synthesizers? Give the reasons.

5.2 Functions

A function is an alternative way to a task to provide a means of reusing a common piece of code. It is used when the procedure has no timing control, (i.e., any statement introduced with #, @, or **wait**), needs to return a single value, has at least one **input** argument, has no **output** or **inout** arguments, or has no nonblocking assignments. A function is usually used to model a combinational circuit. Note that a task can call tasks and functions but a function can only call functions. The features of functions are listed in Table 5.1, as compared with tasks.

5.2.1 Function Definition and Calls

The function definition implicitly declares an internal variable, defaulted to a 1-bit **reg** variable, with the same name as the function name itself. This **reg** variable is used to return the result from the function by explicitly assigning it the function result within the function.

A function is declared with a pair of keywords **function** and **endfunction** and can be one of the following two forms

```
// port-list style
function [automatic] [range_or_type] function_id;
   function_input_declaration{; function_input_declaration};
   {block_item_declaration}   // local variables or constants
   function_statement
endfunction

// port-list declaration style
function [automatic] [range_or_type]
         function_id (function_port_list);
   {block_item_declaration}   // local variables or constants
```

```
    function_statement
endfunction
```

where the keyword **automatic** declares an automatic (recursive) function with all declarations allocated dynamically for each concurrent function entry. `range_or_type` is optional and can be one of the following forms

```
[signed] [range]
return_data_type
```

where `range` is `[msb:lsb]` and `return_data_type` can be any of the set {**integer**, **real**, **realtime**, **time**}. A function returns a 1-bit **reg** value by default if no range or type is specified. The return value of a function may be declared as a signed value by using the keyword **signed** as its return data type is the **reg** variable type. A function definition must be placed within a module definition and it must have at least one **input** argument but cannot have any **output** or **inout** arguments. Values are passed to a function through arguments and the result value is returned through the function name.

The formats of `function_input_declaration` are as follows:

```
input [reg] [signed] [range] port_id{, port_id}
input port_data_type port_id{, port_id}
```

where `port_data_type` can be any of **integer**, **real**, **realtime**, and **time**, and `function_-port_list` has the following format:

```
function_input_declaration{, function_input_declaration}
```

Since a function must be invoked within an **initial** or **always** block or an **assign** continuous assignment, not all data types and constant declarations are allowed to use within a function definition. Hence, the data types and constant declarations that may be included in `block_item_declaration` are variable data types (i.e., **reg**, **integer**, **real**, **realtime**, and **time**), **event**, **localparam**, and **parameter**.

Not all procedural statements can be used in a function definition. The subset of procedural statements (Section 4.1.1) that can be used as `function_statement` is as follows:

```
blocking_assignment;
conditional_statement (if–else)
case_statement (case)
loop_statement
system_task_enable
sequential_block (begin–end)
disable_statement
```

Note that nonblocking assignments cannot be used within a function since they are used to describe sequential circuits by default but the function may merely be used to describe a combinational circuit. In addition, only the sequential (**begin-end**) block can be used within a function; the parallel (**fork-join**) block cannot be used in a function.

5.2.1.1 Function Calls A function call is an operand within an expression and has the following syntax

```
function_id (argument{, argument});
```

where `function_id` is the variable from which the result of the function is returned and the argument list contains one or more input arguments. Consequently, there must have a procedural assignment that assigns the return result of the function to the `function_id` within the function.

Section 5.2 *Functions*

■ Example 5.6: Counting the number of zeros in a byte.

As an example of using the port-list declaration style to declare a function, we rewrite the previous task used to count the number of zeros in a byte as a function.

```
// a module counts the number of zeros in a byte
module zero_count_function(
      input   [7:0] data,
      output reg [3:0] out );
// invoke the function
always @(data)
   out = count_0s_in_byte(data);
// the function declaration starts from here
function [3:0] count_0s_in_byte(input [7:0] data);
integer i;
begin
   count_0s_in_byte = 0;
   for (i = 0; i <= 7; i = i + 1)
      if (data[i] == 0)
         count_0s_in_byte = count_0s_in_byte + 1;
end
endfunction
endmodule
```

Like tasks, functions can only contain procedural statements. No **initial** or **always** block is allowed in a function definition because **initial** and **always** blocks cannot be nested and a function may also be called in an **initial** or **always** block besides an **assign** continuous assignment.

For the same reason as that of using the task definition, the above module can be rewritten as in the following example. This module also illustrates the use of the port-list style of a function.

■ Example 5.7: Counting the number of zeros in a byte.

In this example, we rewrite the function definition described in the preceding example by using the port-list style instead of the port-list declaration style. In addition, we also replace the **if** statement with the statement "count_0s_in_byte = count_0s_in_byte + ~data[i];" in the function definition.

```
// a module to count the number of zeros in a byte
module zero_count_function (
      input   [7:0] data,
      output reg [3:0] out );
always @(data)
    out = count_0s_in_byte(data);
// the function declaration from here
function [3:0] count_0s_in_byte;
input [7:0] data;
integer i;
begin
   count_0s_in_byte = 0;
   for (i = 0; i <= 7; i = i + 1)
      count_0s_in_byte = count_0s_in_byte + ~data[i];
```

```
end
endfunction
endmodule
```

■ Review Questions

5-10 When is a function usually used to describe a piece of code?

5-11 Can timing control be used within functions?

5-12 Can **always** or **initial** blocks be used within functions?

5-13 Can a function contain no arguments?

5-14 Can a function call another function?

5-15 Can a function call tasks?

5.2.2 Types of Functions

Functions can be classified into the following two types: (static) functions and automatic (also called recursive, dynamic) functions. A (static) function is declared without the keyword **automatic** whereas an automatic function is declared with the keyword **automatic**.

The major difference between static and automatic functions is that all declared items are statically allocated in static functions but dynamically allocated for each recursive call in automatic functions. These dynamically allocated items cannot be accessed by their hierarchical path names. Nevertheless, automatic functions, like static functions, can be called through the use of their hierarchical path names. The following example illustrates how an automatic function is declared and called.

■ Example 5.8: An example of using automatic functions.

In this example, an automatic function fact is declared and then recursively called to compute the factorial of an integer n. Since it is similar to the one written in C language, we do not want to further explain it here. Can this module be synthesized? Give your reasons.

```
// the use of an automatic function
module factorial(input [15:0] n, output reg [15:0] result);
// instantiate the fact function
always @(n)
   result = fact(n);

// define the fact function
function automatic [15:0] fact(input [15:0] n);
// the body of the fact function
   if (n == 0) fact = 1;
   else fact = n * fact(n - 1);
endfunction
endmodule
```

5.2.3 Constant Functions

Constant functions are ones that are only evaluated at elaboration time. They are used to support the building of complex calculations of values at elaboration time. The execution of a constant function call has no effect on the initial value of the variable used either at simulation time or among multiple calls of a function at elaboration time.

Section 5.2 Functions

A constant function is associated locally with the calling module where the arguments to the function are constant expressions. Constant functions contain no hierarchical references and can only call constant functions associated locally with the current module. They cannot call system functions. Except for the **$display**, all system tasks within a constant function are ignored. However, the **$display** system task is ignored at elaboration time.

The following module describes how a constant function is used to compute $\log_2(n)$, where n is usually a number of the power of 2. This type of constant function might be useful in writing a parameterizable module. The interested reader is encouraged to synthesize this module and check out the results from the synthesizer.

■ Example 5.9: An example of using constant functions.

This example describes the use of a constant function to help compute the parameter M. Maybe one major purpose of using a constant function is to reduce the number of module parameters. For example, with the help of the constant function count_log_b2(), the zero_count_constant module only needs a single module parameter SIZE instead of two, SIZE and M. From the synthesized result, it can be seen that the constant function is only used to compute the local parameter M, as expected. It does not matter with the synthesized results of the module.

```
// the use of a constant function
module zero_count_constant
      #(parameter SIZE = 8, // set the default size
        parameter M = count_log_b2(SIZE))(
        input  [SIZE-1:0] data,
        output reg [M:0] out);
// the body of the module
integer i;
always @(data) begin
   out = 0;
   for (i = 0; i <= 7; i = i + 1)
      if (data[i] == 0) out = out + 1;
end
// define a constant function for computing log_2 n
function integer count_log_b2(input integer depth);
// the function body
begin
   count_log_b2 = -1;
   while (depth) begin
      count_log_b2 = count_log_b2 + 1;
      depth = depth >> 1;
   end
end
endfunction
endmodule
```

In practice, the above count_log_b2 function can be replaced with the **$clog2**(n) system function, which returns the ceiling function of the $\log_2 n$, where n can be an integer or an arbitrary sized vector value and is treated as an unsigned value. As the argument n is 0, the **$clog2**(n) system function returns a result of 0.

5.2.4 Sharing Tasks and Functions

Although tasks and functions are defined within a module, they may be shared among different modules by either of the following ways:

1. Define tasks and functions to be shared publicly in a separate module, such as `package`, and then use the hierarchical path name to refer to the specific task or function.
2. Define tasks and functions to be shared publicly in a separate text file, say `package.v`, and then use the **'include** compiler directive to include it inside the module which wants to share them.

Next, we demonstrate how these two approaches are used in practice by way of examples that follow. The first example uses a hierarchical path name to refer to the required function or task.

■ Example 5.10: An example of sharing tasks and functions.

In this example, the `fact` function is declared in the `factorial` module and is called in another module, `tasks_function_sharing`. To be properly called, the hierarchical path name, `factorial.fact`, is needed to use.

```verilog
'timescale 1ns/100 ps
// the sharing of tasks and functions through the use of
// the hierarchical path name
module tasks_function_sharing;
reg [15:0] result;

initial begin
   result = factorial.fact(5); // hierarchical path name
   $display ("The result is %d", result);
end
endmodule

// file factorial.v --- an independent module
module factorial;
 // define the fact function
function automatic [15:0] fact(input [15:0] n);
// the body of the fact function
   if   (n == 0) fact = 1;
   else fact = n * fact(n - 1);
endfunction
endmodule
```

The following module shows how the **'include** compiler directive is used to include the required function `fact` into the current module. Note that the **'include** statement must be placed after the keyword **module** because the function definitions in the `package.v` file are not encapsulated by a module declaration and they must be declared within a module. The contents of the `package.v` file are as follows:

```verilog
// file package.v
// define the fact function
function automatic [15:0] fact(input [15:0] n);
// the body of the fact function
   if (n == 0) fact = 1;
   else fact = n * fact(n - 1);
```

```
endfunction
// define other functions
```

■ Example 5.11: Another example of sharing tasks and functions.

As described, the `fact` function is included in the module while compiling. Hence, both the function call and the called function are in the same module. So they proceed their operations as usual.

```
`timescale 1 ns/100 ps
// the sharing of tasks and functions using the
// include compiler directive
module tasks_function_sharing2;
`include "package.v" // in working directory
reg [15:0] result;
initial begin
   result = fact(5);
   $display ("The result is %d", result);
end
endmodule
```

It should be noted that it is also possible to combine the above two approaches into use at the same time. In this situation, the functions to be shared are declared within a separate module declaration. Then, the **`include** compiler directive is used to include the module into the current design. However, at this point, the **`include** compiler directive must be placed outside the module definition because module definitions cannot be nested. Finally, the hierarchical path name should be employed to call the desired functions.

■ Review Questions

5-16 What are the differences between static and automatic functions?
5-17 Describe the features of static functions.
5-18 Describe the features of dynamic functions.
5-19 How could you share tasks and functions in multiple modules within a design?
5-20 What are the features of constant functions?

5.3 System Tasks and Functions

In Verilog HDL, there are many built-in system tasks and functions. These system tasks and functions are predefined in the language and can be used for various applications in writing test benches, describing designs abstractly in behavioral style, and so on. The system tasks and functions provided by Verilog HDL can be subdivided into the following types:

- Display system tasks
- Timescale system tasks
- Simulation time system functions
- Simulation control system tasks
- File I/O system tasks and functions
- String formatting system tasks and functions
- Conversion system functions

- Probabilistic distribution system functions
- Stochastic analysis system tasks and functions
- Math system functions
- Command line arguments
- PLA modeling system tasks

The first four types are often collected as simulation-related system tasks and functions. The last type, PLA modeling system tasks, will be described in Section 10.2.4. All of the other types are addressed in detail in this section.

5.3.1 Simulation-Related System Tasks and Functions

There are four types of system tasks and functions that are widely used during simulation. These are display system tasks, timescale system tasks, simulation time system functions, and simulation control system tasks.

5.3.1.1 Display System Tasks It is very useful to present information about a design visually on the standard output. For this purpose, Verilog HDL provides a group of system tasks for displaying information on the standard output. This group of system tasks is known as **$display** system tasks. It includes the following system tasks:

$display	**$displayb**	**$displayh**	**$displayo**
$write	**$writeb**	**$writeh**	**$writeo**
$monitor	**$monitorb**	**$monitorh**	**$monitoro**
$strobe	**$strobeb**	**$strobeh**	**$strobeo**

For convenience, we further cast them into the following three types:

1. Display and write system tasks
2. Continuous monitoring system tasks
3. Strobed monitoring system tasks

Display and write system tasks. The display and write system tasks include **$display** and **$write**. These two system tasks display values of variables, strings, or expressions in the same order as they appear in the argument list. The difference between them is that after they print out the specified information on the standard output, the **$display** system task also outputs an end-of-line character but the **$write** system task does not. The **$display** and **$write** system tasks have the form as follows

```
display_task_name[(arguments)];
```

where `display_task_name` is one of **$display**, **$displayb**, **$displayo**, **$displayh**, **$write**, **$writeb**, **$writeo**, and **$writeh**, and `arguments` are quoted strings, variables, or expressions. Any null argument produces a single space character in the display. A null argument is characterized by two adjacent commas in the argument list.

To print information with a specified format, an escape sequence is used. Each of the following escape sequences, when included in a string argument, specifies the display format for a subsequent expression.

```
%h or %H   // hexadecimal
%d or %D   // decimal
%o or %O   // octal
%b or %B   // binary
%c or %C   // ASCII character
%l or %L   // library binding information
%v or %V   // net signal strength
```

```
%m or %M  // hierarchical path name
%s or %S  // string
%t or %T  // current time format
```

Any expression with no format specification is displayed using the following default formats:

$display and **$write**   // decimal
$displayb and **$writeb** // binary
$displayo and **$writeo** // octal
$displayh and **$writeh** // hexadecimal

Escape sequences for printing special characters are as follows:

```
\n    // the newline character
\t    // the tab character
\\    // the \ character
\"    // the " character
\ddd  // a character specified by 1 to 3 octal digits
%%    // the % character
```

As invoked without arguments, the **$display** system task simply prints a newline character and the **$write** system task prints nothing at all.

Continuous monitoring system tasks. The continuous monitoring system task, **$monitor**, continuously monitors and displays the values of any variables or expressions. It has the following syntax

```
monitor_task_name[(arguments)];
```

where `monitor_task_name` is one of **$monitor, $monitorb, $monitoro**, and **$monitorh**. The arguments for this system task group are specified exactly in the same way as for the **$display** system task.

Only one **$monitor** system task display list can be activated at a time; however, a new **$monitor** system task with a new display list can be issued any number of times during simulation. In addition to the **$monitor** system task group, there are two related system tasks, **$monitoron** and **$monitoroff**, which are used to enable and disable the monitoring operations, respectively.

Strobed monitoring system tasks. The strobed monitoring system task, **$strobe**, displays the values of any variables or expressions at a specified time but at the end of the time step. It has the following form

```
strobe_task_name[(arguments)];
```

where `strobe_task_name` is one of the following system tasks: **$strobe, $strobeb, $strobeo**, and **$strobeh**. The arguments for this system task group are specified exactly in the same way as for the **$display** system task.

5.3.1.2 Timescale System Tasks Two timescale-related system tasks, **$printtimescale** and **$timeformat**, can be used to display and set the timescale information during simulation. The **$printtimescale** system task displays the time unit and the time precision for a particular module. It has the syntax shown below.

```
$printtimescale [(hierarchical_identifier)];
```

where `hierarchical_identifier` specifies the module whose time unit and time precision are to be displayed. When no argument is specified, the **$printtimescale** system task displays the time unit and the time precision of the current module. The timescale information is displayed with the following format:

```
Time scale of (module_name) is unit / precision
```

The following example demonstrates the use of the **$printtimescale** system task.

■ **Example 5.12: The use of the $printtimescale system task.**

To illustrate the use of the **$printtimescale** system task, consider the following two modules. The `test_a` module calls the **$printtimescale** system task to display the timescale information about another module `test_b`.

`timescale 1 ms/1 us
module test_a;
initial
    $printtimescale(test_b);
endmodule

`timescale 10 ns/1 ns
module test_b;
    ..... ;
endmodule

The timescale information about the `test_b` module is displayed as follows:

```
Time scale of (test_b) is 10 ns/1 ns
```

The **$timeformat** system task specifies how the %t format specification reports the time information. It has the syntax as follows

```
$timeformat [(units_number, precision, suffix,
              numeric_field_width)];
```

where `units_number` is the power of 10^a second, i.e., a, where a is an integer and $-15 \leq a \leq 0$. For instance, to represent 1 μs, the `units_number` is set to -6 because 1 μs $= 10^{-6}$. The default value of the `units_number` is the smallest time precision argument of all **`timescale** compiler directives. The parameter `precision` defaults to 0. The default value of `suffix` is a null and of `numeric_field_width` is 20.

■ **Example 5.13: The use of the $timeformat system task.**

The following call of the **$timeformat** system task

```
$timeformat(-9, 3, " ns.", 5);
$display("The current time is %t", $realtime);
```

will display the %t specifier value in the display task as

```
The current time is 5.102 ns.
```

assuming that the current time is 5.102 ns. If no **$timeformat** system task is specified, the %t format prints the smallest time precision of all timescales in the source.

5.3.1.3 Simulation Time System Functions There are three system functions that provide access to the current simulation time: **$time**, **$stime**, and **$realtime**. The **$time** system function returns a 64-bit integer of time, the **$stime** system function returns an unsigned 32-bit integer of time, and the **$realtime** system function returns a real number of time. All of the returned values from these three system functions are scaled to the time unit of the module that invokes them.

Section 5.3 System Tasks and Functions

■ Example 5.14: The use of the $time system function.

As an illustration of how to use the **$time** system function, consider the following module.

```
'timescale 10 ns/ 1 ns
module time_usage;
reg a;
initial begin
   $monitor($time, "a = ", a);
   #2.55 a = 0;
   #2.55 a = 1;
end
endmodule
```

The simulation result are as follows:

```
#0 a = x
#3 a = 0
#5 a = 1
```

The simulation times are 26 ns and 52 ns and are scaled to 3 and 5 time units, respectively, because the time unit of the module is 10 ns.

The usage of the **$stime** system function is the same as that of the **$time** system function except that it returns a 32-bit unsigned integer rather than a 64-bit integer. It will have the same result for the above example as the **$time** system function is replaced with the **$stime** system function.

The **$realtime** system function returns a real number. Hence, if we replace the first argument **$time** with **$realtime** of the **$monitor** system task in the above module, the simulation results would be as follows:

```
#0    a = x
#2.6  a = 0
#5.2  a = 1
```

5.3.1.4 Simulation Control System Tasks There are two simulation control system tasks: **$stop** and **$finish**. The **$stop**[(n)] system task suspends the simulation and the **$finish**[(n)] system task terminates the simulation, where n (0, 1, or 2) determines the diagnostic messages to be printed before the prompt is issued. The value of 0 means to print nothing, 1 (by default) to print simulation time and location, and 2 to print simulation time, location, and statistics about the memory and central processing unit (CPU) time used in simulation.

■ Review Questions

5-21 What are the differences between **$display** and **$monitor** system tasks?
5-22 What is the distinction between **$realtime** and **$time** system tasks?
5-23 What is the purpose of the use of the **$finish** system task?
5-24 What are the differences between **$finish** and **$stop** system tasks?

5.3.2 File I/O System Tasks and Functions

The file input/output system tasks for file-based operations can be cast into the following categories: the *opening and closing file system function/task*, *file output system tasks*, and *file input system tasks and functions*. We describe their operations in detail in this subsection.

5.3.2.1 Opening and Closing File System Function/Task Before a file can be read, it must be opened by the **$fopen** system function. The file is also needed to close after it has no longer been used. The **$fclose** system task is used for this purpose. In Verilog HDL, there are two types of output files: the *file descriptor* (fd) file and the *multiple channel descriptor* (mcd) file. The fd file is like the files used in most programming languages, such as C/C++ and Java. The mcd file is unique in Verilog HDL; it allows us to write to multiple files the same data in parallel.

The **$fopen** system function and the **$fclose** system task have the following syntax:

```
integer fd = $fopen("file_name", mode);
$fclose(fd);  // close a file opened previously
```

The **$fopen** system function returns an integer value, called a *file descriptor* (fd), after opening the file with the specified mode. The returned file descriptor is 0 if it fails to open the specified file. The file descriptor is a 32-bit value with the MSB (bit 32) set to 1 to distinguish it from a multiple channel descriptor (mcd) file, which will be described later. Like C language, three files are pre-opened: STDIN (32'h8000_0000), STDOUT (32'h8000_0001), and STDERR (32'h8000_0002). The STDIN file is pre-opened for reading, and both STDOUT and STDERR files are pre-opened for appending.

The allowed modes for specifying a file type to be opened are as follows

- `"r"` or `"rb"`: Open a file for reading at the beginning of the file. Return 0 if it fails.
- `"w"` or `"wb"`: Open a file for writing at the beginning of the file or create a file if it does not exist.
- `"a"` or `"ab"`: Open a file for appending at the end of the file or create a file if it does not exist.
- `"r+"`, `"r+b"`, or `"rb+"`: Open a file for update (reading and writing) at the beginning of the file. Issue errors if it does not exist.
- `"w+"`, `"w+b"`, or `"wb+"`: Open a file for update (reading and writing) at the beginning of the file or create a file if it does not exist.
- `"a+"`, `"a+b"`, or `"ab+"`: Open a file for update (reading and writing) at the end of the file or create a file if it does not exist.

where the `"b"` refers to opening binary files.

5.3.2.2 File Output System Tasks The file output system tasks, **$fdisplay, $fwrite, $fmonitor**, and **$fstrobe**, are used to write data to a specific file. They have the same type of arguments with their counterparts, **$display, $write, $monitor**, and **$strobe**, except that the first parameter should be a file descriptor.

$fdisplay	**$fdisplayb**	**$fdisplayh**	**$fdisplayo**
$fwrite	**$fwriteb**	**$fwriteh**	**$fwriteo**
$fmonitor	**$fmonitorb**	**$fmonitorh**	**$fmonitoro**
$fstrobe	**$fstrobeb**	**$fstrobeh**	**$fstrobeo**

$fflush

The file output system tasks have the form

```
file_output_task(fd[, list_of_arguments]);
```

where `file_output_task` can be any of the following system tasks, **$fdisplay, $fwrite, $fmonitor**, and **$fstrobe**, and their variants.

The **$fflush** system task flushes the output buffer to the specified output file. It has the following syntax:

Section 5.3 *System Tasks and Functions*

```
$fflush (fd);
$fflush ();
```

When no argument is specified, the **$fflush** system task flushes any buffered output to all open files.

■ Example 5.15: The use of file output system tasks.

In this example, three file output system function and tasks, **$fopen**, **$fdisplay**, and **$fclose**, are used to completely save the results from the output of the `fact` function.

```verilog
module fact_file;
// test the fact function
integer n, result, fd;
initial begin
   fd = $fopen("fact_result.dat", "w");
   if (fd == 0)
      $display ("File open error !");
   for (n = 0; n <= 12; n = n + 1) begin
      result = fact(n);
      $fdisplay (fd, "%0d factorial = %0d", n, result);
   end
   $fclose (fd);
end
// define the fact function
function automatic integer fact(input [31:0] n);
// the body of the fact function
   if    (n == 0) fact = 1;
   else fact = n * fact(n - 1);
endfunction
endmodule
```

5.3.2.3 Multiple Channel Descriptor (mcd) The mcd is a 32-bit **reg** variable in which each set bit corresponds an opened output file. The MSB (bit 32) is set to 0 to distinguish it from the fd whose MSB is 1. In addition, the LSB (bit 0) of the mcd is reserved for the standard output. As a consequence, there are at most 30 files that can be opened for output through the mcd at the same time. Using the mcd allows us to write multiple files in parallel by ORing together their mcds in a bit-wise fashion and writing to the resultant values.

To open an mcd file, the following syntax is used

```verilog
integer mcd = $fopen("file_name");
```

where mcd is returned from the **$fopen** system function. It has a value of 0 if the **$fopen** system function fails to open the specified file. To close file(s) specified by an mcd, the following syntax may be used.

```
$fclose(mcd);
```

The file output system tasks when mcd files are used have the form

```
file_output_task(mcd[, list_of_arguments]);
```

where file_output_task can be any of the following system tasks: **$fdisplay**, **$fwrite**, **$fmonitor**, and **$fstrobe**, and their variants.

The **$fflush** system task flushes the output buffer to the specified mcd output file(s). It has the following syntax.

```
$fflush (mcd);
$fflush ();
```

When no argument is specified, the **$fflush** system task flushes any buffered output to all open files.

■ Example 5.16: The use of mcd files.

This example demonstrates how to set up and access mcd files. First, two files are opened. Then, their mcds are combined together with a bit-wise OR operation and assigned to the integer variable, `cluster`. In order to output to the standard output as well, a constant 1 is also logically ORed into the `cluster` to form the `all`. The first **$fdisplay** system task writes the current simulation time into all three files, including the standard output, the second **$fdisplay** system task writes to `mcd1` and `mcd2` files, and the third **$fdisplay** system task writes to the `mcd2` file only.

```
integer mcd1, mcd2, cluster, all;
initial begin
   if ((mcd1 = $fopen("cpu.dat")) == 0)
      $display ("File open error !");
   if ((mcd2 = $fopen("dma.dat")) == 0)
      $display ("File open error !");
   cluster = mcd1 | mcd2;
   // include the standard output
   all = cluster | 1'b1; // three files
   // possible uses of opened files
   $fdisplay(all, "System restart at time %d",$time);
   $fdisplay(cluster, "Error occurs at time %d,
             address = %h", $time, addr);
   forever @(posedge clock)
      $fdisplay(mcd2, "dma_mode = %h address = %h",
                dma_mode, address);
end
```

5.3.2.4 File Input System Tasks and Functions File input system tasks and functions provide the capabilities of reading a character or a line at a time, reading formatted data, and reading binary data from a specified file. They also provide a mechanism to control the position to be read of or return I/O status from a specified file. The file input system tasks and functions include the following ones:

$fgetc	$ungetc	$fgets	$fscanf
$fread	$readmemb	$readmemh	$sdf_annotate
$ftell	$fseek	$rewind	$ferror

These system tasks and functions can be classified into the following types: reading a character at a time, reading a line at a time, reading formatted data, reading files, file positioning, and I/O error status.

Reading a character at a time. The **$fgetc** system function reads a character (i.e., a byte) from the file specified by an fd. It has the following form:

```
c = $fgetc(fd);
```

If an error occurs while reading a file, it returns -1 (EOF).

To undo the effect of the **$fgetc** system function, i.e., to insert the character back to the file buffer specified by an fd, the **$ungetc** system function may be used. It has the form as follows:

Section 5.3 System Tasks and Functions

```
code = $ungetc(char, fd);
```

If an error occurs, it returns -1 (EOF).

Reading a line at a time. The **$fgets** system function reads characters from the file specified by an fd. It has the following form:

```
code = $fgets(str, fd);
```

The characters read are stored into str and the operation is terminated when either the buffer is full, a newline character is read, or an end-of-file is encountered. If an error occurs when reading a file, it returns 0 to indicate that no character is read.

Reading formatted data. The **$fscanf** system function reads formatted data from the file specified by an fd. It has the following form

```
code = $fscanf(fd, format, arguments);
```

where the second argument is the format, and the rest are variables to be formatted. The format used in the **$fscanf** system function is exactly the same as the **$display** system task.

■ Example 5.17: Reading formatted data.

As an illustration of how to read formatted data, assume that a file test.dat contains the following data:

```
test_data_1 34 25
```

Then, after the following statements are executed

```
integer code, fd, x, y;
reg [87:0] name;

fd = $fopen("test.dat", "r");
code = $fscanf(fd, "%s %d %d", name, x, y);
```

the **reg** variable name has test_data_1, the integer variable x is 34, and the integer variable y is 25.

Reading files. Three system tasks and functions can be used to read data from a file: **$fread**, **$readmemb**, and **$readmemh**. The **$fread** system function reads binary data from a file specified by an fd into memory or a **reg** variable. It has the following forms

```
code = $fread(memory_name, fd, [start], [count]);
code = $fread(reg, fd);
```

where arguments start and count are optional. If start presents, it specifies the address of the first element in memory_name to be loaded; otherwise, the lowest address of memory_name will be used. If count presents, it specifies the maximum number of locations in memory_name that will be loaded; otherwise, memory_name is filled with whatever are available.

The **$readmemb** and **$readmemh** system tasks read binary and hexadecimal data from a specified file into memory, respectively. They use the same form

```
$readmemb("file_name", memory_name[, start[, end]]);
$readmemh("file_name", memory_name[, start[, end]]);
```

where arguments start and end are used to optionally specify the start and end addresses of memory, respectively. If both start and end arguments are not specified and no address specifications appear within the file, the default start address is the lowest address of memory. If only the start argument is specified, the file is read into memory starting at the specified address and toward the higher address. If both start and end arguments are specified, the file is read into memory beginning at the start address and through the end address, regardless of how the range of memory is declared.

■ Example 5.18: The use of the $readmemb system task.

This example demonstrates the use of the **$readmemb** system task. Here, cache_mem is declared as an array of 256 vectors with 8 bits each.

```
reg   [7:0] cache_mem[255:0];
```

```
$readmemb("cache_mem.data", cache_mem);
$readmemb("cache_mem.data", cache_mem, 24);
$readmemb("cache_mem.data", cache_mem, 156, 1);
```

The first **$readmemb** system task reads data from the cache_mem.data file and stores in cache_mem starting at location 0. The second **$readmemb** system task loads cache_mem starting at address 24 and continues toward address 256. The third **$readmemb** system task loads cache_mem starting at address 156 and continues through address 1.

The text file may contain white space (spaces, new lines, tabs, and form-feeds), comments (including both types of comments), and binary (for **$readmemb**) or hexadecimal (for **$readmemh**) numbers. Numbers may contain x, z, or underscore (_) and are separated by white spaces and/or comments. Moreover, addresses may be specified in the file as hexadecimal numbers using an address specifier, @<hexa_address>. Uninitialized locations default to unknown x. For example,

```
@002    // from location 2
0000 1011
0001 11zz
@008    // from location 8
1111 zz1z
```

Locations 6 and 7 will be initialized to unknown x.

The **$sdf_annotate** system task reads timing data from a standard delay format (SDF) file and put them into a specified region of the design. Its most widely used form is as follows

```
$sdf_annotate("sdf_file", module_instance);
```

where sdf_file is the SDF file path and module_instance specifies the scope to which to annotate the information in the SDF file. Refer to [3] for details.

File positioning. The file positioning system functions include **$ftell**, **$fseek**, and **$rewind**. The **$ftell** system function returns the offset from the beginning of the file specified by an fd. Its form is as follows:

```
position = $ftell(fd);
```

The **$fseek** system function sets the position of the next input or output position on the file specified by an fd. It has the following form:

```
position = $fseek(fd, offset, operation);
```

The offset argument can be counted from the beginning, from the current position, or from the end of the file, according to an operation value of 0, 1, or 2.

The **$rewind** system function is equivalent to the **$fseek**(fd, 0, 0) system function. It has the following form:

Section 5.3 *System Tasks and Functions*

```
code = $rewind(fd);
```

Note that the operations of both **$fseek** and **$rewind** system functions will cancel any **$ungetc** system function operations.

I/O error status. The **$ferror** system function returns more information about the causes of the file I/O operations specified by an fd. It has the following form

```
code = $ferror(fd, str);
```

where the `str` argument must have at least 640 bits to accommondate at least 80 characters. The **$ferror** system function returns 0 and the `str` argument is cleared if the most recent operation did not result in an error.

■ Review Questions

5-25 What are the differences between fd and mcd files?
5-26 Describe the operations of the **$fflush** system task.
5-27 How to read a character from the standard input?
5-28 Describe the operations of the **$fgets** system function.
5-29 Describe the features of the **$readmemb** system task.

5.3.3 String Formatting System Tasks and Functions

The string system tasks and functions provide ways to convert the formats between strings. This group of system tasks includes the following ones:

| **$swrite** | **$swriteb** | **$swriteh** | **$swriteo** |
| **$sformat** | **$sscanf** | | |

The **$swrite** system task provides ways to convert the formats between strings. Their functions are similar to the **$fwrite** system task except that instead of writing to a file, the **$swrite** system task writes to a string of the **reg** variable type, `output_reg`. The **$swrite** system task has the following syntax

```
str_tasks_name(output_reg, arguments);
```

where `str_tasks_name` may be one of the following system tasks: **$swrite**, **$swriteb**, **$swriteh**, and **$swriteo**. If no format specification exists for an argument, the **$swrite** system task defaults to be decimal. The **$swriteb**, **$swriteh**, and **$swriteo** system tasks default to be binary, hexadecimal, and octal, respectively.

The **$sformat** system task is similar to the **$swrite** system task except that the format must be only specified as the second argument. It has the following syntax

```
$sformat(output_reg, format, arguments);
```

where the second argument is the format, and the rest are variables to be formatted. The format used in the **$sformat** system task is exactly the same as the **$display** system task.

The **$sscanf** system function reads a line from a string `input_reg` and stores the values read in the specified arguments. Its syntax is

```
integer code = $sscanf(input_reg, format, arguments);
```

where the second argument is the format, and the rest are variables to be formatted. It returns 0 if it fails to read the string `input_reg`.

■ Example 5.19: The use of file input system functions.

In this example, two file input system functions, **$fgets** and **$sscanf**, are used to read a line of characters from the standard input and then convert it into an integer. Finally, it calls the `fact` function to compute the factorial of the integer and displays the results on the standard output. Note that the maximum valid number is 13 because the `fact` function can process a 32-bit integer only.

```
module fact_file_input;
// an example of using file input system tasks
`define STDIN 32'h8000_0000 // standard input
integer i, n, result, code;
reg [15:0] number;

initial begin
   for (i = 0; i < 10; i = i + 1) begin
      $display ("Please input a number: ");
      code = $fgets(number, `STDIN);
      code= $sscanf (number, "%d", n);
      $display ("The input number is =%d", n);
      result = fact(n);
      $display ("The factorial of %d is %d", n, result);
   end
end

// define the fact function
function automatic integer fact(input [15:0] n);
   // the body of the fact function
   if (n == 0) fact = 1;
   else fact = n * fact(n - 1);
endfunction
endmodule
```

5.3.4 Conversion System Functions

The following utility system functions convert numbers between different types:

| $signed | $unsigned | $rtoi | $itor |
| $realtobits | $bitstoreal | | |

Each of these is described in what follows in more detail.

1. The **$signed** system function interprets the argument as a signed quantity and returns its result as a signed integer.

    $signed(value);

    For example,

    ```
    reg signed [7:0] temp;

    temp = $signed(4'b1101); // temp = -3
    ```

2. The **$unsigned** system function interprets the argument as an unsigned quantity and returns its result as an unsigned integer.

Section 5.3 System Tasks and Functions

```
$unsigned(value);
```

For example,

```
reg [7:0] temp_a, temp_b;

temp_a = $unsigned(-5);      // temp_a = 8'b1111_1011
temp_b = $unsigned(-4'sd3);  // temp_b = 8'b0000_1101
```

3. The **$rtoi** system function converts a real number to an integer by truncating the real value. For example, 324.35 becomes 324.

   ```
   integer int_a;

   int_a = $rtoi(real_value);
   ```

4. The **$itor** system function converts an integer to a real number. For example, 1 becomes 1.0.

   ```
   real real_a;

   real_a = $itor(int_value);
   ```

5. The **$realtobits** system function passes bit patterns across module ports and converts from a real number to the 64-bit vector of that real number.

   ```
   wire [64:1] net_real;

   net_real = $realtobits(real_value);
   ```

6. The **$bitstoreal** system function is the reverse of **$realtobits** system function; it converts from the bit pattern to a real number, rounding the result to the nearest real value.

   ```
   real real_a;

   real_a = $bitstoreal(bit_value);
   ```

The following example illustrates how the **$realtobits** and **$bitstoreal** system functions are used to pass a real value across modules.

■ Example 5.20: $realtobits and $bitstoreal system functions.

As an illustration of using **$realtobits** and **$bitstoreal** system functions to pass a real number between two modules, consider the following two modules. Since a real number cannot be assigned to an output port, it is needed to convert into a bit pattern by using the **$realtobits** system function before it is sent to the output port, net_real. At the receiver, it is converted back to the real number from the bit pattern received by the **$bitstoreal** system function.

```
// module 1: the driver
module driver (output [63:0] net_real);
real r;

assign net_real = $realtobits(r);
endmodule

// module 2: the receiver
module receiver (input [63:0] net_real);
```

```verilog
real r;

initial r = $bitstoreal(net_real);
endmodule
```

■ Review Questions

5-30 Describe the operations of the **$sscanf** system function.
5-31 What are the differences between **$sformat** and **$swrite** system tasks?
5-32 Describe how to pass a real number between two modules.
5-33 Describe the operations of **$signed** and **$unsigned** system functions.

5.3.5 Probabilistic Distribution System Functions

The probabilistic distribution system functions are random number generators that return integer values distributed in accordance with standard probabilistic functions. These functions include the following ones:

$random	**$dist_uniform**	**$dist_normal**
$dist_exponential	**$dist_poisson**	**$dist_chi_square**
$dist_t	**$dist_erlang**	

They can be subdivided into two groups: a random number generator (**$random**) and distribution system functions.

5.3.5.1 The $random System Function The **$random** system function is the most widely used random number generator of the group of probabilistic distribution system functions. It returns a 32-bit random number each time it is called. It has the following form:

$random[(seed)];

The seed argument controls the random numbers returned by the **$random** system function. The seed argument can be either a **reg**, an **integer**, or a **time** variable. The **$random** system function always returns the same value with the same seed.

In general, the **$random** system function returns a 32-bit signed integer. However, we may use the concatenation ({}) operator to interpret the signed integer as an unsigned number. For example,

```verilog
reg [4:0] random_number;
random_number = {$random} % 23;
```

Each random_number is in the set of unsigned integers between 0 and 22.

5.3.5.2 Distribution ($dist_) System Functions Each of these system functions returns a pseudo-random number with characteristics described by the function name. For example, the **$dist_uniform** system function returns random numbers uniformly distributed in the interval specified by its parameters: seed, start, and end. The formats of probabilistic distribution system functions are as follows

```verilog
$dist_uniform(seed, start, end);
$dist_normal(seed, mean, standard_deviation);
$dist_exponential(seed, mean);
$dist_poisson(seed, mean);
$dist_chi_square(seed, degree_of_freedom);
$dist_t(seed, degree_of_freedom);
$dist_erlang(seed, k_stage, mean);
```

Section 5.3 *System Tasks and Functions*

where all parameters to the system functions are integer values. These system functions always return the same value with the same seed. The `seed` argument is an in-out parameter. It should be an **integer** variable that is initialized by the user and only updated by the system functions.

Details of these distribution system functions can be found in the IEEE Language Reference Manual [2, 4].

5.3.6 Stochastic Analysis System Tasks and Functions

Stochastic analysis system tasks and functions are used to manage queues. The set of stochastic analysis system tasks and functions includes:

$q_initialize	$q_add	$q_remove
$q_full	$q_exam	

These system tasks and functions provide a mechanism to implement stochastic queuing models. In what follows, we only describe the functions of these system tasks and functions in brief.

5.3.6.1 The $q_initialize System Task The **$q_initialize** system task creates a new queue and has the following form

$q_initialize(q_id, q_type, max_length, status);

where all parameters are integers. The `q_id` argument is an integer queue identifier. The `q_type` argument specifies which type of the queue will be used: 1 for first-in, first-out (FIFO) queue and 2 for last-in, first out (LIFO) queue, i.e., stack. The `max_length` argument is an integer number specifying the maximum allowed number of entries on the queue.

5.3.6.2 The $q_add System Task The **$q_add** system task adds an entry onto a specified queue. Its syntax is as follows

$q_add(q_id, job_id, inform_id, status);

where the `job_id` argument is an integer job identifier. The `inform_id` argument is an integer input associated with the queue entry. Its meaning is defined by users.

5.3.6.3 The $q_remove System Task The **$q_remove** system task removes an entry from a queue. It has the following form:

$q_remove(q_id, job_id, inform_id, status);

5.3.6.4 The $q_full System Function The **$q_full** system function checks whether the specified queue is full. It returns 1 if the queue is full and returns 0 otherwise. The **$q_full** system function has the form of

code = $q_full(q_id, status);

5.3.6.5 The $q_exam System Task The **$q_exam** system task provides statistical information about activity at the specified queue. It has the syntax of

$q_exam(q_id, q_stat_code, q_stat_value, status);

where the `q_stat_value` argument is the returned value requested by the input request code, `q_stat_code`. The relationship between the `q_stat_code` argument and the `q_stat_value` argument is listed in Table 5.2.

5.3.6.6 Status Codes All of the above queue management tasks and functions return a status code to give some information about the associated operation requested. The status codes and their meanings are listed in Table 5.3.

Table 5.2: The relationship between the `q_stat_code` and the `q_stat_value`.

q_stat_code	q_stat_value
1	Current queue length
2	Mean interarrival time
3	Maximum queue length
4	Shortest wait time ever
5	Longest wait time for jobs still in the queue
6	Average wait time in the queue

Table 5.3: The status codes and their meanings.

Status	Meanings
0	Successful
1	Queue full
2	Undefined q_id
3	Queue empty
4	Unsupported queue type
5	Specified length ≤ 0
6	Duplicate q_id
7	System memory full

5.3.7 Math System Functions

Verilog HDL supports both integer and real math system functions. These math system functions may be used in constant expressions, in particular, in the description of a design in behavioral style and/or in writing a test bench.

5.3.7.1 Integer Math Function The integer math system function only contains one system function **$clog2**, which is often used to compute the minimum address width necessary to address a memory of a given size or the minimum vector width necessary to represent a given number of states. The **$clog2** system function has the syntax

```
result = $clog2(n);
```

which returns the ceiling function of the log base 2 of the argument (the log rounded up to an integer value). The argument n can be an integer or an arbitrary sized vector value. The argument is treated as an unsigned value, and an argument value of 0 will return a result of 0.

5.3.7.2 Real Math Functions The real math system functions in Verilog HDL accept real arguments and return a real result. These system functions are listed in Table 5.4 and have the behavior exactly match the standard math library functions in C language. They are found much useful in writing test benches to verify a design.

5.3.8 Command Line Arguments

In Verilog HDL, two system functions, **$test$plusargs** and **$value$plusargs**, can be used to access to arguments and their values from the command line. These arguments begin with the plus (+) character and hence are referred to as *plusargs*.

5.3.8.1 The $test$plusargs System Function The **$test$plusargs** system function tests whether the plusargs exists. It has the syntax

```
$test$plusargs (string);
```

Section 5.3 System Tasks and Functions

Table 5.4: Real math system functions in Verilog HDL.

Verilog function	Equivalent C function	Description
$ln(x)	log(x)	Natural logarithm
$log10(x)	log10(x)	Decimal logarithm
$exp(x)	exp(x)	Exponential
$sqrt(x)	sqrt(x)	Square root
$pow(x,y)	pow(x,y)	x**y
$floor(x)	floor(x)	Floor
$ceil(x)	ceil(x)	Ceiling
$sin(x)	sin(x)	Sine
$cos(x)	cos(x)	Cosine
$tan(x)	tan(x)	Tangent
$asin(x)	asin(x)	Arc-sine
$acos(x)	acos(x)	Arc-cosine
$atan(x)	atan(x)	Arc-tangent
$atan2(x,y)	atan2(x,y)	Arc-tangent of x/y
$hypot(x,y)	hypot(x,y)	sqrt(x*x+y*y)
$sinh(x)	sinh(x)	Hyperbolic sine
$cosh(x)	cosh(x)	Hyperbolic cosine
$tanh(x)	tanh(x)	Hyperbolic tangent
$asinh(x)	asinh(x)	Arc-hyperbolic sine
$acosh(x)	acosh(x)	Arc-hyperbolic cosine
$atanh(x)	atanh(x)	Arc-hyperbolic tangent

where the string argument is the one to be tested. It returns 0 (false) if no plusarg from the command line matches the string provided and returns 1 (true) otherwise. For example, if the command line contains

    +FINISH=10000

then the following two **$test$plusargs** statements will return 1 (true)

    $test$plusargs ("FINISH");
    $test$plusargs ("FIN");

because the strings FINISH and FIN match the prefixes of the plusarg in the command line.

5.3.8.2 The $value$plusargs System Function The **$value$plusargs** system function returns the value of a specified plusarg, using a format string to access the value in plusargs. The format is much like the **$display** system task. The **$value$plusargs** system function has the syntax

    $value$plusargs (string, variable);

where the format string is the same as the **$display** system task. It returns 0 if no string is found matched and returns a nonzero value otherwise. For example,

    $value$plusargs("FINISH=%d", stop_clock);
    $value$plusargs("START=%d", start_clock);

The variable stop_clock receives the value of 10000 because the string FINISH exactly matches that on the command line. However, the variable start_clock remains unchanged because no +START is specified on the command line.

The aforementioned two system functions can be used to control the execution flow of statements at run time. All statements are compiled but executed conditionally according to the flags set by the command line arguments. In what follows, we give two examples to explore how to use the above two system functions to control the execution of statements conditionally.

■ Example 5.21: Conditional execution with $test$plusargs.

As an illustration of using the **$test$plusargs** system function to control the execution of the **if-else** statement, consider the following module. If the flag DEBUG is set, then the true statement of the **if-else** statement is executed; otherwise, the false statement is executed. The flag DEBUG can be set by specifying the option +DEBUG at run time. (*Hint:* In ModelSim, it can be set through: Simulate → Start Simulation → Verilog → User Defined Arguments [+⟨plusarg⟩]).

```
// an example of the conditional execution with
// $test$plusargs
module test_plusargs;
initial
   if ($test$plusargs("DEBUG"))
      $display("Enter into the debug process !!\n");
   else
      $display("Remain at its normal process. \n");
endmodule
```

Conditional execution of statements can be controlled by using the **$value$plusargs** system function. An illustration of this is explored in the following example.

■ Example 5.22: Conditional execution with $value$plusargs.

As described previously, the **$value$plusargs** system function returns the value of a specified plusarg if the specified string matches the command line arguments and returns a zero value otherwise. As a consequence, after the module is executed the following messages will be displayed if the command line arguments are set with +DEBUG=3 +clk_time=10. (*Hint:* Refer to the preceding example for the details of how to set the command line arguments.)

```
# Enter the debug level   3.
#
# The clock period is 10 ns.

'timescale 1 ns / 100 ps
module value_plusargs;
integer level, clk_period;
reg clk;

initial clk = 1'b0;
initial begin
   if ($value$plusargs("DEBUG=%d", level))
      $display("Enter into debug level =%d.\n", level);
   else
      $display("Remain at normal process. \n");

   // set the clock period
   if ($value$plusargs("clk_time=%d", clk_period)) begin
      $display("The clock period is %d ns\n", clk_period);
      forever #(clk_period/2) clk <= ~clk; end
   else
      $display("Use the default clock period. \n");
end
```

```
initial #200 $finish;
initial
   $monitor($realtime,"ns %h", clk);
endmodule
```

The above two examples reveal how to control the execution of statements conditionally at run time in Verilog HDL environment. In Section 7.4, we will address how to compile statements conditionally at compile time. It is instructive to figure out the differences between these two situations.

■ Review Questions

5-34 Describe the features of the **$random** system function.
5-35 Describe how to generate an unsigned random number in Verilog HDL.
5-36 Describe how to execute statements conditionally at run time.

5.4 User-Defined Primitives

Verilog HDL provides two types of UDPs for users to build their own primitives in addition to the built-in gate and switch primitives. These two UDPs are combinational UDPs and sequential UDPs. Combinational UDPs are defined when the output is uniquely determined by the combination of inputs. Sequential UDPs are defined when the next output is determined by the combination of the current output and inputs. UDPs are instantiated exactly in the same way as gate or switch primitives. In this section, we describe this kind of primitive in detail along with examples.

5.4.1 UDP Basics

UDPs are defined as an entity independent of modules; that is, they are defined at the same level as module definitions in the syntactic hierarchy of Verilog HDL. A UDP is defined with a pair of keywords **primitive** and **endprimitive** and can use either of the following two forms

```
// port-list style
primitive udp_name(port_list);
port_declarations{; port_declarations};
initial output_port = expression; // sequential UDP

// the UDP state table
table     // define the behavior of the UDP
   <table rows>
endtable
endprimitive

// port-list declaration style
primitive udp_name(port_list_declarations);
initial output_port = expression; // sequential UDP

// the UDP state table
table     // define the behavior of the UDP
   <table rows>
endtable
endprimitive
```

where the first argument must be the output port and the other arguments are input ports.

A UDP can have exactly one output port but may have multiple input ports. The maximum number of input ports is limited by implementations, but must have at least 10 input ports for combinational UDPs and 9 input ports for sequential UDPs. All ports of a UDP are scalar; they cannot be bidirectional (**inout**) ports or vector ports. The output can only have a single-bit value with one of three states: 0, 1, or x. The high impedance z is not supported. The z values passed to UDP input ports are treated as x values.

The behavior of a UDP is defined by a *state table*. The state table begins with the keyword **table** and is terminated with the keyword **endtable**. Each row of the table defines the output for a particular combination of the input values and is terminated by a semicolon. The allowed values for inputs and output are 0, 1, and x; the high impedance is not allowed. The order of the input ports of each row must be exactly the same as they are in the port list.

The instantiation of UDPs is exactly the same as that of gate primitives. It has the following syntax

```
udp_name[(strength)][delay2][instance_name[range]](out,in{,in})
        {, [instance_name[range]](out, in{, in})};
```

where delay2 may only use up to two delays because the output of a UDP can never have a value of z. As a consequence, there is no turn-off delay. Because no delay specification can be specified within the definition of a UDP, the only way to use the delay specification is through the instantiation of the UDP.

5.4.2 Combinational UDPs

A combinational UDP is used to describe a combinational circuit, namely, a logic circuit whose output is merely determined by its current inputs. In a combinational UDP, the UDP is evaluated and the output is set to the output field of the table entry that matches all input values whenever an input changes. All combinations of the inputs that are not explicitly specified will drive the output to an unknown x.

As described previously, a UDP can be defined as one of the two forms: the port-list style and the port-list declaration style. They have the following general forms:

```
// port-list style
primitive udp_name(output_port, input_port{, input_port});
output  output_port;
input   input_port{, input_port};

// the UDP state table
table   // define the behavior of the combinational UDP
   <table rows>
endtable
endprimitive

// port-list declaration style
primitive udp_name(
        output output_port,
        input  input_port{, input_port});

// the UDP state table
table   // define the behavior of the combinational UDP
   <table rows>
endtable
endprimitive
```

Section 5.4 User-Defined Primitives

Table 5.5: The shorthand symbols used in the state table of UDPs.

Symbols	Meaning	Comment
?	0, 1, x	Cannot be specified in an output field
b	0, 1	Cannot be specified in an output field
-	No change in state value	Can be only used in the sequential UDP output field
r	(01)	Rising edge of a signal
f	(10)	Falling edge of a signal
p	(01), (0x), or (x1)	Potential rising edge of a signal
n	(10), (1x), or (x0)	Potential falling edge of a signal
*	(??)	Any value change in signals

Each row of the state table has the following form:

<input1><input2> ... <inputn>:<output>;

The <inputi> values appearing in each row of the state table must be in the same order as in the input port list. Inputs and output are separated by a ":". Each row of the state table must end with a ";". All possible combinations of input values must be specified to avoid unknown output values.

Although both the port-list style and the port-list declaration style can be used to define a combinational UDP, the port-list declaration style will be used as an example to further illustrate how combinational UDPs are defined and instantiated.

■ Example 5.23: A simple example of combinational UDPs.

In this simple example, we define a two-input OR gate using a combinational UDP construct. Here, only 0 and 1 are assumed to be the two valid values of inputs. Therefore, only four combinations are needed to consider. The resulting module is as follows:

```
// a simple example of an OR gate UDP
primitive udp_or(
         output f,
         input  a, b);
// define the function of the OR gate
table
// a   b   :   f;
   0   0   :   0;
   0   1   :   1;
   1   0   :   1;
   1   1   :   1;
endtable
endprimitive
```

As a matter of fact, using a combinational UDP to define a combinational circuit corresponds to directly describing the behavior of the logic circuit in truth-table form. To be able to concisely describe a large state table, a set of shorthand symbols is defined, as listed in Table 5.5.

■ Example 5.24: The use of shorthand notation.

As an illustration of using shorthand symbols to shorten the state table, consider an AND gate module. From the behavior of a two-input AND gate, the output is 0 whenever one of its inputs is 0 no matter what the value of the other input is. Hence, the state table can be condensed into three rows. In this example, we assume that the inputs can have three values: 0, 1, and x. The character "?" is used in the state table to denote 0, 1, or x.

```
// an example of an AND gate UDP
primitive udp_and(
          output f,
          input  a, b);
// define the function of the AND gate
table
   //  a   b   :   f;
       1   1   :   1;
       0   ?   :   0;  // ? is expanded to 0, 1, x
       ?   0   :   0;
endtable
endprimitive
```

As mentioned, a combinational UDP is virtually the direct implementation of the truth table of a logic circuit. As an example, consider the following switching expression:

$$f(x,y,z,) = \sim(\sim(x \mid y) \mid \sim x \, \& \, z);$$

To model it using a combinational UDP, we need to compute its truth table on which the state table of a UDP is built.

■ **Example 5.25: A simple example of combinational UDPs.**

Based on the switching expression described above, the truth table is obtained as shown in the state table below. Of course, this example is only intended to illustrate the use of the combinational UDP. In practice, it is more straightforward to describe the switching expression in dataflow style or behavioral style, as stated in the previous two chapters.

```
// a combinational UDP of an arbitrary switching function
primitive my_function(
          output f,
          input  x, y, z);

// define the needed function here
table // truth table for f(x, y, z,) = ~(~(x | y) | ~x & z);
// x y z : f
   0 0 0 : 0;
   0 0 1 : 0;
   0 1 0 : 1;
   0 1 1 : 0;
   1 0 0 : 1;
   1 0 1 : 1;
   1 1 0 : 1;
   1 1 1 : 1;
endtable
endprimitive
```

Recall that UDPs can be instantiated exactly the same way as gate or switch primitives. An illustration of the instantiation of UDPs is demonstrated in the following example. Like a gate or switch primitive, the instance name of a UDP is optional.

Section 5.4 *User-Defined Primitives*   185

■ **Example 5.26: Instantiation of combinational UDPs.**

To illustrate the instantiation of UDPs, assume that we want to construct a full adder. To achieve this, two XOR and two AND gates as well as one OR gate are needed. By properly instantiating these UDPs and connecting them together, a full adder is obtained as in the following module. It is not difficult to verify that the resulting module is indeed a full adder. Note that both AND and OR gates have already been defined previously but the XOR gate is undefined. It is left for the reader as an exercise.

```
// an application of combinational UDPs
module UDP_full_adder(
      input x, y, cin,
      output sum, cout);
wire    s1, c1, c2;
// instantiate udp primitives
   udp_xor (s1, x, y);
   udp_and (c1, x, y);
   udp_xor (sum, s1, cin);
   udp_and (c2, s1, cin);
   udp_or  (cout, c1, c2);
endmodule
```

5.4.3 Sequential UDPs

Sequential UDPs are used to model sequential circuits, such as flip-flops and latches, including both level-sensitive and edge-sensitive behavior. As described, a sequential UDP can be defined by one of the following two forms: the port-list style and the port-list declaration style. They have the following general forms:

```
// port list style
primitive udp_name(output_port, input_port{, input_port});
output output_port;
input  input_port{, input_port};
reg    output_port; // needs to be a reg variable
initial output_port = expression; // optional

// the UDP state table
table    // define the behavior of the sequential UDP
  <table rows>
endtable
endprimitive

// port-list declaration style
primitive udp_name (
         output reg output_port;
         input  input_port{, input_port});
initial output_port = expression; // optional

// the UDP state table
table  // define the behavior of the sequential UDP
  <table rows>
endtable
```

endprimitive

The output port is also needed to declare as a **reg** variable and an **initial** block may be optionally used to initialize the output port. The **initial** block begins with the keyword **initial** and specifies the value of the output port when simulation begins. Unlike in modules, the declaration of the output port in a sequential UDP cannot be combined with the **reg** variable in the port-list style; each of them must be declared in a separate statement. The assignment statement must assign a single-bit value to the output port. In the remainder of this section, we use the port-list declaration style as an example to further illustrate how a sequential UDP is defined and instantiated.

Each row of the state table has the following form:

```
<input1><input2>...<inputn>:<current_state>:<next_state>;
```

The <inputi> values appearing in each row of the state table must be in the same order as in the input port list, input_ports. Inputs, the current state, and the next state are separated by a colon ":". Each row of the state table must end with a ";". The input specifications can be either level-sensitive or edge-sensitive. As in the combinational UDPs, all possible combinations of input values must be specified to avoid unknown output values.

The following simple example models a level-sensitive D latch. As the name implies, the changes of the output value are under the control of the level (high or low) of the enable input.

■ Example 5.27: A level-sensitive D latch.

As described before, the output q has to be declared as a **reg** variable and there is an additional field in each row of the state table. This new field represents the current state of the sequential UDP while the output field in a sequential UDP is denoted by the next state.

```
// a level-sensitive latch defined with a UDP
primitive d_latch(
          output reg q, // needs to be a reg variable
          input   d, gate, clear);
// define the function of the D latch
initial   q = 0; // initialize output q
// state table
table
// d  gate  clear  : q : q+;
   ?   ?     1     : ? : 0 ;  // clear
   1   1     0     : ? : 1 ;  // latch q = 1
   0   1     0     : ? : 0 ;  // latch q = 0
   ?   0     0     : ? : - ;  // no change
endtable
endprimitive
```

In a level-sensitive latch, the output value is determined by the values of the inputs and the current state; in an edge-triggered flip-flop, changes in the output are triggered by the specific transitions of the inputs. To model the transitions of inputs, a pair of values in parenthesis, such as (10), or a symbol, such as f, is used. The shorthand symbols used in UDPs are shown in Table 5.5. Each entry of the state table can have at most a transition specification on each input. Generally, all transitions without affecting the output should be also explicitly specified and otherwise, they will cause the output to change to an unknown x. In addition, all unspecified transitions default to the unknown output value, x.

Section 5.4 *User-Defined Primitives*

■ Example 5.28: A positive-edge-triggered D flip-flop.

As an example of modeling an edge-triggered flip-flop using a sequential UDP, consider a positive-edge-triggered D flip-flop. As the name implies, the changes of the output value are under the control of the edge of the clock signal. To model this flip-flop, the symbols r and f are used in the state table to represent 01 (a positive edge) and 10 (a negative edge), respectively. The resulting module is as follows:

```
// an edge-triggered flip-flop defined with a UDP
primitive d_FF(
         output reg q,
         input clk, d);

// define the function of the D flip-flop
initial q = 1'b1; // initialize output q
table
// clk  d    q    q+
    r   0 :  ? :  0 ; // latch q = 0
    r   1 :  ? :  1 ; // latch q = 1
    f   ? :  ? :  - ; // no change
    ?   * :  ? :  - ; // no change
endtable
endprimitive
```

Mixing level-sensitive with edge-sensitive UDPs. To model a realistic flip-flop with asynchronous clear or reset, the UDP definition allows a combination of both types of level-sensitive and edge-sensitive entries in the same state table. When any input changes, the edge-sensitive cases are processed before the level-sensitive ones. Hence, the level-sensitive entries override the edge-sensitive entries when they specify different output values.

■ Example 5.29: A positive-edge-triggered T flip-flop.

As an illustration of the combination of both types of level-sensitive and edge-sensitive entries in the same state table, consider a positive-edge-triggered T flip-flop with an asynchronous clear input. Because of the inherent features of sequential UDPs, the level-sensitive entries override the edge-sensitive entries. In addition, the first row of the state table overrides the other rows. As a consequence, the output is a logic 0 once the input clear is set to a logic 1, regardless of the status of the clock input clk or the previous output value.

```
// a positive-edge-triggered T flip-flop
primitive T_FF(
         output reg q,
         input clk, clear);

// define the behavior of the edge-triggered T_FF
table
// clk      clear :  q  : q+;
     ?        1   :  ?  : 0 ;  // asynchronous clear
     ?      (10)  :  ?  : - ;  // ignore negative edge of clear
   (01)       0   :  1  : 0 ;  // toggle at positive edge
   (01)       0   :  0  : 1 ;  // of clk
   (1?)       0   :  ?  : - ;  // ignore negative edge of clk
endtable
```

endprimitive

Like combinational UDPs, a number of sequential UDPs are often combined together, even with combinational UDPs, to construct a more useful circuit. To illustrate this, consider the case of constructing a 4-bit ripple counter with four instances of the T flip-flop defined previously. Recall that an important feature of a T flip-flop is that its output is toggled whenever it is triggered each time. Based on this feature, an n-bit ripple counter can be readily constructed by cascading n T flip-flops.

■ Example 5.30: An example of sequential UDP instantiation.

In this example, we instantiate four T flip-flops to construct a 4-bit ripple counter. Since the T flip-flop is positive-edge-triggered, the resulting ripple counter is a down counter. This can be easily seen and verified by plotting the timing diagram of the ripple counter. Because of its intuitive simplicity, the details are left for the reader as an exercise. More details about ripple counters can be referred to in Lin [5].

```
// an example of sequential UDP instantiation
module ripple_counter(
       input clk, clear,
       output [3:0] qout);
// instantiate the T_FFs
   T_FF tff0(qout[0], clk, clear);
   T_FF tff1(qout[1], qout[0], clear);
   T_FF tff2(qout[2], qout[1], clear);
   T_FF tff3(qout[3], qout[2], clear);
endmodule
```

In summary, both combinational and sequential UDPs have the following general features:

1. UDPs model functionality only; they do not model timing or process technology.
2. UDPs have exactly one output terminal and are modeled as a lookup table in memory.
3. UDPs are not applicable to design a logic block because they are usually not accepted by synthesis tools.
4. The UDP state table should be specified as completely as possible.
5. It should use shorthand symbols to combine table entries wherever possible.

Although both types of UDPs have almost the same syntax of declaration, sequential UDPs have the following unique features: First, they must contain a **reg** declaration for the output port. Second, the output port always has the same value as the internal state (next state). Third, the initial value of the output port may be specified with an **initial** block.

■ Review Questions

5-37 What are the two declaration styles of UDPs, regardless of combinational or sequential UDPs?
5-38 Describe the features of combinational UDPs.
5-39 Describe the features of sequential UDPs.
5-40 What is the difference of the state table between combinational and sequential UDPs?
5-41 How would you instantiate a UDP?

5.5 Summary

There are three additional components that can help model a design in behavioral style. These include tasks, functions, and UDPs. Tasks and functions provide the ability to reuse the same piece of code from several places in a design. In addition, they provide a means of breaking up a large design into smaller ones so as to make the source descriptions easier to be read and debugged.

A task is declared with a pair of keywords **task** and **endtask**. A task can contain timing control, has at least one output argument, or has no input arguments. Tasks can invoke tasks and functions. A task can be either static or automatic (also called reentrant, dynamic), depending on whether the keyword **automatic** appears in the declaration or not. Static tasks have the feature that all declared items are statically allocated. Automatic tasks have the feature that all declared items are dynamically allocated for each call.

A function is declared with a pair of keywords **function** and **endfunction**. A function is used when the procedure has no timing control, needs to return a single value, has at least one input argument, has no output or inout arguments, or has no nonblocking assignments. Functions can merely call functions. A function can be either a (static) function, declared without the keyword **automatic**, or an automatic function, declared with the keyword **automatic**. Static functions have the feature that all declared items are statically allocated. Automatic functions have the feature that all function items are dynamically allocated for each (recursive) call.

In addition to user-defined tasks and functions, there is a rich set of predefined system tasks and functions, with identifiers being prefixed with a character "$." System tasks and functions can be used for various applications in writing test benches, describing designs abstractly in behavioral style, and so on. The system tasks and functions can be subdivided into the following types: display system tasks, timescale system tasks, simulation time system functions, simulation control system tasks, file I/O system tasks and functions, string formatting system tasks and functions, conversion system functions, probabilistic distribution system functions, stochastic analysis system tasks and functions, math system functions, command line arguments, and PLA modeling system tasks.

UDPs provide a means to model designs in truth tables and can be cast into combinational UDPs and sequential UDPs. A combinational UDP is used when the output is uniquely determined by the combination of the inputs whereas a sequential UDP is applied when the next output is determined by the combination of the current output and inputs. A UDP is instantiated in exactly the same way as a gate primitive.

References

1. J. Bhasker, *A Verilog HDL Primer*, 3rd ed., Star Galaxy Publishing, 2005.
2. IEEE 1364-2001 Standard, *IEEE Standard Verilog Hardware Description Language*, 2001.
3. IEEE 1497-2001 Standard, *IEEE Standard for Standard Delay Format (SDF) for the Electronic Design Process*, 2001.
4. IEEE 1364-2005 Standard, *IEEE Standard for Verilog Hardware Description Language*, 2006.
5. M. B. Lin, *Digital System Design: Principles, Practices, and Applications*, 4th ed., Chuan Hwa Book Ltd. (Taipei, Taiwan), 2010.
6. S. Palnitkar, *Verilog HDL: A Guide to Digital Design and Synthesis*, 2nd ed., SunSoft Press, 2003.

Problems

5-1 Write a task that counts the number of ones in a byte. In addition, write a test bench to apply stimuli to the task and verify the results.

5-2 Write a task that counts the number of trailing ones in a byte. In addition, write a test bench to apply stimuli to the task and verify the results.

5-3 Write a task that counts the number of leading zeros in a byte. In addition, write a test bench to apply stimuli to the task and verify the results.

5-4 Write a task that counts the number of leading ones in a byte. In addition, write a test bench to apply stimuli to the task and verify the results.

5-5 Write a task to compute the even parity of a 16-bit number. In addition, write a test bench to apply stimuli to the task and verify the results.

5-6 Write a task to generate a reset signal that is raised to 1 for three clock cycles each time it is called. The reset signal starts at the first negative edge of the clock signal after the task is called. Write a test bench to apply stimuli to the task and verify the results.

5-7 Write a task to generate a reset signal that is cleared to 0 for n clock cycles each time the task is called, where n is passed by the calling statement. The reset signal starts at the first negative edge of the clock signal after the task is called. Write a test bench to apply stimuli to the task and verify the results.

5-8 Write a task to dump the contents of a memory block starting from a specified address. In addition, write a test bench to apply stimuli to the task and verify the results.

5-9 Write a function that counts the number of ones in a byte. In addition, write a test bench to apply stimuli to the function and verify the results.

5-10 Write a function that counts the number of trailing ones in a byte. In addition, write a test bench to apply stimuli to the function and verify the results.

5-11 Write a function that counts the number of leading zeros in a byte. In addition, write a test bench to apply stimuli to the function and verify the results.

5-12 Write a function that counts the number of leading ones in a byte. In addition, write a test bench to apply stimuli to the function and verify the results.

5-13 The greatest common divisor (GCD) of two non-negative integer numbers m and n is defined recursively as follows:

$$GCD(m,n) = \begin{cases} m & \text{if } n=0 \\ GCD(n, m \bmod n) & \text{if } m > n \end{cases}$$

(a) Write an automatic function to compute the $GCD(m,n)$.

(b) Use the **$random** system function to generate two random numbers m and n as the inputs and then use the **$display** system task to display the result on the standard output.

(c) Use the **$fgets** and **$sscanf** system functions to read two numbers m and n from the standard input and use the **$display** system task to display the result on the standard output.

5-14 Write a Verilog HDL module to describe an 8-bit adder with two inputs x and y and one output sum and then write a test bench with each of the following specified functions to verify the 8-bit adder.

    (a) Use the **$random** system function to generate two random numbers as the inputs and then use the **$monitor** system task to display the result on the standard output.

    (b) Use the **$fgets** and **$sscanf** system functions to read two numbers from the standard input and use the **$display** and **$monitor** system tasks to display the result on the standard output.

5-15 Write a Verilog HDL module to describe an 8-bit subtracter with two inputs x and y and one output dif and then write a test bench with each of the following specified functions to verify the 8-bit subtracter.

    (a) Use the **$random** system function to generate two random numbers as the inputs and then use the **$monitor** system task to display the result on the standard output.

    (b) Use the **$fgets** and **$sscanf** system functions to read two numbers from the standard input and use the **$display** and **$monitor** system tasks to display the result on the standard output.

5-16 Write a Verilog HDL module to describe a two-digit BCD adder with two inputs x and y and one output sum and then write a test bench with each of the following specified functions to verify the two-digit BCD adder.

    (a) Use the **$random** system function to generate two random numbers as the inputs and then use the **$monitor** system task to display the result on the standard output.

    (b) Use the **$fgets** and **$sscanf** system functions to read two numbers from the standard input and use the **$display** and **$monitor** system tasks to display the result on the standard output.

5-17 Write a Verilog HDL module to describe a two-digit BCD subtracter with two inputs x and y and one output dif and then write a test bench with each of the following specified functions to verify the two-digit BCD subtracter.

    (a) Use the **$random** system function to generate two random numbers as the inputs and then use the **$monitor** system task to display the result on the standard output.

    (b) Use the **$fgets** and **$sscanf** system functions to read two numbers from the standard input and use the **$display** and **$monitor** system tasks to display the result on the standard output.

5-18 Write a function to compute the odd parity of a 16-bit number. In addition, write a test bench to apply stimuli to the function and check the results.

5-19 Write a function that carries out an arithmetic left shift of a 16-bit vector.

5-20 Write a function that carries out an arithmetic right shift of a 16-bit vector.

5-21 Write a combinational UDP to model the logic circuit shown in Figure 5.1.

5-22 Write a combinational UDP to model the logic circuit shown in Figure 5.2.

5-23 Use a combinational UDP to model each of the following switching expressions:

    (a) $f(x,y,z) = \bar{y}\bar{z} + \bar{x}y + xz$
    (b) $f(x,y,z) = \Sigma(0,2,4,5,6)$
    (c) $f(w,x,y,z) = \Sigma(1,3,6,7,11,12,13,15)$

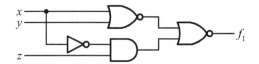

Figure 5.1: The logic circuit for Problem 5-21.

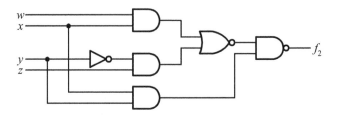

Figure 5.2: The logic circuit for Problem 5-22.

(d) $f(w,x,y,z) = \Sigma(0,1,6,9,10,11,14,15)$

5-24 The operations of an SR flip-flop are as follows. At the positive edge of the clock signal, the output is set to 1 if the input S is 1; the output is cleared to 0 if the input R is 1. The output remains unchanged if both R and S are 0; the output is undetermined if both R and S are 1. Consider each of the following modifications and write a sequential UDP to model the resulting flip-flop:

(a) Suppose that the SR flip-flop is modified as a clear-dominated type; that is, the output is cleared to 0 if both R and S are 1.

(b) Suppose that the SR flip-flop is modified as a set-dominated type; that is, the output is set to 1 if both R and S are 1.

5-25 In at least four ways in Verilog HDL, describe each of the following basic two-input gates by merely writing down the essential statements:

(a) AND
(b) OR
(c) XOR

5-26 In at least four ways in Verilog HDL, describe each of the following basic two-input gates by merely writing down the essential statements:

(a) NAND
(b) NOR
(c) XNOR

6
Hierarchically Structural Modeling

In the previous chapters, we have discussed a variety of modeling styles, including the structural style at the gate level as well as the switch level, the dataflow style, and the behavioral style. In this chapter, we consider three closely related issues of hierarchically structural modeling, including instantiation, **generate** regions, and configurations.

The first issue is the mechanism of instantiation through which a hierarchical structure is formed by embedding modules into others. These modules can be gate and/or switch primitives, UDP primitives, and/or modules. Higher-level modules create instances of lower-level modules and communicate with them through input ports, output ports, and bidirectional ports, each of which can be a scalar or a vector.

The second issue is the **generate** region, which can conditionally generate declarations and instances into a description of a design. Almost what can be put inside a module can also be placed within a **generate** region. The **generate** region can be employed to describe a large regular design, such as $m \times n$ array multipliers, $m \times n$ array dividers, and others.

The final issue is the configuration. A configuration is a set of rules that maps instances to cells, including UDPs, configurations, and modules. By using configurations, we may specify a new set of target libraries so as to change the mapping of a design without having to change the source description of the design.

6.1 Modules

The basic units of Verilog HDL are modules. Like a hardware module or an IC (integrated circuit), each module has two major parts: the *interface* and the *body*. The interface facilitates a way through which the module can communicate with its outside world and the body defines the functionality of the module. In the previous chapters, we have learned how to describe the body of a module in various modeling styles. In this section, we discuss how to define the interface of a module and how to use the interface to communicate with other modules.

6.1.1 Module Definition

A module is defined by using a pair of keywords **module** and **endmodule** and may be one of the following two forms:

```
// port-list style
module module_name [parameter_port_declaration]
                [(port{, port})];   // list of ports
   {port_declaration};
```

```
    {non_port_module_item}
endmodule

// port-list declaration style
module module_name [parameter_port_declaration]
                   [list_of_port_declarations];
    {non_port_module_item}
endmodule
```

Parameters can be declared immediately after module_name and before the port list or list_of_port_declarations. Such parameters constitute the *parameter port* and is often referred to as parameter_port_declaration. Those parameters not present in the parameter port may be declared within the module as parameter_declaration. The syntax of parameter_port_declaration is as follows:

```
#(parameter_declaration{, parameter_declaration})
```

The parameter_declaration will be discussed later in Section 6.1.2.

The port_declaration and list_of_port_declaration parts declare port and data types of all input ports, output ports, and bidirectional ports with the keywords **input**, **output**, and **inout**, respectively. When the port-list declaration style is used, ports cannot be redeclared within the module. The maximum number of ports allowed in a module is limited by implementations but at least 256. The syntax of list_of_port_declaration can be either of the following two formats

```
(port_declaration{, port_declaration})
()
```

where port_declaration will be described later in this section.

The body of the module, i.e., non_port_module_item, can be any of the following ones:

```
module_or_generate_item
parameter_declaration;   // if not in the parameter port
specparam_declaration
specify_block
generate_region
```

where module_or_generate_item defines the functionality of the module and can be any of the following declarations and constructs.

```
data_type_declaration (nets, variables, events, genvars)
local_parameter_declaration
task_declaration
function_declaration
parameter_override
continuous_assignment (assign)
instantiation_statement (gates, udps, modules)
initial_block (also initial_construct)
always_block (also always_construct}
```

The specparam_declaration part will be discussed along with specify_block later in in Section 7.3.2. The generate_region part is to be discussed later in Section 6.2.

A final comment on the definition of a module is as follows. Some synthesis tools allow the keyword **macromodule** to be used interchangeably with the keyword **module** to define a module. However, some other synthesis tools may choose to treat the module definition beginning with the **macromodule** keyword differently. Therefore, in this book, we only consider the use of the keyword **module**.

Section 6.1 Modules

Figure 6.1: Port connection rules.

6.1.1.1 Port Declarations The interface signals (excluding supply and ground) of any Verilog HDL module can be cast into three types: **input**, **output**, and **inout**. In what follows, we describe these three port declarations in detail.

The input port declarations. The keyword **input** is used to declare one or more signals as input ports. It has the following form

    input [net_type] [**signed**] [range] port_id{, port_id}

where net_type can be any of net types and is often omitted if the net type is **wire**. The keyword **signed** is used to declare a signed net. As shown in Figure 6.1, an input port must be a net type but it can be driven by a net port or a variable port. Here, the data type of the variable port may be **reg**, **integer**, or **time**.

The output declarations. The keyword **output** is used to declare one or more signals as output ports. It has the form

    output [net_type] [**signed**] [range] port_id{, port_id}
    output reg [**signed**] [range] port_id[= const_expr]
                      {, port_id[= const_expr]}
    output port_var_type port_id[= const_expr]
                      {, port_id[= const_expr]}

where net_type can be any of net types and is often omitted if the net type is **wire**; port_var_type can be an **integer** or a **time** variable type. As shown in Figure 6.1, an output port can be a net or variable type but it only drives a net port, which of course is an input port of another module. The variable types include **reg**, **integer**, and **time**.

The inout declarations. The keyword **inout** is used to declare one or more signals as bidirectional ports. It has the following syntax

    inout [net_type] [**signed**] [range] port_id{, port_id}

where net_type can be any of net types and is often omitted if the net type is **wire**. As shown in Figure 6.1, an **inout** port can only be a net type and it can drive or be driven by a net port.

6.1.1.2 Port Declaration Examples A complete port declaration of a port must include two components: a port type and its associated data type. The port types include **input**, **output**, or **inout** and the associated data types include nets or variables. A complete port declaration cannot be redeclared again within the module. The following example declares three input ports, x, y, and c_in, as **wire** nets implicitly as well as two output ports, sum and c_out, as **reg** variables.

    input [3:0] x, y;   // net is unsigned
    input c_in;   // net is unsigned
    output reg [3:0] sum;
    output reg c_out;

An implicit net, such as x, y, or c_in, is considered as an unsigned **wire**. A net connected to a port without an explicit net declaration is considered to be unsigned, unless the port is declared to be signed by the keyword **signed**.

```
input    [3:0] x, y;       // unsigned implicit nets
input    signed z;         // signed implicit net
output   reg signed w;     // reg w is signed
output   reg v;            // unsigned reg variable
```

An implicit continuous assignment is assumed to exist between a port and a port expression. Hence, when the bit widths of a port and the port expression are unmatched, the same rules used for the **assign** continuous assignment are applied to the port expression to match the port (see Section 3.1.1 for details).

6.1.2 Parameters

Parameters are constants, which can be used throughout the module defining them, and are often used to specify delays and the bit widths of nets and variables. A parameter can be only defined once. Two types of parameters are *module parameters* and *specify parameters*. The module parameters can be defined by using the keyword **parameter** or **localparam** within a module. The major difference between these two types of module parameters is that a parameter defined by the **parameter** declaration can be modified by using the **defparam** statement or through module instantiation but that defined by the **localparam** declaration cannot. These two types of module parameters combined with the **'define** compiler directive comprise the three ways of defining constants in Verilog HDL. In other words, a constant can be defined in any of the following three methods:

- The **'define** compiler directive
- The **parameter** declaration
- The **localparam** declaration

The details of the **'define** compiler directive will be discussed in Section 7.4.1 and the specify parameters are addressed in Section 7.3.2.

6.1.2.1 Parameter Declarations The **parameter** declaration is the most common approach used to define module parameters that can be overridden by the **defparam** statement or the module instance parameter value assignment. The forms of `parameter_declaration` are as follows

```
parameter [signed] [range] parameter_id = const_expr
                          {, parameter_id = const_expr}
parameter parameter_type parameter_id = const_expr
                          {, parameter_id = const_expr}
```

where `parameter_type` can be one of the following variables: **integer**, **time**, **real**, and **realtime**. `const_expr` can be any constant expression in simple or mintypmax form. Note that no range specification or the keyword **signed** is allowed when the second form is used.

Some examples of the **parameter** declarations are given as follows:

```
parameter SIZE = 7;
parameter WIDTH_BUSA = 24, WIDTH_BUSB = 8;
parameter signed [3:0] mux_selector = 4'b0;
parameter integer CNT_SIZE = 12;
parameter real TWO_PI = 2*3.1415926;
```

6.1.2.2 Local Parameter Declarations The **localparam** declaration is used to define parameters associated locally with a module. It is identical to the **parameter** declaration except that it cannot be overridden by the **defparam** statement or by the module instance parameter value assignment. In addition, the **localparam** declaration cannot be used in the parameter

Section 6.1 *Modules*

port, to be introduced later in this section. The declaration of a local parameter is exactly like that of the module parameter except that the keyword **localparam** is used instead of the keyword **parameter**. The syntax of local_parameter_declaration is as follows

localparam [**signed**] [range] parameter_id = const_expr
{, parameter_id = const_expr}
localparam parameter_type parameter_id = const_expr
{, parameter_id = const_expr}

where parameter_type can be one of the following variables: **integer**, **time**, **real**, and **realtime**. const_expr can be any constant expression in simple or mintypmax form. Note that no range specification or the keyword **signed** is allowed when the second form is used.

Some examples of the **localparam** declarations are given as follows:

localparam SIZE = 7;
localparam WIDTH_BUSA = 24, WIDTH_BUSB = 8;
localparam **signed** [3:0] mux_selector = 4'b0;
localparam **integer** CNT_SIZE = 24;
localparam **real** PI = 3.1415926;

In summary, the **localparam** declaration is used to define parameters when their values are not changed whereas the **parameter** declaration is used to define parameters when their values may be changed at the instant of module instantiation or by using the **defparam** statement.

Parameter dependence. In many applications, it is useful to define a parameter expression containing another parameter so that an update of a parameter, either by the **defparam** statement or in a module instantiation statement for the module defining the parameter, will automatically update another one. For example, in the following parameter declaration, an update of word_size will automatically update memory_size.

parameter word_size = 16,
localparam memory_size = word_size * 512;

Since memory_size depends on the value of word_size, a modification of word_size will change the value of memory_size accordingly.

6.1.2.3 Parameter Ports As described previously, module parameters declared with the keyword **parameter** may be placed after the port list/port-list declarations or between the module name and the port list/port-list declarations. However, when a module is declared in port-list declaration style, the module parameters in the module are needed to declare before the port-list declarations when they are referenced in the port-list declarations.

When a **parameter** declaration is placed between the module name and the port list/port-list declarations, it is called a *parameter port*. Like the regular module parameter, a parameter port can have both type and range specifications and has the following two forms:

module module_name
#(**parameter** [**signed**] [range] parameter_id = const_expr
{, parameter_id = const_expr}
parameter variable_type parameter_id = const_expr
{, parameter_id = const_expr})
[(port list or port—list declarations)]
...
endmodule

Here are some examples of the usage of parameter ports.

module module_name
#(**parameter** SIZE = 7,

```
        parameter WIDTH_BUSA = 24, WIDTH_BUSB = 8,
        parameter signed [3:0] mux_selector = 4'b0,
        parameter integer CNT_SIZE = 12,
        parameter real PERIMETER = 2*3.14 * RADIUS)
    (port list or port-list declarations)
    ...
endmodule
```

Note that the comma (,) is used instead of the semicolon (;) after the **parameter** declarations in the parameter ports. The **parameter** declarations are often used to define constants within *parameterizable modules*. A parameterizable module is one whose features or structural properties are defined with **parameter** declarations and may be modified by using the **defparam** statements or a module instantiation statement for the module.

■ Example 6.1: The use of parameter ports.

In the following parameterizable module, adder_nbit_parameter_port, the module parameter N is declared by using a parameter port. This parameter sets the default value of the constant N, which in turn defines the word width of the n-bit adder and may be modified to any other values through using an appropriate method that will be introduced later in this section. Note that in this example the **parameter** declaration may be placed after the port list instead of using the parameter port.

```
module adder_nbit_parameter_port
        #(parameter N = 4)          // parameter port
        (x, y, c_in, sum, c_out); // port list
// parameter N = 4; --- may be defined here
input   [N-1:0] x, y;
input   c_in;
output  [N-1:0] sum;
output  c_out;
// specify an N-bit adder using a continuous assignment
    assign {c_out, sum} = x + y + c_in;
endmodule
```

Of course, the above port-list style can be rewritten as the port-list declaration style and is given as follows.

■ Example 6.2: Port-list declaration style and parameter ports.

In this example, the port-list declaration style is used. Because of this, the parameter port should be used; otherwise, the module parameter *N* referred to in the port-list declarations will be used before it is defined. As a result, an error message, such as "*Reference to unknown variable N,*" would be raised by simulators or synthesis tools.

```
module adder_nbit_parameter_port2
        #(parameter N = 4)(    // parameter port
          input  [N-1:0] x, y, // port-list declarations
          input  c_in,
          output [N-1:0] sum,
          output c_out);

// specify an N-bit adder using a continuous assignment
```

Section 6.1 Modules

```
assign {c_out, sum} = x + y + c_in;
endmodule
```

■ Review Questions

6-1 What are the two major parts of a module?
6-2 What are the three methods that can be used to define a constant in Verilog HDL?
6-3 What are the differences between **parameter** and **localparam** declarations?
6-4 What is the meaning of the parameter port?
6-5 What is a parameterizable module?

6.1.3 Module Instantiation

In Verilog HDL, module definitions cannot be nested. However, much like hardware modules, a module can incorporate an instance (namely, a copy) of another module into itself through instantiation. To instantiate a module, the following form can be used

```
module_name [#(list_of_parameter_assignments)]
        instance_name [range]([ports])
        {, instance_name [range]([ports])};
```

where #(list_of_parameter_assignments) is the parameter port[1] in which parameters are passed to the instantiated module to override the corresponding parameters defined by the **parameter** declaration within the instantiated module. The range specification is used to create an array of instances through instantiation. In addition, one or more module instances can be created in a single module instantiation statement.

6.1.3.1 Port Connection Rules The port connection rules of Verilog HDL modules are consistent with those of real-world hardware modules. Connecting ports to external signals can be done by one of the following two methods: *named association* and *positional association*.

- *Named association:* In this method, ports are connected by listing their names. Port identifiers and their associated port expressions are explicitly specified. The form is as follows:

    ```
    .port_id1(port_expr1),..., .port_idn(port_exprn)
    ```

    An unconnected port is skipped or placed as an empty character, such as ".port_id()".

- *Positional (or ordered) association:* In this method, ports are connected by the ordered list of the ports, with each corresponding to a port. The form is as follows:

    ```
    port_expr1, ..., port_exprn
    ```

    An unconnected port is just skipped, such as "x, ,y," where a port is skipped between x and y.

Regardless of which method is employed, each port_expr can be any of the following ones:

1. An identifier (a net or a variable)
2. A bit-select of a vector
3. A part-select of a vector
4. A concatenation of the above
5. An expression for input ports

---
[1] The same name is used in both parameter declarations and parameter passing during module instantiation.

Note that the two port connection methods cannot be mixed in the same instantiation of a module. Moreover, as mentioned, Verilog HDL primitives, including built-in (gate and switch primitives) and user-defined primitives (UDPs), can only be connected by positional association.

■ **Example 6.3: Named association.**

An illustration of the use of named association is given in the following module. The instantiated module, ripple_counter_generate, will be described in Section 6.2.2. With named association, the order of the port list is unimportant. It is strongly recommended that named association should be used for port connections at the top-level module because synthesis tools may change the port order in the port list of the module that they generated.

```
`timescale 1 ns / 100 ps
module ripple_counter_generate_tb;
// internal signals declarations:
reg   clk, clear;
wire [0:3] qout;
parameter clk_period = 20;
// Unit Under Test port map
   ripple_counter_generate UUT (
           .clk(clk), .clear(clear), .qout(qout));
initial begin
   repeat (3) @(posedge clk) clear <= 1'b1;
   clear <= 1'b0;
end
initial begin
   clk <= 1'b0;
   forever  #(clk_period/2) clk <= ~clk;
end
initial
   #500 $finish;
initial
   $monitor($realtime,"ns %b %b %h", clk, clear, qout);
endmodule
```

6.1.3.2 Unconnected Ports Unconnected inputs are driven to the "z" state; unconnected outputs are not used. In the instantiation of a module, the unconnected ports can be specified by leaving their port_expr blank.

6.1.3.3 Real Numbers in Port Connections The **real** number is not permitted to pass directly to a port. To pass a real number to a port, the real number is needed to convert into an equivalent bit sequence. At the receiver, the received bit sequence is then converted back to the real number. To help accomplish such operations, two system functions, **$realtobits** and **$bitstoreal**, are provided in Verilog HDL. The **$realtobits** system function converts a real number into a 64-bit vector while the **$bitstoreal** system function carries out the reverse operation; namely, it converts a 64-bit vector into a real number. Refer to in Section 5.3.4 for details.

■ **Review Questions**

6-6 Describe the port connection rules.
6-7 What is the difference between gate instantiation and module instantiation?

Section 6.1 Modules

6-8 What are the values of ports when they are left unconnected?
6-9 What are the differences between **defparam** and **parameter** declarations?
6-10 What are the two methods that can be used to interconnect ports among modules?

6.1.4 Overriding Parameter Values

When a module instantiates other modules, the higher-level module can change the values of parameters defined by the **parameter** declarations within the lower-level modules. This operation is called *parameter overriding*. Parameter overriding provides a mechanism with which a module can be *parameterized*. In other words, the features or structural properties of the module can be changed by modifying the parameters defined within it.

■ **Example 6.4: A parameterizable 2's-complement adder.**

As an example of a parameterizable module, consider the following n-bit 2's-complement adder. The module parameter N of the twos_adder_behavioral_nbit module can be modified from its default value 4 to any other values. Hence, the module is able to function as a new 2's-complement adder with a new word size.

```
module twos_adder_behavioral_nbit
       #(parameter N = 4)(   // set default width
         input   [N-1:0] x, y,
         input   c_in,
         output reg [N-1:0] sum,
         output reg c_out);
reg [N-1:0] t;    // outputs of xor gates

// the function of a 2's-complement adder
always @(x, y, c_in) begin
   t = y ^ {N{c_in}};
   {c_out, sum} = x + t + c_in;
end
endmodule
```

There are two ways that may be used to override parameter values defined within an instantiated module:

1. The **defparam** statement
2. The module instance parameter value assignment

In what follows, we describe these two ways in detail.

6.1.4.1 Using the defparam Statement A **defparam** statement is used to redefine the parameter values defined by the **parameter** declaration in any module instance throughout the design using the hierarchical path names of the parameters and has the syntax

```
defparam hierarchical_path_id = const_expr
        {, hierarchical_path_id = const_expr};
```

where const_expr can be a constant expression involving only numbers and references to parameters. The referenced parameters must be defined in the same module as the **defparam** statement. This approach finds particularly useful for grouping together the parameter overriding assignments in the same module.

Example 6.5: The use of the defparam statement.

As an example of using the **defparam** statement to redefine the parameter values defined within a lower-level module, consider the two_counters module. The default value of N within the counter_nbits module is 4. Through using the **defparam** statement, two instances of counter_nbits with 4 and 8 bits are separately created in the two_counters module.

```
// an example of using the defparam statement
module two_counters(
       input clock, clear,
       output [3:0] qout4b,
       output [7:0] qout8b);
// instantiate two counter modules
defparam cnt_4b.N = 4, cnt_8b.N = 8;
counter_nbits cnt_4b (clock, clear, qout4b);
counter_nbits cnt_8b (clock, clear, qout8b);
endmodule

module counter_nbits
       #(parameter N = 4)(   // set default size
          input clock, clear,
          output reg [N-1:0] qout);

// qout <= (qout + 1) % 2^N;
always @(negedge clock or posedge clear)
   if (clear) qout <= {N{1'b0}};
   else       qout <= (qout + 1);
endmodule
```

6.1.4.2 Module Instance Parameter Value Assignment In this way, the parameters defined by using the **parameter** declaration within a module are overridden by the parameters passed through parameter ports at the time that the module is instantiated. There are two approaches in which parameter values can be specified.

- *Named association:* This approach is similar to connecting module ports by named association. It explicitly links the names specified in the instantiated module and the associated value. One advantage of this approach is that it only needs to specify those parameters desired new values. It is not necessary to specify values for all of the remaining parameters.
- *Positional (or ordered) association:* In this approach, the order of assignments must follow the declaration order of parameters within the module. In practice, it is not often necessary to assign values to all of the parameters declared within a module. It is also not possible to skip over a parameter. As a consequence, for those module parameters that are locally associated with the module are often declared as local parameters with the keyword **localparam**.

The above two parameter assignment approaches cannot be mixed in the same instantiation of a module.

Three examples are given in sequence to illustrate the method of the module instance parameter value assignment. The first one uses named association while the other two use positional association.

■ Example 6.6: An example of named association.

An example of using named association to pass two module parameters to the instantiated module is illustrated in the following module. In this example, an instance of the hazard_-
static module is instantiated and two module parameters are passed to the instantiated module through two parameter ports with named association. The order of parameters to be passed is not important in this case.

```
// an example of named association
module parameter_overriding_example(
      input x, y, z,
      output f);

// instantiate hazard_static module
   hazard_static #(.DELAY2(4), .DELAY1(6)) example(x, y, z, f);
endmodule

module hazard_static
      #(parameter DELAY1 = 2, DELAY2 = 5)(
         input x, y, z,
         output f);
// internal declaration
wire   a, b, c;    // internal nets
// logic circuit body
   and #DELAY2 a1 (b, x, y);
   not #DELAY1 n1 (a, x);
   and #DELAY2 a2 (c, a, z);
    or #DELAY2 o2 (f, b, c);
endmodule
```

■ Example 6.7: An example of positional association.

In this example, two instances of the counter_nbits module are instantiated and the new parameters are passed to the instantiated modules through module instance parameter value assignments with positional association. The resulting two counters are 4 and 8 bits, respectively.

```
// an example of positional association
module two_counters(
      input clock, clear,
      output [3:0] qout4b,
      output [7:0] qout8b);
// instantiate two counter modules
counter_nbits #(4) cnt_4b (clock, clear, qout4b);
counter_nbits #(8) cnt_8b (clock, clear, qout8b);
endmodule
```

■ Example 6.8: Another example of positional association.

An illustration of using positional association to pass two module parameters to the instantiated module with two parameter ports is explored in this example. An instance of the `hazard_static` module is instantiated and two module parameters are passed to the instantiated module through parameter ports with positional association. The order of parameters to be passed is critical in this case.

```
// another example of positional association
module parameter_overriding_example(
      input   x, y, z,
      output  f);
// instantiate hazard_static module
   hazard_static #(4, 8) example (x, y, z, f);
endmodule
```

One advantage of the module instance parameter value assignment with named association is that it can minimize the chance of errors. In addition, we only need to specify those parameters needed to modify. Therefore, it is recommended to use named association in parameter overriding, in particular, when there are many parameters needed to be overridden at a time.

6.1.5 Hierarchical Path Names

In Verilog HDL, an identifier can be defined within one of the following four elements (or components):

1. Modules
2. Tasks
3. Functions
4. Named blocks (See Sections 6.2 and 7.1.3)

Within each element, an identifier must uniquely declare one item (i.e., object). Any identifier can be directly accessed within the elements defined it. To refer to an identifier defined within the other elements, a *hierarchical path name* must be used. The hierarchical path name is formed by using identifiers separated by a period character for each level of the hierarchy and has the form

```
hierarchical_branch[{.simple_hierarchical_branch}]
```

Based on this notation, each identifier in a Verilog HDL description has a unique hierarchical path name. The complete hierarchical path name of any identifier starts from the *top-level module* and can be down to any level in a description. A top-level module is a module that is not instantiated by any other module.

The following example shows some hierarchical path names of Figure 1.6. Each hierarchical path name is formed by appending the lower-level name on the right side of the higher-level name separated by a period. For instance, the 4-bit adder is constructed by four full adders: fa_1, ..., fa_4. Thus, the complete hierarchical path name of fa_1 is 4bit_adder.fa_1. To refer to the net S at the output of the XOR gate within fa_1, the complete hierarchical path name 4bit_adder.fa_1.ha_1.xor1.S is used. Other examples are as follows:

```
4bit_adder                       // top level --- 4bit_adder
4bit_adder.fa_1                  // fa_1 within 4bit_adder
4bit_adder.fa_1.ha_1             // ha_1 within fa_1
4bit_adder.fa_1.ha_1.xor1        // xor1 within ha_1
4bit_adder.fa_1.ha_1.xor1.S      // net s within xor1
```

Section 6.2 *Generate Regions*

■ **Review Questions**

6-11 Describe the two methods that can be used to override module parameter values defined by **parameter**.

6-12 What are the differences between named association and positional association?

6-13 What is the meaning of parameter overriding?

6-14 Explain what a top-level module is.

6-15 Explain the meaning of a hierarchical path name.

6.2 Generate Regions

A **generate** region allows selection and replication of some statements during *elaboration time*. The elaboration time is the time after a design has been parsed but before simulation/logic synthesis begins. A **generate** region is defined by a pair of keywords **generate** and **endgenerate** and has the following general form:

> **generate**
>   generate_if_statement
>   generate_case_statement
>   generate_loop_statement
>   module_or_generate_item
> **endgenerate**

The statements that can be used within a **generate** region are called *generate statements*, and may be one or more of the following ones: declarations (genvars, data types, **event**, and tasks, functions), local parameter declarations (**localparam**) and parameter overriding (**defparam**), instantiation (gates, UDPs, and modules), **assign** continuous assignments, conditional (i.e., **if**) statements, **case** statements, loop statements, **initial** blocks, and **always** blocks.

The power of **generate** regions is that they can conditionally generate declarations and instances of modules into a description of a design. Almost what can be put inside a module can also be placed within a **generate** region except for parameters (declared with the keyword **parameter**), port declarations (declared with keywords: **input**, **output**, and **inout**), and specify blocks.

The usefulness of **generate** regions is centered around three generate-statements: *generate-loop*, *generate-if*, and *generate-case*. Hence, in the rest of this section we address each of these in more detail along with examples.

6.2.1 Generate-Loop Statement

A generate-loop statement is formed by using a **for** statement within a **generate** region. The generate-loop statement allows statements to be duplicated at elaboration time. The generate-loop statement has the following form

> **for** (genvar_initialization; genvar_condition; genvar_update)
>   generate_block

where generate_block can be either of the following two forms:

> generate_statement
>
> **begin**[: block_name]
>   generate_statement
> **end**

To use the **for** statement within a **generate** region, it is necessary to declare an index variable using the keyword **genvar**. This index variable is an integer and often referred to as a *genvar* (generate variable). Since a genvar is used only in the evaluation of a generate-loop statement, it cannot have a negative value. A genvar can be declared either inside or outside the **generate** region using it. The **genvar** declaration is of the form

genvar genvar_id{, genvar_id};

Genvars are only defined during the evaluation of generate blocks. They do not exist during simulation/logic synthesis of the design. In general, generate-loop statements can be nested. However, two generate-loop statements using the same genvar as an index cannot be nested.

The block_name of a block enclosed by a pair of keywords **begin** and **end** is required and used to refer to any instance names that are associated locally with the generate-loop statement. This type of block is called a *named block* and will be dealt with in more detail in Section 7.1.3.

■ **Example 6.9: The generate-loop and assign statements.**

As an illustration of using an **assign** continuous assignment within a generate-loop statement, consider a module that converts an incoming Gray code into its equivalent binary code. To convert a Gray code into its equivalent binary code, we may count the number of "1s" from the MSB to the current position, i. If the result is odd, the current binary bit is 1; otherwise, the current binary bit is 0. Based on this observation, the reduction operator ^ may be applied to an indexed subrange, counting from SIZE-1 to i, of the gray vector and reduces it into a single binary bit. By repeating this process from bit 0 to bit SIZE-1, the conversion is done.

```
// an example of converting the Gray code into the binary code
// using a generate-loop statement
module gray2bin1
       #(parameter SIZE = 8)( // set default width
          input  [SIZE-1:0] gray,
          output [SIZE-1:0] bin);
genvar i;        // define a generate-loop index
generate for (i = 0; i < SIZE; i = i + 1) begin:bit
   assign bin[i] = ^gray[SIZE-1:i];
end endgenerate
endmodule
```

During elaboration time, the generate-loop statement is expanded, an operation referred to as *loop unrolling*, by replicating the body of the **for** statement within a **generate** statement once for each value of the iteration. For instance, the **assign** continuous assignment in the preceding example will be unrolled into the following statements:

```
assign bin[0] = ^gray[SIZE-1:0];
assign bin[1] = ^gray[SIZE-1:1];
assign bin[2] = ^gray[SIZE-1:2];
assign bin[3] = ^gray[SIZE-1:3];
assign bin[4] = ^gray[SIZE-1:4];
assign bin[5] = ^gray[SIZE-1:5];
assign bin[6] = ^gray[SIZE-1:6];
assign bin[7] = ^gray[SIZE-1:7];
```

Of course, you may write the above **assign** continuous assignments by yourself instead of using the generate-loop statement to perform the same operation. Nevertheless, using generate-loop statements allows us to model a design in a more concise and readable way.

■ Example 6.10: The generate-loop and always blocks.

This example simply replaces the **assign** continuous assignment in the preceding example with an **always** block. In this example, it is better to use the wildcard character (*) in the sensitivity list of the **always** block to avoid any possible warning from synthesis tools.

```
// an example of converting the Gray code into the binary code
module gray2bin2
       #(parameter SIZE = 8)( // set default width
          input  [SIZE-1:0] gray,
          output reg [SIZE-1:0] bin);
genvar i;            // define a generate-loop index

// a generate loop
generate for (i = 0; i < SIZE; i = i + 1) begin:bit
   always @(*)      // avoid synthesizer's warning
      bin[i] = ^gray[SIZE - 1: i];
end endgenerate
endmodule
```

6.2.2 Generate-If Statement

A generate-if statement allows modules, gate primitives, UDPs, **assign** continuous assignments, **initial** blocks, **always** blocks, tasks, and functions, to be instantiated into the description of a design based on an **if-else** conditional expression. The generate-if statement can be used alone or inside a generate-loop statement. It has the following forms

```
if (condition) generate_block_or_null
[else generate_block_or_null]

if (condition1) generate_block_or_null
[else if(condition2) generate_block_or_null
    [else generate_block_or_null]]
```

where condition must be a static condition; namely, it must be computable at elaboration time. This implies that condition must only be a function of constants and parameters. generate_block_or_null can be any statement that is allowed within a **generate** region or a null statement (i.e., ;). Based on the condition, an appropriate statement is selected for being expanded during elaboration time.

6.2.2.1 Combining with Modules, assign, and always Blocks In what follows, we give three examples to illustrate the usage of generate-if statements along with modules, **assign** continuous assignments, and **always** blocks. These examples also explore how the powerful generate-if statements can be used to model iterative logic structures. More details about the iterative logic structures can be referred to in Section 11.3.4 and Lin [4].

■ Example 6.11: A parameterizable n-bit ripple-carry adder.

As an illustration of the use of modules within a generate-if statement, consider the n-bit adder depicted in Figure 6.2. In this example, an **if-else** conditional expression is employed to set up the boundary cells, the LSB and the MSB, as well as the other bits. The full_adder module is then created and connected as desired. The LSB cell needs to accept the external carry-in c_in and the MSB cell needs to send the carry-out c_out out of the module. Each of the remaining cells accepts the carry-in from its preceding cell and sends the carry-out to its succeeding cell.

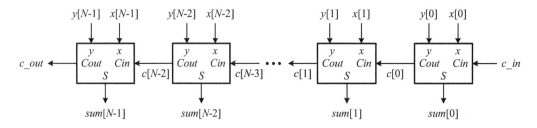

Figure 6.2: The block diagram of an n-bit ripple-carry adder.

```
// using module instantiation inside
// a generate-if statement
module adder_nbit
       #(parameter N = 4)(   // set default width
         input    [N-1:0] x, y,
         input    c_in,
         output   [N-1:0] sum,
         output   c_out);

// specify an N-bit adder using a generate-loop statement
genvar  i;
wire   [N-2:0] c;  // internal carries declared as nets
generate for (i = 0; i < N; i = i + 1) begin: adder
   if (i == 0)         // specify the LSB
      full_adder fa (x[i], y[i], c_in, sum[i], c[i]);
   else if (i == N-1) // specify the MSB
      full_adder fa (x[i], y[i], c[i-1], sum[i], c_out);
   else                // specify other bits
      full_adder fa (x[i], y[i], c[i-1], sum[i], c[i]);
end endgenerate
endmodule
// define a full adder in dataflow style
module full_adder(input x, y, c_in, output sum, c_out);
// specify the function of a full adder
   assign {c_out, sum} = x + y + c_in;
endmodule
```

■ **Example 6.12: A parameterizable n-bit ripple-carry adder.**

This example illustrates the combination of an **if-else** conditional expression and **assign** continuous assignments within a generate-if statement by way of the n-bit adder. In this example, we simply replace the full_adder module of the preceding example with an **assign** continuous assignment. The resulting module is as follows:

```
// using assign statements inside a
// generate-if statement
module adder_nbit
       #(parameter N = 4)(   // set default width
         input    [N-1:0] x, y,
         input    c_in,
```

Section 6.2 Generate Regions

```verilog
           output [N-1:0] sum,
           output c_out);

// specify an N-bit adder using a generate-if
// statement
genvar i;
wire   [N-2:0] c; // internal carries declared as nets
generate for (i = 0; i < N; i = i + 1) begin: adder
   if (i == 0)          // specify the LSB
      assign {c[i], sum[i]} =  x[i] + y[i] + c_in;
   else if (i == N-1) // specify the MSB
      assign {c_out, sum[i]} =  x[i] + y[i] + c[i-1];
   else                 // specify other bits
      assign {c[i], sum[i]} =  x[i] + y[i] + c[i-1];
end endgenerate
endmodule
```

■ Example 6.13: A parameterizable n-bit ripple-carry adder.

As an illustration of the combination of an **if-else** conditional expression and **always** blocks within a generate-if statement, consider the n-bit adder again. In this example, we simply replace the **assign** continuous assignments in the preceding example with **always** blocks. The other operations remain the same and hence we do not explain it furthermore.

```verilog
// using always blocks inside a generate-if
// statement
module adder_nbit
       #(parameter N = 4)(  // set default width
          input  [N-1:0] x, y,
          input  c_in,
          output reg [N-1:0]  sum,
          output reg c_out);

// specify an N-bit adder using a generate-if
// statement
genvar i;
reg    [N-2:0] c; // internal carries declared as nets
generate for (i = 0; i < N; i = i + 1) begin: adder
   if (i == 0)          // specify the LSB
      always @(*) {c[i], sum[i]} =  x[i] + y[i] + c_in;
   else if (i == N-1) // specify the MSB
      always @(*) {c_out, sum[i]} =  x[i] + y[i] + c[i-1];
   else                 // specify other bits
      always @(*) {c[i], sum[i]} =  x[i] + y[i] + c[i-1];
end endgenerate
endmodule
```

The following example demonstrates how an n-bit 2's-complement adder can be constructed by using the powerful generate-if statement. An example of the 4-bit 2's-complement adder can be referred to Figure 1.14.

■ Example 6.14: An n-bit 2's-complement adder.

In this example of an n-bit 2's-complement adder, two generate-loop statements are used to model an n-bit 1's-complement generator and an n-bit ripple-carry adder, respectively. The 2's-complement adder is then performed by using the fact that the 2's complement of a number is equivalent to the 1's complement of the number plus one. Based on this, the 2's-complement adder can then be built by routing the output of the 1's-complement generator to one of the two inputs of the n-bit adder and connecting the control signal of the 1's-complement generator together with the carry-in, named the mode input, of the n-bit adder. The resulting circuit is a subtracter if the mode input is 1 and an adder if the mode input is 0.

```
module twos_adder_nbit
       #(parameter N = 4)(   // set default width
         input   [N-1:0] x, y,
         input   mode, // mode = 1: subtraction; =0: addition
         output  [N-1:0]  sum,
         output  c_out);

// specify the function of an N-bit adder using generate-loop
// statements
genvar i;
wire [N-2:0] c;  // internal carries declared as nets
wire [N-1:0] t;  // true/ones complement outputs
generate for (i = 0; i < N; i = i + 1) begin: ones_cpl_gen
   xor xor_ones_complement (t[i], y[i], mode);
end endgenerate
generate for (i = 0; i < N; i = i + 1) begin: adder
   if (i == 0)       // specify the LSB
      full_adder fa (x[i], t[i], mode, sum[i], c[i]);
   else if (i == N-1) // specify the MSB
      full_adder fa (x[i], t[i], c[i-1], sum[i], c_out);
   else              // specify other bits
      full_adder fa (x[i], t[i], c[i-1], sum[i], c[i]);
end endgenerate
endmodule

// define a full adder in dataflow style
module full_adder(input x, y, c_in, output sum, c_out);
// specify the function of a full adder
   assign {c_out, sum} = x + y + c_in;
endmodule
```

Figure 6.3 shows the synthesized results of the n-bit 2's-complement adder from a synthesis tool with $n=4$. From Figure 6.3(a), we can see that there are four XOR gates and four full-adder modules. Figure 6.3(b) shows the result after dissolving the second and third full-adder modules. Notice that this version of the 2's-complement adder is only for illustrating purpose. In practice, it can be simply modeled in behavioral or dataflow style with vector operands, refer to in Sections 3.3.3 and 4.2.2 for more details.

6.2.2.2 Combining with Tasks, Functions, and UDPs In the rest of this subsection, we use some examples to illustrate how tasks, functions, and UDPs are utilized within generate-if statements. Task and function declarations are allowed within the generate region but not

Section 6.2 Generate Regions

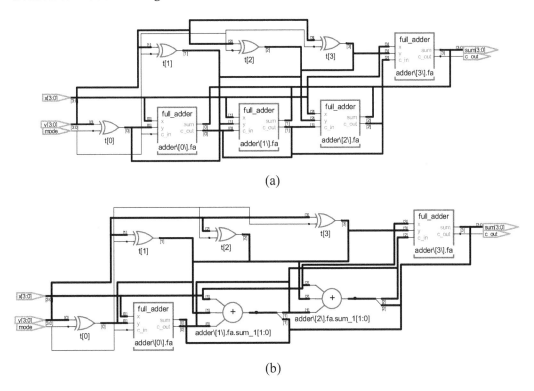

Figure 6.3: An n-bit 2's-complement adder generated from a synthesis tool: (a) RTL schematic and (b) after dissolving the second and the third full adder modules.

within a generate-loop statement. Generated tasks and functions have unique identifiers and can be referred through their hierarchical path names.

■ Example 6.15: The generate-if statement and tasks.

As an example of demonstrating the use of tasks within a generate-if statement, consider the n-bit adder again. To describe an n-bit adder, a task is defined to carry out the operations of a full adder. The full_adder task is then invoked within the **always** blocks inside the generate-if statement to complete the required n-bit adder. The complete module is as follows:

```
// using tasks inside a generate-if statement
module adder_nbit
        #(parameter N = 4)(   // set default width
           input   [N-1:0] x, y,
           input   c_in,
           output reg [N-1:0]  sum,
           output reg c_out);
// specify an N-bit adder using a generate loop
genvar i;
reg    [N-2:0] c;   // internal carries declared as nets
generate for (i = 0; i < N; i = i + 1) begin: adder
   if (i == 0)       // specify the LSB
      always @(*) full_adder(x[i],y[i],c_in,{c[i],sum[i]});
   else if (i == N-1) // specify the MSB
      always @(*) full_adder(x[i],y[i],c[i-1],{c_out,sum[i]});
```

```
         else                  // specify other bits
             always @(*) full_adder(x[i],y[i],c[i−1],{c[i],sum[i]});
end endgenerate

// define a full adder task here
task full_adder(input x, y, c_in, output reg [1:0] sum);
    sum = x + y + c_in;
endtask
endmodule
```

■ **Example 6.16: The generate-if statement.**

As an illustration of the use of functions within a generate-if statement, consider the n-bit adder one more time. To describe an n-bit adder, we first define a function to perform the operations of a full adder. The full_adder function is then called within the **always** blocks inside the generate-if statement to complete the desired n-bit adder.

```
// using functions inside a generate-if statement
module adder_nbit
        #(parameter N = 4)(   // set default width
          input   [N−1:0] x, y,
          input   c_in,
          output reg [N−1:0]  sum,
          output reg c_out);
// specify an N-bit adder using a generate-loop statement
genvar  i;
reg     [N−2:0] c;   // internal carries declared as nets
generate for (i = 0; i < N; i = i + 1) begin: adder
   if (i == 0)         // specify the LSB
      always @(*) {c[i],sum[i]} = full_adder(x[i],y[i],c_in);
   else if (i == N−1) // specify the MSB
      always @(*) {c_out,sum[i]} = full_adder(x[i],y[i],c[i−1]);
   else                // specify other bits
      always @(*) {c[i],sum[i]} = full_adder(x[i],y[i],c[i−1]);
end endgenerate

// define a full adder function here
function [1:0] full_adder(input x, y, c_in);
    full_adder = x + y + c_in;
endfunction
endmodule
```

It is worthwhile comparing the differences between tasks and functions, including their definitions and how they are invoked within a generate-loop statement. In addition to tasks and functions, UDPs can also be used within a generate-if statement. An illustration of this is given in the following example.

■ **Example 6.17: The generate-if statement and UDPs.**

As an illustration of the combination of UDPs and a generate-if statement, assume that we want to model an n-bit ripple counter. To do this, the T_FF UDP defined in Section 5.4.3 is

Section 6.2 Generate Regions

used to carry out the desired function of a T flip-flop. This UDP is then instantiated within a generate-if statement to model an n-bit ripple counter. As compared to the module obtained in Section 5.4.3, this is the better way to model a ripple counter with an arbitrary number of stages. The resulting module is as follows:

```
// an example of sequential UDP instantiation
module ripple_counter_generate
       #(parameter N = 4)(    // set default width
          input   clk, clear,
          output  [N-1:0] qout);

// specify an N-bit ripple counter using a generate-loop
// statement
genvar i;
generate for (i = 0; i < N; i = i + 1)begin:ripple_counter
   if (i == 0) // specify the LSB
      T_FF tff(qout[i], clk, clear);
   else        // specify the other bits
      T_FF tff(qout[i], qout[i-1], clear);
end endgenerate
endmodule
```

6.2.3 Generate-Case Statement

Like the generate-if statement, a generate-case statement allows modules, UDPs, gate primitives, **assign** continuous assignments, **initial** blocks, **always** blocks, tasks, and functions, to be conditionally instantiated into the description of a design. The generate-case statement can be used alone or inside a generate-loop statement. Its general form is as follows

```
case(case_expr)
   case_item1{, case_item1i}: generate_block_or_null
       ...
   case_itemn{, case_itemni}: generate_block_or_null
                 [default: generate_block_or_null]
endcase
```

where case_expr must be computable at elaboration time. This means that case_expr must only be a function of constants and parameters. Based on this value, the statement on a selected branch is expanded during elaboration time. The generate_block_or_null statement associated with each case item can be any statement that is allowed within a **generate** region or a null statement (i.e., ;). The **default** statement is optional and is used to cover unspecified cases.

■ **Example 6.18: An n-bit adder with a generate-case statement.**

As an example of the use of the generate-case statement, consider the n-bit adder again. To model such an adder, the case items 0 and $n-1$ are used to describe the boundary cases of the LSB and MSB, respectively. The other bits are performed by the default case. Based on these considerations, the resulting module is obtained.

```
// using assign continuous assignments inside a generate-case
// statement
module adder_nbit
```

```
        #(parameter N = 4)(   // set default width
          input    [N-1:0] x, y,
          input    c_in,
          output reg [N-1:0]  sum,
          output reg c_out);
// specify an N-bit adder using a generate-loop statement
genvar   i;
wire   [N-2:0] c;   // internal carries declared as nets
generate for (i = 0; i < N; i = i + 1) begin: adder
   case (i)
          0: assign {c[i], sum[i]}   = x[i] + y[i] + c_in;
        N-1: assign {c_out, sum[i]} = x[i] + y[i] + c[i-1];
      default: assign {c[i], sum[i]}   = x[i] + y[i] + c[i-1];
   endcase
end endgenerate
endmodule
```

■ **Example 6.19: The generate-case statement.**

This example demonstrates how the parameter port can be used to control the actual instantiation of the n-bit adder. The n-bit adder is obtained by instantiating the parameterizable module, adder_nbit, when the WIDTH is 4 and 8, and by instantiating the adder_cla module when the WIDTH has other values.

```
module adder_nbit(x, y, c_in, c_out, sum);
// using the parameter port to control the instantiation
// of modules
module adder_nbit
        #(parameter WIDTH = 4)(   // set default width
          input    [WIDTH-1:0] x, y,
          input    c_in,
          output [WIDTH-1:0] sum,
          output c_out);
// specify the function of an N-bit adder
generate
   case (WIDTH)
           4: adder_nbit #(4)   adder4(x, y, c_in, sum, c_out);
           8: adder_nbit #(8)   adder8(x, y, c_in, sum, c_out);
      default: adder_cla  #(WIDTH) cla(x, y, c_in, sum, c_out);
   endcase endgenerate
endmodule
```

■ **Review Questions**

6-16 What statements can be used within a **generate** region?
6-17 Describe the features of the generate-loop statement.
6-18 Describe the features of the generate-if statement.
6-19 Describe the features of the generate-case statement.
6-20 Can the generate-loop statement be nested?
6-21 What elements cannot be used within a **generate** region?

6.3 Configurations

A *configuration* is a set of rules that maps instances to cells. A cell can be a UDP, a configuration, or a module. The collection of compiled cells are stored in a place called a *library*. The operation of mapping an instance to a cell is referred to as *binding*. Hence, a configuration is a set of instance binding rules. The advantage of using configurations is that it may specify a new set of instance binding rules without having to change the source description of a design. In this section, we discuss how to specify a target library for source descriptions and how to use configurations to bind a design to cells associated with different libraries.

6.3.1 Library

A *library* is a logical collection of compiled source descriptions. It can be a logical or a symbolic library. A list of such library declarations (also called library specifications) is specified in a *library map file*. Multiple library map files are allowed. The library map file is read by the Verilog HDL parser before parsing any source code. A tool that is compatible with Verilog HDL must provide a mechanism such as a command line to specify one or more library map files. If multiple map files are specified, then they are read in the order specified.

For convenience, suppose that a file called lib.map already exists in the current working directory, which is readily read by the parser before parsing any source files. A *library declaration* is used to specify a target library into which cells in a source file can be stored after they are compiled. A library declaration uses the keyword **library** and has the following syntax

   library lib_name file_path;

where file_path may use an absolute or a relative path of files. Single character (?) and multiple-character wildcards (*) are allowed. In addition, "..." denotes matching any number of hierarchical directories; ".." specifies the parent directory; "." specifies the current directory, namely, the directory containing the lib.map file. A source file which does not specify a target library is compiled into the work library by default.

The **include** command is used to insert a library map file into another file during parsing. It has the following form:

   include <lib_map_file_path>;

The following is an example of a library map file:

```
// file: my_lib.map
library my_lib    "*.v";
library gate_lib  "./*.vg";
include "../my_lib_definitions";
library homework_lib "./homework*.v";
```

The library map file my_lib.map contains three library declarations and one include statement. The three library declarations create three different libraries: my_lib, gate_lib, and homework_lib. All files in the current directory with a suffix ".v" are compiled into my_lib, all files in the current directory with a suffix ".vg" are compiled into gate_lib, and all files in the current directory with a form of "homework*.v" are compiled into homework_lib.

■ Review Questions

6-22 What is a configuration?
6-23 What is a library?
6-24 What is the meaning of binding?

6-25 What is the meaning of library declarations?
6-26 What is a library map file?
6-27 What is the function of the **include** command in a library mapfile?

6.3.2 Basic Configuration Elements

A configuration is simply a set of rules that are used to search for library cells to which instances bind. Like a module or a UDP, a configuration is a design element at the top level in Verilog HDL. A configuration comprises four basic elements: **design, default, instance**, and **cell** statements. The **design** statement specifies the top-level module in the design and the source description to be used. The **default** statement coupled with the **liblist** statement specifies, by default, which library or libraries, are used to bind instances. The **instance** and **cell** statements override the default rule and specify from which the source description of a particular instance and cell are taken.

A configuration is defined outside a module by using a pair of keywords **config** and **endconfig**. It has the form

```
config config_identifier;
   design    {[lib_name.]cell_id};
   {default   liblist lib_list;}
   {instance instance_name use lib.cell [:config];}
   {instance instance_name liblist lib_list;}
   {cell cell_name use lib.cell [:config];}
   {cell cell_name liblist lib_list;}
endconfig
```

The optional **:config** extension is used explicitly to refer to a configuration that has the same name as a module.

In a configuration, the **design** statement must be unique, but it can include multiple top-level modules. The **design** statement specifies the top-level module or modules and should appear before any statements in the **config**. If lib_name is omitted, then the library which contains the **config** is used to search for the cell.

The **default** statement selects all instances which do not match a more specific selection statement. For simple design configurations, it might be sufficient to specify a **default liblist** statement. The **liblist** clause defines an ordered set of libraries to be searched to find the current instance.

The **instance** statement provides an exception to the default binding rules. It is used to map the specific instance to a cell in lib.cell or in a library searching from lib_list. The instance_name associated with the instance statement is the name of the cell in the design statement.

The **cell** statement specifies the cell to which it applies. Like in the **instance** statement, it can be specified in one of the two forms, the **use** clause and the **liblist** clause. The **use** clause specifies a specific binding for the selected cell. It can only be used in conjunction with an **instance** or a **cell** statement.

In the rest of this section, we give a number of examples to illustrate various configurations. For this purpose, we need to prepare a few libraries with different cells. An illustration for preparing three different libraries is as follows.

■ **Example 6.20: A configuration example.**

This program example includes three files: eight_bit_adder.v, four_bit_adder_gate.v, and four_bit_adder.v. The first file is the top-level module and instantiates two instances

Section 6.3 Configurations

of four_bit_adder. The four_bit_adder module is modeled at both the gate level and the RTL. At the gate level, the four_bit_adder module instantiates full_adder and half_adder in a descendant manner. At the RTL, the four_bit_adder module instantiates full_adder, which is modeled in dataflow style.

```verilog
// top file: eight_bit_adder.v
// RTL modeling: using module instantiation
module eight_bit_adder(
      input   [7:0] x, y,
      input   c_in,
      output  c_out,
      output  [7:0] sum);
wire    c_out_1;  // internal net
// instantiate two instances of a four_bit_adder
   four_bit_adder adder_a
                (x[3:0],y[3:0],c_in,c_out_1,sum[3:0]);
   four_bit_adder adder_b
                (x[7:4],y[7:4],c_out_1,c_out,sum[7:4]);
endmodule
// file: four_bit_adder_gate.v
// a gate-level hierarchical description of a 4-bit adder
module four_bit_adder(
      input   [3:0] x, y,
      input   c_in,
      output  [3:0] sum,
      output  c_out);
wire    c1, c2, c3; // intermediate carries
// the four_bit adder body
// instantiate the full adder
   full_adder fa_1 (x[0], y[0], c_in,   c1, sum[0]);
   full_adder fa_2 (x[1], y[1],   c1,   c2, sum[1]);
   full_adder fa_3 (x[2], y[2],   c2,   c3, sum[2]);
   full_adder fa_4 (x[3], y[3],   c3, c_out, sum[3]);
endmodule
// a gate-level description of a full adder
module full_adder(
      input x, y, c_in,
      output c_out, sum);
wire    s1, c1, c2;  // outputs of both half adders
// the full adder body
// instantiate the half adder
   half_adder ha_1 (x, y, c1, s1);
   half_adder ha_2 (c_in, s1, c2, sum);
   or (c_out, c1, c2);
endmodule
// a gate-level description of a half adder
module half_adder(
      input x, y,
      output c, s);
// the half adder body
   xor (s,x,y);
   and (c,x,y);
```

endmodule

```
// file: four_bit_adder.v
// RTL modeling: using module instantiation inside
// a generate-loop statement
module four_bit_adder(
      input   [3:0] x, y,
      input   c_in,
      output  c_out,
      output  [3:0] sum);
// specify the function of an N-bit adder
genvar  i;
wire    [2:0] c;   // internal carries declared as nets
generate for (i = 0; i < 4; i = i + 1) begin: adder
   if (i == 0)         // specify the LSB
      full_adder fa (x[i], y[i], c_in, c[i], sum[i]);
   else if (i == 3) // specify the MSB
      full_adder fa (x[i], y[i], c[i-1], c_out, sum[i]);
   else                        // specify other bits
      full_adder fa (x[i], y[i], c[i-1], c[i], sum[i]);
end endgenerate
endmodule
// define a full adder in dataflow style
module full_adder(input  x, y, c_in, output c_out, sum);
// specify the function of a full adder
   assign {c_out, sum} = x + y + c_in;
endmodule
```

The library map file `mylib.map` is listed as follows:

```
// file mylib.map
// In ModelSim, the following steps are used to build
// a library:
// step 1: create a new library, e.g., my_lib_top
//         vlib my_lib_top
// step 2: vlog -work my_lib_top
//              ../chapter06/src/eight_bit_adder.v
library my_lib_top
     ../chapter06/src/eight_bit_adder.v
library my_lib_gate
     ../chapter06/src/four_bit_adder_gate.v
library my_lib_rtl
     ../chapter06/src/four_bit_adder.v
```

The relationships of files, modules, and libraries of the underlying example are portrayed in Figure 6.4 for reference later in this section.

In the rest of this section, we discuss how to specify a configuration to meet the specific requirement of a particular application. A configuration can be in the same file as the library map file or as a separate file.

6.3.2.1 Default Configuration The libraries are searched in accordance with the library declaration order in the library map file when no configuration is specified. This means that all

Section 6.3 *Configurations*

```
my_lib_top                      my_lib_gate                      my_lib_rtl
eight_bit_adder.v               four_bit_adder_gate.v            four_bit_adder.v
eight_bit_adder()               four_bit_adder()                 four_bit_adder()
   four_bit_adder --- adder_a      full_adder --- fa_1              full_adder()
   four_bit_adder --- adder_b      full_adder --- fa_2
                                   full_adder --- fa_3
                                   full_adder --- fa_4
                                full_adder()
                                half_adder()
```

Figure 6.4: The relationships between files, modules, and libraries of the underlying example.

instances of the eight_bit_adder module use the cells from the library my_lib_gate (since my_lib_gate is placed before my_lib_rtl).

6.3.2.2 Using the Default Statement As mentioned previously, the **default liblist** statement can be used to change the library search order.

■ **Example 6.21: The use of default liblist statements.**

An illustration of using the **default** statement is given in the following example.

```
// configuration 1: using the default clause
config cfg_1;
   design  my_lib_top.eight_bit_adder
   default liblist my_lib_rtl my_lib_gate;
endconfig
```

The **default liblist** statement overrides the library search order in the mylib.map file, and so the my_lib_rtl library is always searched before the my_lib_gate library. Based on this configuration, the top-level module, eight_bit_adder, uses the following cells:

```
# Loading work.cfg_1
# Loading my_lib_top.eight_bit_adder
# Loading my_lib_rtl.four_bit_adder
# Loading my_lib_rtl.full_adder
```

All descendant cells of the top-level module are taken from the library my_lib_rtl.

6.3.2.3 Using the Cell Statement The **cell** statement selects a specified cell instance and explicitly binds it to a particular library cell or the first cell found from the specified library list.

■ **Example 6.22: The use of cell statements.**

As an example of the use of the **cell** statement, consider the following configuration cfg_2. The **cell** statement selects all cells with a name of full_adder and explicitly binds them to the gate representation in the library my_lib_gate.

```
// configuration 2: using the cell clause
config cfg_2;
   design  my_lib_top.eight_bit_adder
   default liblist my_lib_rtl my_lib_gate;
```

```
      cell full_adder use my_lib_gate.full_adder;
endconfig
```

Using this configuration, the cells used in the top-level module eight_bit_adder are listed as follows:

```
# Loading work.cfg_2
# Loading my_lib_top.eight_bit_adder
# Loading my_lib_rtl.four_bit_adder
# Loading my_lib_gate.full_adder
# Loading my_lib_gate.half_adder
```

Since the **liblist** is inherited, all of the descendants of full_adder inherit its **liblist** from the **cell** selection statement.

6.3.2.4 Using the Instance Statement
The **instance** statement selects a specified instance and explicitly binds it to a particular library cell or the first cell found from the specified library list.

■ **Example 6.23: The use of instance statements.**

As an illustration of the use of the **instance** statement, consider the following configuration cfg_3. The **instance** statement selects the instance with the name of adder_b and explicitly binds it to the gate representation in the library my_lib_gate. All descendants of this instance are taken from the same library. The other instance and its descendant are bound to the library my_lib_rtl specified by the **default liblist** statement.

```
// configuration 3: using the instance clause
config cfg_3;
   design   my_lib_top.eight_bit_adder
   default liblist my_lib_rtl my_lib_gate;
   instance eight_bit_adder.adder_b liblist my_lib_gate;
endconfig
```

Using this configuration, the cells used in the top-level module eight_bit_adder are as follows:

```
# Loading work.cfg_3
# Loading my_lib_top.eight_bit_adder
# Loading my_lib_rtl.four_bit_adder
# Loading my_lib_gate.four_bit_adder
# Loading my_lib_gate.full_adder
# Loading my_lib_gate.half_adder
# Loading my_lib_rtl.full_adder
```

6.3.2.5 Using the Instance And Cell Statements
When both **instance** and **cell** statements are used, the specified instance and cell instance are explicitly bound to a particular library cell or the first cell found from the specified library list.

■ **Example 6.24: The use of instance and cell statements.**

As an illustration of the use of **instance** and **cell** statements, consider the following configuration cfg_4. The instance named adder_b and the cells named full_adder in both instances are explicitly bound to the gate representation in the library my_lib_gate. However, the instance adder_a is bound to the library my_lib_rtl except the cell named full_adder instantiated within it.

Section 6.3 Configurations

```
// configuration 4: using the instance and cell statements
config cfg_4;
   design  my_lib_top.eight_bit_adder
   default liblist my_lib_rtl my_lib_gate;
   instance eight_bit_adder.adder_b liblist my_lib_gate;
   cell full_adder use my_lib_gate.full_adder;
endconfig
```

Based on this configuration, the cells used in the top-level module eight_bit_adder are as follows:

```
# Loading work.cfg_4
# Loading my_lib_top.eight_bit_adder
# Loading my_lib_rtl.four_bit_adder
# Loading my_lib_gate.four_bit_adder
# Loading my_lib_gate.full_adder
# Loading my_lib_gate.half_adder
```

6.3.2.6 Hierarchical Configurations Sometimes, it is desirable to specify a set of configuration rules for a subsection of a design. To see this, consider the following example.

■ Example 6.25: An example of hierarchical configurations.

Suppose that all work has only been on the gate-level module four_bit_adder_gate by itself and we want to use the my_lib_rtl.full_adder cell for fa_3, and the library my_lib_gate for the other cells. Then a possible configuration, called cfg_5, is created as follows:

```
// configuration 5: hierarchical configuration -- part1
config cfg_5;
   design   my_lib_gate.four_bit_adder
   default  liblist my_lib_gate my_lib_rtl;
   instance four_bit_adder.fa_3 liblist my_lib_rtl;
endconfig
```

Based on this configuration, the cells used in the top-level module four_bit_adder are as follows:

```
# Loading work.cfg_5
# Loading my_lib_gate.four_bit_adder
# Loading my_lib_gate.full_adder
# Loading my_lib_gate.half_adder
# Loading my_lib_rtl.full_adder
```

To use this configuration cfg_5 for the adder_b instance of eight_bit_adder and take the full default my_lib_rtl for the adder_a instance, the following configuration cfg_6 can be used:

```
// configuration 6: hierarchical configuration -- part 2
config cfg_6;
   design   my_lib_top.eight_bit_adder_tb  // test bench
            my_lib_top.eight_bit_adder;    // top-level module
   default  liblist my_lib_rtl my_lib_gate;
   instance eight_bit_adder.adder_b use work.cfg_5;
endconfig
```

The **instance** statement specifies the work.cfg_5 configuration and is used to resolve the bindings of the instance adder_b and its descendants. The **design** statement in the configuration cfg_5 defines the exact binding for the adder_b instance itself. The rest of cfg_5 defines the rules to bind the descendants of adder_b. Notice that the **instance** statement in cfg_5 is relative to its own top-level module, four_bit_adder.

```
# Loading work.cfg_6
# Loading my_lib_top.eight_bit_adder_tb
# Loading my_lib_top.eight_bit_adder
# Loading my_lib_rtl.four_bit_adder
# Loading my_lib_top.cfg_5
# Loading my_lib_rtl.full_adder
```

■ Review Questions

6-28 What are the three top-level design elements in Verilog HDL?
6-29 What are the basic elements of a configuration?
6-30 Describe the function of the **design** statement in a configuration.
6-31 Describe the function of the **default** statement in a configuration.
6-32 Describe the function of the **instance** statement in a configuration.
6-33 Describe the function of the **cell** statement in a configuration.

6.4 Summary

In order to build a large design, a mixed style is usually used. Through the use of instantiation, a hierarchical structure is formed. Instantiation is a process that gate and switch primitives, UDP primitives, and modules are embedded into other modules. Higher-level modules create instances of lower-level modules and communicate with them through input, output, and bidirectional ports, which can be scalar or vector.

When a module instantiates another module, the higher-level module can change the values of parameters defined by the **parameter** declaration within the lower-level module. This operation is called parameter overriding. Two ways can be used to override parameter values. These are the **defparam** statement and the module instance parameter value assignment. In the **defparam** statement method, the parameter values defined by the **parameter** declaration in any module are overridden by using the hierarchical path name of the parameter whenever the module is instantiated. In the method of the module instance parameter value assignment, the parameters defined by the **parameter** declaration within a module are overridden by the parameters passed through parameter ports whenever the module is instantiated.

In order to build a large regular design, it is more constructive to use iterative logic modules. To model such a design, a useful and powerful **generate** region, defined by a pair of keywords **generate** and **endgenerate**, is provided by Verilog HDL. A **generate** region allows selection and replication of some statements during elaboration time. By using **generate** regions, one or multiple dimensional iterative logic structures, such as $m \times n$ array multipliers and $m \times n$ array dividers, can be easily described.

The power of **generate** regions is that they can conditionally generate declarations and instances of modules into a description. Almost what can be put inside a module can also be placed within **generate** regions except for parameters, and port declarations. The three most widely used generate-statements are generate-loop, generate-if, and generate-case.

A configuration is a set of rules that map instances to cells. The advantage of using configurations is that it may specify a new set of instance binding rules without having to change the source description of a design. Like a module, a configuration is a design element at the top level. A configuration comprises four basic elements: **design**, **default**, **instance**, and **cell** statements. The **design** statement specifies the top-level module in the design and the source description to be used. The **default** statement along with the **liblist** statement specifies which library or libraries are used to bind instances by default. The **instance** and **cell** statements override the default rule and specify from which the source description of a particular instance and cell are taken.

References

1. J. Bhasker, *A Verilog HDL Primer*, 3rd ed., Star Galaxy Publishing, 2005.
2. IEEE 1364-2001 Standard, *IEEE Standard Verilog Hardware Description Language*, 2001.
3. IEEE 1364-2005 Standard, *IEEE Standard for Verilog Hardware Description Language*, 2006.
4. M. B. Lin, *Digital System Design: Principles, Practices, and Applications*, 4th ed., Chuan Hwa Book Ltd. (Taipei, Taiwan), 2010.
5. S. Palnitkar, *Verilog HDL: A Guide to Digital Design and Synthesis*, 2nd ed., SunSoft Press, 2003.

Problems

6-1 Write a parameterizable module that counts the number of ones in a default byte. Write a test bench to instantiate it with a parameter port value of 16 and verify the results.

6-2 Write a parameterizable module that counts the number of trailing ones in a default byte. Write a test bench to instantiate it with a parameter port value of 16 and verify the results.

6-3 Write a parameterizable module that counts the number of leading zeros in a default byte. Write a test bench to instantiate it with a parameter port value of 16 and verify the results.

6-4 Write a parameterizable module that counts the number of leading ones in a default byte. Write a test bench to instantiate it with a parameter port value of 16 and verify the results.

6-5 Write a parameterizable module that computes the even parity of a default 8-bit number. Write a test bench to instantiate it with a parameter port value of 16 and verify the results.

6-6 Write a parameterizable module that computes the odd parity of a default 8-bit number by using the port-list declaration style. Write a test bench to instantiate it with a parameter port value of 16 and verify the results.

6-7 Use a generate-loop statement to instantiate the **xor** gate primitive and model a parameterizable even-parity checker module with a default size of 8. Assume that a linear structure is used. Write a test bench to instantiate it with a parameter port value of 16 and verify the results.

6-8 Use a generate-loop statement to instantiate the **xor** gate primitive and model a parameterizable odd-parity checker module with a default size of 8. Assume that a linear structure is used. Write a test bench to instantiate it with a parameter port value of 16 and verify the results.

6-9 Use a generate-loop statement to instantiate the **xor** gate primitive and model a parameterizable even-parity checker module with a default size of 8. Assume that a binary-tree structure is used. Write a test bench to instantiate it with a parameter port value of 16 and verify the results.

6-10 Use a generate-loop statement to instantiate the **xor** gate primitive and model a parameterizable odd-parity checker module with a default size of 8. Assume that a binary-tree structure is used. Write a test bench to instantiate it with a parameter port value of 16 and verify the results.

6-11 Assume that an increment-by-1 circuit is required; that is, the circuit always adds a constant 1 to an operand. Answer each of the following questions.
   (a) Derive a 1-bit cell of the circuit.
   (b) Use a generate-if statement to instantiate the 1-bit cell and describe the parameterizable increment-by-1 module. Write a test bench to instantiate it with a parameter port value of 8 and verify the results.
   (c) Redo (b) by using **assign** continuous assignments instead of instantiation.
   (d) Redo (b) by using **always** blocks instead of instantiation.
   (e) Define the 1-bit cell as a task, redo (b) by using **always** blocks to invoke the task.
   (f) Define the 1-bit cell as a function, redo (b) by using **always** blocks to call the function.

6-12 Assume that a decrement-by-1 circuit is required; that is, the circuit always subtracts a constant 1 from an operand. Answer each of the following questions.
   (a) Derive a 1-bit cell of the circuit.
   (b) Use a generate-if statement to instantiate the 1-bit cell and describe the parameterizable decrement-by-1 module. Write a test bench to instantiate it with a parameter port value of 8 and verify the results.
   (c) Redo (b) by using **assign** continuous assignments instead of instantiation.
   (d) Redo (b) by using **always** blocks instead of instantiation.
   (e) Define the 1-bit cell as a task, redo (b) by using **always** blocks to invoke the task.
   (f) Define the 1-bit cell as a function, redo (b) by using **always** blocks to call the function.

6-13 What follows is a module that Professor Idiot wants to describe a parameterizable buffer. Nevertheless, he was told that something is wrong with it from a synthesis tool.

```
module buffer #(parameter numbits = 8)
              (input [numbits-1:0] a,
               output [numbits-1:0] y);

   wire [numbits-1:0] x;

   not #(numbits) i1(a, x);
   not #(numbits) i2(x, y);
endmodule
```

Problems **225**

     (a) What are the problems of the above module?

     (b) Correct the module so that it can work as desired.

6-14 Assume that we want to design an n-digit BCD adder. The details of a single-digit BCD adder can be referred to in Section 3.4.

     (a) Write a module to describe this n-digit BCD adder in behavioral style.

     (b) Write a test bench to verify whether the BCD adder behaves correctly.

6-15 Assume that we want to design an n-digit BCD subtracter. The details of a single-digit BCD subtracter can be referred to Problem 3-16.

     (a) Write a module to describe this n-digit BCD subtracter in behavioral style.

     (b) Write a test bench to verify whether the BCD subtracter behaves correctly.

6-16 Assume that an n-digit BCD adder/subtracter is desired. If the mode selection input `mode` is 0, the BCD adder/subtracter functions as a BCD adder; otherwise, it functions a BCD subtracter.

     (a) Write a module to describe this n-digit BCD adder/subtracter in behavioral style.

     (b) Write a test bench to verify whether the BCD adder/subtracter behaves correctly.

7

Advanced Modeling Techniques

WE have described the basic features of Verilog HDL in the previous chapters. In this chapter, we address a number of additional features of Verilog HDL. These features include block constructs, procedural continuous assignments, specify blocks, timing checks, and compiler directives.

The block constructs, including sequential blocks and parallel blocks, are used to group together two or more statements so that they can act as a single statement. The procedural statements within a sequential block are executed sequentially in the given order but within a parallel block are executed concurrently irrespective of their relative order.

The procedural continuous assignments allow expressions to be driven continuously onto variables or nets. They can only be used within **initial** and **always** blocks. Procedural continuous assignments can be cast into two types: **assign/deassign** and **force/release**. The pair of **assign** and **deassign** procedural continuous assignments assign and deassign values to variables. The pair of **force** and **release** procedural continuous assignments assign and deassign values to nets or variables.

The specify blocks are used to define the module paths accompanied with the propagation delays of cells in ASIC cell libraries. In addition, they provide mechanisms for checking the timing constraints of modules.

Like C language, compiler directives are also defined in Verilog HDL. With the aid of compiler directives, a macro text may be defined, part of a design can be compiled conditionally, default net features can be changed, and unconnected input ports may be pulled up or pulled down.

7.1 Sequential and Parallel Blocks

Block statements are used to group together two or more procedural statements so that they can act as a single (complex) statement. There are two types of block statements: the *sequential block* and the *parallel block*. In this section, we deal with these two block statements in detail.

7.1.1 Sequential Blocks

A sequential block uses a pair of keywords **begin** and **end** to group procedural statements. We often refer to it as a *begin-end block*. The procedural statements within a sequential block are executed sequentially in the given order. Timing control can be used for each procedural statement to control the execution of these procedural statements relative to the simulation time of its previous procedural statement. Once a sequential block completes its execution, the execution will continue to the procedural statements following the sequential block. In other words, the control is only passed out a sequential block after the sequential block completes the execution of its last procedural statement.

The sequential block must be placed within an **initial** or **always** block. The general syntax of a sequential block is as follows

> **begin** [:block_id {block_item_declaration}]
>   {statement}
> **end**

where block_item_declaration can be one of variables, events, local parameters, and parameters. Notice that block_item_declaration can only be used within a named block, meaning that the block has a block name (block_id). In other words, the block needs to have a name before the block_item_declaration statement can be used to declare variables within it.

In a sequential block, timing control (including regular delay control and intra-assignment delay control) can be applied to schedule the executions of procedural statements within the sequential block. To illustrate the effects of regular delay control on procedural statements in a sequential block, let us consider the following two examples. The first example only considers blocking assignments, while the second one deals with only nonblocking assignments.

■ Example 7.1: A sequential block with blocking assignments.

In this example, all procedural statements are blocking assignments. Variable x is assigned to 1 at simulation time 0, and variables y and z are assigned to 1 and 0 at simulation times 12 and 32 (why?), respectively.

```
module sequential_block_blocking;
reg x, y, z;

initial begin
        x = 1'b0;    // execute at time 0
    #12 y = 1'b1;    // execute at time 12
    #20 z = 1'b0;    // execute at time 32
end
endmodule
```

It is interesting to note that the same results will be obtained if nonblocking assignments are used in place of the blocking assignments in the above example. To manifest this, consider the following example, in which only nonblocking assignments are considered.

■ Example 7.2: A sequential block with nonblocking assignments.

The following **initial** block has the same effect as in the preceding example in which blocking assignments are used. The reason is because, as mentioned, regular delay control defers the execution of the entire statement. Hence, in sequential blocks, both blocking and nonblocking assignments yield the same result. The reader is encouraged to run simulations of both examples and compare their simulation results.

```
module sequential_block_nonblocking;
reg x, y, z;

initial begin
        x <= 1'b0;   // execute at time 0
```

Section 7.1 *Sequential and Parallel Blocks*

```
    #12 y <= 1'b1;    // execute at time 12
    #20 z <= 1'b0;    // execute at time 32
end
endmodule
```

From the above two examples, we can realize that for a sequential block, the same results are obtained if regular delay control is used irrespective of whether blocking assignments or nonblocking assignments are considered. However, this is not true under intra-assignment delay control, which will yield different results. Refer to in Section 4.3.1 for more details.

■ Review Questions

7-1 What are the two types of block constructs in Verilog HDL?
7-2 Describe the features of sequential blocks.
7-3 How would you define a sequential block?
7-4 What is the major purpose of the block identifier?
7-5 What data types can be declared within sequential blocks?
7-6 Can intra-assignment delay control be used in sequential blocks?

7.1.2 Parallel Blocks

A parallel block uses a pair of keywords **fork** and **join** to group procedural statements. It is also called a *fork-join block*. The procedural statements within a parallel block are executed concurrently regardless of their relative order. However, timing control can be used to schedule the execution of procedural assignments. The delay for each procedural statement is relative to the simulation time when the block starts its execution. When the last activity, but not necessary the last procedural statement, in a parallel block completes its execution, the execution continues to the procedural statements following the parallel block. In other words, the control is passed out a parallel block only after all procedural statements in the parallel block complete their executions.

Like the sequential block, the parallel block must be placed within an **initial** or **always** block. The general syntax of a parallel block is as follows

    fork [:block_id {block_item_declaration}]
        {statement}
    join

where block_item_declaration can be one of variables, events, local parameters, and parameters. Recall that block_item_declaration can only be used within a named block. In other words, the block must have a name before the block_item_declaration statement may be used to declare variables.

Timing control can be applied in a parallel block to schedule the execution of procedural statements. The following example shows how regular delay control is used to schedule the execution of procedural statements in a parallel block.

■ Example 7.3: A parallel block with blocking assignments.

In this example, all procedural statements are blocking assignments. Variable x is assigned to 0 at simulation time 0, and variables y and z are assigned to 1 and 0 at simulation times 12 and 20, respectively. It is instructive to compare the result of this example with that of the case in the sequential block.

```
module parallel_block_blocking;
reg x, y, z;

initial fork
        x = 1'b0;      // execute at time 0
    #20 z = 1'b0;      // execute at time 20
    #12 y = 1'b1;      // execute at time 12
join
endmodule
```

In the above example, we use blocking assignments. However, the result remains the same if all blocking assignments are replaced with nonblocking assignments. To see this, consider the following example.

■ Example 7.4: A parallel block with nonblocking assignments.

In this example, all procedural statements are nonblocking assignments. Variable x is assigned to 0 at simulation time 0, and variables y and z are assigned to 1 and 0 at simulation times 12 and 20, respectively. It is instructive to compare the result of this example with that of the case in the sequential block.

```
module parallel_block_nonblocking;
reg x, y, z;

initial fork
        x <= 1'b0;     // execute at time 0
    #12 y <= 1'b1;     // execute at time 12
    #20 z <= 1'b0;     // execute at time 20
join
endmodule
```

■ Example 7.5: Parallel block and intra-assignment delay control.

In this example, all procedural statements are blocking assignments. Intra-assignment delay control is used in the parallel block to schedule the execution of three blocking assignments. Variable x is assigned to 0 at simulation time 0, and variables y and z are assigned to 1 and 0 at simulation times 12 and 20, respectively. Note that the result of this example is identical to that of the preceding two examples.

```
module parallel_block_blocking;
reg x, y, z;

initial fork
    x = 1'b0;          // execute at time 0
    z = #20 1'b0;      // execute at time 20
    y = #12 1'b1;      // execute at time 12
join
endmodule
```

In the above example, we use blocking assignments and intra-assignment delay control instead of regular delay control. However, the result remains the same if all blocking assignments are replaced with nonblocking assignments. This is due to the inherent feature of parallel blocks.

Example 7.6: A parallel block with nonblocking assignments.

In this example, all procedural statements are nonblocking assignments. Intra-assignment delay control is used in the parallel block. Variable x is assigned to 0 at simulation time 0, and variables y and z are assigned to 1 and 0 at simulation times 12 and 20, respectively. Note that the result of this example is identical to that of the preceding three examples.

```
// an illustration of a parallel block with
// nonblocking assignments
module parallel_block_nonblocking;
reg x, y, z;
initial fork
    x <= 1'b0;          // execute at time 0
    y <= #12 1'b1;      // execute at time 12
    z <= #20 1'b0;      // execute at time 20
join
endmodule
```

For a parallel block, all statements within it are examined at the simulation time when the block is entered and are scheduled to execute according to their delay control. Based on this, the above four examples yield the same result, even in the case that intra-assignment delay control is used instead of regular delay control. Therefore, we may conclude that the results are the same regardless of whether assignments are blocking or nonblocking and irrespective of whether regular or intra-assignment delay control is used in a parallel block.

Review Questions

7-7 Describe the features of parallel blocks.
7-8 How would you define a parallel block?
7-9 What are the differences between sequential and parallel blocks?
7-10 What types of elements can be declared within a parallel block?
7-11 Can intra-assignment delay control be used in parallel blocks?
7-12 Are there any differences between regular and intra-assignment delay control in parallel blocks?

7.1.3 Special Features of Blocks

In this subsection, we consider two special features of blocks: the *named block* and the *nested block*.

7.1.3.1 Named Blocks As mentioned previously, both sequential and parallel blocks can be given a name by adding :block_id after the keywords **begin** and **fork**, respectively. This type of block is called a named block. A named block has the following features:

1. It allows local variables, parameters, local parameters, and named events to be declared within the block. However, all local variables declared are static; namely, their values remain valid during the entire simulation time.
2. It allows the block to be referenced in procedural statements, such as the **disable** statement.
3. It can be disabled by the **disable** statement.
4. A local variable declared in a named block can be referenced through using its hierarchical path name.

■ Example 7.7: A named block.

In this example, the identifier test is the block name. Variables x, y, and z are locally associated with the block. Since this is a sequential block, the variable z is assigned a logic 1 at simulation time 22. To refer to a **reg** variable associated locally with the **begin-end** block test, such as y, from the outside of the block, a hierarchical path name, test.y, must be used.

```
module named_block;
wire w;

initial begin: test   // test is the block name
reg x, y, z;          // local variables
    x = 1'b0;         // execute at time 0
 #12 y = 1'b1;        // execute at time 12
 #10 z = 1'b1;        // execute at time 22
end
assign w = test.y;
endmodule
```

7.1.3.2 Nested Blocks Blocks can be nested and both sequential and parallel blocks can be mixed. For example, a **begin-end** block may be embedded into a **fork-join** block or another **begin-end** block. Similarly, a **fork-join** block may be encapsulated into another **fork-join** block or a **begin-end** block. Note that in a nested block, the entire inner block is considered as a single (complex) statement from the viewpoint of the outer block.

■ Example 7.8: A nested block.

In this example, a **fork-join** block is embedded into a **begin-end** block. From the viewpoint of the **begin-end** block, the **fork-join** block is regarded as a single statement. Hence, the inner block starts at simulation time 0 and finishes at simulation time 20. As a consequence, the variable x is assigned with a logic 1 at simulation time 45.

```
module nested_block;
reg x, y, z;

initial begin
    x = 1'b0;            // execute at time 0
    fork // parallel block -- enter at time 0
       #12 y <= 1'b1;    // execute at time 12
       #20 z <= 1'b0;    // execute at time 20
    join                 // leave at time 20
    #25 x = 1'b1;        // execute at time 45
end
endmodule
```

■ Review Questions

7-13 Describe the features of named blocks.

7-14 Give an example of a nested block by putting a sequential block inside a parallel block and explain its operations.

7-15 Give an example of a nested block by putting a parallel block inside a sequential block and explain its operations.
7-16 How would you refer to a variable declared within a named block?
7-17 Can intra-assignment delay control be used in nested blocks?

7.1.4 The disable Statement

A **disable** statement is a procedural statement; namely, it can only be used within an **initial** or **always** block. The **disable** statement beginning with the keyword **disable**. It is usually used to terminate a task or a named block, sequential or parallel, before which completes the execution of all its statements. The **disable** statement can be used to get out of a loop, to handle an error condition, to control the execution of a piece of code, and to model a global reset or a hardware interrupt, based on a control signal.

The general form of the **disable** statement is as follows

```
disable hierarchical_block_id;
disable hierarchical_task_id;
```

where hierarchical_block_id and hierarchical_task_id are the hierarchical path names of the named block and the task to be disabled, respectively. After a **disable** statement is executed, the execution continues with the next statement following the block being disabled or the task call.

7.1.4.1 Disabling Named Blocks To illustrate how the **disable** statement can be used to terminate the execution of a named block, consider the following sequential block. According to the operation of the **disable** statement, the third statement within the test block is never executed because it is disabled by the **disable** statement. Nonetheless, the statement "c = b;" following the test block will be executed immediately following the execution of the **disable** statement.

```
begin: test
   a = b;
   disable test;
   c = a; // never execute
end
c = b;   // execute after the disable statement
```

Another illustration of the use of the **disable** statement is given in the following example.

■ Example 7.9: An example of disabling a named block.

In this example, the **disable** statement is controlled by a condition. The **disable** statement disables the test block whenever the flag is true and has no effect on the block otherwise.

```
initial begin: test            // test is the block name
   while (i < 10) begin
      if (flag) disable test;  // the test block is disabled
      #5 i = i + 1;            // if flag is true
   end
end
```

7.1.4.2 Disabling Tasks Although the **disable** statement can be used to disable a task, it is not recommended to use for this purpose, in particular, when the task returns output values, because the values of the output and bidirectional arguments are not specified when a task is disabled. As an illustration of this, consider the following example.

■ Example 7.10: An example of disabling a task.

In this example, the **disable** statement is used as an early return from a task. In this case, the return value x is ensured to be assigned a value of 164 when the "flag == 0" is true.

```
task test (input flag, output reg [7:0] x);
begin: test_block
   x = 164;
   if (flag == 0) disable test_block;
end
endtask
```

However, if we replace the **disable** statement with the following statement

> disable test;

then the return value x is not specified after the **disable** statement is executed.

■ Review Questions

7-18 What are the functions of the **disable** statement?
7-19 What is the next statement to be executed after a **disable** statement is executed?
7-20 Why is it not suggested to disable a task by using the **disable** statement?
7-21 How would you model a hardware interrupt with the **disable** statement?

7.2 Procedural Continuous Assignments

The procedural continuous assignments allow expressions to be driven continuously onto variables or nets. They are procedural statements and hence may only appear within **initial** and **always** blocks. Two types of procedural continuous assignments are **assign/deassign** and **force/release**. In this section, we deal with each of these in detail.

7.2.1 assign and deassign Statements

The **assign** statement overrides the effects of all regular procedural assignments on a variable. Its effect is ended by a **deassign** statement. The variable remains the continuously assigned value after being deassigned until it is changed by a future procedural assignment. If an **assign** statement is applied to a variable which already has an **assign** statement, then it deassigns the variable before making a new procedural continuous assignment to that variable. In short, the **assign** statement behaves like an **assign** continuous assignment except that it is only active a limited period of time set by a pair of the **assign** statement and a **deassign** statement.

The **assign** and **deassign** statements have the syntax as follows:

```
assign    variable_assignment
deassign  variable_lvalue
```

The left-hand side of the **assign** statement must be a variable or a concatenation of variables. It cannot be a memory word (array) or a bit-select or a part-select of a variable. The right-hand side of the **assign** statement can be an expression, treated as a continuous assignment.

Section 7.2 *Procedural Continuous Assignments*

■ **Example 7.11: The use of the assign and deassign statements.**

As an illustration of the use of a pair of **assign** and **deassign** statements, consider a *D* flip-flop with asynchronous reset. The outputs q and qbar are assigned to new values by regular procedural assignments at each negative edge of the clock signal whenever the reset input is low. When the reset input is high, both q and qbar are individually assigned to values 0 and 1 by using two **assign** statements. When the reset is low, both q and qbar are deassigned so that the values of q and qbar are allowed to be changed at the next negative edge of the clock signal.

```
// a negative-edge-triggered D flip-flop with
// asynchronous reset
module edge_dff(input clk, reset, d, output reg q, qbar);
always @(negedge clk) begin // regular assignments
   q <= d; qbar <= ~d; end
always @(reset)
   // override the regular assignments to q and qbar
   if (reset) begin
      assign q = 1'b0;
      assign qbar = 1'b1; end
   else begin // end the effect of assign
      deassign q;
      deassign qbar; end
endmodule
```

Even though the use of the pair of **assign** and **deassign** statements can correctly model a *D* flip-flop, the resulting module is quite cumbersome and difficult to understand, especially for the naive reader. For this reason, the use of the pair of **assign** and **deassign** constructs are now considered as a bad coding style. In Section 9.1, we introduce a better way to describe *D* flip-flops without using these two procedural continuous assignments.

■ **Review Questions**

7-22 What are the two types of procedural continuous assignments?
7-23 Can the **assign** and **deassign** statements assign values to nets?
7-24 What will happen if a variable is assigned by an **assign** statement but does not end with a **deassign** statement?
7-25 Can a concatenation operator be used as the left-hand side of an **assign** statement? What are the constraints in this situation?

7.2.2 force and release Statements

Another type of procedural continuous assignment is the pair of the **force** and **release** statements. They work in a way similar to the pair of **assign** and **deassign** statements except that they can be applied to nets as well. A **force** statement on a variable overrides any procedural assignments or procedural continuous assignments on the variable until the variable is released by a **release** statement. The variable remains the forced value after being released until a future procedural assignment changes it. Note that the pair of **force** and **release** statements overrides the effects of the pair of **assign** and **deassign** statements.

The **force** and **release** statements have the syntax as follows:

```
force    variable_assignment
```

Table 7.1: Comparison of various assignments in Verilog HDL.

Data type	Primitive output	Continuous assignment	Procedural assignment	**assign/ deassign**	**force/ release**
Net	Yes	Yes	No	No	Yes
Variable	Yes (Sequential UDP)	No	Yes	Yes	Yes

```
release variable_lvalue
force   net_assignment
release net_lvalue
```

The left-hand side of the **force** statement can be a variable or a net, a constant bit-select of a vector net, a part-select of a vector net, or a concatenation of the above nets. It cannot be a memory word (array) or a bit-select or a part-select of a vector variable. The right-hand side of the **force** statement can be an expression, which is treated as a continuous assignment.

A **force** statement on a net overrides all continuous assignments until the net is released by a **release** statement. The net will immediately return to its normal driven value when it is released. In summary, the **force** statement on a net behaves like an **assign** continuous assignment except that it is only active a limited period of time set by a pair of the **force** statement and a **release** statement.

Table 7.1 compares various assignments in Verilog HDL, including the primitive output, the **assign** continuous assignment, the procedural assignment, the pair of the **assign** and **deassign** statements, and the pair of the **force** and **release** statements.

The following example simply illustrates the use of the pair of **force** and **release** statements to temporarily force some net to a specific value in debugging a module.

■ Example 7.12: The use of the force and release statements.

As an illustration of using a pair of **force** and **release** statements to debug a design, consider the following module. In this example, an AND gate instance and_1 is replaced by an OR gate implemented by a **force** statement that forces its outputs to the value of its logical OR of inputs between the time units 10 and 20 during simulation. The AND gate functions at the other simulation times.

```
'timescale 1ns/100ps
module force_release_test_tb;
// internal signals declarations
reg  w, x, y;
wire f;
   and and_1 (f, w, x, y);

initial begin
   $monitor("%d, f = %b", $stime, f);
   w = 1;
   x = 0;
   y = 1;
   #10 force f = (w | x | y);
   #10 release f;
   #10 $finish;
end
endmodule
```

Section 7.3 Delay Models and Timing Checks

By using the pair of **force** and **release** statements, an assignment on a net or variable can be temporarily replaced with the other expression a specific amount of time. That is, the pair of the **force** and **release** statements allow us to debug a design without changing the actual contents of the design before we ensure where the bugs are. For this reason, the pair of **force** and **release** statements are often referred to as *test-bench statements*.

■ Review Questions

7-26 Can **force** and **release** statements assign values to variables?
7-27 When would you use **force** and **release** statements?
7-28 What are the differences between **assign** and **force** statements?
7-29 Explain the operations of the **force** statement.
7-30 Explain the operations of the **release** statement.
7-31 Can the right-hand side of a **force** statement be an expression?
7-32 What will happen if a variable is assigned by a **force** statement but is not ended with a **release** statement?
7-33 Can a concatenation operator be used as the left-hand side of a **force** statement? What are the constraints in this case?

7.3 Delay Models and Timing Checks

In a realistic hardware module, propagation delays are often associated with gates and wires. In addition, when accessed by an external module, a module has to verify whether the access timing is confined to its specifications. To this end, a *specify block* is defined. In this section, we address these functions by first introducing the delay model supported by Verilog HDL and then describe the definition of specify blocks as well as timing checks.

7.3.1 Delay models

Three delay models are often used to model the propagation delays in a realistic hardware module: the *distributed delay model*, the *lumped delay model*, and the *path delay model*.

7.3.1.1 Distributed Delay Model The distributed delay model assigns the propagation delays of signal events from inputs to an output gate to each individual gate and net along the path. Recall that a signal event is a value change on a net or variable. In the distributed delay model, delays are associated with individual elements, gates, cells, and wires. Hence, in this model, delays are specified on a per-element basis. An example used to illustrate the distributed delay model of a module is shown in Figure 7.1.

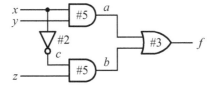

Figure 7.1: An example of the distributed delay model.

The distributed delays can be modeled by assigning one delay value to each individual gate or by assigning a delay value in each **assign** continuous assignment describing a gate. Two examples follow to illustrate these modeling styles.

■ Example 7.13: The distributed delay model in structural style.

As an example of modeling a gate-level logic circuit in structural style with the distributed delay model, consider the circuit in Figure 7.1. The details are revealed in the following module. Each gate instance associates with a specified delay value.

```
module test_structural(input x, y, z, output f);
wire a, b, c;
// modeling distributed delays in structural style
    and  #5  a1 (a, x, y);
    not  #2  n1 (c, x);
    and  #5  a2 (b, c, z);
    or   #3  o1 (f, a, b);
endmodule
```

■ Example 7.14: The distributed delay model in dataflow style.

In this example, we simply rewrite the preceding gate-level module with **assign** continuous assignments, one for each gate. Each **assign** continuous assignment associates with a specified delay value. This results in a module in dataflow style.

```
module test_dataflow(input x, y, z, output f);
wire a, b, c;
// modeling distributed delays in dataflow style
    assign  #5  a = x & y;
    assign  #2  c = ~x;
    assign  #5  b = c & z;
    assign  #3  f = a | b;
endmodule
```

7.3.1.2 Lumped Delay Model The lumped delay model assigns the propagation delays of signal events from inputs passing through all possible paths to an output gate. In other words, delays are associated with the entire module. In the lumped delay model, delays are specified on a per-output basis; namely, a single delay is assigned to each output. To do this, the worst-case cumulative delays of all possible paths destined to the same output gate from different inputs are calculated and then lumped at that output gate. An example used to illustrate the lumped delay model of a module is shown in Figure 7.2.

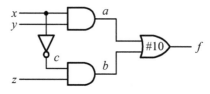

Figure 7.2: An example of the lumped delay model.

The lumped delay can be modeled by assigning a single delay value to the output gate or by using a delay value only in the **assign** continuous assignment describing the output gate. Two examples follow to illustrate these modeling styles.

Section 7.3 *Delay Models and Timing Checks*

■ Example 7.15: The lumped delay model in structural style.

An illustration of modeling the gate-level logic circuit in Figure 7.2 in structural style with the lumped delay model is concerned in this example. Only the output gate instance associates with a specified delay value. This delay is the worst-case value of all possible paths from any input to the output.

```
module test_structural(input x, y, z, output f);
wire a, b, c;
// modeling lumped delays in structural style
    and     a1 (a, x, y);
    not     n1 (c, x);
    and     a2 (b, c, z);
    or  #10 o1 (f, a, b);
endmodule
```

■ Example 7.16: The lumped delay model in dataflow style.

As an illustration of modeling the logic circuit shown in Figure 7.2 in dataflow style with the lumped delay model, we rewrite the preceding example with **assign** continuous assignments, one for each gate. Only the **assign** continuous assignment for the output gate associates with a delay value.

```
module test_dataflow(input x, y, z, output f);
wire a, b, c;
// modeling lumped delays in dataflow style
    assign      a = x & y;
    assign      c = ~x;
    assign      b = c & z;
    assign #10  f = a | b;
endmodule
```

7.3.1.3 Path Delay Model In the path delay model, the propagation delays of signal events from inputs to an output are individually assigned to each path from an input to the output. The input may be an input port or a bidirectional (inout) port and the output may be an output port or a bidirectional port. Hence, a path delay is specified on a pin-to-pin (or port-to-port) basis. This type of path will be called the *module path* later to distinguish it from the other types of paths.

A module path specifies the time it takes signal events to propagate from an input port to an output port along all possible paths. It is often the case that each module path may contain many internal paths. In this case, the worst-case internal path delay is taken as the delay of the module path. An example used to illustrate the path delays of a module is shown in Figure 7.3. The path delay can be modeled in Verilog HDL by using a specify block, which will be introduced in the next subsection.

In summary, the lumped delay model is the easiest one to use among the three delay models when a module is considered as a unit. Nevertheless, it only provides the worst-case delay of all possible path combinations of input and output ports. The path delay model is easier to use than the distributed delay model in that it need not know the details of the internal structure of a module. This feature makes the delay calculation much easier than the other two delay models so that the path delay model is the most popular one, in particular, in describing the delays of ASIC cells.

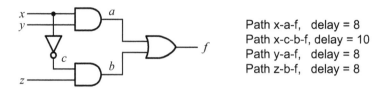

Figure 7.3: An example of the path delay model.

■ Review Questions

7-34 What are the delay models often used to model the delays of a hardware module?
7-35 What are the features of the distributed delay model? How would you model this in dataflow style?
7-36 What are the features of the lumped delay model? How would you model this in dataflow style?
7-37 What are the features of the path delay model?
7-38 What is the meaning of the module path?
7-39 Why does it need to use the worst-case path delay from all paths in the lumped delay model?

7.3.2 Specify Blocks

Once we have known how to model delays in a hardware module, in this subsection we deal with how to describe the module delays in the specify block of Verilog HDL. The specify block is a mechanism in Verilog HDL for providing the following functions:

1. Describe various paths across a module
2. Assign delays to these paths
3. Perform the necessary timing checks

The first two functions of specify blocks are discussed in this subsection and the timing checks will be discussed in the next subsection in more detail.

7.3.2.1 Declaration of Specify Blocks A specify block is defined by a pair of keywords **specify** and **endspecify** and must be placed within a module. The general form of a specify block is as follows:

```
specify
    {path_declaration}
    {specparam_declaration}
    {system_timing_check}
    {pulsestyle_declaration}
    {showcancelled_declaration}
endspecify
```

■ Example 7.17: A specify block.

As an example to explain how a specify block can be used to model module path delays, consider the logic circuit in Figure 7.3. We can see from the figure that there are three module paths: x-f, y-f, and z-f. The path x-f indeed contains two paths: x-a-f and x-c-b-f, with delays of 8 and 10 time units, respectively. Hence, the worst-case path x-c-b-f is used for

Section 7.3 *Delay Models and Timing Checks*

the specified module path. All three module paths are specified in a **specify** block along with their delays. The resulting module is as follows:

```
// an illustration of the use of specify blocks
module test_gate_specify(input x, y, z, output f);
wire a, b, c;
// gate instances
   and  a1 (a, x, y);
   not  n1 (c, x);
   and  a2 (b, c, z);
   or   o1 (f, a, b);
// a specify block with path delay statements
specify
   (x => f) = 10;
   (y => f) = 8;
   (z => f) = 8;
endspecify
endmodule
```

It is worth noting that the specify block can also be used to describe the delay of a single gate, although it is usually specified by using a delay specifier when it is instantiated. Refer to Section 2.2.2 for more details about the delay specifier.

7.3.2.2 Path Declarations
Three types of path declarations can be declared within a specify block:

1. Simple path
2. Edge-sensitive path
3. State-dependent path

Simple path. A simple path means that it simply constructs a path from a path source to a path destination. The path source is an input port or an inout port and the path destination is an output port or an inout port. A simple path can be declared in one of the following two forms:

1. parallel connection (using source => destination)
2. full connection (using source *> destination)

Both connections are illustrated graphically in Figure 7.4.

1. Parallel connection. Parallel connection uses the symbol "=>" and every path source connects to exactly one path destination. Hence, both the source and destination must have the same size. The syntax of parallel connection is as follows

   (source [polarity] => destination) = delay_value;

where `destination` can be any scalar output or inout port, or the bit-select of a vector output or inout port. The `polarity` (may be + or -) parameter arbitrarily denotes whether a signal is inverted when it propagates from `source` to `destination`. However, it does not matter with the actual propagation, depending on the internal logic, of signals through the module. The positive polarity (+) is used when the transitions at both the source and destination are identical while the negative polarity (-) is employed when the transitions at both the source and destination are different. The default is unknown polarity (no specification), which means both transitions at the source and destination cannot be predicted. An example of parallel connection is as follows

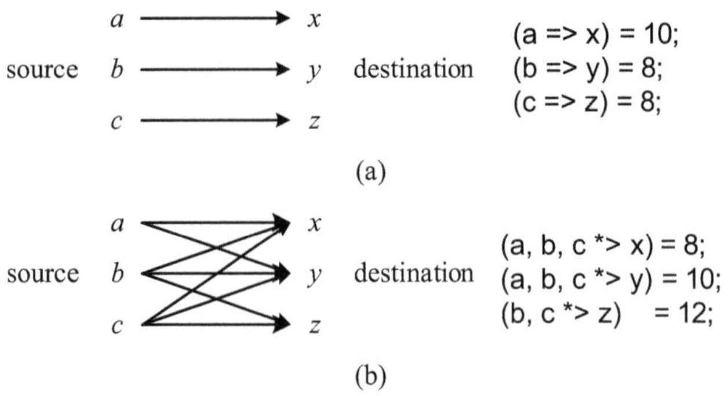

Figure 7.4: An example of simple path connections: (a) parallel connection and (b) full connection.

```
specify
    (a => x) = 10;
    (b => y) = 8;
    (c => z) = 8;
endspecify
```

which is illustrated in Figure 7.4(a).

2. Full connection. Full connection uses "*>" and every path source connects to all path destinations. The source need not have the same size as the destination. It is used to describe a module path between a vector and a scalar, between vectors of different sizes, and with multiple sources or multiple destinations in a single statement. The syntax of full connection is as follows:

```
(source [polarity] *> destination) = delay_value;
```

The destination can be a list of one or more of the vector or scalar output and inout ports, bit-selects or part-selects of vector output, and inout ports. An example of full connection is as follows

```
specify
    (a, b, c *> x) = 8;
    (a, b, c *> y) = 10;
    (b, c *> z) = 12;
endspecify
```

which is depicted in Figure 7.4(b).

■ Example 7.18: An example of multiple module paths.

This example considers how to describe multiple module paths in a single statement. As an illustration, consider the following specify block.

```
specify
    (a, b, c *> x, y) = 8;
endspecify
```

It is equivalent to

```
specify
```

```
    (a, b, c *> x) = 8;
    (a, b, c *> y) = 8;
endspecify
```

This can be further expanded into the following six individual module path assignments:

```
specify
    (a => x) = 8;
    (b => x) = 8;
    (c => x) = 8;
    (a => y) = 8;
    (b => y) = 8;
    (c => y) = 8;
endspecify
```

Edge-sensitive path. In an edge-sensitive path, the module path is described based on an edge transition of a specified input.

```
([edge_id] input_descriptor =>
    (destination [polarity]:source)) = delay_value;
([edge_id] input_descriptor *>
    (destination [polarity]:source)) = delay_value;
```

where edge_id can be either **posedge** or **negedge**. input_descriptor may be any input port or inout port. When input_descriptor is a vector port, the edge transition is detected on the LSB. If edge_id is not specified, it is active on any transition.

The following example demonstrates an edge-sensitive path declaration along with a positive polarity operator.

```
(posedge clock => (out +: in)) = (8, 6);
```

It means that at the positive edge of clock, a module path extends from clock to out using a rise time and a fall time of 8 and 6 time units, respectively. The data path is from in to out and the signal is not inverted.

The next example addresses an edge-sensitive path declaration coupled with a negative polarity operator.

```
(negedge clock => (out -: in)) = (8, 6);
```

It denotes that at the negative edge of clock, a module path extends from clock to out using a rise time and a fall time of 8 and 6 time units, respectively. The data path is from in to out and the signal is inverted.

Of course, when the path is active on any transition at the input port, it is required to leave the edge_id unspecified. For example,

```
(clock => (out : in)) = (8, 6);
```

It means that at any changes in clock, a module path extends from clock to out using a rise time and a fall time of 8 and 6 time units, respectively. The data path is from in to out and the signal transition is unknown.

State-dependent path. In a realistic hardware module, the module path delays might be changed when the states of input signals to a circuit change. To reflect this situation, the module path delay of the circuit should be assigned conditionally, based on the value of the signals in the circuit. Such a module path is referred to as a *state-dependent path*. The general forms of state-dependent paths are as follows

```
if (cond_expr) simple_path_declaration
if (cond_expr) edge_sensitive_path_declaration
ifnone simple_path_declaration
```

where `cond_expr` controls the module path. The `cond_expr` expression can contain any logical, bitwise, reduction, concatenation, or conditional operators. The operands in the `cond_expr` expression can be one of the following options: scalar or vector input ports or inout ports or their bit-selects or part-selects, locally defined variables or nets or their bit-selects or part-selects, constants, and specify parameters. The **ifnone** statement is a default state-dependent path when all other conditions for the path are false.

■ Example 7.19: The use of state-dependent paths.

This example illustrates the use of state-dependent paths. As shown in the following specify block, two module paths are individually described by a simple state-dependent path.

```
specify
    if (x)  (x => f) = 10;
    if (~x) (y => f) = 8;
endspecify
```

If the input x is true, a module path extends from x to f with a delay of 10 time units; if the input x is false, a module path extends from y to f with a delay of 8 time units.

■ Example 7.20: The timing of a NOR gate.

In this example, the delays of a NOR gate are specified with state-dependent paths. The first two state-dependent paths describe a pair of output rise and fall times when the NOR gate (nor1) outputs a 0 independent of the other input. The last two state-dependent paths describe another pair of output rise and fall times when the NOR gate inverts a changing input.

```
// the use of the specify block
module my_nor(input a, b, output out);

// the body of the module
    nor nor1 (out, a, b);

// the specify block
specify
    if (a)  (b => out) = (1, 2);
    if (b)  (a => out) = (1, 2);
    if (~a) (b => out) = (2, 3);
    if (~b) (a => out) = (2, 3);
endspecify
endmodule
```

The **if** (`cond_expr`) statement can also be combined with an edge-sensitive path. This results in the *state-dependent edge-sensitive path*. An illustration is given in the following example:

```
specify
    if (reset_n && clear_n)
        (negedge clock => (out +: in)) = (8, 6);
endspecify
```

Section 7.3 Delay Models and Timing Checks

If the negative edge of clock occurs when reset_n and clear_n inputs are high, a module path extends from clock to out using a rise time and a fall time of 8 and 6 time units, respectively. The data path is from in to out and the signal is not inverted.

The ifnone condition. As mentioned before, the **ifnone** statement is used to specify a default state-dependent path when no other conditions are true. The source and destination used in the **ifnone** statement should be the same as the state-dependent module paths.

■ Example 7.21: The use of the ifnone condition.

This example illustrates how to use the **ifnone** statement. To see this, consider the following specify block. The state-dependent path descriptions in this specify block are valid.

```
specify
   if (reset_n && clear_n)
       (posedge clk => (q +: d)) = (3, 4);
   ifnone (clk => q) = (2, 5);
endspecify
```

However, the following state-dependent path descriptions are invalid because the **ifnone** state-dependent path is the same as an unconditional path below it rather than the state-dependent path above it.

```
specify   // invalid state-dependent path descriptions
   if (~a)(b => out) = (3,3);
   ifnone (a => out) = (2,2); // the same module paths
          (a => out) = (3,3);
endspecify
```

■ Review Questions

7-40 What are the functions of specify blocks?
7-41 What can be declared inside a specify block?
7-42 Define a simple path, an edge-sensitive path, and a state-dependent path.
7-43 What are the differences between parallel connection and full connection?
7-44 When multiple module paths want to be declared as concisely as possible, which type of connection, parallel or full, should be used?
7-45 What is the meaning of the **ifnone** statement? When would you use it?

7.3.2.3 Delay Specifications As we have encountered many times in the previous examples, each module path can be assigned a delay with a delay specification, which may be one, two, three, six, or twelve delays in accordance with the actual requirements. The delays must be constant expressions containing literals or specify parameters, and can be a delay expression of the min:typ:max form. If the path delay expression results in a negative value, it is treated as zero. The general forms of a delay specification are as follows:

```
tpath_expr                         // one delay
trise_expr, tfall_expr             // two delays
trise_expr, tfall_expr, tz_expr    // three delays

t01_expr, t10_expr, t0z_expr,      // six delays
tz1_expr, t1z_expr, tz0_expr
```

```
t01_expr, t10_expr, t0z_expr,    // twelve delays
tz1_expr, t1z_expr, tz0_expr,
t0x_expr, tx1_expr, t1x_expr,
tx0_expr, txz_expr, tzx_expr
```

For one delay, the specified delay is the propagation time of the path. For two delays, the specified delays are the rise time and fall time of the path in order. For three delays, an additional delay called the *turn-off time* of the path is appended to the two delays. The six and twelve delays include all transition combinations of the value set, {0, 1, z} and {0, 1, x, z}, respectively.

7.3.2.4 specparam Declarations The **specparam** (a specify parameter) declaration is used to define specify parameters, which are intended to provide timing and delays. It may be placed within or outside a specify block but within a module. When a **specparam** declaration is placed outside a specify block, it must be declared before it is used. The syntax of specparam_declaration is as follows

> **specparam** [range] specparam_id = const_expr
>   {, specparam_id = const_expr};

where specparam_id is an identifier and const_expr can be a constant expression or a mintypmax constant expression, which may be composed of the other specify parameters or module parameters (i.e., **parameter** and **localparam**). A specify parameter can be used as part of a constant expression for a subsequent specify parameter declaration. However, it can only be changed through SDF annotation. An example to illustrate the use of **specparam** and specify parameters is as follows:

```
specify
   // define parameters inside the specify block
   specparam   d_to_q = 10;         // a constant
   specparam   clk_to_q = 13:15:16; // a mintypmax constant
     ( d  => q) = d_to_q;
     (clk => q) = clk_to_q;
endspecify
```

■ Example 7.22: The timing of a NOR gate.

This example illustrates the combined use of the **specparam** declaration and state-dependent paths. It specifies the delays of a NOR gate. The first two state-dependent paths describe a pair of output rise and fall times when the NOR gate (nor1) outputs a 0 independent of the other input. The last two state-dependent paths describe another pair of output rise and fall times when the NOR gate inverts a changing input.

```
// the use of the specparam statement
module my_nor(input a, b, output out);
// the body of the module
   nor nor1 (out, a, b);
// the specify block
specify
   specparam trise = 1, tfall = 2;
   specparam trise_n = 2, tfall_n = 3;
   if (a) (b => out) = (trise, tfall);
   if (b) (a => out) = (trise, tfall);
   if (~a)(b => out) = (trise_n, tfall_n);
```

Section 7.3 *Delay Models and Timing Checks*

```
      if (~b)(a => out) = (trise_n, tfall_n);
endspecify
endmodule
```

7.3.2.5 Pulse-Width Limit Control Because the inertial delay model is used to model module path delays, any signal with a pulse width (the time between two consecutive transitions) less than the specified module path delay will be filtered out and not present at the output of the module path. However, three ways in Verilog HDL can be used to change the pulse-width limits from their default values. They are summarized as follows:

1. *The specify block control approach*: The **specparam PATHPULSE$** statement is provided by Verilog HDL to modify the pulse-width limits from their default values.
2. *The global control approach*: Invocation options (command lines) can specify percentages that apply to all module path delays to form the corresponding pulse-width limits.
3. *The SDF annotation approach*: SDF annotation can individually annotate the pulse-width limits of each module path.

The pulse-width ranges associated with each module path delay are defined by two limit values: *the error limit* and *the rejection limit*. The error limit is always at least as large as the rejection limit. Pulses with a width greater than or equal to the error limit pass unfiltered. Pulses with a width less than the error limit, but greater than or equal to the rejection limit, are filtered to an unknown value x. Pulses with a width less than the rejection limit are rejected. By default, both the error limit and rejection limit are set equal to the delay.

1. Specify block control approach. Pulse-width limits may be set with the **specparam PATHPULSE$** statement within the specify block by using the generic syntax of

```
specparam pulse_control_specparam
       {, pulse_control_specparam};
```

where `pulse_control_specparam` can be in either of the following two forms:

```
PATHPULSE$ = (reject_limit[, error_limit])
PATHPULSE$source$destination=(reject_limit[,error_limit])
```

If only the reject limit is specified, it applies to both pulse-width limits. The pulse-width limits may be specified for a specific module path by using the second form to specify both the source and destination. As no module path is specified, the pulse-width limits apply to all module paths defined in the module containing the specify block. The path-specific **PATHPULSE$** statement has higher precedence than the non-path-specific **PATHPULSE$** statement.

■ Example 7.23: The use of specparam PATHPULSE$.

Two path-specific **PATHPULSE$** statements and one non-path-specific **PATHPULSE$** statement are used to declare the pulse-width limits. The path (clk => q) has a reject limit of 2 and an error limit of 3. The path (reset *> q) has a reject limit of 0 and an error limit of 3. The path (data *> q) receives both reject and error limits of 3, as defined by the non-path-specific **PATHPULSE$** statement.

```
specify
   // define parameters inside the specify block
   specparam    d_to_q   = 4:5:6;
   specparam    clk_to_q = 4:5:6;
   specparam    reset_to_q = 3:4:5;
      (clk   => q) = clk_to_q;
      (data  *> q) = d_to_q;
```

```
        (reset *> q) = reset_to_q;
    specparam
        PATHPULSE$clk$q   = (2, 3),
        PATHPULSE$reset$q = (0, 3),
        PATHPULSE$        = 3;
endspecify
```

2. Global control approach. The reject limit and error limit invocation options can specify percentages applying globally to all module path delays. The percentages are integers between 0 and 100. The default value is 100%. The **PATHPULSE$** values have higher precedence than global pulse limit invocation options. What follows is a general form of invocation options.

>vsim +pulse_int_e/<percent> +pulse_int_r/<percent>

For example, consider an interconnect delay of 10 along with a +pulse_int_e/80 option. The error limit is 80% of 10 and the rejection limit defaults to 80% of 10. This results in the propagation of pulses greater than or equal to 8, and all other pulses are filtered out.

3. SDF annotation approach. SDF annotation can be used to specify the pulse-width limits of module path delays.

Detailed pulse control capabilities. The above-mentioned default style of pulse filtering behavior reveals two shortcomings: First, pulse filtering to the x state may result in an output pulse with an x-state duration too short to be useful. Second, unequal delays can result in pulse rejection whenever the trailing edge precedes the leading edge, thereby leaving no indication that a pulse was rejected.

The first difficulty can be solved by replacing the default on-event method with the on-detect method. With the on-detect method, the specified output will transition to x immediately upon detection of a pulse at the input rather than transition to x at the scheduled time of the leading edge of the pulse used in the default on-event method. To specify which method is to use, the following syntax can be employed.

pulsestyle_ondetect output_id[range]{, output_id[range]};
pulsestyle_onevent output_id[range]{, output_id[range]};

where output_id can be either an **output** or **inout** port identifier. The keyword **pulsestyle_onevent** represents the default, unchanged behavior.

This functionality can also be controlled globally using the vsim command line switch

>vsim +pulse_e_style_onevent

The command line switches take precedence over the specify block declarations.

The second difficulty can be solved by using the specify block declaration, **showcancelled**, for an output to schedule a transition to and from x over the duration of the negative pulse (trailing edge to leading edge). The keyword **noshowcancelled** denotes the default, unchanged behavior. The **showcancelled** and **noshowcancelled** specify block declarations have the following syntax.

showcancelled output_id[range]{, output_id[range]};
noshowcancelled output_id[range]{, output_id[range]};

The above functionality can also be controlled globally using the vsim command line switches.

>vsim +show_cancelled_e
>vsim +no_show_cancelled_e

The command line switches take precedence over the specify block declarations.

Section 7.3 Delay Models and Timing Checks

As an illustration of the on-detect method and the use of the **showcancelled** specify block declaration, consider the following example.

■ Example 7.24: The on-detect method.

Consider the following my_buf_on_detect module, where the buffer has the rise and fall times of 4 and 6, respectively. As an input pulse with a width of 2 time units is applied to the input a of the buffer, the output b of the buffer will present different results, depending on whether the on-detect method or the default on-event method is used, as illustrated in Figure 7.5. To exhibit the cancelled pulse, the **showcancelled** specify block declaration must be used.

```
// an illustration of the on-detect method and
// the use of the showcancelled declaration
module my_buf_on_detect(output b, input a);

// the body of the module
   buf buf1(b,a);

// the specify block
specify
   pulsestyle_ondetect b;
   showcancelled b;
   (a => b) = (4, 6);
endspecify
endmodule
```

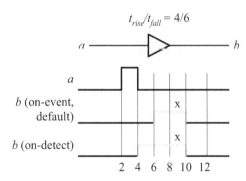

Figure 7.5: An illustration of the one-detect versus on-event methods.

■ Review Questions

7-46 What are the differences between specify and module parameters?
7-47 List the six transitions from the value set: {0, 1, z}.
7-48 List the twelve transitions from the value set: {0, 1, x, z}.
7-49 What is the error limit when only the reject limit is specified to 5 time units?
7-50 What are specified in the three delays?
7-51 How would you specify a specific module path in **PATHPULSE$**?
7-52 How would you change the limits of a pulse width?

7.3.3 Timing Checks

In Verilog HDL, a set of functions is provided for timing checks. All timing checks must be inside the specify blocks. Although they begin with the character "$," these timing checks are not system tasks. It should be memorized that no system task can appear in a specify block and no timing check can appear in procedural constructs: **initial** and **always** blocks.

The timing checks can be roughly cast into two types: *window-related timing checks* as well as *clock and signal timing checks*.

- Window-related timing checks are used to check the *setup time*, *hold time*, *recovery time*, and *removal time* of signals, and include **$setup, $hold, $setuphold, $recovery, $removal**, and **$recrem**.
- Clock and signal timing checks are used to check the *width*, *period*, *skew*, and *change* of signals, and include **$width, $period, $skew, $timeskew, $fullskew**, and **$nochange**.

Every timing check may include an optional *notifier* which toggles its value whenever a violation is detected. In addition, all timing checks have a reference event and a data event, and may associate with a boolean condition.

7.3.3.1 Conditioned Events In timing checks, every event can be associated with a boolean condition that controls the occurrence of the event conditionally. This type of event is called a *conditioned event*. The event used in timing checks has the following general form

```
[timing_event] specify_terminal_descriptor
               [&&& check_condition]
```

where `timing_event` can be either of **posedge**, **negedge**, and an edge-control specifier. The edge-control specifier contains the keyword **edge** and has the following form

edge [edge_descriptor[, edge_descriptor]]

where `edge_descriptor` can be one or more transitions from the signal value set $\{0, 1, x\}$, separated by commas. These transitions are listed as follows

```
01  Transition from 0 to 1
0x  Transition from 0 to x
10  Transition from 1 to 0
1x  Transition from 1 to x
x0  Transition from x to 0
x1  Transition from x to 1
```

where z is treated as x. The **posedge** is equivalent to **edge** [01, 0x, x1] and the **negedge** is equivalent to **edge** [10, x0, 1x].

The `check_condition` can be one of the following expressions:

```
expression, or ~expression
expression == scalar_constant
expression === scalar_constant
expression != scalar_constant
expression !== scalar_constant
```

The `scalar_constant` value is only allowed to use 0 (or 1'b0) and 1 (1'b1). For example,

$hold(**posedge** clk &&& (~clr), data, 3);
$hold(**posedge** clk &&& (clr === 0), data, 3);

show two ways to trigger on the positive edge of clk only when clr is low.

7.3.3.2 Window-Related Timing Checks

For window-related timing checks, two signals, named the *reference event* and the *data event*, are accepted and a window is defined with respect to one signal while checking the transition time of the other with respect to the window. When the data event occurs within the window, timing violation is reported.

Setup time check. For a latch- or flip-flop-based circuit, the setup time is defined as the amount of time that data must be stable before they are sampled, as shown in Figure 7.6. The setup time can be checked by the **$setup** timing check, which has the following form

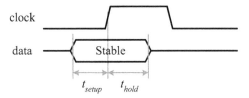

Figure 7.6: A timing diagram showing the setup and hold times.

```
$setup (data_event, reference_event, limit[, [notifier]]);
```

where `data_event` is usually a data signal and `reference_event` is a clock signal. `limit` is a non-negative constant expression and `notifier` is an optional **reg** variable. Violation is reported when

```
t_reference_event - t_data_event < limit
```

When the limit is zero, the **$setup** timing check never reports a violation.

An illustration of how to use the **$setup** timing check is given as follows. A violation occurs when the `data` change values within 5 time units before the positive edge of `clock`.

```
specify
    $setup (data, posedge clock, 5);
endspecify
```

Hold time check. For a latch- or flip-flop-based circuit, the hold time is defined as the amount of time that data must continually remain stable after they have been sampled, as shown in Figure 7.6. The hold time can be checked by the **$hold** timing check, which has the following form

```
$hold (reference_event, data_event, limit[, [notifier]]);
```

where `data_event` is usually a data signal and `reference_event` is a clock signal. `limit` is a non-negative constant expression. Violation is reported when

```
t_data_event - t_reference_event < limit
```

When the limit is zero, the **$hold** timing check never reports a violation.

An illustration of the use of the **$hold** timing check is given as follows. A violation occurs when the `data` change values within 3 time units after the positive edge of `clock`.

```
specify
    $hold (posedge clock, data, 3);
endspecify
```

The following example explains how the **$setup** and **$hold** timing checks are used in an actual case to check whether the input signals confine to the timing requirements of the instantiated module.

■ Example 7.25: Setup and hold-time timing checks.

As an illustration of the use of **$setup** and **$hold** timing checks, consider a positive-edge-triggered D flip-flop with asynchronous active-low reset. In this example, we use **specparam** to define the required delays. The propagation delays of clock and reset signals are represented in min:typ:max format. When no specification is given, the simulator defaults to use the typical values. The simulator will report errors when the timing relationship between inputs clk, d, and reset_n violates any specification on the timing checks: **$setup** and **$hold**. The reference events in both timing checks are conditioned events, which are triggered only when the reser_n signal is high. In addition, two module paths are specified and each module path associates with a two-delay specification.

```
module dff_setuphold_check(
      input clk, d, reset_n,
      output reg q);
specify
   // define timing specparam values
   specparam tSU = 12, tHD = 5;
   // define module path delays with min:typ:max
   specparam tPLHclk = 12:14:16, tPHLclk = 12:13:16;
   specparam tPLHreset = 13:15:16, tPHLreset = 11:12:15;
   // specify module path delays
   (clk *> q)     = (tPLHclk, tPHLclk);
   (reset_n *> q) = (tPLHreset, tPHLreset);
   // setup time: data to clock, only when reset_n is 1
   $setup(d, posedge clk &&& reset_n, tSU);
   // hold time: clock to data, only when reset_n is 1
   $hold(posedge clk &&& reset_n, d, tHD);
endspecify

// the body of the D flip-flop
always @(posedge clk or negedge reset_n)
   if (!reset_n) q <= 0; // active-low reset
   else          q <= d;
endmodule
```

Remember that synthesis tools generally ignore the specify block in synthesizing a module. Thus, the above module is acceptable for synthesis tools.

Setup and hold times check. Both setup and hold times are often checked together and can be performed by using a single **$setuphold** timing check. When positive limits are used, the **$setuphold** timing check is a combination of both **$setup** and **$hold** timing checks. The **$setuphold** timing check with positive limits has the following form

```
$setuphold (reference_event, data_event, setup_limit,
            hold_limit [, [notifier]]);
```

where reference_event is a clock signal and data_event is usually a data signal. Both setup_limit and hold_limit are constant expressions and their values may be less than 0. We will return to this case and discuss it in more detail later in this section. Violation is reported when

```
t_reference_event - t_data_event < setup_limit
t_data_event - t_reference_event < hold_limit
```

Section 7.3 *Delay Models and Timing Checks*

When both limits are zero, the **$setuphold** timing check never reports a violation.

An illustration of the use of the **$setuphold** timing check is as follows

specify
    $setuphold (**posedge** clock, data, 10, 5);
endspecify

which is equivalent to

specify
    $setup(data, **posedge** clock, 10);
    $hold (**posedge** clock, data, 5);
endspecify

Recovery timing check. This timing check specifies the minimum time that an asynchronous input must not be asserted prior to the active edge of the clock signal, as shown in Figure 7.7(a). The **$recovery** timing check has the following form

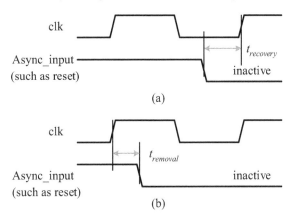

Figure 7.7: A timing diagram showing the (a) recovery and (b) removal times.

    $recovery (reference_event, data_event, limit[, [notifier]]);

where reference_event is usually an asynchronous control signal, such as clear, reset, or preset, and data_event is a clock signal. limit is a non-negative constant expression. Violation is reported when

    t_data_event − t_reference_event < limit

When the limit is zero, the **$recovery** timing check never reports a violation.

The use of the **$recovery** timing check is shown in the following example.

specify
    $recovery (**negedge** aysyn_input, **posedge** clk, 5);
endspecify

Removal timing check. This timing check specifies the minimum time that an asynchronous input, such as clear and preset, must remain asserted after the active edge of the clock signal, as shown in Figure 7.7(b). The **$removal** timing check has the following form

    $removal (reference_event, data_event, limit[, [notifier]]);

where reference_event is usually an asynchronous control signal, such as clear, reset, or preset, and data_event is a clock signal. limit is a non-negative constant expression. Violation is reported when

$$t\_reference\_event - t\_data\_event < \text{limit}$$

When the limit is zero, the **$removal** timing check never reports a violation.

The use of the **$removal** timing check is shown in the following example.

specify
    $removal (**negedge** aysyn_input, **posedge** clk, 5);
endspecify

Recovery and removal timing checks. Both removal and recovery times are often checked together. To this end, the **$recrem** timing check can be used. When limits are positive or zero, the **$recrem** timing check is a combination of both **$removal** and **$recovery** timing checks. The **$recrem** timing check with positive limits has the following form

$recrem (reference_event, data_event, recovery_limit,
        removal_limit [, [notifier]]);

where reference_event is usually an asynchronous input signal and data_event is a clock signal. Both recovery_limit and removal_limit are constant expressions and their values may be less than 0. Violation is reported when

$$t\_data\_event - t\_reference\_event < \text{recovery\_limit}$$
$$t\_reference\_event - t\_data\_event < \text{removal\_limit}$$

When both limits are zero, the **$recrem** timing check never reports a violation.

An illustration of the use of the **$recrem** timing check is as follows

specify
    $recrem (**negedge** aysyn_input, **posedge** clk, 5, 5);
endspecify

which is equivalent to

specify
    $recovery (**negedge** aysyn_input, **posedge** clk, 5);
    $removal (**negedge** aysyn_input, **posedge** clk, 5);
endspecify

7.3.3.3 Negative Timing Checks As stated before, both **$setuphold** and **$recrem** timing checks can accept negative limit values. Their behavior is identical with respect to negative limit values. Hence, the following descriptions are special for the **$setuphold** timing check but may apply equally well to the **$recrem** timing check. The general forms of **$setuphold** and **$recrem** timing checks are as follows

$setuphold (reference_event, data_event, setup_limit,
        hold_limit[, [notifier][, [stamptime_cond]
        [, [checktime_cond][, [delayed_ref]
        [, [delayed_data]]]]]]);
$recrem (reference_event, data_event, recovery_limit,
        removal_limit[, [notifier][, [stamptime_cond]
        [, [checktime_cond][, [delayed_ref]
        [, [delayed_data]]]]]]);

where stamptime_cond and checktime_cond are conditions for negative timing checks, and delayed_ref and delayed_data are delayed reference and data signals for negative timing checks, respectively.

The setup and hold times define a *violation window* with respect to the reference event during which the data must remain stable. Any change of the data during the specified window

Section 7.3 Delay Models and Timing Checks

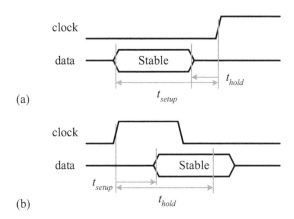

Figure 7.8: A timing diagram showing the negative (a) hold and (b) setup times.

causes a timing violation. A positive value for setup and hold times means that the violation window straddles the reference signal while a negative value for hold or setup time implies the violation window is shifted to either left or right of the reference edge, as shown in Figure 7.8.

Because of unequal delays of internal signal paths, a hardware module can have a negative setup or hold time, but can never have both to be negative simultaneously. In order to accurately model negative-value timing constraints, delayed copies of the data and reference signals are generated and used internally for timing check evaluation at runtime. The setup and hold times used internally are adjusted so that the violation window can overlap the reference signal. Delayed data and reference signals can also be explicitly declared within the timing check to ensure accurate simulation. If no delayed signals are declared and a negative setup or hold value is used, then implicit delayed signals are created instead.

For instance, in the following example, implicit delayed signals are created for clk and data because a setup time with a negative value of 5 is used. The data will not be properly clocked in if it does not transition after the positive edge of clk in 5 time units.

$setuphold (**posedge** clk, data, −5, 10);

The following example explicitly declares the delayed signals dclk, and ddata:

$setuphold (**posedge** clk, data, −5, 10,,,, dclk, ddata);

The delayed timing check signals are delayed only when negative limit values are present.

As mentioned previously, conditions can be associated with both reference and data signals by using the &&& operator. The following example illustrates the use of conditions.

$setup (data, clk &&& condition, tsetup, notifier);
$hold (clk, data &&& condition, thold, notifier);

is equivalent to a single **$setuphold** timing check:

$setuphold(clk, data, tsetup, thold, notifier, ,condition);

where stamptime_cond is null, and checktime_cond is the condition.

■ Review Questions

7-53 Explain the basic operations of window-related timing checks.
7-54 What is the conditioned event?
7-55 Define the following two terms: setup time and hold time.

7-56 What is the function of the keyword **edge**?

7-57 What is the meaning of a violation window?

7.3.3.4 Clock and Signal Timing Checks The set of clock and signal timing checks accepts one or two signals and verifies transitions on them that are never separated by more than the limit. This set of timing checks is used to check the width, period, skew, and change of signals.

Pulse width check. The **$width** timing check tests whether the width of a pulse meets the minimum width requirement, as shown in Figure 7.9. The **$width** timing check has the following general form

Figure 7.9: A timing diagram showing the pulse width.

$width (reference_event, limit[,threshold[, notifier]]);

where data_event is derived from reference_event. data_event is the next opposite edge of reference_event. In other words, the **$width** timing check examines the time between two consecutive transitions of a signal. Both limit and threshold are non-negative constant expressions. If the comma before the threshold is present, the comma before the notifier must also be present, even though both arguments are optional. Violation is reported when

threshold < t_data_event − t_reference_event < limit

Note that no violation is reported for glitches smaller than the threshold. In order to avoid a timing violation, the pulse width must be greater than or equal to the specified limit. For example, the following specify block specifies the pulse width of the reset signal must be at least 10 time units.

   specify
      $width (**posedge** reset, 10);
   endspecify

Period check. This is used to check whether the period of a signal meets the minimum period requirement, as shown in Figure 7.10. The **$period** timing check has the following general form

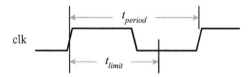

Figure 7.10: A timing diagram showing the period.

$period (reference_event, limit[, [notifier]]);

where data_event is derived from reference_event. It is the next same polarity edge of the reference_event signal. In other words, the **$period** timing check examines the time between two adjacent transitions of the same polarity of a signal. limit is a non-negative constant expression. Violation is reported when

Section 7.3 Delay Models and Timing Checks

```
t_data_event − t_reference_event < limit
```

For example, the following specify block specifies that the clock signal clk must have a period of 25 time units at least.

specify
    $period (**posedge** clk, 25);
endspecify

■ Example 7.26: A more complex timing check example.

In this example, we add three timing checks, **$period**, **$width**, and **$recovery**, to the preceding example. These timing checks are used to check the timing constraints of the period of the clock signal, the pulse width of the reset signal, and the recovery time of the reset signal, respectively.

```verilog
module dff_timing_check(
      input clk, reset_n, d,
      output reg q);
specify
   // define timing specparam values
   specparam tSU=10, tHD=5, tPW=25, tWPC=10, tREC=5;
   // define module path delays with min:typ:max
   specparam tPLHclk = 12:14:16, tPHLclk = 12:13:16;
   specparam tPLHreset = 13:15:16, tPHLreset = 11:12:15;
   // specify module path delays
   (clk *> q)     = (tPLHclk, tPHLclk);
   (reset_n *> q) = (tPLHreset, tPHLreset);
   // setup time: data to clock, only when reset_n is 1
   $setup(d, posedge clk &&& reset_n, tSU);
   // hold time: clock to data, only when reset_n is 1
   $hold(posedge clk &&& reset_n, d, tHD);
   // clock period check
   $period(posedge clk, tPW);
    // pulse width: reset_n
   $width(negedge reset_n, tWPC, 0);
    // recovery time: reset_n to clock
   $recovery(posedge reset_n, posedge clk, tREC);
endspecify

// the body of the D flip-flop
always @(posedge clk or negedge reset_n)
   if (!reset_n) q <= 0; // active-low reset
   else          q <= d;
endmodule
```

Skew Timing checks. Skew timing checks examine whether the time difference between two signals meets the specified timing constraint. Two timing violation detection mechanisms of skew timing checks are *event-based* and *timer-based*. The event-based skew timing check relies on signal transitions whereas the timer-based skew timing check employs the simulation time.

There are three skew timing checks: **$skew**, **$timeskew**, and **$fullskew**. The **$skew** timing check is event-based whereas the **$timeskew** and **$fullskew** timing checks are timer-based by

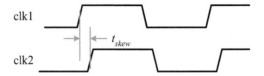

Figure 7.11: A timing diagram showing the clock skew.

default and can be altered to be event-based. The timing diagram showing the skew between two signals is depicted in Figure 7.11.

$skew timing check. The **$skew** timing check examines whether the time skew between two signals meets the maximum requirement based on events. It has the following form

```
$skew (reference_event, data_event, limit[, [notifier]]);
```

where `limit` is a non-negative constant expression. Violation is reported when

```
t_data_event - t_reference_event > limit
```

A simple example given below shows that the skew between two clock signals clk1 and clk2 must be within 5 time units; otherwise, a timing violation is reported.

specify
    $skew (**posedge** clk1, **posedge** clk2, 5);
endspecify

$timeskew timing check. The **$timeskew** timing check examines whether the time skew between two signals meets the maximum requirement based on events or the simulation time. It has the following general form

```
$timeskew (reference_event, data_event, limit[, [notifier]
           [, [event_based_flag][, [remain_active_flag]]]]);
```

where `limit` is a non-negative constant expression. Both `event_based_flag` and `remain_active_flag` are optional and are constant expressions. Violation is reported when

```
t_data_event - t_reference_event > limit
```

After reporting the violation, the **$timeskew** timing check becomes dormant and reports no more violations until the next reference event occurs.

A simple example is demonstrated below:

specify
    $timeskew (**posedge** clk1, **posedge** clk2, 5);
endspecify

The operation of the **$timeskew** timing check defaults to be timer-based. However, it can be altered to be event-based using `event_based_flag`. As only `event_based_flag` is set, the **$timeskew** timing check behaves like the **$skew** timing check except that it becomes dormant after reporting the first violation. To make it behave exactly like the **$skew** timing check, both `event_based_flag` and `remain_active_flag` have to be set.

$fullskew timing check. The **$fullskew** timing check is the same as the **$timeskew** timing check except that the reference and data events can transition in either order. It has the following general form

```
$fullskew (reference_event, data_event, limit1, limit2
           [,[ notifier][,[ event_based_flag][,
           [ remain_active_flag]]]]);
```

Section 7.3 *Delay Models and Timing Checks*

where limit1 and limit2 are non-negative constant expressions. Violation is reported when either of the following two conditions occurs:

```
t_data_event - t_reference_event > limit1
t_reference_event - t_data_event > limit2
```

An illustration of how to use the **$fullskew** timing check is as follows:

specify
    $fullskew (**posedge** clk1, **negedge** clk2, 5, 7);
endspecify

Timing violation will be reported when the negative edge of clk2 is 5 time units lagged behind the positive edge of clk1 or the positive edge of clk1 is 7 time units lagged behind the negative edge of clk2.

$nochange timing check. The **$nochange** timing check examines whether a signal remains unchanged in a time window defined by the reference event. The time window is defined as the time interval between the leading edge and trailing edge of the reference event. In other words, it reports a violation when a data event occurs during the specified level (high or low) of the reference event. It has the following general form

$nochange (reference_event, data_event, start_offset,
        end_offset[, [notifier]]);

where start_offset and end_offset are used to expand or shrink the timing violation region. Both are constant expressions relative to reference_event. reference_event must be an edge-triggered event and specified with the keyword **posedge** or **negedge**.

A violation arises if the data event occurs anytime within the time window. Violation is reported when

```
t_start_window =
        t_leading_edge_reference_event - start_offeset
t_end_window   =
        t_trailing_edge_reference_event + end_offeset
t_start_window < t_data_event < t_end_window
```

For example, the following **$nochange** timing check will report a timing violation if clear changes while reset_n is high. The time window is the time interval between the positive edge and the negative edge of the reset_n signal.

specify
    $nochange (**posedge** reset_n, clear, 0, 0);
endspecify

■ Review Questions

7-58 What is the basic principle of checking the width of a pulse? Which timing check can be used for this purpose?

7-59 What is the basic principle of checking the period of a periodic signal? Which timing check can be used for this purpose?

7-60 What are the differences between event-based and time-based skew detection mechanisms?

7-61 What are the differences among **$skew**, **$timeskew**, and **$fullskew** timing checks?

7-62 Explain the operation of the **$nochange** timing check. Why is it so named?

7.4 Compiler Directives

Like C language, Verilog HDL provides a rich set of compiler directives. All compiler directives begin with the character "`" (i.e., accent grave) and has the form: `<keyword>`. A compiler directive, when compiled, remains in effect across all files being processed until another compiler directive specifies otherwise. The set of compiler directives provided by Verilog HDL is as follows:

1. **`define** and **`undef** compiler directives
2. **`include** compiler directive
3. **`ifdef**, **`else**, **`elsif**, **`endif**, and **`ifndef** compiler directives
4. **`timescale** compiler directive
5. **`celldefine** and **`endcelldefine** compiler directives
6. **`line** compiler directive
7. **`default_nettype** compiler directive
8. **`unconnected_drive** and **`nounconnected_drive** compiler directives
9. **`resetall** compiler directive

7.4.1 `define and `undef Compiler Directives

The **`define** compiler directive is used to create a macro for text substitution. It is similar to the #define construct in C language. Two major features of the **`define** compiler directive are as follows: (1) it is usually placed at the beginning of a file or in a separated file, and (2) it can be used both inside and outside module definitions. The **`define** compiler directive has the general form

`define macro_name [(formal_arguments)] macro_text

Once a macro text is defined with macro_name, every occurrence of the `macro_name in the source file will be substituted by the macro text, macro_text. For example, if we define

`define BUS_WIDTH 8

then any place where a `BUS_WIDTH appears will be substituted by 8.

In order to allow a macro to be customized individually for each use, a text macro may also be defined with arguments. A text macro with argument(s) is expanded by substituting each formal argument with the actual argument whenever it is called. The general form of using a text macro is as follows:

`macro_name [(actual_argument{, actual_argument})]

When a text macro is expanded, each formal argument is substituted for the corresponding actual argument literal by literal. Therefore, when an expression is used as an actual argument, the entire expression is substituted into the macro text. For example,

`define min(a, b)((a) < (b) ? (a) : (b))
y = `min(p+q, r+s);   // macro call

will be expanded into:

y = ((p+q) < (r+s) ? (p+q) : (r+s));

The **`undef** compiler directive removes a previously defined text macro. It has the form

`undef macro_name

Any attempt to remove an undefined text macro using the **`define** compiler directive will result in a warning message.

Section 7.4 *Compiler Directives*

7.4.2 'include Compiler Directive

The **'include** compiler directive allows us to insert the entire contents of a source file in another file during compilation. It is usually used to include commonly used definitions, which are often placed in a separate file. The work of the **'include** compiler directive is similar to that of the #include construct in C language. The **'include** compiler directive can be specified anywhere within a source file. The syntax of the **'include** compiler directive is as follows

```
'include "filename"
```

where filename is the name of the file to be included. The filename can be a full (absolute) or relative path name. Only white spaces or a comment may appear on the same line as the **'include** compiler directive. In addition, the **'include** compiler directive may be nested. That is, a file included in the source using the **'include** compiler directive may contain the other **'include** compiler directives. The following two examples illustrate the usage of the **'include** compiler directive.

```
'include "count.v"
'include "../program/chapter06/count.v"
```

7.4.3 'ifdef, 'else, 'elsif, 'endif, and 'ifndef Compiler Directives

In Verilog HDL, the optional parts of a source file can be compiled conditionally based on some macro definitions controlled by some conditionally compiled directives. These conditionally compiled directives can be classified into two types: check the existence of a macro definition (**'ifdef**) and check the lack of a macro definition (**'ifndef**).

The **'ifdef** compiler directive has the following form

```
'ifdef macro_name
   ifdef_group_of_lines
{'elsif macro_name elsif_group_of_lines}
['else   else_group_of_lines]
'endif
```

where the **'elsif** and **'else** compiler directives are optional with the **'ifdef** compiler directive. The following example illustrates the use of the **'ifdef** compiler directive to select WORD_SIZE by using a parameter, DWORD.

■ **Example 7.27: The use of the 'ifdef compiler directive.**

During compilation, the WORD_SIZE parameter is set to 32 if the text macro name DWORD is defined and set to 16 otherwise.

```
'ifdef DWORD
   parameter WORD_SIZE = 32;
'else
   parameter WORD_SIZE = 16;
'endif
```

The following example demonstrates how to use an **'ifdef** compiler directive in a realistic case to control the actual parameter used in a design.

■ Example 7.28: Another example of the 'ifdef compiler directive.

In the following module, the **'ifdef** compiler directive is used to control the word width of the n-bit adder. The adder_nbit_ifdef module will be synthesized into a 32-bit adder if the DWORD macro is defined and into a 16-bit adder otherwise. The DWORD macro can be defined somewhere before it is used in the design and may have any value.

```
// a simple use of ifdef compiler directive
`define DWORD 1  // set switch DWORD to true
// module definition
module adder_nbit_ifdef(x, y, c_in, sum, c_out);
`ifdef DWORD      // set default value conditionally
   parameter N = 32;
`else
   parameter N = 16;
`endif
// I/O port declarations
input   [N-1:0] x, y;
input   c_in;
output  [N-1:0] sum;
output  c_out;

// specify an N-bit adder using a continuous assignment
   assign {c_out, sum} = x + y + c_in;
endmodule
```

The **'ifndef** compiler directive has the following form

```
`ifndef macro_name
   ifndef_group_of_lines
{`elsif macro_name elsif_group_of_lines}
[`else  else_group_of_lines]
`endif
```

where the **'elsif** and **'else** compiler directives are optional with the **'ifndef** compiler directive. The following example illustrates the usage of the **'ifndef** compiler directive.

■ Example 7.29: The use of the 'ifndef compiler directive.

During compilation, if the text macro named DATAFLOW is not defined, an **xor** gate is instantiated; otherwise, a net declaration assignment is compiled instead.

```
module xor_op(output a, input b, c);
`ifndef DATAFLOW
   xor a1 (a, b, c);
`else
   wire a = b ^ c;
`endif
endmodule
```

Section 7.4 *Compiler Directives*

■ **Review Questions**

7-63 Explain the meaning and usage of the **'define** compiler directive.
7-64 Explain the meaning and usage of the **'include** compiler directive.
7-65 Explain the meaning and usage of the **'ifdef** compiler directive.
7-66 Explain the meaning and usage of the **'ifndef** compiler directive.

7.4.4 'timescale Compiler Directive

The **'timescale** compiler directive is used to specify both the physical measure units or the time scale of a numerical delay and the resolution of the time scale, i.e., the minimum step size of the scale during simulation. The format of the **'timescale** compiler directive is as follows

'timescale time_unit/time_precision

where time_unit specifies the measure units of simulation times and propagation delays whereas time_precision specifies how delays are rounded during simulation. Only the integers 1, 10, and 100 may be used to specify both time_unit and time_precision; the valid units are s, ms, us, ns, ps, and fs.

■ **Example 7.30: A use of the 'timescale compiler directive.**

As an illustration of the use of the **'timescale** compiler directive, consider the following statements.

'timescale 10 ns/1 ns

#2.55 a = b + 1; // corresponds to 26 ns
#2.52 b = c + 1; // corresponds to 25 ns

The **'timescale** 10 ns/1 ns specifies that the time unit is 10 ns and the time precision is 1 ns. Based on this, the time values in the module are the multiples of 10 ns, rounded to the nearest 1 ns. Therefore, the delay 2.55 and 2.52 are scaled and rounded to 26 ns and 25 ns, respectively.

■ **Example 7.31: Another use of the 'timescale compiler directive.**

As another example of the use of the **'timescale** compiler directive, consider the following statements.

'timescale 1 ns /10 ps

#2.55 a = b + 1; // corresponds to 2.55 ns
#2.52 b = c + 1; // corresponds to 2.52 ns

In this example, the time unit is 1 ns and the time precision is 10 ps. Hence, all time values are the multiples of 1 ns because the time unit is 1 ns and are rounded to real numbers with two decimal places because the time precision is 10 ps.

The time_precision value must not exceed the time_unit value. Hence, **'timescale** 100 ps /1 ns is illegal. In addition, the effect of the **'timescale** compiler directive lasts for all modules that follow this compiler directive until another **'timescale** compiler directive specifies otherwise.

■ Review Questions

7-67 Explain the meaning of the **'timescale** compiler directive.

7-68 Explain the meaning of the `time_unit` in the **'timescale** compiler directive.

7-69 Explain the meaning of the `time_precision` in the **'timescale** compiler directive.

7.4.5 Miscellaneous Compiler Directives

In this subsection, we describe each of the following compiler directives:

1. **'celldefine** and **'endcelldefine**.
2. **'line**.
3. **'unconnected_drive** and **'nounconnected_drive**.
4. **'default_nettype**.
5. **'resetall**.

'celldefine and 'endcelldefine compiler directives. These two compiler directives are used to mark a module as a cell module. That is, their combination defines the cell boundary. These two compiler directives may appear anywhere in the source file, but it is recommended that these two compiler directives be specified outside the module definition.

■ Example 7.32: The use of the 'celldefine compiler directive.

This example demonstrates the use of a pair of **'celldefine** and **'endcelldefine** compiler directives. These two compiler directives are used to mark the AND2 module as a cell module.

```
'timescale 1ns / 10ps

'celldefine
module AND2(input a1, a2, output f);
   and   (f, a1, a2);
specify
   (a1 => f)=(2, 2);
   (a2 => f)=(2, 2);
endspecify
endmodule
'endcelldefine
```

Cell modules are used by some PLI routines, such as delay calculations. They typically come along with a specify block, as shown in the above example, which specifies the delays of the cell modules.

'line compiler directive. The **'line** compiler directive is used to reset the line number and the file name of the current file to that as specified. It is of the form

```
'line number "file_name" level
```

where `number` is the new line number of the next line, `file_name` is the new name of the file, and the value of `level` indicates whether an include file has been entered if it is 1, an include file is exited if it is 2, or neither has been done if it is 0. The **'line** compiler directive can be specified anywhere within the Verilog HDL source file. The results of this compiler directive are not affected by the **'resetall** compiler directive.

'unconnected_drive and 'nounconnected_drive compiler directives. These two compiler directives are employed to pull up or pull down any unconnected input ports of a module.

Section 7.5 Summary

More precisely, any unconnected input ports of a module appearing between **'unconnected_drive** and **'nounconnected_drive** compiler directives are pulled up or pulled down dependent on what value is specified: **pull1** or **pull0**. The general form is as follows:

```
// the first example
`unconnected_drive pull0
// all unconnected input ports are pulled down
`nounconnected_drive
// the second example
`unconnected_drive pull1
// all unconnected input ports are pulled up
`nounconnected_drive
```

These two compiler directives must be specified in pairs and placed outside the module declarations.

'default_nettype compiler directive. The default net type is **wire** for implicit net declarations. The **'default_nettype** compiler directive changes the default net type as specified. It can only be used outside module definitions. Multiple **'default_nettype** compiler directives are allowed. It has the form

```
`default_nettype net_type
```

where net_type is one of the following net types: **wire, tri, tri0, tri1, wand, triand, wor, trior, trireg, uwire,** or **none**. When none is specified, all nets must be explicitly declared.

'resetall compiler directive. The **'resetall** compiler directive resets all compiler directives to their default values. It is of the form

```
`resetall
```

It is recommended to place the **'resetall** compiler directive at the beginning of a source file before any compiler directives so as to reset all compiler directives to their default values.

∎ Review Questions

7-70 Explain the meaning and usage of the **'celldefine** compiler directive.
7-71 Explain the meaning and usage of the **'unconnected_drive** compiler directive.
7-72 Explain the meaning and usage of the **'default_nettype** compiler directive.
7-73 Explain the meaning and usage of the **'resetall** compiler directive.

7.5 Summary

In this chapter, we discussed a number of additional features of Verilog HDL. These features include block statements, procedural continuous assignments, specify blocks, and timing checks.

The block statements are used to group together two or more procedural statements so that they can act as a single procedural statement. They can be subdivided into sequential and parallel blocks. The procedural statements within sequential blocks are executed sequentially in the order they appear but within parallel blocks are executed concurrently regardless of their relative order.

The procedural continuous assignments allow expressions to be driven continuously onto variables or nets. They are procedural statements and hence must be used within an **initial** or **always** block. The two types of procedural continuous assignments are **assign/deassign** and **force/release**. The pair of **assign** and **deassign** statements assign values to and release variables, respectively. The pair of **force** and **release** statements assign values to and release nets or variables, respectively.

The specify block is a mechanism of Verilog HDL for describing various paths across a module, assigning delays to these paths, and performing necessary timing checks. The specify blocks are often used accompanied with the definition of the cells in ASIC cell libraries.

The timing checks provided in Verilog HDL are used to check various timing constraints of a module. All timing checks must be inside the specify blocks. The timing checks can be roughly subdivided into two sets: window-related timing checks as well as clock and signal timing checks. Window-related timing checks are used to check the setup time, hold time, recovery time, and removal time of signals. Clock and signal timing checks are employed to check the width, period, skew, and change of signals.

Like C language, Verilog HDL equips a rich set of compiler directives. A compiler directive, when compiled, remains its effect across all files being processed until another compiler directive specifies otherwise. With the aid of compiler directives, a macro text may be defined, part of a design can be compiled conditionally, default net features can be changed, and unconnected input ports may be pulled up or pulled down.

References

1. J. Bhasker, *A Verilog HDL Primer*, 3rd ed., Star Galaxy Publishing, 2005.
2. IEEE 1364-2001 Standard, *IEEE Standard Verilog Hardware Description Language*, 2001.
3. IEEE 1364-2005 Standard, *IEEE Standard for Verilog Hardware Description Language*, 2006.
4. S. Palnitkar, *Verilog HDL: A Guide to Digital Design and Synthesis*, 2nd ed., SunSoft Press, 2003.

Problems

7-1 Explain the operations of the following two parallel blocks:

```
// block a: cannot swap a and b
fork: race
   #5 a = b;
   #5 b = a;
join
// block b: swap a and b properly
fork: swap
   a = #5 b;
   b = #5 a;
join
```

7-2 Explain the operations of the following two **always** blocks:

```
// block a: cannot swap a and b
always @(*)
   a = b;
always @(*)
   b = a;
// block b: swap a and b properly
always @(*)
```

```
        a <= b;
    always @(*)
        b <= a;
```

7-3 What are the differences between the following two sequential blocks:

```
// block a
begin: block_a
    a = #5 b;
    b = #5 a;
end
// block b
begin: block_b
    a <= #5 b;
    b <= #5 a;
end
```

7-4 What are the differences between the following two parallel blocks:

```
// block a
fork: block_a
    a = #5 b;
    b = #5 a;
join
// block b
fork: block_b
    a <= #5 b;
    b <= #5 a;
join
```

7-5 Explain the operations of the following nested block and give the time when each statement executes.

```
// block a
fork: block_a
    a = #5 b;
    begin: block_b
        d <= #15 c;
        c <= #15 d;
    end
    b = #5 a;
join
```

7-6 Explain the operations of the following nested block and give the time when each statement executes.

```
// block a
fork: block_a
    a = #5 b;
    begin: block_b
        #15 d <= c;
        #15 c <= d;
    end
```

```
      b = #5 a;
join
```

7-7 Explain the operations of the following nested block and give the time when each statement executes.

```
// block a
fork: block_a
    #5 a = b;
    begin: block_b
        d <= #15 c;
        c <= #15 d;
    end
    #5 b = a;
join
```

7-8 Explain the operations of the following nested block and give the time when each statement executes.

```
// block a
fork: block_a
    #5 a = b;
    begin: block_b
        #15 d <= c;
        #15 c <= d;
    end
    #5 b = a;
join
```

7-9 Explain the operations of the following nested block and give the time when each statement executes.

```
// block a
begin: block_a
    a <= #5 b;
    fork: block_b
        d = #15 c;
        c = #15 d;
    join
    b <= #5 a;
end
```

7-10 Explain the operations of the following nested block and give the time when each statement executes.

```
// block a
begin: block_a
    a <= #5 b;
    fork: block_b
        #15 d = c;
        #15 c = d;
    join
    b <= #5 a;
```

end

7-11 Use a pair of **assign** and **deassign** procedural continuous assignments to define a positive-edge-triggered D flip-flop with asynchronous preset and clear. The output is set to 1 when the preset input is active and cleared to 0 when the clear input is active.

7-12 Use a pair of **force** and **release** procedural continuous assignments to define a negative-edge-triggered D flip-flop with asynchronous preset and clear. The output is set to 1 when the preset input is active and cleared to 0 when the clear input is active.

7-13 Considering the bidirectional buffer shown in Figure 2.9(a), answer each of the following questions.
   (a) Write a Verilog HDL module to describe the bidirectional buffer.
   (b) Write a test bench to verify the module of the bidirectional buffer.

7-14 Assume that an n-bit counter with a bidirectional data bus is required. As the write (rd) signal is activated, the counter is preset to the value from the data bus; as the read (rd) signal is asserted, the counter value is read to the data bus. In addition, a gate (gate) input is employed to enable the operations of the counter.
   (a) Write a Verilog HDL module to describe the n-bit counter.
   (b) Write a test bench to verify the module of the n-bit counter.

7-15 Professor Bob writes a 4-bit 2's-complement adder, which is as follows:

```
module twos_adder_behavioral(
        input   [3:0] x, y,      // 4-bit array
        input   c_in,
        output reg [3:0] sum,    // 4-bit array
        output reg c_out);
reg     [3:0] t;           // outputs of xor gates
// specify the function of the 2's-complement adder
always @(x, y, c_in) begin
   t = y ^ c_in;
   {c_out, sum} = x + t + c_in;
end
endmodule
```

   (a) Write a test bench to verify the functionality of the above module. Is it correct? Give your reasons.
   (b) Use a pair of **force** and **release** procedural continuous assignments to tentatively correct the bug and verify the functionality of the module again.

7-16 Explain the operations of the following parallel block:

```
fork
   begin: blk1
      @start;
      repeat (3) @trigger;
      #d1 f = x ^ y;
   end
   @reset disable blk1;
join
```

7-17 Explain the operations of the following parallel block:

```
always begin: monostable
   #120 qout = 0;
end
always @trigger begin
   disable monostable;
   qout = 1;
end
```

7-18 Referring to Figure 7.5 and assuming that $t_{rise}/t_{fall} = 6/2$, answer each of the following questions.
   (a) Draw a timing diagram showing the waveforms of output b as the on-event method and the on-detect method are used, respectively.
   (b) Write a Verilog HDL module to describe the buffer depicted in Figure 7.5 along with the specifications of rise and fall times.
   (c) Write a test bench to verify the timing diagram that you plotted in (a).

7-19 Explain the operations of the following specify block:

```
specify
   if (reset)
      (posedge clk => (q[3:0]-:data)) = (4,5);
endspecify
```

7-20 Explain why the following two state-dependent path declarations are invalid.

```
specify
   if (reset)
      (posedge clk => (q[7:0]:data)) = (5,8);
   if (!reset)
      (posedge clk => (q[2]:data)) = (10,6);
endspecify
```

7-21 Which of the following statements are legal?

```
$width (posedge clear, limit);
$width (posedge clear, limit, thresh, notifier);
$width (negedge clear, limit, notifier);
```

7-22 Which of the following statements are illegal?

```
$width (negedge clear, limit, 0, notifier);
$width (posedge clear, limit, , notifier);
$width (posedge clear, limit, notifier);
```

7-23 Up to now, there are three approaches that can be used to define a constant within the source description of a design. These are the **'define** compiler directive, the **parameter** declaration, and the **localparam** declaration. Describe their meanings and explain the differences when they are used to define a constant.

7-24 Explain the meaning of the following timing check:

$setup(data, **posedge** clk, 5);

Rewrite the above statement so that the **$setup** timing check is triggered only when the positive edge of the clk signal is coming and the clear signal is high.

7-25 Considering the figure shown in Figure 7.12, answer each of the following questions.

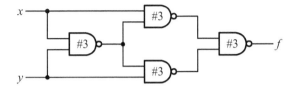

Figure 7.12: The figure for Problem 7-25.

(a) Using the distributed delay model, write a module at the gate level to model the circuit in structural style.
(b) Using the distributed delay model, write a module at the gate level to model the circuit in dataflow style.
(c) Using the lumped delay model, write a module at the gate level to model the circuit in structural style.
(d) Using the lumped delay model, write a module at the gate level to model the circuit in dataflow style.
(e) Using the path delay model, write a module at the gate level to model the circuit in structural style.

7-26 Considering the positive-edge-triggered D flip-flop shown in Figure 7.13(a), answer each of the following questions.

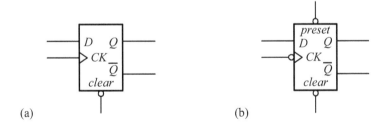

Figure 7.13: The figures for Problems 7-26 and 7-27.

(a) Write a module to describe the D flip-flop with timing parameters. Suppose that paths CK-Q and CK-Qbar have the rise and fall times of 6 ns and 6.8 ns, respectively. Paths clear-Q and clear-Qbar have the propagation delays of 3 ns.
(b) Write a test bench to verify the module in both functionality and timing. Your test bench must be able to verify all the above-mentioned timing values.

7-27 Considering the negative-edge-triggered D flip-flop shown in Figure 7.13(b), answer each of the following questions.

(a) Write a module to describe the D flip-flop with timing parameters. Suppose that paths CK-Q and CK-Qbar have the rise and fall times of 6 ns and 6.8 ns, respectively. Paths clear-Q and clear-Qbar have the propagation delays of 3 ns. Paths preset-Q and preset-Qbar have the propagation delays of 3.6 ns.

(b) Write a test bench to verify the module in both functionality and timing. Your test bench must be able to verify all the above-mentioned timing values.

8

Combinational Logic Modules

In the previous chapters, we have described most features of Verilog HDL. In this and the next chapters, we examine some basic combinational and sequential modules, which are often used as basic building blocks to construct a complex design. In particular, when the datapath and controller approach in a complex design is used, these modules are the basic building blocks of datapaths to be described in more detail in Section 11.2.3.

In this chapter, we are concerned with the most commonly used combinational logic modules. These modules include decoders and encoders, multiplexers and demultiplexers, and magnitude comparators. In addition, a multiplexed seven-segment light-emitting diode (LED) display system combining the use of decoders and a multiplexer is discussed in detail. The other widely used combinational logic modules, including carry-look-ahead (CLA) adders and subtractors, along with other arithmetic logic circuits, such as multipliers and dividers, are considered in Chapter 14.

As we have learned, many options may be used to model combinational circuits. These options include Verilog HDL primitives, the **assign** continuous assignment, behavioral statements, functions, tasks without delay or event timing control, combinational UDPs, and interconnected combinational modules. We will use these whenever they are appropriate.

8.1 Decoders

For convenience, we often refer to the numeric code used in an underlying digital system as the *reference code*. When the reference code differs from that used in the outside world of the underlying system, a conversion process is needed to convert the reference code into the outside-world code and vice versa, as illustrated in Figure 8.1. The circuit used to perform the conversion process is known as a *code converter*. The conversion process is referred to as an *encoding process* if it transforms an outside-world code into the reference code. In contrast, the reverse process, which transforms the reference code into the outside-world code, is called the *decoding process*. The corresponding circuit is then known as an *encoder* and a *decoder*, respectively. In this section, we assume that the reference code is an n-bit binary code and the outside-world code is a 1-out-of-m code, where n and m are positive integers.

8.1.1 Decoders

A decoder is an n-input, m-output combinational circuit that accepts an n-bit binary code and generates a 1-out-of-m code on the output such that the physical position of the activated

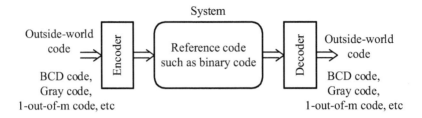

Figure 8.1: The concept of encoding and decoding.

output bit corresponds to the input binary code. Usually, $m \leq 2^n$. The decoder is said to be *fully decoded* if $m = 2^n$ and *partially decoded* if $m < 2^n$.

The general block diagrams of an n-to-m decoder are shown in Figure 8.2. The output may be noninverting or inverting, depending on the actual requirement. A decoder usually associates with one or more optional enable inputs, which may be either active-high or active-low, determined by the actual need. The enable inputs control the operations of the decoder. The output of a decoder is a 1-out-of-m code if all its enable inputs are asserted and is all 1s, all 0s, or even high impedances, depending on the actual requirement and design, if all its enable inputs are deasserted. One advantage of using enable is that it makes easier the cascade of two or more decoders to form a bigger one.

An example of a 2-to-4 active-low decoder with inverting output is shown in Figure 8.3. Figures 8.3(a) and (b) show the logic symbol and function table of the circuit, respectively. From the function table, it is easy to derive the logic circuit, as depicted in Figure 8.3(c). To model this simple 2-to-4 decoder in Verilog HDL, any modeling style described in Section 1.3 may be used. In what follows, we show an example to model it in behavioral style with a **case** statement.

■ Example 8.1: A 2-to-4 active-low decoder.

As an example of using a **case** statement to describe a 2-to-4 active-low decoder with inverting output, consider the function table shown in Figure 8.3(b). The output vector y is all 1s if the enable input is high and outputs a 1-out-of-4 code in accordance with the value of the input vector x otherwise.

```
// a 2-to-4 active-low decoder with inverting output
```

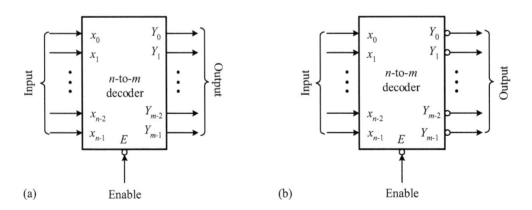

Figure 8.2: The general block diagrams of an n-to-m decoder: (a) noinverting output; (b) inverting output.

Section 8.1 Decoders

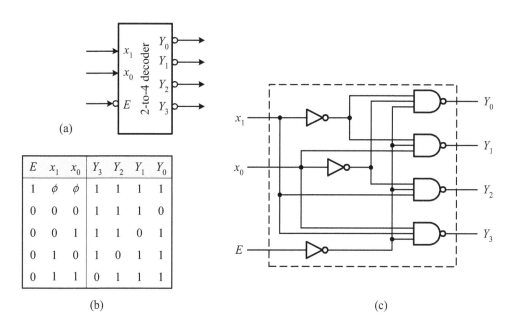

Figure 8.3: A 2-to-4 active-low decoder with inverting output: (a) logic symbol; (b) function table; (c) logic circuit.

```
module decoder_2to4_low(
       input   [1:0] x,
       input   enable_n,
       output reg [3:0] y);
// the body of the 2-to-4 decoder
always @(x or enable_n)
   if (enable_n) y = 4'b1111; else
      case (x)
         2'b00 : y = 4'b1110;
         2'b01 : y = 4'b1101;
         2'b10 : y = 4'b1011;
         2'b11 : y = 4'b0111;
      endcase
endmodule
```

Another type of 2-to-4 decoder with active-high enable and noninverting output is described in the following example.

■ Example 8.2: A 2-to-4 active-high decoder.

As in the preceding example, a **case** statement is used to describe the decoding operation of the decoder. In fact, this example almost has the same structure as the preceding one, except that now the enable input is active-high and the output is noninverting.

```
// a 2-to-4 active-high decoder with noninverting output
module decoder_2to4_high(
       input   [1:0] x,
       input   enable,
       output reg [3:0] y);
```

```verilog
// the body of the 2-to-4 decoder
always @(x or enable)
   if (!enable) y = 4'b0000; else
      case (x)
         2'b00 : y = 4'b0001;
         2'b01 : y = 4'b0010;
         2'b10 : y = 4'b0100;
         2'b11 : y = 4'b1000;
      endcase
endmodule
```

Both of the above two modules simply use a **case** statement to implement the desired operations and cannot be parameterized. Recall that a module is said to be parameterizable if its features or structural properties can be specified whenever it is instantiated. Although using a **case** statement to model a decoder is straightforward, it is difficult to make the resulting module parameterizable. In order to model a parameterizable decoder, it is necessary to explore the basic operation of decoders in greater depth. As we can see from Figure 8.3(b), the essential operation of a decoder with inverting output is to place a logic 0 at the output bit specified by the input code. More precisely, if the input $\{x_1, x_0\}$ is equal to i then a logic 0 is placed in the ith physical position of the output bits, which are numbered from 0 to $n-1$. For example, the decoder puts a logic 0 to the output bit Y_1 when the input code x is 2'b01. Based on this idea, a parameterizable decoder with inverting output can then be modeled. An illustration of this is revealed in the following example.

■ Example 8.3: A parameterizable n-to-m decoder.

As mentioned, the essential operation of a decoder with noninverting output is to put a logic 1 to the output bit specified by the input code x. To realize this idea, we first put a logic 1 in the LSB and a logic 0 in all other bits. Next, these values are shifted left logically the number of x positions under the control of a **for** loop and then assigned to the output vector y. This results in the following module.

```verilog
// a parameterizable N-to-M decoder with noninverting output
module decoder_n2m_high
       #(parameter  N = 2,  // the number of input lines
         parameter  M = 4)( // the number of output lines
         input    [N-1:0] x,
         input    enable,
         output reg [M-1:0] y);
// the body of the n-to-m decoder
integer i;
always @(x or enable)
   for (i = 0; i < M; i = i + 1)
      if   (!enable) y = {M{1'b0}};
      else if (x == i) y = {{M-1{1'b0}},1'b1} << i;
endmodule
```

The reader is encouraged to synthesize the above module and see what happens. In addition, it is recommended to instantiate the above parameterizable module with different parameters and observe their synthesized results. For instance, you may try to design a 2-to-4 decoder and a 3-to-6 decoder by instantiating the above module separately with parameters #(2, 4) and #(3, 6) as follows:

Section 8.1 Decoders

```
decoder_n2m_high #(2, 4) decoder1(x, enable, y);
decoder_n2m_high #(3, 6) decoder2(x, enable, y);
```

It is instructive to note that the condition (x == i) can be removed from the **else-if** part of the **if-else** statement in the above module. In addition, the **for** loop can be removed. That is, we could simply write the **always** statement as follows:

```
always @(x or enable)
    if  (!enable) y = {M{1'b0}};
    else y = {{M-1{1'b0}},1'b1} << x;
```

Try to verify and synthesize it and explain what you obtained from the synthesis tool you used. It is interesting to note that the above **always** statement is indeed a parameterizable expansion of the **case** statement used to describe the truth table of an n-to-m decoder. With most current synthesis tools, the **always** statement can be further simplified into the following statement:

```
always @(x or enable)
    if  (!enable) y = {M{1'b0}};
    else begin
        y = {M{1'b0}};   // avoid inferring a latch
        y[x] = 1'b1;
    end
```

Try to verify and synthesize it and explain what you obtained from the synthesis tool you used. It should be noted that this simplified statement is much like the software-minded style rather than hardware-minded one. However, it still works well and is synthesized into a well-structured n-to-m decoder.

8.1.2 Expansion of Decoders

In a number of practical applications, a big decoder, such as an 8-to-256 decoder or even a bigger one, is often required. To model such a big decoder, at least two ways can be employed. The most straightforward approach is like what we have done in the case of a 2-to-4 decoder to use a **case** statement. This would result in a very large source file which may not be easy to follow and is prone to errors. Another approach is by instantiating a parameterizable decoder. This would result in a very concise module, which essentially has the same source file for all decoders of different numbers of inputs. Nonetheless, regardless of which approach is used, the synthesized result is a single big decoder.

The most widely used approach to constructing a big decoder in practice is by properly cascading a number of small decoder modules in a hierarchical manner. For this reason, decoders are usually designed as modules with an appropriate number of inputs and one or more enable inputs so that they can be cascaded hierarchically to form a bigger one. For instance, we can combine two 2-to-4 decoders with one 1-to-2 decoder to form a 3-to-8 decoder, as shown in Figure 8.4. An illustration of how this 3-to-8 decoder works is explored in the following example.

■ **Example 8.4: A 3-to-8 decoder built with two 2-to-4 decoders.**

As shown in Figure 8.4, the input x_2 is directly connected to the upper decoder and connected to the lower decoder through an inverter. Therefore, the upper decoder is enabled if the input x_2 is 0 and the lower decoder is enabled if the input x_2 is 1. The resulting module is a 3-to-8 decoder. Note that the inverter acts as a 1-to-2 decoder here.

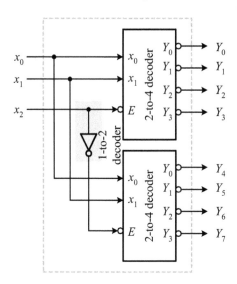

Figure 8.4: A 3-to-8 decoder constructed by cascading two 2-to-4 decoders.

■ Review Questions

8-1 Define an n-to-m decoder.
8-2 Describe a 2-to-4 active-high decoder with inverting output.
8-3 Describe a 2-to-4 active-low decoder with noninverting output.
8-4 Describe an n-to-m active-low decoder with inverting output.
8-5 Write a Verilog HDL module to model the logic circuit shown in Figure 8.4.

8.2 Encoders

Remember that the encoding process is the reverse of the decoding process and transforms an outside-world code into the reference code. The logic circuit used to perform the encoding process is known as an *encoder*. In this section, we begin to describe the principles of general encoders and then introduce priority encoders.

8.2.1 Encoders

An encoder has $m = 2^n$ (or fewer) inputs and n outputs. All inputs are arranged and fixed in their physical positions in a way such that all inputs are numbered from 0 up to $m-1$ in sequence. The output generates a binary code corresponding to the physical position of the activated input.

The general block diagrams of an m-to-n encoder are shown in Figure 8.5, where Figure 8.5(a) is noninverting output while Figure 8.5(b) is inverting output. Like decoders, an encoder usually associates with one or more optional enable inputs, which may be either active-high or active-low, depending on the actual requirement. The enable inputs control the operations of the encoder. The output of an encoder is a binary code if all its enable inputs are asserted and is all 1s, all 0s, or even high impedances, determined by the actual requirement and design, if all its enable inputs are deasserted. One advantage of using enable is that it makes easier the cascade of two or more encoders to form a bigger one.

A simple 4-to-2 encoder is shown in Figure 8.6. From the function table shown in Figure 8.6(a), it is easy to derive the logic circuit as shown in Figure 8.6(b). To model an encoder,

Section 8.2 Encoders

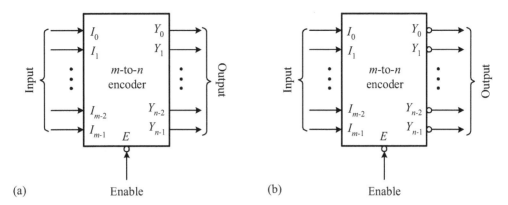

Figure 8.5: The general block diagrams of an m-to-n encoder with: (a) noninverting output; (b) inverting output.

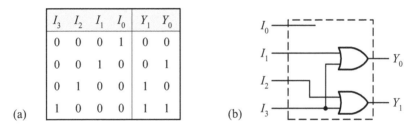

Figure 8.6: The (a) function table and (b) logic diagram of a simple 4-to-2 encoder.

it proves useful to use an **if-else** or a **case** statement. An illustration of using a nested **if-else** statement to describe a 4-to-2 encoder is explored in the following example.

■ Example 8.5: Modeling 4-to-2 encoder with if-else statement.

As an illustration of using a nested **if-else** statement to model an encoder, consider the 4-to-2 encoder shown in Figure 8.6. From the function table shown in Figure 8.6(a), we can use a nested **if-else** statement to model such an encoder by comparing the input code sequentially with a constant which determines the corresponding output value. The resulting module is as follows:

```
// a 4-to-2 encoder using an if-else statement
module encoder_4to2_ifelse(
       input  [3:0] in,
       output reg [1:0] y);
// the body of the 4-to-2 enecoder
always @(in) begin
   if (in == 4'b0001) y = 2'b00; else
   if (in == 4'b0010) y = 2'b01; else
   if (in == 4'b0100) y = 2'b10; else
   if (in == 4'b1000) y = 2'b11; else
       y = 2'bx;
end
endmodule
```

Although a nested **if-else** statement and a **case** statement can be used interchangeably, in most practical applications we often prefer a **case** statement to a nested **if-else** statement when describing an encoder because a **case** statement is much easier to follow then an **if-else** statement. This will be more apparent in comparing the following example with the preceding one.

■ Example 8.6: Modeling 4-to-2 encoder with case statement.

Using a **case** statement to model a 4-to-2 encoder is similar to using an **if-else** statement. In this case, the constants that determine the output values are used as the case items. It should be noted that a **default** statement is necessary in this module to avoid the latch inferred by synthesis tools.

```
// a 4-to-2 encoder using a case statement
module encoder_4to2_case(
      input   [3:0] in,
      output reg [1:0] y);

// the body of the 4-to-2 enecoder
always @(in)
   case (in)
      4'b0001 : y = 2'b00;
      4'b0010 : y = 2'b01;
      4'b0100 : y = 2'b10;
      4'b1000 : y = 2'b11;
      default : y = 2'bx;
   endcase
endmodule
```

In fact, the above two examples cannot exactly model the 4-to-2 encoder depicted in Figure 8.6(b) because a priority is inherent in both **if-else** and **case** statements and we set the output y to an unknown for all inputs that do not belong to the specified set of input constants. More details about this will be discussed in the next subsection. For now, a better way to exactly model this 4-to-2 encoder is by describing the logic circuit in Figure 8.6(b) directly in any modeling style.

8.2.2 Priority Encoders

One disadvantage of the simple encoder discussed above is that the input code must be a 1-out-of-m code, namely, only one input being activated at a time; otherwise, the encoded output code might be meaningless. To see this, consider the logic circuit in Figure 8.6(b) again. As both inputs I_1 and I_2 are 1, the output code Y_1Y_0 is 11, which does not represent any actual input I_1 or I_2, but is I_3 instead.

To avoid the above situation when multiple inputs are activated, a *priority* is often associated with each input. For example, as shown in Figure 8.7, the priority is associated with the index values of the inputs; namely, the priority of the inputs are $I_3 > I_2 > I_1 > I_0$. Based on this, the output Y_1Y_0 is 11 when the input I_3 is 1, independent of the values of the other inputs. An encoder constructed in such a way that each input associates with a priority is called a *priority encoder*. It should be noted that the *Valid* output indicating that at least one input is being activated is necessary to distinguish the encoded output A_1A_0 of a value of 00 from the situation that there is no input is activated.

Section 8.2 Encoders

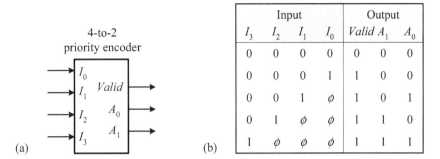

Figure 8.7: The (a) block diagram and (b) function table of a 4-to-2 priority encoder.

Recall that a nested **if-else** statement has an inherent priority, which can be readily used to describe an m-to-n priority encoder. An illustration of using a nested **if-else** statement to model a 4-to-2 priority encoder is explored in the following example.

■ Example 8.7: A 4-to-2 priority encoder.

Because the priority may be set by the order of statements associated with the nested **if-else** statement, the input I_3 with the highest priority is put in the first statement and hence it is examined first. If it fails, then the input I_2 is checked, and so on. By this way, the description of a 4-to-2 priority encoder is obtained.

```
// a 4-to-2 priority encoder using an if-else statement
module priority_encoder_4to2_ifelse(
       input   [3:0] in,
       output  valid_in,
       output reg [1:0] y);
// the body of the 4-to-2 priority encoder
assign valid_in = |in;
always @(in) begin
   if (in[3]) y = 3; else
   if (in[2]) y = 2; else
   if (in[1]) y = 1; else
   if (in[0]) y = 0; else
   y = 2'bx;  // avoid inferring a latch
end
endmodule
```

Recall that the **casex** statement only compares non-x positions in the case_expr and case_item_expr. Based on this feature, a **casex** statement can be used to model a priority encoder. An illustration of this is explored in the following example.

■ Example 8.8: Another 4-to-2 priority encoder.

To illustrate how a **casex** statement can be used to model a 4-to-2 priority encoder, consider the function table shown in Figure 8.7(b). By using the feature of the **casex** statement, each entry of the function table can be coded into a case_item_expr expression directly. Nevertheless, it should be noted that because the **casex** statement is incompletely specified, a **default** statement is needed to avoid inferring a latch. The reader is encouraged to synthesize it with and without the **default** statement and see what happens.

```verilog
// a 4-to-2 priority encoder using a casex statement
module priority_encoder_4to2_casex(
       input   [3:0] in,
       output  valid_in,
       output reg [1:0] y);
// the body of the 4-to-2 priority encoder
assign valid_in = |in;
always @(in) casex (in)
   4'b1xxx: y = 3;
   4'b01xx: y = 2;
   4'b001x: y = 1;
   4'b0001: y = 0;
   default: y = 2'bx;
endcase
endmodule
```

Like decoders, a parameterizable module of a priority encoder is often preferred in practical applications. A parameterizable priority encoder can be readily modeled by exploring the basic operation of a general priority encoder. To see this, consider the function table shown in Figure 8.7(b). The output code of a priority encoder is corresponding to the index of the leftmost bit with a value 1. Hence, it suffices to detect this bit and output its index as the desired code. Illustrations to explore this idea are revealed in the following two examples.

■ **Example 8.9: Modeling m-to-n priority encoder with a for loop.**

Because the basic operation of a priority encoder is that it outputs as its output code the index of the leftmost bit with a value 1, it is sufficient to use a loop running from $M-1$ to 0 to detect such a bit. Once this bit is detected, the loop breaks and sets the corresponding index as the output code. A straightforward way to realize this idea is summarized in the following module.

```verilog
// a parameterizable M-to-N priority encoder
module priencoder_m2n
       #(parameter M = 8,  // the number of inputs
         parameter N = 3)( // the number of outputs
         input   [M-1:0] x,
         output  valid_in,  // indicate the input x is valid
         output reg [N-1:0] y);
// the body of the M-to-N priority encoder
integer i;
assign valid_in = |x;
always @(*) begin: check_for_1
   for (i = M - 1; i >= 0; i = i - 1)
       if (x[i] == 1) begin
          y = i;
          disable check_for_1; end // break for loop
       else y = 0;  // why need else part?
end
endmodule
```

Even though the above module works and is synthesizable, it is not good practice to use the **disable** statement in a synthesizable module. In fact, it is often possible to write a module without the need of a **disable** statement if we properly arrange the logical sequence of the operations to be performed. As an illustration, consider the following example which uses a **while** loop in place of the **for** loop to manage the required check of conditions and hence the **disable** statement is removed.

■ Example 8.10: An m-to-n priority encoder with a while loop.

This example uses the same idea as the preceding one except that a **while** statement is used instead of the **for** statement. The reader is encouraged to compare the differences between these two examples and explore why the **disable** statement may be removed.

```
// a parameterizable M-to-N priority encoder
module priencoder_m2n_while
       #(parameter M = 4,  // define the number of inputs
         parameter N = 2)(  // define the number of outputs
         input    [M-1:0] x,
         output valid_in,  // indicate the input x is valid
         output reg [N-1:0] y);
// the body of the M-to-N priority encoder
integer i;
assign valid_in = |x;
always @(*) begin
   i = M - 1 ;
   while(x[i] == 0 && i >= 0 ) i = i - 1;
   y = i;
end
endmodule
```

Of course, the **while** loop in the above module can be replaced by a **for** loop. Nevertheless, we would like to leave this as an exercise for the reader. (Refer to Section 4.5 for more details about iteration (loop) statements.)

■ Review Questions

8-6 Define an m-to-n encoder.
8-7 Explain the difficulty of an encoder when multiple inputs are active at the same time.
8-8 Define an m-to-n priority encoder.
8-9 Describe a 4-to-2 priority encoder with inverting output.

8.3 Multiplexers

Multiplexing and demultiplexing are two widely used techniques in communications. The former selects one of incoming channels and routes it to the destination and the latter routes the incoming channel to its selected destination channel, as illustrated in Figure 8.8. In fact, demultiplexing is the reverse process of multiplexing and vice versa. In this section, we address devices used to perform multiplexing, known as *multiplexers*. The devices used to carry out demultiplexing are called *demultiplexers*, which will be discussed in the next section. A multiplexer is also referred to as a *data selector* whereas a demultiplexer as a *data distributor*.

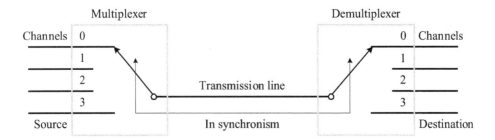

Figure 8.8: The concept of multiplexing and demultiplexing.

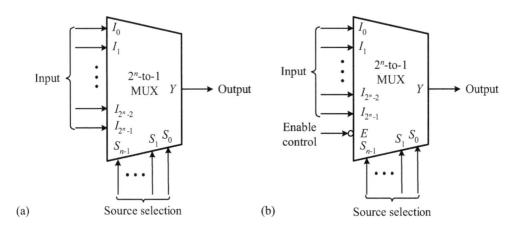

Figure 8.9: The general block diagrams of a 2^n-to-1 multiplexer: (a) without enable; (b) with enable.

8.3.1 Multiplexers

An m-to-1 ($m = 2^n$) multiplexer (Mux or MUX) has m inputs, one output, and n source selection inputs. The input I_i selected by the binary combination of n source selection inputs is routed to the output Y. In other words, the input I_i is routed to the output Y if the value of source selection inputs is i.

The general block diagrams of a 2^n-to-1 multiplexer are shown in Figure 8.9. Figure 8.9(a) does not contain an enable input while Figure 8.9(b) includes an active-low enable input. Like decoders and encoders, a multiplexer usually associates with one or more enable inputs, which may be either active-high or active-low, depending on the actual need. The enable inputs control the operations of the multiplexer. The output of a multiplexer is the value of the selected input if all its enable inputs are asserted and is 1, 0, or even a high impedance, depending on the actual requirement and design, if all its enable inputs are deasserted.

An example of a 4-to-1 multiplexer without enable is depicted in Figure 8.10. Figure 8.10(a) is the logic symbol and Figure 8.10(b) is the function table. From the function table, it is easy to derive the logic circuit, as shown in Figure 8.10(c). There are a number of ways that can be used to model a multiplexer. The most straightforward approach is based on the structural style. It instantiates gate primitives and connects them together according to the logic circuit shown in Figure 8.10(c). The details of this approach can be found in Section 2.1.1.

An illustration of modeling an n-bit 4-to-1 multiplexer without enable in dataflow style using a nested conditional operator is explored in the following example.

Section 8.3 Multiplexers

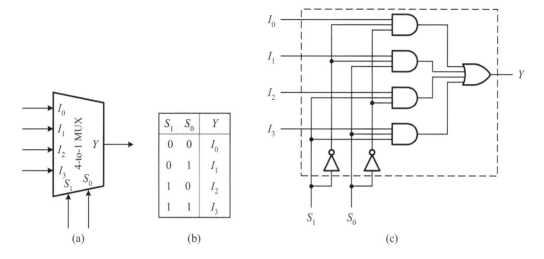

Figure 8.10: A simple 4-to-1 multiplexer without enable: (a) logic symbol; (b) function table; (c) logic circuit.

■ Example 8.11: An n-bit 4-to-1 multiplexer.

A single **assign** continuous assignment with a nested conditional operator is used in this example. If select[1] is 1, the input in3 or in2 is selected; otherwise, the input in1 or in0 is selected. The input in3 or in2 is further selected by the value of select[0]. Similarly, the input in1 or in0 is also further selected by the value of select[0]. The resulting behavior is an n-bit 4-to-1 multiplexer.

```
// an N-bit 4-to-1 multiplexer using a conditional operator
module mux_nbit_4to1
       #(parameter N = 4)(  // set default bit width
          input   [N-1:0] in3, in2, in1, in0,
          input   [1:0] select,
          output  [N-1:0] y);
// the body of the N-bit 4-to-1 multiplexer
assign y = select[1] ?
          (select[0] ? in3 : in2) :
          (select[0] ? in1 : in0) ;
endmodule
```

A multiplexer usually associates with an enable input. To describe such a multiplexer, the combination of an **if-else** statement with a nested conditional operator (? :) is often used.

■ Example 8.12: An n-bit 4-to-1 active-high multiplexer.

In this example, besides the fact that an active-high enable input is added, the description of the n-bit 4-to-1 multiplexer is also changed into the behavioral style. The resulting module is a combination of an **if-else** statement and a nested conditional operator. Of course, it is possible and not difficult to replace the nested conditional operator with a nested **if-else** statement. However, we would like to leave this for the reader as an exercise.

```
// an N-bit 4-to-1 active-high multiplexer
module mux_nbit_4to1_en
```

```verilog
        #(parameter N = 4)( // set default bit width
          input  [N−1:0] in3, in2, in1, in0,
          input  [1:0] select,
          input  enable,
          output reg [N−1:0] y);
// the body of the N-bit 4-to-1 multiplexer
always @(select or enable or in0 or in1 or in2 or in3)
   if (!enable) y = {N{1'b0}};
   else         y = select[1] ? (select[0] ? in3 : in2) :
                                (select[0] ? in1 : in0) ;
endmodule
```

Although the above example can model an n-bit 4-to-1 active-high multiplexer properly, it is hard to read, especially for naive readers. Like decoders and priority encoders, the more understandable way to describe a multiplexer is by using a **case** statement to list all pairs of each combination of source selection inputs and its associated output. An illustration is given as follows.

■ Example 8.13: An n-bit 4-to-1 multiplexer.

In this example, a **case** statement is employed to model a multiplexer. Using this approach, we only need to use the source selection inputs as case_expr and list each combination of source selection inputs as case_item_expr and its associated output as the case_item_expr statement. The resulting module is as follows:

```verilog
// an N-bit 4-to-1 multiplexer using a case statement
module mux_nbit_4to1_case
        #(parameter N = 8)( // set default bit width
          input  [N−1:0] in3, in2, in1, in0,
          input  [1:0] select,
          output reg [N−1:0] y);

// the body of the N-bit 4-to-1 multiplexer
always @(*)
   case (select)
      2'b11: y = in3 ;
      2'b10: y = in2 ;
      2'b01: y = in1 ;
      2'b00: y = in0 ;
   endcase
endmodule
```

Although using a **case** statement to model a multiplexer is easy to understand, it may not be an efficient and useful way to describe a big multiplexer. In addition, as we want to model a multiplexer in a parameterizable way, the method of using a **case** statement may not be a good choice. Like the case of decoders and priority encoders, the way to describe a parameterizable multiplexer is to explore the basic operation of a general multiplexer: The input I_i selected by the source selection inputs with a value i is routed to the output Y. As an illustration, consider Figure 8.10(b), where each row of the function table is read as if $\{S_1, S_0\} = i$, then $Y = I_i$. Based on this operation, a parameterizable m-to-1 multiplexer can then be described by the combination of a **for** loop and an **if** statement. An illustration is demonstrated in the following example.

Section 8.3 *Multiplexers*

■ Example 8.14: A parameterizable m-to-1 multiplexer.

As stated above, the essential operation of a multiplexer is that the input I_i selected by the source selection inputs of a value i is routed to the output Y. Based on this, we use a loop running from 0 to $M - 1$ to provide the values of current source selection inputs, which are then detected by the **if** statement associated with the **for** loop in the **always** block. The input in[i] is assigned to the output y if the value of current source selection inputs exactly matches the value i. The resulting module is as follows:

```
// a parameterizable M-to-1 multiplexer
module mux_mto1
       #(parameter M = 5,   // default size
         parameter K = 3)(  // K = log2 M
          input   [M-1:0] in,
          input   [K-1:0] select,
          output reg y);
// the body of the m-to-1 multiplexer
integer i;
always @(*) begin: mux_body
   y = 1'b0;  // avoid inferring a latch when M < 2**K
   for (i = 0; i < M; i = i + 1)
      if (select == i) y = in[i];
end
endmodule
```

You are encouraged to synthesize the above module and try to explain the rationale behind it. In particular, it is worth paying attention to the function of the statement y = 1'b0;. Why is it needed? If we remove it, what will happen? (Actually, it still works properly if $M = 2^K$.) It is instructive to note that the above **always** statement is indeed a parameterizable expansion of the **case** statement used to describe the truth table of an m to 1 multiplexer. With most current synthesis tools, the above **always** statement can be further simplified as follows:

```
always @(*) begin
   y = 1'b0;  // a redundant expression, why?
   y = in[select];
end
```

Try to verify and synthesize it and explain what you obtained from the synthesis tool you used. It should be noted that this simplified statement is much like the software-minded style rather than hardware-minded one. However, it still works well and is synthesized into a well-structured m-to-1 multiplexer.

8.3.2 Expansion of Multiplexers

Like decoders, in practical applications, a big multiplexer, such as a 256-to-1 multiplexer or an even bigger one, is often required. To describe such a big multiplexer, at least two ways can be used. The most straightforward approach is like what we have done for the case of a 4-to-1 multiplexer to use a **case** statement to write down each combination of source selection inputs one by one. This would result in a very large source file which may not be easy to follow and is prone to errors. Another approach is by instantiating a parameterizable multiplexer. This would yield a very concise module, which essentially has the same source file for all multiplexers of different numbers of inputs. Nevertheless, regardless of which approach is used, the result is a single big multiplexer module.

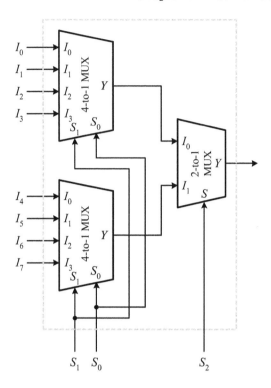

Figure 8.11: An 8-to-1 multiplexer consisting of two 4-to-1 and one 2-to-1 multiplexers.

The most widely used approach to building a big multiplexer is by appropriately cascading a number of small multiplexer modules hierarchically. The resulting bigger multiplexer is known as *a multiplexer tree*. As an illustration, consider the 8-to-1 multiplexer depicted in Figure 8.11, which is a combination of two 4-to-1 multiplexers and one 2-to-1 multiplexer. The operation of this 8-to-1 multiplexer is illustrated in the following example.

■ Example 8.15: A multiplexer tree — an 8-to-1 multiplexer.

As shown in Figure 8.11, the source selection inputs S_1 and S_0 of both 4-to-1 multiplexers are connected together. Both outputs of the two 4-to-1 multiplexers are separately connected to the inputs I_0 and I_1 of the output 2-to-1 multiplexer. The source selection input of the output 2-to-1 multiplexer is denoted by the source selection input S_2. The upper 4-to-1 multiplexer is selected when S_2 is 0 and the lower 4-to-1 multiplexer is selected when S_2 is 1. Therefore, the resulting module is an 8-to-1 multiplexer.

■ Review Questions

8-10 Describe the operation of multiplexers.
8-11 How many 4-to-1 multiplexers are needed to construct a 16-to-1 multiplexer?
8-12 Describe the meaning of a multiplexer tree.
8-13 How many source selection inputs are required in an m-to-1 multiplexer?
8-14 Model a 4-to-1 multiplexer with inverting output.

Section 8.4 *Demultiplexers*

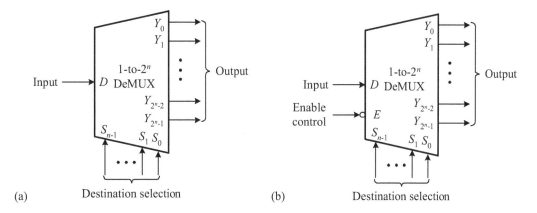

Figure 8.12: The general block diagrams of a 1-to-2^n demultiplexer: (a) without enable; (b) with enable.

8.4 Demultiplexers

Remember that demultiplexing is the reverse process of multiplexing and routes the input to a selected output. The logic circuits used to perform demultiplexing are known as demultiplexers. In this section, we define the general demultiplexers and provide an introduction to the modeling of demultiplexers in Verilog HDL.

8.4.1 Demultiplexers

A 1-to-m ($m = 2^n$) demultiplexer (DeMux or DeMUX) has one input, m outputs, and n destination selection inputs. The input D is routed to the output Y_i selected by the binary combination of n destination selection inputs. In other words, the output Y_i is equal to the input D if the value of destination selection inputs is i.

The general block diagrams of a 1-to-2^n demultiplexer are shown in Figure 8.12. Figure 8.12(a) does not contain an enable input while Figure 8.12(b) includes an active-low enable input. Like multiplexers, a demultiplexer usually associates with one or more enable inputs, which may be either active-high or active-low, depending on the actual requirement. The enable inputs control the operations of the demultiplexer. The output selected by the destination selection inputs has the same value as the input if all its enable inputs are asserted and all outputs are 1s, 0s, or even high impedances, determined by the actual requirement and design, if all its enable inputs are deasserted.

An example of a 1-to-4 demultiplexer is shown in Figure 8.13. Figure 8.13(a) is the logic symbol and Figure 8.13(b) is its function table. From the function table, it is easy to derive the logic circuit, as shown in Figure 8.13(c). There are many ways that can be used to model a demultiplexer. The most straightforward approach is based on the structural style. It instantiates gate primitives and connects them together according to the logic circuit shown in Figure 8.13(c). Because of its intuitive simplicity, we would not like to describe it furthermore here.

An illustration of using four **if-else** statements to model an n-bit 1-to-4 demultiplexer without enable in behavioral style is explored in the following example.

■ Example 8.16: An n-bit 1-to-4 demultiplexer.

This example uses four **if-else** statements to describe an n-bit 1-to-4 demultiplexer. Because of its intuitive simplicity, we will not further explain it here.

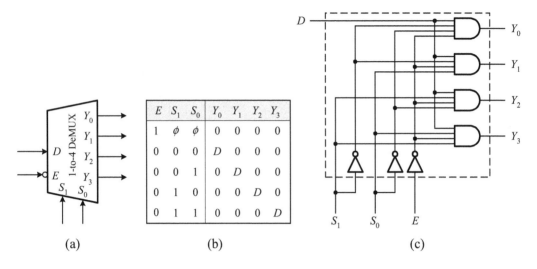

Figure 8.13: A simple 1-to-4 active-low demultiplexer with noninverting output: (a) logic symbol; (b) function table; (c) logic circuit.

```
// an N-bit 1-to-4 demultiplexer using if-else statements
module demux_1to4_ifelse
       #(parameter N = 4)(         // set default width
          input       [N-1:0] in,  // input
          input       [1:0] select, // dest. selection inputs
          output reg  [N-1:0] y3, y2, y1, y0); // outputs

// the body of the N-bit 1-to-4 demultiplexer
always @(select or in) begin
   if (select == 3) y3 = in; else y3 = {N{1'b0}};
   if (select == 2) y2 = in; else y2 = {N{1'b0}};
   if (select == 1) y1 = in; else y1 = {N{1'b0}};
   if (select == 0) y0 = in; else y0 = {N{1'b0}};
end
endmodule
```

It is instructive to note that the function of the **else** part of each **if-else** statement in the above module. What will happen if the **else** part is removed? Like multiplexers, a demultiplexer usually associates with an enable input to enable or inhibit its operations. An illustration of this is given below.

■ **Example 8.17: An n-bit 1-to-4 active-high demultiplexer.**

This example is exactly the same as the preceding one except that an active-high enable input is added to it. It is worth paying attention to the need of the **else** part of each **if-else** statement in the first **begin-end** block and the second **begin-end** block. Explain the reason why this block is necessary. You may answer this question with the aid of the synthesized result from a synthesis tool available to you.

```
// an N-bit 1-to-4 active-high demultiplexer
module demux_1to4_ifelse_en
       #(parameter N = 4)(         // set default width
```

Section 8.4 Demultiplexers

```verilog
        input      [N-1:0] in,      // input
        input      enable,          // enable input
        input      [1:0] select,    // dest. selection inputs
        output reg [N-1:0] y3, y2, y1, y0); // outputs

// the body of the N-bit 1-to-4 demultiplexer
always @(select or in or enable) begin
   if (enable) begin
      if (select == 3) y3 = in; else y3 = {N{1'b0}};
      if (select == 2) y2 = in; else y2 = {N{1'b0}};
      if (select == 1) y1 = in; else y1 = {N{1'b0}};
      if (select == 0) y0 = in; else y0 = {N{1'b0}};
   end
   else begin
      y3 = {N{1'b0}}; y2 = {N{1'b0}};
      y1 = {N{1'b0}}; y0 = {N{1'b0}};
   end
end
endmodule
```

Although the above two examples can model a 1-to-4 demultiplexer properly, they may be hard to read for naive readers. Like encoders and multiplexers, a more understandable way to describe a demultiplexer is to use a **case** statement to list all pairs of each combination of the destination selection inputs and its associated output. An illustration of this is revealed in the following example.

■ Example 8.18: Another n-bit 1-to-4 demultiplexer.

Like the situation of multiplexers, using a **case** statement to model a demultiplexer may be the most straightforward method. In this approach, we only need to use the destination selection inputs as case_expr and list each combination of the destination selection inputs as case_-item_expr and its associated output as the case_item_expr statement. The resulting module is as follows:

```verilog
// an N-bit 1-to-4 demultiplexer using a case statement
module demux_1to4_case
       #(parameter N = 4)(           // set default width
         input      [N-1:0] in,      // input
         input      [1:0] select,    // dest. selection inputs
         output reg [N-1:0] y3, y2, y1, y0); // outputs

// the body of the N-bit 1-to-4 demultiplexer
always @(select or in) begin
   y3 = {N{1'b0}}; y2 = {N{1'b0}}; // avoid latch inference
   y1 = {N{1'b0}}; y0 = {N{1'b0}};
   case (select)
      2'b11: y3 = in;
      2'b10: y2 = in;
      2'b01: y1 = in;
      2'b00: y0 = in;
   endcase
end
```

endmodule

Although using a **case** statement to model a demultiplexer is easy to understand, it may not be an efficient and useful way to describe a big demultiplexer. In addition, to model a demultiplexer in a parameterizable way, the method of using a **case** statement may not be a good choice. In this situation, like the descriptions of a parameterizable decoder and a parameterizable multiplexer, we may resort to the basic operation of a general demultiplexer: The input D is routed to the output Y_i if the value of destination selection inputs is i. For instance, consider the function table shown in Figure 8.13(b), where $Y_i = D$ if $\{S_1, S_0\} = i$ and $E = 0$. Based on this idea, the description of a parameterizable 1-to-m demultiplexer can be done as illustrated in the following example.

■ Example 8.19: A parameterizable 1-to-m demultiplexer.

As described above, the essential operation of a demultiplexer is that the input D is routed to the output Y_i if the value of destination selection inputs is i. Based on this, we may use a loop running from 0 to $M - 1$ to provide the values of current destination selection inputs which are then detected by the **if-else** statement associated with the **for** loop in the **always** block. The output y[i] is assigned with the input in if the value of destination selection inputs is equal to i. The resulting module is as follows:

```
// a parameterizable 1-to-m demultiplexer
module demux_1tom
       #(parameter M = 4,   // set the default size of outputs
         parameter K = 2)(  // K = log2 M
         input    in,            // input
         input    [K-1:0] select, // destination selection inputs
         output reg [M-1:0] y);  // outputs
integer i;

// the body of the 1-to-m demultiplexer
always @(*)
   for (i = 0; i < M; i = i + 1) begin
      if (select == i) y[i] = in;
      else             y[i] = 1'b0;
   end
endmodule
```

You are encouraged to synthesize the above module and try to explain the rationale behind it. It is instructive to note that the above **always** statement is indeed a parameterizable expansion of the **case** statement used to describe the truth table of a 1-to-m demultiplexer. With most current synthesis tools, the **always** statement can be further simplified as follows:

```
always @(*) begin
   y = {M{1'b0}};   // avoid inferring a latch
   y[select] = in;
end
```

Try to verify and synthesize it and explain what you obtained from the synthesis tool you used. Like multiplexers, this modeling style is much like the software-minded style rather than the hardware-minded one. However, it still works well and is synthesized into a well-structured 1-to-m demultiplexer.

Section 8.4 *Demultiplexers*

8.4.1.1 Demultiplexers versus Decoders Both decoders and demultiplexers have the same logic structure; namely, they both contain AND gates as their output stages with each gate detecting a combination of the primary input values. As a consequence, some relationship must exist between these two types of logic circuits. To explore the relationship between decoders and demultiplexers, observe the 2-to-4 active-low decoder with inverting output shown in Figure 8.3(b). As the enable input (E) is used as the data input (D), it can be used as a 1-to-4 demultiplexer with inverting output. Generally speaking, this observation is true: For an n-to-2^n enable-controlled decoder can function as a 1-to-2^n demultiplexer if the enable input (E) is used as the data input (D). More specifically, we have

- An n-to-2^n active-low decoder with inverting output can function as a 1-to-2^n demultiplexer with inverting output if the enable input (E) is used as the data input (D) of the demultiplexer.
- An n-to-2^n active-high decoder with inverting output can function as a 1-to-2^n demultiplexer with inverting output if the complemented enable input ($\bar{E}$) is used as the data input (D) of the demultiplexer.
- An n-to-2^n active-low decoder with noninverting output can function as a 1-to-2^n demultiplexer with noninverting output if the complemented enable input ($\bar{E}$) is used as the data input (D) of the demultiplexer.
- An n-to-2^n active-high decoder with noninverting output can function as a 1-to-2^n demultiplexer with noninverting output if the enable input (E) is used as the data input (D) of the demultiplexer.

Similarly, as shown in Figure 8.13(b), the 1-to-4 active-low demultiplexer with noninverting output may function as a 2-to-4 active-low decoder with noninverting output if its data input (D) is set to 1. In general, a 1-to-2^n (enable-controlled) demultiplexer can function as an n-to-2^n (enable-controlled) decoder when its data input (D) is set to an appropriate value, 1 or 0, depending on whether the output is noninverting or inverting. More precisely, we have

- A 1-to-2^n active-low (active-high) demultiplexer with noninverting output is able to function as an n-to-2^n active-low (active-high) decoder with noninverting output when its data input (D) is set to 1.
- A 1-to-2^n active-low (active-high) demultiplexer with inverting output is able to function as an n-to-2^n active-low (active-high) decoder with inverting output when its data input (D) is set to 0.
- A 1-to-2^n demultiplexer with noninverting output is able to function as an n-to-2^n decoder with noninverting output when its data input (D) is set to 1.
- A 1-to-2^n demultiplexer with inverting output is able to function as an n-to-2^n decoder with inverting output when its data input (D) is set to 0.

8.4.2 Expansion of Demultiplexers

In many practical applications, a big demultiplexer, such as a 1-to-128 demultiplexer or an even bigger one, is often required. To describe such a big demultiplexer, a number of ways may be used. Among these, the most straightforward approach is to use a **case** or an **if-else** statement as we have done in the preceding examples. Another approach is by instantiating a parameterizable demultiplexer. Nevertheless, regardless of which approach is used, the result is a single big demultiplexer module.

The most widely used approach to building a big demultiplexer is by properly cascading a number of small demultiplexer modules in a hierarchical way. The resulting bigger demultiplexer is known as *a demultiplexer tree*. As an illustration, consider the 1-to-8 demultiplexer depicted in Figure 8.14, which combines two 1-to-4 demultiplexers with one 1-to-2 demultiplexer. The operation of this 1-to-8 demultiplexer is illustrated in the following example.

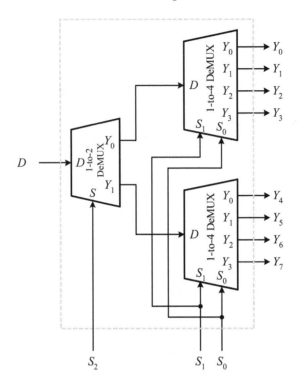

Figure 8.14: A 1-to-8 demultiplexer constructed by cascading two 1-to-4 and one 1-to-2 demultiplexers.

■ **Example 8.20: A demultiplexer tree — a 1-to-8 demultiplexer.**

As shown in Figure 8.14, the destination selection inputs S_1 and S_0 of both 1-to-4 demultiplexers are connected together in the way shown in the figure. Their inputs are separately connected to the outputs Y_0 and Y_1 of the input 1-to-2 demultiplexer. The destination selection input of the input 1-to-2 multiplexer is denoted by the destination selection input S_2. Therefore, the upper 1-to-4 demultiplexer is selected if S_2 is 0 and the lower 1-to-4 demultiplexer is selected if S_2 is 1. The resulting module is a 1-to-8 demultiplexer.

■ **Review Questions**

8-15 Describe the operation of demultiplexers.
8-16 Explain the differences between a 2-to-4 decoder and a 1-to-4 demultiplexer.
8-17 Describe the meaning of a demultiplexer tree.
8-18 How many 1-to-4 demultiplexers are needed to construct a 1-to-16 demultiplexer?
8-19 How many destination selection inputs are required in a 1-to-m demultiplexer?
8-20 Model a 1-to-4 demultiplexer with inverting output.

8.5 Magnitude Comparators

It is often needed to compare two numbers in magnitude for equality in digital systems. A circuit that compares two numbers in magnitude and indicates whether they are equal is called

Section 8.5 Magnitude Comparators

a *comparator*. When a comparator not only compares two numbers in magnitude for testing their equality but also indicates the arithmetic relationship between them, it is called a *magnitude comparator*. The numbers input to a magnitude comparator can be signed or unsigned generally.

The structure of an n-bit comparator is quite simple. It is only an n-bit XOR gate and can be modeled in much the same way as that of the parity generator introduced in Sections 2.1.1 and 3.3.4. As for magnitude comparators, two types of circuits widely used in practical applications are magnitude comparators and *cascadable magnitude comparators*.

8.5.1 Magnitude Comparators

A magnitude comparator is used to compare two numbers in magnitude, signed or unsigned, in order to determine and indicate their arithmetic relationship, namely, less than, equal to, or greater than. An n-bit magnitude comparator can be easily modeled by using relational operators. An illustration is given in the following example, where both inputs are assumed to be unsigned numbers.

■ **Example 8.21: An n-bit magnitude comparator.**

Assuming that both two input numbers are unsigned numbers, all inputs and outputs are declared as unsigned **wire** nets and **reg** variables, respectively. By using relational operators, the desired magnitude comparator can then be modeled as follows:

```
// an N-bit unsigned comparator
module comparator_simple
       #(parameter N = 4)(   // set default size
          input    [N-1:0] a, b,
          output reg cgt, clt, ceq);

// the body of the N-bit comparator
always @(*) begin
   cgt = (a > b); // greater than
   clt = (a < b); // less than
   ceq = (a == b); // equality
end
endmodule
```

The synthesized result of the above example is shown in Figure 8.15. It is instructive to change the input numbers of the preceding module to be signed and see what happens after synthesizing it.

8.5.2 Cascadable Magnitude Comparators

Like other combinational logic modules, a magnitude comparator is often designed as a cascadable module. To achieve this, the magnitude comparator module needs to receive the compared results from its preceding stage and to combine them with the current inputs of the module to determine the combined results up to this stage.

The block diagram of a cascadable 4-bit magnitude comparator is shown in Figure 8.16. The inputs, $I_{A>B}$, $I_{A=B}$, and $I_{A<B}$, are the compared results from its preceding stage, namely, the lower significant digit. These inputs are referred to as *input relationships*. The outputs, $O_{A>B}$, $O_{A=B}$, and $O_{A<B}$, known as *output relationships*, indicate the combined results of the current input numbers of this stage and the input relationships. The output result is an equality if both current inputs are equal and the input relationship is also an equality. The

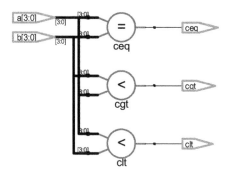

Figure 8.15: The synthesized result of a 4-bit unsigned magnitude comparator.

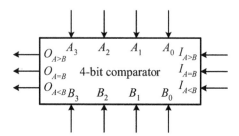

Figure 8.16: The block diagram of a cascadable 4-bit comparator.

output result is greater than if the relationship of current inputs is greater than or the relationship of current inputs is an equality but the input relationship is greater than. The output result is less than if the relationship of current inputs is less than or the relationship of current inputs is an equality but the input relationship is less than.

Based on the above discussion, a cascadable n-bit magnitude comparator can be readily described by using relational and logical equality operators. An illustration is given in the following example.

■ Example 8.22: A cascadable n-bit magnitude comparator.

Because the operation of a magnitude comparator is to compare and indicate the relative magnitude of the two input numbers, logical equality (==) and relational operators, greater than (>) and less than (<), are suitable for describing such a circuit. To make the circuit cascadable, the three inputs used to indicate the input relationships from its preceding stage are also needed. Based on these considerations, the resulting module is as follows and its synthesized result is shown in Figure 8.17.

```
// a cascadable N-bit comparator example
module comparator_cascadable
       #(parameter N = 4)(              // set default size
         input  Iagtb, Iaeqb, Ialtb,   // input relationships
         input  [N-1:0] a, b,           // input numbers
         output reg Oagtb, Oaeqb, Oaltb);

// use relational and logical equality operators
always @(*) begin
   Oaeqb = (a == b) && (Iaeqb == 1);                // equality
```

Section 8.5 *Magnitude Comparators*

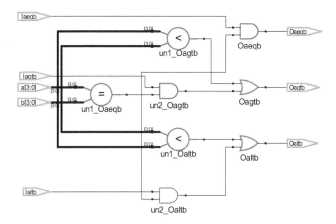

Figure 8.17: The synthesized result of a cascadable 4-bit magnitude comparator.

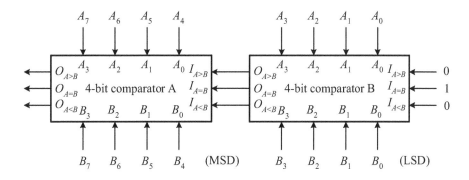

Figure 8.18: The block diagram of an 8-bit magnitude comparator constructed by cascading two 4-bit magnitude comparators.

```
    Oagtb = (a > b) || ((a == b) && (Iagtb == 1)); // greater than
    Oaltb = (a < b) || ((a == b) && (Ialtb == 1)); // less than
end
endmodule
```

Figure 8.18 illustrates how to cascade two 4-bit magnitude comparators into an 8-bit magnitude comparator. As described in the preceding example, the most significant digit (MSD) magnitude comparator may solely determine the final arithmetic relationship from its two input numbers unless for the case where both input numbers are equal. In this situation, the final arithmetic relationship is determined by its preceding magnitude comparator, namely, the least significant digit (LSD).

■ Review Questions

8-21 Design a logic circuit to compare two 1-bit numbers.
8-22 Design a logic circuit to compare two 2-bit numbers.
8-23 Explain the differences between a comparator and a magnitude comparator.
8-24 Describe the meaning of a cascadable magnitude comparator.
8-25 What will happen when we set the input relationships of Figure 8.18 to 000?

8.6 Case Study — Seven-Segment LED Displays

Light-emitting diodes (LEDs) are widely used in most digital or mixed-signal systems as indicators or to present information visually. By appropriately combining a number of LEDs into patterns, such as seven-segment digits or dot-matrix characters, digits and characters may be displayed as required. In this section, we are concerned with the issues of basic LED devices, seven-segment LED displays, a BCD-to-seven-segment decoder, and a typical multiplexed seven-segment display system.

8.6.1 Seven-Segment LED Displays

An LED is a device that can emit visible or invisible light when an appropriate voltage is applied across it. There are a variety of visible or invisible LEDs. In this book, we only consider visible LEDs. Nowadays, many visible LEDs with a wide variety of colors are available. The most popular colors used in digital systems are green, orange, and red. Regardless of what colors they are, all LEDs have a forward voltage dropped between their terminals when they are turned on. The magnitude of the forward voltage of an LED is determined by the material used to manufacture the device. As summarized in Table 8.1, the forward voltages of LEDs are about 2.0 V for green, orange, and red colors, and are much higher than 2 V for blue and white colors, which are 5 and 4 V, respectively. Nevertheless, regardless of what color an LED is, the light intensity of an LED will increase with an increasing forward current up to a point where any further increase in current will not significantly increase the level of illumination. This point is called the *saturation point* and the corresponding current is called the *saturation current*. Based on this observation, the forward current of an LED must be limited in the range between 0 and the saturation current. To control the forward current of an LED, a resistor is often connected in-series with the LED.

Table 8.1: Various commercial light-emitting diodes.

Color	Material	Forward voltage (V)
Amber	AlInGaP	2.0
Blue	GaN	5.0
Green	GaP	2.2
Orange	GaAsP	2.0
Red	GaAsP	1.8
White	GaN	4.0
Yellow	AlInGaP	2.0

LEDs can be individually used as indicators or combined into a particular pattern, such as the one shown in Figure 8.19(a), which is patterned as a digit and is known as a *seven-segment display* because it consists of seven LED devices. In practice, there are two types of seven-segment displays available for use, as shown in Figures 8.19(b) and (c), respectively. A seven-segment display is referred to as a *common-anode structure* if all anodes of the seven LEDs are connected together and is known as a *common-cathode structure* if all cathodes of the seven LEDs are connected together.

The LED devices shown in Figure 8.20(a) are used as individual indicators. Recall that to turn on an LED, we need to apply a forward bias on the device. For instance, to turn on the leftmost LED of Figure 8.20(a), we have to activate the input signal a in a way such that the output of the buffer is lowered down to the ground level, thereby making the LED forward biased. To protect the LED from being damaged, the current flowing through it must be limited in the range of 10 to 20 mA. This is done by a resistor with the resistance of 220 Ω connected in series between the LED and the power supply.

Section 8.6 Case Study — Seven-Segment LED Displays

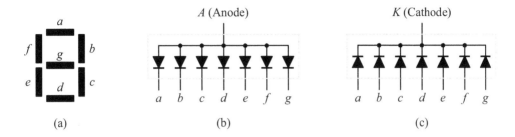

Figure 8.19: The (a) pattern, (b) common-anode structure, and (c) common-cathode structure of the seven-segment LED display [ignoring the decimal-point (dp) LED].

A common-anode seven-segment LED display is shown in Figure 8.20(b). To display a digit on the seven-segment display, it needs to turn on an appropriate subset of LEDs according to the digit being displayed, referring to Figure 2.32 for more details about digital patterns.

The on-off status of all LEDs for displaying a specific digit pattern is called the *display codeword*. The set of display codewords is referred to as a *display code*. In order to specify the display codeword for a digit pattern, we need to define a display codeword format, as shown in Figure 8.21(b). Based on this format, we can derive the display codeword for each digit to be displayed. For instance, we can see from Figure 8.21(a) that to display the digit "0," all LEDs except the one labeled with g must be turned on. This means that all values of a to f must be 0s and g must be 1. As a result, the display codeword for the digit "0" is 40 in hexadecimal. The other digits can be derived in a similar way. The complete display code is summarized in Figure 8.21(c). It is worth noting that this display code is special for use with common-anode seven-segment LED displays. For the devices with a common cathode, each display codeword of the display code is needed to complement. In addition, when the input is only limited to digits, "0" to "9," we can only use the first 10 display codewords of the display code and ignore the remaining six ones.

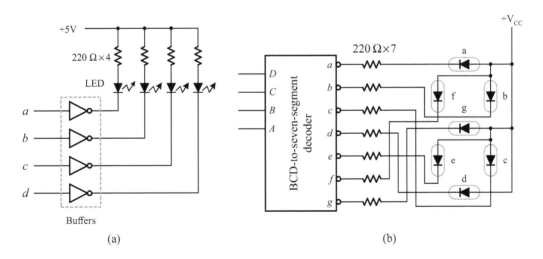

Figure 8.20: Examples of (a) LED indicators and (b) a seven-segment LED display circuit.

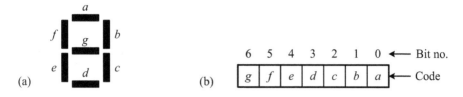

(a) (b)

(c)

Figure 8.21: The (a) seven-segment LED display, (b) display codeword format, and (c) display code of the common-anode seven-segment LED display.

■ Example 8.23: A BCD-to-seven-segment decoder.

As an example of illustrating how to convert a BCD input into a common-anode seven-segment LED display codeword, suppose that the input is a BCD code. According to the display code exhibited in Figure 8.21(c), the first ten display codewords can be described by a **case** statement as shown in the following module. The remaining six display codewords are treated by the **default** statement, which has the display codeword of 7'b111_1111. Hence, the seven-segment LED display will be blank when the input is not a valid BCD code. The circuit associated with this decoder is shown in Figure 8.20(b).

```
// converting input data into seven_segment_display
// code with inverting outputs
// the input data is assumed in the range of 0 to 9
module BCD_to_seven_segment_decoder(
       input     [3:0] data_in,
       output reg [6:0] data_out);
// the default case is assumed to be blank
always @(data_in)
   case (data_in)
      0: data_out = 7'b100_0000;   // h40
      1: data_out = 7'b111_1001;   // h79
      2: data_out = 7'b010_0100;   // h24
      3: data_out = 7'b011_0000;   // h30
      4: data_out = 7'b001_1001;   // h19
      5: data_out = 7'b001_0010;   // h12
      6: data_out = 7'b000_0010;   // h02
      7: data_out = 7'b111_1000;   // h78
      8: data_out = 7'b000_0000;   // h00
      9: data_out = 7'b001_0000;   // h10
```

```
        default: data_out = 7'b111_1111;
    endcase
endmodule
```

8.6.1.1 Displaying Multiple Digits The above discussion is only for displaying one digit on a seven-segment LED display. In most realistic applications, it is desirable to display many digits on seven-segment LED displays at the same time. For this purpose, there are two widely used methods: the *direct-driven approach* and the *multiplexed approach*. The direct-driven approach is the most straightforward way to display multiple digits. This approach is simply to duplicate the circuit in Figure 8.20(b) the desired number of times. For instance, as four digits to be displayed at a time is desired, four copies of the circuit of Figure 8.20(b) are needed, with each for one digit. Based on this idea, a seven-segment LED display with any number of digits can be readily constructed.

One major drawback of the direct-driven approach is that it consumes too much power. To see this, recall that each illuminated LED has to consume a current of about 10 to 20 mA. If we assume that the illuminated current of an LED is 20 mA, then the average current is up to 392 mA for a 4-digit seven-segment LED display. This is quite tremendous and often intolerable in most real-world applications. Therefore, an alternative known as the multiplexed approach is used instead to circumvent this drawback. In what follows, we focus on this approach.

8.6.2 Multiplexed Seven-Segment LED Display Module

It can be shown that we are quite satisfied with the brightness (illumination) generating by an LED as it is illuminated for 1 ms at a current level of 10 mA and refreshed every 10 to 16 ms at least. Based on this, we can arrange a multiple-digit seven-segment LED display in a way such that all digit patterns on the display are illuminated in a time-division-multiplexed manner, with each digit being displayed for about 1 ms and repeated every 10 to 16 ms. This results in a multiplexed approach and the seven-segment LED display using this approach along with its associated logic circuitry is often referred to as a *multiplexed seven-segment LED display module*.

A widely used multiplexed common-anode seven-segment LED display is shown in Figure 8.22(a). In a multiplexed seven-segment LED display, all seven-segment LED displays are connected in a way such that all segments with the same label are connected together and each anode is driven by an individual transistor, as shown in Figure 8.22(a). The base of each transistor serves as a control input. To display a digit pattern, the corresponding control input is lowered down to the ground level to turn on its associated transistor so as to power LEDs and meanwhile the corresponding display codeword is applied to the seven-segment LED display.

Figure 8.22(b) shows the timing diagram that controls the turned-on period of all four transistors and hence the four seven-segment LED displays. It can be seen from Figure 8.22(b) that each seven-segment LED display is enabled periodically to display its own digit for a short time. In other words, all four seven-segment LED displays share a common BCD-to-seven-segment decoder and are illuminated in a time-division multiplexing manner.

An important feature of a multiplexed seven-segment LED display is that the amount of current consumed is only equal to that of one seven-segment LED display using the direct-driven approach. Therefore, the power dissipation is very low in comparison with the direct-driven approach, in particular, when the number of digits is large.

A complete 4-digit multiplexed common-anode seven-segment LED display module is revealed in Figure 8.23. It consists of four parts: a 2-bit binary counter and a 2-to-4 decoder, a 4-bit 4-to-1 multiplexer, a BCD-to-seven-segment decoder, and the 4-digit seven-segment LED display module, namely, Figure 8.22(a). The 2-bit binary counter and the 2-to-4 decoder generate the selection signals, S1 and S0, for the 4-bit 4-to-1 multiplexer and the control signals, A3, A2, A1, and A0, for the 4-digit seven-segment LED display module, as shown in

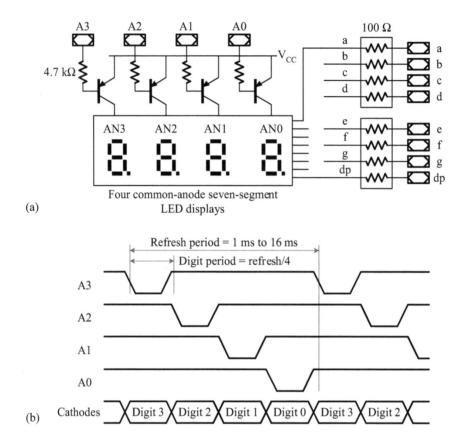

Figure 8.22: The (a) logic circuit and (b) timing diagram of a 4-digit multiplexed seven-segment LED display.

Figure 8.22(b). Both selection signals, S1 and S0, and control signals, A3, A2, A1, and A0, have to be in synchronism with each other. This can be easily accomplished by decoding the selection signals directly into the control signals. The 4-bit 4-to-1 multiplexer multiplexes the 4-digit data source into the input of the BCD-to-seven-segment decoder through which a BCD input is converted into a seven-segment display codeword.

■ Example 8.24: A multiplexer timing generator.

The function of the 2-bit (modulo-4) binary counter is to generate the selection signals for the 4-bit 4-to-1 multiplexer and the control input signals for the multiplexed seven-segment LED display. These two sets of signals must be in synchronism with one another. To achieve this, the outputs of the 2-bit binary counter are served as the selection signals and at the same time decoded into the control input signals with a 2-to-4 decoder. The 2-bit binary counter is described by the first **always** block and the decoder by the second **always** block. More details about counters will be explored in the next chapter.

```
// a binary counter with a decoder serve as a timing generator
module mux_timing_generator(
       input  clk, reset,
       output [1:0] mux_sel,
       output reg [3:0] addr);// generates A3 to A0 signals
// the body of the binary counter
```

Section 8.6 Case Study — Seven-Segment LED Displays

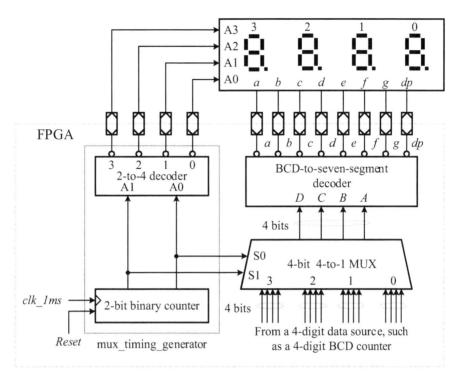

Figure 8.23: The complete logic diagram of a 4-digit multiplexed seven-segment LED display module.

```
reg     [1:0] qout;
always @(posedge clk or posedge reset)
   if (reset) qout <= 0;
   else       qout <= qout + 1;
assign mux_sel = qout;
// decode the output of the binary counter
always @(qout)
   case(qout)
      2'b00:   addr = 4'b1110;
      2'b01:   addr = 4'b1101;
      2'b10:   addr = 4'b1011;
      2'b11:   addr = 4'b0111;
   endcase
endmodule
```

■ Example 8.25: A 4-bit 4-to-1 multiplexer.

The 4-bit 4-to-1 multiplexer used in the seven-segment LED display module shown in Figure 8.23(a) is described as in the following module. The data width in this multiplexer is 4 bits. Since it is the typical one that can be described with a **case** statement, we will not further discuss it here.

```
// a 4-bit 4-to-1 multiplexer
module four_bit_mux_4to1(
```

```verilog
        input     [1:0] mux_sel,
        input     [3:0] data_in3, data_in2, data_in1, data_in0,
        output reg [3:0] data_out);
// the body of the multiplexer
always @(*)
    case (mux_sel)
        2'b00: data_out = data_in3;
        2'b01: data_out = data_in2;
        2'b10: data_out = data_in1;
        2'b11: data_out = data_in0;
    endcase
endmodule
```

■ Example 8.26: Top-level module of complete display system.

The top-level module of the 4-digit multiplexed seven-segment LED display module receives a clock signal, clk_1ms, a reset signal, reset, and four 4-bit data words, and generates a 4-bit control input signal and a 7-bit display codeword output, as shown in Figure 8.23 with a dashed-line box. The top-level module instantiates three modules, mux_timing_generator, four_bit_4_to_1_mux, and BCD_to_seven_segment_decoder. Its details are as follows:

```verilog
// the top-level module of the 4-digit multiplexed
// seven-segment LED display system
module four_digit_LED_display(
        input     clk_1ms, reset,
        input     [3:0] data3, data2, data1, data0,
        output    [3:0] addr,
        output    [6:0] DB_out);
// the body of the module
wire [3:0] current_data_out;
wire [1:0] mux_sel;
mux_timing_generator mux_select
                    (clk_1ms, reset, mux_sel, addr);
four_bit_mux_4to1 mux_BCD (mux_sel, data3, data2,
                    data1, data0, current_data_out);
BCD_to_seven_segment_decoder segment_seg_LED
                    (current_data_out, DB_out);
endmodule
```

■ Review Questions

8-26 What is the forward voltage of a typical blue LED device?

8-27 What are the forward voltages of the most widely used green, orange, and red LED devices?

8-28 What is the major disadvantage of a direct-driven seven-segment LED display?

8-29 What is the rationale behind the multiplexed seven-segment LED display?

8-30 Compare the features of the direct-driven and multiplexed seven-segment LED displays?

8.7 Summary

In this chapter, we have examined a number of widely used combinational modules, including encoders and decoders, multiplexers and demultiplexers, and magnitude comparators. In addition, a multiplexed seven-segment LED display module combining the use of a binary counter, a decoder, a multiplexer, and a BCD-to-seven-segment decoder was discussed in detail.

A decoder is a logic circuit that receives an n-bit input code and outputs a 1-out-of-m code in a way such that the physical position of the activated output bit corresponds to the input binary code. A decoder can be fully decoded or partially decoded, depending on the actual requirement. An encoder is a logic circuit with an m-bit ($m = 2^n$, or fewer) input and an n-bit output. It outputs a binary code corresponding to the physical position of the activated input. Encoders may be either a simple encoder in which only one of its inputs can be activated at a time or a priority encoder in which each input has a unique priority. Nevertheless, priority encoders are usually used in practical applications.

An m-to-1 ($m = 2^n$) multiplexer is a logic circuit that has m inputs, one output, and n source selection inputs. The input I_i selected by the binary combination of n source selection inputs is routed to the output Y. A 1-to-m ($m = 2^n$) demultiplexer is a circuit that has one input, m outputs, and n destination selection inputs. The input D is routed to the output Y_i selected by the binary combination of n destination selection inputs.

A comparator is a logic circuit that compares two numbers in magnitude and indicates whether they are equal. When a comparator not only compares two numbers in magnitude to test their equality but also indicates the arithmetic relationship between them, it is called a magnitude comparator. The input numbers of a magnitude comparator can be signed or unsigned.

LEDs are widely used in most digital or mixed-signal systems as indicators or to present information visually. By appropriately combining a number of LEDs into patterns, such as seven-segment digits or dot-matrix characters, digits and characters may be displayed as desired. There are two ways that can be used to display multiple digits on seven-segment LED displays: the direct-driven approach and the multiplexed approach. In the former, each seven segment LED display associates an individual BCD-to-seven-segment decoder and is always turned on. In the latter, all seven-segment LED displays share a common BCD-to-seven-segment decoder and are turned on in a time-division multiplexing manner. The direct-driven approach has a simpler control circuit but consumes much more power and needs more hardware than the multiplexed approach. In contrast, the multiplexed approach consumes less power but needs a much more complicated control circuit than the direct-driven approach.

References

1. J. Bhasker, *A Verilog HDL Primer*, 3rd ed., Star Galaxy Publishing, 2005.
2. R. L. Boylestad and L. Nashelsky, *Electronic Devices and Circuit Theory*, 9th ed., Upper Saddle River, New Jersey: Prentice-Hall, 2006.
3. M. B. Lin, *Digital System Design: Principles, Practices, and Applications*, 4th ed., Chuan Hwa Book Ltd. (Taipei, Taiwan), 2010.
4. M. B. Lin and S. T. Lin, *Basic Principles and Applications of Microprocessors: MCS-51 Embedded Microcomputer System, Software, and Hardware*, 3rd ed., Chuan Hwa Book Ltd. (Taipei, Taiwan), 2013.
5. J. F. Wakerly, *Digital Design Principles & Practices*, 3rd ed., Upper Saddle River, New Jersey: Prentice-Hall, 2001.

Problems

8-1 Design a 3-to-8 decoder with each of the following specified output polarities and then write Verilog HDL modules to describe them.
  (a) Noninverting output
  (b) Inverting output

8-2 Design a 4-to-16 decoder using at most two enable-controlled 3-to-8 decoders and write a Verilog HDL module to describe it.

8-3 Design a 4-to-16 decoder using at most five enable-controlled 2-to-4 decoders and write a Verilog HDL module to describe it.

8-4 Design a 5-to-32 decoder using at most four enable-controlled 3-to-8 decoders and a 2-to-4 decoder. Write a Verilog HDL module to describe it.

8-5 Design a 4-to-2 priority encoder and write a Verilog HDL module to describe it. The priority encoder must have an enable input to enable the encoder and a valid output to indicate that at least an input is active.

8-6 Design a 4-to-1 active-low multiplexer and write a Verilog HDL module to describe it. The output is at a high impedance when the enable input is deasserted.

8-7 Use each of the following specified approaches to design a 32-to-1 multiplexer and write Verilog HDL modules to describe them.
  (a) Two 16-to-1 multiplexers and one 2-to-1 multiplexer.
  (b) Four 8-to-1 multiplexers and one 4-to-1 multiplexer.

8-8 Use two 2-to-1 multiplexers to design a 3-to-1 multiplexer and write a Verilog HDL module to describe it.

8-9 Write a Verilog HDL module to describe a parameterizable n-bit m-to-1 multiplexer. Also write a test bench to verify its functionality.

8-10 How many 1-to-4 demultiplexers are required when we use them to design each of the following specified demultiplexers?
  (a) A 1-to-32 demultiplexer
  (b) A 1-to-64 demultiplexer
  (c) A 1-to-128 demultiplexer
  (d) A 1-to-256 demultiplexer

8-11 Write a Verilog HDL module to describe a parameterizable n-bit 1-to-m demultiplexer. Also write a test bench to verify its functionality.

8-12 Use 4-bit magnitude comparators to design each of the following specified magnitude comparators. Also write Verilog HDL modules to describe them.
  (a) An 8-bit magnitude comparator
  (b) A 12-bit magnitude comparator

8-13 Considering a common-cathode seven-segment LED display, answer each of the following questions.
  (a) Define the seven-segment LED display code.

(b) Write a BCD-to-seven-segment decoder module in Verilog HDL.

8-14 Write a test bench to test the functionality of the multiplexed seven-segment LED display module described in Figure 8.23.

8-15 Design an eight-digit seven-segment LED display system, supposing that common-cathode seven-segment LED displays are used.
   (a) Draw the block diagram and control timing of the seven-segment LED display system.
   (b) Draw a complete seven-segment LED display system.
   (c) Write various Verilog HDL modules required for the seven-segment LED display system.
   (d) Write a test bench to verify the functionality of the seven-segment LED display system.

8-16 Supposing that the current passing through a turned-on LED is 10 mA and each decimal digit appears equally likely on the seven-segment LED display, calculate the average current consumed in an eight-digit seven-segment LED display system designed based on each of the following specified approaches.
   (a) The direct-driven approach
   (b) The multiplexed approach

8-17 When designing a big decoder, we may use either a single-module approach or a hierarchical approach. Using an 8-to-256 decoder as an example, study the performance of both approaches in terms of the hardware cost and propagation delay of the resulting decoder.
   (a) Construct a single 8-to-256 decoder using any approach you like.
   (b) Construct a decoder tree using enable-controlled 3-to-8 decoders as the major building blocks.
   (c) Synthesize the above two modules and compare their hardware cost in terms of the number of LUTs.
   (d) Compare the propagation delays of both modules.

8-18 When designing a big encoder, we may use either a single-module approach or a hierarchical approach. Using a 256-to-8 encoder as an example, study the performance of both approaches in terms of the hardware cost and propagation delay of the resulting encoder.
   (a) Construct a single 256-to-8 encoder using any approach you like.
   (b) Construct an encoder tree using enable-controlled 8-to-3 encoders as the major building blocks.
   (c) Synthesize the above two modules and compare their hardware cost in terms of the number of LUTs.
   (d) Compare the propagation delays of both modules.

8-19 When designing a big multiplexer, we may use either a single-module approach or a hierarchical approach. Using a 256-to-1 multiplexer as an example, study the performance of both approaches in terms of the hardware cost and propagation delay of the resulting multiplexer.
   (a) Construct a single 256-to-1 multiplexer using any approach you like.

(b) Construct a multiplexer tree using enable-controlled 8-to-1 multiplexers as the major building blocks.

(c) Synthesize the above two modules and compare their hardware cost in terms of the number of LUTs.

(d) Compare the propagation delays of both modules.

8-20 When designing a big demultiplexer, we may use either a single-module approach or a hierarchical approach. Using a 1-to-256 demultiplexer as an example, study the performance of both approaches in terms of the hardware cost and propagation delay of the resulting demultiplexer.

(a) Construct a single 1-to-256 demultiplexer using any approach you like.

(b) Construct a demultiplexer tree using enable-controlled 1-to-8 demultiplexers as the major building blocks.

(c) Synthesize the above two modules and compare their hardware cost in terms of the number of LUTs.

(d) Compare the propagation delays of both modules.

8-21 When designing a big magnitude comparator, we may use either a single-module approach or a hierarchical approach. Using a 64-bit magnitude comparator as an example, study the performance of both approaches in terms of the hardware cost and propagation delay of the resulting magnitude comparator.

(a) Construct a single 64-bit magnitude comparator using any approach you like.

(b) Construct a comparator tree using 4-bit comparators as the major building blocks.

(c) Synthesize the above two modules and compare their hardware cost in terms of the number of LUTs.

(d) Compare the propagation delays of both modules.

8-22 A majority circuit is a device that outputs 1 whenever more than half of its inputs are 1. Answer each of the following questions.

(a) Describe the behavior of an n-input majority circuit.

(b) Model the n-input majority circuit in behavioral style.

(c) Synthesize the above module and notice its hardware cost in terms of the number of LUTs.

(d) What is the propagation delay of the resulting module?

8-23 A minority circuit is a device that outputs 1 whenever less than half of its inputs are 1. Answer each of the following questions.

(a) Describe the behavior of an n-input minority circuit.

(b) Model the n-input minority circuit in behavioral style.

(c) Synthesize the above module and notice its hardware cost in terms of the number of LUTs.

(d) What is the propagation delay of the resulting module?

9

Sequential Logic Modules

In this chapter, we examine a number of basic sequential modules that are widely used in digital systems. These include flip-flops, synchronizers, a switch-debounce circuit, registers, data registers, register files, shift registers, counters (binary, BCD, ring, and Johnson counters), CRC code generators and detectors, clock generators, pulse generators, and timing generators.

The most basic building blocks of any sequential module are flip-flops and latches. Hence, in this chapter, we first consider the flip-flop in detail and give some of its basic applications. Synchronizers are circuits used to sample an external asynchronous signal (which may occur at any time) in synchronism with an internal system clock signal. Switch debouncers are circuits that remove the bouncing effects of mechanical switches due to the mechanical inertia of switches.

After introducing flip-flops and latches, we consider a broad variety of applications of flip-flops, including registers, shift registers, and counters. Registers usually contain basic registers and register files. Shift registers are often used to perform arithmetic operations, such as divide-by-2, multiply-by-2, and data format conversions. Counters that we consider include both asynchronous and synchronous types. In addition, shift-register-based counters are also considered in great detail. Furthermore, we discuss a variety of clock and timing generators widely used in digital systems. Finally, we use a clock/timer case study to comprehensively explore the combination of modules introduced in the previous and this chapters into a useful and practical system.

Many options can be used to model sequential logic. These options include behavioral statements, tasks with delay or event timing control, sequential UDPs, instantiated library register cells, and interconnected sequential logic modules. We will use these whenever they are appropriate.

9.1 Flip-Flops

Flip-flops are the basic building blocks of sequential circuits. They can be used to construct registers, data registers, register files, shift registers, counters, clock generators, timing generators, and others. In this section, we introduce the fundamentals of flip-flops, distinguish flip-flops from latches, and describe a number of basic applications of flip-flops, including synchronizers and a switch-debounce circuit.

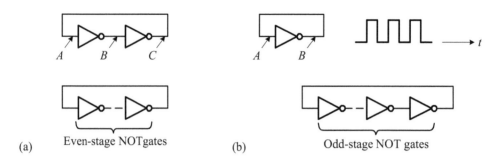

Figure 9.1: The basic structures of (a) a bistable circuit and (b) an oscillator circuit.

9.1.1 Flip-Flops

A bistable device is a circuit with two stable states, as shown in Figure 9.1(a). A bistable device is usually built from an even number of NOT gates or inverters. For instance, the bistable circuit exhibited in Figure 9.1(a) is constructed from two NOT gates cascaded together into a loop structure. The signal A propagates to the output of the first NOT gate as an inverted version, denoted by B, which then propagates to the output of the second NOT gate as an inverted signal again, denoted by C. Finally, the signal C is fed back to the input of the first NOT gate, with a consistent phase with the signal A. Thus, the circuit is bistable.

To change the state of a bistable device electronically, either two-input NOR controlled gates or NAND controlled gates can be used to replace the inverters in the bistable device, as illustrated in Figures 9.2(a) and (b), respectively. A bistable device is called a NOR-gate (or an active-high) SR latch if both the inverters of the bistable device are replaced with NOR controlled gates and a NAND-gate (or an active-low) SR latch if both the inverters of the bistable device are replaced with NAND controlled gates. It should be noted that the inputs R (reset) and S (set) are at different places in the NOR-gate SR latch and in the NAND-gate SR latch.

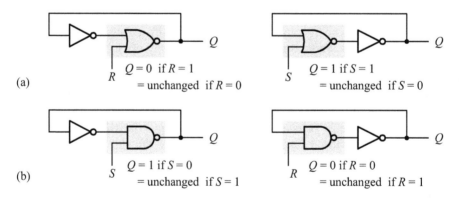

Figure 9.2: The use of (a) NOR controlled gates and (b) NAND controlled gates to change the state of bistable devices.

Figure 9.1(b) shows an unstable circuit, which consists of an odd number of NOT gates. A conceptual unstable circuit is simply a one-stage loop-connected NOT gate. A signal appearing at its input is inverted and appeared at its output as an inverted version after a finite delay, which is then fed back to its input. Because the polarities of both original and feedback signals are different, the feedback signal is to be continuously inverted and fed back forever. Hence, the

Section 9.1 Flip-Flops

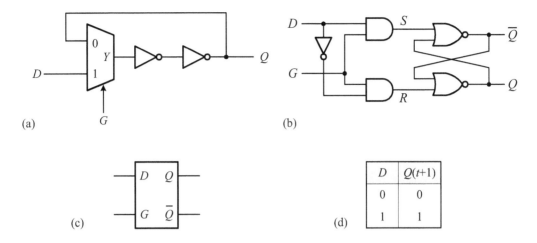

Figure 9.3: D latches: (a) multiplexed D latch; (b) SR-latch D latch; (c) logic symbol; (d) characteristic table.

resulting circuit is unstable, thereby, leading to an oscillator. For CMOS technology, it proves safe to construct an oscillator by cascading three or more odd number of stages of NOT gates.

9.1.1.1 Latches The fundamental applications of bistable devices are used to form latches and flip-flops. A latch is a bistable device capable of latching or storing the external data. To achieve this, some mechanism must be provided to route the external data into the bistable device without causing a conflict with the internal feedback data. According to the mechanism used to route the external data into the latch, two basic techniques can be used to construct a latch, which are described as follows:

- *Multiplexing technique:* In this technique, a multiplexer is employed to break the feedback path inherent in the bistable device and to route the external data into the latch. The resulting latch is known as a *multiplexed latch*.
- *Controlled-gate technique:* In the controlled-gate technique, controlled gates, such as NOR gates or NAND gates, are used to introduce the external data into a bistable device. The resulting latch is called a NOR-gate (or active-high) SR latch or NAND-gate (or active-low) SR latch, depending on whether NOR gates or NAND gates are utilized to construct the device.

An example of multiplexed D latches is shown in Figure 9.3(a), which accepts and passes the external data D to the output when the source selection input G of the 2-to-1 multiplexer is set to 1 and latches the sampled data otherwise. The second structure is based on a NOR-gate SR latch, as shown in Figure 9.3(b), where a NOR-gate SR latch is combined with two AND gates and one inverter to form a D latch. This type of latch is known as an SR-latch D latch. The logic symbol of the D latch and its characteristic table are shown in Figures 9.3(c) and (d), respectively. Refer to Lin [3] for more details.

A *positive latch* is a latch that passes the external data through it to its output when its enable input is high and a *negative latch* is one that passes the external data through it to its output when its enable input is low. Such a property is called the *transparent property*. In addition to being used as latches alone, latches are often employed to form flip-flops, known as *master-slave flip-flops*, each being constructed by combining a positive latch with a negative latch or vice versa.

9.1.1.2 Flip-Flops Like latches, a flip-flop is a bistable device capable of storing the external data. In CMOS technology, flip-flops are usually built from two latches with different enabling polarities: a positive latch and a negative latch. The resulting type of flip-flop is called a *master-slave flip-flop* and can function as an edge-triggered flip-flop. To see this, consider the logic circuit depicted in Figure 9.4(a), which consists of two latches, called the master and the slave latches. The *master latch* is a negative latch, accepting the external data when the clock signal (*clk*) is low and storing the sampled data otherwise. The *slave latch* is a positive latch, receiving the output data from the master latch when its clock signal *clk* is high and storing the sampled data otherwise. As a result, the slave latch presents at its output the external data sampled at the positive edge of the clock signal and hence the flip-flop functions as a positive-edge-triggered flip-flop. The logic symbol, characteristic table, and state diagram, of the positive-edge-triggered D flip-flop are shown in Figures 9.4(b), (c), and (d), respectively.

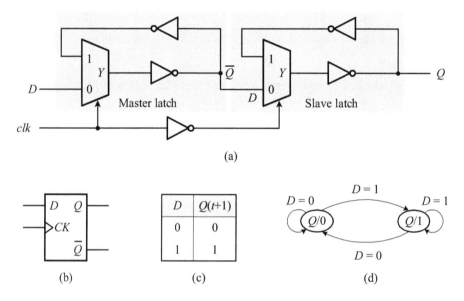

Figure 9.4: A master-slave D flip-flop: (a) logic circuit; (b) logic symbol; (c) characteristic table; (d) state diagram.

9.1.1.3 Transparent Property Although both latches and flip-flops are widely used in digital systems, there exists an essential difference between them. Recall that the output of a latch will follow its input data whenever its enable input G is high; that is, it owns the transparent property. However, the output of a flip-flop is only the sampled value of the input data at the specified (positive or negative) edge of the clock signal. Hence, in the context of the transparent property, a latch is a bistable device with the transparent property whereas a flip-flop is a bistable device without the transparent property.

To get more insight into the difference between latches and flip-flops, let us consider the timing diagrams depicted in Figure 9.5. As we can see, the output of the (positive) latch depicted in Figure 9.3(a) follows the input data when its enable input G is high and holds the value sampled before the enable input G goes to low. In contrast, the output of the positive-edge-triggered D flip-flop described in Figure 9.4(a) holds the value sampled at each positive edge of the clock signal and retains that value during the rest of the whole clock cycle. In a number of applications, the transparent property may cause troubles, such as inducing noises into the system. In addition, in designing digital systems based on the synthesis flow, it is more difficult to handle latches than flip-flops even though using latches with care may achieve higher performance. Therefore, in this book we only consider flip-flops.

Section 9.1 Flip-Flops

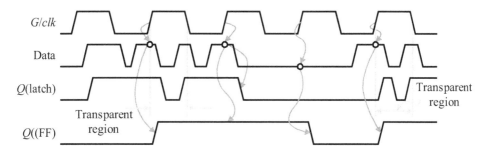

Figure 9.5: An illustration of the difference between a D latch and a D flip-flop.

9.1.1.4 Modeling Flip-Flops Although there are many types of flip-flops, including SR, JK, and T flip-flops, nowadays the D flip-flop is the most widely used bistable device in sequential logic. A practical D flip-flop usually has a *reset* (or *clear*) input, which clears or resets the output of the flip-flop asynchronously or synchronously. The following two examples illustrate how to model such a practical D flip-flop.

■ Example 9.1: A D flip-flop with asynchronous reset.

The word "asynchronous" means randomness, i.e., beyond the control of the clock signal. In other words, an asynchronous signal occurs in a random and unpredictable way. As a consequence, to model an asynchronous reset signal of a D flip-flop, it needs to put the reset signal within the sensitivity list of the **always** block being used to describe the D flip-flop so that the activities of the **always** block can also be triggered by the reset signal. The resulting module is as follows. The output q is cleared to 0 whenever the reset signal reset_n is active, i.e., 0, regardless of the status of the clock signal.

```
// a D flip-flop with asynchronous reset
module DFF_async_reset(
       input   clk, reset_n, d,
       output reg q);

// the body of the flip-flop
always @(posedge clk or negedge reset_n)
   if (!reset_n) q <= 0; // active-low reset
   else          q <= d;
endmodule
```

Of course, the reset operation can also be performed in synchronism with the clock signal. A flip-flop operated in this manner is said to be synchronously resettable. An illustration of this type of flip-flop is explored in the following example.

■ Example 9.2: A D flip-flop with synchronous reset.

The word "synchronous" means under the control of the clock signal. In other words, a synchronous signal occurs in a predictable way. As a consequence, to model a synchronous reset signal of a D flip-flop, it needs to put the reset signal outside the sensitivity list of the **always** block being used to describe the D flip-flop so that the reset signal can merely be tested after the **always** block is triggered by the clock signal. This results in the following module. At each positive edge of the clock signal clk, the reset signal reset is checked to see if it is active. If yes, then the output q is cleared to 0; otherwise, the input data d is sampled into the flip-flop.

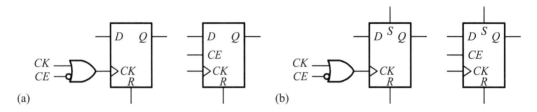

Figure 9.6: Symbols for the D flip-flops of CPLD and FPGA devices with: (a) reset and (b) reset and preset.

```
// a D flip-flop with synchronous reset
module DFF_sync_reset (
      input   clk, reset, d,
      output reg  q);

// the body of the flip-flop
always @(posedge clk)
   if (reset)  q <= 0; // active-high reset
   else        q <= d;
endmodule
```

9.1.1.5 General Considerations of Flip-Flop Modeling Most field-programmable devices, such as CPLDs and FPGAs (see Chapter 10), have flip-flops with a clock-enable (CE) input in addition to the asynchronous reset (R), as shown in Figure 9.6. Moreover, sometimes an asynchronous preset (i.e., set, S) input is also added as an optional feature. Both reset and preset inputs can be active-low or active-high. Like the clock-enable input, the reset and preset inputs can also be either synchronous or asynchronous. To illustrate how to model such flip-flops, we use positive-edge-triggered D flip-flops as examples to explore a number of useful modeling paradigms.

- A flip-flop output qout is inferred and updated with the value of the input data data_in on each rising clock signal, clock.

    always @(**posedge** clock)
       qout <= data_in;

- A flip-flop output qout with synchronous reset and clock enable is inferred and updated with the value of the input data data_in on each rising clock signal, clock, under the control of the clock-enable input, clock_en, if the synchronous reset signal is not asserted.

    always @(**posedge** clock)
       if (reset) qout <= 0;
       else if (clock_en) qout <= data_in;

- A flip-flop output qout with asynchronous reset and clock enable is inferred and updated with the value of the input data data_in on each rising clock signal, clock, under the control of the clock-enable input, clock_en, if the asynchronous reset signal is not asserted.

    always @(**posedge** clock **or posedge** reset)
       if (reset) qout <= 0;
       else if (clock_en) qout <= data_in;

Section 9.1 Flip-Flops

- A flip-flop output `qout` with both asynchronous reset and preset inputs is inferred and updated with the value of the input data `data_in` on each rising clock signal, `clock`, if both asynchronous `reset` and `preset` signals are not asserted.

   always @(**posedge** clock **or posedge** reset **or posedge** preset)
      if (reset) qout <= 0;
      else if (preset) qout <= 1;
         else qout <= data_in;

   However, this paradigm can merely be used with a flip-flop having both asynchronous reset and preset inputs.

- A flip-flop output `qout` with both asynchronous reset and preset inputs along with clock enable is inferred and updated with the value of the input data `data_in` on each rising clock signal, `clock`, under the control of the clock-enable input, `clock_en`, if both asynchronous `reset` and `preset` signals are not asserted.

   always @(**posedge** clock **or posedge** reset **or posedge** preset)
      if (reset) qout <= 0;
      else if (preset) qout <= 1;
         else if (clock_en) qout <= data_in;

   Similarly, this paradigm can merely be used with a flip-flop that has both asynchronous reset and preset inputs.

■ **Review Questions**

9-1 What are the differences between a bistable device and an oscillator?
9-2 Explain the operation of the circuit shown in Figure 9.3(b).
9-3 What is the meaning of the transparent property?
9-4 Distinguish the synchronous reset from the asynchronous one of flip flops.
9-5 What is the basic difference between latches and flip-flops?
9-6 Design a negative-edge-triggered D flip-flop by properly combining two latches.
9-7 Describe the basic features of D flip-flops.

9.1.2 Basic Timing and Metastable State

Remember that a flip-flop uses the clock signal being applied to it to sample the data appearing at its data input D and then stores the sampled value for the rest of the whole clock cycle. To make a flip-flop properly sample the input data, the input data must be stable at the data input D of the flip-flop for a period of time, referred to as the *sampling window*, being the sum of the *setup time* and *hold time* [3] of the flip-flop. The setup and hold times are constrained by the *max-delay constraints* and *min-delay constraints*, respectively. In what follows, we first address these two types of constraints and then deal with in depth clock skew and its effects, and the metastable state of flip-flops.

9.1.2.1 Max-Delay Constraints The max-delay constraints consider the worst-case delays of both combinational logic and flip-flops. These delays determine the maximum setup time of flip-flops that can be used in a system. Generally speaking, as we use flip-flops, D or other types, to design a digital system, we need to consider the following three basic timing parameters associated with flip-flops.

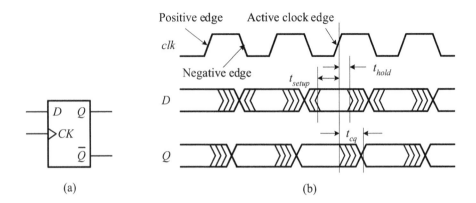

Figure 9.7: The (a) logic symbol and (b) basic timing parameters of D flip-flops.

- The *setup time* (t_{setup}) of a flip-flop is the amount of time that the input data D of the flip-flop must be stable before the active clock edge (the sampling point). The active clock edge may be positive or negative, depending on whether the underlying flip-flop is positive-edge-triggered or negative-edge-triggered.
- The *hold time* (t_{hold}) of a flip-flop is the amount of time that the input data D of the flip-flop must remain stable after the active clock edge (the sampling point).
- The *clock-to-Q delay* (t_{cq}) of a flip-flop is the maximum propagation delay from the active clock edge (the sampling edge) to the time that a new value appears at the Q output of the flip-flop.

An illustration of the definition of these three timing parameters of D flip-flops is revealed in Figure 9.7(b). To further illustrate the relationship among these timing parameters, the clock period, and the sampling window, consider the logic circuit shown in Figure 9.8(a).

The setup time (t_{setup}) and clock-to-Q delay (t_{cq}) combining with the propagation delay (t_{pd}) of the combinational circuit between two D flip-flops determine the minimum period of the clock signal, as shown in Figure 9.8(b). More specifically, this timing relationship can be represented as in the following equation

$$T_{clk} \geq t_{cq} + t_{pd} + t_{setup} \tag{9.1}$$

where T_{clk} is the clock period. This timing constraint is known as the *max-delay constraint*. As it is violated, the flip-flop B will miss its setup time and sample the wrong data, even enter the metastable state, to be described later. This situation is often called a *setup-time failure* or *max-delay failure*.

The max-delay constraint can be stated in another way. Since both clock-to-Q delay (t_{cq}) and the setup time (t_{setup}) are the inherent features of flip-flops and hence are the overhead of a digital system, for a fixed clock period T_{clk} the maximum available time in a clock cycle for the combinational logic to carry out useful computation can be expressed as follows:

$$t_{pd} \leq T_{clk} - (t_{cq} + t_{setup}) \tag{9.2}$$

Consequently, the overhead ($t_{cq} + t_{setup}$) of flip-flops must be as small as possible to maximize the available time for the combinational circuit to carry out more complicated functions. Another benefit of a small setup time will manifest itself in the design of synchronizers, to be discussed in the next subsection.

9.1.2.2 Min-Delay Constraints Min-delay constraints involve the best-case delays of both combinational logic and flip-flops. These constraints determine the maximum hold time of

Section 9.1 Flip-Flops

flip-flops that can be used in a system. To see this, observe the following facts from a practical digital system: First, in a general combinational circuit, there always exist many different paths with unequal propagation delays from the input to the output of the combinational circuit (see Section 2.3.2). Second, the propagation delays of a combinational circuit and a flip-flop are determined by both the underlying logic gates and input data profiles. Third, the propagation delays of a combinational circuit and a flip-flop may vary in a broad range due to process, voltage, and temperature (PVT) variations. As a result, besides the worst-case delays a combinational circuit and a flip-flop may also exist the best-case propagation delays.

Based on the above observations, the best-case propagation delays of both combinational logic and flip-flops are separately defined as the following two terms:

- The *contamination delay* ($t_{pd(\min)}$) of a combinational circuit is the minimum propagation delay of the combinational circuit.
- The *minimum clock-to-Q delay* or *contamination of clock-to-Q delay* ($t_{cq(\min)}$) of a flip-flop is the minimum propagation delay from the active clock edge (sampling edge) to the new value appearing at the output Q of the flip-flop.

The effects of minimum propagation delays of both combinational logic and flip-flops can be illustrated as the timing diagram depicted in Figure 9.8(c). The new data, which is supposed to appear at the later time of the current clock cycle, may corrupt the current data appearing at flip-flop B and make flip-flop B enter the metastable state if the sum of the contamination delays of flip-flop A and combinational logic is smaller than the hold-time requirement of flip-flop B. For this reason, the contamination delay is usually given to mean the *minimum delay*.

In order to guarantee that a system properly works, the hold time of flip-flop B must be shorter than the sum of the minimum clock-to-Q delay of flip-flop A and the minimum propagation delay of the combinational circuit between flip-flop A and flip-flop B. That is, the

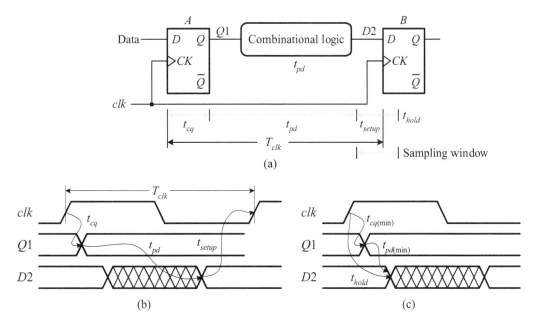

Figure 9.8: The basic timing and parameters of digital systems: (a) logic circuit and timing term definitions; (b) max-delay constraint; (c) min-delay constraint.

318   Chapter 9   Sequential Logic Modules

min-delay constraint is given as

$$t_{hold} \leq t_{cq(\min)} + t_{pd(\min)} \tag{9.3}$$

As it is violated, the data in flip-flop B will be corrupted by the new data, which is supposed to be arrived at later in the current cycle. This situation is called a *race condition, hold-time failure*, or *min-delay failure*.

In summary, the setup-time failure can be eliminated by elongating the clock period, namely, slowing the operating clock signal or by using flip-flops with a shorter setup time and/or clock-to-Q delay. Nevertheless, the hold-time failure can only be fixed by redesigning the logic circuit; it cannot be simply fixed by slowing the operating clock signal. It is good practice to design a system very conservatively to avoid such failures because redesigning or modifying a system or a chip is very expensive and time-consuming.

■ Review Questions

9-8 Define the following two terms: the setup time and hold time.
9-9 What is the meaning of the clock-to-Q delay of a flip-flop?
9-10 Describe the basic timing relationship of a flip-flop-based system.
9-11 What is the contamination delay of a combinational circuit?
9-12 How would you fix the hold-time failure? How about the setup-time failure?

9.1.2.3 Effects of Clock Skew Ideally, the clock period, T_{clk}, is determined by the three factors defined by Equation (9.1). In practice, it is also affected by two other factors: *clock jitter* and *clock skew*. They are defined as follows:

- *Clock jitter* is the temporal variations of a clock signal at a given point on a digital system. In other words, the clock period may shrink or expand on a cycle-by-cycle basis.
- *Clock skew* is the misalignment of clock edges in a synchronous system or circuit. Clock skew reduces the timing margin between the data and the clock signal at the destination register.

Ideally, a clock signal should arrive at every flip-flop at exactly the same time. In reality, the clock signal will arrive at different flip-flops at different times due to the above-mentioned factors. This time difference is called the *clock skew* [3] of flip-flops. If flip-flop i is the destination, while flip-flop j is the source, then the clock skew between two flip-flops i and j on a circuit is given by $t_{skew} = t_i - t_j$. Depending on whether the directions of the clock signal and data flow are the same or not, clock skew can be positive or negative, as shown in Figures 9.9 and 9.10, respectively. Because clock skew is caused by static mismatches in different clock paths and differences in the loading of clock paths, it is constant from cycle to cycle. The design of a low-skew clock network is essential for modern high-speed digital systems.

Positive clock skew. Positive clock skew may arise when a clock signal is routed in the same direction as the data flow through a sequential structure, as shown in Figure 9.9. Flip-flop A receives its clock signal earlier than flip-flop B. As a result, the positive clock skew occurs, that is, $t_{skew} > 0$. The constraints on the timing of a digital system including the clock skew are as follows:

$$T_{clk} \geq t_{cq} + t_{pd} + t_{setup} - t_{skew}$$
$$t_{hold} \leq t_{cq(\min)} + t_{pd(\min)} - t_{skew}$$
$$t_{skew} > 0 \tag{9.4}$$

This means that the positive clock skew effectively reduces the minimum clock period required for a given system and hence promises to improve system performance. However,

Section 9.1 Flip-Flops

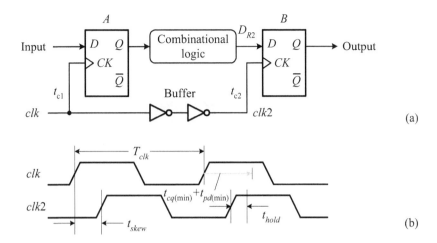

Figure 9.9: The effect of positive clock skew on a digital system: (a) logic circuit; (b) timing diagram.

positive clock skew may result in a hold-time constraint violation. For a system to be reliably operated, the hold-time constraint stated above must be satisfied. In addition, if $t_{skew} \geq t_{cq(\min)} + t_{pd(\min)}$, then flip-flop B will sample the new data instead of the old one from flip-flop A, as indicated in Figure 9.9(b).

Negative clock skew. Negative clock skew may occur when a clock signal is routed in the opposite direction as the data flow through a sequential structure, as shown in Figure 9.10. Flip-flop A receives its clock signal later than flip-flop B. This results in a negative clock skew, namely, $t_{skew} < 0$. The constraints on the timing of a digital system can be represented as follows:

$$T_{clk} \geq t_{cq} + t_{pd} + t_{setup} - t_{skew}$$
$$t_{hold} \leq t_{cq(\min)} + t_{pd(\min)} - t_{skew}$$
$$t_{skew} < 0 \tag{9.5}$$

This means that the clock skew reduces the time available for actual computation so that the clock period has to be elongated by an amount of $|t_{skew}|$ to maintain the validation of the setup time if the other two factors are not changed. However, the hold-time constraint in this situation is unconditionally satisfied.

From the above discussion, we may conclude that although positive clock skew promises to improve system performance while negative clock skew can eliminate races and make hold-time constraint unconditionally satisfied, they do not always properly work because a generic logic circuit can generally have data flow in both directions relative to the clock signal due to feedback paths that may exist in the system, in particular, in the datapath. Hence, it is necessary to control the amount of clock skew regardless of whether it is positive or negative in an acceptable range when designing a system.

■ Review Questions

9-13 What is clock skew? Define it.
9-14 What is the meaning of positive clock skew and negative clock skew?
9-15 Explain why positive clock skew may cause the hold-time constraint to be violated.
9-16 Explain why the control of clock skew is so important in a digital system.

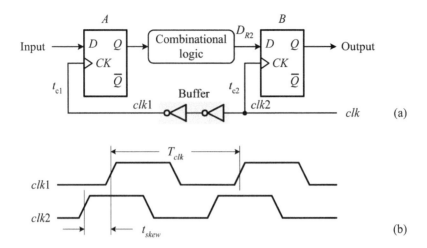

Figure 9.10: The effect of negative clock skew on a digital system: (a) logic circuit; (b) timing diagram.

9.1.2.4 Metastable State Recall that the output of a flip-flop will present the value sampled at the data input D after a clock-to-Q delay if both timing constraints, setup time and hold time (their sum is combined into the *sampling window*), are satisfied, and may output a value between a logic 1 and a logic 0 for a period of time otherwise. The latter situation is known as a *metastable* state. More precisely, a metastable state is the one between two stable states, as shown in Figure 9.11(a). It is temporarily stable at a point between two stable states. Any disturbance will let it go down to either stable state: a logic 0 or logic 1. It is worth noting that a metastable state can occur in any bistable device and any device stemmed from it. Hence, any latch and flip-flop may have opportunities to enter the metastable state if the sampling window is not satisfied.

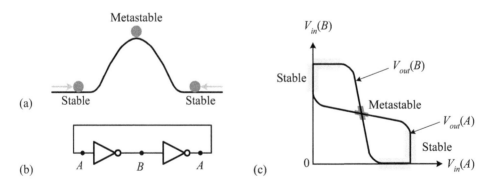

Figure 9.11: The (a) metastable state, (b) bistable circuit, and (c) voltage-transfer characteristics of a bistable device.

To get more insight into the occurrence of the metastable state, consider the bistable device in Figure 9.11(b). When the input signal A does not sustain long enough, both NOT gates may leave their outputs at the midpoints of their voltage-transfer characteristics, as shown in Figure 9.11(c), thereby taking the bistable device to the metastable state. Fortunately, the bistable device will come back to either stable state, a logic 0 or logic 1, after a short time due

Section 9.1 Flip-Flops

to the positive feedback existing in the bistable device and both NOT gates residing in their high voltage gain regions.

Once a bistable device entered the metastable state, the only way that we can do is to wait for it to leave the metastable state by itself. The amount of time required for a bistable device to move out the metastable state is referred to as the *recovery time*, t_{rec}, which is the amount of time elapsed from the time when the bistable device samples the input data to the point when the bistable device leaves the metastable state and goes back to its normal state, i.e., a stable state 0 or 1.

■ Review Questions

9-17 What is the meaning of the sampling window of a flip-flop?
9-18 What is the meaning of the metastable state of a flip-flop?
9-19 How would you tackle the metastable state of a flip-flop?
9-20 Describe the meaning of the recovery time.

9.1.3 Synchronizers

In a number of applications, it is often necessary to sample an asynchronous external signal and then to carry out the desired operations accordingly. Recall that an asynchronous external signal can be simply sampled by using a D flip-flop, as shown in Figure 9.12(a). The D flip-flop used for this purpose is known as a *synchronizer* because it synchronizes the asynchronous external signal with the internal system clock signal.

In some applications, this simple synchronizer could actually do its job quite well. Nonetheless, in many other applications, this single-stage synchronizer may cause a situation called the *synchronization failure* due to the synchronizer entering the metastable state. As illustrated in Figure 9.12(a), when the synchronizer entering the metastable state, both its succeeding stages B and C may generate different output values for the same output w from the synchronizer. For example, the output of the D flip-flop B may be a logic 1 and of the D flip-flop C may be an indeterminate value, or other combinations of a logic 0, a logic 1, and an indeterminate value, as illustrated in Figure 9.12(b). This would result in a "chaos" because of two different recognitions of the same signal. Synchronization failure may cause a "disaster" for a system and is usually not allowed in any practical system.

Remember that once a D flip-flop entered the metastable state, it needs a recovery time (t_{rec}) to go back to its normal state. Hence, as depicted in Figure 9.13, the timing relationship among the setup time, recovery time, and clock period can be expressed as follows

$$T_{clk} \geq t_{rec} + t_{pd} + t_{setup} \qquad (9.6)$$

where t_{rec} is the recovery time of the D flip-flop used as the synchronizer. For a given system, the clock period, T_{clk}, is determined by the required performance of the underlying system. For most practical systems, there is no combinational circuit associated with the synchronizer and hence the propagation delay of the combinational circuit t_{pd} is 0. The available recovery time is then equal to

$$t_{rec} \leq T_{clk} - t_{setup} \qquad (9.7)$$

which is dependent on the setup time of the D flip-flop that follows the synchronizer. To maximize the available recovery time for a specific clock period T_{clk}, we need to use a D flip-flop with a setup time as small as possible.

Because whether a synchronizer (a D flip-flop) enters the metastable state is a statistical process, a widely used measure for estimating the probability, namely, the *mean time between*

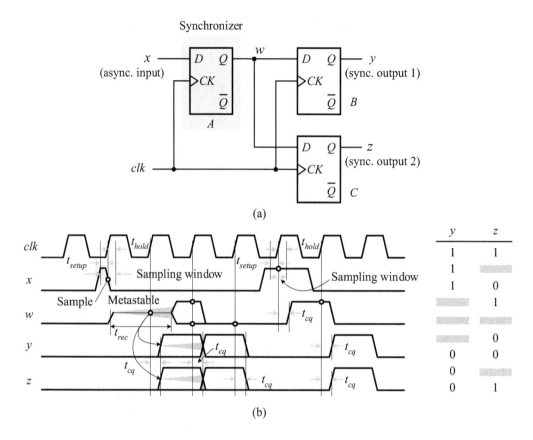

Figure 9.12: An illustration of the metastable state and synchronization failure: (a) logic circuit and (b) timing diagram.

Table 9.1: Parameters T_0 and τ of some sample devices [4].

Device	τ(ns)	T_0(s)
74LSxx	1.35	4.8×10^{-3}
74Sxx	2.80	1.3×10^{-9}
74ALSxx	1.00	8.7×10^{-6}
74ASxx	0.25	1.4×10^{3}
74Fxx	0.11	1.8×10^{8}
74HCxx	1.82	1.5×10^{-6}

failures (MTBF), of a given synchronizer entering the metastable state is given as follows

$$\text{MTBF}(t_{rec}) = \frac{1}{T_0 \cdot f \cdot a} \exp\left(\frac{t_{rec}}{\tau}\right) \tag{9.8}$$

where f is the clock operating frequency of the synchronizer, a is the number of asynchronous input changes per second, and T_0 and τ are constants determined by the electrical characteristics of the synchronizer. Some sample values of T_0 and τ are given in Table 9.1. To get more insight into the meanings of the MTBF, two illustrations are given in the following examples.

Section 9.1 Flip-Flops

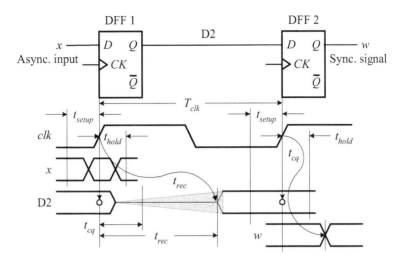

Figure 9.13: An illustration of the synchronizer timing for showing the relationship among the recovery time, set-up time, and the clock period.

■ Example 9.3: A numerical example of MTBF(t_{rec}).

Suppose that $f = 10$ MHz, $a = 100$ kHz, and a 74LS74 D flip-flop is used as a synchronizer. Calculate both MTBF(20 ns) and MTBF(40 ns).

Solution: Because T_0 and τ of the 74LS74 are 4.8×10^{-3} sec and 1.35 ns, respectively, the MTBF can be computed as follows:

$$\text{MTBF}(20 \text{ ns}) = \frac{1}{4.8 \times 10^{-3} \cdot 10 \times 10^6 \cdot 100 \times 10^3} \exp\left(\frac{20 \text{ ns}}{1.35 \text{ ns}}\right)$$
$$= 5.66 \times 10^{-4} \text{ sec}$$

Consequently, the circuit is almost entirely in the metastable state when it is operating in this environment.

However, if we relax the recovery time to 40 ns, then the MTBF is increased to a value calculated as follows

$$\text{MTBF}(40 \text{ ns}) = \frac{1}{4.8 \times 10^{-3} \cdot 10 \times 10^6 \cdot 100 \times 10^3} \exp\left(\frac{40 \text{ ns}}{1.35 \text{ ns}}\right)$$
$$= 1537.25 \text{ sec}$$

which is a much better result than the case of the 20-ns recovery time. Nevertheless, it still cannot be used in practical systems.

The approaches to increasing the MTBF of a synchronizer are to use a flip-flop that follows the synchronizer with a smaller setup time and to increase the clock period of the system. The former has to use more advanced technology whereas the latter needs to slow down the system clock signal and hence the system performance. The following example gives some feelings about the effects of the operating frequency on the MTBF.

■ Example 9.4: Another numerical example of MTBF(t_{rec}).

Suppose that $a = 100$ kHz, and a 74ALS74 D flip-flop is used as a synchronizer. The setup time, t_{setup}, of the flip-flop that follows the synchronizer is 10 ns. Calculate the MTBF(t_{rec}) at $f = 20$ MHz and 50 MHz, respectively.

Solution: T_0 and τ of the 74ALS74 are 8.7×10^{-6} sec and 1.0 ns, respectively. At $f = 20$ MHz, $t_{clk} = 50$ ns, and $t_{rec} = t_{clk} - t_{setup} = 40$ ns. Hence, the MTBF is computed as follows:

$$\text{MTBF}(40 \text{ ns}) = \frac{1}{8.7 \times 10^{-6} \cdot 20 \times 10^6 \cdot 100 \times 10^3} \exp\left(\frac{40 \text{ ns}}{1.0 \text{ ns}}\right)$$
$$= 1.35 \times 10^{10} \text{ sec} \approx 429 \text{ years}$$

As a result, the mean time between two metastable states is approximately 429 years.

As the synchronizer operates at $f = 50$ MHz, $t_{clk} = 20$ ns, and $t_{rec} = t_{clk} - t_{setup} = 10$ ns.

$$\text{MTBF}(10 \text{ ns}) = \frac{1}{8.7 \times 10^{-6} \cdot 50 \times 10^6 \cdot 100 \times 10^3} \exp\left(\frac{10 \text{ ns}}{1.0 \text{ ns}}\right)$$
$$= 506 \times 10^{-6} \text{ sec}$$

Consequently, it is a high probability to enter a metastable state when the synchronizer circuit is operating in this environment.

The two approaches widely used in practice to attack metastable states are the *cascaded synchronizer* and *frequency-divided synchronizer*, as shown in Figure 9.14. The cascaded synchronizer is merely an n-stage shift register, as shown in Figure 9.14(a), where n is typically 2 or 3. The resulting synchronizer is often referred to as a two-stage or three-stage synchronizer. The rationale behind the cascaded synchronizer is that every stage of the synchronizer has the same probability of entering the metastable state. Hence, the probability of an n-stage synchronizer is the product of the individual probabilities of all stages. This probability is quite small when a D flip-flop with a good MTBF is used as the basic building block. An illustration of such a cascaded synchronizer with the capability of asynchronous reset is described in the following example.

■ Example 9.5: A cascaded synchronizer with asynchronous reset.

As illustrated in Figure 9.14(a), a cascaded synchronizer is basically an n-stage shift register and hence may be readily modeled by using the same style of shift registers. The resulting module is given as follows. The output qout is cleared to 0 whenever the active-low reset signal reset_n is active, i.e., 0, regardless of the status of the clock signal, and is the shifted input, otherwise.

```
// a cascaded synchronizer with asynchronous reset
module cascaded_synchronizer
       #(parameter N = 2)(// set default number of stages
         input   clk, reset_n, d,
         output  qout);
reg [N-1:0] q; // define an N-stage shift register
// the body of the cascaded synchronizer
assign qout = q[N-1];

// the operation of an N-stage shift register
always @(posedge clk or negedge reset_n)
   if (!reset_n) q <= 0; // active-low reset signal
```

Section 9.1 Flip-Flops

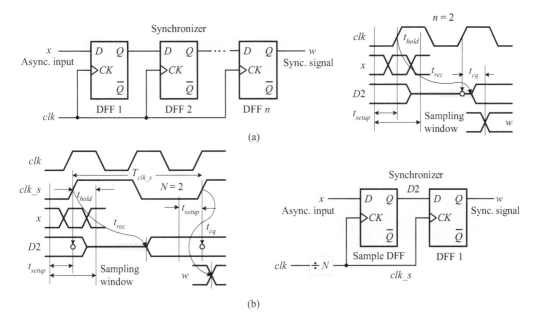

Figure 9.14: Two basic types of synchronizers: (a) a cascaded synchronizer and (b) a frequency-divided synchronizer (Note that in the running example $T_{clk\_s} = 2T_{clk}$).

```
        else         q <= {q[N-2:0], d};
endmodule
```

The frequency-divided synchronizer is shown in Figure 9.14(b). In such an approach, the system clock signal is first divided by N with a frequency divider and then applied to the two-stage synchronizer, thereby increasing the clock period of the clock signal applied to the two-stage synchronizer by a factor of N. The available recovery time is then increased to $t_{rec} = N \times T_{clk} - t_{setup}$. As a result, the MTBF is improved and a flip-flop (used as DFF1) with a larger setup time may be used to construct the synchronizer. An illustration of the frequency-divided synchronizer with the capability of asynchronous reset is explored in the following example.

■ Example 9.6: A frequency-divided synchronizer.

The rationale behind the frequency-divided synchronizer is the use of a modulo-n frequency divider by which the system clock signal is divided by $n\ (=2^N)$ before being applied to the synchronizer. An **always** block is used for this purpose. The second **always** block describes a two-stage shift register. The output qout is cleared to 0 whenever the active-low reset signal reset_n is active, i.e., 0, regardless of the status of the clock signal, and is the shifted input, otherwise.

```
// a frequency-divided synchronizer with asynchronous reset
module frequency_divided_synchronizer
       #(parameter N = 1)(   // default to modulo 2
         input  clk, reset_n, d,
         output qout);

reg  [1:0] q;         // default to 2-stage D flip-flops
```

```
reg  [N-1:0] clk_q; // define a divide-by-2^N counter
wire clk_sync;
// the body of the frequency-divided synchronizer
// a modulo-2^N frequency divider
always @(posedge clk or negedge reset_n)
   if   (!reset_n) clk_q <= 0;
   else clk_q <= clk_q + 1;
assign clk_sync = clk_q[N-1];
// the body of the synchronizer --- a two-stage shift register
always @(posedge clk_sync or negedge reset_n)
   if (!reset_n) q <= 0; // active-low reset signal
   else          q <= {q[0], d};
assign qout = q[1];
endmodule
```

■ Review Questions

9-21 Describe the meaning of synchronizers.
9-22 What is the meaning of the synchronization failure?
9-23 Describe the meaning of the mean time between failures (MTBF).
9-24 Describe the rationale behind the cascaded synchronizer.
9-25 Describe the rationale behind the frequency-divided synchronizer.

9.1.4 A Switch-Debounce Circuit

Another basic application of flip-flops is found in constructing a very useful circuit called the *switch-debounce circuit*, which is used to generate a single pulse each time a mechanical switch is pressed and then released. Because of the inertia of mechanical switches, an inherent feature of a mechanical switch is that it would bounce back and forth several times during the period of 5 ms to 20 ms after each time it is pressed or released. As a result, each press or release of the mechanical switch would be recognized as many pulses instead of only one because the speed of any electronic circuit is much faster than that of the mechanical switch.

Although many approaches can be used to overcome the switch-bouncing effect, in this section, we introduce a synthesizable scheme, which is based on the idea of a technique often found in software. The idea is that *a valid key press or release for a switch is recognized whenever two samples of the switch status separated 10 ms apart are the same*. Based on this principle, a simple switch-debounce circuit can be constructed with two D flip-flops and one JK flip-flop along with three basic gates, as illustrated in Figure 9.15. The first two D flip-flops serve as a two-stage synchronizer and sample the switch status 10 ms apart, controlled by the clock signal, clk_10ms. The switch status is recognized as a valid switch press if both samples have the same value of a logic 1 and is an invalid one, otherwise. Once a valid switch press is detected, the JK flip-flop generates a high-level output until both D flip-flops sample a low-level input, which clears the JK flip-flop and terminates the output pulse. Consequently, only one pulse is generated each time the switch is pressed and then released. This switch-debounce circuit is described as the following module in behavioral style.

Section 9.1 Flip-Flops

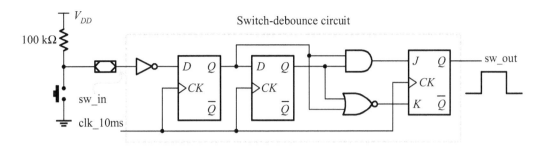

Figure 9.15: A simple but synthesizable switch-debounce circuit.

■ Example 9.7: A switch-debounce circuit.

As illustrated in Figure 9.15, the switch-debounce circuit consists of two parts: a two-stage shift register and a JK flip-flop. The shift register comprises two D flip-flops and is employed to detect the switch status. The JK flip-flop generates a single pulse once a valid key press is detected and is implemented with a D flip-flop by writing its excitation equation: $Q(t+1) = J\bar{Q}(t) + \bar{K}Q(t)$. In the module, the first **always** block describes the two-stage shift register and the second **always** block models the JK flip-flop with a D flip-flop. The excitation equation of the D flip-flop accompanied with AND and OR gates are described by an **assign** continuous assignment. The output of the JK flip-flop is assigned as the output sw_out.

```
// a single switch-debounce logic circuit
module switch_debounce(
        input   clk_10ms, reset, sw_in,
        output wire sw_out);
wire    d;
reg     [1:0] q_sample;
reg     q_jk;

// remove the bouncing effect of a mechanical switch
always @(posedge clk_10ms or posedge reset)
   if (reset) q_sample <= 0; // active-high reset signal
   else begin
      q_sample[1] <= q_sample[0];
      q_sample[0] <= ~sw_in;
   end
// using a JK flip-flop to generate a single pulse
// each time the switch is pressed
assign d = (((q_sample[0] & q_sample[1]) & ~q_jk) |
         (~(~q_sample[0] & ~q_sample[1]) & q_jk));
always @(posedge clk_10ms or posedge reset)
   if (reset)    q_jk <= 1'b0;
   else          q_jk <= d;
assign sw_out = q_jk;
endmodule
```

■ Review Questions

9-26 What is the purpose of the switch-debounce circuit?

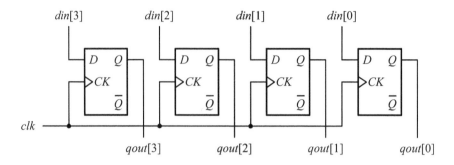

Figure 9.16: The logic diagram of a 4-bit data register.

9-27 Describe the rationale behind the switch-debounce circuit.
9-28 What would happen if we remove the NOR gate from Figure 9.15?
9-29 What is the function of the AND gate in Figure 9.15?

9.2 Memory Elements

In any large digital system, about one-third or more hardware is contributed by memory modules, which are used to store the intermediate information of the system. A memory element can be a data register or a random-access memory (RAM) module. In synthesizable design, the three most popular types of memory elements are *data registers*, *register files*, and *synchronous RAM* modules. In addition, asynchronous RAM is also often used externally for storing the data required for the system.

9.2.1 Registers

Registers are usually used to store small amount of information in digital systems. A data register (or a register for short) is a set of flip-flops or latches in which each flip-flop or latch is capable of storing 1-bit information. A single-bit register is just a single flip-flop or a latch. For a synthesizable design, it usually uses flip-flops rather than latches when constructing a memory element to make easy the design. An n-bit register is a set of n flip-flops placed in parallel with a common clock signal and a clear (or reset) signal.

Figure 9.16 shows a 4-bit data register consisting of four D flip-flops with a common clock signal clk connected to all flip-flops. An illustration of modeling this register is revealed in the following example.

■ Example 9.8: An n-bit data register.

It is quite intuitive to model an n-bit data register. The resulting module is as follows. At each positive edge of the clock signal, clk, the input data, din, are sampled and pass to the output, qout, of the data register.

```
// an N-bit data register
module register
    #(parameter N = 4)(   // set default size
      input   clk,
      input   [N-1:0] din,
      output reg [N-1:0] qout);
// the body of an N-bit data register
```

Section 9.2 Memory Elements

```
always @(posedge clk) qout <= din;
endmodule
```

In most practical applications, a data register often needs to have the capability of clearing its outputs asynchronously. An illustration to explore how to model such a data register is treated as follows.

∎ Example 9.9: An n-bit data register with asynchronous reset.

This example simply expands the D flip-flop with asynchronous reset to an n-bit register by declaring both din and qout as an individual n-bit vector. The resulting module is as follows:

```
// an N-bit data register with asynchronous reset
module register_reset
     #(parameter N = 4)(   // default number of bits
        input   clk, reset_n,
        input   [N-1:0] din,
        output reg [N-1:0] qout);
// the body of an N-bit data register
always @(posedge clk or negedge reset_n)
   if (!reset_n) qout <= {N{1'b0}};
   else          qout <= din;
endmodule
```

A more practical data register often needs to have the capability of loading new data in synchronism with the clock signal and the capability of clearing its outputs asynchronously. As an illustration, consider the following n-bit data register.

∎ Example 9.10: An n-bit resettable, parallel-load data register.

The output, qout, is cleared whenever the active-low reset signal, reset_n, is asserted. As the reset signal, reset_n, is inactive, new input data, din, are sampled and loaded into the data register, qout, at each positive edge of the clock signal, clk, if the load input, load, is enabled. The output remains unchanged if the load input, load, is disabled.

```
// an N-bit data register with synchronous load and
// asynchronous reset
module register_load_reset
     #(parameter N = 4)(   // set default size
        input   clk, load, reset_n,
        input   [N-1:0] din,
        output reg [N-1:0] qout);
// the body of an N-bit data register
always @(posedge clk or negedge reset_n)
   if (!reset_n)   qout <= {N{1'b0}};
   else if (load)  qout <= din;
   //   else       qout <= qout; // a redundant expression
endmodule
```

Notice that the **else** part is a redundant expression for sequential circuits if it merely keeps the output of the **if** statement unchanged. It is instructive to explain it. (*Hint:* It is proved helpful to consider the fundamental difference between combinational and sequential logic.)

■ Coding Style

1. *We should avoid using latches in synthesis-flow-based design as possible as we can because they are more difficult to be tested than flip-flops.*
2. *For an active-low signal, it is good practice to end the signal name with an underscore followed by a lowercase letter* b *or* n, *such as* reset_n *or* reset_b.

9.2.2 Register Files

Register files often find their broad use in the datapaths of most digital systems. A register file is a set of registers with multiple access ports and the capability of random access. Here, the random access means that the access time of any register in the register file is the same. A register file is usually characterized by the following parameters: the number of registers, the width of each register, and the number of ports. Depending on the actual requirements of practical systems, the most widely used register files have 16 or 32 registers with each containing 16 or 32 bits. As for the number of ports, one read port and one write port as well as two read ports and one write port are the most common. Of course, other combinations of port types and the number of ports are also possible and widely used in particular designs.

Register files are usually built from registers, which are in turn composed of flip-flops. However, they can also be built by using synchronous RAM. Synchronous RAM will be introduced in the next subsection. Here, we give an example to illustrate how to model an n-word register file, with each register having w bits, with one synchronous write port and two asynchronous read ports.

■ Example 9.11: An n-word register file.

In most register-file-based datapaths, three operands are often required for carrying out an operation, and thus the register file needs to have two read ports and one write port. In the module that follows, suppose that the access to each read port is unconditional and asynchronous, and the access of the write port is controlled by a write enable signal, wr_enable, in synchronism with the clock signal. Each port regardless of read or write has its own access address.

```
// an N-word register file with one write port and two read ports
module register_file
       #(parameter M = 4,  // default number of address bits
         parameter N = 16, // default number of words,N=2**M
         parameter W = 8)( // default word width
         input   clk, wr_enable,
         input   [W-1:0] din,
         input   [M-1:0] rd_addra, rd_addrb, wr_addr,
         output  [W-1:0] douta, doutb);
reg [W-1:0] reg_file [N-1:0];
// the body of the N-word register file
assign douta = reg_file[rd_addra],
       doutb = reg_file[rd_addrb];
always @(posedge clk)
   if (wr_enable) reg_file[wr_addr] <= din;
endmodule
```

The reader is invited to synthesize the above module and take a closer look at the synthesized results. In particular, examine very carefully how the synthesis tool generates the register file and how much hardware is needed.

9.2.3 Synchronous RAM

Synchronous RAM is also widely used for storing a moderate amount of information within digital systems. A synchronous RAM is a random-access memory in which the data access is in synchronism with a clock signal. In general, random-access memory modules can be either asynchronous or synchronous. However, synchronous RAM modules are particularly popular in digital system designs based on the synthesis flow due to the ease of use.

A synchronous RAM module may be constructed from flip-flops, compiled RAM cells, or a memory macro. However, a flip-flop may take up 10 to 20 times the area of a 6-transistor static RAM cell. Hence, for a synchronous RAM module of a moderate capacity, it is better to use the compiled RAM cell in cell-based design or the memory macro built into the device in FPGA-based design. Generally speaking, there are two types of memory macros of modern FPGA devices: *distributed memories* and *block memories*. Distributed memories are formed by reconfiguring some logic cells and hence reduce the available logic cells for implementing logic functions. Block memories are pre-fabricated along with logic cells on FPGA devices. They have much faster access time than distributed memories and are often referred to as *hardwired memory blocks*.

Synchronous RAM modules can be characterized by the following parameters: the word width, the number of words (i.e., the capacity), and the number of access ports. In addition, the access time of a synchronous RAM module is also an important factor as the synchronous RAM module is used in a design. An example of synchronous RAM modules with one read port and one write port is considered and modeled as follows.

■ Example 9.12: An example of synchronous RAM modules.

This example explores how to model a synchronous RAM module with one read port and one write port. Since both read and write operations are in synchronism with the clock signal, they are placed inside the **always** block under the event control of the clock edge. A chip select (cs) input is also used to control the operations of the whole module.

```
// a synchronous RAM module
module syn_ram
       #(parameter N = 16,// default number of words
         parameter A = 4, // default number of address bits
         parameter W = 4)(// default word width in bits
         input   [A−1:0] addr,
         input   [W−1:0] din,
         // chip select, read-write control, and clock
         input   cs, wr, clk,
         output reg [W−1:0] dout);
reg    [W−1:0] ram [N−1:0]; // declare an N*W memory array
// the body of the synchronous RAM
always @(posedge clk)
   if (cs) if (wr) ram[addr] <= din;       // write
           else         dout <= ram[addr]; // read
endmodule
```

The reader is encouraged to synthesize the above module and check the synthesized result. It is instructive to compare the result with that of the preceding example. Also, it is strongly recommended that you should familiarize yourself with the generating methods of memory blocks provided by cell-library or FPGA vendors before using memory blocks in your designs.

■ Review Questions

9-30 What is the relationship between an n-bit register and n flip-flops?

9-31 What are the features of register files?

9-32 Describe the features of synchronous RAMs.

9-33 What are the two implementations of synchronous RAMs usually found in synthesizable designs?

9-34 What are the two types of memory macros usually found in FPGA devices?

9-35 What are the possible parameters in describing a register file?

9.2.4 Asynchronous RAMs

Asynchronous static random-access memories (SRAMs, often called RAMs for short) are still widely used in today's low- to middle-end products. Consequently, in this section, we introduce the basic features and access timing of typical RAM devices. In addition, we give an example to illustrate how to model these RAM devices in Verilog HDL.

9.2.4.1 General Block Diagram The block diagram of a typical 32-kB RAM device is shown in Figure 9.17, where the memory cells are configured into a 512×512 memory matrix. As a result, the row address needs nine bits and the column address requires six bits because the data bus is eight bits. Besides the row and column decoders, there are a timing generator and a read/write control logic circuit. The timing generator produces all timing required for the memory device to perform the read and write operations. The read/write control logic circuit determines the data flow direction and enables the read or write operation.

In general, when there is only one access control signal in a RAM device, the control input is denoted by R/$\overline{\text{W}}$ (Read/Write). The RAM device is in the read mode as R/$\overline{\text{W}}$ = 1 and in the write mode as R/$\overline{\text{W}}$ = 0, assuming that in both cases the chip is enabled (i.e., $\overline{\text{CE}}$ = 0). When a RAM device has two access control inputs, one of them is often named the write enable ($\overline{\text{WE}}$) input and the other is called the output enable ($\overline{\text{OE}}$) input. The chip is in the write mode when $\overline{\text{WE}}$ = 0 and in the read mode when $\overline{\text{OE}}$ = 0 and $\overline{\text{WE}}$ = 1, assuming that the chip is enabled. The output bus is at a high impedance when both $\overline{\text{WE}}$ and $\overline{\text{OE}}$ inputs are high.

The pin assignments of the two most widely used RAM devices, 6264 (8 kB) and 62256 (32 kB), are shown in Figure 9.18. Both devices 6264 and 62256 have two access control inputs: a write enable ($\overline{\text{WE}}$) input and an output enable ($\overline{\text{OE}}$) input. The write enable ($\overline{\text{WE}}$) input asserts the write operation and the output enable ($\overline{\text{OE}}$) input asserts the read operation by enabling the output buffers. The 6264 device has two chip select inputs, $\overline{\text{CE1}}$ and CE2, and the 62256 device has only one chip select input, $\overline{\text{CE}}$. The chip select input controls the operations of the entire chip. The data bus is in a high impedance whenever the chip enable input is deasserted, regardless of the values of both write enable ($\overline{\text{WE}}$) and output enable ($\overline{\text{OE}}$) inputs.

9.2.4.2 Read-Cycle Timing In order to read data from a RAM device, the address and access control inputs, $\overline{\text{OE}}$ and $\overline{\text{CE}}$, must be applied to the device in accordance with the timing shown in Figure 9.19. When both $\overline{\text{OE}}$ and $\overline{\text{CE}}$ inputs are low, the valid data will appear on the output data bus after all three timing constraints, t_{AA}, t_{CE}, and t_{OE}, are satisfied. In other words, the amount of time required for the valid data to appear on the output data bus is determined by the above three timing constraints which is satisfied last.

In general, the read-cycle timing of a RAM device are governed by the following four types of access times:

- *Address access time* (t_{AA}) is the time interval from the time when the address is applied to the address bus to the point when stable data appear on the output data bus.

Section 9.2 Memory Elements

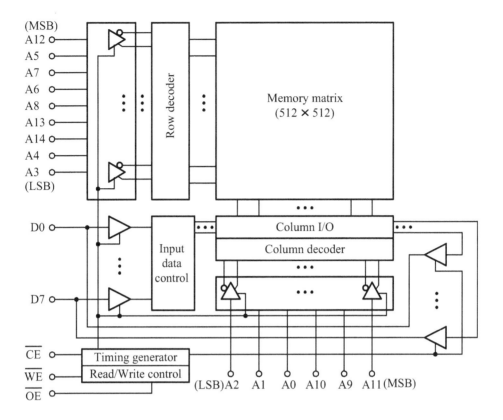

Figure 9.17: The block diagram of a typical 32-kB RAM device.

- $\overline{CE}$ *access time* (t_{CE}) is the time interval from the time when the chip enable ($\overline{CE}$) input is asserted to the point when stable data appear on the output data bus.
- $\overline{OE}$ *access time* (t_{OE}) is the time interval from the time when the output enable ($\overline{OE}$) input is asserted to the point when stable data appear on the output data bus.
- *Read cycle time* (t_{RC}) is the minimum amount of time required to separate two consecutive random read operations.

The output data bus begins to be active in a time of t_{CLZ} and t_{OLZ} after the chip enable ($\overline{CE}$) and output enable ($\overline{OE}$) inputs are asserted, respectively. For consecutive read operations, both chip enable ($\overline{CE}$) and output enable ($\overline{OE}$) inputs may remain in the asserted status while only addresses are changed. In this case, the data appearing on the output data bus will continually sustain an amount of time known as the *hold time*, t_{OH}. The output data bus goes to a high impedance in a time of t_{CHZ} or t_{OHZ} after the chip enable ($\overline{CE}$) or output enable ($\overline{OE}$) signal is deasserted.

The parameters related to the read cycle of typical RAM devices are listed in Table 9.2. For most RAM devices, both t_{AA} and t_{CE} have the same value. The value of t_{OE} is generally less than that of t_{AA} or t_{CE}. In addition, the read-cycle time is usually equal to both the address access time (t_{AA}) and the $\overline{CE}$ access time (t_{CE}).

9.2.4.3 Write-Cycle Timing
Generally speaking, writing data into a RAM device can be controlled by either the $\overline{WE}$ or $\overline{CE}$ input, as shown in Figure 9.20. Figure 9.20(a) shows the $\overline{WE}$-controlled write operation and Figure 9.20(b) presents the $\overline{CE}$-controlled write operation. In fact, it can be seen that the write operation is accomplished by the control input $\overline{WE}$ or $\overline{CE}$ that is deasserted earlier.

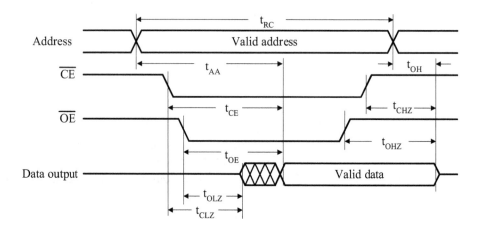

Figure 9.18: The pin assignment of two widely used commercial RAM devices: 6264 and 62256.

Figure 9.19: The read-cycle timing diagram of a typical SRAM device.

In order to correctly write data into a RAM device, the write timing must meet the write-cycle specification of the device, especially, the *data setup time* (t_{DW}) and *data hold time* (t_{DH}). Like a flip-flop or latch, the data setup time is defined as the amount of time that data must be stable before it is sampled and the data hold time is the amount of time that data must remain stable after it is sampled. Here, the sampling point (i.e., the timing reference point) is the positive edge of the $\overline{WE}$ or $\overline{CE}$ signal, depending on whether the underlying write operation is $\overline{WE}$-controlled or $\overline{CE}$-controlled.

In the $\overline{WE}$ ($\overline{CE}$)-controlled write operation, the address must be stable for an amount of time known as the *address setup time* (t_{AS}) before the $\overline{WE}$ ($\overline{CE}$) signal is asserted, and must be still stable for an amount of time known as the *address hold time* after the $\overline{WE}$ ($\overline{CE}$) signal is deasserted. The *write recovery time* (t_{WR}) is virtually the address hold time. Similarly, the write data must be stable for an amount of time known as the *data setup time* (t_{DW}) before the

Section 9.2 Memory Elements

Table 9.2: The read-cycle timing parameters of typical SRAM devices (in units of ns).

Symbol	Parameter	HM6264B-10L Min	HM6264B-10L Max	MCM60L256A-C Min	MCM60L256A-C Max	HM62256B-5 Min	HM62256B-5 Max
t_{RC}	Read cycle time	100	-	100	-	55	-
t_{AA}	Address access time	-	100	-	100	-	55
t_{CE}	Chip select access time	-	100	-	100	-	55
t_{OE}	Output enable to output valid	-	50	-	50	-	35
t_{CLZ}	Chip select to output in low-Z	10	-	10	-	5	-
t_{OLZ}	Output enable to output in low-Z	5	-	5	-	5	-
t_{OH}	Output hold from address change	10	-	10	-	5	-
t_{OHZ}	Output disable to output in high-Z	0	35	0	35	0	20
t_{CHZ}	Chip deselect to output in high-Z	0	35	0	35	0	20

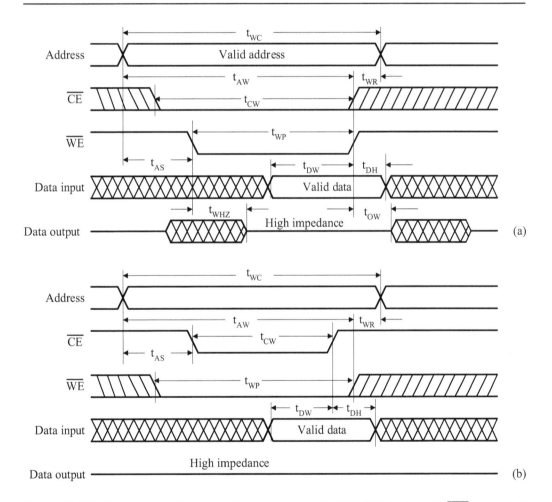

Figure 9.20: The write-cycle timing diagram of a typical SRAM device: (a) $\overline{WE}$-controlled write and (b) $\overline{CE}$-controlled write.

$\overline{WE}$ ($\overline{CE}$) signal is deasserted, and must still remain stable for an amount of time known as the *data hold time* (t_{DH}) after the $\overline{WE}$ ($\overline{CE}$) signal is deasserted.

In the $\overline{WE}$-controlled write operation, when the $\overline{OE}$ signal is deasserted before the write cycle begins, the output data bus enters a high impedance in a time of t_{OHZ} after the $\overline{OE}$ signal

Table 9.3: The write-cycle timing parameters of typical SRAM devices (in units of ns).

Symbol	Parameter	HM6264B-10L Min	HM6264B-10L Max	MCM60L256A-C Min	MCM60L256A-C Max	HM62256B-5 Min	HM62256B-5 Max
t_{WC}	Write cycle time	100	-	100	-	55	-
t_{AS}	Address setup time	0	-	0	-	0	-
t_{AW}	Address valid to end of write	80	-	80	-	40	-
t_{WP}	Write pulse width	60	-	60	-	35	-
t_{DS}	Data setup time	40	-	35	-	25	-
t_{DH}	Data hold time	0	-	0	-	0	-
t_{OHZ}	Output disable to output in high-Z	0	35	0	35	0	20
t_{WHZ}	Write to output in high-Z	0	35	0	25	0	20
t_{OW}	Output active from end of write	5	-	10	-	5	-
t_{WR}	Write recovery time	0	-	0	-	0	-
t_{CW}	Chip select to end of write	80	-	80	-	40	-

is deasserted. When the $\overline{OE}$ signal is asserted before the write cycle begins, the output data bus enters a high impedances in a time of t_{WHZ} after the $\overline{WE}$ signal is asserted and stays there until a time of t_{OW} after the $\overline{WE}$ signal is deasserted. In the $\overline{CE}$-controlled write operation, the output data bus remains at a high impedance.

Regardless of either the $\overline{WE}$ or $\overline{CE}$-controlled write operation, both $\overline{WE}$ and $\overline{CE}$ signals must sustain an amount of time t_{WP} and t_{CW}, respectively. In addition, the address must at least sustain an amount of time t_{WC}, known as the *write-cycle time*. The parameters related to the write cycle of typical RAM devices are summarized in Table 9.3.

9.2.4.4 A Modeling Example In the rest of this section, we demonstrate how to model an asynchronous static RAM module in Verilog HDL by way of an example.

■ Example 9.13: An asynchronous static RAM module.

In this example, we demonstrate how to model a static RAM module with timing checks. The details of timing checks can be referred to in Section 7.3.3 again.

```
'timescale 1 ns / 1 ns
// an example of modeling a 32k * 8 asynchronous
// sram with timing checks
module sram_timing_check(
       input CE_b, WE_b, OE_b,
       input [14:0] addr,
       inout [7:0] data);
reg [7:0] data_int;
reg [7:0] mem [2**15-1:0];
specify
   // parameters for the read cycle
   specparam t_RC  = 55;   // read cycle time
   specparam t_AA  = 55;   // address access time
   specparam t_CE  = 55;   // chip select access time
   specparam t_OE  = 35;   // output enable to output valid
   specparam t_CLZ = 5;    // chip select to output in low-Z
   specparam t_OLZ = 5;    // output enable to output in low-Z
   specparam t_OH  = 5;    // output hold from address change
   specparam t_OHZ = 0;    // output disable to output in high-Z
   specparam t_CHZ = 0;    // chip deselect to output in high-Z
```

```verilog
        // read cycle time
        $width(negedge addr, t_RC);
        // module path timing specifications
        (addr *> data) = t_AA;
        (CE_b *> data) = (t_CE, t_CE, t_CHZ);
        (OE_b *> data) = (t_OE, t_OE, t_OHZ);
        // write cycle timing checks
        // parameters for write cycle
        specparam t_WC  = 55;   // write cycle time
        specparam t_AS  = 0;    // address setup time
        specparam t_AW  = 40;   // address valid to end of write
        specparam t_WP  = 35;   // write pulse width
        specparam t_DW  = 25;   // data setup time
        specparam t_DH  = 0;    // data hold time
        specparam t_WHZ = 0;    // write to output in high-Z
        specparam t_OW  = 5;    // output active from end of write
        specparam t_WR  = 0;    // write recovery time
        specparam t_CW  = 40;   // chip selection to end of write
        // write cycle time
        $width(negedge addr, t_WC);
        // address valid to end of write
        $setup (addr, posedge WE_b &&& CE_b == 0, t_AW);
        $setup (addr, posedge CE_b &&& WE_b == 0, t_AW);
        // address setup time
        $setup (addr, negedge WE_b &&& CE_b == 0, t_AS);
        $setup (addr, negedge CE_b &&& WE_b == 0, t_AS);
        // write pulse width
        $width(negedge WE_b, t_WP);
        // data setup time
        $setup (data, posedge WE_b &&& CE_b == 0, t_DW);
        $setup (data, posedge CE_b &&& WE_b == 0, t_DW);
        // data hold time
        $hold (data, posedge WE_b &&& CE_b == 0, t_DH);
        $hold (data, posedge CE_b &&& WE_b == 0, t_DH);
        // chip select to end of write
        $setup (CE_b, posedge WE_b &&& CE_b == 0, t_CW);
        $width (negedge CE_b &&& WE_b == 0, t_CW);
endspecify
// process the memory access cycle
assign data = ((CE_b == 0)&& (WE_b == 1)&&(OE_b == 0))
            ? data_int: 8'bzzzz_zzzz;
always @(*) // wait for chip enable
    if (!CE_b)
        if (!WE_b) mem[addr] = data; // write cycle
        else if (WE_b) // read cycle
                if (!OE_b) data_int = mem[addr];
                else data_int = 8'bzzzz_zzzz;
            else data_int = 8'bzzzz_zzzz;
    else    // chip not selected, disable output
        data_int <= 8'bzzzz_zzzz;
endmodule
```

■ Review Questions

9-36 Define the data setup and data hold times.

9-37 What are the three timing constraints that must be satisfied for valid data to appear on the output data bus when reading a RAM device?

9-38 Define address access time, $\overline{CE}$ access time, and $\overline{OE}$ access time.

9-39 What are the two ways that can be used to write data into a RAM device?

9-40 Describe the functions of the $\overline{CE}$ and $\overline{OE}$ inputs of RAM devices.

9.3 Shift Registers

Another type of register widely used in digital systems is *shift registers*. As the name implies, a shift register is a register that may shift its contents left or right a specified number of bit positions. Moreover, shift registers are often employed in digital systems to perform various data format conversions. In this section, we begin with the introduction of simple shift registers and then deal with universal shift registers, which combines both serial load and parallel load with both left and right shifts into the same logic circuit.

9.3.1 Shift Registers

The basic structure of a 4-bit shift register is shown in Figure 9.21(a). An n-bit shift register is generally composed of n D flip-flops connected in series along with a common clock signal. From the timing diagram depicted in Figure 9.21(b), the serial input data are sampled at each positive edge of the clock signal *clk* and shifted to the next stage on the next clock cycle. An n-bit shift register with active-low asynchronous reset is described in the following example.

■ Example 9.14: An n-bit shift register.

From Figure 9.21(a), the output, qout, is cleared to 0 if the active-low reset signal, reset_n, is asserted and assigned the concatenation of the serial input, sin, and a portion of qout, i.e., qout[N-1:1], otherwise. Based on this, the following module is obtained.

```
// a right-shift register module
module shift_register
       #(parameter N = 4)( // set default number of bits
          input   clk, reset_n, sin,
          output reg [N-1:0] qout);

// the body of an N-bit right-shift register
always @(posedge clk or negedge reset_n)
   if (!reset_n) qout <= {N{1'b0}};
   else          qout <= {sin, qout[N-1:1]};
endmodule
```

As another way of modeling a shift register, consider the shift register shown in Figure 9.21(a) again. By observing the operations of the shift register, we can see that each flip-flop actually receives the output value from its preceding stage in the previous cycle except the first stage which receives the serial input data, sin. Based on this, a shift register associated with some desired computation (namely, a trivial pipeline) can be modeled as in the following example.

Section 9.3 Shift Registers

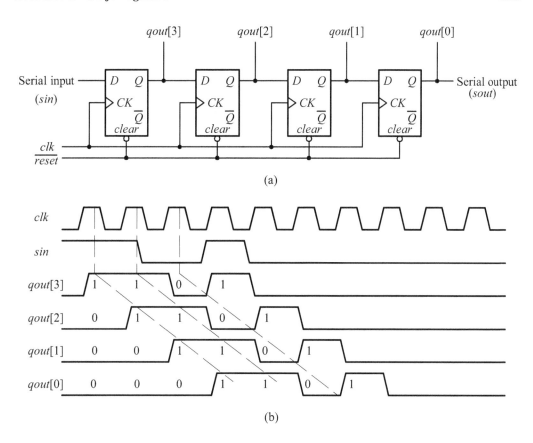

Figure 9.21: The (a) logic diagram and (b) timing of a 4-bit shift register.

■ Example 9.15: A 4-bit shift register/a trivial pipeline structure.

Consider the module shown below. If the active-low reset signal, reset_n, is activated, the output, qout, is cleared to 0; otherwise, each individual output is assigned the output value of its preceding stage along with the desired computation except the first stage which receives the serial input data, sin. Hence, the result is a right shift with computation. It should be noted that nonblocking assignments should be used here; otherwise, the result will be wrong.

```
// a right-shift register as a trivial pipeline
module shift_register_pipeline(
        input  clk, reset_n, sin,
        output reg [3:0] qout);
// the body of a 4-bit right-shift register
always @(posedge clk or negedge reset_n)
   if (!reset_n) qout <= 0;
   else begin
      qout[3] <= sin;
      qout[2] <= qout[3];
      qout[1] <= qout[2] ^ qout[3];
      qout[0] <= qout & qout[1];
   end
endmodule
```

The above example would not work properly if all nonblocking assignments are replaced with blocking assignments (see Problem 9-12). The reader is invited to synthesize it with both types of assignments and compare their synthesized results very carefully. Also, it is worth noting that the combination of both blocking and nonblocking assignments in the same **always** block is usually not allowed in most synthesis tools. Another insight into this example is that it illustrates how to model a pipeline operation. As a matter of fact, a shift register may be regarded as a pipeline structure with a trivial combinational circuit inserted in between any two stages.

In addition to inputting data through the serial input, it is not uncommon to load data into a shift register in parallel. As an illustration, consider the following example.

■ Example 9.16: An n-bit shift register with parallel load.

In this module, the shift register, qout, is cleared to 0 if the active-low reset signal, reset_n, is asserted, is loaded with the external data, din, if the synchronous load input, load, is enabled, or is assigned the concatenation of the serial input, sin, and a portion of qout, qout[N-1:1], if the load input, load, is not enabled. Hence, the result is a right-shift register with asynchronous reset and synchronous parallel load.

```verilog
// a right-shift register with parallel load
module shift_register_parallel_load
       #(parameter N = 4)( // set default number of bits
         input   clk, load, reset_n, sin,
         input   [N-1:0] din,
         output reg [N-1:0] qout);
// the body of an N-bit right-shift register
always @(posedge clk or negedge reset_n)
   if (!reset_n)    qout <= {N{1'b0}};
   else if (load)   qout <= din;
   else             qout <= {sin, qout[N-1:1]};
endmodule
```

9.3.2 Universal Shift Registers

Another use of shift registers is to perform data format conversions: serial to parallel and vice versa. Depending on the specific requirements of practical systems, a data format conversion may perform one of the following operations:

- Serial in, serial out (SISO)
- Serial in, parallel out (SIPO)
- Parallel in, serial out (PISO)
- Parallel in, parallel out (PIPO)

A register capable of facilitating the above four operations is called a *universal shift register* because the shift register must be able to carry out the following operations:

- Parallel load
- Serial in and serial out
- Left shift and right shift

An illustration of an n-bit universal shift register modeled in behavioral style is explored in the following example.

Section 9.3 *Shift Registers*

■ **Example 9.17: An n-bit universal shift register.**

Suppose that the serial data inputs for the right and left shifts are rsi and lsi, respectively. Since the shift register has four operating modes, no change, right shift, left shift, and parallel load, two mode selection signals s1 and s0 are required. With these in mind, the shift register can be readily modeled with an **always** block containing a **case** statement to select the desired operating mode. The resulting module is as follows:

```
// a universal shift register module
module universal_shift_register
       #(parameter N = 4)( // define the default size
          input   clk, reset_n, s1, s0, lsi, rsi,
          input   [N-1:0] din,
          output reg [N-1:0] qout);

// the shift register body
always @(posedge clk or negedge reset_n)
   if (!reset_n) qout <= {N{1'b0}};
   else case ({s1,s0})
      2'b00: ;// qout <= qout;              // no change
      2'b01: qout <= {rsi, qout[N-1:1]};    // right shift
      2'b10: qout <= {qout[N-2:0], lsi};    // left shift
      2'b11: qout <= din;                   // parallel load
   endcase
endmodule
```

Another universal shift register often encountered in digital logic textbooks [2] is shown in Figure 9.22(a), where an explicit 4-bit 4-to-1 multiplexer is used to implement the **case** statement in the preceding example. The logic symbol of such a universal shift register is portrayed in Figure 9.22(b). Figure 9.22(c) summarizes the operating modes of the universal shift register. To model such a universal shift register in a parameterizable fashion, it proves useful to utilize a generate-loop statement. An illustration of this is revealed in the following example.

■ **Example 9.18: Another n-bit universal shift register.**

To model the circuit in Figure 9.22(a), we use an **assign** continuous assignment to describe the 4-to-1 multiplexer and an **always** block to model a *D* flip-flop with asynchronous reset. Three cases are needed to process separately, including two boundary conditions, the LSB and the MSB, and the rest of bits. For the LSB, it needs to take into account the left-shift serial input, and for the MSB, it needs to consider the right-shift serial input. For the remaining bits, they take the outputs from their preceding or succeeding stages depending on whether the shift is left or right.

```
// a universal shift register module using a generate-loop
// statement
module universal_shift_register_generate
       #(parameter N = 4)(   // set default size
          input   clk, reset_n, s1, s0, lsi, rsi,
          input   [N-1:0] din,
          output reg [N-1:0] qout);
wire [N-1:0] y;
// the shift register body
```

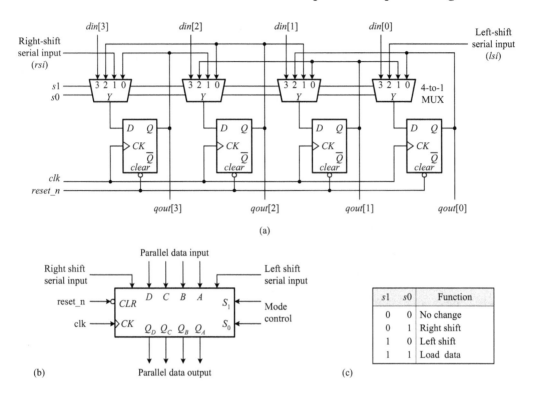

Figure 9.22: The (a) logic diagram, (b) logic symbol, and (c) function table of a 4-bit universal shift register.

```
genvar   i;
generate for (i = 0; i < N; i = i + 1) begin: universal_SR
   if (i == 0) begin: lsb    // specify the LSB
      assign y[i] = (~s1 & ~s0 & qout[i])  |
                    (~s1 &  s0 & qout[i+1])|
                    ( s1 & ~s0 & lsi)      |
                    ( s1 &  s0 & din[i]);
      always @(posedge clk or negedge reset_n)
         if (!reset_n) qout[i] <= 1'b0;
         else   qout[i] <= y[i];  end
   else if (i == N−1) begin: msb   // specify the MSB
      assign y[i] = (~s1 & ~s0 & qout[i])  |
                    (~s1 &  s0 & rsi)|
                    ( s1 & ~s0 & qout[i−1])|
                    ( s1 &  s0 & din[i]);
      always @(posedge clk or negedge reset_n)
         if (!reset_n) qout[i] <= 1'b0;
         else   qout[i] <= y[i];  end
   else begin: rest_bits   // specify the remaining bits
      assign y[i] = (~s1 & ~s0 & qout[i])  |
                    (~s1 &  s0 & qout[i+1]) |
                    ( s1 & ~s0 & qout[i−1])|
                    ( s1 &  s0 & din[i]);
      always @(posedge clk or negedge reset_n)
```

Section 9.4 *Counters*

```
        if (!reset_n) qout[i] <= 1'b0;
        else     qout[i] <= y[i]; end
endgenerate // the universal_shift_register
endmodule
```

■ Review Questions

9-41 What are the four possible operations when a shift register is used as a parallel/serial format conversion?

9-42 What does a universal shift register mean?

9-43 Model an n-bit shift register with active-high synchronous reset.

9-44 Model an n-bit shift register with active-high asynchronous reset.

9.4 Counters

A counter is a device that counts the input events, such as input pulses or clock pulses. In terms of whether they are in synchronism with a clock signal, counters can be asynchronous or synchronous. In practical applications, the most common asynchronous (ripple) counters are binary counters and the most widely used synchronous counters include binary counters and BCD counters as well as Gray counters. It is worth noting that a counter is called a *timer* when its input or clock source is from a standard or known timing source with a fixed, known period, such as the system clock signal.

9.4.1 Ripple Counters

Like ripple-carry adders, ripple counters are one of the most widely known types of counters used in digital systems. An n-bit binary ripple counter consists of n T flip-flops cascaded together; namely, the output of each flip-flop is connected to the clock input of its succeeding flip-flop. Figure 9.23(a) shows an example of a 3-bit binary ripple counter. Here each T flip-flop is realized with a JK flip-flop by connecting its J and K inputs to a logic 1. As we can see from the figure, the essential feature of ripple counters is that each flip-flop is triggered by the output of its preceding stage except the first stage which is triggered by an external signal known as the clock signal. A simple illustration of binary ripple counters described in behavioral style is given in the following example.

■ Example 9.19: A 3-bit binary ripple counter.

This example intuitively models the binary ripple counter shown in Figure 9.23(a). To describe such a counter, three **always** blocks are needed, with each describing a T flip-flop. However, instead of describing a T flip-flop directly, we model the T flip-flop in behavioral style by noting that the toggle operations inherently associated with the T flip-flop—the output value of the flip-flop is toggled each time it is triggered. The complete module is as follows:

```
// a 3-bit ripple counter module
module ripple_counter(
        input clk,
        output reg [2:0] qout);

// the body of the 3-bit ripple counter
always @(negedge clk)
```

344 Chapter 9   Sequential Logic Modules

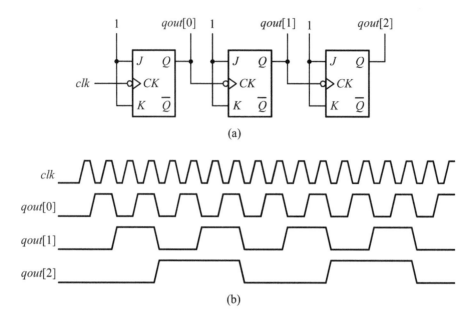

Figure 9.23: The (a) logic diagram and (b) timing of a 3-bit binary ripple counter.

```
   qout[0] <= ~qout[0];
always @(negedge qout[0])
   qout[1] <= ~qout[1];
always @(negedge qout[1])
   qout[2] <= ~qout[2];
endmodule
```

The reader is encouraged to synthesize the above module and see what happens. It should be noted that the outputs of the above module cannot be observed from simulators due to the lack of initial values of qout; thereby, their values remain unknown at all times. In most practical applications, a useful counter is often needed to be resettable, such as the one shown in the following example.

■ Example 9.20: A 3-bit resettable binary ripple counter.

As an illustration of resettable binary ripple counters, consider a 3-bit ripple counter with asynchronous reset. The counter is cleared to 0 if the active-low reset signal, reset_n, is asserted and functions as a normal ripple counter, otherwise. The operation of each flip-flop is described by an **if-else** statement.

```
// a 3-bit ripple counter with an enable input
module ripple_counter_enable(
     input  clk, enable, reset_n,
     output reg [2:0] qout);

// the body of the 3-bit ripple counter
always @(posedge clk or negedge reset_n)
   if (!reset_n) qout[0] <= 1'b0; else
   if (enable)   qout[0] <= ~qout[0];
always @(posedge qout[0] or negedge reset_n)
```

Section 9.4 Counters

```
      if (!reset_n) qout[1] <= 1'b0; else
      if (enable)   qout[1] <= ~qout[1];
   always @(posedge qout[1] or negedge reset_n)
      if (!reset_n) qout[2] <= 1'b0; else
      if (enable)   qout[2] <= ~qout[2];
endmodule
```

The drawback of the above two examples is the lack of parameterizable capability. To make a binary ripple counter parameterizable, it proves useful to employ a generate-loop statement to expand the required number of stages of the binary ripple counter. An illustration of this technique is explored in the following example.

■ Example 9.21: A parameterizable binary ripple counter.

As it can be seen from Figure 9.23(a), except for the first stage that receives the external clock signal, all other stages accept the outputs from their preceding stages as their triggered signals. Hence, to model such an n-bit binary ripple counter, we only need to consider these two cases. The resulting module is as follows:

```
// an N-bit ripple counter using a generate-loop statement
module ripple_counter_generate
       #(parameter N = 4)( // set default size
          input  clk, reset_n,
          output reg [N-1:0] qout);

// the body of the N-bit ripple counter
genvar i;
generate for (i = 0; i < N; i = i + 1) begin:ripple_counter
   if (i == 0) // specify the LSB
      always @(negedge clk or negedge reset_n)
         if (!reset_n) qout[0] <= 1'b0;
         else          qout[0] <= ~qout[0];
   else         // specify the remaining bits
      always @(negedge qout[i-1] or negedge reset_n)
         if (!reset_n) qout[i] <= 1'b0;
         else          qout[i] <= ~qout[i];
end endgenerate
endmodule
```

Notice that the above three binary ripple counters may also be modified into down counters by simply changing the triggered means from the negative edge to the positive edge. In addition, a binary ripple counter can be constructed as an up/down counter under the control of a mode input. For example, it is an up counter when the mode input is 0 and a down counter when the mode input is 1. Furthermore, a binary ripple counter is not necessary to be modulo-2^n in nature, where n is the number of flip-flops. It can be designed as one with a modulus other than 2^n. Here, the *modulus* of a binary ripple counter means the number of states that the counter may have during its normal operation. Actually, it can be any modulus if you like. However, these issues are left for the reader as exercises. More details about binary ripple counters can be found in Lin [2].

9.4.2 Synchronous Counters

Remember that a synchronous counter means that it operates in synchronism with the clock signal. In other words, all flip-flops of a synchronous counter change their states at the same time under the control of the clock signal, namely, at the positive or the negative edge of the clock signal. Like ripple counters, a synchronous counter may be designed to count with a modulus of m, where m is any positive integer, equal to 2^n or smaller than 2^n, where n is the number of flip-flops. In addition, a synchronous counter may be designed as an up counter, a down counter, or an up/down counter under the control of one or more mode inputs.

■ Example 9.22: An n-bit binary counter.

The following module is a synchronous modulo-2^n binary counter with synchronous reset and enable. It consists of n flip-flops. In order to be cascadable, this counter generates a carry-out signal, cout, with a duration of one clock cycle. The carry-out signal, cout, can be used as an enable input to enable the next higher module each time it is generated.

```
// an N-bit binary counter with synchronous reset and enable
module binary_counter
       #(parameter N = 4)( // set default size
         input  clk, enable, reset,
         output reg [N-1:0] qout,
         output cout);      // carry-out

// the body of the N-bit binary counter
always @(posedge clk)
   if (reset)        qout <= 0;
   else if (enable)  qout <= qout + 1;
// generate a carry-out signal
assign #2 cout = &qout; // why #2 is required ?
endmodule
```

To make this counter work well at the behavioral level, it is necessary to include a delay with some non-zero value in the **assign** continuous assignment for generating the cout signal. Otherwise, the next higher module would not count up at all. However, the gate-level simulation after place and route works very well because all real-world gates have definite propagation delays. This can be readily justified by simulating the above module with and without a delay and then checking their results.

The above synchronous binary counter may be modified into an up/down counter by adding a mode selection input to control its operations. An illustration of such a synchronous binary up/down counter is explored in the following example.

■ Example 9.23: An n-bit binary up/down counter (Version 1).

This example simply uses a mode selection signal (upcnt) to control the up/down operations of a synchronous binary counter. Both the reset and enable inputs of the counter are assumed to be synchronous. At each positive edge of the clock signal, the counter is cleared to 0 if the active-high reset signal is asserted. Otherwise, the counter counts up if the mode selection signal, upcnt, is 1 or down if the mode selection signal, upcnt, is 0 as the counter is enabled. The counter remains unchanged as the enable input, enable, is 0.

```
// an N-bit binary up/down counter with synchronous
// reset and enable inputs
```

Section 9.4 Counters

```verilog
module binary_up_down_counter_reset
        #(parameter N = 4)(  // set default size
           input   clk, enable, reset, upcnt,
           output reg [N-1:0] qout,
           output  cout, bout);  // carry and borrow outputs
// the body of the N-bit up/down binary counter
always @(posedge clk)
   if (reset) qout <= 0;
   else if (enable) begin
        if (upcnt) qout <= qout + 1;
        else       qout <= qout - 1;
   end
// generate carry and borrow outputs
assign #2 cout = &qout;  // why #2 is required ?
assign #2 bout = ~|qout;
endmodule
```

In order to provide the capability of cascading multiple modules together to form a bigger counter, it proves much helpful to combine together the function of both the enable input and the mode selection input and then to partition the result into two mode control signals: the enable count up (eup) and the enable count down (edn) inputs. Such an idea is illustrated in the following example.

■ Example 9.24: An n-bit binary up/down counter (Version 2).

This example still uses the synchronous reset and enable signals. At each positive edge of the clock signal, the counter is cleared to 0 if the active-high reset signal is active; otherwise, it may count up, count down, or remain unchanged, depending on the values of the enable count up (eup) and the enable count down (edn) signals. The counter counts up if eup is 1 and down if edn is 1. The counter would not count down or up if both count enable signals, eup and edn, are active at the same time.

```verilog
// an N-bit up/down binary counter with synchronous
// reset and enable inputs
module up_dn_bin_counter
        #(parameter N = 4)( // set default size
           input   clk, reset, eup, edn,
           output reg [N-1:0] qout,
           output  cout, bout);
// the body of the N-bit binary counter
always @(posedge clk) begin
   if (reset) qout <= 0; // synchronous reset
   else begin
        if (eup) qout <= qout + 1;
        if (edn) qout <= qout - 1;
   end
end
// generate carry and borrow outputs
assign #1 cout = (&qout) & eup;  // generate carry-out
assign #1 bout = (~|qout) & edn; // generate borrow-out
endmodule
```

Although the above up/down counter can be expanded to any bits by simply setting the parameter N to a desired value, a practical discipline is to cascade two or more counters together to form a bigger one. An illustration of this is given in the following example by way of the up/down counter introduced in the preceding example.

■ Example 9.25: A cascade of two up/down counters (Version 2).

In this example, we instantiate two 4-bit up/down counters and cascade them together to form an 8-bit up/down counter. From this example, we can see the importance of the delay associated with the **assign** continuous assignments which generate both carry-out and borrow-out signals during simulation. Without them, the next higher-order counter module would not work properly. However, the synthesized gate-level module is able to do very well.

```
// the cascading of two up/down counters
module up_dn_bin_counter_cascaded
        #(parameter N = 4)(      // set default size
          input   clk, reset, eup, edn,
          output  [2*N-1:0] qout,
          output  cout, bout);

// declare internal nets for cascading both counters
wire    cout1, bout1;
// the body of the cascaded up/down counter
   up_dn_bin_counter #(4) up_dn_cnt1
        (clk, reset,eup, edn, qout[3:0], cout1, bout1);
   up_dn_bin_counter #(4) up_dn_cnt2
        (clk, reset,cout1, bout1, qout[7:4], cout, bout);
endmodule
```

It is quite often to have a counter with a modulus other than 2^n, where n is the number of flip-flops. For instance, a counter with a modulus of 6 or 10 is often desired in practical applications. The following example demonstrates how to model a synchronous counter with an arbitrary modulus.

■ Example 9.26: A modulo-r binary counter.

To construct a synchronous counter with a modulus of r, it proves useful to first construct an n-bit synchronous binary counter with $2^n \geq r$ and then a combinational circuit is used to detect the output value of the counter. If the output value of the counter is $r-1$ then the counter is cleared at the next clock cycle; otherwise, the counter continues to count up as usual [2]. Consequently, by setting the appropriate value of the modulus r, a synchronous binary counter with an arbitrary modulus can be readily built. The module describes such a counter is revealed as follows:

```
// a modulo-R binary counter with synchronous reset
module modulo_r_counter
        #(parameter R = 10,   // default modulus
          parameter N = 4)(   // N = log2 R
          input   clk, reset,
          output  cout,       // carry-out
          output  reg [N-1:0] qout);

// the body of the modulo-r binary counter
```

```
assign cout = (qout == R - 1);
always @(posedge clk)
   if (reset) qout <= 0;
   else if  (cout) qout <= 0;
       else        qout <= qout + 1;
endmodule
```

It is worth mentioning that you may simply describe a modulo-r counter in behavioral style by using the following statement

```
qout <= (qout + 1) % r;
```

Unfortunately, it is generally unsynthesizable. This is because the modulus operator (%) is usually synthesizable only if its right operand is an integer power of 2.

■ Review Questions

9-45 Distinguish between counters and timers.

9-46 What does an asynchronous counter mean?

9-47 What does a synchronous counter mean?

9-48 Explain the meaning of ripple counters.

9-49 What are the differences between using a single statement: `{cout, qout} <= qout + 1`, and using two statements: `qout <= qout + 1` and `assign cout = &qout`?

9-50 Write a parameterizable modulo-r counter with an active-high enable input and an asynchronous active-low reset input.

9.5 Sequence Generators

In addition to being used as data registers and data format converters, shift registers find their widespread use as *sequence generators*. In this section, we focus our attention on the following four circuits: pseudo-random sequence (PR-sequence) generators, cyclic-redundancy check (CRC code) generators/detectors, ring counters, and Johnson counters.

9.5.1 PR-Sequence Generators

The general logic block diagram of sequence generators consisting of an n-bit shift register and a combinational circuit is shown in Figure 9.24, where the outputs of all D flip-flops of the shift register are combined together through the combinational circuit and the result of the combinational circuit is then fed back into the input of the shift register. The resulting sequence generator is called a *linear feedback shift register* (LFSR) if the combinational circuit is a network of XOR gates. An LFSR is called a *pseudo-random sequence generator* (PRSG) if it is able to generate a maximum-length sequence; namely, the sequence has a period of $2^n - 1$, excluding the case of all 0s, where n is the number of flip-flops. Such a logic circuit is also known as an *autonomous linear feedback shift register* (ALFSR) and the generated sequence is often referred to as a *pseudo-random sequence* (PR-sequence) or *pseudo-random binary sequence* (PRBS). PR sequences are widely used in data networks, communication systems, and VLSI testing.

Although the general paradigm shown in Figure 9.24 can also be used to generate sequences other than PR-sequences, in the rest of this section, we only focus our attention on PR-sequences. The reader interested in the issues of generating other possible sequences can be referred to in Lin [2] for more details.

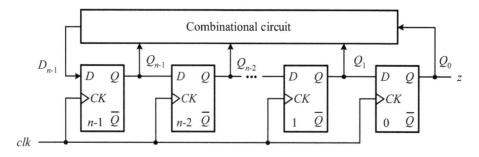

Figure 9.24: The general logic block diagram of sequence generators.

Table 9.4: Sample primitive polynomials for n from 1 to 60.

n	$f(x)$	n	$f(x)$	n	$f(x)$
1,2,3,4,6	$1+x+x^n$	24	$1+x+x^2+x^7+x^n$	43	$1+x+x^5+x^6+x^n$
7,15,22,60		26	$1+x+x^2+x^6+x^n$	44,50	$1+x+x^{26}+x^{27}+x^n$
5,11,21,29	$1+x^2+x^n$	30	$1+x+x^2+x^{23}+x^n$	45	$1+x+x^3+x^4+x^n$
10,17,20,25	$1+x^3+x^n$	32	$1+x+x^2+x^{22}+x^n$	46	$1+x+x^{20}+x^{21}+x^n$
28,31,41,52		33	$1+x^{13}+x^n$	48	$1+x+x^{27}+x^{28}+x^n$
9	$1+x^4+x^n$	34	$1+x+x^{14}+x^{15}+x^n$	49	$1+x^9+x^n$
23, 47	$1+x^5+x^n$	35	$1+x^2+x^n$	51,53	$1+x+x^{15}+x^{16}+x^n$
18	$1+x^7+x^n$	36	$1+x^{11}+x^n$	54	$1+x+x^{36}+x^{37}+x^n$
8	$1+x^2+x^3+x^4+x^n$	37	$1+x^2+x^{10}+x^{12}+x^n$	55	$1+x^{24}+x^n$
12	$1+x+x^4+x^6+x^n$	38	$1+x+x^5+x^6+x^n$	56,59	$1+x+x^{21}+x^{22}+x_n$
13	$1+x+x^3+x^4+x^n$	39	$1+x^4+x^n$	57	$1+x^7+x^n$
14,16	$1+x^3+x^4+x^5+x^n$	40	$1+x^2+x^{19}+x^{21}+x^n$	58	$1+x^{19}+x^n$
19,27	$1+x+x^2+x^5+x^n$	42	$1+x+x^{22}+x^{23}+x^n$		

A PR-sequence is a polynomial code which represents a bit string as a polynomial with coefficients of 0s and 1s only. Let x denote a unit delay, corresponding to a D flip-flop, and x^k denote a k-unit delays. Then, an n-stage LFSR can be expressed as the following polynomial

$$f(x) = a_n x^n + a_{n-1} x^{n-1} + \cdots + a_1 x + a_0 \qquad (9.9)$$

where coefficients $a_i \in \{0,1\}$, for all $0 \leq i \leq n$. Each combination of coefficients a_i for a given n corresponds to a function $f(x)$. However, only a few such functions can yield a maximum-length sequence. As a polynomial $f(x)$ generates a maximum-length sequence, it is called a *primitive polynomial*. In other words, a PR-sequence is generated by a primitive polynomial. Sample primitive polynomials for n from 1 to 60 are listed in Table 9.4.

Any primitive polynomial of a given n can be implemented by either the *standard format* or the *modular format*, as shown in Figure 9.25. The standard format is a simple application of the general paradigm exhibited in Figure 9.24 in which all XOR gates are lumped together. In contrast, the modular format distributes the XOR gates into each stage and hence it is so named.

Illustrated in Figure 9.26 is an example of a 4-bit PR-sequence generator, which has a primitive polynomial, $1+x+x^4$, and is realized in standard format. Here, the coefficients a_2 and a_3 are zero and hence their corresponding XOR gates are not necessary and omitted.

Modeling an n-bit PR-sequence generator in standard format is rather simple because the feedback XOR network can be simply computed by using an xor (^) reduction operator. An illustration of this is explored in the following example.

Section 9.5 Sequence Generators

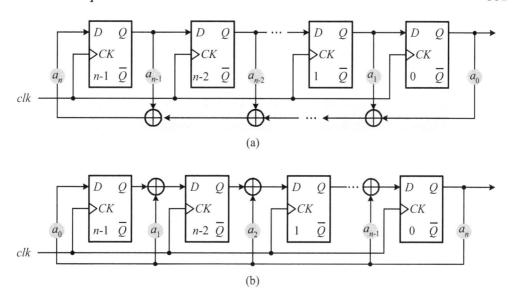

Figure 9.25: The two standard paradigms for implementing primitive polynomials: (a) the standard format and (b) the modular format.

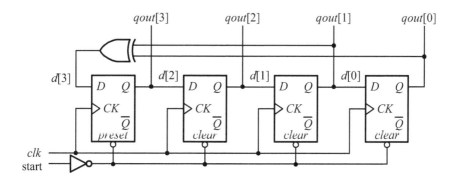

Figure 9.26: A 4-bit PR-sequence generator with a primitive polynomial: $1 + x + x^4$.

■ Example 9.27: An n-bit PRSG in standard format.

The basic idea of implementing a primitive polynomial with the standard format is to use an xor (^) reduction operator to reduce the XOR network into a single-bit result which is then fed back to the first stage as the serial input. To achieve this, it proves very helpful to declare an $(n+1)$-bit parameter tap to store the coefficients of the given primitive polynomial. The lowest n bits of tap and the outputs of all D flip-flops are ANDed together and then the xor reduction operator is applied to reduce the ANDed result into a single bit value. This value is fed back to the first stage as the serial input data.

```
// an N-bit PR-sequence generator --- in standard format
module pr_sequence_generator
       #(parameter N = 4,   // define the default size
         parameter [N:0]tap = 5'b10011)(  // x^4 + x + 1
         input  clk,
         output reg [N-1:0] qout);
wire   d;
```

```verilog
// the body of the PR-sequence generator
assign d = ^(tap[N-1:0] & qout[N-1:0]);
always @(posedge clk)
    qout <= {d, qout[N-1:1]};
endmodule
```

The drawback of the above module is that it cannot be simulated due to the lack of the initial value of qout and hence simulators cannot calculate qout; thereby, we could not observe the qout values, i.e., only unknown values. To overcome this difficulty, an initial value, say, qout = 4'b1000, may be used. In practical applications, a PR-sequence generator is needed to start automatically. To this end, a start-up circuit must be added to a PR-sequence generator to preset its qout with an initial value, say, 4'b1000, as illustrated in Figure 9.26. An illustration of this idea is explored as follows.

■ Example 9.28: An n-bit PRSG in standard format.

This example is an extension of the previous one with the addition of a start-up circuit to preset the output, qout, with an initial value of N'b10...0. It is also possible to preset the PR-sequence generator with other initial values. In fact, any initial value other than all 0s can be used. The start-up circuit used in this example is controlled by a start signal, start. After the start signal is applied, the PR-sequence generator will begin from the state N'b10...0.

```verilog
// an N-bit PR-sequence generator --- in standard format
module pr_sequence_generator
       #(parameter N = 4,     // define the default size
         parameter [N:0]tap = 5'b10011)(
         input  clk, start,
         output reg [N-1:0] qout );
wire   d;

// the body of the PR-sequence generator
assign d = ^(tap[N-1:0] & qout[N-1:0]);
always @(posedge clk or posedge start)
    if (start) qout <= {1'b1, {N-1{1'b0}}};
    else       qout <= {d, qout[N-1:1]};
endmodule
```

Recall that a PR-sequence generator will go through every possible state except the all-zero state, which will lock out the circuit forever. Due to ubiquitous noise in the environment, it is always possible for a PR-sequence generator to enter the all-zero state. Hence, some mechanisms must be provided to prevent such a situation from occurring or to make the PR-sequence generator get out the all-zero state once it entered this state inadvertently. The simplest way to get out the all-zero state is to add an all-zero detector to the circuit, as shown in Figure 9.27. The all-zero detector is an n-input AND gate with each input connected to the complemented output of each D flip-flop of the PR-sequence generator. The output of the AND gate is combined with the output from the network of XOR gates through an OR gate and is then fed into the input of the shift register. Consequently, as the outputs of all flip-flops are 0s, the output of the all-zero detector is 1, thereby causing the shift register to start with the state N'b10...0 at the next clock cycle. An illustration to reveal how to model this self-start PR-sequence generator is explored in the following example.

Section 9.5 Sequence Generators

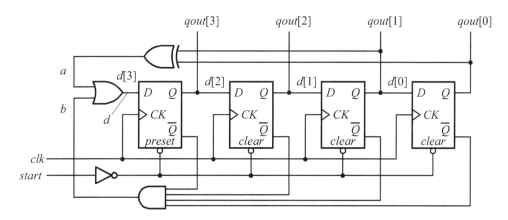

Figure 9.27: A 4-bit PR-sequence generator with a self-start circuit.

■ Example 9.29: An n-bit PRSG in standard format.

This circuit is basically the same as the preceding example, except that it adds an all-zero detector (AND gate) to detect the situation that the outputs of all flip-flops are zeros so as to generate an output value of 1. This value of 1 will transfer the shift register back to its normal loop at the next clock cycle. After the reset signal is activated, the circuit starts up from the state N'b10...0.

```
// an N-bit PR_sequence generator --- in standard format
module pr_sequence_generator_self_start
       #(parameter N = 4,  // define the default size
         parameter [N:0]tap = 5'b10011)(
         input   clk, start,
         output reg [N-1:0] qout);
wire    a, b, d;

// the body of the PR-sequence generator
assign a = ^(tap[N-1:0] & qout[N-1:0]),
       b = ~(|qout[N-1:0]),
       d = a | b;
always @(posedge clk or posedge start)
   if (start) qout <= {N{1'b0}};
   else       qout <= {d, qout[N-1:1]};
endmodule
```

■ Review Questions

9-51 What is an LFSR?
9-52 Model an n-bit PR-sequence generator in modular format.
9-53 What is a pseudo-random sequence?
9-54 What is a primitive polynomial?
9-55 Explain the function of the AND gate in the circuit of Figure 9.27.

9.5.2 CRC Code Generators/Detectors

Another important application of the LFSR is to generate and check (detect) the CRC code of a message block, which is widely used to ensure data integrity in disk devices as well as wired and wireless communication networks. Like PR-sequences, a CRC code is also a polynomial code and can be represented as a polynomial of the x operator. An important property of a CRC code with n check bits is that it has the capability of detecting all burst errors of lengths less than or equal to n and all burst errors with odd number of bits.

9.5.2.1 Principles of CRC The use of a CRC code to ensure data integrity is based on the following principle. At the transmitter, the message to be sent is first divided by a CRC polynomial (also called a *CRC generator*) and the remainder (called the CRC code) is then appended to the message polynomial and sent to the destination along with the message polynomial. At the receiver, the received polynomial is divided by the same CRC polynomial again. If the remainder is 0, it is correct; otherwise, it contains errors.

To get more insight into the generation of a CRC code, suppose that the message to be sent is represented as a polynomial of x and denoted by $D(x)$, the CRC polynomial is used as the divisor polynomial and denoted by $G(x)$, and the remainder polynomial is denoted by $R(x)$. For convenience, all above three polynomials are expressed as coefficient representation. The resulting CRC polynomial to be sent is $T(x) = \{D(x), R(x)\}$, where $\{,\}$ denotes the concatenation of two strings, and the received CRC polynomial is denoted by $T'(x)$, which may contain erroneous bits.

The detailed operations of CRC code generation is described as the following algorithm.

■ Algorithm 9-1: CRC code generator

Input: The message $D(x)$ to be sent.
Output: The CRC polynomial to be sent, $T(x) = \{D(x), R(x)\}$.

1. Supposing that $G(x)$ is a degree-n polynomial, append n 0s to the message polynomial $D(x)$ and denote the resulting polynomial as $D'(x)$.
2. Divide the resulting polynomial $D'(x)$ by $G(x)$ using modulo-2 operation.
3. Append the remainder $R(x)$ (i.e., the CRC code) obtained from the above step to the original message $D(x)$ to form the CRC polynomial $T(x)$ to be sent; namely, $T(x) = \{D(x), R(x)\}$, which corresponds to $D'(x) - R(x)$.

■ Example 9.30: A numeric example of the CRC code generator.

The operation of the above algorithm is illustrated in Figure 9.28. Here, we assume that $G(x) = x^4 + x + 1$ and the message to be sent is 1101101010. Because the degree of $G(x)$ is 4, four 0s are appended to the message polynomial. After completing the division, the remainder (i.e., the CRC code) is found to be 0010. Therefore, the CRC polynomial to be sent is 11011010100010. It is easy to see that the remainder will be 0 when this CRC polynomial is divided by $G(x)$ again.

To check (or detect) a CRC polynomial, the received CRC polynomial is divided by the same CRC generator $G(x)$. If the remainder is 0, the received CRC polynomial is error free; otherwise, it is erroneous. The logic circuit used to compute the remainder is called a *CRC code generator* and the logic circuit used to check/detect whether a CRC polynomial is error free is called a *CRC code detector*. Because both circuits are identical, they are indeed a single circuit and is often referred to as a *CRC code generator/detector*.

Section 9.5 Sequence Generators

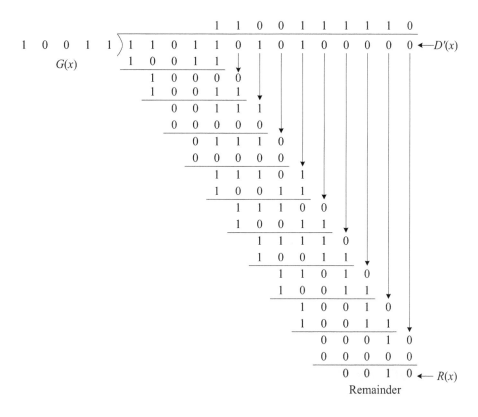

Figure 9.28: A numerical example of a 4-bit CRC code generator.

Before going into the implementations of CRC code generators/detectors, we summarize below a number of CRC polynomials commonly used in data communications.

$$\text{CRC-8} = x^8 + x^7 + x^6 + x^4 + x^2 + 1$$
$$\text{CRC-12} = x^{12} + x^{11} + x^3 + x^2 + x + 1$$
$$\text{CRC-16} = x^{16} + x^{15} + x^2 + 1$$
$$\text{CRC-CCITT} = x^{16} + x^{12} + x^5 + 1$$
$$\text{CRC-32a} = x^{32} + x^{30} + x^{22} + x^{15} + x^{12} + x^{11} + x^7 + x^6 + x^5 + x + 1$$
$$\text{CRC-32b} = x^{32} + x^{26} + x^{23} + x^{22} + x^{16} + x^{12} + x^{11} + x^{10} + x^8 + x^7 +$$
$$x^5 + x^4 + x^2 + x + 1$$

9.5.2.2 Implementations of CRC Code Generators/Detectors The paradigms for implementing CRC code generators/detectors are basically the same as the modular format of the PR-sequence generator described in Figure 9.25(b) except that here we need to introduce the external data into the circuit. Because of the inherent bit-serial nature of CRC code generators/detectors, the external data may be introduced into the circuit from the leftmost (LSB) or rightmost (MSB) end, as shown in Figures 9.29(a) and (b), respectively.

The approach to feeding the external data into the circuit from the LSB is illustrated in Figure 9.29(a). In this approach, the input of each flip-flop receives the combined result from the output of the last stage with the input data $D(x)$ or with the output from its previous stage under the control of the coefficients of $G(x)$. In fact, this method realizes the above algorithm directly; that is, the actual message that enters the CRC code generator is a combination of the

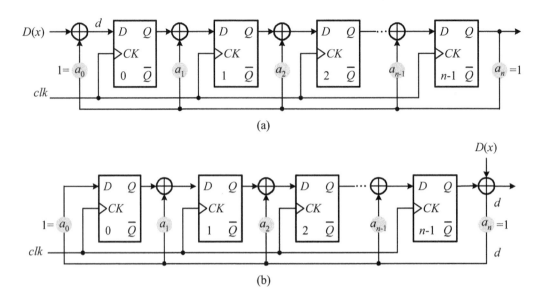

Figure 9.29: The general paradigms of n-bit CRC code generators/detectors: (a) feeding data from the LSB and (b) feeding data from the MSB.

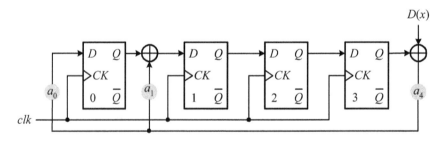

Figure 9.30: A 4-bit CRC code generator with $G(x) = x^4 + x + 1$.

message to be sent and n 0s. In other words, after entering the entire message into the circuit, it still needs to fill another n 0s, one 0 for each clock cycle, in order to generate the remainder (i.e., CRC code). This may cause some troubles in practical applications because it needs to wait n cycles for generating the desired remainder after the message has been entered the CRC code generator.

To overcome the above difficulty, an alternative method is to feed the data into the circuit from the MSB, as shown in Figure 9.29(b), where the input data $D(x)$ is combined with the output from the last stage and the result is then fed back to the preceding stages under the control of the coefficients of $G(x)$. Based on this, the remainder is left in the circuit right after the message is entirely entered the circuit. Consequently, the remainder may be sent after the message immediately. Because of this important feature, this method is most widely used in practical applications. It is worth noting again that the same circuit is used for both CRC code generation and detection.

An example of implementing $G(x) = x^4 + x + 1$ is shown in Figure 9.30. Because only coefficients a_4, a_1, and a_0 are not zero, only the connections indicated by these three coefficients are needed. The following example shows how to describe an n-bit CRC code generator using a generate-loop statement.

Section 9.5 *Sequence Generators* 357

■ Example 9.31: An n-bit CRC code generator.

This example indeed illustrates the implementation of the modular format shown in Figure 9.25(b) except that the external data $D(x)$ is needed to introduce into the circuit. Observing Figure 9.29(b), we know that two cases are needed to consider, the LSB and the other bits. For convenience, suppose that d is the result obtained from data XORed with qout[N-1]. For the LSB, the data d is directly input into the flip-flop when the active-low reset signal, reset_n, is not asserted. For the remaining bits, the input of a flip-flop directly comes from the output of its preceding stage if the associated coefficient of $G(x)$ is 0 or is a combined result of d and the output of its preceding stage if the associated coefficient of $G(x)$ is 1.

```verilog
// a parameterizable CRC generator/detector module
// using generate-for statement --- feeding data from MSB
module CRC_MSB_generate
       #(parameter N = 16,    // set default size
         parameter [N:0]tap = 17'b11000000000000101)(
         input   clk, reset_n, data,
         output reg [N-1:0] qout);
wire d;
// the CRC code generator/detector body
assign d = data ^ qout[N-1];
genvar i;
generate for(i = 0; i < N ; i = i + 1) begin: crc_generator
   if (i == 0) // specify the LSB
      always @(posedge clk or negedge reset_n)
         if (!reset_n) qout[i] <= 1'b0;
         else qout[i] <= d;
   else        // the remaining bits
      always @(posedge clk or negedge reset_n)
         if (!reset_n) qout[i] <= 1'b0;
         else if (tap[i] == 1) qout[i] <= qout[i-1] ^ d;
         else qout[i] <= qout[i-1];
end endgenerate
endmodule
```

As a matter of fact, the generate-loop statement within the above module may also be replaced by the following **always** block:

```verilog
always @(posedge clk or negedge reset_n)
   if (!reset_n) qout <= 1'b0;
   else qout <= {qout[N-2:0], 1'b0} ^ (tap[N-1:0] & {N-1{d}});
```

The reader is encouraged to explain how this works. It is instructive to write a complete module by using the above statement and write a test bench to verify the functionality of the resulting module.

9.5.3 Ring Counters

There are two basic types of ring counters: the *standard ring counter* and the *twisted ring counter*. A modulo-n standard ring counter is an n-bit (or n-stage) shift register with the serial output fed back to the serial input. An essential feature of an n-bit ring counter is that it outputs a 1-out-of-n code directly from its flip-flop outputs.

A 4-bit ring counter is shown in Figure 9.31, where the output $qout[3]$ is connected backward to the input $d[0]$. In addition, a start-up circuit is used to set the initial value of the ring counter to 4'b1000.

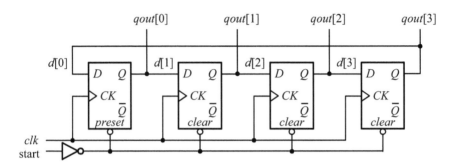

Figure 9.31: An example of a 4-bit ring counter.

■ Example 9.32: An n-bit ring counter with an initial value.

As an illustration of how to model an n-bit ring counter, consider the 4-bit ring counter in Figure 9.31. In order to start up the circuit, an initial value other than all 0s is required. This is done in this example by using a start signal, start. After the start signal is applied, the circuit will start from the state N'b10...00. The complete module is as follows:

```
// a ring counter with an initial value
module ring_counter
       #(parameter N = 4)(   // set default size
          input   clk, start,
          output reg  [0:N-1] qout );
// the body of the ring counter
always @(posedge clk or posedge start)
  if (start)   qout <= {1'b1, {N-1{1'b0}}};
  else         qout <= {qout[N-1], qout[0:N-2]};
endmodule
```

In general, a standard ring counter has n valid states and $2^n - n$ invalid states. Once the ring counter enters one of these invalid-state loops, it will be locked in the loop and cannot go back to its normal valid-state loop. To prevent a ring counter from being locked inside these invalid-state loops, some extra logic circuit must be added to instruct the ring counter to return to its normal valid-state loop from the invalid-state loops. To illustrate this, consider the state diagram of the 4-bit standard ring counter in Figure 9.31, which, except a normal valid-state loop, has five independent invalid-state loops, as plotted in Figure 9.32. To instruct the 4-bit standard ring counter to return to the normal valid-state loop, a widely used approach is to replace the input $d[0]$ with the following switching function:

$$d[0] = \overline{(qout[0] + qout[1] + qout[2])} \tag{9.10}$$

It can be shown that once the ring counter enters any invalid-state loop, it will go back to its normal valid-state loop after four clock cycles at most (see Problem 9-21). A standard ring counter designed in such a way is called a *self-correcting ring counter*.

Section 9.5 Sequence Generators

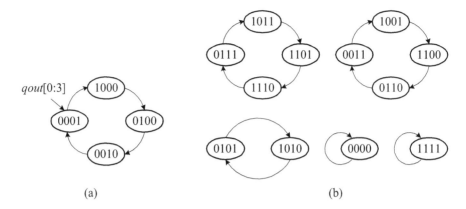

Figure 9.32: The state diagram of the 4-bit ring counter shown in Figure 9.31: (a) valid states and (b) invalid states.

9.5.4 Johnson Counters

A twisted-ring counter is also known as a *Johnson counter*, a Moebius, or a *switched-tail counter*. A modulo-$2n$ Johnson counter is an n-bit shift register with the complemented serial output fed back to the serial input. The essential feature of Johnson counters is that they only need half the number of flip-flops needed in ring counters to achieve the same number of counting states.

A 4-bit Johnson counter is shown in Figure 9.33, where the complement of the last-stage output is fed back to the input of the first stage. Because its structure is similar to the shift register, the same technique for modeling a shift register can be applied equally well here to describe an n-bit Johnson counter. An example to illustrate this is given as follows.

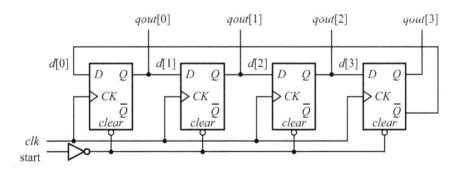

Figure 9.33: An example of a 4-bit Johnson counter.

■ Example 9.33: An n-bit Johnson counter with an initial value.

Like a standard ring counter, the essential structure of Johnson counters is also an n-bit shift register, except that the complemented output of the last stage is fed back to the first stage. In contrast to standard ring counters, the initial value set in the Johnson counter is all 0s rather than a 1-out-of-n code.

```
// Johnson counter with an initial value
module Johnson_counter
       #(parameter N = 4)(   // set default size
```

```
          input  clk, start,
          output reg [0:N-1] qout);
// the body of the Johnson counter
always @(posedge clk or posedge start)
   if (start) qout <= {N{1'b0}};
   else       qout <= {~qout[N-1], qout[0:N-2]};
endmodule
```

An n-bit Johnson counter has $2n$ valid states and $2^n - 2n$ invalid states. Hence, a Johnson counter is also subjected to the same problem as standard ring counters; namely, it may be locked in one of the invalid-state loops and cannot return to its normal valid-state loop. To make a Johnson counter self-correctable, some extra logic circuit must be added to instruct the Johnson counter to return to its normal valid-state loop from the invalid-state loops. A widely used approach for the 4-bit Johnson counter shown in Figure 9.33 is to replace the input $d[2]$ with the following switching function:

$$d[2] = (qout[0] + qout[2])qout[1] \tag{9.11}$$

The reader is encouraged to show that the resulting Johnson counter is a self-correcting counter (see Problem 9-22).

■ Review Questions

9-56 Describe the basic principles of applying the CRC code to ensure data integrity.
9-57 Describe how to compute the CRC code.
9-58 Describe the basic structure of an n-bit ring counter.
9-59 Describe the basic structure of an n-bit Johnson counter.
9-60 What is the basic feature of an n-bit ring counter?
9-61 What is the meaning of a "self-correcting" counter?
9-62 Discuss the reason why a counter is needed to be self-correctable.

9.6 Timing Generators

A timing generator is a logic circuit that generates the timing required for a specific application. Two types of timing generators widely used in designing digital systems are *multiphase clock generators* and *digital monostable circuits*. A multiphase clock generator can be constructed using a ring counter, a Johnson counter, or a binary counter with a decoder. A digital monostable circuit may be either retriggerable or non-retriggerable.

9.6.1 Multiphase Clock Generators

A typical multiphase clock signal is shown in Figure 9.34. Each clock phase lasts for a clock cycle and is repeated after a given number of clock cycles. In other words, an n-phase clock generator yields a 1-out-of-n code at every clock cycle. This exactly matches the feature of an n-bit ring counter. Hence, it is quite natural to use an n-bit ring counter to generate an n-phase clock signal.

Another widely used approach to generating an n-phase clock signal is to use a $\log_2 n$-bit binary counter with a $(\log_2 n)$-to-n decoder. As depicted in Figure 9.35 is an 8-phase clock generator using the binary counter with a decoder approach. The 3-bit binary counter generates the outputs, 0 to 7, in sequence, for the decoder to yield the corresponding 1-out-of-8 code. An illustration of modeling such a circuit as a parameterizable module is explored in the following example.

Section 9.6 Timing Generators

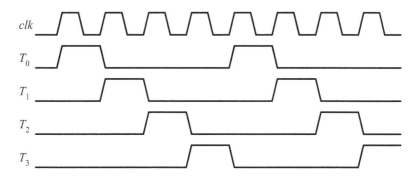

Figure 9.34: An example of a multiphase clock signal.

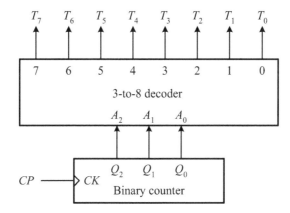

Figure 9.35: A multiphase clock generator based on the binary counter with a decoder approach.

■ Example 9.34: A multiphase clock signal generator.

There are two major components in the binary counter with a decoder approach. The first component is an m-bit binary counter, which is described by the first **always** block. The second component is an m-to-n decoder, which is modeled by the second **always** block. Note that here we assume that $n = 2^m$.

```
// a timing generator constructed using the approach of
// a binary counter with a decoder
module binary_counter_timing_generator
       #(parameter N = 8, // set the number of phases
         parameter M = 3)(// set the number of bits
         input  clk, reset, enable,
         output reg [N-1:0] qout);

reg    [M-1:0] bcnt_out; // an M-bit binary counter
// the body of an M-bit binary counter
always @(posedge clk or posedge reset)
   if (reset)  bcnt_out <= {M{1'b0}};
   else if (enable) bcnt_out <= bcnt_out + 1;

// decode the output of the binary counter
```

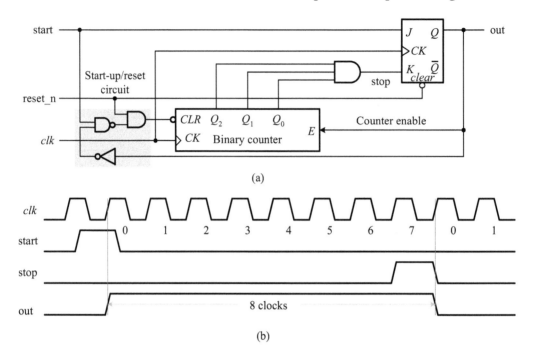

Figure 9.36: A digital non-retriggerable monostable circuit: (a) logic circuit and (b) timing diagram.

```
always @(bcnt_out)
    qout = {{N-1{1'b0}}, 1'b1} << bcnt_out;
endmodule
```

The reader is encouraged to synthesize the above module and observe the results. It is also instructive to change the parameters, M and N, where $N = 2^M$, and repeat the above operations.

9.6.2 Digital Monostable Circuits

A monostable circuit, as the name implies, has only one stable state. Normally, a monostable circuit is in its stable state. Whenever it is triggered, it temporarily leaves the stable state and goes into a transient state where it outputs a pulse signal and then returns to its stable state after a period set by an external circuit. A monostable circuit can be either *retriggerable* or *non-retriggerable*. A retriggerable monostable circuit is one that could be retriggered before it ends its previous transient state and restarts another. A non-retriggerable monostable circuit is one that ignores any triggered signal before it completes its current transient state.

Although most monostable circuits are based on some kind of RC circuit, a digital monostable circuit is usually desired in designing modern digital systems because it facilitates a more testable and robust circuit than an RC-based circuit. An example of a non-retriggerable digital monostable circuit is shown in Figure 9.36, which consists of a JK flip-flop, a binary counter, an AND gate, and a start-up/reset circuit.

The digital monostable circuit works as follows. Each time the trigger signal *start* is asserted, the output, *out*, of the JK flip-flop rises to high at the next positive edge of the clock signal. This high-level output signal enables the binary counter to count up. Once the binary counter reaches a specified value (namely, 7 in the current example) set by the AND gate, the output of the JK flip-flop resets to 0 and the *out* pulse terminates, which in turn inhibits the

Section 9.7 *Summary*

operation of the binary counter. Consequently, it generates an output pulse with a width equal to a specified number of clock cycles, namely, 8 clock cycles in this example.

Based on the above idea and circuit structure, it is easy to construct a parameterizable non-retriggerable digital monostable circuit. An illustration is demonstrated in the following example.

■ **Example 9.35: A non-retriggerable digital monostable example.**

This example describes a parameterizable non-retriggerable digital monostable circuit with a maximum output pulse width of 2^m clock cycles. The pulse output signal out will indeed go to high for PW clock cycles once the start signal is activated. The first **always** block models the binary counter and its associated logic. The **assign** continuous assignment and the second **always** block describe the JK flip-flop and its related logic circuits.

```
// a non-retriggerable digital monostable circuit --- the out
// goes high for PW cycles once the start signal is asserted
module digital_monostable
      #(parameter M = 3,  // define maximum pulse width = 2**M
        parameter PW = 5)( // define the actual pulse width
        input   clk, reset_n, start,
        output reg out);
reg   [M-1:0] bcnt_out;
wire  d, bcnt_reset;

// the body of the binary counter
assign bcnt_reset = !reset_n || (start && !out);
always @(posedge clk or posedge bcnt_reset)
   if  (bcnt_reset) bcnt_out <= {M{1'b0}};
   else if (out) bcnt_out <= bcnt_out + 1;

// describe the JK flip-flop and its related circuits
assign d = (start & ~out) | (~(bcnt_out == (PW-1)) & out);
always @(posedge clk or negedge reset_n)
   if (!reset_n) out <= 1'b0;
   else          out <= d;
endmodule
```

■ **Review Questions**

9-63 What is a multiphase clock generator?
9-64 What is a digital monostable circuit?
9-65 What are the retriggerable and non-retriggerable monostable circuits?
9-66 Can we use a Johnson counter to construct a multiphase clock generator?
9-67 Modify the circuit shown in Figure 9.36 so that it can be retriggered.

9.7 Summary

In this chapter, we examined many widely used sequential modules, including flip-flops, synchronizers, a switch-debounce circuit, registers, data registers, register files, shift registers, counters (binary, BCD, ring, and Johnson), CRC code generators and detectors, clock generators, pulse generators, and timing generators.

Flip-flops and latches are the basic building blocks for any sequential module. The flip-flops and latches are basically bistable devices. The fundamental difference between them is that latches have the transparent property whereas flip-flops do not. Hence, the output of a latch follows the input if its enable is active whereas a flip-flop only samples its input data at some specific time instant, such as the positive or negative edge of the clock signal.

The basic applications of flip-flops at least include synchronizers and switch-debounce circuits. The objective of synchronizers is to sample an external asynchronous signal in synchronism with the internal system clock signal and the aim of switch debouncers is to remove the switch-bouncing effects of mechanical switches, caused by the mechanical inertia of switches.

Flip-flops also find their widespread use in constructing registers and shift registers as well as counters. Registers include basic registers and register files. Shift registers are often used to perform arithmetic operations, such as divide-by-2 and multiply-by-2, and data format conversions. Counters that we have considered include both asynchronous and synchronous types.

Another important application of shift registers is to serve as sequence generators, including PR-sequence generators, CRC code generators/detectors, ring counters, and Johnson counters. The general scheme of sequence generators is an n-stage LFSR in which the outputs of all D flip-flops of the shift register are combined together through a combinational circuit and the result is then fed back into the input of the shift register. An LFSR is called a PR-sequence generator if it generates a maximum-length sequence; namely, the sequence has a period of $2^n - 1$, excluding the case of all 0s, where n is the number of flip-flops. Ring counters and Johnson counters are the two basic applications of LFSR-based sequence generators.

A timing generator is a device that generates timing required for specific applications. Multiphase clock generators and digital monostable circuits are the two types of widely used timing generators in digital systems. A multiphase clock generator can be constructed from a ring counter, a Johnson counter, or a binary counter with a decoder. For a typical multiphase clock signal, each clock phase lasts for one clock cycle and is repeated after a given number of clock cycles.

A monostable circuit is a circuit that can generate an output pulse with a specified duration whenever it is triggered. The digital monostable circuit can be either retriggerable or non-retriggerable. A retriggerable monostable circuit can be retriggered before it ends its previous transient state and restarts another. A non-retriggerable monostable circuit ignores any triggered signal before it completes its current transient state.

A case study was given at the end of this chapter. This case study of a 12-hr clock/timer system combines the results from the previous chapter with those from this chapter into a useful and practical system. In this case study, we reveal the combination of the 4-digit multiplexed seven-segment LED display module, the switch-debounce circuit, and counters, into a useful and practical 12-hr clock/timer system.

References

1. J. Bhasker, *A Verilog HDL Primer*, 3rd ed., Star Galaxy Publishing, 2005.

2. M. B. Lin, *Digital System Design: Principles, Practices, and Applications*, 4th ed., Chuan Hwa Book Ltd. (Taipei, Taiwan), 2010.

3. M. B. Lin, *Introduction to VLSI Systems: A Logic, Circuit, and System Perspective*, CRC Press, 2012.

4. J. F. Wakerly, *Digital Design Principles & Practices*, 3rd ed., Upper Saddle River, New Jersey: Prentice-Hall, 2001.

Problems

9-1 Write a Verilog HDL module to describe a D flip-flop with both asynchronous clear and preset.

9-2 Write a Verilog HDL module to describe a D flip-flop with both synchronous clear and preset.

9-3 Considering the simple logic circuit shown in Figure 9.37, answer each of the following questions.

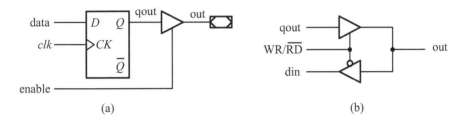

Figure 9.37: Figure for Problem 9-3.

(a) Write a Verilog HDL module to describe the circuit shown in Figure 9.37(a).

(b) Replace the output tristate buffer of the circuit shown in Figure 9.37(a) with the bidirectional buffer shown in Figure 9.37(b) and redo (a).

9-4 This problem concerns a synchronizer consisting of a single-stage D flip-flop. Supposing that $f = 50$ MHz, $a = 100$ kHz, and 74LS74 D flip-flops with the setup time of 10 ns are used in the entire system, answer each of the following questions.

(a) Calculate the MTBF.

(b) What is the maximum operating frequency allowed so that the MTBF can be greater than 100 years?

9-5 This problem involves the design of synchronizers. Suppose that $f = 50$ MHz, $a = 200$ kHz, and 74ALS74 D flip-flops are used in the entire system. The setup time of flip-flops is 10 ns.

(a) Calculate the MTBF if a single-stage D flip-flop is used as the synchronizer.

(b) Calculate the MTBF when the frequency-divided synchronizer with $N = 2$, as shown in Figure 9.14(b), is used.

9-6 Considering the circuits shown in Figure 9.38, answer each of the following problems:

(a) For the logic circuit shown in Figure 9.38(a), as the sampler flip-flop enters the metastable state, what are the possible values of the output a?

(b) For the case of the logic circuit shown in Figure 9.38(b), what are the possible values of the outputs a, b, and c?

(c) What are the conclusions you make from the above cases?

9-7 Design a circuit that can generate a pulse with a width of 10 ms each time a switch is pressed. Write a Verilog HDL module to describe this circuit. Of course, you have to take care of the switch-bounce problem.

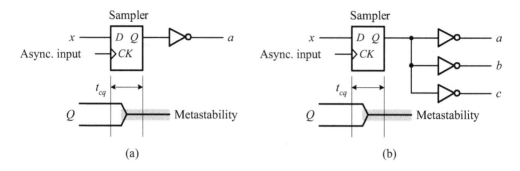

Figure 9.38: Figures for Problem 9-6.

9-8 Design a circuit to generate a single-cycle pulse whenever the input x is set to high, regardless of how long it remains in the high state. Write a Verilog HDL module to describe this circuit.

9-9 Referring to Figure 9.8(a), assume that the D flip-flop has the following parameters: $t_{cq,rise}$ = 5:6:8 ns, $t_{cq,fall}$ = 4.6:5.6:7.6 ns, t_{clearq} = 2.6:3:3.8 ns, and $t_{setup} = t_{hold}$ = 7:8:9 ns. For simplicity, the combinational logic is assumed to be merely an inverter (a NOT gate) with the propagation delay of 2:3:4 ns. All delays are specified in min:typ:max format.

   (a) Write a Verilog module for the D flip-flop with a specify block to include the timing parameters and the **$setup** and **$hold** timing checks to check the setup and hold times of the D flip-flop.

   (b) Assume that the external data are toggled to the first D flip-flop 1 ns after the negative edge of the clock signal clk. If the clock signal clk has a period (T_{clk}) of 20 ns with a duty cycle of 50%, what would happen under the typical delays? Write a test bench and carry out the required simulation to check this.

   (c) Rerun (b) with the worst-case delays of the logic circuit and check to see what happen. The worst-case delays can be specified with the **+maxdelays** command option. (*Hint:* In ModelSim, it can be set through: Simulate → Start Simulation → Verilog → Delay Selection → max.)

   (d) Rerun (b) with the best-case delays of the logic circuit and check to see what happen. The best-case delays can be specified with the **+mindelays** command option. (*Hint:* In ModelSim, it can be set through: Simulate → Start Simulation → Verilog → Delay Selection → min.)

9-10 Write a test bench to verify the sram_timing_check module described in Section 9.2.4.

9-11 Suppose that a special-purpose memory module with the layout and access features shown in Figure 9.39 is required.

   (a) Model the special-purpose memory module in behavioral style.
   (b) Write a test bench to verify the functionality of your module.

9-12 Considering the shift register with computation module shown below, answer each of the following questions.

```
// a right-shift register with computation
module shift_reg_blocking(clk, sin, qout);
input   clk;
```

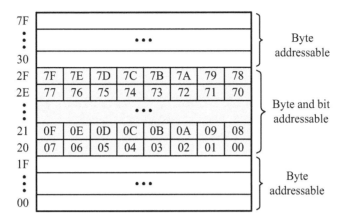

Figure 9.39: The layout of a special-purpose memory module considered in Problem 9-11.

```
input  sin;  // serial data input
output reg [3:0] qout;
// a 4-bit right-shift register with computation
always @(posedge clk) begin
   qout[3] = sin;
   qout[2] = qout[3];
   qout[1] = qout[2] ^ qout[3];
   qout[0] = qout[2] & qout[1];
end
endmodule
```

  (a) Can it be operated as a 4-bit left-shift register with the required computation?
  (b) Synthesize it and examine the result.
  (c) Explain why the synthesized result is like what you have obtained.
  (d) Correct the statements within the module so that it can function correctly.

9-13 Design an n-bit program counter (PC) with the following functions and write a Verilog HDL module to describe it.
  (a) PC is cleared to 0 when the asynchronous reset input is asserted.
  (b) PC is loaded a new value in parallel when the PCload input is asserted.
  (c) PC is incremented by 1 when the PCinc input is asserted.

9-14 Suppose that a 4-bit programmable counter is needed. The modulus of the counter can be set to n, where n may be set from 0 to 15 externally by the control signals $C3$ to $C0$. Write a Verilog HDL module to describe this counter and write a test bench to verify it.

9-15 Suppose that an 8-bit programmable counter is required. The modulus of the counter can be set to $4n$, where n may be set from 1 to 15 externally by the control signals $C3$ to $C0$. Write a Verilog HDL module to describe this counter and write a test bench to verify it.

9-16 Considering the CRC-16 code generator shown in Figure 9.40, which is found in a commercial device, answer each of the following questions.

(a) Write a Verilog HDL module to describe the circuit shown in Figure 9.40(a). Write a test bench to verify its functionality.

(b) Write a Verilog HDL module to describe the circuit shown in Figure 9.40(b). Write a test bench to verify its functionality.

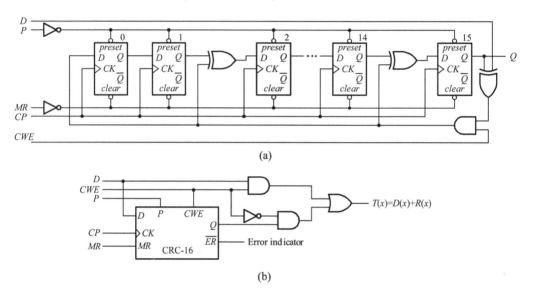

Figure 9.40: The (a) logic circuit and (b) an application of the CRC-16 code generator/detector.

9-17 Considering the general paradigms shown in Figures 9.29(a) and (b), respectively, answer each of the following questions.

(a) Use $G(x) = x^4 + x + 1$ and message $M(x) = 1101001110$ to show that both the paradigms generate the same CRC code.

(b) What are the differences between these two paradigms?

(c) If the CRC code must follow immediately after the message, give an approach to accomplishing this by using the paradigm shown in Figure 9.29(a).

9-18 Considering the general paradigms shown in Figure 9.29(a), answer each of the following questions.

(a) Write a parameterizable CRC code generator/detector module according to the paradigm shown in Figure 9.29(a) by using a generate-loop statement. Write a test bench to verify the functionality of the resulting module.

(b) Write a parameterizable CRC code generator/detector module according to the paradigm shown in Figure 9.29(a) without using the generate-loop statement. Write a test bench to verify the functionality of the resulting module.

9-19 Write Verilog HDL modules to describe each of the following CRC code generators and detectors:

(a) CRC-8 $= x^8 + x^7 + x^6 + x^4 + x^2 + 1$

(b) CRC-12 $= x^{12} + x^{11} + x^3 + x^2 + x + 1$

(c) CRC-16 $= x^{16} + x^{15} + x^2 + 1$

9-20 Write Verilog HDL modules to describe each of the following CRC code generators and detectors:

(a) CRC-32a = $x^{32} + x^{30} + x^{22} + x^{15} + x^{12} + x^{11} + x^7 + x^6 + x^5 + x + 1$
(b) CRC-32b = $x^{32} + x^{26} + x^{23} + x^{22} + x^{16} + x^{12} + x^{11} + x^{10} + x^8 + x^7 + x^5 + x^4 + x^2 + x + 1$

9-21 Show that the 4-bit standard ring counter shown in Figure 9.31 is self-correctable if its $d[0]$ function is replaced with the following switching function:

$$d[0] = \overline{(qout[0] + qout[1] + qout[2])}$$

In other words, once the ring counter entered any invalid-state loop, it only needs to take at most four clock cycles to go back to its normal valid-state loop.

9-22 Show that the 4-bit Johnson counter shown in Figure 9.33 is self-correctable if its $d[2]$ function is replaced with the following switching function:

$$d[2] = (qout[0] + qout[2])qout[1]$$

In other words, once the Johnson counter entered any invalid-state loop, it only needs to take at most five clock cycles to go back to its normal valid-state loop.

9-23 Using the binary counter with a decoder approach, design a timing generator with the following characteristics:

(a) The timing generator generates an eight-phase timing signal, T_i, where $0 \leq i \leq 7$.
(b) Each phase sustains one clock cycle except phases T_3 and T_5.
(c) Phase T_3 sustains two clock cycles and phase T_5 sustains four clock cycles.

9-24 This problem involves the design of a retriggerable digital monostable circuit that has the following features:

(a) Both the maximum pulse and actual pulse widths can be specified as module parameters.
(b) The output of the monostable circuit is cleared whenever the reset_n signal is asserted.
(c) The pulse output signal out goes to high for a specified number of clock cycles each time the start signal is activated.

Write a Verilog HDL module to describe the circuit you designed and write a test bench to verify its functionality.

9-25 Using counters, write a synthesizable Verilog HDL module to generate a periodic waveform with a 5-μs high duration and a 4-μs low duration. Suppose that a clock signal with a period of 100 ns is available as an input.

(a) Using two counters and a JK flip-flop, design and draw the block diagram of the waveform generator.
(b) Write a Verilog HDL module to describe the waveform generator in behavioral style.
(c) Write a test bench to verify the waveform generator.

9-26 Design a single-cycle pulse generator that generates a single pulse each time a push button is pressed. The generated pulse has a duration of one clock cycle.

(a) Draw the logic diagram and explain its operation.
(b) Write a Verilog HDL module to describe the single-cycle pulse generator in behavioral style.

(c) Write a test bench to verify the single-cycle pulse generator.

9-27 Design a logic circuit that outputs a high-level pulse with a width of four clock cycles each time the trigger signal, Tr, is asserted.
(a) Draw the logic diagram and explain its operation.
(b) Write a Verilog HDL module to describe it in behavioral style.
(c) Write a test bench to verify it.

9-28 A special shift register. Assume that four 4-bit left-shift registers are cascaded together in a way such that only the rightmost shift register can receive the external data in parallel and all shift registers can output their data at the same time. The combined shift register works as follows. Each time when a new data item is coming, a data valid signal is asserted. Once having received this asserted signal, the contents of the combined shifter register are shifted left 4-bit positions so as to accommodate the new data item in the rightmost shifter register.
(a) Draw the block diagram and label each component appropriately to indicate its functionality and size.
(b) Write a Verilog HDL module to describe it in behavioral style.
(c) Write a test bench to verify it.

9-29 A two-digit scoreboard. Suppose we want to design a simple scoreboard, which can display scores from 0 to 99 in decimal. The inputs to the system include a reset (or clear) signal and two control signals. The reset signal is used to clear the score and the two control signals are employed to increment and decrement the score, respectively. The score is reset to 0 whenever the reset signal is activated for at least three clock cycles. The score gets incremented by one if the increment signal is asserted and is decremented by 1 if the decrement signal is asserted. Using two seven-segment LED displays and the direct-driven approach, design this scoreboard system.
(a) Draw a block diagram, label each component properly to indicate its functionality and size, and explain its operation.
(b) Write a Verilog HDL module to describe the scoreboard system in behavioral style.
(c) Write a test bench to verify the scoreboard system.

9-30 A key module with four push buttons. Supposing that we have four push buttons, B_0, B_1, B_2, and B_3, design a logic circuit that outputs a code i to indicate that the push button B_i is pushed, where $0 \leq i \leq 3$. For example, the code is 00 when the push button B_0 is pushed and is 11 when the push button B_3 is pushed. The logic circuit you would design should also include the switch-debounce capability in order to eliminate the switch-bouncing effect.
(a) Draw a block diagram and explain its operation.
(b) Write a Verilog HDL module to describe the push-button system in behavioral style.
(c) Write a test bench to verify the push-button system.

10

Implementation Options of Digital Systems

RECALL that a design can often have many different implementations with the same or different technologies. In this chapter, a great variety of implementation options for digital systems with gate counts ranging from tens of hundreds to tens of millions are introduced. These options include application-specific integrated circuits (ASICs), field-programmable devices, and platforms. ASICs are devices that must be fabricated in IC foundries and can be designed with one of the following approaches: full-custom, cell-based, and gate-array-based. Field-programmable devices are ones that can be personalized in laboratories and include programmable logic devices (PLDs), complex PLDs (CPLDs), and field-programmable gate arrays (FPGAs). Platforms are general-purpose microprocessors or microcomputers, digital-signal processing processors, or their combinations, and include μP/DSP systems, platform-IP, and platform-FPGAs.

For a system composed of a variety of devices with different logic levels and power-supply voltages, two important issues related to the interface between devices must be taken into account: I/O standards and voltage tolerance. An I/O standard defines a set of electrical rules that connect two devices so as to transfer information through electrical signals properly. A device is said to be voltage-tolerant if it can withstand a voltage greater than its V_{DD} on its I/O pins.

This chapter is concluded with a case study, which demonstrates how to design a keypad scanner along with the related FPGA implementation details. Keypads are widely used in digital systems to enter data required in the systems from the outside world. The design of a keypad scanner is fundamentally important but is not an easy task because it involves the key-scan operation, the switch-debounce operations, and the conversion from the scan code to the key code.

10.1 Implementation Options of Digital Systems

Currently, ASICs, platforms, and field-programmable devices are ubiquitous in digital systems. From the acronym of the word "ASIC," which is *application specific integrated circuit*, any integrated circuit not in a standard catalog may be referred to as an ASIC; namely, it is a customized integrated circuit and may also include all field-programmable devices. Nevertheless, in industry the word "ASIC" is in particular reserved for the devices that must be directly fabricated from IC foundries only; namely, an ASIC is one that is designed with one of the

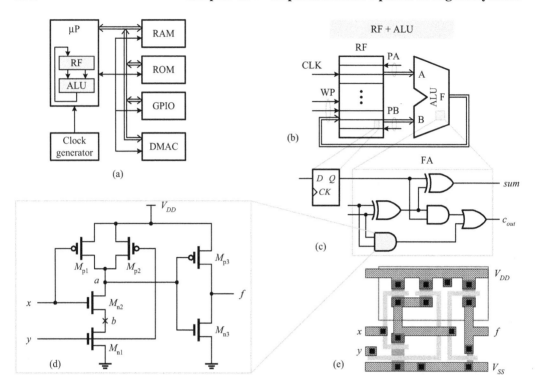

Figure 10.1: The design hierarchy of general digital systems: (a) system level; (b) register-transfer level; (c) gate level; (d) circuit level; (e) physical level.

following methods: full-custom, cell-based, and gate-array-based. In this section, we first introduce the design hierarchy of digital systems and then examine a wide variety of available options used to implement digital systems.

10.1.1 Hierarchical System Design

Nowadays, digital system designers widely use the so-called *hierarchical structure* or known as the *modular design* technique to design digital systems. The essence of the hierarchical structure is to partition the system into many smaller independent subsystems which are combined to perform the same functionality of the original system.

Basically, the design hierarchy of any digital system can be classified into four levels: the *system level*, the *register-transfer level*, the *gate level*, and the *circuit level*. The conceptual differences of these four design levels are illustrated in Figure 10.1.

10.1.1.1 System Level The system level is the topmost level. At this level, the data units processed by digital systems are usually bytes or blocks of bytes. The processing time is in the range between ms and μs. The logic devices used at this level are microprocessors/microcomputers (μPs/μCs), memory devices, and peripherals, such as timers, the general-purpose input and output (GPIO), the universal asynchronous receiver and transmitter (UART), the universal serial bus (USB) controller, the inter-integrated circuit (IIC) controller, the Ethernet MAC, and so on. In addition, a real-time operating system, C language, and others, are usually employed as a tool chain to write the desired application software.

Because of the highly developed manufacturing techniques of integrated circuits, nowadays a system-level design can be built on one printed-circuit board (PCB) or a single silicon chip. A PCB-based design consists of discrete standard components, such as μPs/μCs, periph-

Section 10.1 *Implementation Options of Digital Systems*

erals, and memory devices. The features of this design are as follows. The fixed cost is low but the product may cost too much to be accepted by the end user because too many devices are required. In addition, the actual size and power dissipation of the product are large.

A silicon-based system-level design can be accomplished by either a *programmable system chip* (PSC) or a cell-based platform IP. The PSC is a platform FPGA and also called a *system on a programmable chip* (SoPC) or a silicon-on-a-chip (SoC) FPGA. A PSC-based design consists of system cells, such as MicroBlaze and NIOS—both are 32-bit RISC CPUs associated some commonly used peripherals—much like the ones used in PCB-based design. The entire system can be implemented on the same FPGA chip. PSCs will be increasingly popular in designing most digital systems in the near future. A cell-based approach comprises soft cells (synthesizable modules) that can be configured into an optimized system hardware module. An example is Tensilica's Xtensa. At present, this approach proves most applicable to the fields of digital-signal processing (DSP) and multimedia.

10.1.1.2 Register-Transfer Level (RTL) The RTL components operate on bits or bytes on a time scale of 10^{-8} to 10^{-9} seconds. An RTL design consists of combinational logic modules and sequential logic modules. The combinational logic modules include decoders, encoders, multiplexers, demultiplexers, arithmetic circuits, magnitude comparators, and others. The sequential logic modules include registers, shift registers, counters, sequential logic modules, and so on. Like the system level, modules constructed using RTL components may exist some unavoidable overhead because some functions in a specific component may be undesired but its associated hardware cannot be removed. However, this overhead is much less than that encountered at the system level because the component granularities at the RTL are much finer than those at the system level.

10.1.1.3 Gate Level The gate-level components operate on bits on a time scale of 10^{-9} to 10^{-11} seconds. A gate-level design uses basic logic gates, flip-flops, and latches to construct an RTL module. These basic gates, flip-flops, and latches are in turn built from transistors, bipolar junction transistors (BJTs) or MOS transistors. Like the RTL, modules constructed using gate-level components may exist some unavoidable overhead because some function in a specific component may be undesired but its associated hardware cannot be removed. However, this overhead is much less than that encountered at the RTL because the component granularities at the gate level are much finer than those at the RTL.

10.1.1.4 Circuit Level The circuit-level components operate on bits on a time scale of 10^{-10} to 10^{-12} seconds. A circuit-level design consists of MOS transistors, BJTs, and MESFETs. The physical level is just another view of the circuit level. Modules constructed using circuit-level components have the best performance and no hardware overhead because only the desired hardware is employed to construct the modules.

■ Review Questions

10-1 Describe the concept of the design hierarchy of digital systems.
10-2 What are the four levels of the design hierarchy of digital systems?
10-3 What are the three major system-level implementation options?
10-4 Give the major features of PSC-based platforms.
10-5 What are the major features of the platform-based design?

10.1.2 Implementation Options of Digital Systems

With the maturity of very large scale integrated circuit (VLSI) technology, nowadays a wide variety of options are available for implementing a given digital system. These options can

374 Chapter 10 Implementation Options of Digital Systems

be cast into the following three classes: *application-specific integrated circuits* (ASICs), *field-programmable devices*, and *platforms*.

10.1.2.1 Implementation Options The three classes of implementation options for digital systems along with their detailed ingredients are summarized in Figure 10.2. ASICs are devices that must be fabricated in IC foundries and include full-custom, cell-based, and gate-array-based devices. Field-programmable devices are ones that can be personalized in laboratories, and cover programmable logic devices (PLDs), complex PLDs (CPLDs), and field-programmable gate arrays (FPGAs). Platforms are general-purpose microprocessors or microcomputers, digital-signal processing processors, or their combinations, and cover μP/DSP systems, platform-IP, and platform-FPGAs.

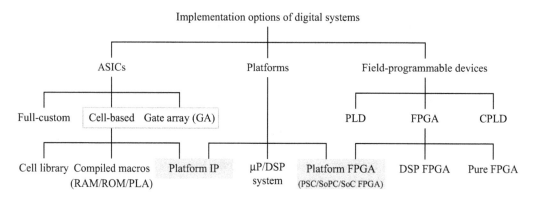

Figure 10.2: Implementation options of digital systems.

ASICs. Approaches to designing an ASIC include the *full-custom approach*, the *cell-based approach*, and the *gate-array-based approach*. The essential feature of ASICs is that they all are needed to process in IC foundries by using a partial or full set of masks (or photomasks). In the full-custom and cell-based approaches, the designer or the design team must provide all masks required for the chip to be manufactured. The major difference between the full-custom approach and cell-based approach is that the former creates a design from scratch while the latter uses a standard cell library, in which the layout of each cell has been done. In the gate-array-based approach, the wafer to be used to implement a digital system has been processed partially by the IC foundry up to the point where only the interconnect between cells or transistors is left to be defined by the designer. Hence, it is often called a *semi-custom approach*. Using gate arrays to implement a design, the designer only needs to specify the required interconnect with CAD tools in order to define or commit the specified function of the chip. Both cell-based and gate-array-based approaches use an HDL synthesis flow along with cell or macro libraries to create an ASIC and hence they have the features of fast prototyping and shorter time to market as compared to the full-custom approach.

In summary, the full-custom approach starts from scratch, i.e., from the zero to an IC. The cell-based approach incorporates a standard cell library into the synthesis flow so that it can cut the development time of products profoundly. The gate-array-based approach is also based on the synthesis flow but targets to prefabricated logic devices, known as *gate arrays*, which are arrays of prefabricated transistors on a wafer and only metal masks are left to be defined by users. Hence, it takes the advantage of the cell-based approach and speeds up the prototyping relative to the full-custom approach.

Field-programmable devices. Field-programmable devices, as their names imply, are ones that can be personalized in laboratories and include PLDs, CPLDs, and FPGAs. PLDs are devices built on the two-level AND-OR logic structure and include three different types: read-only memories (ROMs), programmable logic arrays (PLAs), and programmable array logic (PAL). CPLDs are devices that embody many PALs onto the same chip and interconnect them together through a programmable interconnect network. FPGAs use programmable logic blocks (PLBs) to implement switching functions and can be divided into two structures, the matrix type and the row type, in terms of the arrangement of PLBs on the chip. It is worth noting that gate arrays are mask-programmable while FPGAs are field-programmable. For this reason, gate arrays are often called *mask-programmable gate arrays* (MPGAs).

Platforms. The platform option uses system-level components, such as $\mu C/\mu P$ and/or digital signal processing (DSP) processors, memory devices, and peripherals, to implement a desired system. The platform approach includes the following three options: the $\mu P/DSP$ system, platform IP, and platform FPGA. The platform FPGA is also called the *programmable system chip* (PSC), *system on a programmable chip* (SoPC), or system-on-chip (SoC) FPGA (i.e., SoC FPGA).

The $\mu P/DSP$ system uses system-level standard ICs and is built on a single or multiple printed-circuit boards (PCBs) [8, 10, 9]. Because $\mu P/DSP$ (computer) systems are used as components to design desired systems, they are often called *embedded systems*. Instead of using standard ICs in the $\mu C/\mu P$-based system, both platform-IP design and platform-FPGA design use system-level cells, such as $\mu P/DSP$ modules, memory modules, and peripherals, to construct the desired systems on silicon chips rather than on PCBs. The system cells used in both the platform-IP design and platform-FPGA design are known as *intellectual property* (IP). In general, an IP module is a predesigned component that can be used in a larger design. IP can be further classified into two types: *hard IP* and *soft IP*. A hard IP module comes as a predesigned layout with an accurately measured block's size, performance, and power dissipation. A soft IP module is synthesizable and described in a hardware description language, Verilog HDL, SystemVerilog, or VHDL. It has features of being flexible to new technologies but harder to be characterized, and is generally larger and slower as compared to the hard IP module. In FPGA devices, another type of IP, called *hardwired IP*, is widely used to mean a module, such as a CPU core, a multiplier, an so on, has already fabricated on the device along with FPGA fabrics. Refer to Section 10.3.2 for more about FPGA fabrics.

The essential difference between the platform-IP design and the platform-FPGA design is that the former uses cell-based libraries to build the desired system while the latter uses FPGA devices. Like $\mu P/DSP$ systems, the $\mu P/DSP$ systems used in these two types of designs are also called *embedded systems*. In addition, systems implemented with these two approaches are founded on a silicon chip regardless of using cell-based libraries or FPGA devices. Hence, such systems are often called the *system-on-chip* (SoC). In a word, an embedded system is a computer system used as a component in a digital system and an SoC is a digital system containing one or more embedded systems on a single silicon chip.

In summary, the major difference among all three platform-based implementation options is that the playgrounds for $\mu P/DSP$ systems are PCBs while for the platform-IP and platform-FPGA designs are silicon chips. Nonetheless, all three platform options use computers as components to implement desired systems. Because all these three options employ system-level components and involve hardware and software co-design at the system level, they are beyond the scope of this book and hence we will not further discuss them.

■ Review Questions

10-6 What are the three classes of implementation options of digital systems?
10-7 What is the meaning of the platform-based design?
10-8 What is the meaning of the PSC-based design?
10-9 Define the terms: embedded system, SoC, and IP.
10-10 What are the distinctions between the hard IP and soft IP?

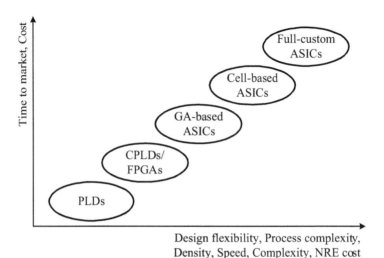

Figure 10.3: Comparison of various implementation options of digital systems in terms of the time to market and cost.

10.1.2.2 Comparison of Implementation Options Figure 10.3 compares the features of various implementation options of digital systems in terms of the time to market and cost. The time to market and cost of a product are represented as a function of design flexibility, process complexity, (integration) density, speed, complexity, and non-recurring engineering (NRE) cost. The time to market is the time elapsed from the time when the design specifications are created to the point when the desired product goes to the market. Roughly speaking, the NRE cost is the fixed cost needed to develop a product. It generally includes the engineering cost, such as engineer salaries, training fees, and software costs, the cost of prototyping the product, and the expenses for marketing the product. The NRE cost is independent of the sales volume and amortized over all units of a given product. Consequently, for a product with a high NRE cost, the expected sales volume of the product should be large enough to lower down the end price of the product to a point that are acceptable by users.

Because the design flexibility, density, speed, complexity, process complexity, NRE cost, and time to market are the highest among all implementation options, the full-custom approach is only aimed at high-performance, high-speed, and high-volume products, such as CPUs, memory devices (including RAM, DRAM, Flash, etc), and field-programmable devices (including PLDs and CPLDs as well as FPGAs).

The advantages of using an ASIC to implement a system are as follows: First, both the size and power dissipation of the system can be reduced considerably because an ASIC may replace many standard ICs. It is shown that about one-third or more power dissipation of an IC is on the I/O-related circuits. As many chips are integrated into an ASIC, this amount of power dissipation can be removed due to the removal of unnecessary I/O circuits. Second, the reliability and performance of the system are increased due to the reduced number of devices needed in the system.

Figure 10.4 shows the comparison of various implementation options of digital systems in terms of cost and design flexibility, respectively. In general, the cost of a product is a combination of the NRE cost and the recurring cost. Although the NRE cost only needs once during the entire lifetime of a particular product, it is often the highest one among all costs. The recurring cost is the portion that needs to be paid each time the product is produced. The amount of recurring cost is proportional to the sales volume of the underlying product.

Section 10.1 *Implementation Options of Digital Systems*   377

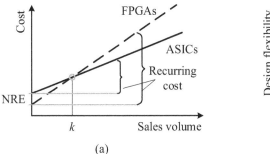
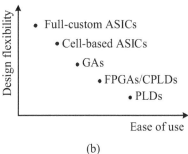

(a)   (b)

Figure 10.4: Comparison of various implementation options of digital systems in terms of (a) cost versus sales volume and (b) design flexibility versus ease of use.

In general, the cost of a product using a given implementation option can be represented as a function of the sales volume, as shown in Figure 10.4(a), where the intercept of the vertical coordinate is the NRE cost of the product and the slope denotes the recurring cost per unit of the product. The cost above the intercept point is the recurring cost of the product. In general, field-programming devices have a lower NRE cost but a higher recurring cost than ASICs due to its high overheads caused by the highly flexible interconnect and programmability. Consequently, the slope of the cost function of FPGAs is greater than that of ASICs. Because of the lower NRE cost of FPGAs than that of ASICs, there must exist a cross point, denoted by the k point, of their curves. This cross point determines which implementation option should be used for a given sales volume. Above it, the ASIC design is a better choice; below it, using FPGAs is more appropriate. The same approach can be applied equally well to compare the other implementation options.

Figure 10.4(b) compares the relative design flexibility with the ease of use of various implementation options. Here, the *design flexibility* means the capability of designing a circuit with an arbitrary function. The easier to use, the less design flexibility. For example, PLDs are easiest to use but have the least design flexibility. The full-custom approach is very difficult to use but has the highest design flexibility.

■ **Review Questions**

10-11 Compare the features of ASICs and field-programmable devices.
10-12 What are the advantages of using ASICs to design digital systems?
10-13 Compare the features of gate arrays and FPGAs.
10-14 Describe the meaning of the k point in Figure 10.4(a).
10-15 How would you decide to use an ASIC or an FPGA device to implement a system?

10.1.3 ASIC Approaches

As mentioned previously, an ASIC can be created by any of full-custom, cell-based, and gate-array-based approaches. In this section, we describe each of these in more detail.

10.1.3.1 Full-Custom Approach In the full-custom approach, each transistor and its physical layout are delicately designed by the designer in order to achieve the best performance. However, even though it has the best performance among the three approaches, the full-custom approach has very low throughput because it needs much more time to create an ASIC from scratch. In addition, it takes much more time than the other implementation options to prototype a design. Furthermore, because nowadays the lifetimes of electronic products become shorter and shorter, the design time and the time to market of the products are needed to cut accordingly. As a result, in those applications needed to fast prototype, the cell-based approach, the gate-array-based approach, and the approach based-on field-programmable devices are much more popular than the full-custom approach.

In summary, because of the much shorter lifetimes of today's electronic products than before and the increasing NRE cost of the full-custom approach with the reduction in feature size, nowadays the full-custom approach is only used for the cases when high-volume and/or high-speed products, such as CPUs, memory devices, gate arrays, and field-programmable devices, are required. The rationale behind this is that the high NRE cost can be amortized over a large number of units so that the cost of each unit is low enough for realizing a cost-effective product. Please refer to Figure 10.4(a) again for gaining a more insightful explanation of this claim.

10.1.3.2 Cell-Based Approach The essence of the cell-based approach is that a design is composed of a set of predefined cells, known as *standard cells*, as shown in Table 10.1. The mask layouts of these cells have already been done in advance; only the higher-layer metal wires are left to be defined and connected on demand [7].

Table 10.1: The basic cell types of a typical standard cell library.

Types	Variations
Inverter/buffer/tristate buffer	1X, 2X, 4X, 8X, 16X
NAND/AND gate	2 to 8 inputs
NOR/OR gate	2 to 8 inputs
XOR/XNOR gate	2 to 8 inputs
MUX/DeMUX	2 to 8 inputs (inverted/noninverted output)
Encoder/Decoder	4 to 16 inputs (inverted/noninverted output)
Schmitt trigger circuit	Inverted/noninverted output
Latch/register/counter	DFF/JKFF (Sync./Async., clear/reset)
I/O pad circuits	Input/Output (tristate/bidirectional)

A typical standard cell library usually includes the following ones:

- *Basic logic gates*: NAND, NOR, XOR, AOI, OAI, AND, OR, buffers, and inverters.
- *Combinational logic modules*: decoders, encoders, priority encoders, multiplexers, demultiplexers, parity checkers, adders, subtractors, and shifters.
- *Sequential logic modules*: D flip-flops, registers, counters, timing generators, memory (ROM, RAM) modules.
- *System building blocks*: multipliers, arithmetic-and-logic units (ALUs), CPUs, UARTs, IIC controllers, GPIO modules, USB controllers, Ethernet MAC modules, and so on.

By and large, cells in a standard cell library with the similar order of complexity usually have a fixed height, and their power-supply rail and ground are separately routed at the top and bottom of the cells so that they can be abutted end-to-end and have their power-supply rails connected. The width is allowed to vary to accommodate different functions. For instance, a NOT gate has merely two transistors but a 2-input NAND gate contains four transistors. As a result, they must have different widths if their heights are set to be identical. Figure 10.5 shows an example of using the above-mentioned NAND and NOT gates to construct a D flip-flop. All gates have the same height (note that the figure has been rotated by $90°$) and hence they can be abutted together. Both left and right parts contain four NAND gates and one NOT gate. The left part constitutes the master latch of the D flip-flop and the right part comprises the slave latch. Between them is the routing channel used to connect together NAND gates and

Section 10.1 Implementation Options of Digital Systems

NOT gates. For modern multiple-layer metallization processes, the routing channel is virtually no longer needed and hence the complex cell can be built more compactly.

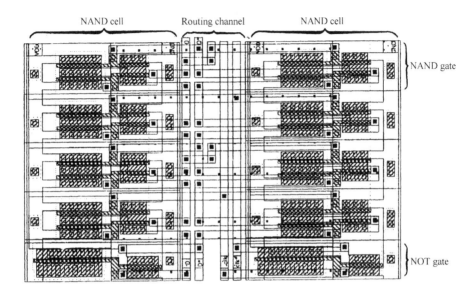

Figure 10.5: A D flip-flop constructed from a standard cell library.

Compared with the full-custom approach, the cell-based approach uses predesigned cells with layouts. The size of each cell in a typical standard cell library is generally much larger than the one created by the full-custom approach. Consequently, an ASIC created using a standard cell library often needs much more area and has worse performance than the one created by the full-custom approach. This implies that as the sales volume of a product is large enough and thus the NRE cost of the product is of little importance after being amortized over all product units, the full-custom approach would be more appropriate than the cell-based approach.

The main advantage of using the cell-based approach to create an ASIC is to improve productivity because the cell-based approach is basically an HDL synthesis flow. However, like the full-custom approach, the prototype designed with the cell-based approach is still needed to fabricate in IC foundries with a full set of masks. Therefore, in light of this viewpoint the cell-based approach is indeed a type of the full-custom approach.

10.1.3.3 Gate-Array-Based Approach Gate arrays (GAs) are also known as *uncommitted logic arrays* (ULAs) because their functions are left to be defined by users and are a type of semi-custom design of ASICs. Here, the semi-custom means that only a partial set of masks is needed to process in IC foundries after logic functions have been committed. The essential features of gate arrays are that standard transistors have been fabricated in advance using standard masks and only the metallization (i.e., interconnect) masks are left to users to commit their final functions [7]. The basic elements of gate arrays can be either NOR gates or NAND gates in CMOS technology. A particular subclass of gate arrays is known as *sea-of-gates* (SoGs). Sea-of-gates differ from gate arrays in that the array of transistors is continuous rather than segmented.

A typical structure of CMOS gate arrays is shown in Figure 10.6, which consists of a set of pairs of CMOS transistors and routing channels. If the rows of nMOS and pMOS transistors are broken into segments with each having two or three transistors, such as the one shown in Figure 10.7, the result is the gate-array structure; otherwise, it is the sea-of-gates structure.

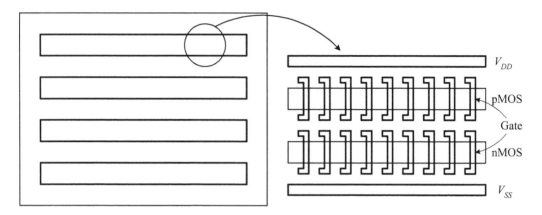

Figure 10.6: The basic structure of gate arrays and sea-of-gates.

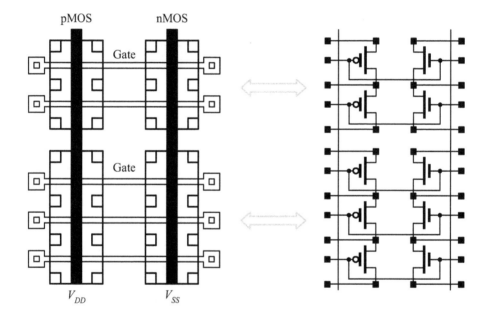

Figure 10.7: The basic structure of gate-array cells.

The major feature of sea-of-gates is that they utilize multiple-layer metallization to remove the routing channels. Hence, they have a higher integration density than gate arrays.

The basic circuit structure of gate arrays is shown in Figure 10.7, where two segments, one with two transistors and the other with three transistors, along with their equivalent circuits, are depicted. Because the structure of a MOS (pMOS or nMOS) transistor is symmetric in the sense that its drain and source can be interchanged without deteriorating its performance, and in a CMOS logic gate, both the gates of an nMOS transistor and a pMOS transistor must be connected together, the gates of both nMOS and pMOS transistors are closely placed like the one shown in Figure 10.6 or connected in pairs like the one shown in Figure 10.7. In addition, all transistors are intimately placed to increase the density of integration.

When using the basic devices of a gate array to design logic gates, we only need to connect the drain to the power supply V_{DD}, to connect the source to ground GND, or to connect them together. As an illustration of how to apply this idea to construct a NOT gate and a NAND

Section 10.1 *Implementation Options of Digital Systems*

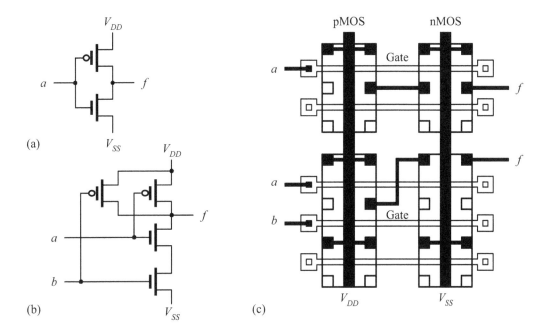

Figure 10.8: An example of interconnected gate-array cells: (a) NOT gate; (b) NAND gate; (c) gate array implementation.

gate from a gate array, consider the following example.

■ Example 10.1: A NOT gate and a NAND gate

Assume that the gate array shown in Figure 10.7 is used in this example to construct a NOT gate and a NAND gate.

To construct a NOT gate, one pMOS transistor and one nMOS transistor are required, as shown in Figure 10.8(a). The gates of both transistors are connected together to serve as the input as they have already been done. The drains of both transistors are connected together via a metal wire to serve as the output. The source of the nMOS transistor is connected to V_{SS} while the source of the pMOS transistor is connected to V_{DD}. The resulting gate-array layout along with connection is portrayed in the upper part of Figure 10.8(c).

The NAND gate is constructed in the way depicted in the Figure 10.8(b). The gates of both pMOS and nMOS transistors are connected together in pairs and serve as inputs, denoted by a and b, respectively. Two nMOS transistors are connected in series whereas two pMOS transistors are connected in parallel, as shown in Figure 10.8(b). The source of the series nMOS chain is grounded and the drain of the series nMOS chain is connected with the drains of the parallel-connected pMOS transistors to form the output. The source of both pMOS transistors are connected to V_{DD}. The resulting gate-array layout along with connection is portrayed in the lower part of Figure 10.8(c).

Like the full-custom approach, it is quite clumsy and time-consuming to start an ASIC implementation of a design with gate arrays from scratch. Just as the case of the cell-based approach, macro libraries are usually provided by the gate-array vendors to save the developing time of a product. A typical macro library associated with gate arrays is similar to the standard-cell library with only one essential difference in that it contains only the metallization masks because the transistors on gate arrays are already fabricated.

Like the cell-based approach, using gate arrays to create an ASIC also encounters the overhead problem in the chip area. The area overhead is due to the unused logic components associated with macros, which cannot be removed when these macros are used. To see this, consider the case that if only an 8-bit adder macro is available for designing a 6-bit adder, the area overhead is 25% because the highest or the lowest two-bit adders are left unused but still occupy the chip area. In contrast, this area overhead can be completely removed without any penalty in the full-custom approach.

As compared with the full-custom and cell-based approaches, the gate-array-based approach has less functionality per unit area due to the overheads of interconnect and those unused prefabricated transistors. However, unlike the full-custom and the cell-based approaches which need to process every mask, the gate-array-based approach only needs to process the metallization masks. In addition, like the cell-based approach, the gate-array-based approach also uses the HDL synthesis flow. Thus, using the gate-array-based approach takes much less time to create an ASIC and prototype the product.

Since all full-custom, cell-based, and gate-array-based approaches need their resulting ASIC designs to be fabricated in IC foundries, these approaches are time-consuming processes and need a great amount of time to prototype a product. To solve this difficulty and provide a way to faster prototype a product, field-programmable devices are emerged into the market as an alternative. Field-programmable devices will be focused in more detail in the next subsection.

■ Review Questions

10-16 Which approaches can be used to create an ASIC?
10-17 What are the features of the full-custom approach?
10-18 What are the features of the cell-based approach?
10-19 What are the features of gate arrays?
10-20 Describe the meaning of uncommitted logic arrays (ULAs).
10-21 Describe the meaning of SoGs.

10.1.4 Field-Programmable Devices

Field-programmable devices can be cast into two types, PLDs/CPLDs and FPGAs, in terms of their logic structures. The common features of PLDs and CPLDs are that they use the two-level AND-OR logic structure to realize (or implement) switching functions and can be on-site programmed. FPGA devices combine the features of gate arrays with the on-site programmability of PLDs/CPLDs.

PLDs/CPLDs and FPGAs can be cast into two types in terms of the programmability: *mask-programmable* and *field-programmable*. Like the case of gate arrays, when using mask-programmable devices, the designer needs to provide the vendor the designed interconnect pattern of the given device for preparing the required masks to fabricate the desired IC. Field-programmable devices can be personalized on the scene or in the laboratory by the designer through using appropriate *programming equipment*, referred to as a *programmer*. Field-programmable devices can be further subdivided into two types: *one-time programmable* (OTP) and *erasable*. OTP devices can only be programmed one time while erasable devices can be reprogrammed as many times as needed. In summary, field-programmable devices are usually used at the start-up time of a design to gain the flexibility or in low-volume production to save the NRE cost of the design while mask-programmable devices are used in high-volume production to reduce the cost.

Section 10.1 *Implementation Options of Digital Systems* 383

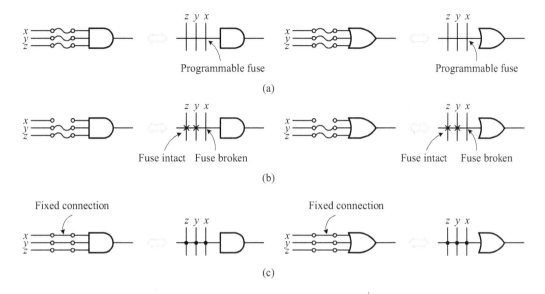

Figure 10.9: The shorthand notation used to describe the structures of PLDs: (a) before programming; (b) after programming; (c) fixed connection.

10.1.4.1 Programmable Logic Devices In designing a digital system, a PLD can replace a number of small-scale integration (SSI) and/or medium-scale integration (MSI) devices. Therefore, using PLDs to design digital systems can significantly reduce the number of wires, the number of devices used, the area of the printed circuit board (PCB), and the number of connectors, thereby cutting the hardware cost of the system considerably.

To represent the structures of various PLDs, it is convenient to use the shorthand notation shown in Figure 10.9. Because CMOS technology is the most popular today, the "fuse" shown in the figure is a RAM cell, an EEPROM (Flash) cell, or an antifuse cell when the PLD is field-programmable. In mask-programmable devices, there is a MOS transistor if a "fuse" is present and there is no MOS transistor otherwise.

Recall that PLDs can be classified into three types: PLA, ROM, and PAL. As shown in Figure 10.10, all these three types of devices have a similar two-level AND-OR logic structure but their programmability of AND and OR arrays is different.

- For ROM devices, the AND array generates all minterms of inputs and hence is fixed but the OR array is programmable in order to implement the desired functions.
- For PLA devices, both AND and OR arrays are programmable. As a consequence, they provides the maximum flexibility among the three types of PLDs.
- For PAL devices, the AND array is programmable but the OR array is fixed to connect to some specified AND gates.

Nowadays, the discrete PLA devices have become obsolete. However, PLA structures and ROMs are often used in the full-custom and cell-based approaches to take advantage of their regular structures. Both PAL and ROM are still widely used in digital systems and hence have commercial discrete devices.

10.1.4.2 Programmable Interconnect (PIC) Structures The programmable interconnect structures of field-programmable devices can be classified into the following three types: static RAM (SRAM) cells, Flash cells, and antifuse cells, as shown in Figure 10.11. The interconnect structure based on the SRAM cell is an nMOS switch or a transmission gate controlled

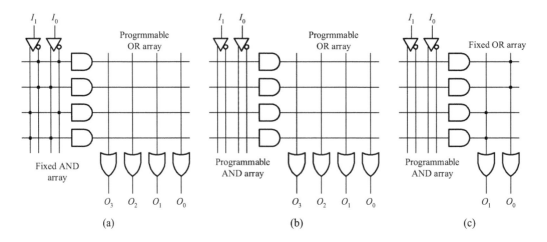

Figure 10.10: The basic structures of PLDs: (a) ROM; (b) PLA; (c) PAL.

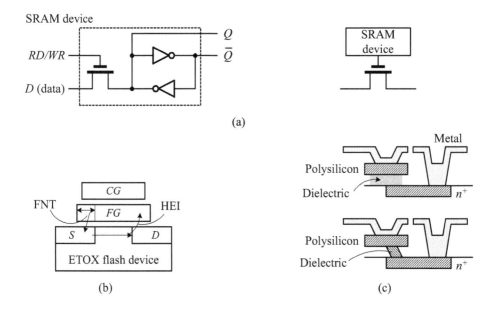

Figure 10.11: The basic structures of programmable interconnect: (a) SRAM cell; (b) Flash cell; (c) antifuse cell.

by an SRAM cell. The basic structure of an SRAM cell is a bistable circuit, as shown in Figure 10.11(a). Once programmed, the SRAM cell retains its state until it is reprogrammed or the power supply applied to it is removed. The Flash cell is like the cell used in Flash memory devices, as shown in Figure 10.11(b). Its basic structure is a floating-gate transistor, which can be programmed to store a logic 1 or a logic 0. Once programmed, the Flash cell retains its state permanently, even when the power supply applied to it is removed; however, it can be reprogrammed many times as needed. As shown in Figure 10.11(c), the antifuse cell is a device that operates in the reverse direction of a normal fuse; namely, it has a high resistance in normal condition but is changed to a low resistance permanently when an appropriate voltage is applied to it.

The basic characteristics of the above-mentioned three types of programmable interconnect structures are summarized in Table 10.2. From the table, we can see that the antifuse structure has the best performance because it has the lowest resistance and capacitance. The other two structures have almost the same performance. However, the SRAM structure is volatile while Flash and antifuse structures are involatile.

Table 10.2: The basic features of programmable interconnect structures.

	SRAM	Flash	Antifuse
Process technology	Standard CMOS	Standard two-level ploysilicon	New type polysilicon
Programming approach	Shift register	FAMOS	Avalanche
Cell area	Very large	Large	Small
Resistance	$\approx 2\ k\Omega$	$\approx 2\ k\Omega$	$\approx 500\ \Omega$
Capacitance	50 fF	50 fF	10 fF

10.1.4.3 Complex Programmable Logic Devices (CPLDs) Because of the popularity of PAL devices along with the maturity of VLSI technology, combining many PALs with a programmable interconnect structure into the same chip is feasible. Such a device is known as a complex PLD (CPLD). A CPLD comprises a number of macrocells, with each consisting of an output function of the PAL device. As mentioned, each output function of PAL is composed of an OR gate and a number of AND gates, ranging from 5 to 8 gates. Such a type of circuit when used as the building blocks of CPLDs is usually referred to as a *PAL macro* or a *macrocell* for short.

The basic structures of CPLDs consist of *PAL macros, interconnect,* and *input/output blocks* (I/O blocks or IOBs for short). There are two basic types of CPLDs, which are classified according to the arrangements of PAL macros and the interconnect structures, as shown in Figures 10.12(a) and (b), respectively. Figure 10.12(a) shows the first type and is most widely used in commercial CPLDs, where PAL macros are placed on both sides of the programmable interconnect. Another type is depicted in Figure 10.12(b), where PAL macros are placed on all four sides and a programmable interconnect is placed at the center region, called the *global routing area*. In addition, an *output routing area* is deployed between the PAL macros and input/output blocks.

10.1.4.4 Field-Programmable Gate Arrays (FPGAs) The basic structures of FPGAs are composed of *programmable logic blocks* (PLBs), *interconnect,* and *input/output blocks* (I/O blocks or IOBs for short). These three components are known as the *fabrics* of FPGAs. PLBs are referred to as configurable logic blocks (CLBs) in Xilinx terminology and logic elements (LEs) in Altera terminology. PLBs are usually used to implement combinational logic and sequential logic functions.

According to the arrangements of PLBs on the chip, the basic structures of FPGA devices can be partitioned into two types: the *matrix type* and the *row type*. The matrix-type FPGA is shown in Figure 10.13(a), where PLBs are placed in the 2-D matrix way. Between PLBs, there are two types of interconnect, called *horizontal routing channels* and *vertical routing channels*, respectively. Figure 10.13(b) is a row-type FPGA, where PLBs are placed intimately in a row fashion. The spaces between two rows are the routing channels.

■ Review Questions

10-22 What types can field-programmable devices be classified into?
10-23 What is the meaning of OTP devices?
10-24 Distinguish the differences between the three types of PLDs.
10-25 What are the three types of PICs?

386   Chapter 10 ■ Implementation Options of Digital Systems

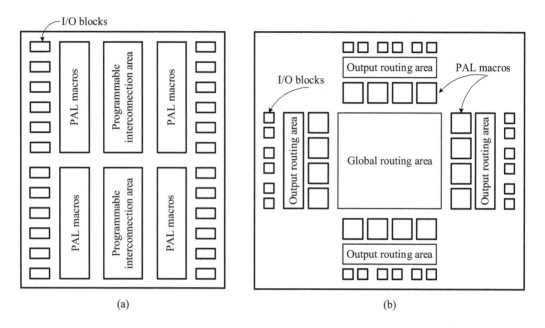

Figure 10.12: The basic structures of CPLDs: (a) CPLD basic structure and (b) pLSI basic structure.

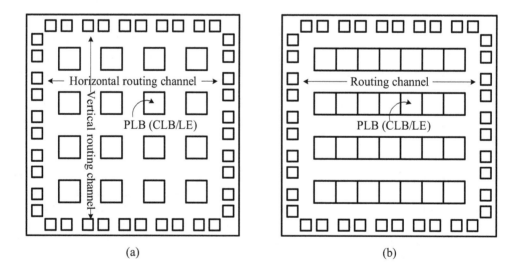

Figure 10.13: The basic structures of FPGAs: (a) the matrix type and (b) the row type.

10-26 Describe the feature of the antifuse cell.
10-27 Describe the basic structures of CPLD devices.
10-28 Describe the basic structures of FPGA devices.

10.2 PLD Structure and Modeling

Recall that programmable logic devices (PLDs) are built on the two-level AND-OR logic structure and include three types of structures, according to the programmability of AND

Section 10.2 PLD Structure and Modeling

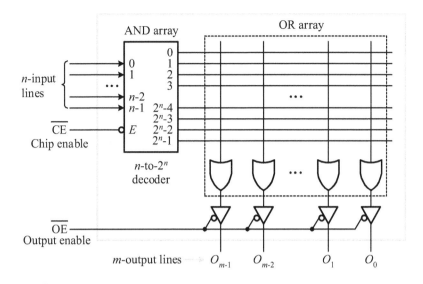

Figure 10.14: The general structure of ROM devices.

and OR arrays. In this section, we begin with the discussion of these structures and their related devices in more detail and then use PLAs as examples to illustrate how to model these structures in Verilog HDL.

10.2.1 Read-Only Memory

As mentioned previously, the basic structure of ROM is a two-level AND-OR logic structure. A $2^n \times m$ ROM device is a circuit that includes an n-to-2^n (full) decoder and an array of m programmable 2^n-input OR gates, as shown in Figure 10.14. Each combination of n input lines is called an *address* and the corresponding m-bit output is called a *word*. For a ROM device with n input lines and m output lines, there are 2^n m-bit words. The number of bits, namely, $2^n \times m$, contained in a ROM device is known as *capacity*. To specify the capacity of a ROM device, the form of $2^n \times m$ is generally used. For example, 16×4 ROM and 256×8 ROM. Sometimes, we also informally use the word "size" to mean the capacity.

For commercial ROM devices, there are two control inputs: the chip enable ($\overline{CE}$) and the output enable ($\overline{OE}$). The chip enable input allows multiple ROM devices to be connected to a common bus so that each device may be accessed by its uniquely designated address space. The output enable input controls the operations of the output tristate buffers connected at the outputs of the OR array. The output tristate buffers are used to either pass the values of the OR array to the output lines or place the output lines in their high impedances, depending on whether the output enable ($\overline{OE}$) input is asserted or not.

The operation of writing data into a ROM device is referred to as *program*. According to the programming mechanisms of memory cells, ROM devices can be cast into three types: *mask ROMs*, *programmable ROMs* (PROMs), and *erasable programmable ROMs* (EPROMs). The contents of mask ROMs are set by the vendor according to the patterns provided by the designer and cannot be changed later after the device is fabricated. A programmable ROM device is an OTP device; that is, it can only be programmed one time. An erasable programmable ROM device can be reprogrammed a number of times. Currently, most EPROM devices are designed with CMOS floating-gate technology and known as *electrically erasable PROMs* (EEPROMs) because their contents can be erased electrically.

The most widely used EEPROMs nowadays are referred to as *Flash memories*. They can be accessed just like SRAM devices without concerning the programming process, which is automatically taken care by an internal controller to generate all timing signals required for programming a specified block of cells.

Perhaps, the most common use of ROMs in digital systems is to implement switching functions. To see this, consider the general structure of ROM devices shown in Figure 10.14, which is simply a two-level AND-OR logic structure. The AND array generates all minterms of n input lines and the OR array connects the required minterms to its m output lines. In other words, it is a two-level logic circuit for multiple-output functions in standard SOP form. As a result, a $2^n \times m$ ROM device is able to implement m output functions with n input variables at most.

■ Example 10.2: ROM — Implementing switching functions.

Figure 10.15 shows an example of using a $2^2 \times 2$ ROM device to implement a multiple-output switching function. Since the input AND array generates all minterms of input variables, it only needs to connect the minterms with a logic-1 output value to the programmable OR array, as shown in Figure 10.15(b). It is instructive to compare both the truth table and the logic circuit shown in Figures 10.15(a) and (b), respectively.

Input		Output	
A_1	A_0	O_1	O_0
0	0	0	1
0	1	1	0
1	0	0	0
1	1	1	1

(a)

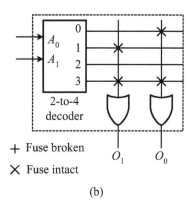

+ Fuse broken
× Fuse intact

(b)

Figure 10.15: A $2^2 \times 4$ ROM of implementing two switching functions: (a) truth table and (b) logic circuit.

The following example illustrates how to model a ROM device without timing checks and demonstrates how to set the initial values by using procedural assignments.

■ Example 10.3: Modeling a ROM device.

In this example, we use a $2^2 \times 4$ ROM device to implement the switching functions shown in Figure 10.15. To model a ROM device without timing checks, it only needs to declare a memory array with the desired number of words and the desired word width. Then, the memory array is initialized by using procedural assignments or reading from a file through the use of system tasks: **$readmemb** or **$readmemh**. In this example, procedural assignments are used. Using system tasks to fill the memory array is deferred to the PLA modeling subsection.

```
module ROM_example(input a1, a0, output out1, out0);
reg [1:0] mem[3:0];

// using procedural assignment statements to
// initialize the contents of ROM
```

Section 10.2 PLD Structure and Modeling

```
initial begin
   mem[0] = 2'b01;
   mem[1] = 2'b10;
   mem[2] = 2'b00;
   mem[3] = 2'b11;
end
assign {out1, out0} = mem[{a1, a0}];
endmodule
```

■ Review Questions

10-29 Describe the features of ROM devices in terms of the programmability of the AND-OR logic structure.

10-30 Describe the following terms of ROM devices: address, program, capacity, and word.

10-31 What are the types of ROM devices according to the programming mechanisms of memory cells?

10-32 Is the AND array of ROM devices fully or partially decoding the input lines?

10-33 How many functions can a $2^n \times m$ ROM device implement?

10-34 Write a test bench to verify the ROM_example module.

10.2.2 Programmable Logic Arrays

One drawback of using ROM devices to implement switching functions is that some minterms may be unused, such as the minterm 2 in Figure 10.15. These unused minterms are indeed wasted and may be removed. To this end, the AND array must be made to be programmable so that it may only accommodate the desired minterms or product terms, thereby leading to a structure that makes both AND and OR arrays programmable. The resulting structure is known as the *programmable logic array* (PLA).

For a typical $n \times k \times m$ PLA, as shown in Figure 10.16, there are n inputs and buffers/NOT gates, k AND gates, and m OR gates, as well as m XOR gates. Between the inputs and the AND gates, there are $2n \times k$ programmable points, known as *fuses* for a historical reason that the programming points of earlier PLA devices were composed of fuses; between the AND gates and the OR gates, there are $k \times m$ programmable points. At the output XOR gates, there are m programming points, which are used to program the outputs as inverting or noninverting.

The features of PLA devices are that both AND and OR arrays are programmable and all AND gates may be shared by all OR gates. As a result, when a PLA device is used to implement switching functions, a multiple-output minimization process is usually utilized to simplify the switching functions in order to obtain the shared product terms and hence to reduce the required number of product terms.

The following example illustrates how to implement a multiple-output switching function with a typical PLA device. Nonetheless, the multiple-output minimization process is beyond the scope of this book and hence we would like to omit it here. The interested reader may be referred to Lin [6].

■ Example 10.4: A PLA — Implementing switching functions.

In this example, we use a PLA device to implement three switching functions, as shown in Figure 10.17. These three switching functions are as follows:

$f_1(w, x, y, z) = \Sigma(2, 3, 5, 7, 8, 9, 10, 11, 13, 15)$
$f_2(w, x, y, z) = \Sigma(2, 3, 5, 6, 7, 10, 11, 14, 15)$

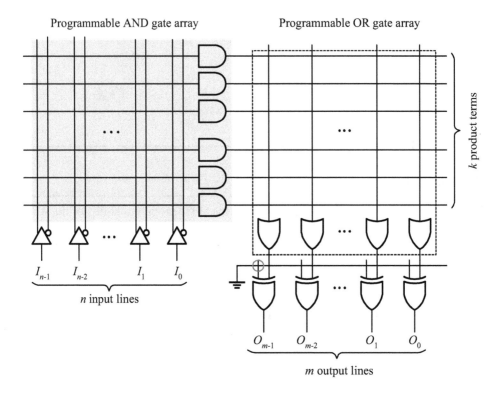

Figure 10.16: The general structure of PLA devices.

$$f_3(w,x,y,z) = \Sigma(6,7,8,9,13,14,15)$$

Simplifying by using a multiple-output minimization process, such as the one introduced in Lin [6], we obtain the following simplified switching expressions:

$$f_1(w,x,y,z) = \bar{x}y + w\bar{x}\bar{y} + \bar{w}xz + wxz$$
$$f_2(w,x,y,z) = \bar{w}xz + y$$
$$f_3(w,x,y,z) = xy + w\bar{x}\bar{y} + wxz$$

These three switching functions can be represented as the PLA programming table shown in Figure 10.17(a). There are six product terms, numbered from 0 to 5. The product terms $w\bar{x}\bar{y}$ and wxz are shared by functions f_1 and f_3, and $\bar{w}xz$ is shared by functions f_1 and f_2. The function f_1 consists of four product terms: 2, 3, 4, and 5. The function f_2 consists of two product terms: 0 and 4. The function f_3 consists of three product terms: 1, 3, and 5. The logic circuit that realizes the programming table shown in Figure 10.17(a) is depicted in Figure 10.17(b).

■ **Review Questions**

10-35 Describe the features of PLA devices in terms of the programmability of the AND-OR logic structure.

10-36 How many programmable points can an $n \times k \times m$ PLA have?

10-37 Can a product term be shared among multiple switching functions in a PLA device?

10-38 Is the AND array of PLA devices fully or partially decoding the input combinations?

Section 10.2 PLD Structure and Modeling

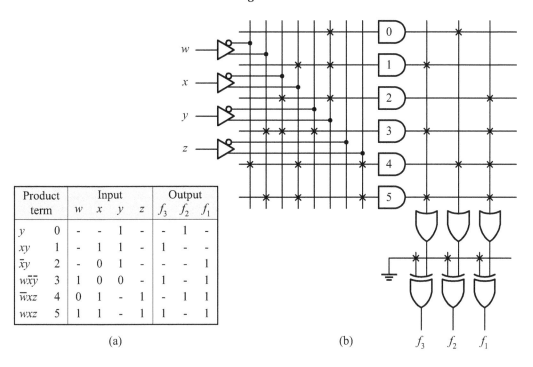

Figure 10.17: A PLA device of implementing a multiple-output switching function: (a) PLA programming table and (b) logic circuit.

10.2.3 Programmable Array Logic

A major disadvantage of PLA devices is that they consume too much area in fuses and hence for a fixed area budget, it is not possible to make a PLA device with the enough number of product terms for practical applications. One way to overcome this is to trade off the fuse area for product terms. It is shown that most switching functions in practical applications contain less than five product terms. As a result, it is unnecessary to allow all product terms to be connected to all OR gates. Based on this consideration, the third type of the two-level AND-OR logic structure, known as *programmable array logic* (PAL), is obtained.

Remember that the essential structure of PAL devices is that its AND array is programmable and its OR array is connected to a fixed number of AND gates. This results in that each OR gate along with its associated AND array comprises a logic module suitable for implementing a switching function with a limited number of product terms, ranging from 5 to 8. Based on this feature, in using a PAL device to implement a multiple-output function, it is sufficient to simplify each output function independently because no AND gates can be shared by different OR gates.

Nowadays, discrete PAL devices [6] are still on the market, although they are only used as the glue logic between large logic modules. These PAL devices have the standard features as listed in Table 10.3. They can be generally grouped into two series: PAL16xx and PAL20xx. The former has 16 inputs for each AND gate and the latter has 20 inputs for each AND gate. The next character "L" or "R" distinguishes the combinational logic type from the registered output type. The last digit denotes the number of outputs. An example of a PAL device, 16R8, is shown in Figure 10.18.

The important features of PAL devices are summarized as follows: First, each OR gate associated with an AND array forms a two-level AND-OR logic structure. This AND-OR logic structure can be used to implement a switching function represented in the sum-of-product

Table 10.3: The standard features of PAL devices.

PAL devices	Package pins	AND-gate inputs	Primary inputs	Bidirectional I/Os	Registered outputs	Combinational outputs
PAL16L8	20	16	10	6	0	2
PAL16R4	20	16	8	4	4	0
PAL16R6	20	16	8	2	6	0
PAL16R8	20	16	8	0	8	0
PAL20L8	24	20	14	6	0	2
PAL20R4	24	20	12	4	4	0
PAL20R6	24	20	12	2	6	0
PAL20R8	24	20	12	0	8	0

(SOP) form. Second, since there are no shared product terms in a PAL device, a single-output minimization process is sufficient in minimizing switching functions. An illustration of how to implement a multiple-output switching function with a typical PAL device is demonstrated in the following example.

■ Example 10.5: PAL — Implementing switching functions.

In this example, a multiple-output switching function consisting of four switching functions is implemented by using a PAL device. These four switching functions are as follows:

$f_1(w,x,y,z) = \Sigma(2,12,13)$
$f_2(w,x,y,z) = \Sigma(7,8,9,10,11,12,13,14,15)$
$f_3(w,x,y,z) = \Sigma(0,2,3,4,5,6,7,8,10,11,15)$
$f_4(w,x,y,z) = \Sigma(1,2,8,12,13)$

By simplifying them separately using Karnaugh maps or the other minimization process, we obtain the following simplified switching expressions:

$f_1(w,x,y,z) = wx\bar{y} + \bar{w}\bar{x}y\bar{z}$
$f_2(w,x,y,z) = w + xyz$
$f_3(w,x,y,z) = \bar{w}x + yz + \bar{x}\bar{z}$
$f_4(w,x,y,z) = wx\bar{y} + \bar{w}\bar{x}y\bar{z} + w\bar{y}\bar{z} + \bar{w}\bar{x}\bar{y}z$

The resulting PAL programming table and logic diagram are shown in Figures 10.19(a) and (b), respectively.

■ Review Questions

10-39 Describe the features of PAL devices in terms of the programmability of the AND-OR logic structure.
10-40 Can a product term be shared among multiple switching functions in a PAL device?
10-41 Is the AND array of PAL devices fully or partially decoding the input combinations?

10.2.4 PLA Modeling

Recall that all PLD devices have a similar two-level AND-OR structure and are categorized into three different types in terms of the programmability of both AND and OR arrays. As an AND or OR array is being programmed with a pattern, it is said to be personalized and the

Section 10.2 PLD Structure and Modeling

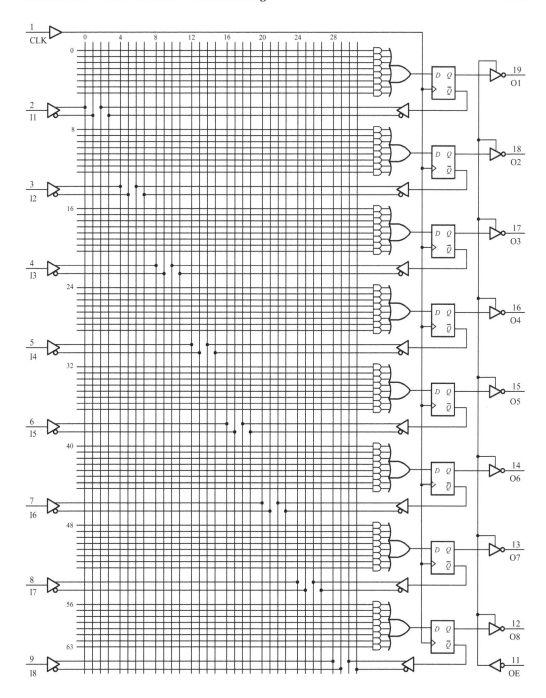

Figure 10.18: The structure of the PAL device 16R8.

resulting array is denoted by an AND or OR array *personality*. For convenience, in this section, we will use the *logic array* to denote an array of AND, OR, or other logic gates and the *logic array personality* to mean a logic array that has been personalized.

In addition to the two-level AND-OR structure, the two-level OR-AND structure is also widely used in ASIC devices to construct a PLA module. Hence, to model these two possible structures, Verilog HDL provides a flexible mechanism with which we can model one of the

Product term		Input w x y z	Output function
$wx\bar{y}$	16	1 1 0 -	
$\bar{w}\bar{x}yz$	17	0 0 1 0	f_1
w	24	1 - - -	
xyz	25	- 1 1 1	f_2
$\bar{w}x$	32	0 1 - -	
yz	33	- - 1 1	f_3
$\bar{x}\bar{z}$	34	- 0 - 0	
$wx\bar{y}$	40	1 1 0 -	
$\bar{w}\bar{x}yz$	41	0 0 1 0	
$w\bar{y}\bar{z}$	42	1 - 0 0	f_4
$\bar{w}\bar{x}\bar{y}z$	43	0 0 0 1	

(a)

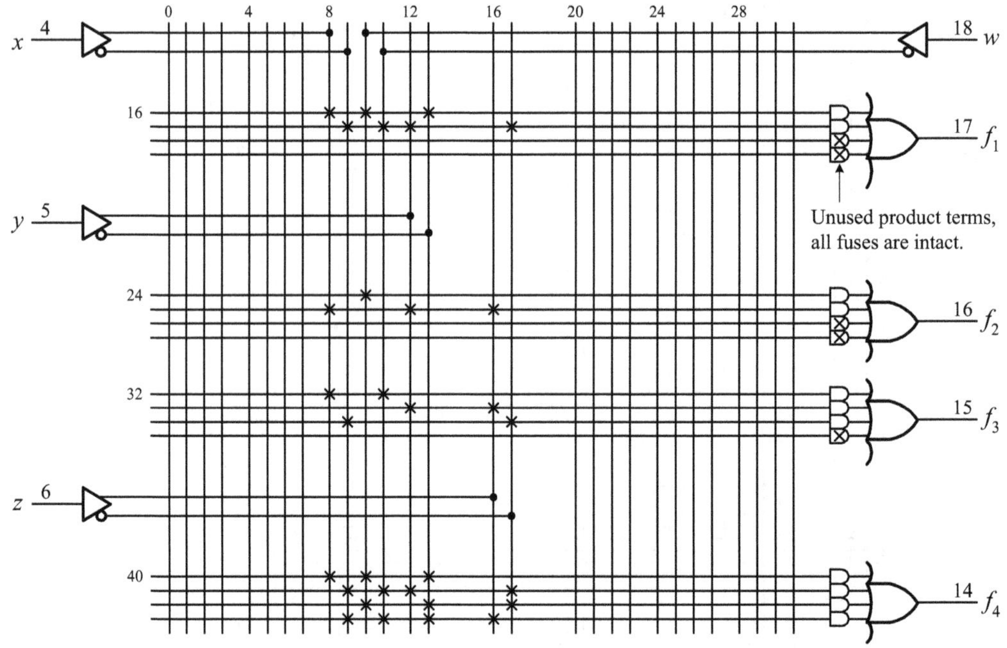

(b)

Figure 10.19: A PAL device of implementing a multiple-output switching function: (a) PAL programming table and (b) logic circuit.

following two types of structures:

1. *Sum of product* (SOP): An SOP expression comprises an AND array (also called an *AND plane*) and an OR array (an *OR plane*). Through using DeMorgan's law, an SOP expression can be equivalently represented as a NAND-NAND structure, constructed by cascading two NAND arrays.
2. *Product of sum* (POS): A POS expression consists of an OR array and an AND array. By using DeMorgan's law, a POS expression can be equivalently expressed as a NOR-NOR structure, constructed by cascading two NOR arrays.

Section 10.2 PLD Structure and Modeling

From the above two structures, we know that there are four types of logic arrays, AND, OR, NAND, and NOR, that are needed to model. Each array is personalized according to the actual specifications of the desired logic structure. During simulation, each logic array personality must be loaded into its *personality memory* to evaluate the values of the logic array.

Since the PLA structure is most widely used in HDL-based design and takes both ROM and PAL structures as special cases, in the rest of this section, we will only focus our attention on the modeling of PLAs. Nonetheless, the same technique can be applied equally well to PAL and ROM devices.

10.2.4.1 System Tasks for PLA Modeling A group of system tasks is provided by Verilog HDL with which any type of PLA device can be modeled as desired through an appropriate combination of two system tasks. The general form of PLA modeling system tasks is as follows

`$array_type$logic$format(memory_type, input_terms, output_terms);`

where `array_type` can be either `async` (asynchronous) or `sync` (synchronous), `logic` can be any of `and`, `or`, `nand`, and `nor`, and `format` can be either `array` or `plane`. As a consequence, there are sixteen system tasks that can be obtained through these combinations, as shown in Table 10.4. Note that `memory_type` is an array of **reg** vectors, `input_terms` can be nets or variables, and `output_terms` may only be variables.

Each PLA modeling system task can be either asynchronous or synchronous. For an asynchronous system task, the evaluation of the logic array is executed whenever any input changes or any word of the personality memory is modified. For a synchronous system task, the time to evaluate the logic array can be controlled by the clock signal. However, regardless of which type of system task is used, the output is updated without any delay.

Table 10.4: PLA modeling system tasks in Verilog HDL.

Array format		Plane format	
Asynchronous	Synchronous	Asynchronous	Synchronous
$async$and$array	$sync$and$array	$async$and$plane	$sync$and$plane
$async$nand$array	$sync$nand$array	$async$nand$plane	$sync$nand$plane
$async$or$array	$sync$or$array	$async$or$plane	$sync$or$plane
$async$nor$array	$sync$nor$array	$async$nor$plane	$sync$nor$plane

When used to model a PLA structure, a logic array personality can be specified by either of the following two formats:

- **Array format:** In this format, only the values of 1 and 0 are used to denote whether the input value is taken. Consequently, the inputs to a logic array personality must be literals rather than variables. Recall that a variable x and its complement $\bar{x}$ are two different literals but is the same variable. Hence, when modeling a logic array personality with the array format, we have to generate the complemented signals if necessary.
- **Plane format:** The plane format (compliant with Espresso) allows us to specify a logic array personality with the true or complemented input value and don't care (namely, not taken).
  1. Values 1 and 0 separately denote the true and complemented input values.
  2. Symbol x denotes the worst-case input value.
  3. Symbol ? or z denotes a don't care condition (i.e., the input value is of no significance).

396   Chapter 10   Implementation Options of Digital Systems

Since the above two formats use different rules to specify the values of a logic array personality, each system task indicates explicitly in its name the actual format used. For instance, the **$async$or$array** system task uses the array format and the **$async$or$plane** system task employs the plane format.

As an illustration of how to use PLA modeling system tasks, let us consider the following simple example, which reveals the declarations of two asynchronous AND arrays and one synchronous AND array.

■ Example 10.6: An example of using PLA system tasks.

This example demonstrates the use of PLA modeling system tasks. The AND-array personality memory, `mem_and`, is declared as an array of 7-bit vectors with a depth of three words because there are seven input terms and three output terms. In order to simulate the function of the AND array, the AND-array personality is needed to load into its personality memory, `mem_and`, through the use of the **$readmemb** or **$readmemh** system task, or by using procedural assignments, as illustrated in the succeeding two examples.

```
wire a1, a2, a3, a4, a5, a6, a7;
reg  b1, b2, b3;
wire [1:7] awire;
reg  [1:3] breg;
reg  [1:7] mem_and [1:3]; // define personality memory

   // an asynchronous AND-array PLA modeling system task
   $async$and$array(mem_and,{a1,a2,a3,a4,a5,a6,a7},{b1,b2,b3});
   // or using the following statement
   $async$and$array(mem_and, awire, breg);
   // a synchronous AND-array PLA modeling system task
   always @(posedge clock)
      $sync$and$array(mem_and,{a1,a2,a3,a4,a5,a6,a7},{b1,b2,b3});
```

Generally, the personality memory of a logic array personality is declared as an array of **reg** vectors with the vector width equal to the number of input terms and the array depth equal to the number of output terms. The contents of a personality memory can be loaded from a file by using either the **$readmemb** or **$readmemh** system task, or be filled using procedural assignments. The personality memory can be changed dynamically during simulation simply by changing the contents of the memory array. An illustration of how to load the AND-array personality into its personality memory from a file is given in the following example.

■ Example 10.7: Loading the personality memory from a file.

In this example, only one AND array is considered. The contents of the AND-array personality memory, `mem_and`, are loaded from the `array.dat` file by using the **$readmemb** system task.

```
// an example of the usage of the array format.
module async_array(
      input  a1, a2, a3, a4, a5, a6, a7,
      output reg b1, b2, b3);
reg [1:7] mem_and[1:3]; // for array personality

initial begin
   // setup the personality from the array.dat file
   $readmemb ("array.dat", mem_and);
```

Section 10.2 PLD Structure and Modeling

```
   // setup an asynchronous logic array with the input
   // and output terms expressed as concatenations
   $async$and$array (mem_and,{a1,a2,a3,a4,a5,a6,a7},{b1,b2,b3});
end
endmodule
```

The output functions are as follows:

```
b1 = a1 & a2
b2 = a3 & a4 & a5
b3 = a5 & a6 & a7
```

Hence, the AND array personality is defined as follows:

```
1100000
0011100
0000111
```

The following example demonstrates how to set the personality memory of the AND-array personality through using procedural assignments.

■ Example 10.8: Loading the personality memory by assignments.

In this example, only one AND array is considered. The contents of the AND-array personality memory, mem_and, are assigned by using procedural assignments. Three assignments are needed in this example because there are three output terms.

```
module pla(input   a0, a1, a2, a3, a4, a5, a6, a7,
           output reg b0, b1, b2);
reg [7:0] mem_and[0:2];

// an example of the usage of array format
initial begin    // using procedure assignment statements
   mem_and[0] = 8'b11001100;
   mem_and[1] = 8'b00110011;
   mem_and[2] = 8'b00001111;
   $async$and$array(mem_and,{a0,a1,a2,a3,a4,a5,a6,a7},{b0,b1,b2});
end
endmodule
```

What follows shows the simulation results from a test bench:

```
A = 11001100 -> B = 100
A = 00110011 -> B = 010
A = 00001111 -> B = 001
A = 10101010 -> B = 000
A = 01010101 -> B = 000
A = 11000000 -> B = 000
A = 00111111 -> B = 011
```

10.2.4.2 PLA Modeling Examples To explain how PLA modeling system tasks may be used to model a practical PLA device, consider the PLA example shown in Figure 10.20. Here, two switching functions with three input variables, x, y, and z, are implemented with a PLA device. After minimization, these two switching functions are represented as the PLA programming table shown in Figure 10.20(a). There are four product terms, numbered from 0

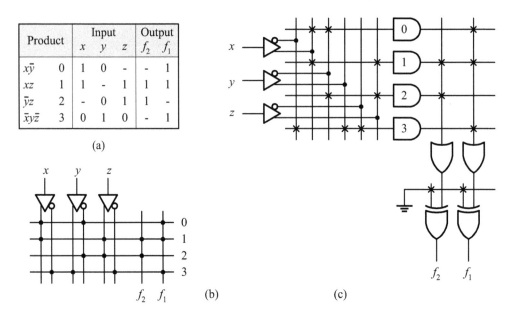

Figure 10.20: An example of modeling a PLA device: (a) PLA programming table; (b) PLA symbolic diagram; (c) logic circuit.

to 3, in which the product term 1, xz, is shared by two functions. The function f_1 consists of three product terms: 0, 1, and 3. The function f_2 comprises two product terms: 1 and 2. The programming table shown in Figure 10.20(a) can be expressed by a shorthand notation, named the *PLA symbolic diagram*, as shown in Figure 10.20(b). Figure 10.20(c) is the logic circuit that realizes the programming table shown in Figure 10.20(a).

In what follows, we give two examples to illustrate how this PLA device can be separately modeled with the array and the plane formats. We begin to consider the case of using the array format and then deal with the situation of using the plane format.

■ Example 10.9: A PLA modeling example with the array format.

As stated before, when the array format is used, only two values of 1 and 0 are allowed in the logic array personality to denote whether the input variable is taken or not taken. Hence, we need to derive the complemented signals from the inputs by using **assign** continuous assignments. The resulting module is as follows:

```
// an example of the usage of the array format
module pla_example(input x, y, z, output reg f1, f2);
reg  p0, p1, p2, p3;    // internal product terms
reg  [0:5] mem_and[0:3];
reg  [0:3] mem_or[1:2];
wire x_n, y_n, z_n;
assign x_n = ~x, y_n = ~y, z_n = ~z;
initial begin: pla_model_array
   mem_and[0] = 6'b100100;  // define the AND array
   mem_and[1] = 6'b100010;
   mem_and[2] = 6'b000110;
   mem_and[3] = 6'b011001;
```

Section 10.2 PLD Structure and Modeling

```
      mem_or[1] = 4'b1101;     // define the OR array
      mem_or[2] = 4'b0110;
   end
   always @(*) begin
      $async$and$array(mem_and, {x, x_n, y, y_n, z, z_n},
                                {p0, p1, p2, p3});
      $async$or$array(mem_or, {p0, p1, p2, p3}, {f1, f2});
   end
endmodule
```

Note that when the array format is used, both AND-array and OR-array personality memories indeed store the contents of the PLA symbolic diagram. Justify this by comparing the contents of personality memories of both AND and OR arrays with the PLA symbolic diagram shown in Figure 10.20(b). The simulation results from a test bench are as follows:

```
# 0 ns   x x x x x
# 5 ns   0 0 0 0 0
# 10 ns  0 0 1 0 1
# 15 ns  0 1 0 1 0
# 20 ns  0 1 1 0 0
# 25 ns  1 0 0 1 0
# 30 ns  1 0 1 1 1
# 35 ns  1 1 0 0 0
# 40 ns  1 1 1 1 1
```

■ Example 10.10: A PLA modeling example with plane format.

As mentioned above, when the plane format is used, values 1 and 0 are used in the logic array personality to denote the true input and the complemented input, respectively. In addition, the character ? or z is used to denote the don't care condition. Hence, the AND-array personality memory only needs three bits and four words because there are three input variables and four product terms. The contents of the AND-array personality memory directly correspond to the PLA programming table shown in Figure 10.20(a). Similarly, the OR-array personality memory needs four bits and two words because there are four product terms and two switching functions. The contents of the OR-array personality memory can also be obtained from the PLA programming table of Figure 10.20(a). The results are shown in the following module.

```
// an example of the usage of the plane format
module pla_example(input x, y, z, output reg f1, f2);
reg p0, p1, p2, p3;      // internal product terms
reg [0:2] mem_and[0:3];
reg [0:3] mem_or[1:2];
initial begin: pla_model
   mem_and[0] = 3'b10?;    // define AND array
   mem_and[1] = 3'b1?1;
   mem_and[2] = 3'b?01;
   mem_and[3] = 3'b010;

   mem_or[1] = 4'b11?1;    // define OR array
   mem_or[2] = 4'b?11?;
end
always @(*) begin
```

```
        $async$and$plane(mem_and, {x, y, z}, {p0, p1, p2, p3});
        $async$or$plane(mem_or, {p0, p1, p2, p3}, {f1, f2});
    end
endmodule
```

Note that when the plane format is used, both AND-array and OR-array personality memories indeed store the contents of the PLA programming table. Justify this by comparing the contents of both AND-array and OR-array personality memories with the PLA programming table shown in Figure 10.20(a).

■ Review Questions

10-42 Describe the general form of PLA modeling system tasks.
10-43 Describe the meaning of the logic array personality.
10-44 Describe the differences between the asynchronous and synchronous PLAs.
10-45 Describe the features of the array format used in modeling PLAs.
10-46 Describe the features of the plane format used in modeling PLAs.

10.3 CPLDs and FPGAs

Recall that both CPLDs and FPGAs comprise three types of components: *function blocks* (FBs), the *interconnect structure*, and *input/output blocks* (I/O blocks or IOBs for short). Function blocks are used to implement the desired logic functions; the interconnect structure provides programmable communicating paths between function blocks as well as between function blocks and I/O blocks; I/O blocks are I/O pins with related logic circuits and can configure their associated I/O pins into the input, output, or bidirectional mode as needed. I/O blocks also support other features, such as low-power or high-speed connections.

10.3.1 CPLDs

The basic structure of a typical CPLD is composed of three major parts: function blocks (i.e., a set of macrocells), an interconnect structure (i.e., one or more switch matrices and related interconnect), and I/O blocks, as shown in Figure 10.21. Each function block contains a number of macrocells, with each responsible for implementing a logic function with m input variables. The interconnect (switch matrix) structure provides communicating paths between function blocks as well as between function blocks and I/O blocks. I/O blocks provide buffers for inputs and outputs. The complexity of CPLDs can be achieved by adding more function blocks and I/O blocks, and expanding the interconnect structure accordingly.

10.3.1.1 Function Blocks Each function block contains a number of macrocells. As depicted in Figure 10.21, each function block (FB) accepts m input signals from and outputs n signals to the interconnect structure, through which it connects to other macrocells. A simplified structure of macrocells is plotted in Figure 10.22. Each macrocell contains 5 to 8 product terms along with a programmable D/T flip-flop and is responsible for implementing a combinational or sequential logic function. The product terms act as an AND array that can be directly used as an input source of the OR and XOR gates to implement the desired function or as the clock signal, the clock enable, or the set/reset control signal for the programmable D/T flip-flop within the macrocell, or as the output enable signal to control the output buffer within its associated I/O block. The output from a macrocell can serve as an input to both the interconnect structure and I/O blocks at the same time.

Section 10.3 CPLDs and FPGAs

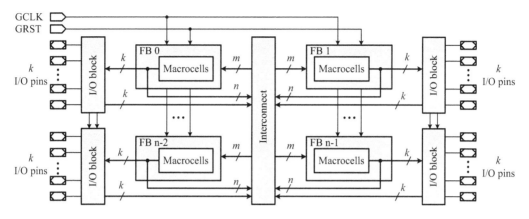

Figure 10.21: The conceptual block diagram of typical CPLDs.

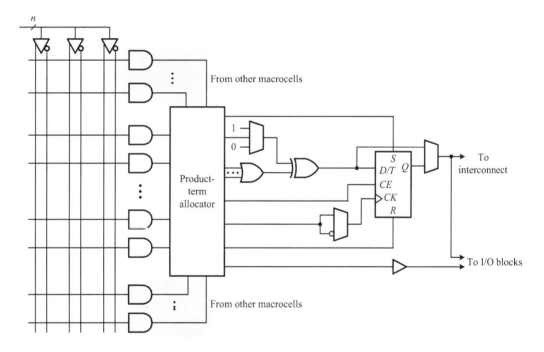

Figure 10.22: A simplified macrocell of the typical CPLD.

Since each macrocell is a form of the two-level AND-OR logic structure, the switching function realized by each macrocell is in SOP form. The product-term allocator determines the actual usage of product terms. It can accept the product terms from other macrocells, only use basic product terms, or allocate the unused product terms to other macrocells. Thus, although each macrocell only has a few (5 to 8) product terms, it can borrow from or lend to other macrocells within the same function block the desired number of product terms.

The programmable D/T flip-flop within each macrocell may be programmed as either a D or T flip-flop to fit the actual requirements of the underlying system. To implement a JK flip-flop, some extra logic is needed from the product-term allocator. Of course, it is bypassed when the macrocell is used to realize a combinational logic function. Each programmable D/T flip-flop can be set or cleared asynchronously through its asynchronous set and reset control inputs, which can be either the global set/reset control signal (not shown in the figure) or the

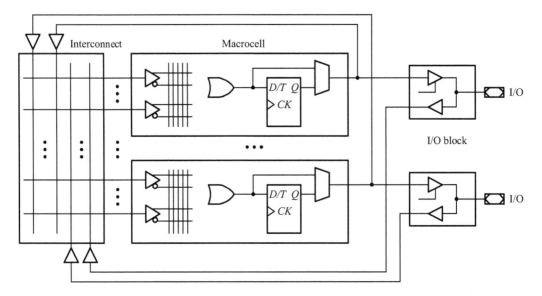

Figure 10.23: The interconnect structure of the typical CPLD.

product-term set and the product-term reset control signals. When the power supply starts up, all programmable D/T flip-flops are set to the default value of 0 unless they are set to the values specified otherwise by users. The clock input of the programmable D/T flip-flop may be programmed to be positive- or negative-edge-triggered. The clock signal can be obtained either from the global clock signal (not shown in the figure) or from the product-term allocator. The use of the global clock signal achieves the fastest clock-to-output performance while the use of the product-term clock signal has the feature that the programmable D/T flip-flop can be clocked by signals from macrocells or I/O pins.

10.3.1.2 Interconnect Structure The interconnect structure of a typical CPLD is usually a programmable switch matrix or a combination of several programmable switch matrices. A conceptual illustration of using a single switch matrix to connect macrocells and I/O blocks is shown in Figure 10.23. The switch matrix functions as an interconnect network used to provide communicating paths between I/O blocks and macrocells as well as between macrocells and macrocells. Each crosspoint may be an SRAM bit, an antifuse cell, or a Flash cell, as we have described before.

10.3.1.3 Input/Output Blocks The I/O block is used to route the signal either from the output of a macrocell to an I/O pin or from an I/O pin to the input of macrocells through the interconnect structure. The structure of a typical I/O block is depicted in Figure 10.24. Each I/O block contains an input buffer, an output driver, an output enable select multiplexer, a slew-rate control circuit, a pull-up resistor, and a programmable ground.

To make CPLDs versatile, the input buffers of CPLDs are usually made to be compatible with 5-V CMOS, 5-V TTL, 3.3-V CMOS, and 2.5-V CMOS, referring to Table 10.5. To achieve this, a constant input threshold voltage must be provided to avoid the threshold voltage varying with the I/O power-supply voltage (V_{CCIO}). For this purpose, an internal +3.3-V power-supply voltage V_{CCINT} is used.

Similarly, all output drivers of CPLDs are made to be tolerant either to 3.3-V CMOS and 5-V TTL as $V_{CCIO} = 3.3$ V or to 2.5-V CMOS as $V_{CCIO} = 2.5$ V. In addition, each output driver has a slew-rate control circuit used to control both the rise and fall times of the output signal. The output enable signal can be set to one of three options: always "1", always "0", and the output enable signal generated from the associated macrocell. Finally, each I/O block also has

Section 10.3 *CPLDs and FPGAs*

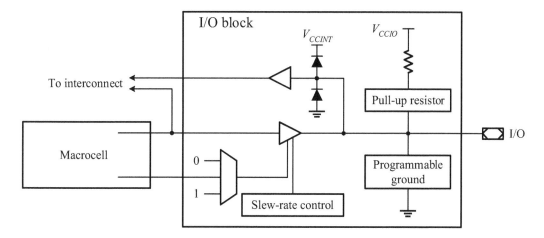

Figure 10.24: A simplified I/O block of the typical CPLD.

the capability of programmable ground to ground its unused I/O pin so as to reduce power dissipation and switching noise.

■ Review Questions

10-47 Describe the basic structure of a typical CPLD.
10-48 Describe the structure of function blocks.
10-49 Describe the structure of macrocells.
10-50 Describe the structure of the switch matrix of CPLDs.
10-51 Describe the structure of I/O blocks.
10-52 Could a switching function implemented by CPLDs contain more than five product terms?

10.3.2 FPGAs

Like CPLDs, a typical FPGA also consists of three basic types of components: function blocks (namely, PLBs), the interconnect structure, and input/output blocks (I/O blocks or IOBs for short), as depicted in Figure 10.13. PLBs provide the functional elements for implementing the desired logic functions, the interconnect structure connects together PLBs and I/O blocks, and I/O blocks provide the interface between I/O pins and internal signals.

10.3.2.1 Function Blocks Each PLB usually consists of one or more k-input *function generators* (namely, *universal logic modules*) with each associated with a programmable output-stage logic circuit, where k is usually set from 3 to 8. A logic circuit is known as a k-input function generator if it is capable of implementing any switching function with k input variables at most. Conceptually, a k-input function generator is a lookup table (LUT), which can be implemented by an SRAM device, a Flash memory device, a multiplexer, or other combinational logic. The output-stage logic circuit consists of a number of multiplexers and a flip-flop. As a result, each PLB is capable of implementing one or more combinational or sequential logic functions.

A simplified PLB is shown in Figure 10.25. Each PLB contains two groups, with each consisting of a 4-input function generator (a 4-input LUT) and a D flip-flop. The 4-input function generator can realize any switching function with 4 input variables at most. It can also be used to implement a sequential logic function when combining with the output D

flip-flop. The D flip-flop is a universal flip-flop, associated with clock-enable control as well as asynchronous set and clear. The 2-to-1 multiplexer at the output of each group is used to bypass the D flip-flop when the group realizes a combinational logic function. The multiplexer is configured by the M bit, which may be an SRAM bit, an antifuse cell, or a Flash cell, as we have described before.

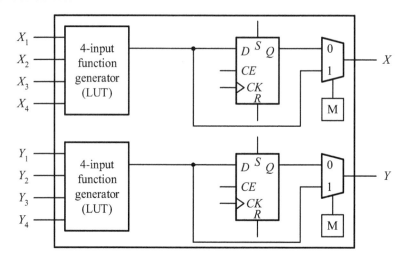

Figure 10.25: A simplified PLB showing the essential elements of typical PLBs.

The k-input function generator can be realized by either a $2^k \times 1$ RAM or Flash memory, or a 2^k-to-1 multiplexer. Regardless of which way is used, the essential idea is first to represent the switching function to be realized as a truth table and then the truth table is stored in the RAM/Flash memory or used to configure the multiplexer (or a multiplexer tree) for any lookup operation later. The following example illustrates how this idea works.

■ **Example 10.11: Implementing a switching function using a LUT.**

Assume that we want to realize the majority function with three variables

$$f(x,y,z) = xy + yz + xz$$

then we need to represent it as a truth table. Because there are three variables, eight combinations exist and the function values corresponding to the eight combinations of x, y, and z in order are as follows: 0, 0, 0, 1, 0, 1, 1, and 1. As a result, to realize this function with a 3-input LUT, we only need to store the function values in the LUT, as depicted in Figure 10.26(a). Similarly, to realize this function with an 8-to-1 multiplexer, we only need to set each of its inputs with the corresponding function value, as shown in Figure 10.26(b). More about the use of multiplexers to realize switching functions can be referred to Lin [6].

10.3.2.2 Interconnect Structure The interconnect structure of typical FPGAs is shown in Figure 10.27(a). It mainly consists of interconnect lines and switch matrices (SMs). Vertical and horizontal lines are used to connect two switch matrices.

The switch matrix (SM) has the structure shown in Figure 10.27(b). Each cross-point of the switch matrix is composed of a switch circuit that contains six nMOS pass transistors; that is, it forms a fully connected switch in all directions. As a result, the horizontal and vertical lines that intersect with an SM can establish the connections between them. For example, a signal entering on the left side of the SM can be routed to a line on the top, the right, or the bottom side, or any combination of them, if multiple branches are required.

Section 10.3 CPLDs and FPGAs

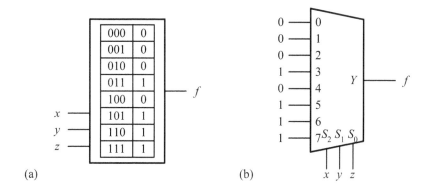

Figure 10.26: An example illustrating the use of (a) a 3-input LUT and (b) an 8-to-1 MUX to implement a three-input switching function.

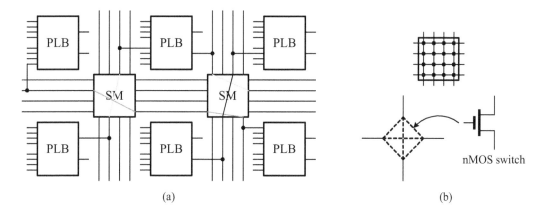

Figure 10.27: The programmable interconnect structure of typical FPGA devices: (a) a general-purpose interconnect and (b) a switch matrix.

10.3.2.3 Input/Output Blocks

The I/O blocks provide the interface between external I/O pins and the internal logic. Each I/O block controls one I/O pin and can configure the I/O pin into the input, output, or bidirectional mode. To make FPGA devices versatile, inputs are often made to be compatible with both TTL and 3.3-V CMOS and outputs are pulled up to a 3.3-V power-supply voltage. Likewise, all outputs of FPGA devices are often made to be fully 5-V tolerant (Section 10.4.2) even though the V_{CC} is 3.3 V. A simplified I/O block structure of typical FPGA devices is depicted in Figure 10.28, which consists of many multiplexers, buffers, a D latch/flip-flop, and a D flip-flop.

The output D flip-flop can be only used as a flip-flop and the input D flip-flop can be programmed as either a flip-flop or a latch. Both input and output D flip-flops have their own clock signals (CLK_{in} and CLK_{out}) but with a common clock enable input (CE). Each flip-flop can be programmed into the positive- or negative-edge-triggered mode, through a 2-to-1 multiplexer at the clock input of the flip-flop.

The input signal is connected to the input D flip-flop that can be programmed as either an edge-triggered flip-flop or a level-sensitive latch. The input signal may be delayed a few nanoseconds ($\Delta \tau$) before entering the input D flip-flop in order to compensate for the hold time insufficiency of the external signal due to the delay of the clock signal. Signal I_{pin} is from the I/O pin whereas signal I_Q is from the output of the input D flip-flop/latch.

The output signal can be from the OUT end or from the output of the output D flip-flop.

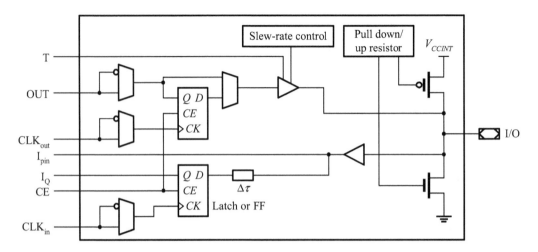

Figure 10.28: A simplified I/O block of the typical FPGA devices.

Besides the output enable (T) signal, there is a slew-rate control circuit used to control the slew rate of the output buffer, namely, the rise and fall times of the output signal. The function of the slew-rate control circuit is to control the current sunk or sourced by the output buffer. During switching noncritical signals, the slew rate of the output buffer is reduced to minimize the power-supply transients and during switching critical signals, it is increased to promote performance. As usual, the unused I/O pins are pulled up through pull-up resistors or grounded in order to reduce unnecessary power dissipation and switching noise.

10.3.3 Advanced Structures of FPGAs

Nowadays, many modern FPGAs have also incorporated into specialized features for specific applications in the fields of communications, digital-signal processing, multimedia, or consumer products. The following ones are the most common.

10.3.3.1 Carry and Cascade Chains In order to provide high-speed arithmetic functions, modern FPGAs provide dedicated datapaths, *carry chains* and *cascade chains*, through which adjacent PLBs (i.e., LEs) can be connected without using local interconnect paths, as illustrated in Figure 10.29. A carry chain supports arithmetic functions, such as counters and adders, and a cascade chain implements wide-input functions, such as equality comparators.

Through a carry chain, the carry-in signal from a lower-order bit drives forward into the higher-order bit, and feeds into both the LUT and the next portion of the carry chain so as to allow to implement counters, adders, and comparators of an arbitrary width. With the cascade chain, adjacent LUTs can compute portions of a function in parallel. The cascade chain can use a logical AND or OR to connect the outputs of adjacent PLBs.

10.3.3.2 Distributed Memory In the PLB shown in Figure 10.25, if each of the two 4-input function generators is individually implemented by a 16×1 RAM circuit, then it indeed realizes a switching function by storing the truth table of the switching function in the RAM and then uses the input variables (X_i and Y_i, for $1 \leq i \leq 4$) as the address signals to read the contents of the RAM later. When both X and Y function generator circuits are a RAM circuit, by properly adding some extra control signals, they both can be configured as a RAM module to be used otherwise.

As a PLB composed of m k-input LUTs is built on the basis of SRAMs, it usually can be configured as either a $(2^k \times 1)$-, $(m \times 2^k \times 1)$-, or $(2^k \times m)$-bit memory array. For instance, as $m=2$ and $k=4$, then the possible memory arrays are 16×1 bits, 32×1 bits, and 16×2 bits.

Section 10.3 CPLDs and FPGAs

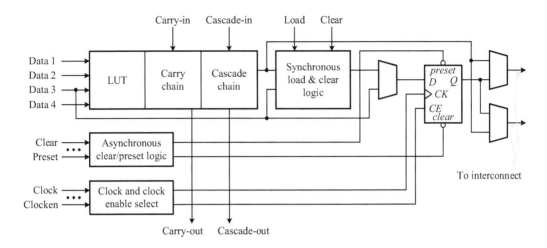

Figure 10.29: The PLB with carry and cascade chains in some FPGA devices.

Such a type of memory is often referred to as the *distributed memory* due to the inherently distributed feature of PLBs. Furthermore, the distributed memory can often be configured as either an edge-triggered (synchronous) or a dual-port RAM. Edge-triggered RAMs simplify system timing while dual-port (such as one read and one write) RAMs double the effective throughput of FIFO applications. It is instructive to note that as a PLB is employed as a distributed memory, it is no longer used as a logic element. An example to illustrate such an application is as follows.

■ Example 10.12: Single-port RAM: the data port is bidirectional.

In this example, we design a synchronous memory with the same port for both write and read accesses; that is, we need to combine both ports into a bidirectional port. Thus, tristate buffers are needed to use at the read port before they are combined with the write port. This can be readily modeled by an **assign** continuous assignment as demonstrated in the following module. Note that the write port is clocked whereas the read port is not.

```
// an N-word single-port ram
module single_port_ram
       #(parameter M = 6,  // number of address bits
         parameter N = 64, // number of words, N=2**M
         parameter W = 8)( // number of bits in a word
         input clk, write,
         inout [W-1:0] data,
         input [M-1:0] addr);
reg    [W-1:0] ram [N-1:0];

// the body of the N-word single-port ram
assign data = (!write) ? ram[addr] : {W{1'bz}};
always @(posedge clk)
   if (write) ram[addr] <= data;
endmodule
```

It is instructive to note that tristate buffers are used in the data bus within the above module. This means that the data bus in the above module will be directly routed to the outside of the system if an FPGA device is used since only the I/O blocks of FPGA devices may have tristate buffers. In addition, depending on the synthesis tool used, the memory array, `ram`, may be synthesized by using distributed memory or block memory.

10.3.3.3 Memory Blocks A key feature of modern FPGAs is to embed memory blocks (i.e., hardwired IP) onto the chips. Memory blocks can be used to implement filter delay lines, small first-in first-out (FIFO) buffers, shift registers, processor code storage, packet buffers, video frame buffers, dual-port memory, and general-purpose applications. The word width of each embedded memory block may often be varied. For instance, a 16-kb memory block may be configured into a $16k \times 1$, $8k \times 2$, $4k \times 4$, or $2k \times 8$. To make the embedded memory blocks more flexible for use in various applications, memory blocks with a variety of amounts of capacity are often embedded in the same FPGA device.

To guarantee that the desired memory array is implemented with memory blocks, it is often necessary to use the *macro generator* provided by the vender to instantiate a memory module manually rather than to entirely rely on the capability of the synthesis tool to generate the memory array automatically.

10.3.3.4 DSP Blocks For consumer applications, modern FPGAs are equipped with a variety of DSP blocks (i.e., hardwired IP), with each capable of implementing the following functions: multiplication, multiply-add, multiply-accumulate (MAC), and dynamic shifts. The multipliers generally have word lengths from 9 to 36 bits and may work in a fully registered, pipelined fashion. Along with soft logic fabrics and memory blocks, DSP blocks may be configured to carry out sophisticated fixed-point and floating-point arithmetic functions for applications in the DSP fields, including finite impulse response (FIR) filters, complex FIR filters, infinite impulse response (IIR) filters, fast Fourier transform (FFT) functions, and discrete cosine transform (DCT) functions.

10.3.3.5 Clock Networks and PLLs The hierarchical clock structure and multiple PLLs are often equipped in high-performance FPGAs. A hierarchical clock structure means that many clock networks are provided in the same device and these clock networks can be combined in a variety of ways. As an example, consider a clock structure with three clock networks: *dedicated global clock networks* (GCLKs), *regional clock networks* (RCLKs), and *periphery clock networks* (PCLKs). These three clock networks may be combined to provide up to a few hundred unique *clock domains* (CDs). Another feature of modern FPGAs is to provide a large number of on-chip PLLs. Each PLL can even support many outputs simultaneously, with each allowing to be independently programmed to create a uniquely customized clock frequency.

10.3.3.6 High-Speed Transceivers Modern high-end FPGAs usually equip a broad variety of transceivers for applications ranging from low-cost consumer products to high-end networking systems. By incorporating adaptive equalization techniques, such transceivers can achieve the performance up to 30 Gb/s, even 100-Gb/s, and can support the requirements of various I/O buses, such as *peripheral component interconnect* (PCI) *express* (PCIe), *serial advanced technology attachment* (SATA), and *universal serial bus* (USB).

10.3.3.7 Platform/SoC FPGAs FPGAs equipped with hardwired CPU cores, such as PowerPC and ARM Cortex processors along with various peripherals (including ADCs and DACs), are also widely found in many high gate-count devices. These FPGA devices are called platform (or SoC) FPGAs (or SoPCs), referred to Section 10.1.2 for more details. Each CPU core may also have an *auxiliary processor unit* (APU) for supporting hardware acceleration and an integrated crossbar switch for high data throughput. Besides, all FPGA devices with enough gate count (or the number of PLBs) are allowed to embed soft CPU cores, such as the MicroBlaze, PicoBlaze, or NIOS, along with peripherals for specific applications. Although using soft IP cores, they are still referred to as platform FPGAs. In summary, a platform FPGA

Table 10.5: The most widely used standard logic levels.

	5-V CMOS	5-V TTL	3.3-V LVTTL	2.5-V CMOS	1.8-V CMOS	1.5-V CMOS	1.2-V CMOS
V_{CC}	5.0 V	5.0 V	3.3 V	2.5 V	1.80 V	1.50 V	1.20 V
V_{OH}	4.44 V	2.4 V	2.4 V	2.0 V	1.35 V	1.15 V	0.84 V
V_{IH}	3.5 V	2.0 V	2.0 V	1.7 V	1.17 V	0.975 V	0.78 V
V_{th}	2.5 V	1.5 V	1.5 V	1.2 V	0.90 V	0.75 V	0.60 V
V_{OL}	0.5 V	0.4 V	0.4 V	0.4 V	0.45 V	0.35 V	0.36 V
V_{IL}	1.5 V	0.8 V	0.8 V	0.7 V	0.63 V	0.525 V	0.42 V

(or SoC FPGA) means that an FPGA device embeds one or more CPU cores and related peripherals in any form of hardwired, soft IP, and hard IP.

■ **Review Questions**

10-53 Describe the basic structure of PLBs.
10-54 What is the meaning of a k-input universal logic module?
10-55 Describe the programmable-interconnect structure of FPGAs.
10-56 Describe the SM structure of FPGAs.
10-57 Describe the I/O block structure of FPGAs.

10.4 Practical Issues

Because of the evolution of VLSI technology, a wide variety of ASICs and CPLDs/FPGAs may coexist in the same system. These devices may have different power-supply voltages and logic voltage levels. As a consequence, the interface between them becomes an important issue in designing practical systems. In this section, we consider the two most important issues related to the interface between two devices: *I/O standards* and *voltage tolerance*. An I/O standard defines a set of electrical rules that governs information transfer between two devices through electrical signals. Voltage tolerance deals with the possible problems associated with the interface between two devices operating at different power-supply voltages.

10.4.1 I/O Standards

Because of the fast evolution of VLSI technology, the feature sizes of MOS transistors have been reduced from several micrometers down to tens of nanometers (7 nm and beyond). This evolution has led the operating power-supply voltages of chips to move toward lower values. Nowadays, it is not uncommon that many power-supply voltages coexist in the same PCB system, even on the same chip. For instance, the core logic operates at a lower power-supply voltage, say, 1.2 V, but the I/O circuits operate at a higher power-supply voltage, say, 3.3 V, for being compatible with the other legacy devices in the system.

10.4.1.1 I/O Standards For the ease of interfacing chips with different power-supply voltages, an IC industry standard group, called *Joint Electron Device Engineering Council* (JEDEC), defines the following standard power-supply voltages for logic circuits: 3.3 V ± 0.3 V, 2.5 V ± 0.2 V, 1.8 V ± 0.15 V, 1.5 V ± 0.1 V, and 1.2 V ± 0.1 V. The logic levels associated with the most widely used standard power-supply voltages are listed in Table 10.5.

The JEDEC standard for 3.3-V logic circuits indeed contains two sets of logic levels: low-voltage CMOS (LVCMOS) and low-voltage TTL (LVTTL). LVCMOS levels are used in CMOS applications where the outputs have light loads (currents), usually less than 100 μA, so that V_{OL} and V_{OH} are maintained within 0.2 V of the power-supply rails, namely, $V_{OL} = 0.2$

V and $V_{OH} = 3.1$ V. LVTTL levels are used in TTL applications where the outputs have significant loads (currents). The $V_{OL} = 0.4$ V and $V_{OH} = 2.4$ V, which are compatible with traditional 5-V TTL levels.

10.4.1.2 LVDS Although a lot of I/O standards are defined and used in various application environments, they can be classified into the following three categories: the chip-to-chip interface, backplane interface, and high-speed memory interface. In what follows, we only focus on the two most widely used chip-to-chip interfaces: low-voltage TTL (LVTTL) and *low-voltage differential signal* (LVDS). The interested reader may be referred to Granberg [5] for the details of these I/O standards. Figure 10.30 shows the single-ended signal transfer circuit and its related logic levels. The voltage swing between the high and low levels is 1.2 V.

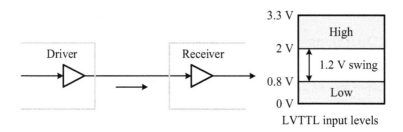

Figure 10.30: LVTTL chip-to-chip interfaces.

LVDS, also known as TIA/EIA-644 and introduced in 1994, is a technical standard that specifies *electrical characteristics* of a differential, serial communications protocol. The simple termination, low power, low noise generation, and tolerance to common-mode shifts between the driver and receiver generally make LVDS the technology of choice for data rates from tens of Mbps (0 to 655 Mbit/s, specified in the standard) up to 3.125 Gbit/s and beyond (dependent on circuit design). LVDS is a physical layer specification only; many data communication standards and applications use it and add a data link layer on top of it. The LVDS drivers are capable of transmitting signals over twisted-pair cables up to 10 to 15 m.

A typical LVDS driver and receiver pair in a point-to-point topology is depicted in Figure 10.31(a). The key points for consideration here are to properly control the impedance of the cable, proper driver load, and cable termination when designing for low-jitter signal transmission. Figures 10.31(b) and (c) reveal the single-ended and differential signal waveforms, respectively. The low-swing current-mode driver outputs (nominal 3.5 mA) create low noise and yield ± 350 mV (depending on the current direction) typical signal swings with a 100-Ω terminator and hence consume very low power. Changing the current direction results in the same voltage amplitude but in opposite polarity at the receiver. Logic 1s and 0s are generated in this manner. The receiver threshold voltage is $\pm$ 100 mV and is able to accommodate a common mode input voltage of 1.2 ± 1.0 V (i.e., from 0.2 V to 2.2 V).

Except for the point-to-point topology revealed in Figure 10.31(a), LVDS can also be applied to multipoint topologies, where multiple signal drivers and receivers all share a single cable. Figure 10.32 exhibits two examples. Figure 10.32(a) is a typical multidrop topology with the signal bus being terminated on the far receiver side. Such a scheme is advisable only when the signal driver is on the opposite end of the bus from the terminated receiver. In all other cases, such as the driver being connected to the middle of the bus, the bus needs to be terminated at both ends of the bus. Figure 10.32(b) illustrates a half-duplex topology, which consists of two driver/receiver pairs that transmit and receive signals between two points over a single cable.

Broadly speaking, data transfer in a bus (i.e., data communications) can be divided into the following three schemes: *simplex*, *half-duplex*, and *full-duplex* [9, 10, 11]. The *simplex*

Section 10.4 Practical Issues

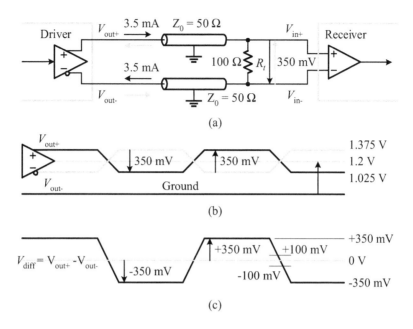

Figure 10.31: LVDS specification: (a) LVDS point-to-point connection; (b) single-ended signal waveform; (c) differential signal waveform.

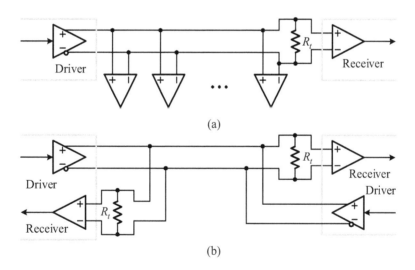

Figure 10.32: Typical common LVDS topologies: (a) multidrop topology; (b) half-duplex topology.

scheme means that data transfer can merely proceed in one direction at a time. If a bidirectional signal line (or channel) is connected between two devices but data transfer can only proceed in one direction at a time, the communication scheme is called a *half-duplex* scheme. If the data path (or channel) between two devices is bidirectional and data transfer can proceed in both directions at the same time, the communication scheme is called a *full-duplex* scheme.

In summary, the common advantages of differential signaling technologies are as follows. First, the current source is always on and routed in different directions to drive logic 1s and 0s, thereby eliminating the switching-noise spikes and *electromagnetic interference* (EMI)

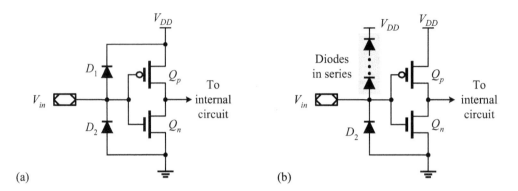

Figure 10.33: The input stages of typical CMOS logic circuits: (a) non-5V tolerant input and (b) 5-V tolerant input.

resulting from turning high-current transistors on and off. Second, since the two lines of the differential pair are closely adjacent to each other, common-mode noise that appears on both lines of the pair will cancel at the receiver. Third, as the two closely adjacent lines carry an equal amount of current but in opposite directions, EMI generation is minimized.

10.4.2 Voltage Tolerance

A device is *voltage compliance* if it is capable of receiving and transmitting signals from and to a device with a higher power-supply voltage than itself. Both input and output stages have the voltage-tolerant problems. A device is said to be *voltage-tolerant* if it can withstand a voltage greater than its V_{DD} on its I/O pins. If the device is capable of receiving a voltage greater than its V_{DD} on its input pins, it is said to have *input voltage tolerance*. If the device is able to drive a load with a power-supply voltage greater than its V_{DD} on its output pins, it is said to have *output voltage tolerance*.

The input voltage-tolerant problem is needed to consider when an input of a logic circuit with a lower power-supply voltage is driven by an output from a logic circuit with a higher power-supply voltage. The output voltage-tolerant problem is needed to take into account when two or more outputs of logic circuits with different power-supply voltages are connected together. Since in many practical applications the interface between two devices operating at 5-V TTL and 3.3-V LVTTL logic levels, respectively, is often encountered, in what follows we will focus on the input voltage-tolerant and output voltage-tolerant problems associated with these two voltages.

10.4.2.1 Input Voltage Tolerance

As we have shown in Table 10.5, even though both LVTTL and TTL families have the same logic levels, it might not be able to work properly if we ignore the potential voltage-tolerant problem between them. To see this, consider the circuit shown in Figure 10.33(a), where a standard dual-diode input protection network is used to protect the input buffer from being damaged by *electrostatic discharge* (ESD) [7]. Because of the use of clamped diodes $D1$ and $D2$, the input voltage will be clamped in between -0.7 V and $+0.7 + V_{DD}$, where 0.7 V is the turn-on voltage of diodes. Nevertheless, a problem will arise when this circuit operates at the power-supply voltage V_{DD} of 3.3 V and the input is connected to the output of a CMOS device operating at a 5-V power-supply voltage. From Table 10.5, the V_{OH} is greater than 4.44 V, which makes the diode $D1$ turn on and hence a low-resistance path is formed between the input and the power-supply rail V_{DD}, thereby leading to unnecessary power dissipation

Section 10.4 Practical Issues

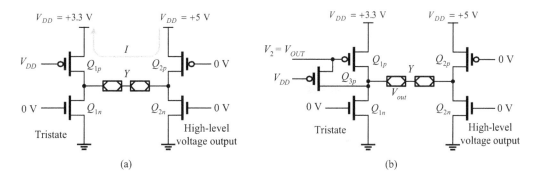

Figure 10.34: The typical output stage of CMOS logic circuits: (a) non-5V-tolerant output and (b) 5-V-tolerant output.

The above problem can be simply solved by a cascade of several diodes instead of just one diode $D1$, but still to keep $D2$ for clamping the negative voltage. The resulting circuit is shown in Figure 10.33(b). The needed number of diodes to be cascaded is determined by the voltage difference between the input and the power-supply voltage V_{DD}. As a result, the diode chain remains cutoff under the application of V_{OH} (4.44 V) and hence the input of the device is said to be 5-V tolerant.

10.4.2.2 Output Voltage Tolerance When the outputs of two logic circuits with two different power-supply voltages of 3.3 V and 5.0 V, respectively, are connected together, there may exist the 5-V output voltage-tolerant problem. To see this, consider the circuit shown in Figure 10.34(a), where two output stages with different power-supply voltages are connected together. Under normal condition without considering the output voltage at Y, both nMOS transistors Q_{1n} and Q_{2n}, and pMOS transistor Q_{1p}, are off. However, the voltage at node Y is 5 V because the gate of pMOS transistor Q_{2p} is at the ground level and hence it is turned on. Hence, the pMOS transistor Q_{1p} is also turned on by the high voltage appearing at the output Y, which is 5 V and greater than the 3.3 V appearing at the gate of Q_{1p} by at least an amount that can turn on the transistor, i.e., the threshold voltage. As a result, there exists a low-resistance path between the two power-supply rails and a small direct current is passing through both pMOS transistors. This current will continually consume power and generate heat on the I/O circuits and even degrade the performance of both devices.

Figure 10.34(b) shows a solution [12] of the problem encountered in Figure 10.34(a) by adding a pMOS transistor Q_{3p} to the output buffer of the device with the lower power-supply voltage. This transistor is connected across the gate and drain of the pMOS transistor Q_{1p} and its gate is connected to V_{DD} (3.3 V) so that it can be turned on and make the gate and drain of the pMOS transistor Q_{1p} at the same potential when the output Y is at the high voltage of 5 V. Therefore, the pMOS transistor Q_{1p} remains off. In other words, the output of the 3.3-V device is 5-V tolerant.

■ Review Questions

10-58 Explain the meaning of voltage tolerance of inputs and outputs.
10-59 Explain the meaning of voltage compliance of inputs and outputs.
10-60 Describe the features of the LVDS scheme.
10-61 Can an LVTTL device be driven by a standard TTL device?

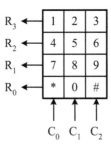

Figure 10.35: The layout of a 3 × 4 keypad.

10.5 Case Study — A Keypad Scanner and Decoder

Keypads are widely used in digital systems to enter desired data into the systems. A keypad is often composed of an array of keys, with each being a push-button or other types of switches. Like a single push button, any key in a keypad is also needed to debounce as it is pressed or pushed. However, it is definitely not necessary to associate with each key a debounce circuit because this will make the resulting *keypad scanner* not only cost too much hardware but also become too complicated. Here, the keypad scanner means a logic circuit used to detect which key in the keypad is being pressed and to output its corresponding code correctly. In this section, we study how to design a cost-effective keypad scanner and its associated decoder. In addition, we demonstrate how to implement the resulting design with an FPGA device, in particular, including the pin assignment problem of FPGA devices that we must face in practical applications.

10.5.1 A Keypad Scanner

Assume that we would like to design a scanner for a keypad with three columns and four rows (namely, a 3 × 4 keypad), as shown in Figure 10.35. The keypad is wired in matrix form with a push button served as the key at each intersection of each column and each row. Pressing a key establishes the connection between a column and a row. For simplicity, suppose that only one key is detected and decoded each time. The design problem of a keypad scanner is to determine which key has been stroked and to output a binary number, corresponding to the key number labeled on the figure. For example, pressing key 9 must output 1001, pressing key * must output 1010, and pressing key # must output 1011.

The keypad circuit and its scanning circuit are shown in Figure 10.36. To provide the correct operations of the keypad, each of R_i and C_i, where $0 \leq i \leq 3$, has a 10-kΩ pull-up resistor connected to V_{DD}. To detect whether a key is being pressed, the keypad is scanned one by one in a given sequence known as the *scan code*. The scan code is then converted into a *key code* by a logic circuit, referred to as the *keypad decoder*. The scan code is determined by the scanning circuit while the key code is determined by how people recognize the key. For example, the key labeled 0 has the key code 0 (4'b0000) but has the scan code 4'b0100.

The basic principle of the keypad scanner depicted in Figure 10.36 is as follows. We begin to output a 4'b0111 on the $C_0C_1C_2C_3$ (C_3 is unused here) to scan the first column. Then, the inputs from R_0 to R_3 are detected by the 4-to-1 multiplexer one by one being scanned by a modulo-4 counter. Once a key has been detected — some R_i is zero — the output of the multiplexer is 0, which in turn stops the clock signal and freezes both modulo-4 counters and hence the scan code is frozen to be ready for being latched. This event also triggers the switch-debounce circuit, which produces a valid-key pulse, Kv, after a debouncing time interval. This valid-key signal latches the scan code into a 4-bit register and restarts both counters to

Section 10.5 Case Study — A Keypad Scanner and Decoder

continue the scanning operation of the keypad at the same time. Through a keypad decoder, the scan code is converted into its key code.

Some problems associated with the above keypad scanner are as follows: First, the first key being pressed must be the key * (its scan code is 4'b0000) after the circuit is reset. Second, the keypad might be locked due to the lack of a scanning clock signal. This might happen when a key is only pressed in a very short duration in which a glitch (a short pulse) is generated. This glitch may stop the scanning clock signal by way of clearing the D flip-flop but cannot generate the valid-key signal because it is too short to be recognized as a valid key. As a consequence, the scanning clock signal is stopped but could not be resumed any more. To compensate for the effects caused by these two potential problems, an error-recovery counter is used to generate a pulse every four clock cycles whenever there is no valid-key signal. This pulse is then used to clock the D flip-flop and hence to restart the scanning clock signal, as illustrated in Figure 10.36.

■ Example 10.13: A 3×4 keypad scanner.

The following module describes the 3×4 keypad scanner depicted in Figure 10.36. The first two **always** blocks comprise the modulo-4 counter and the 2-to-4 decoder for providing the column scanning code. The next two **always** blocks comprise the modulo-4 counter and the 4-to-1 multiplexer for generating the row scanning code. The fifth **always** block models the D flip-flop. The sixth **always** block describes the error recovery counter. The last two **always**

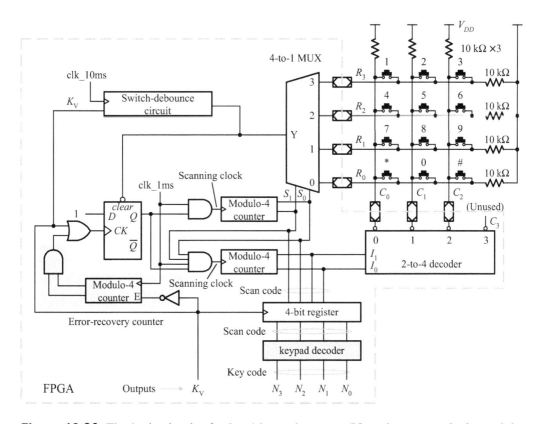

Figure 10.36: The logic circuit of a 3 × 4 keypad scanner (Note that except the keypad decoder, all other components inside the lightly dashed line comprise the keypad scanner module.)

blocks separately describe the 4-bit register and the keypad decoder, which are combined to latch and map the scan code to its equivalent key code.

```verilog
// a 3-by-4 keypad scanner --- detects a key being pressed
// and returns its corresponding code
module keypad_3by4(
       input  clk_1ms, clk_10ms, reset,
       input  [3:0] row,
       output reg [2:0] column,
       output reg [3:0] N,    // the key code
       output Kv);   // code is valid
reg  [1:0] col_qout, row_qout; // a modulo-3/4 counter
reg  qout, Y;
reg  [3:0] scan_code;// the scan code
reg  [1:0] rec_qout; // a modulo-4 error recovery counter

wire    rec_reset, qout_out;
// detects any valid key and debounces it
switch_debouncer switch_kv (clk_10ms, reset, Y, Kv);
// a modulo-4 counter used to generate the column outputs
always @(posedge clk_1ms or posedge reset)
   if (reset) col_qout <= 0;
   else if (qout && (row_qout == 3)) col_qout <= col_qout + 1;
// a 2-to-3 decoder --- outputs the column scanning code
always @(*)
   case (col_qout)
      2'b00: column = 3'b011;
      2'b01: column = 3'b101;
      2'b10: column = 3'b110;
      default: column = 3'b111;
   endcase
// a modulo-4 counter used to scan the row inputs
always @(posedge clk_1ms or posedge reset)
   if (reset) row_qout <= 0;
   else if (qout) row_qout <= row_qout + 1;
// a 4-to-1 multiplexer --- receives a row
always @(*)
   case (row_qout)
      2'b00: Y = row[0];
      2'b01: Y = row[1];
      2'b10: Y = row[2];
      2'b11: Y = row[3];
   endcase
// a flip-flop used to enable the counters col_qout and
// row_qout; qout is set to 1 whenever there is a Kv
// pulse or rec_qout = 3
assign qout_out = Kv || (rec_qout == 3);
always @(negedge qout_out or negedge Y)
   if (!Y) qout <= 1'b0;
   else qout <= 1'b1;
// an error recovery counter --- being used to set qout
// whenever there exists a Y pulse but no Kv pulse
```

Section 10.5 Case Study — A Keypad Scanner and Decoder

```verilog
// because the Y pulse is too short (less than 10 ms)
assign rec_reset = reset || Kv;
always @(posedge clk_10ms or posedge rec_reset)
   if (rec_reset) rec_qout <= 0;
   else if (!Kv) rec_qout <= rec_qout + 1;
// latch scan_code at the rising edge of Kv
always @(posedge Kv or posedge reset)
   if (reset) scan_code <= 4'b1111;
   else scan_code <= {col_qout, row_qout};
// decode the scan code into the key code
always @(*)
   case (scan_code)
      4'b0000: N = 4'b1010;   // *
      4'b0001: N = 4'b0111;   // 7
      4'b0010: N = 4'b0100;   // 4
      4'b0011: N = 4'b0001;   // 1
      4'b0100: N = 4'b0000;   // 0
      4'b0101: N = 4'b1000;   // 8
      4'b0110: N = 4'b0101;   // 5
      4'b0111: N = 4'b0010;   // 2
      4'b1000: N = 4'b1011;   // #
      4'b1001: N = 4'b1001;   // 9
      4'b1010: N = 4'b0110;   // 6
      4'b1011: N = 4'b0011;   // 3
      default: N = 4'b1111;   // otherwise
   endcase
endmodule
```

The salient features of the keypad scanner shown in Figure 10.36 are as follows: First, it is designed in a very straightforward way without needing a finite-state machine, which will be addressed in Chapter 11. Second, it applies the switch-debounce principle described in Section 9.1.4 intuitively. Third, it is easily adapted into the other matrix-type keypads of any size simply by adjusting the size of related components without changing its structure virtually. As a matter of fact, it is able to be modeled as a parameterizable module. However, we would like to leave this as an exercise for the reader.

10.5.2 A Keypad Scanner and LED Display System

In this section, we combine the keypad scanner with the 4-digit LED display module described in Section 8.6.2 into one simple system. As illustrated in Figure 10.37, to connect the keypad scanner to the 4-digit LED display module, we need a 4-digit register module, with each digit having a 4-bit shift register (SR), to store up to four key strokes because the 4-digit LED display module cannot retain its inputs.

Conceptually, as a new key stroke arrives, the 4-digit register is shifted left one-digit position to vacate the least significant digit (LSD) for the new coming stroke data (i.e., the key code). Physically, this operation can be implemented by at least either of the following two ways: *a shift register* and *a parallel-load register*. Regardless of which approach is used, we will call this resulting module a 4-digit shift and load module.

10.5.2.1 Shift-Register Approach Since the new key code enters the 4-digit register from the rightmost (namely, the LSD) end, the 4-digit register must be a left-shift type and the rightmost digit of the 4-digit register must also have the parallel-load capability. As illustrated

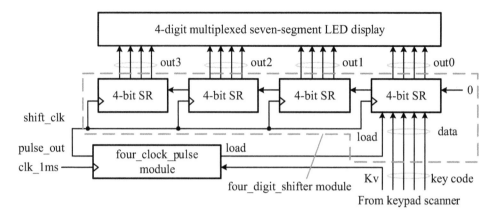

Figure 10.37: The logic circuit showing the 4-digit shift and load module.

in Figure 10.37, each time a new key code is incoming, the contents of the 4-digit register should be shifted left four bits (one digit) to vacate the LSD for the new coming key code. To provide this capability, a four_clock_pulse module is used to generate a pulse chain of four clock cycles, triggered by the valid-key signal, Kv, coming along with the new key code. This pulse chain is then applied to the 4-digit register to shift left its content in a way of 1-bit position per pulse. At the last clock cycle, the new key code from the keypad scanner is also loaded into the LSD register.

The following two examples describe the four-clock pulse generator and the 4-digit register, respectively. Because of the aforementioned reason, the 4-digit register is called a *4-digit shift register.*

■ Example 10.14: A four-clock pulse generator.

In fact, this is a simple non-retriggerable monostable circuit. As stated, the four-clock pulse generator mainly consists of a JK flip-flop and a modulo-4 binary counter. However, to avoid repeatedly enabling the modulo-4 counter and hence outputting pulses, a single-cycle pulse generator is first employed to reshape the valid-key signal into one single-cycle pulse before it is applied to enable the JK flip-flop. The four-clock pulse generator works as follows. When the reshaped valid-key signal is applied to the J input of the JK flip-flop, the output Q of the JK flip-flop rises high. This in turn enables the modulo-4 counter and the generator starts to output the clock pulses. As the counter reaches 3, a logic 1 is applied to the K input of the JK flip-flop, which in turn clears the output Q of the JK flip-flop in the next cycle. As a result, the counter is disabled and the output pulses are terminated. To enable the 4-digit shift register to load the new key code, a load signal is also generated when the modulo-4 counter reaches 3.

```
// a four-clock pulse generator --- generates a pulse
// with four-clock cycles each time the trigger signal
// Kv is asserted
module four_clock_pulse(
        input   clk_1ms, reset, Kv,
        output  load, pulse_out );
reg     [1:0] qout4;    // a modulo-4 counter
reg     qout, Kv_qout;  // a JK flip-flop
wire    Kv_out;
```

Section 10.5 *Case Study — A Keypad Scanner and Decoder*

```
// the modulo-4 counter
always @(posedge clk_1ms or posedge reset)
   if (reset) qout4 <= 0;
   else if (qout) qout4 <= qout4 + 1;
assign load = (qout4 == 3);
// reshape the triggered signal Kv into one cycle
always @(posedge clk_1ms or posedge reset)
   if (reset) Kv_qout <= 1'b0;
   else       Kv_qout <= Kv;
assign Kv_out = Kv & (~Kv_qout );
// using a JK flip-flop to generate a four-clock pulse
always @(posedge clk_1ms or posedge reset)
   if (reset) qout <= 1'b0;
   else qout <= (Kv_out)&(~qout) | (~load)&(qout);
// generate the pulse_out signal
assign pulse_out = clk_1ms & qout;
endmodule
```

■ Example 10.15: A 4-digit shift and load module.

This module describes the logic circuit shown in Figure 10.37. It consists of two modules: four_clock_pulse and four_digit_shifter. The four_clock_pulse module has been described in the previous example and hence we omit it here. The four_digit_shifter module is simply a shift register with parallel load. Because of its intuitive simplicity, we will not further discuss it here.

```
// a 4-digit shift and load LSD module
module four_digit_shift_load (
      input   clk, reset, Kv,
      input   [3:0] data,
      output  [3:0] out3, out2, out1,out0);

wire  load;
// instantiate a 4-clock cycle module
four_clock_pulse shift_clock
     (clk, reset, Kv, load, pulse_out);
four_digit_shifter shift_reg
     (pulse_out,reset,load,out3,out2,out1,out0,data);
endmodule

// a 4-digit shift register
module four_digit_shifter (
      input   shift_clk, reset, load,
      input   [3:0] data,
      output reg   [3:0] out3, out2, out1,out0);
// model a 16-bit shift register
always @(posedge shift_clk or posedge reset)
   if (reset) {out3, out2, out1, out0} <= 16'b0;
   else if (load) out0 <= data;
   else {out3,out2,out1,out0}<={out3,out2,out1 out0}<<1;
endmodule
```

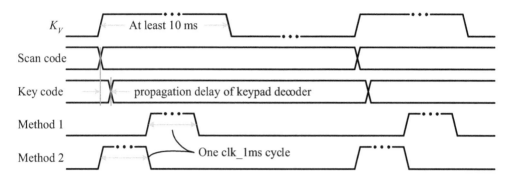

Figure 10.38: The timing diagram showing the relationships of various signals: K_v, the scan code, the key code, method1 and method2 (Not drawn in scale).

10.5.2.2 Parallel-Load Register Approach Another implementation of the 4-digit register is based on the following concern. Since the shift-by-4 operation is only to reposition the data lines a fixed number of positions, it can be simply modeled by using the following statement:

{out3, out2, out1, out0} <= {out2, out1, out0, data};

However, from Figure 10.37, we can see the valid-key signal K_v cannot be simply employed to sample the new coming data because it is also utilized to sample the scan code, which is in turn decoded as the key code. Hence, if this is done, the 4-digit LED display will present the previous key stroke rather than the current one. To solve this problem, we should sample the scan code a small amount of time later after the K_v has been asserted to allow the keypad decoder to complete its operation. In addition, because the pulse width of K_v is much wider than one cycle of the clock signal, clk_1ms, a reshaping circuit is needed to reshape the K_v signal into a single clock cycle in order to prevent the new coming data from being sampled several times. Figure 10.38 shows the timing diagram of the relationships of various signals, K_v, the scan code, the key code, and the sample signals.

As a result, the above 4-digit shift and load module can be modeled in a simpler way, as in the following example, without using the four-clock pulse generator and hence can be completed on a single clock cycle.

■ Example 10.16: Another 4-digit shift and load module.

In this example, the method 1 is employed to load the new coming data. As stated, the valid-key signal, K_v, is first reshaped into one single-cycle signal of a period of clk_1ms with a delay of one clk_1ms clock cycle by using the first **always** block. The actual shift and load operation is done by the second **always** block.

```
// file --- four_digit_shift_loadc.v
// a 4-digit shift and load LSD module ---
// without using the four-clock pulse generator
module four_digit_shift_load (
       input   clk_1ms, reset, Kv,
       input   [3:0] data,
       output reg [3:0] out3, out2, out1,out0);
reg    [1:0] Kv_qout;
// reshape the triggered signal Kv into a single-cycle pulse
always @(posedge clk_1ms or posedge reset)
   if (reset) Kv_qout <= 2'b00;
   else begin Kv_qout[0] <= Kv;
```

Section 10.5 Case Study — A Keypad Scanner and Decoder

```
                    Kv_qout[1] <= Kv & (~Kv_qout[0]); end
// perform the shift and load operations
always @(posedge Kv_qout[1] or posedge reset)
   if   (reset) {out3, out2, out1, out0} <= 16'b0;
   else {out3,out2,out1,out0} <= {out2,out1,out0,data};
endmodule
```

The above code is supposed that only the positive edges of the clock signal are allowed throughout the entire system for the purpose of easily testing. If this constraint may be relaxed, the negative edge of the single-cycle signal may be used to load the new coming data into the 4-digit register without the need of shifting it one more `clk_1ms` clock cycle; namely, the method 2 may be used, as shown in Figure 10.38. Because of its intuitive simplicity, we would like to leave it as an exercise for the reader. Nonetheless, it is worth noting that we could not use the positive edge of the pulse in method 2 to load the new coming data. The reader is encouraged to explain the reason for it.

10.5.2.3 Comparison of Both Approaches The parallel-load register approach needs less hardware than the shift-register approach because it directly loads into its least three significant digits concatenated with the new key-stroked data. In contrast, the shift-register module uses a four-clock pulse generator for generating four clock cycles to shift its content in a way of one bit per clock cycle to vacate its LSD for the new coming key data. As for the speed, although the shift-register approach needs four clock cycles while the parallel-load register approach only needs one clock cycle, both approaches have the same performance because the valid-key signal, K_v, has a duration much wider than four clock cycles. A final remark about these two approaches is that the shift-register approach is a multicycle implementation while the parallel-load register approach is a single-cycle implementation of the conceptual operations of the 4-digit register. More about multicycle and single-cycle implementations of a design will be dealt with in the next chapter.

10.5.3 FPGA Implementation

In this subsection, we put together all modules described in this case study. As illustrated in Figure 10.37, three modules are needed to be instantiated, including the keypad scanner, the 4-digit shift and load module, and the 4-digit multiplexed seven-segment LED display module. In addition, a timing generator is also needed to generate the required clock signals, `clk_1ms` and `clk_10ms`, from a 40-MHz clock source.

■ Example 10.17: The top-level module of the underlying system.

The top-level module of the resulting keypad and display system is as follows. It instantiates a `timing_base_generator` module to generate all clock signals required in the system, a `keypad_3by4` module to scan the keypad and return key codes, a `four_digit_shift_load` register to hold the last four key strokes, and a `four_digit_LED_display` module to display the key strokes being held in the shift register.

```
// the top-level module of the keypad scanner plus LED display
// module using a common-anode multiplexed seven-segment
// LED display module
module keypad_LED_display_top (
        input   clk_40MHz, reset,
        input   [3:0] row,
        output  [2:0] column,
        output  [3:0] addr,
```

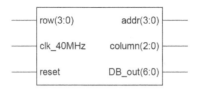

Figure 10.39: The logic symbol of the keypad scanner and LED display system after it is synthesized.

```
        output [6:0] DB_out );

wire    [3:0] out3, out2, out1, out0, N;
wire    clk_1ms, clk_10ms, Kv;
wire    reset_top;
assign  reset_top = ~reset; // reset is an active-low signal
// a timing generator generates clk_1ms and clk_10ms from
// a 40-MHz source
timing_base_generator basic_timing
        (clk_40MHz, reset_top, clk_1ms, clk_10ms);
// the 3-by-4 keypad module
keypad_3by4 keypad_12keys
        (clk_1ms, clk_10ms, reset_top, row, column, N, Kv);
// the 4-digit shift register with the LSD entry
four_digit_shift_load shift_load
        (clk_1ms, reset_top, Kv, N, out3, out2, out1, out0);
// the 4-digit multiplexed seven-segment LED display
four_digit_LED_display LED_display
        (clk_1ms, reset_top, out3, out2, out1, out0, addr,
         DB_out );
endmodule
```

The synthesized symbol of the top-level module is shown in Figure 10.39. From this figure, we can see all ports of the top-level module. In order to perform the real-world test, we need to implement the top-level module using an available FPGA device. Here, we assume that a Xilinx's FPGA device is used. To accomplish this, a user-constrained file (*.ucf) is also needed to map each port of the top-level module to an input/output pin. This file is given in the following example.

■ **Example 10.18: The user-constrained file.**

What follows is the user-constrained file (UCF) used in the project. In this file, only the pin assignments are given. Different FPGA devices have their own pin assignments. Refer to the related data sheet of the device you want to use for details.

```
## file—— keypad_LED_display_top.ucf
NET "clk_40MHz" LOC = P9;
NET "reset" LOC = K5;

## pins for 3-by-4 keypad
NET "row<0>" LOC = L10; # row 0 —— upmost
NET "row<1>" LOC = L9;
```

Section 10.5 *Case Study — A Keypad Scanner and Decoder* 423

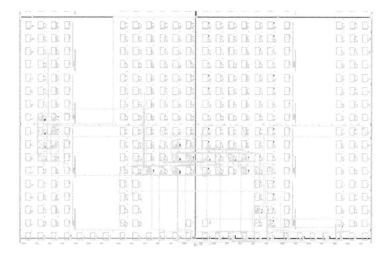

Figure 10.40: The chip layout of the keypad scanner and LED display system on a Xilinx's FPGA device (only showing the related half part).

```
NET "row<2>" LOC = L8;
NET "row<3>" LOC = L7;  # row 3 —— bottom
NET "column<2>" LOC = N12; # column outputs
NET "column<1>" LOC = P13;
NET "column<0>" LOC = T13;

## for the 4-digit LED display module
NET "DB_out<0>" LOC = T11;
NET "DB_out<1>" LOC = T10;
NET "DB_out<2>" LOC = P10;
NET "DB_out<3>" LOC = N10;
NET "DB_out<4>" LOC = M10;
NET "DB_out<5>" LOC = R9;
NET "DB_out<6>" LOC = T9;
NET "addr<0>" LOC = N11; # AN4
NET "addr<1>" LOC = P11; # AN3
NET "addr<2>" LOC = P12; # AN2
NET "addr<3>" LOC = R13; # AN1
```

After the completion of the place-and-route process, we can see the resulting chip layout of the keypad scanner and LED display system on an FPGA device, as portrayed in Figure 10.40. To reduce the space of presentation, we only show here the related half part. From this figure, it can be seen how the design is mapped to the CLBs of an FPGA device and the related routing information. The more details of the chip layout can be viewed by using the zoom-in operation provided by the CAD tool.

∎ Review Questions

10-62 What are the differences between the scan code and the key code.
10-63 What is the function of keypad scanners?
10-64 Why is an error recovery counter needed in the keypad scanner?

10-65 Refer to Figure 10.38, explain why we could not use the positive edge of the pulse in method 2 to load the new data.

10.6 Summary

In this chapter, we have examined the hierarchical structure and various implementation options of digital systems. The hierarchical structure, also known as the modular design technique, is often used to design digital systems. The essence of the hierarchical structure is to partition a system into many smaller independent subsystems which combine to perform the same functionality of the original system.

The implementation options of digital systems include field-programmable devices and ASICs in addition to platforms. Field-programmable devices are ones that can be programmed in laboratories and include PLDs, CPLDs, and FPGAs. PLDs are devices built on the two-level AND-OR logic structure. PLDs contain three types: ROM, PLA, and PAL. PLDs are used in low-end applications or as glue logic in large systems.

Both types of CPLD and FPGA devices are used in more complex digital systems and dominate the current market of digital systems. The basic functional structure of CPLDs is the same as that of PLDs; namely, it is still a two-level AND-OR logic structure. A CPLD virtually embeds many PLDs onto the same chip and interconnects them through a programmable interconnect network. As a consequence, a complex digital system may be created by using a single or a few CPLDs. FPGAs have a quite different structure from CPLDs. They use programmable logic blocks (PLBs), such as configurable logic blocks (CLBs) or logic elements (LEs), to implement switching functions and can be roughly divided into two structures: the matrix type and the row type, according to the arrangements of PLBs on the chip.

ASICs mean those devices that are needed to manufacture in IC foundries. An ASIC may be created with the full-custom, cell-based, or gate-array-based approach. The full-custom approach starts from scratch, namely, from zero to the final IC. The cell-based approach incorporates a standard cell library into a synthesizable design flow so that it can reduce both the design time and the time to market considerably. However, the standard cell library is still needed to design with the full-custom approach and the final prototypes using it are also needed to fabricate in IC foundries. Gate arrays are arrays of prefabricated transistors with the metallization masks left undefined. Design with gate arrays is still a synthesizable design flow by incorporating into cells from a macro library supplied by the vendor of gate arrays. Consequently, as compared with the full-custom approach, it not only can shorten the design time like the cell-based approach but can also reduce the prototyping time of the design. In addition, the time to market of a product implemented with the gate-array-based approach is shorter than that with the cell-based and full-custom approaches because gate arrays are on prefabricated wafers and hence save a lot of time-consuming steps.

Because of the evolution of VLSI technology, it is not uncommon that a wide variety of devices, including ASICs and field-programmable devices, coexist in the same system. These devices have different logic levels and power-supply voltages because they are manufactured with different process technologies. Consequently, the interface between them becomes an important issue. The two most important issues related to the interface between two devices are I/O standards and voltage tolerance. An I/O standard defines a set of electrical rules that governs information transfer between two devices through electrical signals. A device is said to be voltage-tolerant if it can withstand a voltage greater than its V_{DD} on its I/O pins.

A case study concluded this chapter. This case study explores the design of a keypad scanner and its implementation with an FPGA device. Keypads are widely used in digital systems to enter data required in digital systems. Therefore, the design of keypad scanners is fundamentally important in digital systems. In this case study, we showed how to design a cost-effective keypad scanner and its associated decoder to convert the scan code into the key code. In addition, we demonstrated how to implement the resulting design with an FPGA device, in particular, including the pin assignment problem of the FPGA device that we have to face in practical applications.

References

1. Altera Corp., *APEX 20K Programmable Logic Device Family*, ver. 5.1, 2004. (http://www.altera.com)
2. Altera Corp., *MAX 7000 Programmable Logic Device Family*, ver. 6.7, 2005.
3. Altera Corp., *Cyclone III Device Handbook*, Vol. 1, 2008.
4. Altera Corp., *Stratix IV Device Handbook*, Vol. 1, 2008.
5. T. Granberg, *Handbook of Digital Techniques for High-Speed Design*, Upper Saddle River, New Jersey: Prentice-Hall, 2004.
6. M. B. Lin, *Digital System Design: Principles, Practices, and Applications*, 4th ed., Taipei, Taiwan: Chuan Hwa Book Ltd, 2010.
7. M. B. Lin, *Introduction to VLSI Systems: A Logic, Circuit, and System Perspective*, CRC Press, 2012.
8. M. B. Lin and S. T. Lin, *8051 Microcomputer Principles and Applications*, Taipei, Taiwan: Chuan Hwa Book Ltd., 2012.
9. M. B. Lin and S. T. Lin, *Basic Principles and Applications of Microprocessors: MCS-51 Embedded Microcomputer System, Software, and Hardware*, 3rd ed., Taipei, Taiwan: Chuan Hwa Book Ltd., 2013.
10. M. B. Lin, *Microprocessor Principles and Applications: x86/x64 Family Software, Hardware, Interfacing, and Systems*, 6th ed., Taipei, Taiwan: Chuan Hwa Book Ltd., 2018.
11. M. B. Lin, *An Introduction to Cortex-M4-Based Embedded Systems — TM4C123 Microcontroller Principles and Applications*, CreateSpace Independent Publishing Platform, 2019.
12. J. F. Wakerly, *Digital Design Principles & Practices*, 3rd ed., Upper Saddle River, New Jersey: Prentice-Hall, 2001.
13. Xilinx Inc., *XC4000E and XC4000X Series Field Programmable Gate Arrays*, v1.6, 1999. (http://www.xilinx.com)
14. Xilinx Inc., *FastFLASH XC9500XV: High-Performance, Low-Cost CPLD Family*, v1.1, February 2000.
15. Xilinx Inc., *Spartan-3 FPGA Family: Complete Data Sheet*, 2005.
16. Xilinx Inc., *Virtex-5 FPGA User Guide*, V4.4, 2008.

Problems

10-1 Considering the switching functions shown in Figure 10.15, answer each of the following questions.
   (a) Use a PLA device to implement the switching functions.
   (b) Write a Verilog HDL module to describe the PLA device you used.

10-2 Considering the switching functions shown in Figure 10.15, answer each of the following questions.

426 Chapter 10 Implementation Options of Digital Systems

  (a) Use a PAL device to implement the switching functions.
  (b) Write a Verilog HDL module to describe the PAL device you used.

10-3 Considering the switching functions shown in Figure 10.17, answer each of the following questions.
  (a) Use a ROM device to implement the switching functions.
  (b) Write a Verilog HDL module to describe the ROM device you used.

10-4 Considering the switching functions shown in Figure 10.17, answer each of the following questions.
  (a) Use a PAL device to implement the switching functions.
  (b) Write a Verilog HDL module to describe the PAL device you used.

10-5 Considering the switching functions shown in Figure 10.19, answer each of the following questions.
  (a) Use a ROM device to implement the switching functions.
  (b) Write a Verilog HDL module to describe the ROM device you used.

10-6 Considering the switching functions shown in Figure 10.19, answer each of the following questions.
  (a) Use a PLA device to implement the switching functions.
  (b) Write a Verilog HDL module to describe the PLA device you used.

10-7 Considering the switching functions shown in Figure 10.20, answer each of the following questions.
  (a) Use a ROM device to implement the switching functions.
  (b) Write a Verilog HDL module to describe the ROM device you used.

10-8 Considering the switching functions shown in Figure 10.20, answer each of the following questions.
  (a) Use a PAL device to implement the switching functions.
  (b) Write a Verilog HDL module to describe the PAL device you used.

10-9 Using the simplified PLB described in Figure 10.25, draw the block diagram when realizing the 4-bit shift register depicted in Figure 9.16 (Section 9.2.1).

10-10 Using the simplified PLB described in Figure 10.25, draw the block diagram when realizing the hierarchical 4-bit adder depicted in Figure 1.6 (Section 1.3.2).

10-11 A 4×4 **keypad module.** Suppose that we have a keypad with four columns and four rows (namely, a 4×4 keypad), as shown in Figure 10.41. The keypad structure and its operations are the same as those of the 3×4 keypad described in Section 10.5.1. Design a keypad scanner to determine which key is being pressed and output a key code, corresponding to the key number labeled on the figure.
  (a) Draw the block diagram of the keypad scanner, including the key matrix, and label each component properly to indicate its functionality and size.
  (b) Write a Verilog HDL module to describe the keypad scanner in behavioral style.
  (c) Write a test bench to verify the keypad scanner.

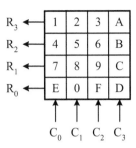

Figure 10.41: The layout of a 4×4 keypad.

10-12 A keypad plus display system. Combining the 4×4 keypad with the 4-digit multiplexed seven-segment LED display module described in Section 8.6.2, design a keypad and display system. The digit is entered from the rightmost digit and the display is shifted one digit left each time a new digit is entered.

  (a) Draw the block diagram of the keypad and display system.
  (b) Write a Verilog HDL module to describe the keypad and display system in behavioral style.
  (c) Write a test bench to verify the keypad and display system.

10-13 Considering the block diagram shown in Figure 10.37, draw a timing diagram to show the timing relationships among the key code, K_v, the pulse_out, and the load signals.

10-14 Consider the `four_digit_shift_load` module that uses the parallel-load register approach. Assuming that the negative rather than the positive edge of the single-cycle pulse signal is used to load the new coming data into the 4-digit register, rewrite the module to accommodate this situation.

11

System Design Methodologies

In the previous chapters, we have dealt with various combinational and sequential logic modules widely used in digital systems. In this chapter, we introduce two useful techniques by which a system can be designed. These techniques include the finite-state machine (FSM) and register-transfer-level (RTL) approaches. The former may be described by a state diagram or an algorithmic state machine (ASM) chart; the latter may be described by an ASM chart or a finite-state machine with datapath (FSMD).

The ASM chart is also known as a state-machine (SM) chart and only composed of three types of building blocks: the state box, decision box, and conditional output box. The important features of ASM charts are that they precisely define the operations on every cycle and clearly show the control flow from state to state. In this chapter, we are concerned with how to model both state diagrams and ASM charts in Verilog HDL.

An RTL approach may be achieved by either coding an ASM chart directly or using the datapath-and-controller (DP+CU) paradigm based on the FSMD model. For simple systems, the datapath and the controller of a design can be derived from its ASM chart directly. The datapath portion corresponds to the registers and function units in the ASM chart, and the controller portion corresponds to the generation of control signals needed in the datapath. To reveal this technique, we outline a three-step paradigm by which the datapath and the controller of a design could be derived from its ASM chart or a variant of the ASM chart, called an algorithmic state machine with datapath (ASMD) chart. For complex systems, the datapath and the controller of systems are often derived from specifications in a state-of-the-art fashion. An example of displaying a 4-digit data on a commercial dot-matrix liquid-crystal display (LCD) module is used to illustrate this approach.

11.1 Finite-State Machines

A *finite-state machine* (FSM) is usually used to model a circuit with memory such as a sequential circuit or the controller in a small digital system. A sequential circuit is one that its outputs are determined not only by its present inputs but also by its past history. Hence, it is a circuit with memory. In this section, we focus our attention on finite-state machines and their related issues often encountered in designing digital systems.

11.1.1 Types of Sequential Circuits

A finite-state machine (FSM) $\mathcal{M}$ is a quintuple $\mathcal{M} = (\mathcal{I}, \mathcal{O}, \mathcal{S}, \delta, \lambda)$, where $\mathcal{I}$, $\mathcal{O}$, and $\mathcal{S}$ are the finite, nonempty sets of input, output, and state symbols, respectively. Here, a symbol in each

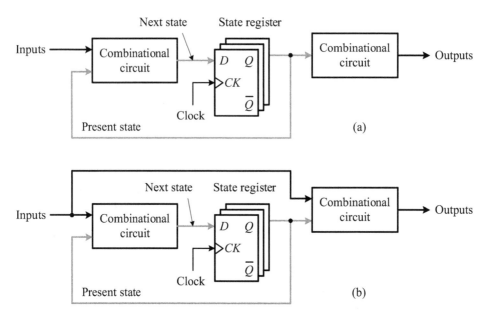

Figure 11.1: Types of sequential circuits: (a) Moore machines and (b) Mealy machines.

set is a combination of variables in that set. δ and λ are the state transition function and the output function, respectively.

The state transition function δ is defined as follows

$$\delta : \mathcal{I} \times \mathcal{S} \to \mathcal{S}$$

which means that the δ function is determined by both input and state symbols.

The output function λ is defined as follows

$$\lambda : \begin{cases} \mathcal{S} \to (\mathcal{O} \cup A) & \text{(Moore machine)} \\ \mathcal{S} \times (\mathcal{I} \cup S) \to (\mathcal{O} \cup A) & \text{(Mealy machine)} \end{cases}$$

where in a Moore machine the output function is solely determined by the present state, whereas in a Mealy machine, the output function is determined by both present inputs and present state.

11.1.1.1 Paradigms of Sequential Circuits The general paradigms of Moore and Mealy machines in modeling sequential circuits are depicted in Figure 11.1. Regardless of using the Moore or Mealy machine to model a sequential circuit, three components are common: a combinational circuit for computing the state transition function, a combinational circuit for computing the output function, and one set of flip-flops (memory) for storing the set of states. This set of flip-flops is called the *state register*.

The design of a sequential circuit based on a finite-state machine is to derive the set of state symbols, $\mathcal{S}$, the state transition function, δ, and the output function, λ, from the known sets of input and output symbols, $\mathcal{I}$ and $\mathcal{O}$. In other words, we need to derive a state diagram or table from the relationship between the inputs and outputs of the sequential circuit and then design the combinational circuits of the state transition and output functions, respectively, from both the current inputs and the present state based on the derived state diagram or table. More specifically, the design procedure can be described as the following algorithm [5].

Table 11.1: Some state encoding examples.

State	Binary	Gray	One-hot
A	00	00	1000
B	01	01	0100
C	10	11	0010
D	11	10	0001

■ Algorithm 11-1: Design Procedure of Sequential Circuits

Input: The sets of input and output symbols, $\mathcal{I}$ and $\mathcal{O}$.
Output: The output and state transition functions, λ and δ.
Begin

1. Derive the state diagram (or state table) of the sequential circuit from the relationships between $\mathcal{I}$ and $\mathcal{O}$.
2. Simplify the number of states in the state diagram (or state table).
3. Assign an appropriate code to each state in the state diagram (or state table).
4. Derive the transition function (assuming that D flip-flops are used) and implement it.
5. Derive the output function and implement it.

End

11.1.1.2 State Encoding In practice, as we derive the state diagram (or state table) for a sequential circuit, it proves much more convenient to use symbolic names for the states. However, to realize this state diagram in a logic circuit, we need to replace the symbolic names of states with an appropriate code from which the output and state transition functions can be derived and implemented. The most common FSM state encoding options include the following ones:

- *One-hot encoding* sets one of the flip-flops of the state register to 1 at each state; namely, it uses the 1-out-of-n encoding scheme. One-hot encoding is used in the situation where performance is critical. One advantage of the 1-out-of-n encoding scheme is that it is easier to debug than the other encoding schemes because at any time only one flip-flop of the state register is active.
- *Binary encoding* assigns a binary code to the state register according to the state sequence. Binary encoding is used for area optimization because it needs fewer flip-flops for the state register.
- *Gray encoding* assigns a Gray code to the state register according to the state sequence. Gray encoding is used for hazard and glitch minimization because only one bit is different between two adjacent states.
- *Random encoding* assigns a random code to the state register for each state.

One-hot encoding is usually used in FPGA-based designs due to a lot of flip-flops available for use in theses devices while the other encoding schemes are often found in cell-based designs to save area. An example of state encoding using binary, Gray, and one-hot encoding schemes is shown in Table 11.1.

■ Review Questions

11-1 Define a finite-state machine.
11-2 What are the differences between a Moore machine and a Mealy machine?
11-3 What is the meaning of the state register?

11-4 What are the features of one-hot encoding?
11-5 In what situation, is binary encoding preferred to use?
11-6 In what situation, is Gray encoding preferred to use?

11.1.2 FSM Modeling Styles

Now that we know the definition and circuit types of FSMs, the next issue naturally arising is how to model (or describe) an FSM in Verilog HDL. Two major modeling issues of an FSM are (1) the declaration and update of the state register, and (2) the computation of both next-state and output functions. The state register can be declared as either one register, such as state, or two registers, such as present_state and next_state. The one-register style is harder to follow and model. In contrast, the two-register style is easier to follow and model. Both output and next-state functions can be computed by one of the following approaches or their combinations: **assign** continuous assignments, functions, and **always** blocks.

11.1.2.1 Modeling Styles As mentioned above, to model an FSM explicitly two styles can be used depending on whether one or two state registers are used. As a consequence, in what follows we classify the modeling styles of an FSM into two types according to whether the state register is declared as one or two state registers.

Style 1: One state register.

In this modeling style, only one state register is declared. The description can be partitioned into three parts.

Part 1: *Initialize and update the state register.* Usually, an **always** block is used for this purpose. The next state is computed by a next_state() function.

```
always @(posedge clk or negedge reset_n)
    if (!reset_n) state <= A; // A is the start state
    else state <= next_state(state, x) ;
```

Part 2: *Compute the next state with a function.* A function is usually used to compute the next-state function from both the current inputs and the present state.

```
function next_state (input present_state, x)
    case (present_state)
        ...
    endcase
endfunction
```

Part 3: *Compute the output function.* The output function can be computed by an **always** block or a number of **assign** continuous assignments. When an **always** block is used, a **case** statement is often used within the **always** block to compute the output function associated with each state.

```
always @(state or x)
    case (state)... endcase

assign continuous assignments
```

Style 2: Two state registers.

In this modeling style, two state registers, present_state and next_state, are declared. The description can also be divided into three parts.

Part 1: *Initialize and update the state register.* Usually, an **always** block is used for this purpose. Inside the **always** block, an **if-else** statement is used to select whether the state assigned to present_state is the start state or the next state.

Section 11.1 Finite-State Machines

```
always @(posedge clk or negedge reset_n)
if (!reset_n) present_state <= A;// A is the start state
else present_state <= next_state;
```

Part 2: *Compute the next state.* It usually uses an **always** block to compute the next-state function from both the current inputs and the present state. Sometimes **assign** continuous assignments are used instead.

```
always @(present_state or x)
   case (present_state)... endcase
```

assign continuous assignments

Part 3: *Compute the output function.* The output function can be computed by an **always** block or a number of **assign** continuous assignments. When an **always** block is used, a **case** statement is often employed inside the **always** block to compute the output function associated with each state.

```
always @(present_state or x)
   case (present_state) ... endcase
```

assign continuous assignments

11.1.2.2 Examples In the rest of this section, we use a number of examples to illustrate how to describe an FSM in Verilog HDL using the above modeling styles.

Problem description. Suppose that a sequential circuit has an input x and an output z. The input x receives a sequence of 0s and 1s. The output z outputs 1 whenever the circuit detects an occurrence of the 0101 pattern in its input stream x. An example of this 0101 sequence detector is shown in Figure 11.2. Figure 11.2(a) shows the block diagram of the 0101 sequence detector. The state diagram of the 0101 sequence detector is shown in Figure 11.2(b), which can be derived by the approach described in most digital logic textbooks, such as Gajski [3] and Lin [5]. Figure 11.2(c) gives an instance of an input stream along with the output of the 0101 sequence detector in terms of a timing diagram.

A number of examples are given next to model the 0101 sequence detector in behavioral style with one and two state registers, respectively. We first consider the case with one register and one **always** block.

■ Example 11.1: The 0101 sequence detector — style 1a.

This example declares a state register, state, and uses an **always** block to initialize and update the state register. The next-state function is computed by a function fsm_next_state, which uses a **case** statement with each state associating with an **if-else** statement to describe the operations associated with the state. The output function is simply computed by an **assign** continuous assignment.

```
// a 0101 sequence detector in behavioral style --- style 1a
// a Mealy machine example --- one state register
// and one always block
module sequence_detector_mealy(
      input  clk, reset_n, x,
      output z);
// declare the state register
reg  [1:0] state;   // for both ps and ns
localparam A = 2'b00, B = 2'b01, C = 2'b10, D = 2'b11;
```

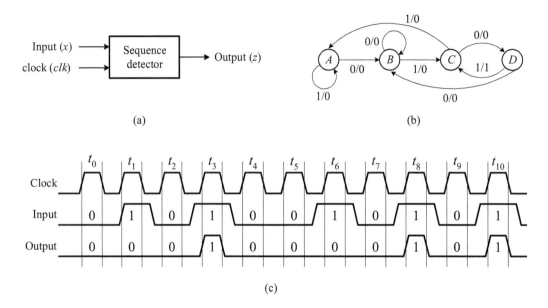

Figure 11.2: A 0101 sequence detector: (a) block diagram, (b) state diagram, and (c) timing diagram.

```
// part 1: update the state register
always @(posedge clk or negedge reset_n) begin
   if (!reset_n) state <= A; // A is the atart state
   else state <= fsm_next_state(state, x);
end
// part 2: compute the next-state function
function [1:0] fsm_next_state (input [1:0] present_state,
                               input x);
reg [1:0] next_state;
begin
   case (present_state)
      A: if (x) next_state = A;  else next_state = B;
      B: if (x) next_state = C;  else next_state = B;
      C: if (x) next_state = A;  else next_state = D;
      D: if (x) next_state = C;  else next_state = B;
   endcase
   fsm_next_state = next_state;
end endfunction
// part 3: evaluate the output function
assign z = (x == 1 && state == D);
endmodule
```

Another illustration of the 0101 sequence detector is given in the following example with style 1. It uses two **always** blocks, one for initializing and updating the state register and the other for computing the output function.

Section 11.1 *Finite-State Machines*

■ **Example 11.2: The 0101 sequence detector — style 1b.**

This example declares a state register, `state`, and uses an **always** block to initialize and update the state register. The next-state function is computed by a function `fsm_next_state` and its associated **case** and **if-else** statements. The output function is simply computed by another **always** block and its associated **case** and **if-else** statements.

```
// a 0101 sequence detector in behavioral style --- style 1b
// a Mealy machine example ---
// using only one state register and two always blocks
module sequence_detector_mealy(
      input clk, reset_n, x,
      output reg z );
// declare a state register
reg [1:0] state;  // for both ps and ns
localparam A = 2'b00, B = 2'b01, C = 2'b10, D = 2'b11;
// part 1: update the state register
always @(posedge clk or negedge reset_n) begin
   if (!reset_n) state <= A;  // A is the atart state
   else state <= fsm_next_state(state, x);
end
// part 2: compute the next-state function
function [1:0] fsm_next_state (input [1:0] present_state,
                               input x);
reg [1:0] next_state;
begin
   case (present_state)
      A: if (x) next_state = A;   else next_state = B;
      B: if (x) next_state = C;   else next_state = B;
      C: if (x) next_state = A;   else next_state = D;
      D: if (x) next_state = C;   else next_state = B;
   endcase
   fsm_next_state = next_state;
end endfunction
// part 3: evaluate output function z
always @(state or x)
   case (state)
      A: if (x) z = 1'b0;  else z = 1'b0;
      B: if (x) z = 1'b0;  else z = 1'b0;
      C: if (x) z = 1'b0;  else z = 1'b0;
      D: if (x) z = 1'b1;  else z = 1'b0;
   endcase
endmodule
```

The third illustration of the 0101 sequence detector is given in the following example with style 2. It uses three **always** blocks, for initializing and updating the state register, computing the next-state function, and computing the output function.

■ **Example 11.3: The 0101 sequence detector — style 2.**

This example declares two state registers, `present_state` and `next_state`, and uses an **always** block to initialize and update the state register, `present_state`. The next-state function is computed by another **always** block, which updates the state register, next_state, by using a **case** statement and its associated **if-else** statements. The output function is simply computed by the third **always** block, with a **case** statement and its associated **if-else** statements.

```
// a 0101 sequence detector in behavioral syle --- style 2
// a Mealy machine example ---
// using two state registers and three always blocks
module sequence_detector_mealy(
       input   clk, reset_n, x,
       output reg z);
// declare state registers: present state and next state
reg [1:0] present_state, next_state;
localparam A = 2'b00, B = 2'b01, C = 2'b10, D = 2'b11;
// part 1: Initialize to state A and update the state register
always @(posedge clk or negedge reset_n)
   if (!reset_n) present_state <= A;// A is the atart state
   else present_state <= next_state;// update the present state
// part 2: compute the next-state function
always @(present_state or x)
   case (present_state)
      A: if (x) next_state = A; else next_state = B;
      B: if (x) next_state = C; else next_state = B;
      C: if (x) next_state = A; else next_state = D;
      D: if (x) next_state = C; else next_state = B;
   endcase
// part 3: evaluate the output function
always @(present_state or x)
   case (present_state)
      A: if (x) z = 1'b0; else z = 1'b0;
      B: if (x) z = 1'b0; else z = 1'b0;
      C: if (x) z = 1'b0; else z = 1'b0;
      D: if (x) z = 1'b1; else z = 1'b0;
   endcase
endmodule
```

From the above three examples, it is worth noting that only part 1 is sequential logic; the other two parts are combinational logic. In addition, both modeling styles for this simple 0101 sequence detector yield the same synthesized result, as shown in Figure 11.3, regardless of whether one state register or two state registers are used. Moreover, the declaration of using two registers, present_state and next_state, does not mean the synthesized result will create two state registers. It only provides a convenient way to write the RTL code. As compared with Figure 11.1, we can realize that style 2 is indeed the approach that is most closely related to the sequential circuit between the two modeling styles.

Guidelines for writing finite-state machines that may help in optimizing the logic of the underlying circuit are listed as follows.

■ Coding Style

1. *State names should be described using parameters,* **parameter** *or* **localparam**.
2. *Combinational logic for computing the next-state function should be separated from the state registers, i.e., using its own* **always** *block or function.*

Section 11.1 *Finite-State Machines*   437

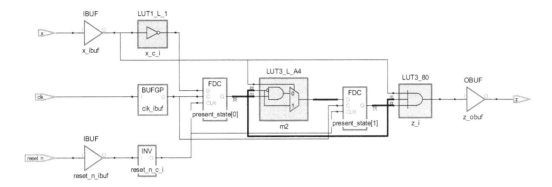

Figure 11.3: The synthesized result of the 0101 sequence detector.

3. *The combinational logic of the next-state function should be implemented with a* **case** *statement for clarity.*

4. *The combinational logic of the output function can be implemented with a* **case** *statement for clarity or implemented by* **assign** *continuous assignments for simplicity.*

11.1.3 Implicit versus Explicit Finite-State Machines

Recall that Verilog HDL is only a hardware description and verification language. It is simply used to model or describe the hardware module that we have designed. As mentioned, coding style always determines synthesis results. For a sequential circuit, there are two types of finite-state machines that could be resulted from the descriptions of Verilog HDL in accordance with different coding styles: *implicit finite-state machines* (implicit FSMs) and *explicit finite-state machines* (explicit FSMs).

11.1.3.1 Implicit Finite-State Machines As the name implies, an implicit FSM means that no FSM is designed by the designer explicitly but the FSM is implied by the activities within the cyclic behavior of procedural statements, such as **always** @(**posedge** clk...), and can be extracted by synthesis tools. In order to make this idea clear, let us consider the following example, where a finite-state machine will be inferred by synthesis tools.

■ Example 11.4: An implicit FSM example.

Supposing that we want to accumulate three data inputs, a natural way is to write a module as shown below. An **always** block containing three procedural assignments with event timing control is used for carrying out the desired operations. Since only one data input is accumulated in total per clock cycle, an implicit FSM is inferred by synthesis tools to schedule the operations. Three clock cycles are required to complete the desired operations. The synthesized result of this module is shown in Figure 11.4.

```
// an implicit FSM example
module sum_3data_implicit
       #(parameter N = 8)(
         input  clk,
         input  [N-1:0] data_a, data_b, data_c,
         output reg [N-1:0] total);

// we do not declare state registers
```

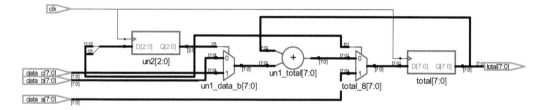

Figure 11.4: The synthesized result of an implicit FSM.

```
always begin
    @(posedge clk) total <= data_a;
    @(posedge clk) total <= total + data_b;
    @(posedge clk) total <= total + data_c;
end
endmodule
```

Although the above module can be correctly synthesized by synthesis tools, it is strongly recommended that an implicit FSM should be used only as the data flow is in a linear fashion such as the one shown in the above example. For more complex situations, the designer should explicitly explore and design the finite-state machine by himself. The major drawbacks of implicit FSMs are as follows: First, they are rather hard to write a source code clearly and correctly in this approach. Second, not every synthesis tool supports this type of FSM. As a result, it is not good practice to use this method when an FSM circuit is needed.

11.1.3.2 Explicit Finite-State Machines In contrast to the implicit FSM, an explicit FSM (or FSM for short) is one that is designed explicitly by the designer in accordance with the desired behavior of the sequential circuit in question. As an illustration, consider the following example which rewrites the above module in explicit FSM style.

■ Example 11.5: An explicit FSM example.

In this example, an FSM with three states is used to schedule the accumulation operations. Because of its inherent simplicity, only one **always** block is used here to describe the FSM along with all desired computations.

```
// an explicit FSM example
module sum_3data_explicit
        #(parameter N = 8)(
            input   clk, reset,
            input   [N-1:0] data_a, data_b, data_c,
            output reg [N-1:0] total);

reg      [1:0]    state;
localparam A = 2'b00, B =2'b01, C = 2'b10;
// the FSM used to schedule the sequential operations
always @(posedge clk or posedge reset)
    if (reset) state <= A; else
    case (state)
        A: begin total <= data_a; state <= B; end
        B: begin total <= total + data_b; state <= C; end
        C: begin total <= total + data_c; state <= A; end
```

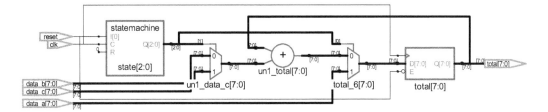

Figure 11.5: The synthesized result of an explicit FSM.

```
        endcase
endmodule
```

The synthesized result of the above module is shown in Figure 11.5. Compared to the previous example, we can see that even though both examples virtually yield the same synthesized result, the example with an explicit FSM has the total control of the desired operations in general. As a consequence, it is good practice to use an *explicit FSM* in practical digital systems.

In contrast to implicit FSMs, explicit FSMs have the following advantages: First, because an explicit FSM is designed and all states are declared clearly, it is rather easy to write a source code in a clear and correct manner. Second, every synthesis tool supports this type of FSM. As a consequence, almost all FSMs are written in this type and an explicit FSM is usually called an FSM for short.

■ Coding Style

1. *It should avoid mixing positive- with negative-edge-triggered flip-flops for the purpose of easily testing.*
2. *It is good practice to use an explicit FSM instead of an implicit FSM for clarity and easy control.*
3. *It should isolate the finite-state machine from the other logic.*
4. *It should avoid multiple clock signals within a block.*
5. *Each register can only be updated in one **always** block.*

■ Review Questions

11-7 What is the one-register style in describing an FSM?
11-8 What is the two-register style in describing an FSM?
11-9 Which statements are suitable for describing a state diagram?
11-10 Explain what an explicit FSM is.
11-11 Explain what an implicit FSM is.

11.2 RTL Design

The register-transfer-level (RTL) approach is often used for designing a large system at the algorithmic level. The rationale behind the RTL approach is that *any digital system can be considered as a system of interconnected registers with a combinational circuit being placed in between two registers to perform the desired functions*. The RTL approach has the following important features: First, an RTL design is fundamentally the design of an FSM. Second, an RTL design is structural in nature because a complex combination of FSMs may not be easily described solely by a large state diagram or an ASM chart. Third, an RTL design concentrates only on functionality, not on the details of logic designed. Nowadays, there are two widely used RTL design approaches: *ASM charts* and the *datapath-and-controller approach*. In this section, we will address these two approaches in sequence.

11.2.1 ASM Charts

The ASM chart, sometimes called a *state machine* (SM) chart, is often used to describe the design of a digital system at the algorithmic level. Two essential features of ASM charts are as follows: First, they specify RTL operations because they define what happens on every clock cycle. Second, they show clearly the control flow from state to state.

11.2.1.1 RTL Statements The operations of a digital system consist of a sequence of RTL statements [5, 7]. Each RTL statement starts to read data from a set of registers, computes the data (namely, changes the property of the data), and transfers (or stores) the results to the same set or another set of registers.

An RTL statement has the following general format

```
[control_statement:] operation_statement
```

where `control_statement` is optional and may be any form of boolean expressions when it exists. Sometimes, **if** (`control_statement`) is used instead of `control_statement`. The `operation_statement` may be either a simple statement or a complex statement containing two or more simple statements separated with a comma "," and ends with a semicolon ";". An RTL statement without `control_statement` is referred to as an *unconditional statement*. It is executed unconditionally whenever it is encountered. In contrast, an RTL statement with `control_statement` is called a *conditional statement*. It is executed under the control of `control_statement`.

The `operation_statement` in an RTL statement must be a valid RTL operation; that is, it combines registers or variables with Verilog HDL operators listed in Table 3.1. An RTL operation usually uses ← to denote the direction of data transfer, capital characters to denote registers, and lower-case characters to denote variables. The four types of RTL operations most often encountered in digital systems are *data transfer operations, arithmetic operations, logical operations,* and *shifts*.

11.2.1.2 Building Blocks An ASM chart consists of three basic types of building blocks: the *state box, conditional output box,* and *decision* (also called *selection* or *conditional*) *box*, as shown in Figure 11.6.

A state box, as shown in Figure 11.6(a), specifies a machine state and a set of unconditional RTL operations associated with the state. It may execute as many actions as desired and all actions in a state box occur in parallel. Each state, along with its related operations, occupies one clock period. In an ASM chart, state boxes are executed sequentially in a way of one state box per clock cycle. In addition, a register can only be assigned once in a state box. This is known as a *single-assignment rule*.

A conditional output box, as shown in Figure 11.6(b), describes the RTL operations that are executed under the conditions specified by one or more decision boxes. The input of a conditional output box must be from the output of a decision box or the other conditional output boxes. A conditional output box can only execute its RTL operations based on the present state and/or primary input values on the present cycle.

As shown in Figure 11.6(c), a decision box describes the condition by which specific actions will be executed. It can be drawn in either a two-way selection or a multi-way selection.

Section 11.2 RTL Design

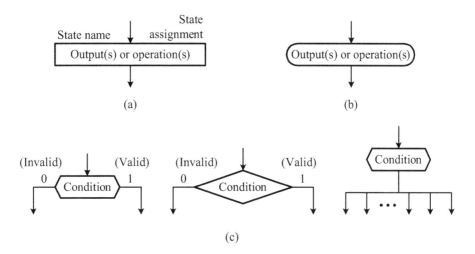

Figure 11.6: The ASM building blocks: (a) state box; (b) conditional output box; (c) decision box.

11.2.1.3 Operations on ASM Charts Recall that ASM charts can be used to describe finite-state machines (Mealy or Moore machines) or algorithms at the gate level or at the RTL. Hence, as revealed in Figure 11.7, the operations (also called *commands*) in the state or conditional output boxes of ASM charts are quite diverse but can be cast into the following three types:

- *Single-bit signal*: A signal is a bit that conveys information and is transmitted via a wire. As a signal appears in a state box or a conditional output box, such as z (meaning that $z=1$) or z' (meaning that $z=0$), the signal is asserted in the state or conditional output box in question. In other state or conditional output boxes, where the signal is not mentioned (namely, asserted), the signal takes its default value. Hence, *the single-bit signal will take its effect immediately as it is activated.*

  For simplicity of notation, in practical applications we usually only list the single-bit outputs whose values are asserted. In other words, the z' ($z=0$) is not put in the ASM chart if its asserted value is z ($z=1$), and the z ($z=1$) is not put in the ASM chart if its asserted value is z' ($z=0$).

- *Multiple-bit signal* (=): As a multiple-bit signal appears in a state or conditional output box with a value different from its default one, an assignment with the equal sign (=) is employed to give its value. The right-hand side of the equal sign can be a constant or an expression consisting of operators and operands. In other state or conditional output boxes,

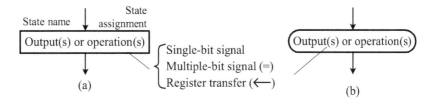

Figure 11.7: The three basic operations used in the state and conditional output boxes of ASM charts.

442   Chapter 11   System Design Methodologies

where the multiple-bit signal is not mentioned, the multiple-bit signal takes its default value. Thus, like the single-bit signal *the multiple-bit signal will take its effect immediately as it is activated*.

> In practice, the notation of multiple-bit signals is often used to denote the single-bit signal whenever it is convenient or more understandable. For example, $z=0$ and z' are often used interchangeably.

- *Register transfer* ($\leftarrow$): To transfer the contents of registers from one to another, the notation of $A \leftarrow B$ is used, where registers A and B are the destination and source registers, respectively. In general, the register-transfer notation has the form of $A \leftarrow f(.)$, where $f(.)$ denotes a function and may be a constant, a register, or an expression consisting of operators and operands. *The effect of the assignment ($\leftarrow$) is delayed in the sense that it takes the effect in the next state*. The destination register will keep its value until it receives another value later.

> In practice, the other output/operation notations may be useful and employed as long as you clearly define your terminology.

11.2.1.4 ASM Blocks An ASM block contains exactly one state box, together with the possible decision boxes and conditional output boxes associated with that state. Each ASM block describes the operations executed in one state. An ASM block has the following two features: First, it has exactly one entrance path but can have one or more exit paths. Second, it contains one state box and a possible serial-parallel network of decision boxes and conditional output boxes. The following example gives two instances of valid ASM blocks.

■ Example 11.6: Two examples of valid ASM blocks.

Both ASM blocks shown in Figure 11.8(a) and (b) are valid. Figure 11.8(a) shows a serial test but Figure 11.8(b) displays a parallel test. In Figure 11.8(a), we first test condition x_1 and then condition x_2. Conditional output box $c=0$ is executed whenever $x_1 = 0$ and conditional output box $d=5$ is executed whenever $x_2 = 1$. The reader can check that all four combinations of x_1 and x_2 are consistent without any conflicts. In Figure 11.8(b), both conditions x_1 and x_2 are independently tested in parallel. Conditional output box $c=0$ is executed whenever $x_1 = 0$ and conditional output box $d=5$ is executed whenever $x_2 = 1$. It is worth noting that although both ASM blocks seem to be different, they indeed perform the same function.

Even though the construction of an ASM chart is more or less an art, the following two basic constructing rules are found helpful:

> **Rule 1:** Each state box and its associated serial-parallel network of decision and conditional output boxes must define a unique next state at any time. In other words, each ASM block could not go to two or more next states at the same time.

> **Rule 2:** Every path of the serial-parallel network of decision and conditional output boxes must terminate at a next state; namely, there cannot exist any loop in the serial-parallel network.

To further illustrate the meanings of these two rules, consider the following two examples.

■ Example 11.7: Examples of invalid ASM blocks.

The ASM block shown in Figure 11.9(a) is invalid because the next state is not unique. This can be seen from the figure that when both x_1 and x_2 are 0, the next state will be at states B and C at the same time, not a unique state. Hence, it violates **rule 1**. Another invalid ASM

Section 11.2 RTL Design

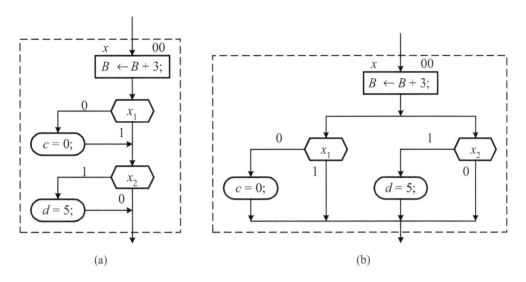

Figure 11.8: Two examples of valid ASM blocks: (a) serial test and (b) parallel test.

block is revealed in Figure 11.9(b). In this ASM block, there exists a loop around two decision boxes and one conditional output box, namely, the path of decision boxes, x_1 and x_2, with their values equal to 1 and conditional output box $c = 0$. Consequently, it violates **rule 2.**

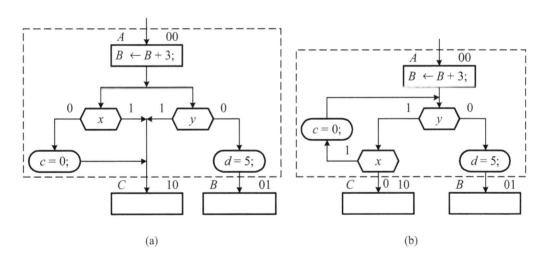

Figure 11.9: Two examples of invalid ASM blocks: (a) undefined next state and (b) undefined exit path.

11.2.1.5 ASM Charts versus State Diagrams
The ASM chart can also be employed to describe a Moore or Mealy machine. To describe a Moore (sequential) machine, only state and decision boxes are required. In such a case, the state boxes also need assignments to perform the output functions. To describe a Mealy (sequential) machine, conditional output boxes (i.e., outputs) are also needed to carry out the output functions in addition to both state and decision boxes. Nevertheless, in such a case, the state boxes do not need to incorporate into any assignment because in Mealy machines outputs are only associated with state transitions and hence conditional output boxes.

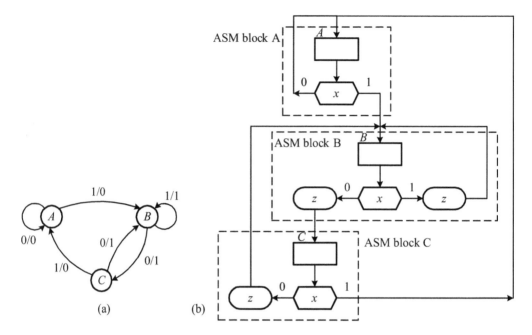

Figure 11.10: An example of the equivalence between (a) a state diagram and (b) an ASM chart.

According to the above discussion, the state diagram of a Mealy or Moore machine can be converted into an equivalent ASM chart, with each state corresponding to an ASM block. An illustration is explored in the following example.

■ Example 11.8: Converting a state diagram into an ASM chart.

Figure 11.10 shows an example of converting the state diagram of a Mealy machine into an equivalent ASM chart. Each state of the state diagram corresponds to an ASM block, as indicated in the figure with the dashed-line block. For example, state A in the state diagram tests input signal x and outputs 0, regardless of what value of x is. The next state of state A is A if the input signal x is 0 and is state B if the input signal x is 1. The other two ASM blocks corresponding to states B and C, respectively, can be derived in a similar way. The complete ASM chart is depicted in Figure 11.10(b).

■ Review Questions

11-12 Describe the rationale behind the RTL design.
11-13 What are the major features of ASM charts?
11-14 What are the three types of building blocks of ASM charts?
11-15 When would you use conditional output boxes in an ASM chart?
11-16 How would you describe a Moore machine with an ASM chart?
11-17 How would you describe a Mealy machine with an ASM chart?

Section 11.2 *RTL Design*

11.2.2 ASM Modeling Styles

In what follows, we first describe the general modeling styles of ASM charts and then using the Booth algorithm as an example to illustrate how to code these modeling styles in Verilog HDL.

11.2.2.1 Modeling Styles Much like FSMs, an ASM chart may be modeled in either of two styles, depending on whether one or two state registers are used. Based on this, the modeling styles of an ASM chart can be cast into two types as described below.

Style 1a: One state register and one always block (suitable for small ASM charts).

In this modeling style, only one state register and one **always** block are used to describe the ASM chart.

Part 1: *Initialize, compute, update the state register, and execute RTL operations.* This part usually uses an **always** block to initialize, compute, and update the state register, as well as to execute the desired RTL operations. An **if-else** statement is used to separate the initial operations from the rest in the ASM chart.

```
always @(posedge clk or negedge reset_n)
   if (!reset_n) begin
      state <= A;
         ...      // set initial values
   end
   else case      // perform RTL operations and
         ...;     // determine the next state
   endcase
```

Style 1b: One state register and two always blocks (suitable for small ASM charts).

In this modeling style, one state register and two **always** blocks are used to describe the ASM chart. It can be subdivided into two parts.

Part 1: *Initialize, compute, and update the state register.* This part usually uses an **always** block to compute the next-state function from both the current inputs and the present state. An **if-else** statement is used to distinguish whether the state assigned to the state register, state, is the start state or the next state.

```
always @(posedge clk or negedge reset_n)
   if (!reset_n) state <= A;
   else          state <= ...;
```

Part 2: *Execute RTL operations.* This part often uses an **always** block accompanied with a **case** statement to perform the desired RTL operations. Sometimes, **assign** continuous assignments are used instead.

```
always @(posedge clk)
   case (state)
      ...
   endcase

   assign continuous assignments
```

Style 2: Two state registers and three always blocks (suitable for large ASM charts).

In this modeling style, two state registers and three **always** blocks are used to describe the ASM chart. This style is much like the style 2 of FSM modeling. It also consists of three parts with each being described by an **always** block.

Part 1: *Initialize and update the state register.* An **always** block is used for this purpose. Inside the **always** block, an **if-else** statement is used to select whether the state assigned to the state register, present_state, is the start state or the next state.

```
always @(posedge clk or negedge reset_n)
    if (!reset_n) present_state <= A;
    else          present_state <= next_state;
```

Part 2: *Compute the next state.* This part usually uses an **always** block to compute the next-state function from both the current inputs and the present state. Sometimes, **assign** continuous assignments are used instead.

```
always @(present_state or x)
    case (present_state)
        ...
    endcase

assign continuous assignments
```

Part 3: *Execute RTL operations.* The RTL operations can be computed by an **always** block or **assign** continuous assignments. Inside the **always** block, a **case** statement is usually utilized to compute the RTL operations associated with each state.

```
always @(posedge clk)
    case (present_state)
        ...
    endcase

assign continuous assignments
```

11.2.2.2 Modeling Examples of the Booth Algorithm To illustrate the aforementioned ASM modeling styles, a well-known 2's-complement multiplication algorithm, called the Booth algorithm, is employed as an example. The Booth algorithm receives two signed inputs and produces a signed output. The details of the Booth algorithm is described as follows.

Booth Algorithm. Assume that the two signed inputs of the Booth algorithm are $X = x_{n-1}x_{n-2}\cdots x_2 x_1 x_0$ and $Y = y_{n-1}y_{n-2}\cdots y_2 y_1 y_0$, respectively. Then, the operations of the Booth algorithm can be described as follows. At the ith step, where $0 \leq i \leq n-1$, one of the following operations is performed according to the values of $x_i x_{i-1}$.

- Add 0 to the partial product (P) if $x_i x_{i-1} = 00$;
- Add Y to the partial product (P) if $x_i x_{i-1} = 01$;
- Subtract Y from the partial product (P) if $x_i x_{i-1} = 10$;
- Add 0 to the partial product (P) if $x_i x_{i-1} = 11$.

The initial value of x_{-1} is assumed to be 0.

The above operations can be described as the following algorithm.

■ Algorithm 11-1: Booth algorithm

Input: An n-bit multiplicand and an n-bit multiplier in 2's-complement form.
Output: The product left in $acc:mp[n:1]$ in 2's-complement form.
Begin

1. while (start == 0) **do** ; // wait for start to become 1

Section 11.2 RTL Design

2. Load the multiplicand and multiplier into registers $mcand$ and $mp[n:1]$, respectively; clear the accumulator acc and the bit $mp[0]$, and set the loop count (cnt) equal to $n-1$;
3. **repeat**

    3.1 **if** $mp[1:0] = 01$ **then** $acc \leftarrow acc + mcand$;

    3.2 **if** $mp[1:0] = 10$ **then** $acc \leftarrow acc - mcand$;

    3.3 Shift register pair $acc:mp$ right one bit; $cnt \leftarrow cnt - 1$;

    until ($cnt == 0$);
4. Set the *finish* flag to 1;

End

An ASM chart. In order to derive an ASM chart for the Booth algorithm, suppose that multiplicand Y and multiplier X are stored in the $mcand$ and mp registers, respectively. The mp register has an extra bit in the rightmost position to accommodate the value of $x_{-1} = 0$ for the 0th step. The multiplicand is added to or subtracted from the accumulator, acc, at each step when the values of the two least significant bits of mp are 01 and 10, respectively. The result in acc, concatenated with the contents of mp, is then shifted right arithmetically 1-bit position. The accumulator, acc, concatenated with the leftmost $i+1$ bits of the mp register, forms an $(n+i+1)$-bit partial product register at the ith step. To control the desired number of iterations of the above operations, a counter, cnt, is used. Based on this description, an ASM chart can be derived, as shown in Figure 11.11.

To further illustrate the detailed operations of the Booth algorithm, a numerical example is given in Figure 11.12. You are encouraged to trace each step of this example according to the ASM chart shown in Figure 11.11. This process is essentially needed when an algorithm is initially developed in order to ensure that you have certainly understood the specification of a design and developed a correct algorithm.

Once we have understood the detailed operations of the Booth algorithm, we are in a position to describe it in Verilog HDL. In what follows, three examples are given to illustrate how the aforementioned three different modeling styles can be used to describe an ASM chart. We begin with the modeling of the Booth algorithm in style 1a.

■ Example 11.9: The Booth algorithm: style 1a.

When using style 1a to model a hardware module, we only need to use one **always** block to initialize, compute, and update the state register, as well as execute RTL operations. An **if-else** statement is used to separate the initial operations from the rest of the ASM chart. Based on this, the ASM chart shown in Figure 11.11 can be described as follows. If the `reset_n` signal is active, the system enters the idle state. In the idle state, if the `start` signal is active, the system loads the `mcand` and `mp` registers, and loads the counter, `cnt`, as well as clears the accumulator, `acc`, and the `finish` flag; otherwise, the system remains in the idle state. Each ASM block is modeled by a case item selected by the `state` variable, which is used as the conditional expression of the **case** statement.

```
// a Booth multiplier: style 1a --- realizing directly its ASM
module booth_step1
       #(parameter W = 8,   // the default word size
         parameter N = 3)(  // N = log2(W)
         input   clk, reset_n, start,
         input   [W-1:0] multiplier, multiplicand,
         output  [2*W-1:0] product,
         output reg finish);   // the finish flag
```

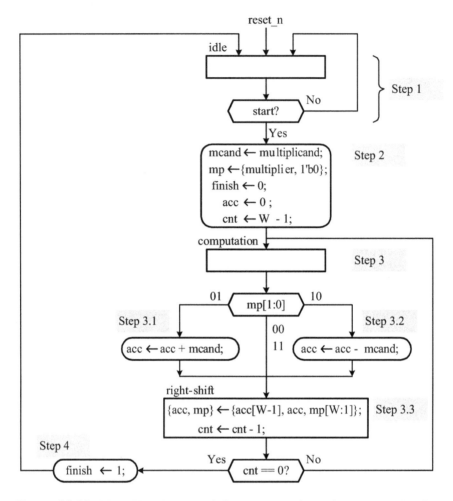

Figure 11.11: The ASM chart describing the operations of the Booth algorithm.

```
reg    [W-1:0] mcand, acc;
reg    [N-1:0] cnt;
reg    [W:0]   mp;       // one extra bit
reg    [1:0]   state;
localparam    idle = 2'b00, computation = 2'b01,
              right_shift = 2'b10;

// the body of the w-bit booth multiplier
always @(posedge clk or negedge reset_n)
   if (!reset_n) state <= idle;
   else case (state)
      idle: if (!start) state <= idle;
         else begin
            mp    <= {multiplier,1'b0};
            mcand <= multiplicand;
            acc <= 0; cnt <= W -1;
            finish <= 0;
            state <= computation;
```

Section 11.2 RTL Design

	acc	mp	mp(0)	cnt	
mp = 77H	0 0 0 0 0 0 0 0	0 1 1 1 0 1 1 1	0	7	$acc \leftarrow acc - mcand$
$mcand$ = 13H	- 0 0 0 1 0 0 1 1				
	1 1 1 0 1 1 0 1				
	1 1 1 1 0 1 1 0	1 0 1 1 1 0 1 1	1	7	shift right $acc{:}mp$
	1 1 1 1 1 0 1 1	0 1 0 1 1 1 0 1	1	6	shift right $acc{:}mp$
	1 1 1 1 1 1 0 1	1 0 1 0 1 1 1 0	1	5	shift right $acc{:}mp$
	+ 0 0 0 1 0 0 1 1				$acc \leftarrow acc + mcand$
	0 0 0 1 0 0 0 0				
	0 0 0 0 1 0 0 0	0 1 0 1 0 1 1 1	0	4	shift right $acc{:}mp$
	- 0 0 0 1 0 0 1 1				
	1 1 1 1 0 1 0 1				
	1 1 1 1 1 0 1 0	1 0 1 0 1 0 1 1	1	3	shift right $acc{:}mp$
	1 1 1 1 1 1 0 1	0 1 0 1 0 1 0 1	1	2	shift right $acc{:}mp$
	1 1 1 1 1 1 1 0	1 0 1 0 1 0 1 0	1	1	shift right $acc{:}mp$
	+ 0 0 0 1 0 0 1 1				$acc \leftarrow acc + mcand$
	0 0 0 1 0 0 0 1				
08D5H →	0 0 0 0 1 0 0 0	1 1 0 1 0 1 0 1	0	0	shift right $acc{:}mp$

Figure 11.12: A numerical example illustrates the operations of the Booth algorithm.

```
      end
   computation: begin
      case (mp[1:0])
         2'b01: acc <= acc + mcand;
         2'b10: acc <= acc - mcand;
         default: ; // do nothing
      endcase
      state <= right_shift; end
   right_shift: begin
      {acc, mp} <= {acc[W-1], acc, mp[W:1]};
      cnt <= cnt - 1;
      if (cnt == 0) begin
         finish <= 1; state <= idle; end
      else state <= computation; end
   default: ; // do nothing
   endcase
assign product = {acc, mp[W:1]};
endmodule
```

The modeling of the Booth algorithm described in the ASM chart given in Figure 11.11 in style 1b is explored in the following example.

■ Example 11.10: The Booth multiplier: style 1b.

When using style 1b, we employ one state register and two **always** blocks to describe the Booth algorithm. It can be subdivided into two parts. The first **always** block accompanied with a function fsm_next_state initializes, computes, and updates the state register. The second **always** block performs the desired RTL operations.

```verilog
// a Booth multiplier: style 1b --- using two always blocks
module booth_step1
        #(parameter W = 8, // the default word size
          parameter N = 3)( // N = log2(W)
          input clk, reset_n, start,
          input [W-1:0] multiplier, multiplicand,
          output [2*W-1:0] product,
          output reg  finish); // the finish flag

// the body of the w-bit booth multiplier
reg [W-1:0] mcand, acc;
reg [N:0]   cnt;
reg [W:0]   mp;     // one extra bit
reg [1:0]   state;  // only declare one state register
localparam  idle = 2'b00, computation = 2'b01,
            right_shift = 2'b10;

// part 1: initialize to the idle state, determine the
//         next state, and update the state register
always @(posedge clk or negedge reset_n)
   if (!reset_n) state <= idle;
   else  state <= fsm_next_state(state, cnt, start);
// the function to compute the next state
function [1:0] fsm_next_state;
input [1:0] fsm_present_state;
input [N:0] fsm_cnt;
input fsm_start;
reg  [1:0] next_state;
begin
   case (fsm_present_state)
     idle:    if (!fsm_start) next_state = idle;
              else            next_state = computation;
     computation: next_state = right_shift;
     right_shift: if (fsm_cnt == 0) next_state = idle;
              else            next_state = computation;
     default: next_state = idle; // return the idle state
   endcase
   fsm_next_state = next_state;
end endfunction

// part 2: execute RTL operations
always @(posedge clk)
   case (state)
     idle: if (start) begin
              mp <= {multiplier,1'b0};
              mcand <= multiplicand;
              finish <= 0;
              acc <= 0;
              cnt <= W -1; end
```

Section 11.2 RTL Design

```verilog
         computation: begin case (mp[1:0])
                     2'b01: acc <= acc + mcand;
                     2'b10: acc <= acc - mcand;
                     default: ; // do nothing
                  endcase end
         right_shift: begin
            {acc, mp} <= {acc[W-1], acc, mp[W:1]};
            cnt <= cnt - 1;
            if (cnt == 0) finish <= 1; end
         default: ; // do nothing
      endcase
   assign product = {acc, mp[W:1]};
endmodule
```

Like the case of FSMs, an ASM chart can also be described by using two state registers and three **always** blocks. This style is much like the style 2 of FSM modeling. To illustrate this, the Booth algorithm is employed as an example again.

■ Example 11.11: The Booth multiplier: style 2.

In this style, three **always** blocks are used. The first **always** block initializes and updates the state register. The second **always** block computes the next-state function from the present state, the current inputs, and the condition computed. The third **always** block performs the desired RTL operations.

```verilog
// a Booth multiplier: style 2 --- using three always blocks
module booth_style2_step1
       #(parameter W = 8,  // the default word size
         parameter N = 3)( // N = log2(W)
         input    clk, reset_n, start,
         input    [W-1:0] multiplier, multiplicand,
         output   [2*W-1:0] product,
         output reg finish);   // the finish flag
// the body of the W-bit booth multiplier
reg    [W-1:0]   mcand, acc;
reg    [N-1:0]   cnt;
reg    [W:0]     mp;   // one extra bit
reg    [1:0]     ps, ns;
localparam   idle = 2'b00, computation = 2'b01,
             right_shift = 2'b10;
// part 1: initialize and update the state registers
always @(posedge clk or negedge reset_n)
   if (!reset_n) ps <= idle;
   else    ps <= ns;
// part 2: compute the next state
always @(*)
   case (ps)
      idle: if (!start) ns = idle;
            else        ns = computation;
      computation: ns = right_shift;
      right_shift: if (cnt == 0) ns = idle;
                   else          ns = computation;
```

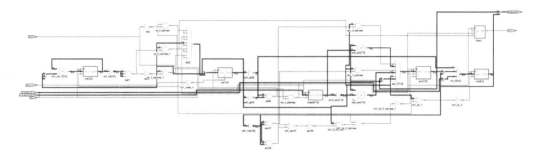

Figure 11.13: The synthesized result of the Booth algorithm from a synthesis tool.

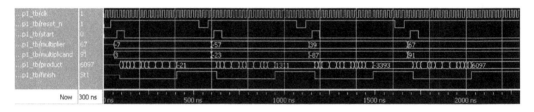

Figure 11.14: The simulation results of the Booth multiplier algorithm.

```
         default: ns = idle;
      endcase
// part 3: execute RTL operations
always @(posedge clk)
   case (ps)
      idle: if (start) begin
         mp      <= {multiplier,1'b0};
         mcand   <= multiplicand;
         finish  <= 0;
         acc     <= 0;
         cnt     <= W - 1; end
      computation: begin
         case (mp[1:0])
            2'b01: acc <= acc + mcand;
            2'b10: acc <= acc - mcand;
            default: ; // do nothing
         endcase end
      right_shift: begin
         {acc, mp} <= {acc[W-1], acc, mp[W:1]};
         cnt <= cnt - 1;
         if (cnt == 0) finish <= 1; end
      default: ; // do nothing
   endcase
assign product = {acc, mp[W:1]};
endmodule
```

Almost all of the above three modeling styles yield the same synthesized result, as shown in Figure 11.13. The simulation results of the Booth algorithm are portrayed in Figure 11.14.

Section 11.2 RTL Design

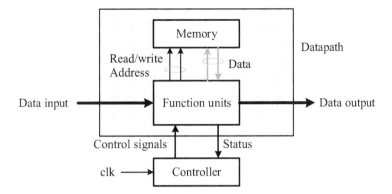

Figure 11.15: The general block diagram of the datapath and controller architecture of digital systems.

11.2.3 Datapath-and-Controller Approach

In this section, we first introduce the basic design paradigm of the datapath-and-controller approach and then address the first one of the following two most common approaches.

- In the *three-step paradigm*, the datapath and the controller of a digital system are derived from its ASM chart. The datapath corresponds to the registers and the function units in the ASM chart and the controller corresponds to the logic used to generate all needed control signals in the datapath. To help partition the digital system, a variant of the ASM chart, called an *algorithmic state machine with datapath* (ASMD) chart is often used instead. This approach is suitable for simple digital systems.

- In the *heuristic approach*, the datapath and the controller of a digital system are derived from the specifications in a state-of-the-art fashion. The controller is then modeled by either a *finite-state machine with datapath* (FSMD) diagram or an ASM chart or its variant, an ASMD chart. This approach is often used in designing a complex system. Refer to Section 11.4 for more details of this approach.

11.2.3.1 The Paradigm of Datapath and Controller Approach The general circuit model of the *datapath and controller architecture* of digital systems is shown in Figure 11.15. It is also referred to as the *control-point structure* (usually called *the DP+CU structure* for short). The related design approach is called the *datapath and controller approach* or the *control-point approach* (usually called *the DP+CU approach* for short).

General speaking, a digital system at the RTL can be considered as a system composed of three major parts, a *datapath*, a *memory* module, and a *controller*, as shown in Figure 11.15. The datapath performs all operations desired in the system. The memory module temporarily stores intermediate data used and generated by the datapath. The controller controls and schedules all operations to be performed by the datapath. In order to adaptively generate the control signals, a set of status signals is often fed back to the controller to indicate the status of the datapath.

A datapath is a logical and physical structure, which usually comprises arithmetic units, such as an adder/subtractor, a multiplier, a shifter, a comparator, registers, an ALU, and some other function units. The memory may be one or more of the following components, RAMs, CAMs, ROMs, FIFOs (buffers), shift registers, and registers, depending on the actual requirements. A controller is a finite-state machine, being described by a state diagram or an ASM chart and implemented by using PLA, ROM, or random logic. In addition, an interconnect network is often needed to pass data among the above three portions. The most widely used interconnect network is crossbar switches, arbiters, buses, and multiplexers. In large digital systems, such as CPUs, the two most important design issues of datapaths are the memory (registers and/or register files) and the interconnect network because they severely affect the system performance and cost.

11.2.3.2 Finite-State Machine with a Datapath (FSMD) A finite-state machine with datapath (FSMD) [2, 3] is an extension of the FSM introduced in Section 11.1 and is popular for use in high-level synthesis to generate an RTL result. However, it is also widely used in designing a complex system at the RTL.

An FSMD $\mathcal{M}_\mathcal{D}$ is a quintuple $\mathcal{M}_\mathcal{D} = \langle \mathcal{S}, \mathcal{I} \cup S, \mathcal{O} \cup A, \delta, \lambda \rangle$, where $\mathcal{S} = \{s_0, \ldots, s_{n-1}\}$ is the set of n finite states, s_0 is the reset state, $\mathcal{I}$ is the set of input variables, $\mathcal{O} = \{o_k\}$ is the set of primary output values, $S = \{Rel(a,b) : a,b \in E\}$ is the set of statements specifying relations between two expressions from the set E, $E = \{f(x,y,z,\ldots) : x,y,z,\ldots, \in V\}$ is the set of expressions, V is the set of storage variables, and $A = \{x \Leftarrow e : x \in V, e \in E\}$ is the set of storage assignments.

The state transition function is defined as follows

$$\delta : \mathcal{S} \times (\mathcal{I} \cup S) \to \mathcal{S}$$

which means that the δ function is determined by inputs, statements, and the present state.

The output function λ is defined as follows

$\lambda : \mathcal{S} \times (\mathcal{I} \cup S) \to (\mathcal{O} \cup A)$ (Mealy machine)
$\lambda : \mathcal{S} \to (\mathcal{O} \cup A)$ (Moore machine)

where in a Mealy machine, the output function is determined by the present inputs, statements, and the present state whereas in a Moore machine the output function is solely determined by the present state.

In summary, an FSMD is the extension of an FSM in which the RTL operations of the datapath are annotated beside the states and/or state transitions along with the state transition conditions. Since the FSMD method is not widely used as the ASM chart, in the rest of this section, we will focus on the ASM chart and its variant, the ASMD chart.

11.2.3.3 The Three-Step Paradigm To explore how the three-step paradigm can be used to derive the datapath and the controller [1] from an ASM chart of a digital system, in what follows we first illustrate the paradigm by way of a simple example and then apply it to the Booth algorithm.

Transforming an ASM chart into the datapath and controller architecture generally follows the following three steps:

1. *Model the design* (usually described by an ASM chart) as a single module using any modeling style described above.
2. *Extract the datapath* from the module and construct it as an independent module. The datapath module is then instantiated by the original module, which now contains only the portion of control signal generation.
3. *Extract the controller module and construct a top-level module.* Rewrite the portion of control signal generation as an independent module in parallel to the datapath module. Add a *top-level module* that instantiates both datapath and controller modules. Here, a top-level module means a module that is not instantiated by the other modules.

To illustrate this three-step paradigm in a step-by-step way, in what follows, we use a very simple problem that accumulates an input number in the number of times determined by the number of 1s in another input number. The ASM chart is portrayed in Figure 11.16. Indeed, this problem can be done in one cycle simply using a combinational circuit.

Section 11.2 RTL Design

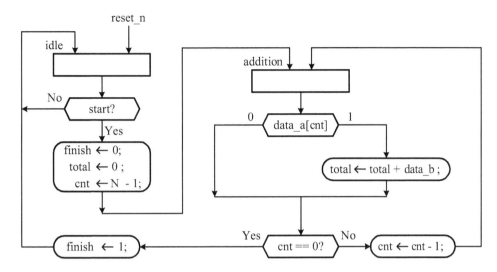

Figure 11.16: The ASM chart of the three-step example.

■ Example 11.12: Step 1: Model the design.

Suppose that a bit counter, cnt, is used to control the desired number of iterations of addition, where a sum, total, accumulates the specified input number, data_b, under the control of another input, data_a. The data_b is added to total when the bit value of the input number, data_a[cnt], is 1; otherwise, the data_b need not be added to total at all. In order to implement this idea, an FSM with two states is used. If the reset_n signal is active, the system enters the idle state. In the idle state, if the start signal is active, the system clears total and finish as well as loads the bit numbers, $n-1$, into cnt; otherwise, the system remains in the idle state. In the addition state, the input number, data_a, is examined bit by bit to determine whether the input number, data_b, is needed to add to total or not. Besides, in this state, the counter, cnt, is decremented by 1 and the FSM remains in this state until the counter, cnt, reaches 0. At this point, the operation is finished and the finish flag is set to 1.

```
// an example illustrates the three-step design paradigm
module three_steps_step1
      #(parameter N = 8,   // the default word size
        parameter M = 3)(  // M = log_2(N)
         input   clk, reset_n, start,
         input   [N-1:0] data_a, data_b,
         output reg [N-1:0] total,
         output reg  finish);  // the finish flag
// the body of the module
reg next_state, present_state;
reg [M-1:0] cnt;
localparam idle = 1'b0, addition = 1'b1;
// part 1: initialize and update the state register
always @(posedge clk or negedge reset_n)
   if  (!reset_n) present_state <= idle;
   else           present_state <= next_state;
// part 2: compute the next state
always @(present_state or start or cnt)
   case (present_state)
```

```verilog
      idle:      if (start)   next_state = addition;
                 else         next_state = idle;
      addition:  if (cnt == 0) next_state = idle;
                 else         next_state = addition;
   endcase
// part 3: compute RTL operations
always @(posedge clk)
   case (present_state)
      idle: if (start) begin
            finish <= 0; total <= 0;
            cnt <= N - 1; end
      addition: begin
            if (data_a[cnt] == 1) total <= total + data_b;
            if (cnt == 0) finish <= 1;
            else cnt <= cnt - 1; end
   endcase
endmodule
```

The second step is to extract the datapath operations from the Verilog HDL description. The following example explains the details of this.

■ Example 11.13: Step 2: Extract the datapath.

The datapath structure is extracted from the part 3 of the above Verilog HDL description. In this example, only one `total` register is used to accumulate the sum of the input number and two operations are performed: clear and load the accumulated sum. Hence, the datapath is quite simple and only composed of an accumulator, as shown in the datapath module. What is left in part 3 are those signals without relating to the datapath, and the control signals for the datapath. In this example, only one control signal is required, namely, `acc_load`, which loads the accumulator.

```verilog
// an example illustrates the three-step design paradigm
module three_steps_step2
      #(parameter N = 8,  // the default word size
        parameter M = 3)( // M = log_2(N)
        input  clk, reset_n, start,
        input  [N-1:0] data_a, data_b,
        output [N-1:0] total,
        output reg finish);  // the finish flag
// the body of the module
reg next_state, present_state;
reg [M-1:0] cnt;
wire load;
localparam idle = 1'b0, addition = 1'b1;
// datapath --- consisting of a single accumulator
datapath #(8) dp (.clk(clk), .acc_reset_n(reset_n),
               .acc_load(load), .acc_data_b(data_b),
               .acc_total(total));
// part 1: initialize and update the state register
always @(posedge clk or negedge reset_n)
   if   (!reset_n) present_state <= idle;
   else            present_state <= next_state;
```

Section 11.2 RTL Design

```verilog
// part 2: compute the next state
always @(present_state or start or cnt)
   case (present_state)
      idle: if (start)   next_state = addition;
            else         next_state = idle;
      addition: if (cnt == 0) next_state = idle;
                else          next_state = addition;
   endcase
// part 3: execute RTL operations
always @(posedge clk)
   case (present_state)
      idle: if (start) begin
            finish <= 0;
            cnt <= N - 1; end
      addition: begin
            if (cnt == 0) finish <= 1;
            else cnt <= cnt -1; end
   endcase
assign load = (present_state==addition )&&(data_a[cnt]==1);
endmodule

// define the datapath module
module datapath
      #(parameter N = 8)( // the default word size
         input  clk, acc_reset_n, acc_load,
         input  [N-1:0] acc_data_b,
         output reg [N-1:0] acc_total);
// the body of the datapath
always @(posedge clk or negedge acc_reset_n)
   if (!acc_reset_n)  acc_total <= 0;
   else if (acc_load) acc_total <= acc_total + acc_data_b;
endmodule
```

The third step is to rewrite the module as a controller module and construct another module, called the top-level module, to include both datapath and controller modules as its instances.

■ Example 11.14: Step 3: Extract the controller module.

This part is simply to rewrite the module, except for the instance of the datapath, as a controller module. In order to make the notation more apparent, in this example, we prefix cu_ to all ports of the controller module so that they can be distinguished from the top-level module. Of course, it is not necessary to do this. However, it is good practice to make signals distinguishable.

```verilog
// an example illustrates the three-step design paradigm
module three_steps_step3
      #(parameter N = 8,  // the default word size
        parameter M = 3)( // M = log_2(N)
         input  clk, reset_n, start,
         input  [N-1:0] data_a, data_b,
         output [N-1:0] total,
         output finish);  // the finish flag
```

```verilog
// the body of the module
wire load;
localparam idle = 1'b0, addition = 1'b1;
// datapath --- consisting of a single accumulator.
datapath #(8) dp (.clk(clk), .acc_reset_n(reset_n),
              .acc_load(load), .acc_data_b(data_b),
              .acc_total(total));
controller cu (.cu_clk(clk), .cu_reset_n(reset_n), .cu_start(start),
            .cu_data_a(data_a), .cu_load(load),
            .cu_finish(finish));
endmodule

// the controller module
module controller
     #(parameter N = 8,  // the default word size
       parameter M = 3)( // M = log_2(N)
       input cu_clk, cu_reset_n, cu_start,
       input [N-1:0] cu_data_a,
       output wire cu_load,
       output reg  cu_finish);  // the finish flag
// the body of the controller
reg next_state, present_state;
reg [M-1:0] cnt;
localparam idle = 1'b0, addition = 1'b1;
// part 1: initialize and update the state register
always @(posedge cu_clk or negedge cu_reset_n)
   if (!cu_reset_n) present_state <= idle;
   else             present_state <= next_state;
// part 2: compute the next state
always @(present_state or cu_start or cnt)
   case (present_state)
      idle:     if (cu_start) next_state = addition;
                else          next_state = idle;
      addition: if (cnt == 0) next_state = idle;
                else          next_state = addition;
   endcase
// part 3: execute RTL operations
always @(posedge cu_clk)
   case (present_state)
      idle: if (cu_start) begin
         cu_finish <= 0;
         cnt <= N - 1; end
      addition: begin
         if (cnt == 0) cu_finish <= 1;
         else cnt <= cnt - 1; end
   endcase
assign cu_load = (present_state==addition)&&(cu_data_a[cnt]==1);
endmodule

// define the datapath module
```

Section 11.2 RTL Design

```
module datapath
      #(parameter N = 8)( // the default word size
        input   clk, acc_reset_n, acc_load,
        input   [N-1:0] acc_data_b,
        output reg [N-1:0] acc_total);
// the body of the datapath
always @(posedge clk or negedge acc_reset_n)
   if (!acc_reset_n)   acc_total <= 0;
   else if (acc_load) acc_total <= acc_total + acc_data_b;
endmodule
```

In summary, the datapath and controller architecture generally not only has a simpler architecture but also consumes less area than the ASM chart approach. In the rest of this section, we further illustrate this three-step paradigm by another more complex example, i.e., the Booth algorithm.

11.2.3.4 The Booth Algorithm The datapath of the Booth algorithm can be derived from the ASM chart given in Figure 11.11 and is shown in Figure 11.17. The datapath has an n-bit adder and subtractor, *add_sub*, one accumulator, *acc*, and two registers, *mp* and *mcand*. The accumulator, *acc*, and the register, *mp*, constitute a shift register and serve as the partial product register.

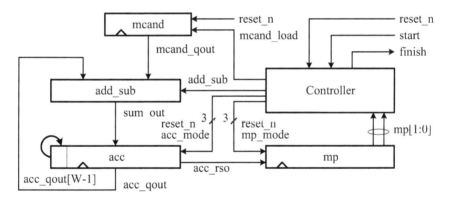

Figure 11.17: The datapath of the Booth algorithm derived from the ASM chart given in Figure 11.11.

The controller generates all control signals needed in the datapath. It reads the two least significant bits of the *mp* register, namely, *mp*[1:0], and generates all control signals accordingly. In addition, a reset signal, *reset_n*, is input to both the datapath and the controller to reset all modules to their initial states and a *start* signal is utilized to start the operations of the algorithm. The control signal *mcand_load* loads the multiplicand into the *mcand* register. The control signal *acc_sub* determines whether the operation is addition or subtraction. The control signal *acc_mode* is a two-bit mode selection signal, which controls the operations of the accumulator. There are three modes associated with the accumulator *acc*: 00—does nothing, 10—loads the accumulator, and 01—shifts the contents of the accumulator right 1-bit position. Similarly, the control signal *mp_mode* is also a two-bit mode selection signal, which controls the operations of the *mp* register. The *mp* register has three modes: 00—does nothing, 10—loads the *mp* register, and 01—shifts the contents of the *mp* register right 1-bit position.

In the following two examples, we use the modeling style 2 of the Booth algorithm as an example to illustrate how to transfer an ASM chart or a Verilog HDL description into a datapath and controller architecture.

■ Example 11.15: The Booth multiplier: style 2 — step 2.

Recall that the second step of the three-step paradigm is to extract the datapath structure from the RTL operations. The datapath has the structure as shown in Figure 11.17 along with related control signals. After the datapath is extracted, the ASM chart is only left with the state transition operations and datapath-related control signals.

```verilog
// a Booth multiplier --- style 2 : step 2
module booth_style2_step2
      #(parameter W = 8,  // the default word size
        parameter N = 3)( // N = log2(W)
        input   clk, reset_n, start,
        input   [W-1:0] multiplier, multiplicand,
        output  [2*W-1:0] product,
        output reg finish);  // the finish flag
// the body of the W-bit booth multiplier
reg  [1:0] acc_mode, mp_mode;
reg  mcand_load, add_sub;
wire [1:0] mp_qb10;
localparam  idle = 2'b00, computation = 2'b01,
            right_shift = 2'b10;
// the datapath part --- a structural description
booth_datapath #(W)  bdatapath
      (.clk(clk), .reset_n(reset_n), .add_sub(add_sub),
       .mcand_load(mcand_load), .acc_mode(acc_mode),
       .mp_mode(mp_mode), .multiplier(multiplier),
       .multiplicand(multiplicand), .mp_qb10(mp_qb10),
       .product(product));
// the controller --- a behavioral description
// local variable declarations
reg [1:0] ns, ps;
reg [N-1:0] cnt;
// part 1: initialize and update the state registers
always @(posedge clk or negedge reset_n)
   if (!reset_n) ps <= idle;
   else          ps <= ns;
// part 2: compute the next state
always @(*)
   case (ps)
      idle: if (!start) ns = idle;
            else        ns = computation;
      computation: ns = right_shift;
      right_shift: if (cnt == 0) ns = idle;
                   else          ns = computation;
      default: ns = idle;
   endcase
// part 3: generate control signals
// define the finish signal
```

Section 11.2 RTL Design

```verilog
always @(posedge clk)
   if (start) finish <= 0; else
   if ((ps == right_shift) && (cnt == 0)) finish <= 1;
// define the counter block
always @(posedge clk or negedge reset_n)
   if (!reset_n) cnt <= W - 1;
   else if (ps == idle && finish == 0) cnt <= W - 1 ;
       else if (ps == right_shift)     cnt <= cnt - 1;
// all signals are initially set to their initial
// values and then set to be active accordingly
always @(*) begin begin
   mcand_load = 1'b0; mp_mode = 2'b00;
   acc_mode = 2'b00;  add_sub = 1'b0;    end
   case (ps)
      idle: if (start) begin
         mcand_load = 1'b1;  mp_mode = 2'b10; end
      computation: case (mp_qb10)
         2'b01: begin
            add_sub = 1'b0;
            acc_mode = 2'b10; end // addition
         2'b10: begin
            add_sub = 1'b1;
            acc_mode = 2'b10; end // subtraction
         default: ; endcase
      right_shift: begin
         acc_mode = 2'b01;
         mp_mode  = 2'b01;  end // shift right 1-bit position
      default: ;  // do nothing
   endcase
end // the end of the always block
endmodule

//  the Booth datapath is described in a structural way
module booth_datapath #(parameter W = 8)( // the default word size
     input   clk, reset_n, add_sub, mcand_load,
     input   [1:0]   acc_mode, mp_mode,
     input   [W-1:0] multiplier, multiplicand,
     output wire [1:0] mp_qb10,
     output wire [2*W-1:0] product);
// the body of the datapath
wire   [W-1:0] acc_qout, mcand_qout, sum_out;
wire   acc_rso, mp_rso;
wire   [W:0]   mp_qout;

// instantiate data path components
addsub    #(W)    addsub
     (.a(acc_qout), .b(mcand_qout), .mode(add_sub),
       .sum(sum_out));
shift_reg #(W)    acc
     (.clk(clk), .reset_n(reset_n), .lsi(acc_qout[W-1]),
       .mode(acc_mode), .data(sum_out), .qout(acc_qout),
```

```verilog
              .rso(acc_rso));
shift_reg #(W+1) mp
        (.clk(clk), .reset_n(reset_n), .lsi(acc_rso),
         .mode(mp_mode), .data({multiplier, 1'b0}),
         .qout(mp_qout), .rso(mp_rso));
register   #(W)   mcand
        (.clk(clk), .reset_n(reset_n), .load(mcand_load),
         .data(multiplicand), .qout(mcand_qout));
assign product = {acc_qout, mp_qout[W:1]};
assign mp_qb10 = mp_qout[1:0];
endmodule

// the shift register module
module shift_reg #(parameter N = 8)(
        input   clk, reset_n, lsi,
        input   [1:0] mode,
        input   [N-1:0] data,
        output reg [N-1:0] qout,
        output wire rso);

// the shift register body
assign rso = qout[0];
always @(posedge clk or negedge reset_n)
   if (!reset_n) qout <= {N{1'b0}};
   else case (mode)
      2'b00: ;                             // no change
      2'b01: qout <= {lsi, qout[N-1:1]};   // shift right
      2'b10: qout <= data;                 // parallel load
      default:; // no operation
   endcase
endmodule

// the register module
module register #(parameter N = 8)(
        input   clk, reset_n, load,
        input   [N-1:0] data,
        output reg [N-1:0] qout);
// the register body
always @(posedge clk or negedge reset_n)
   if (!reset_n)  qout <= {N{1'b0}};
   else if (load) qout <= data;
endmodule

// the addition and subtraction module
module addsub #(parameter N = 8)(
        input   [N-1:0] a, b,
        input   mode, // define addition or subtraction
        output reg [N-1:0] sum);
// the adder-subtractor body
always @(a or b or mode)
    if  (mode) sum = a - b; // mode = 1 --> subtraction
```

Section 11.2 RTL Design

```
        else        sum = a + b; // mode = 0 --> addition
endmodule
```

The third step of the three-step paradigm is to rewrite the module left in the second step as a controller module and construct another module, called the top-level module, to include both the datapath and the controller modules as its instances. The following example illustrates this.

■ Example 11.16: The Booth multiplier: style 2 — step 3.

This part is simply to rewrite the module left in step 2 as a controller module and then construct a top-level module to instantiate both the datapath and the controller modules. This results in the following module.

```
// a Booth multiplier: the datapath and controller approach
module booth_style2_step3
        #(parameter W = 8,  // the default word size
          parameter N = 3)( // N = log2(W)
          input  clk, reset_n, start,
          input  [W-1:0] multiplier, multiplicand,
          output [2*W-1:0] product,
          output finish);   // the finish flag
// the body of the W-bit booth multiplier
wire    [1:0] acc_mode, mp_mode;
wire    mcand_load, add_sub;
wire    [1:0] mp_qb10;
// instantiate the datapath and the controller
booth_datapath #(W)  bdatapath
        (.clk(clk), .reset_n(reset_n), .add_sub(add_sub),
         .mcand_load(mcand_load), .acc_mode(acc_mode),
         .mp_mode(mp_mode), .multiplier(multiplier),
         .multiplicand(multiplicand), .mp_qb10(mp_qb10),
         .product(product));
booth_controller #(W,N) bcontroller
        (.clk(clk), .reset_n(reset_n), .start(start),
         .mp_qout(mp_qb10), .add_sub(add_sub),
         .mcand_load(mcand_load), .acc_mode(acc_mode),
         .mp_mode(mp_mode), .finish(finish));
endmodule
// the Booth datapath is described in a structural way
module booth_datapath #(parameter W = 8)( // the default word size
          input  clk, reset_n, add_sub, mcand_load,
          input  [1:0]   acc_mode, mp_mode,
          input  [W-1:0] multiplier, multiplicand,
          output wire [1:0] mp_qb10,
          output wire [2*W-1:0] product);
// the body of the datapath
wire    [W-1:0] acc_qout, mcand_qout, sum_out;
wire    acc_rso, mp_rso;
wire    [W:0] mp_qout;
// instantiate data path components
addsub    #(W)    addsub
        (.a(acc_qout), .b(mcand_qout), .mode(add_sub),
```

```verilog
         .sum(sum_out));
shift_reg #(W)     acc
      (.clk(clk), .reset_n(reset_n), .lsi(acc_qout[W-1]),
       .mode(acc_mode), .data(sum_out), .qout(acc_qout),
       .rso(acc_rso));
shift_reg #(W+1) mp
      (.clk(clk), .reset_n(reset_n), .lsi(acc_rso),
       .mode(mp_mode), .data({multiplier, 1'b0}),
       .qout(mp_qout), .rso(mp_rso));
register  #(W)    mcand
      (.clk(clk), .reset_n(reset_n), .load(mcand_load),
       .data(multiplicand), .qout(mcand_qout));
assign product = {acc_qout, mp_qout[W:1]};
assign mp_qb10 = mp_qout[1:0];
endmodule
//
// the controller --- a behavioral description
module booth_controller
       #(parameter W = 8,
         parameter N = 3)(
         input  clk, reset_n, start,
         input  [1:0] mp_qout,
         output reg add_sub, mcand_load,
         output reg [1:0] acc_mode, mp_mode,
         output reg finish);    // the finish flag
// the body of the Booth multiplier
localparam idle = 2'b00, computation = 2'b10,
           right_shift = 2'b11;
// local variable declarations
reg [1:0] ns, ps;
reg [N-1:0] cnt;
// part 1: initialize and update state registers
always @(posedge clk or negedge reset_n)
   if (!reset_n) ps <= idle;
   else          ps <= ns;
// part 2: compute the next state
always @(*)
   case (ps)
      idle: if (!start) ns = idle;
            else        ns = computation;
      computation: ns = right_shift;
      right_shift: if (cnt == 0) ns = idle;
                   else          ns = computation;
      default: ns = idle;
   endcase
// part 3: generate control signals
// define the finish signal
always @(posedge clk)
   if (start) finish <= 0; else
   if ((ps == right_shift) && (cnt == 0)) finish <= 1;
// define the counter block
```

Section 11.2 RTL Design

```verilog
   always @(posedge clk or negedge reset_n)
      if (!reset_n) cnt <= W - 1;
      else if (ps == idle && finish == 0) cnt <= W - 1 ;
         else if (ps == right_shift)    cnt <= cnt - 1;
// all signals are initially set to their initial
// values and then set to active accordingly
   always @(*) begin begin
      mcand_load = 1'b0;  mp_mode = 2'b00;
      acc_mode = 2'b00;   add_sub = 1'b0;     end
      case (ps)
         idle: if (start) begin
            mcand_load = 1'b1;  mp_mode = 2'b10; end
         computation: case (mp_qout[1:0])
            2'b01: begin
               add_sub = 1'b0;
               acc_mode = 2'b10; end // addition
            2'b10: begin
               add_sub = 1'b1;
               acc_mode = 2'b10; end // subtraction
            default: ; endcase
         right_shift: begin
            acc_mode = 2'b01;
            mp_mode  = 2'b01;  end // shift right 1-bit position
         default: ; // do nothing
      endcase
end // end of the always block
endmodule

// the shift register module
module shift_reg #(parameter N = 8)(
        input  clk, reset_n, lsi,
        input  [1:0] mode,
        input  [N-1:0] data,
        output reg [N-1:0] qout,
        output wire rso);
// the shift register body
assign rso = qout[0];
always @(posedge clk or negedge reset_n)
   if (!reset_n) qout <= {N{1'b0}};
   else case (mode)
      2'b00: ;  // no change
      2'b01: qout <= {lsi, qout[N-1:1]}; // shift right
      2'b10: qout <= data;               // parallel load
      default:;  // no operation
   endcase
endmodule
// the register module
module register #(parameter N = 8)(
        input  clk, reset_n, load,
        input  [N-1:0] data,
        output reg [N-1:0] qout);
```

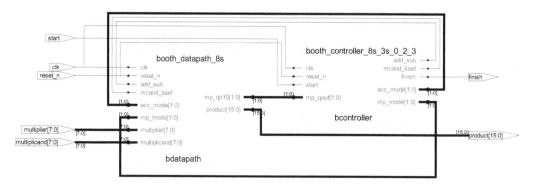

Figure 11.18: The synthesized result of the Booth algorithm based on the DP+CP architecture.

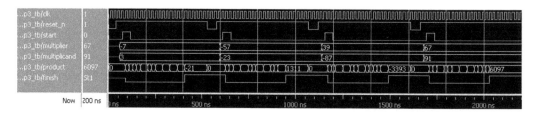

Figure 11.19: The simulation results of the Booth algorithm.

```
// the register body
always @(posedge clk or negedge reset_n)
    if (!reset_n)  qout <= {N{1'b0}};
    else if (load) qout <= data;
endmodule
// the addition and subtraction module
module addsub #(parameter N = 8)(
        input   [N-1:0] a, b,
        input   mode, // define addition or subtraction
        output reg [N-1:0] sum);
// the adder-subtractor body
always @(a or b or mode)
    if    (mode) sum = a - b; // mode = 1 --> subtraction
    else         sum = a + b; // mode = 0 --> addition
endmodule
```

The synthesized result is given in Figure 11.18. From the figure, it is apparent that there are two modules: the datapath and the controller. The details of each module can be seen by dissolving it. Generally, the design using the datapath and controller approach yields a result with an area less than the ASM chart approach. The reader is encouraged to justify this by comparing various modeling approaches described in this section. The simulation results the Booth algorithm are shown in Figure 11.19.

11.2.3.5 ASMD Charts An *algorithmic state machine with datapath* (ASMD) chart [8] is indeed an ASM chart for modeling the controller in the controller-datapath architecture. It can also be regarded as an ASM chart with the following modification:

Section 11.3 RTL Implementation Options

- RTL operations are moved out rather than listed within the state and conditional output boxes to optionally annotate their associated state and conditional output boxes. The state and conditional output boxes identify signals to control the corresponding RTL operations performed by the datapath.

In other words, an ASMD chart with annotation partitions the ASM chart and hence the complex digital system that it represents into the controller and the datapath, and clearly indicates the relationship between them. The RTL operations used to annotate the state and conditional output boxes comprise the datapath whereas the ASMD chart itself is the controller.

With the above modification in mind, an ASMD chart can be generally obtained from the design specifications directly in a way similar to the ASM chart with the following three steps: First, derive and draw an ASMD chart that shows only inputs. Second, annotate optionally state and conditional output boxes with their associated RTL operations. Third, identify control signals that control the corresponding RTL operations of the datapath and fill them into the state and conditional output boxes. Alternatively, an ASMD chart can also be obtained from an ASM chart by using the following three steps:

1. Derive an ASM chart from the design specifications.
2. Move out RTL operations from and optionally put beside to annotate their associated state and conditional output boxes, if any, of the ASM chart.
3. Fill state and conditional output boxes with control signals to control the related datapath operations.

As an illustration of the above procedure, consider the following example that converts the ASM chart shown in Figure 11.11 into an equivalent ASMD chart.

■ Example 11.17: An ASMD chart of the Booth algorithm.

From the above procedure, to convert the ASM chart shown in Figure 11.11 into an equivalent ASMD chart, all RTL operations from both the state and conditional output boxes of the ASM chart are moved out and put beside their corresponding state and conditional output boxes. Then a simple phrase is written into each individual state box and conditional output box to denote the desired control signals. Finally, the control signals corresponding to state and conditional output boxes are filled in place. The resulting ASMD chart is depicted in Figure 11.20.

Because an ASMD chart provides the same information and has the same coding style as an ASM chart, in this book we will not further address the ASMD chart.

■ Review Questions

11-18 What are the three basic components in the datapath and controller architecture?
11-19 Describe how to derive the datapath from an ASM chart?
11-20 Describe how to derive the controller from an ASM chart?
11-21 Describe the operations of the three-step paradigm.
11-22 What is an ASMD chart?

11.3 RTL Implementation Options

There are many options that can be used to implement an RTL design. The rationale behind these options is a trade-off among performance (throughput, operating frequency, or propagation delays), cost (the number of LUTs/LEs, hardware area), and power dissipation (energy).

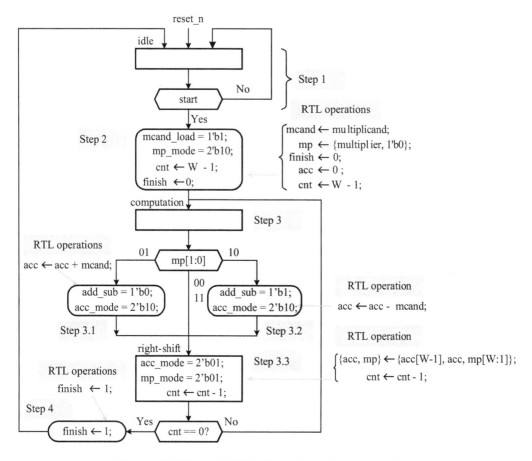

Figure 11.20: An ASMD chart of the Booth algorithm.

The most common options widely used in practical systems are: *single-cycle structure*, *multicycle structure*, and *pipeline structure*. Basic single-cycle structure and multicycle structure are depicted in Figure 11.21. In this section, we discuss these options in detail.

11.3.1 Single-Cycle Structure

Generally speaking, single-cycle structure only uses combinational logic to realize the desired functions as shown in Figure 11.21(a). This may require a quite long propagation delay to finish the desired computations. A simple example of using single-cycle structure is an n-bit adder, which has appeared a number of times in this book.

As another illustration of using single-cycle structure, consider the following example of computing an arithmetic expression with a combination of addition and subtraction.

■ **Example 11.18: An example of single-cycle structure.**

In this example, an arithmetic expression $total = data\_c - (data\_a + data\_b)$ is to be computed in one clock cycle. The propagation delay is the sum of the propagation delays of the adder and the subtracter.

```
// a single-cycle example
module single_cycle_example
       #(parameter N = 8)(
```

Section 11.3 RTL Implementation Options

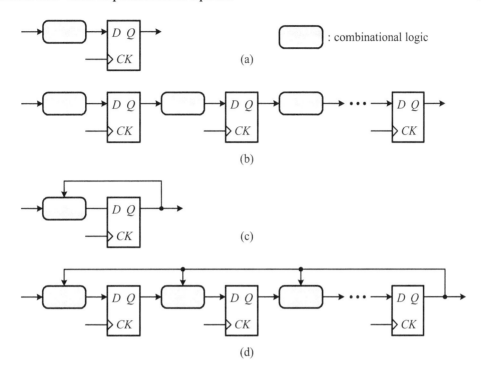

Figure 11.21: The RTL implementation options of digital systems: (a) single-cycle structure; (b) linear multicycle structure; (c) nonlinear single-stage multicycle structure; (d) nonlinear multiple-stage multicycle structure.

```
         input  clk,
         input  [N-1:0] data_a, data_b, data_c,
         output reg [N-1:0] total);

// compute total = data_c - (data_a + data_b)
always @(posedge clk) begin
   total  <= data_c - (data_a + data_b);
end
endmodule
```

The synthesized result of the above single-cycle module is revealed in Figure 11.22. From the figure, we can see that all input data are added together using addition and subtraction units. The result is then stored in the register, `total`.

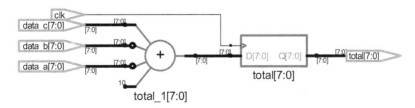

Figure 11.22: A single-cycle structure used to compute the expression: $total = data\_c - (data\_a + data\_b)$.

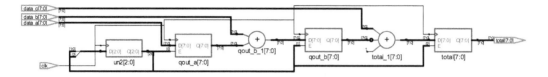

Figure 11.23: A multicycle structure used to compute the expression: $total = data\_c - (data\_a + data\_b)$.

11.3.2 Multicycle Structure

Multicycle structure executes the desired functions in consecutive clock cycles. The actual number of clock cycles needed in a multicycle structure is determined by the specific functions to be computed. Multicycle structure may be further classified into two basic types: *linear multicycle structure* and *nonlinear multicycle structure*.

11.3.2.1 Linear Multicycle Structure A linear multicycle structure simply cascades a number of stages together and performs the desired functions without sharing resources, as shown in Figure 11.21(b). In this structure, multiple copies of the same combinational circuit are needed if two or more stages carry out the same function. As an example of multicycle structure, let us rewrite the preceding example as follows.

■ **Example 11.19: An example of linear multicycle structure.**

This example is to restructure the preceding example as a multicycle structure. The operations of this module is as follows. At the first positive edge of the clock signal, the input data, data_a, is latched into a register, qout_a. The next clock cycle computes the sum of qout_a and data_b and stores the result into register qout_b. The third clock cycle computes the final result and stores the result into register total. As a result, a complete computation of the arithmetic expression requires three clock cycles.

```
// a multiple cycle example
module multiple_cycle_example
       #(parameter N = 8)(
          input   clk,
          input   [N-1:0] data_a, data_b, data_c,
          output reg [N-1:0] total);
reg [N-1:0] qout_a, qout_b;

// compute total = data_c - (data_a + data_b)
always @(posedge clk) begin
   qout_a <= data_a;
   @(posedge clk) qout_b <= qout_a + data_b;
   @(posedge clk)  total <= data_c - qout_b;
end
endmodule
```

The synthesized result of the above example is shown in Figure 11.23. From this figure, we can see that a controller is inferred from the **always** block to schedule the operations for computing the arithmetic expression. It is apparent that this is an example of the implicit FSM. See section 11.1.3 for more details of the implicit FSM.

11.3.2.2 Nonlinear Multicycle Structure Nonlinear multicycle structure can be further cast into single-stage structure and multiple-stage structure, as shown in Figures 11.21(c) and

Section 11.3 RTL Implementation Options

(d), respectively. Nonlinear single-stage multicycle structure performs the desired functions with resource sharing through the use of feedback connection. Nonlinear multiple-stage multicycle structure carries out the required functions with resource sharing through the use of feedback and/or feed-forward connection; that is, it may be a feedback or feed-forward structure or a combination of both. As a consequence, one important feature of nonlinear multicycle structure is that it may reuse the same hardware stage many times. In contrast, the important feature of linear multicycle structure is that it does not contain any feedback or feed-forward paths.

The concept of nonlinear single-stage multicycle structure is illustrated in Figure 11.24 by way of an n-bit serial adder, where the 1-bit adder (i.e., full adder) is reused n times to perform the computation of the desired n-bit addition. At each clock cycle, an individual bit from each operand along with the carry-in, c_in, for the bit are added by the full adder (FA). The resulting sum is then stored in the output register, and the carry-out, c_out, from the full adder is saved in the D flip-flop, which is in turn fed back to the carry-in of the full adder for the addition of the next bit. A controller (not shown) is needed to control the entire operation of the n-bit adder.

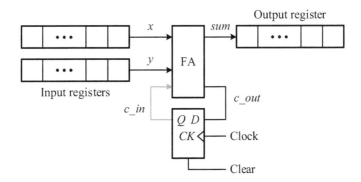

Figure 11.24: An n-bit serial adder.

11.3.2.3 Features of Multicycle Structure
The features of multicycle structure are as follows. At one extreme, the use of nonlinear single-stage multicycle structure needs the minimum hardware resources because it explores the features of resource sharing to an extreme limit. At the other extreme, the use of linear (multiple-stage) multicycle structure needs the maximum hardware resources because it does not explore the features of resource sharing. Nonetheless, the use of linear multiple-stage (including nonlinear too) multicycle structure has a feature of being ready to pipeline the operations that it performs.

As a final comment, a nonlinear single-stage multicycle structure can be generally transformed into a linear multicycle structure by duplicating both the combinational circuit and the register the number of times equal to the required cycles, and then cascading together all stages in a linear manner. This may also result in iterative logic if all registers are removed, to be described later in this section.

11.3.3 Pipeline Structure

A pipeline structure is a multiple-stage multicycle structure in which new data can be fed into the structure at a regular rate (usually, every clock cycle). Therefore, it may output a result regularly (usually, per clock cycle) after the pipeline is fully filled. Pipeline structure may also have linear and nonlinear types. In fact, the difference between the regular multiple-stage multicycle structure and the pipeline structure is on the fact whether the new data are fed into the structure at a regular rate or not. Because of the operations of nonlinear pipeline structure are quite complicated and beyond the scope of this book, we would like to omit them here. In the rest of this section, we only concern the linear pipeline structure.

11.3.3.1 Linear Pipeline Structure The general block diagram of (linear) pipeline structure is depicted in Figure 11.25. From this figure, we can see that a pipeline is composed of a number of stages, with each consisting of a combinational circuit and a register, called a *pipeline register*. In every clock cycle, the data in the pipeline are forwarded to the next stage through which the combinational logic performs the desired computations.

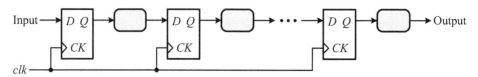

Figure 11.25: The general block diagram of a pipeline structure.

The performance of a pipeline can be measured by two metrics: *pipeline latency* and the *pipeline clock frequency*. Pipeline latency is the number of clock cycles required between the presentation of the input data and the appearance of its associated output. Sometimes, the *pipeline depth* is used to denote the number of stages in a pipeline. The *pipeline clock frequency*, f_{pipe}, is set by the slowest stage and is usually much higher than that in the original single-cycle system.

A number of observations can be obtained from the relationships among the pipeline depth, the pipeline clock period, and pipeline latency. As shown in Figure 11.26, the pipeline clock period is inversely proportional to the pipeline depth because for a given computation as the pipeline depth is increased, the combinational circuit between two pipeline registers becomes less complexity and hence has a shorter propagation delay due to only a simpler logic function needed to execute. In addition, pipeline latency increases with the pipeline depth because each stage of the pipeline consumes one clock cycle and hence virtually both the pipeline latency and the pipeline depth are synonymous in the context of the number of clock cycles.

In order to determine the pipeline clock frequency, we need to estimate the propagation delay of each stage quantitatively. For the ith-stage, the smallest allowable clock period, T_i, is determined by the following condition

$$T_i = t_{cq} + t_{setup} + t_{pd,i} - t_{skew,i+1} \tag{11.1}$$

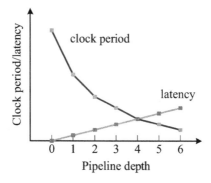

Figure 11.26: The pipeline clock period and pipeline latency versus the pipeline depth.

Section 11.3 RTL Implementation Options

where the first two terms, t_{cq} and t_{setup}, are the clock-to-Q delay and the setup time of flip-flops, respectively. The third term, $t_{pd,i}$, is the propagation delay of the ith-stage combinational logic. The final term, $t_{skew,i+1}$ is the clock skew. Based on these factors, the pipeline clock period for an m-stage pipeline is then chosen to be

$$T_{pipe} = \max\{T_i | i = 1, 2, \ldots, m\}. \tag{11.2}$$

That is, the pipeline clock period is determined by the slowest stage, which has the longest combinational propagation delay. Once the pipeline clock period is determined, the pipeline clock frequency is simply the reciprocal of the pipeline clock period, i.e., $f_{pipe} = 1/T_{pipe}$.

11.3.3.2 Pipeline Modeling From Figure 11.25, we can see that a pipeline structure is virtually a shift register with a combinational circuit placed in between two adjacent registers. In other words, we may regard the shift register as a special case of the pipeline structure in which there is no combinational logic existing in between two adjacent registers. Based on these observations, a pipeline structure can be readily modeled with the same method for modeling shift registers, namely, using nonblocking assignments, as we have described in Chapters 4 and 9.

To concretely demonstrate the modeling of a pipeline structure, we use the arithmetic expression of the preceding multicycle example as an instance. In the examples that follow, we begin with the introduction of a straightforward but improper way of modeling a pipeline structure and then give a correct one.

■ Example 11.20: A simple but not good pipeline example.

As described, a pipeline structure is like a shift register in which the output of a register is connected to the input of its succeeding stage in a linear fashion and all registers are steered by a common clock signal. Based on this, we may simply use nonblocking assignments to describe a pipeline structure. Therefore, the following module is resulted.

```
// a simple pipeline example --- not a good one
module simple_pipeline
       #(parameter N = 8)(
          input   clk,
          input   [N-1:0] data_a, data_b, data_c,
          output reg [N-1:0] total);
// local declaration
reg [N-1:0] qout_a, qout_b;
// compute total = data_c - (data_a + data_b)
always @(posedge clk) begin
   qout_a <= data_a;
   qout_b <= qout_a + data_b;
   total  <= data_c - qout_b;
end
endmodule
```

The synthesized result of the above simple pipeline module is shown in Figure 11.27. The reader is encouraged to synthesize this module and explore its features and problems. Indeed, the above module is incorrect for modeling a pipeline structure because both data_b and data_c are not forwarded to their next stages properly; namely, they are not forwarded in a hop-by-hop fashion. To remedy this, some extra registers are necessary to be added, as illustrated in the following example.

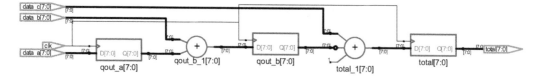

Figure 11.27: An improper pipeline structure used to compute $total = data\_c - (data\_a + data\_b)$.

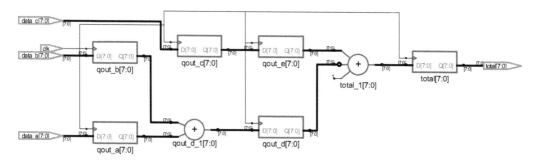

Figure 11.28: A correct pipeline structure used to compute the expression: $total = data\_c - (data\_a + data\_b)$.

■ Example 11.21: A correct pipeline example.

In order to perform a correct pipeline operation, the data in the system must be forwarded in a hop-by-hop way; namely, they must be forwarded to their next stage at the next clock cycle. Based on this fact, the preceding example is modified as follows. The reader is encouraged to compare the **always** block of this example with that of the previous one and find out their differences.

```
// a simple pipeline example
module pipeline_example
       #(parameter N = 8)(
         input   clk,
         input   [N-1:0] data_a, data_b, data_c,
         output reg [N-1:0] total);
reg [N-1:0] qout_a, qout_b, qout_c, qout_d, qout_e;
// compute total = data_c - (data_a + data_b)
always @(posedge clk) begin
   qout_a <= data_a; qout_b <= data_b; qout_c <= data_c;
   qout_d <= qout_a + qout_b;
   qout_e <= qout_c;
   total  <= qout_e - qout_d;
end
endmodule
```

The synthesized result of the above pipeline module is shown in Figure 11.28. It is easy to justify that the result is indeed a pipeline structure exactly like the one shown in Figure 11.25.

■ Review Questions

11-23 Describe the basic features of single-cycle structure.

Section 11.3 *RTL Implementation Options*

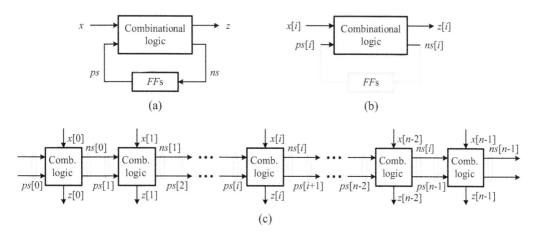

Figure 11.29: The relationships between FSM and iterative logic: (a) the FSM structure; (b) the combinational logic portion; (c) the 1-D iterative logic.

11-24 Describe the basic features of multicycle structure.
11-25 Describe the basic features of pipeline structure.
11-26 How would you determine the pipeline clock period?
11-27 Define the pipeline latency and pipeline depth.

11.3.4 FSM versus Iterative Logic

An *iterative logic* circuit is a circuit composed of a set of basic circuits, referred to as *cells*, in a way such that the set of basic circuits are repeatedly used as many times as needed. An iterative logic circuit can be regarded as an FSM being expanded along the temporal dimension. At each time step, the combinational logic portion of the FSM is duplicated and the memory portion is ignored, as shown in Figure 11.29. Figure 11.29(a) shows the general structure of an FSM, where the next-state function is stored in the state register. The combinational logic portion of the FSM is revealed in Figure 11.29(b). As an FSM is expanded along the temporal dimension n times, an n-stage one-dimensional iterative logic circuit is obtained, as depicted in Figure 11.29(c), where each cell corresponds to the combinational logic portion of the FSM in Figure 11.29(a).

In principle, any sequential circuit implemented by an FSM can be transformed into an iterative logic circuit by duplicating and cascading the combinational logic portion of the FSM in a number of times equal to the length of the input sequence if the storage elements of the FSM are D flip-flops. In what follows, we first use the 0101 sequence detector discussed in Section 11.1.2 as an example to illustrate the above concept of transforming an FSM into one-dimensional (1-D) iterative logic. Then, we employ the Booth algorithm as an illustration to explore the features of two-dimensional (2-D) iterative logic.

11.3.4.1 A 0101-Sequence Detector
The implementation of a 0101 sequence detector using iterative logic is given in Figure 11.30. As can be seen from Figure 11.30(a), each cell receives the present state and current inputs, and produces the next state and output. The function of each cell is described by a state table identical to that of its FSM. Indeed, the design of the ith cell is exactly the same as that of an FSM [5]. Hence, we can derive the transition table and output table from the state table. With the help of Karnaugh maps shown in Figure 11.30(d), we obtain the state transition function and output function and have the combinational logic depicted in Figure 11.30(e).

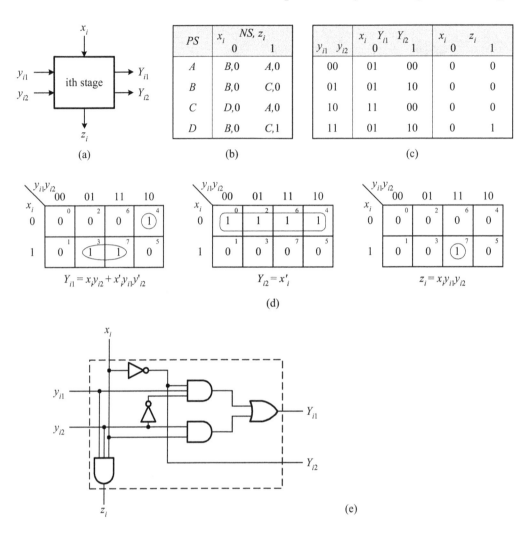

Figure 11.30: The iterative logic implementation of the 0101 sequence detector: (a) the ith-stage block diagram; (b) the ith-stage state table; (c) the transition table and output table; (d) Karnaugh maps; (e) the logic circuit.

An illustration of how to model iterative logic is given in the following example by way of the 0101 sequence detector. In the example, we begin to define the basic cell in accordance with the state diagram from Figure 11.2(b) and then we use a generate-loop statement to instantiate the desired number of stages and cascade them together to form the required iterative logic circuit. To make this module more flexible, we describe it as a parameterizable module.

■ Example 11.22: An iterative logic of 0101 sequence detector.

Because of the inherently structural features of iterative logic, it is convenient to model an iterative logic circuit with a generate-loop statement. The description of the 0101 sequence detector can be subdivided into two types of cells: the LSB and the remaining bits. The LSB cell receives the initial state, namely, A, and an input x, and then produces an output z and a next state. The remaining bit cells have almost the same operations as the LSB cell except that they receive the next state from their previous cells instead of the initial state, A. The

Section 11.3 RTL Implementation Options

definition of each cell contains only the computation of both the next-state function and the output function. The resulting module is as follows:

```verilog
// the behavioral description of 0101 sequence detector
// an iterative logic version
module sequence_detector_iterative
       #(parameter N = 6)( // the default size
           input  [N-1:0] x,
           output [N-1:0] z);
// local declarations
wire   [1:0] next_state[N-1:0];
localparam A = 2'b00, B = 2'b01, C = 2'b10, D = 2'b11;
// produce an N-bit 0101 sequence detector
genvar i;
generate for(i = 0; i < N; i = i + 1) begin: detector_0101
   if (i == 0)begin: LSB  // the LSB cell
      basic_cell bc (x[i], A,  next_state[i], z[i]); end
   else begin: rest_bits  // the remaining bits
      basic_cell bc (x[i], next_state[i-1], next_state[i],
                    z[i]); end
end endgenerate
endmodule

// define the basic cell
module basic_cell(
      input x,
      input [1:0] present_state,
      output reg [1:0] next_state,
      output reg output_z);
localparam A = 2'b00, B = 2'b01, C = 2'b10, D = 2'b11;
// determine the next state
always @(present_state or x)
   case (present_state)
      A: if (x) next_state = A; else next_state = B;
      B: if (x) next_state = C; else next_state = B;
      C: if (x) next_state = A; else next_state = D;
      D: if (x) next_state = C; else next_state = B;
   endcase
// determine the output value
always @(present_state or x)
   case (present_state)
      A: if (x) output_z = 1'b0; else output_z = 1'b0;
      B: if (x) output_z = 1'b0; else output_z = 1'b0;
      C: if (x) output_z = 1'b0; else output_z = 1'b0;
      D: if (x) output_z = 1'b1; else output_z = 1'b0;
   endcase
endmodule
```

An instance of the 0101 sequence detector is given in Figure 11.31, where n is set to 6. Generally, an n-bit 0101 sequence detector is implemented by simply cascading n basic cells together.

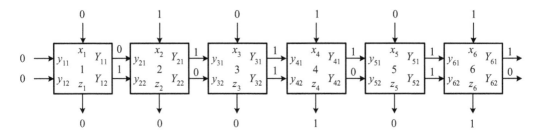

Figure 11.31: An example of a 6-bit 0101 sequence detector implemented by iterative logic.

The above example is one instance of 1-D iterative logic. In a number of practical applications, 2-D iterative logic is also found its widespread use in realizing logic circuits. Such logic circuits are indeed an array structure. As an illustration, the Booth array multiplier is considered in the rest of this section. More examples will be addressed in Chapter 14.

11.3.4.2 A Booth Array Multiplier Recall that the Booth algorithm is implemented as a sequential circuit in Section 11.2.2, namely, using a single-stage multicycle structure. In what follows, we explore how to implement the Booth algorithm with a 2-D iterative logic structure, referred to as an *array structure*. Note that iterative logic usually means a single-cycle structure because it is a combinational circuit in nature.

The array structure of the Booth algorithm is indeed to expand the operations of the single-stage multicycle structure described previously along the temporal dimension. To see this, recall that at each step of the Booth algorithm, two bits of the multiplier are examined from the LSB to the MSB, and then an appropriate operation is taken accordingly. The partial product is then shifted right 1-bit position, equivalent to shifting the multiplicand left 1-bit position. Based on these observations, a 4×4 Booth array multiplier can be derived and shown in Figure 11.32(d). It requires 4×4 complementer and subtractor (CAS) cells and 4 controllers (CTRLs). The function and logic circuit of the CAS cell are revealed in Figure 11.32(a) and (c), respectively. The CTRL controls the operations of CAS cells and has the logic circuit shown in Figure 11.32(b), which can be easily derived from the encoding table shown in Figure 11.32(a).

The key points to successfully model a 2-D array structure are as follows: First, draw a sample array structure with a small size and label each of its components with indexes properly to indicate all related nets and signals. Second, declare a number of 2-D net arrays for intermediate nets, with one array for each type of net. Third, indicate both indexes of x and y directions to be used in generate-loop statements clearly on the sample array structure. As an illustration of how these key points can be applied, an $m \times n$ Booth array multiplier with generate-loop statements is considered in the following example.

■ **Example 11.23: An m×n Booth array multiplier.**

This module consists of two parts: CTRL cells and a multiplier array. The CTRL cells are described in accordance with the logic circuit shown in Figure 11.32(b) except that for the LSB, which can be further simplified. The multiplier array can be partitioned into two cases: the first row and the other rows. In the first row, the operand P_{in} of each CAS cell is zero and in the remaining rows, the operand P_{in} of each CAS cell comes from its previous row. The other operands, y_i, of all rows are the multiplicand. As shown in Figure 11.32(d), the P_{in} of the last CAS cell of all rows except the first row, is from its previous cell at the same row. Based on these observations, the multiplier array can then be modeled with generate-loop statements as in the following module.

Section 11.3 *RTL Implementation Options*

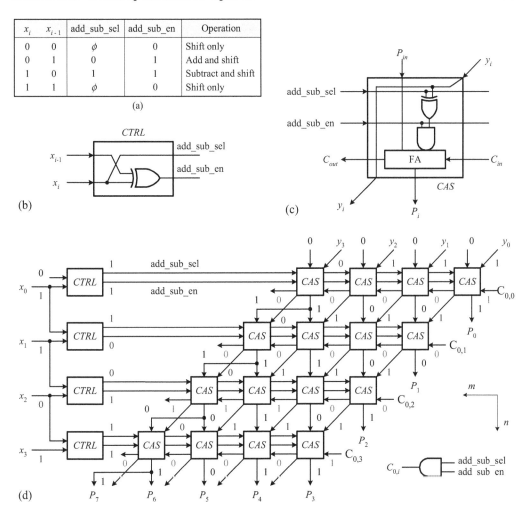

Figure 11.32: A 4×4 Booth array multiplier: (a) the encoding table; (b) CTRL logic; (c) CAS logic; (d) a 4×4 Booth array multiplier.

```
// an M-by-N radix-2 Booth array multiplier
module booth_array_multiplier
       #(parameter M = 4,
         parameter N = 2)(
         input  [N-1:0] x,   // multiplier
         input  [M-1:0] y,   // multiplicand
         output [M+N-1:0] product);
// internal wires
wire sum    [M-1:0][N-1:0]; // internal nets
wire carry  [M-1:0][N-1:0];
wire add_sub_en[N-1:0], add_sub_sel[N-1:0];
genvar i, j;
generate for (i = 0; i < N; i = i + 1) begin: booth_encoder
   // generate add_sub_en and add_sub_sel control signals
   if (i == 0) begin
      assign add_sub_en[i] = x[0];
```

```verilog
            assign add_sub_sel[i] = x[0]; end
        else begin
            assign add_sub_en[i] = x[i]^x[i-1];
            assign add_sub_sel[i] = x[i]; end
end endgenerate
// generate the multiplication array
generate for (i = 0; i < N; i = i + 1) begin: booth_array
    if (i == 0) begin    // describe the first row
        for (j = 0; j < M; j = j + 1) begin: first_row
            if (j == 0) // the LSB of the first row
                assign {carry[j][i],sum[j][i]}=(add_sub_en[i]&
                       (add_sub_sel[i] ^ y[j])) +
                       (add_sub_sel[i] & add_sub_en[i]);
            else   // the remaining bits of the first row
                assign {carry[j][i], sum[j][i]}=(add_sub_en[i]&
                       (add_sub_sel[i] ^ y[j])) + carry[j-1][i];
        end
    end
    else begin: rest_rows // the rest of rows
        for (j = 0; j < M; j = j + 1) begin: rest_rows
            if (j == 0) // describe the LSB of each row
                assign {carry[j][i],sum[j][i]}=(add_sub_en[i]&
                       (add_sub_sel[i] ^ y[j])) + sum[j+1][i-1]+
                       (add_sub_sel[i] & add_sub_en[i]);
            else if (j == M - 1) // describe the MSB of each row
                assign {carry[j][i],sum[j][i]}=(add_sub_en[i]&
                       (add_sub_sel[i]^y[j])) + sum[j][i-1] +
                       carry[j-1][i];
            else     // the remaining bits of each row
                assign {carry[j][i],sum[j][i]}=(add_sub_en[i]&
                       (add_sub_sel[i]^y[j]))+sum[j+1][i-1] +
                       carry[j-1][i];
        end
    end
end endgenerate
// generate product bits
generate for (i = 0; i < N ; i = i + 1)begin: lower_product
    assign product[i] = sum[0][i];
end endgenerate
generate for (i = 1; i < M ; i = i + 1)begin: higher_product
    assign product[N-1+i] = sum[i][N-1];
end endgenerate
    assign product[M+N-1] = sum[M-1][N-1];
endmodule
```

11.3.4.3 Single-Cycle versus Multicycle Structure Up to now, we have described several design examples with both single-cycle and multicycle implementations. A design can be generally implemented by either a single-cycle or a multicycle structure, depending on the performance required and the amount of hardware allowed. To see this, consider the 0101 sequence detector. When it is implemented with a multicycle structure, only one 2-bit register and one combinational circuit are required. Nevertheless, it needs to iterate n

Section 11.4 Case Study — Liquid-Crystal Displays

times when the input sequence is n bits. As a consequence, the total running time is equal to $n \times (t_{comb} + t_{cq} + t_{setup})$. When it is implemented with iterative logic (namely, a single-cycle structure), n copies of the combinational circuit without flip-flops are required. The total running time in this case only equals $n \times t_{comb}$. As a result, iterative logic runs faster at the expense of more hardware resources.

Similarly, for the case of the $m \times n$ Booth algorithm, when implemented with a multicycle structure, it requires three registers for the multiplicand, the multiplier, and the partial product, respectively. In addition, an m-bit adder is needed to add the multiplicand to the partial product. The total running time is at least $2nt_{clk}$, where t_{clk} is determined by the propagation delay of the m-bit adder as well as the t_{cq} and t_{setup} of the registers used. The constant factor 2 counts both the computation and right-shift steps of the ASM chart shown in Figure 11.11. However, when implemented with iterative logic (namely, a single-cycle structure), it requires $m \times n$ CAS cells and n CTRL cells. The total running time is $[2(m-1) + n]t_{CAS}$, where t_{CAS} is the propagation delay of the CAS cell. Of course, this is one more example of a trade-off between the performance and the cost of hardware.

■ Review Questions

11-28 Describe the basic features of iterative logic.
11-29 Describe what 1-D iterative logic is.
11-30 Describe what 2-D iterative logic is.
11-31 What is the relationship between FSMs and iterative logic?

11.4 Case Study — Liquid-Crystal Displays

The *liquid-crystal display* (LCD) is another widely used display device in addition to seven-segment LED display devices. It has ultra-low power dissipation on the order of microwatts, compared to the order of milliwatts for LED devices. In this section, we first consider the basic principles of LCD devices and then use a commercial reflective field-effect LCD module as an example to illustrate how to interface to and utilize an LCD module to display information. Through this case study, the reader should be also able to understand how the datapath and controller approach can be applied to design a digital system heuristically.

11.4.1 Principles of LCDs

A broad variety of LCD devices have been constructed during the past decades. The types of major interest are field-effect and dynamic scattering devices. Both types of LCD devices can operate in the *reflective mode* or the *transmissive mode*. In this book, we only consider the field-effect devices because they are most widely used today. The discussion will include both reflective and transmissive modes.

A field-effect liquid crystal is a material that flows like a liquid and the individual molecules have a rodlike appearance. A useful feature of a liquid crystal is that its optical characteristics can be influenced by an externally applied electric field. Under normal condition (namely, when no applied bias exists), all of rodlike molecules are aligned in a spiral way such that the light passing through it is shifted by 90°. However, when a bias is applied, the rodlike molecules align themselves with the electric field and the light passes directly through it without a 90° shift.

Based on the above property of field-effect liquid crystals, two widely used types of display devices are constructed on the basis of whether the light source is provided internally or externally. A liquid crystal display (LCD) device is in the reflective mode as an external light source is used and in the transmissive mode as an internal light source is employed.

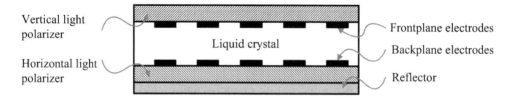

Figure 11.33: The structure of a typical reflective field-effect LCD device.

11.4.1.1 Reflective Field-Effect LCD Devices The basic structure of a reflective field-effect LCD device is shown in Figure 11.33, which consists of six planes: a *vertical light polarizer, frontplane electrodes*, a *liquid crystal, backplane electrodes*, a *horizontal light polarizer*, and a *reflector*.

As described above, the property of a liquid crystal is that the light passing through it will be shifted by 90° when no bias exists and have no shift when a bias is applied. As a result, when no bias is applied to the liquid crystal, the vertically polarized light passing through the vertical light polarizer is shifted by 90° and transferred into the horizontally polarized light. This horizontally polarized light encounters a horizontal light polarizer and passes through it to reach the reflector, where it is reflected back into the liquid crystal, bent back to the other vertical polarization, and returned to the observer. A transparent area appears on the liquid crystal.

However, when a bias is applied, the rodlike molecules align themselves with the electric field and the light passes directly through them without a 90° shift. The vertically incident light cannot pass through the horizontal light polarizer and reach the reflector. A dark area yields on the liquid crystal. Based on these two observations, by designing the frontplane and backplane electrodes into a desired pattern, the resulting device is a liquid-crystal display (LCD) device. A variety of such display devices are widely used in various consumer products and industry controllers as data indicators or signal monitors.

11.4.1.2 Transmissive Field-Effect LCD Devices An important feature of liquid crystals is that the degree of polarization can be controlled by the amount of applied bias. This implies that we are able to control the amount of light passes through the liquid crystal by controlling the magnitude of applied bias. Based on this feature, an LCD device with gray levels can be built. In addition, by properly using a color filter array and controlling the active pixels (picture element), a color LCD display device is resulted.

In order to display images, the basic pixel cells are arranged in a 2-*D* array. Each pixel cell controls the amount of light passing through it. By grouping three basic cells and activating each individual cell with a variable bias, a color image dot may be formed if a color filter is used appropriately. To achieve this and provide high integration density, the bias for the basic cell is applied by using an active element, usually a *thin-film transistor* (TFT). The active element allows the transfer of the signal at each pixel to the liquid crystal cell controlled by a proper timing signal. Because active elements are used and the pixel cells are arranged into a 2-*D* array, the resulting LCD device is known as an *active-matrix LCD* device or a *TFT LCD panel*. Sometimes we call it an *LCD panel* for short. Active-matrix LCD devices are widely used in TV sets, computer systems, instruments, consumers, cell phones, and so on.

Figure 11.34 shows the basic structure of an active-matrix LCD device. Like the reflective LCD device, an active-matrix LCD device is still composed of polarizers, electrodes, and a liquid crystal. However, an internal light source known as the *backlight* is also needed. During the past two decades, the dominant backlighting technology for LCD devices is the *cold cathode fluorescent lamp* (CCFL), which uses a low-voltage DC to high-voltage AC converter as the driver. This driver consumes the largest amount of power in the display system. For this reason, white LEDs have become the major light sources for the active-matrix LCD panels

Section 11.4 Case Study — Liquid-Crystal Displays

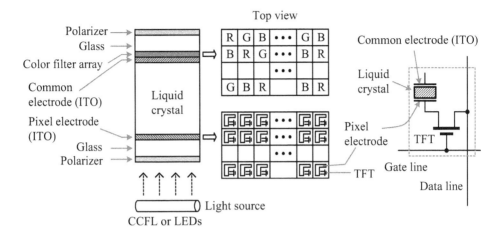

Figure 11.34: The structure of a typical active-matrix LCD device.

nowadays. Compared with the CCFL approach, the LED backlight approach may reduce up to 30% power dissipation.

One major feature distinguishing it from the above mentioned reflective LCD device is that each pixel cell of the active-matrix LCD device can be controlled in such a way that the amount of light passing through it is determined by the bias voltage applied to it. To achieve this, both electrodes of each pixel cell are made of *indium tin oxide* (ITO), a transparent conducting material.

Although active-matrix LCD devices find their widespread use in consumer products, their controllers are too complicated to be discussed and beyond the scope of this book. Hence, we will not further discuss them here. In the rest of this section, we consider the most common dot-matrix LCD modules, which are based on the reflective field-effect LCD device and often find their broad use in various consumer products except the large-scale display devices. These LCD devices are usually made into various patterns, such as seven-segment digits and dot-matrices, to display digits and characters.

■ Review Questions

11-32 Distinguish between the reflective mode and transmissive mode of LCD devices.
11-33 Describe the operation of field-effect liquid crystals.
11-34 Describe the principle of reflective field-effect LCD devices.
11-35 Describe the principle of transmissive field-effect LCD devices.
11-36 What is the meaning of active-matrix LCD devices?

11.4.2 Commercial Dot-Matrix LCD Modules

A widely used commercial dot-matrix LCD module is one that uses a Hitachi 44780 or its equivalent as a controller, as shown in Figure 11.35. In what follows, we address the controller architecture, access timing, and access commands of this dot-matrix LCD module.

11.4.2.1 Controller Architecture The controller contains a command register, an address counter (AC), a character generator (CG) ROM, a character generator (CG) RAM, a display data (DD) RAM, and a busy flag (BF).

The CG ROM is accessed by the ASCII code and is able to generate 192 character patterns, including 160 5×7 character patterns and 32 5×10 character patterns. The CG RAM allows

484   Chapter 11   System Design Methodologies

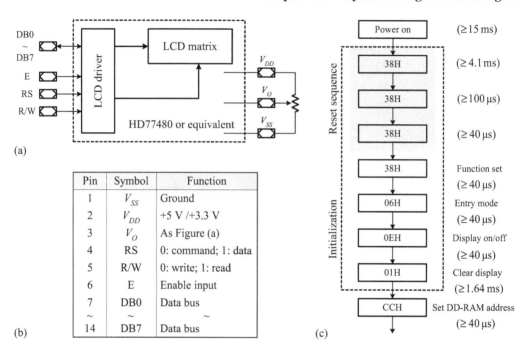

Figure 11.35: The (a) block diagram, (b) pin assignments and functions, and (c) initialization process of a typical commercial reflected LCD device.

the user to create up eight custom 5×7 characters plus a 5×8 cursor. The display data (DD) RAM is capable of buffering up to 80 or even more bytes, depending on the types of LCD modules. The command register stores the command to be executed. The address counter (AC) stores the address of CG ROM or DD RAM. Because of only one address counter being available, only either CG ROM or DD RAM can be accessed at a time. The busy flag (BF) indicates whether the LCD module is working on a command.

The standard electrical interface of the LCD module is shown in Figure 11.35(a). The pin assignments and associated functions are revealed in Figure 11.35(b). The functions of pins are as follows. DB7 to DB0 are the eight-bit data bus. There are two modes: 4 bits and 8 bits. In 4-bit mode, only DB7 to DB4 are required and each command or data are sent twice, one nibble at a time. The upper nibble is transferred first and then the lower nibble follows. This mode is useful for interfacing with 4-bit microcomputers, which are widely used in low-end consumer products. The other three control signals, enable (E), register select (RS), and read/write (R/W), are used to control the access of the LCD module. The enable signal E is to enable a read or write cycle. The RS selects the register to be accessed. When RS = 1, the data buffer is selected; when RS = 0, the command register is selected. The R/W signal controls the access mode. When R/W = 1, the access mode is read; when R/W = 0, the access mode is write.

11.4.2.2 Access Timing To access the LCD module properly, the above interface signals must be applied to the LCD module following the timing specification shown in Figure 11.36, including both the read and write timing of the LCD module. In the read cycle, both RS and R/W must be stable for a period of time t_{AS} before the rising edge of E and must remain stable for another period of time t_{AH} after the falling edge of E. The data DB7 to DB0 appear at the data bus within a period of time t_{DDR} after the rising edge of E and remain stable for a period of time t_{DHR} after the falling edge of E. The pulse width of E must be at least t_{WEP} and the period of E must be at least t_{CYC}. Detailed values of these parameters are listed in Figure 11.36(c).

Section 11.4 Case Study — Liquid-Crystal Displays

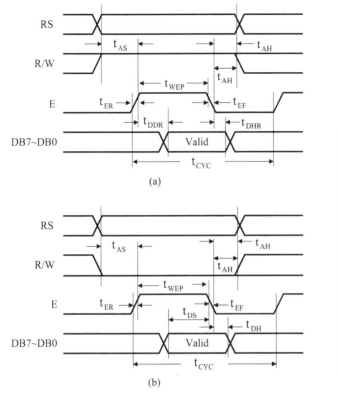

Figure 11.36: The (a) read cycle, (b) write cycle, and (c) timing parameters of typical LCD modules.

In the write cycle, both RS and R/W have exactly the same timing relationship as in the read cycle. The data on the data bus is latched into the LCD module on the falling edge of E. The data DB7 to DB0 to be written must be stable at the data bus for a period of time t_{DS} before the falling edge of E and must remain stable for another period of time t_{DH} after the falling edge of E. Detailed values of the various parameters are listed in Figure 11.36(c).

11.4.2.3 Access Commands In the LCD module, there are two registers: the *command register* and the *data register*. These two registers are selected by the RS signal, as described before. Another register, called the address counter (AC), specifies an address when accessing CG RAM or DD RAM.

There are a number of commands associated with the LCD module for controlling and accessing the module, as summarized in Table 11.2. In what follows, we describe each command in detail.

- **Clear display.** This command clears all display memory and returns the cursor to the home position.
- **Cursor home.** It returns the cursor to the home position.
- **Entry mode set.** This command sets the entry mode and makes the cursor automatically move to the right when incremented by one or to the left when decremented by one. The I/D bit sets the address counter (AC) to increment (I/D = 1) or decrement (I/D = 0) by one when a character is written to or read from DD RAM or CG RAM. After each data is written to DD RAM, the entire display can be shifted either right or left (S = 1). When

Table 11.2: The command summary of typical commercial dot-matrix LCD devices.

RS	RW	DB7	DB6	DB5	DB4	DB3	DB2	DB1	DB0	Function
Clear display										Execution time: 82 μs to 1.64 ms.
0	0	0	0	0	0	0	0	0	1	Clear all display memory and place the cursor to home location (i.e., 0).
Cursor home										Execution time: 40 μs to 1.6 ms.
0	0	0	0	0	0	0	0	1	*	Return the cursor to home location (i.e., 0). DD RAM contents remain unchanged.
Entry mode set										Execution time: 40 μs to 1.64 ms.
0	0	0	0	0	0	0	1	I/D	S	I/D=1: Increment; I/D=1: Decrement. S=1: Display shift; S=0: No display shift.
Display and cursor on/off control										Execution time: 40 μs.
0	0	0	0	0	0	1	D	C	B	D=1: Display ON; D=0: Display OFF. C=1: Cursor ON; C=0: Cursor OFF. B=1: Blink ON; B=0: Blink OFF.
Cursor or display shift										Execution time: 40 μs.
0	0	0	0	0	1	S/C	R/L	*	*	S/C=1: Display shift; S/C=0: cursor movement. R/L=1: Right shift; R/L=0: left shift. DD RAM remains unchanged.
Function set										Execution time: 40 μs.
0	0	0	0	1	DL	N	F	*	*	DL=1: 8 bits; DL=0: 4 bits. N=1: 2 lines; F=0: 5*7 dot matrix.
CG-RAM address set (CG RAM: Character generator RAM)										Execution time: 40 μs.
0	0	0	1	A_{CG} (CG-RAM address)						
DD-RAM address set (DD RAM: Display data RAM)										Execution time: 40 μs.
0	0	1	A_{DD} (DD-RAM address)							
Busy flag and address read										Execution time: 1 μs.
0	1	BF	A_C (Address counter for DD- and CG-RAM address)							
Data write to CG or DD RAM										Execution time: 40 μs.
1	0	Data to be written								
Data read from CG or DD RAM										Execution time: 40 μs.
1	1	Data read								

S = 1 and I/D = 1, the display is shifted left one character position; when S = 1 and I/D = 0, the display is shifted right one character position.

- **Display and cursor on/off control.** This command controls the display and cursor operations. It includes three control bits: D, C, and B. The D bit controls whether the display is turned on (D = 1) or off (D = 0). The display data remain unchanged when the display is turned off. The C bit controls whether the cursor is displayed (C = 1) or not displayed (C = 0). The B bit controls whether the cursor is blinking (B = 1) or not (B = 0).
- **Cursor or display shift.** This command moves the cursor or shifts the display without

Section 11.4 Case Study — Liquid-Crystal Displays

changing DD RAM contents. It consists of two bits: S/C and R/L. The four operation modes are as follows:

1. **S/C R/L = 00:** The cursor position is shifted left (AC decrements one).
2. **S/C R/L = 01:** The cursor position is shifted right (AC increments one).
3. **S/C R/L = 10:** The entire display is shifted left with the cursor.
4. **S/C R/L = 11:** The entire display is shifted right with the cursor.

- **Function set.** This command sets the bus width and the number of display lines as well as the character font. It consists of three bits: DL, N, and F. When DL = 1, the data bus width is set to 8 bits; when DL = 0, the data bus width is set to 4 bits (DB7 to DB4). The upper nibble is transferred first and then the lower nibble follows. The N bit specifies the number of display lines. When N = 1, the display is two lines. The F bit sets the character font. When F = 0, the character is a 5×7 dot-matrix.
- **CG-RAM address set.** This command sets the CG-RAM address to the contents of AC before the CG RAM can be accessed.
- **DD-RAM address set.** This command sets the DD-RAM address to the contents of AC before the DD RAM can be accessed.
- **Busy flag/address read.** This command reads the busy flag and the contents of AC. The busy flag indicates whether the controller is working on a current command (BF = 1) or not (BF = 0). When BF = 1, the module is working internally and cannot accept a new command. Therefore, it is necessary to ensure BF = 0 before writing the next command to the LCD module.
- **Data write to CG RAM or DD RAM.** This command writes data to either CG RAM or DD RAM, which is set by the CG-RAM address set command or the DD-RAM address set command before this command is executed.
- **Data read from CG RAM or DD RAM.** This command reads data from either CG RAM or DD RAM, which is set by the CG-RAM address set command or the DD-RAM address set command before this command is executed.

11.4.2.4 Cursor Addresses
From the DD-RAM address set command, we know that

```
RS R/W DB7 DB6 DB5 DB4 DB3 DB2 DB1 DB0
 0   0   1   A   A   A   A   A   A   A
```

where AAAAAAA = 0000000 to 0100111 for line 1 and AAAAAAA = 1000000 to 1100111 for line 2. The upper address limit can be up to 0100111 (39) for the 40-character-wide LCD and up to 0010011 (19) for the 20-character-wide LCD. Based on this, some widely used address ranges of commercial LCD devices are summarized in Table 11.3.

11.4.2.5 Initialization Process
When the power supply is turned on and before any access to the LCD module can proceed, an initialization process must be carried out. This initialization process is a sequence of commands used to setup the modes of the LCD module, as revealed in Figure 11.35(c). The details of the initialization sequence are stated in words as follows:

■ Algorithm 11-1: Initialization of an LCD module

Begin

1. Wait for at least 15 ms after V_{DD} has risen to 4.5 V.
2. Write command 8'h38 and wait for at least 4.1 ms
3. Write command 8'h38 and wait for at least 100 μs

Table 11.3: Cursor addresses for some typical commercial dot-matrix LCD devices.

Type	Address range								
16 × 2 LCD	8'h80	8'h81	8'h82	8'h83	...	8'h8c	8'h8d	8'h8e	8'h8f
	8'hc0	8'hc1	8'hc2	8'hc3	...	8'hcc	8'hcd	8'hce	8'hcf
20 × 1 LCD	8'h80	8'h81	8'h82	8'h83	...	8'h90	8'h91	8'h92	8'h93
20 × 2 LCD	8'h80	8'h81	8'h82	8'h83	...	8'h90	8'h91	8'h92	8'h93
	8'hc0	8'hc1	8'hc2	8'hc3	...	8'hd0	8'hd1	8'hd2	8'hd3
20 × 4 LCD	8'h80	8'h81	8'h82	8'h83	...	8'h90	8'h91	8'h92	8'h93
	8'hc0	8'hc1	8'hc2	8'hc3	...	8'hd0	8'hd1	8'hd2	8'hd3
	8'h94	8'h95	8'h96	8'h97	...	8'ha4	8'ha5	8'ha6	8'ha7
	8'hd4	8'hd5	8'hd6	8'hd7	...	8'he4	8'he5	8'he6	8'he7
40 × 2 LCD	8'h80	8'h81	8'h82	8'h83	...	8'ha4	8'ha5	8'ha6	8'ha7
	8'hc0	8'hc1	8'hc2	8'hc3	...	8'he4	8'he5	8'he6	8'he7

4. Write command 8'h38 and wait for at least 40 μs
5. Write the following commands with the desired execution time in sequence:

```
function set:    8'b0011_NFxx
entry mode:      8'b0000_0110
display on/off:  8'b0000_1110
clear display:   8'b0000_0001
```

End

■ Review Questions

11-37 What is the function of CG ROM? Can we access CG ROM directly?
11-38 What is the function of DD RAM?
11-39 Can we access both DD RAM and CG RAM at the same time?
11-40 Describe the write cycle of the dot-matrix LCD module.
11-41 Describe the initialization process of the dot-matrix LCD module.

11.4.3 Datapath Design

Now that we know the details of the widely used commercial LCD module, our focus in the rest of this section is on the design of a hardware module to interface the commercial LCD module. Upon understanding this hardware module, the reader can easily modify the underlying hardware module to his or her own applications.

For simplicity, suppose that we want to design an LCD driver that displays four decimal digits on the LCD module starting at the address of 8'hcc (from 8'hcc to 8'hcf.) The digits may come from a wide variety of sources, such as the output of a 4-digit decimal counter, the output of a clock/timer, the output of a keypad scanner, and other data sources. A clock signal with a period of 1 ms is employed in all modules to simplify the design.

As described before, the essential operations of using an LCD module for displaying information are to write/read commands and data to/from the LCD module through the electrical interface of the module properly. In the initialization process, many constants used. These constants along with the starting address may be gathered together with the 4-digit input as a unit,

Section 11.4 Case Study — Liquid-Crystal Displays

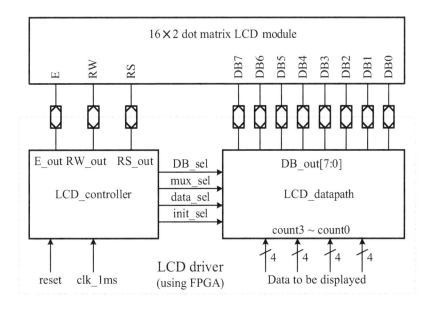

Figure 11.37: The block diagram of a 4-digit LCD driver.

called the datapath, because these data are required to be sent to the LCD module through the data bus, as shown in Figure 11.37.

Once the datapath has been constructed, the next problem to be solved is how to schedule the constants and the 4-digit input to be sent to the LCD module in a way such that all data must be written into the LCD module following the timing constraints of the write cycle revealed in Figure 11.36(b). This problem is solved by a circuit called the controller. As a result, the LCD driver is composed of two parts, a datapath and a controller, as shown in Figure 11.37. In this figure, we also show the details of the interface between the LCD driver realized in an FPGA device and the LCD module.

As mentioned above, the datapath of the LCD driver contains the constants used in the initialization process and the starting address of DD RAM as well as the 4-digit data to be displayed. However, only one constant or one digit data is allowed to route to the data bus at a time. Consequently, a multiplexer tree is used to route these data one by one to the data bus DB7 to DB0 in a way scheduled by the controller. Details of the datapath of the LCD driver are depicted in Figure 11.38.

As shown in Figure 11.38, the datapath consists of two 4-to-1 multiplexers and two 2-to-1 multiplexers. The initialization multiplexer (`init_reset_mux`) is a 6-bit 4-to-1 multiplexer because the two most significant bits of all initialization constants are 0s and hence they can be concatenated at the output of the multiplexer. Similarly, the multiplexer used for routing the 4-digit input is a 4-bit 4-to-1 multiplexer (`counter_mux`). At the output of this multiplexer, a 4-bit constant 4'b0011 is prefixed to convert the input decimal digits into their ASCII codes. The 8-bit 2-to-1 data_select multiplexer (`data_out`) chooses either the initialization constants or the input data to be sent to the LCD module. It should be noted that this 8-bit 2-to-1 multiplexer in fact can be replaced with a 6-bit 2-to-1 multiplexer by cascading the two most significant bits at the output instead of at the input. However, it would be much clearer to leave it as it is now. Another 8-bit 2-to-1 multiplexer (`DB_out`) is used to route another constant, namely, the starting address, to the LCD module. As a result, the datapath has four sets of control signals that are under the control of the controller. This datapath can be modeled as in the following example.

■ Example 11.24: The datapath of the 4-digit LCD driver.

For simplicity, in this example, we use **case** statements to describe various multiplexers used in the datapath shown in Figure 11.38. Each **case** statement within an **always** block implements a multiplexer.

```verilog
// an LCD datapath --- a multiplexer used to select
// appropriate data for displaying on a two-line
// dot-matrix LCD panel
module LCD_datapath (
        input  [1:0] mux_sel, init_sel,
        input  data_sel, DB_sel,
        input  [3:0] count3, count2, count1, count0,
        output reg [7:0] DB_out );
reg    [3:0] counter_mux;
reg    [5:0] init_reset_mux;
reg    [7:0] data_out;
// the counter_mux multiplexer
always @(*)
   case (mux_sel)
      2'b00: counter_mux = count3;
      2'b01: counter_mux = count2;
      2'b10: counter_mux = count1;
      2'b11: counter_mux = count0;
   endcase
// the initialization and reset process
always @(*) // the init_reset_mux multiplexer
   case (init_sel)
      2'b00: init_reset_mux = 6'b000001;// 01H--00_000001
      2'b01: init_reset_mux = 6'b001110;// 0EH--00_001110
```

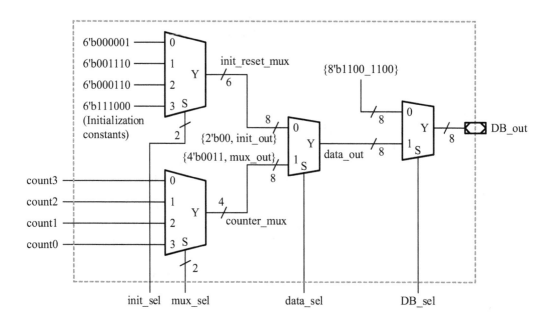

Figure 11.38: The datapath of the 4-digit LCD driver.

Section 11.4 *Case Study — Liquid-Crystal Displays*

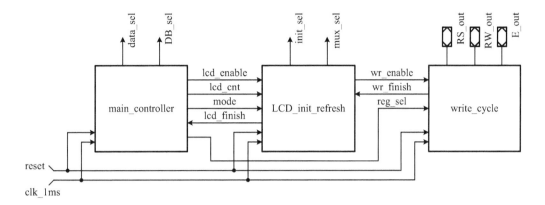

Figure 11.39: The controller of the 4-digit LCD driver.

```
         2'b10: init_reset_mux = 6'b000110;// 06H--00_000110
         2'b11: init_reset_mux = 6'b111000;// 38H--00_111000
      endcase
   always @(*) // select the initialization or normal data
      case (data_sel)
         1'b0: data_out = {2'b00, init_reset_mux};
         1'b1: data_out = {4'b0011, counter_mux};
      endcase
   always @(*) // select the starting address or normal data
      case (DB_sel)
         1'b0: DB_out = 8'b1100_1100;    // address at CCH
         1'b1: DB_out = data_out;
      endcase
endmodule
```

11.4.4 Controller Design

As described before, the controller has to properly schedule the constants or input data sent to the LCD module and generate a correct timing relationship between various control signals according to the timing specifications of the write cycle revealed in Figure 11.36(b). For convenience, the controller is partitioned into three modules, the main controller, the LCD initialization and refresh module, and the write-cycle module, as shown in Figure 11.39.

The main controller module determines when to initialize and when to refresh the LCD module. The LCD initialization and refresh module performs the initialization process and refresh operations of the LCD module. The write-cycle module generates the required write-cycle timing for writing data to the LCD module. To design the controller, we assume that a clock signal with a period of 1 ms is used. This assumption not only simplifies the design considerably but also conforms to the write-cycle timing specifications of the LCD module.

In the rest of this section, we describe each module in more detail. Before proceeding, the top-level module of the controller of the 4-digit LCD driver is described first as follows.

■ Example 11.25: The controller of the 4-digit LCD driver.

The top-level module of the controller of the 4-digit LCD driver under consideration instantiates three lower-level modules: `main_controller`, `LCD_init_refresh`, and `write_cycle`.

```verilog
// an LCD driver --- initialize and write data to the LCD module
module LCD_controller(
        input   clk_1ms, reset,
        output  [1:0] mux_sel, init_sel,
        output  data_sel, DB_sel, RS_out, E_out, RW_out);
wire mode,lcd_enable,lcd_finish,reg_sel,wr_enable,wr_finish;
wire [1:0] lcd_cnt;

// instantiate various modules
main_controller  main_contl
    (clk_1ms, reset, lcd_finish, lcd_enable, mode,
     reg_sel, data_sel, DB_sel, lcd_cnt);
LCD_init_refresh LCD_wr_data
    (clk_1ms, reset, lcd_enable, wr_finish, mode,
     lcd_cnt, mux_sel, init_sel, lcd_finish, wr_enable);
write_cycle      write_cycle
    (clk_1ms, reset, wr_enable, reg_sel, wr_finish,
     RS_out, E_out, RW_out);
endmodule
```

11.4.4.1 Write-Cycle Module The write-cycle module generates the required write-cycle timing conforming to the timing specifications shown in Figure 11.36(b). According to the timing specifications, it is easy to partition the timing into four states: outputs RS and R/W signals (wr_idle), activates the E signal for two states (wr_init and wr_Eout), and deactivates the E signal (wr_end). More details of the write cycle are shown in Figure 11.40 represented as an ASM chart.

Because the write-cycle module is required for writing each input data or a constant to the LCD module, a handshaking mechanism is needed to coordinate the operations between the write-cycle module and the `LCD_init_refresh` module. For this purpose, two control signals are used: `wr_enable` and `wr_finish`. The `wr_enable` signal enables the operations of the write-cycle module. When the write-cycle module completes its assigned operations, it asserts the `wr_finish` signal to notify the `LCD_init_refresh` module of this fact. The ASM chart shown in Figure 11.40 is modeled as follows.

■ Example 11.26: The write-cycle module of the controller.

This module generates the required timing for writing data or a constant to the LCD module. The detailed operations of the module are exactly the same as those described in the ASM chart shown in Figure 11.40. Hence, we will not further explain them.

```verilog
// a write-timing generator -- a four-state FSM
// it is enabled by wr_enable and asserts the wr_finish
// signal when it completes its operations
module write_cycle(
        input  clk_1ms, reset, wr_enable, reg_sel,
        output reg wr_finish, RS_out, E_out,
        output RW_out);
reg [1:0] wr_state;
```

Section 11.4 Case Study — Liquid-Crystal Displays

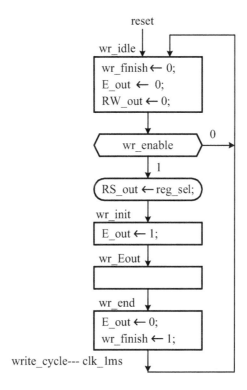

Figure 11.40: The ASM chart for the write-cycle operations of the 4-digit LCD driver.

```
// define local parameters
localparam wr_idle = 2'b00, wr_init = 2'b01,
           wr_Eout = 2'b10, wr_end  = 2'b11;

// in the write cycle RW_out is always 0
assign RW_out = 1'b0;
always @(posedge clk_1ms or posedge reset)
   if (reset) wr_state <= wr_idle;
   else case (wr_state)
      wr_idle: begin
         wr_finish <= 0; E_out <= 1'b0;
         if (wr_enable) begin
            wr_state <= wr_init; RS_out <= reg_sel; end
         else wr_state <= wr_idle; end
      wr_init: begin
         wr_state <= wr_Eout; E_out <= 1'b1; end
      wr_Eout: wr_state <= wr_end;
      wr_end: begin
         wr_state <= wr_idle; E_out <= 1'b0;
         wr_finish <= 1; end
   endcase
endmodule
```

11.4.4.2 LCD Initialization and Refresh Module Because of the identical operations of both initialization and (data) refresh processes, both processes can be combined into a single module, called the LCD initialization and refresh module, `LCD_init_refresh`. The goal of the `LCD_init_refresh` module is to schedule the initialization constants and the input data to be sent to the LCD module through the data bus. Hence, it has to generate the desired selection signals, `mux_sel` and `init_sel`, for counter_mux and init_reset_mux multiplexers, as shown in Figure 11.41. After this, it has to enable the write-cycle module to generate the required write-cycle control signals. In addition, the `LCD_init_refresh` module is enabled in either the initialization process or the refresh process. The initialization process is only executed once when the reset signal is asserted, whereas the refresh process is executed repeatedly to update the LCD module with the data. Like the way that the write-cycle module is enabled by the LCD initialization and refresh module, the LCD initialization and refresh module is enabled by the main controller module, which determines when to initialize the LCD module and when to refresh the LCD module with the input data. The control mechanism used in this module comprises both `lcd_enable` and `lcd_finish` signals. The detailed operations of this module are revealed in the ASM chart shown in Figure 11.41. This ASM chart can be described as a Verilog HDL module given in the following example.

■ **Example 11.27: The LCD initialization and refresh module.**

This module receives the `lcd_enable`, `lcd_cnt`, and `mode` control signals from the main controller module, `main_controller`, and generates the required control signals, `init_sel` and `mux_sel`, to the LCD module. Besides, it enables the write-cycle module, `write_cycle`, to generate the required timing for writing initialization constants or data onto the LCD module. The detailed operations of the module are identical to those described in the ASM chart shown in Figure 11.41 and hence we will not further explain them.

```verilog
// the LCD initialization and refresh module --- a 4-state FSM
module LCD_init_refresh(
       input   clk_1ms, reset, lcd_enable, wr_finish, mode,
       input   [1:0] lcd_cnt,
       output reg [1:0] mux_sel, init_sel,
       output reg lcd_finish,wr_enable );
reg [1:0] state;
localparam LCD_INIT = 0;
localparam lcd_idle  = 2'b00, lcd_data = 2'b01,
           lcd_data1 = 2'b10, lcd_end  = 2'b11;

always @(posedge clk_1ms or posedge reset)
   if (reset) state <= lcd_idle;
   else case (state)
      lcd_idle: begin
         lcd_finish <= 0; mux_sel <= 0;
         init_sel <= 0; wr_enable <= 0;
         if (lcd_enable) begin
            state <= lcd_data;
            if (mode == LCD_INIT) init_sel <= lcd_cnt;
            else mux_sel <= lcd_cnt; end
         else state <= lcd_idle; end
      lcd_data: begin
         wr_enable <= 1; state <= lcd_data1; end
      lcd_data1:  begin
```

Section 11.4 Case Study — Liquid-Crystal Displays

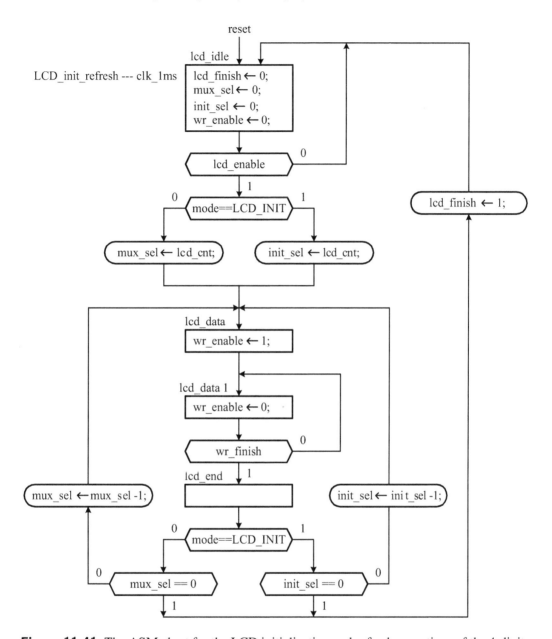

Figure 11.41: The ASM chart for the LCD initialization and refresh operations of the 4-digit LCD driver.

```
           wr_enable <= 0;
           if (wr_finish == 1) state <= lcd_end;
           else state <= lcd_data1; end
    lcd_end:
           if (mode == LCD_INIT) begin
              if (init_sel == 0) begin
                 lcd_finish <= 1; state <= lcd_idle; end
              else begin
                 init_sel <= init_sel - 1;
```

```
                    state <= lcd_data; end end
         else begin
            if (mux_sel == 0) begin
               lcd_finish <= 1; state <= lcd_idle; end
            else begin
               mux_sel <= mux_sel - 1;
               state <= lcd_data; end end
   endcase
endmodule
```

11.4.4.3 Main Controller Module The main controller module determines when to initialize the LCD module and when to update the LCD module with the input data, and then routes the required data to the LCD module through the rightmost two 8-bit 2-to-1 multiplexers, as shown in Figure 11.38. It enables the LCD initialization and refresh module through a handshaking mechanism with two signals: lcd_enable and lcd_finish. The detailed operations of the main controller module are revealed in the ASM chart shown in Figure 11.42.

The ASM chart can be divided into three parts, with each containing two states. The first part performs the initialization process by enabling the LCD_init_refresh module to write the initialization constants onto the LCD module. The second part sets the starting address of DD RAM. This address must be set each time the refresh operation is executed so that the data may be displayed on the LCD module at the same place because the cursor of the LCD display is moved right one-digit position each time a digit is written. The third part carries out the refresh operations.

■ Example 11.28: The main controller module of the controller.

This module generates both data_sel and DB_sel control signals for the datapath shown in Figure 11.38 in order to route initialization constants, data to be displayed, or the starting address to the LCD module. It also generates lcd_enable, lcd_cnt, and mode signals to control the appropriate operations of the LCD_init_refresh module. The detailed operations of the module are identical to those described in the ASM chart shown in Figure 11.42 and thus we will not further explain them.

```
// the main controller --- schedule the initialization
// and refresh module
module main_controller(
        input    clk_1ms, reset, lcd_finish,
        output reg lcd_enable, mode, reg_sel, data_sel, DB_sel,
        output reg [1:0] lcd_cnt);

reg [2:0] main_state;
localparam INIT_CONST_NO = 4,
          REF_DATA_NO   = 4;
localparam LCD_INIT = 0, LCD_REF = 1;

// the main-loop operation --- a seven-state FSM
localparam main_idle = 3'b000, main_init  = 3'b001,
           main_addr = 3'b010, main_addr1 = 3'b011,
           main_ref  = 3'b100, main_ref1  = 3'b101;

always @(posedge clk_1ms or posedge reset)
   if (reset) main_state <= main_idle;
```

Section 11.4 Case Study — Liquid-Crystal Displays

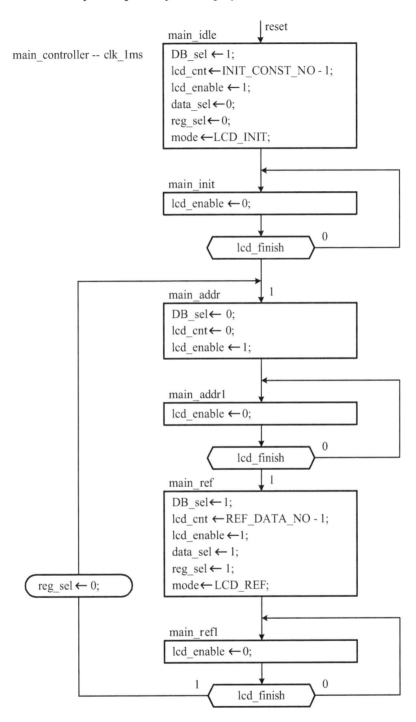

Figure 11.42: The ASM chart for the main controller operations of the 4-digit LCD driver.

```
else case (main_state)
   main_idle: begin // begin initialization
      DB_sel <= 1'b1; lcd_cnt <= INIT_CONST_NO - 1;
      lcd_enable <= 1; data_sel <= 1'b0;
```

```verilog
         reg_sel <= 1'b0; mode <= LCD_INIT;
         main_state <= main_init;   end
   main_init: begin  // inializing ...
         lcd_enable <= 0;
         if (lcd_finish) main_state <= main_addr;
         else main_state <= main_init; end
   main_addr: begin // set up the starting address
         DB_sel <= 1'b0; lcd_enable <= 1;
         lcd_cnt <= 0;   // send only one data
         main_state <= main_addr1; end
   main_addr1: begin
         lcd_enable <= 0;
         if (lcd_finish) main_state <= main_ref;
         else main_state <= main_addr1; end
   main_ref:  begin   // begin to refresh
         DB_sel <= 1'b1;  lcd_cnt <= REF_DATA_NO - 1;
         lcd_enable <= 1; data_sel <= 1'b1;
         reg_sel <= 1'b1; mode <= LCD_REF;
         main_state <= main_ref1;  end
   main_ref1: begin
         lcd_enable <= 0;
         if (lcd_finish) begin
            reg_sel <= 1'b0; main_state <= main_addr; end
         else main_state <= main_ref1; end
   endcase
endmodule
```

■ Review Questions

11-42 Describe the structure of the datapath module of the 4-digit LCD driver.

11-43 Describe the structure of the controller module of the 4-digit LCD driver.

11-44 Describe the operations of the write-cycle module of the 4-digit LCD driver.

11-45 Describe the initialization process of the 4-digit LCD driver.

11-46 Describe the refresh operations of the 4-digit LCD driver.

11-47 Describe the main controller operations of the 4-digit LCD driver.

11-48 Explain why the starting address of DD RAM must be set each time the refresh operation is executed.

11.5 Summary

In this chapter, we have introduced two major methods, the FSM approach and the RTL design, for designing digital systems. The FSM approach is employed in designing small systems whereas the RTL design is used for designing large systems.

An FSM is widely used to model a circuit with memory, such as a sequential circuit or a controller in a digital system. An FSM may be described by a state diagram or an ASM chart. Two widely used FSM models, the Mealy machine and the Moore machine, are introduced to model sequential circuits. In the Mealy machine, the output function is determined by both the current inputs and the present state, whereas in the Moore machine the output function is solely determined by the present state.

Two widely used RTL design methodologies are the ASM chart and the datapath and controller approach. The ASM chart, also known as a state machine (SM) chart, is often used to describe a digital system at the algorithmic level. An ASM chart is merely composed of three types of building blocks: state boxes, decision boxes, and conditional output boxes. It is much like a flow chart but also takes timing into consideration. ASM charts have the following features that they specify RTL operations because they precisely define the operations on every cycle and clearly show the control flow from state to state.

Using the datapath and controller approach, a digital system can be designed with either the three-step paradigm or the heuristic approach. In the three-step paradigm, an ASM (or ASMD) chart is first derived and then both the datapath and the controller are in turn derived from this ASM chart. The datapath part corresponds to the registers and function units in the ASM chart and the controller part corresponds to the control signals in the ASM chart. With the heuristic approach, the datapath and the controller are derived from the specifications in a state-of-the-art fashion. The controller is then modeled by either a FSMD diagram, an ASM chart, or an ASMD chart. To illustrate this heuristic approach, an example of displaying data on an LCD module is given as a case study.

An RTL design can often be implemented by a number of ways. The choice of these options is a trade-off among performance (throughput), space (area or hardware cost), and time (operating frequency or propagation delays). The most common options widely used in practical systems are single-cycle structure, multicycle structure, and pipeline structure. In single-cycle structure, combinational logic is used to perform the desired computation. This structure often runs faster than the other two at the cost of a large amount of hardware. Multicycle structure uses much less hardware than single-cycle structure but has worse performance. Pipeline structure is often used in the case of a stream input to improve throughput at the expense of both hardware and data latency.

References

1. M. G. Arnold, *Verilog Digital Computer Design: Algorithms into Hardware*, Upper Saddle River, New Jersey: Prentice-Hall, 1999.

2. D. Gajski, N. Dutt, A. Wu, and S. Lin, *High-Level Synthesis Introduction to Chip and System Design*, Kluwer Academic Publisher, Boston, 1992.

3. D. D. Gajski, *Principles of Digital Design*, Upper Saddle River, New Jersey: Prentice-Hall, 1997.

4. H. C. Lee, *Introduction to Color Imaging Science*, New-York: Cambridge University Press, 2005.

5. M. B. Lin, *Digital System Design: Principles, Practices, and Applications*, 4th ed., Chuan Hwa Book Ltd. (Taipei, Taiwan), 2010.

6. M. B. Lin, *Introduction to VLSI Systems: A Logic, Circuit, and System Perspective*, CRC Press, 2012.

7. M. B. Lin and S. T. Lin, *Basic Principles and Applications of Microprocessors: MCS-51 Embedded Microcomputer System, Software, and Hardware*, 3rd ed., Chuan Hwa Book Ltd. (Taipei, Taiwan), 2013.

8. M. M. Mano and M. D. Ciletti, *Digital Design: With An Introduction to the Verilog HDL*, 5th ed., Upper Saddle River, New Jersey: Prentice-Hall, 2013.

9. C. H. Roth, Jr., *Fundamentals of Logic Design*, 5th ed., Boston, Massachusetts: PWS, 2004.

Problems

11-1 A 4-bit ECC memory. This problem considers the design of an *error-correcting code* (ECC) memory module in which the Hamming code is used to detect and locate the error bit. The located error bit can then be corrected by inverting its value. The rationale behind the Hamming code is to insert k parity bits into an m-bit message to form a $(k+m)$-bit codeword. The parity bits are placed at the positions of 2^i, where $0 \leq i < k$. In order to correct a single-bit error, including the m-bit message and k parity bits, m and k must satisfy the equation: $2^k \geq m+k+1$. For instance, when the message is 4 bits, at least three parity bits are required to satisfy the inequality and they are placed at positions of $2^0 = 1$, $2^1 = 2$, and $2^2 = 4$, respectively. As a result, the new codeword is $m_3 m_2 m_1 p_2 m_0 p_1 p_0$, where p_0 is at position 1 and m_3 is at position 7. As an m-bit message is written to the memory, the parity bits are computed by a criterion that each parity bit combined with some specific message bits must form an even parity; namely, all bits indicated by a digit 1 in each row of Table 11.4 have to form an even parity. For instance, p_0 combined with m_3, m_1, and m_0 have to form an even parity and hence $p_0 = m_3 \oplus m_1 \oplus m_0$. The other two parity bits p_1 and p_2 can be computed similarly. These parity bits along with the m-bit message, referred to as a codeword, are written into memory. Whenever the codeword is read out from memory, its parity status is checked and the result, called a syndrome or a positional number, is represented as $c_2 c_1 c_0$. For instance, c_0 checks the parity status established by m_3, m_1, m_0, and p_0, and hence $c_0 = m_3 \oplus m_1 \oplus m_0 \oplus p_0$. The other two check bits c_1 and c_2 can be computed similarly. If $c_2 c_1 c_0 = 0$, the read-out codeword is error free; otherwise, the bit indicated by $c_2 c_1 c_0$ is erroneous.

Table 11.4: The positional number of the Hamming (7, 4) code.

	7	6	5	4	3	2	1
	m_3	m_2	m_1	p_2	m_0	p_1	p_0
c_0	1	0	1	0	1	0	1
c_1	1	1	0	0	1	1	0
c_2	1	1	1	1	0	0	0

(a) Design a logic circuit to implement the above parity check generator as well as the error check and correction. Label each component appropriately to indicate its functionality and size.

(b) Describe your design in Verilog HDL as an ECC memory module.

(c) Write a test bench to verify the functionality of your ECC memory module.

11-2 An 8-bit ECC memory. This problem is an extension of the above 4-bit ECC memory and the result is referred to as the Hamming (12, 8) code. Its positional number is shown in Table 11.5.

(a) Design a logic circuit to implement the parity check generator as well as the error check and correction. Label each component appropriately to indicate its functionality and size.

(b) Describe your design in Verilog HDL as an ECC memory module.

(c) Write a test bench to verify the functionality of your ECC memory module.

11-3 An 8-bit SECMED memory. In order to provide the capability of multiple-bit error detection in addition to single-bit error correction, an extra parity check bit has to be added. The resulting memory module is referred to as a *single-error correction and*

Table 11.5: The positional number of the Hamming (12, 8) code.

	12	11	10	9	8	7	6	5	4	3	2	1
	m_7	m_6	m_5	m_4	p_3	m_3	m_2	m_1	p_2	m_0	p_1	p_0
c_0	0	1	0	1	0	1	0	1	0	1	0	1
c_1	0	1	1	0	0	1	1	0	0	1	1	0
c_2	1	0	0	0	0	1	1	1	1	0	0	0
c_3	1	1	1	1	1	0	0	0	0	0	0	0

Table 11.6: The parity check generation for an SECMED code.

	m_7	m_6	m_5	m_4	m_3	m_2	m_1	m_0
c_0	×	×		×	×			
c_1	×		×	×		×	×	
c_2		×	×		×	×		×
c_3	×	×	×				×	×
c_4				×	×	×	×	×

multiple-error detection (SECMED) memory. The parity check generation is listed as Table 11.6. Each data bit is checked by three parity bits. Each parity bit, c_i, along with its associated data bits (marked by "×") in the same row of the table, forms an even parity. For instance, c_0, and data bits m_7, m_6, m_4, and m_3, form an even parity. To locate a possible single-bit error, it is necessary to determine whether the even parity property of the codeword stored in memory is still valid. This is achieved by calculating the syndrome (i.e., the combined parity values of all rows) and the result is denoted by new c_i, This new c_i is then used to locate the single error bit and correct the erroneous bit accordingly. The relationships between syndromes and bit locations are shown in Table 11.7.

(a) Design a logic circuit to implement the parity check generator as well as the error check and correction. Label each component appropriately to indicate its functionality and size.

(b) Describe your design in Verilog HDL as an ECC memory module.

(c) Write a test bench to verify the functionality of your ECC memory module.

11-4 ASM Charts. Consider the state diagram in Figure 11.2(b) and answer each of the following questions.

(a) Draw an equivalent ASM chart for this state diagram.

(b) Model the ASM chart in style 1a. Write a test bench to verify the functionality of your module.

(c) Model the ASM chart in style 1b. Write a test bench to verify the functionality of your module.

Table 11.7: Locating a single error in the SECMED code shown in Table 11.6.

	No error	m_7	m_6	m_5	m_4	m_3	m_2	m_1	m_0	c_4	c_3	c_2	c_1	c_0
c_0	1	0	0	1	0	0	1	1	1	1	1	1	1	0
c_1	1	0	1	0	0	1	0	0	1	1	1	1	0	1
c_2	1	1	0	0	1	0	0	1	0	1	1	0	1	1
c_3	1	0	0	0	1	1	1	0	0	1	1	1	1	1
c_4	1	1	1	1	0	0	0	0	0	0	0	1	1	1

(d) Model the ASM chart in style 2. Write a test bench to verify the functionality of your module.

11-5 RTL implementation options. Consider an n-bit 2's-complement adder which can be implemented by using a single-cycle structure or a multicycle structure.

- **(a)** Suppose that both inputs x and y are stored in each individual n-bit register; the output sum is also stored in an n-bit register. Design a circuit using only one full adder along with an XOR gate and a D flip-flop to add these two inputs and generate the sum. Write a test bench to verify the functionality of the module.
- **(b)** Using an iterative logic circuit, redesign the above module. Write a test bench to verify the functionality of the module.

11-6 An LCD module. Modify the LCD display system described in Section 11.4 so that it is capable of displaying eight digits starting from address 8'hC8.

- **(a)** Draw the datapath module. Label each component properly to indicate its functionality and size.
- **(b)** Construct an ASM chart of the controller.
- **(c)** Write a test bench to verify the functionality of your system.

11-7 Synchronization between two different clock domains. This problem explores the synchronization between two different clock domains. In order to examine this, suppose that the `write_cycle` module shown in Figure 11.40 uses a clock signal with a period of 50 μs rather than 1 ms. Redesign the `write_cycle` module and write a test bench to ensure that the resulting LCD display system still functions correctly.

11-8 A start/stop watch. This problem considers the design of a start/stop watch with a time resolution of 10 ms and a maximum count of 99.99 seconds. The start/stop watch has a start/stop push button that starts and stops the operations of the start/stop watch. In addition, a clear push button is provided to reset the start/stop watch. The clear operation can only be carried out during which the start/stop watch is stopped.

- **(a)** Design this start/stop watch module and draw its block diagram. Label each component properly to indicate its functionality and size.
- **(b)** Write a Verilog HDL module to model the start/stop watch module in behavioral style.
- **(c)** Write a test bench to verify its functionality.

11-9 A clock. This problem involves the design of a clock in 24-hr display format. The clock can display hours, minutes, and seconds. Three push buttons, corresponding to hours, minutes, and seconds, respectively, are used to set the time of the clock whenever another push button, clock-set, is pushing and holding. Assume that the master clock signal available is 4 MHz.

- **(a)** Design this 24-hr clock module and draw its block diagram. Label each component appropriately to indicate its functionality and size.
- **(b)** Write a Verilog HDL module to describe the 24-hr clock module.
- **(c)** Write a test bench to verify the functionality of the 24-hr clock module.

11-10 A timer. In this problem, we consider the design of a timer with a maximum duration of 24 hours. The timer can display hours, minutes, and seconds. The timer will output an alarm signal for 30 seconds once it counts down to 0. In addition, the timer can be started and stopped in a toggled manner by a push button, start/stop. When stopped, it is able to be set or cleared. To set the timer, three push buttons corresponding to hours,

Problems

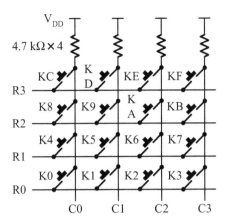

Figure 11.43: The layout of a typical 16-key keypad.

minutes, and seconds are used; to clear the timer, just push the clear button. Assume that the master clock signal available is 4 MHz.

(a) Design this 24-hr timer module and draw its block diagram. Label each component appropriately to indicate its functionality and size.

(b) Write a Verilog HDL module to model the 24-hr timer module.

(c) Write a test bench to verify the functionality of the 24-hr timer module.

11-11 A FIFO. This problem concerns the design of a FIFO having the depth of 16 words, with each containing 16 bits. The front end (read port) and rear end (write port) have a separate set of control signals: clock and access enable (read or write). In addition, two flags, empty and full, are used to indicate the status of the FIFO.

(a) Design this FIFO module and draw its block diagram. Label each component appropriately to indicate its functionality and size.

(b) Derive an ASM chart of the controller unit.

(c) Write a Verilog HDL module to model the FIFO module.

(d) Write a test bench to verify the functionality of the FIFO module.

11-12 A stack. In this problem, we concern the design of a stack having the depth of 16 words, with each containing 16 bits. The top end (read and write port) has the following control signals: clock, push, and pop. In addition, two flags, empty and full, are used to indicate the status of the stack.

(a) Design this stack module and draw its block diagram. Label each component appropriately to indicate its functionality and size.

(b) Derive an ASM chart of the controller unit.

(c) Write a Verilog HDL module to model the stack module.

(d) Write a test bench to verify the functionality of the stack module.

11-13 A 16-key keypad. This problem involves the design of a keypad scanning circuit that detects a key being stroked and outputs its key code. The keypad and its key code are shown in Figure 11.43. Assume that R_i are outputs and C_i are inputs, for all $0 \leq i \leq 3$. In addition, each time only one of R_i is low.

(a) Design a scanning circuit that scans the keypad in sequence. The switch-debounce problem must be taken into account. Draw the circuit at the RTL.

- **(b)** Derive the ASM chart of the controller unit.
- **(c)** Describe the keypad scanning circuit in Verilog HDL.
- **(d)** Write a test bench to verify the functionality of the resulting module.

11-14 A programmable lock. Suppose that we want to design a programmable lock with the capability of password verification. The length of the password is limited to 6 digits and the initial password is all 0s. In order to change the password, the following steps are proceeded: First, we stroke the key denoted by "Change Password" and then enter the old password and the new password twice. After entering the password each time, we need to stroke the key denoted by "End" to notify the lock that the password has been completely entered. To open the lock, we just enter the password and the lock is open (sends an open signal) if the password exactly matches the one stored. Otherwise, it does nothing.

- **(a)** Design this lock module and draw its block diagram. Label each component appropriately to indicate its functionality and size.
- **(b)** Derive an ASM chart of the controller unit.
- **(c)** Write a Verilog HDL module to describe the lock module.
- **(d)** Write a test bench to verify the functionality of the lock module.

11-15 A keypad plus LED module. By combining the LED module described in Section 8.6.2 with the keypad module from Problem 11-13, use an available FPGA device to construct a keypad plus an LED module. You may limit the maximum number of digits that may be entered and design the system accordingly.

- **(a)** Design this module and draw its block diagram. Label each component appropriately to indicate its functionality and size.
- **(b)** Write a Verilog HDL module to model the resulting module.
- **(c)** Write a test bench to verify the functionality of the resulting module.
- **(d)** Using an available FPGA device and a keypad to perform the real-world test of the resulting module.

11-16 A keypad plus LCD module. By combining the LCD module described in Section 11.4 with the keypad module from Problem 11-13, use an available FPGA device to construct a keypad plus an LCD module. You may limit the maximum number of digits that allows to be entered and design the system accordingly.

- **(a)** Design this module and draw its block diagram. Label each component appropriately to indicate its functionality and size.
- **(b)** Write a Verilog HDL module to describe the resulting module.
- **(c)** Write a test bench to verify the functionality of the resulting module.
- **(d)** Using an available FPGA device and a keypad to perform the real-world test of the resulting module.

11-17 An FPGA-based calculator. In this problem, we are concerned with the design of a simple calculator. The calculator is capable of performing the basic arithmetic operations: +, -, *, and /, according to the entry sequence without the capability of parentheses. Combining the 16-key keypad with an LCD display module, use an available FPGA device to construct the system.

- **(a)** Design this module and draw its block diagram. Label each component appropriately to indicate its functionality and size.
- **(b)** Derive an ASM chart of the controller unit.

Problems 505

(c) Write a Verilog HDL module to model the resulting module.

(d) Write a test bench to verify the functionality of the resulting module.

(e) Using an available FPGA device and a keypad to perform the real-world test of the resulting module.

11-18 A simple traffic light controller. This problem involves the design of a simple traffic light controller for the intersection of streets A and B. Assume that the red light is turned on for 40 seconds, the green light is on for 32 seconds, and the yellow light blinks for 5 seconds. Both red lights in streets A and B overlap 3 seconds.

(a) Plot a timing diagram to show the timing relationships of traffic lights between streets A and B.

(b) Draw the block diagram of the traffic light controller and explain its operations.

(c) Write a Verilog HDL module to describe the traffic light controller in behavioral style.

(d) Write a test bench to verify the functionality of the traffic light controller.

11-19 A BCD-to-excess-3-code converter. Supposing that the incoming BCD code is input in series, i.e., in a bit-by-bit fashion, design a converter that converts the BCD code into its equivalent excess-3 code. Of course, you may first use a 4-bit serial-in, parallel-out (SIPO) shift register to receive the incoming BCD code and then use a combinational BCD-to-excess-3-code converter to convert it into the desired excess-3 code. Finally, the excess-3 code is output in series through a parallel-in, serial-output (PISO) shift register. Another way is to use a finite-state machine to directly convert the incoming BCD code to the excess-3 code on the fly.

(a) Draw the block diagram of the first approach and label each component appropriately to indicate its functionality and size.

(b) Write a Verilog HDL module to describe the above block diagram. Also write a test bench to verify it.

(c) Draw the state diagram of the second approach. Write a Verilog HDL module to describe the state diagram. Also write a test bench to verify it.

(d) Compare both approaches in terms of the required number of clock cycles and the number of LUTs (or area).

11-20 An excess-3-to-BCD-code converter. Supposing that the incoming excess-3 code is input in series, i.e., in a bit-by-bit manner, design a converter that converts the excess-3 code into its equivalent BCD code. Of course, you may first use a 4-bit serial-in, parallel-out (SIPO) shift register to receive the incoming excess-3 code and then use a combinational excess-3-to-BCD-code converter to convert it into the desired BCD code. Finally, the BCD code is output in series through a parallel-in, serial-output (PISO) shift register. Another way is to use a finite-state machine to directly convert the incoming excess-3 code to the BCD code on the fly.

(a) Draw the block diagram of the first approach and label each component appropriately to indicate its functionality and size.

(b) Write a Verilog HDL module to describe the above block diagram. Also write a test bench to verify it.

(c) Draw the state diagram of the second approach. Write a Verilog HDL module to describe the state diagram. Also write a test bench to verify it.

(d) Compare both approaches in terms of the required number of clock cycles and the number of LUTs (or area).

11-21 A BCD-to-binary converter. This problem concerns the design of a BCD-to-binary converter that converts an incoming 10-bit BCD number with a maximum value of 255 into its equivalent 8-bit binary number. The rationale behind the conversion is based on the following observations: First, for any number the least significant bit (LSB) must be the same regardless of whether the number is represented as binary or BCD. Second, the right shift of 1-bit position corresponds to a divide-by-2 operation for binary numbers but is not true for BCD numbers. However, if we subtract 4'b0011 from each right-shifted BCD digit being greater than or equal to 4'b1000, the resulting BCD number is a divide-by-2 number. Assuming that the incoming 10-bit BCD number and the resulting 8-bit binary number are stored in registers D and B, respectively, design a BCD-to-binary converter.

  (a) Illustrate the algorithm starting with the BCD number 179, showing the details of registers D and B at each step.
  (b) Draw the block diagram of the BCD-to-binary converter.
  (c) Derive an ASM chart of the controller unit.
  (d) Write a Verilog HDL module to describe the BCD-to-binary converter.
  (e) Write a test bench to verify the functionality of the resulting module.

11-22 A binary-to-BCD converter. This problem considers the design of a binary-to-BCD converter that converts an incoming 8-bit binary number into its equivalent 10-bit BCD number. The essence of conversion is based on the following observation. Suppose that the incoming 8-bit binary is represented as: $(b_7 b_6 \cdots b_1 b_0)$, then the equivalent BCD number can be obtained by applying Horner's rule as follows

$$\begin{aligned} \text{BCDNO} &= b_7 \times 2^7 + b_6 \times 2^6 + \cdots + b_1 \times 2^1 + b_0 \times 2^0 \\ &= [[[[[[[0 + b_7] \times 2 + b_6] \times 2 + b_5] \times 2 + b_4] \times 2 + b_3] \times \\ &\quad 2 + b_2] \times 2 + b_1] \times 2 + b_0 \end{aligned}$$

where the addition must be carried out in decimal. Assuming that the incoming 8-bit binary number and the resulting 10-bit BCD number are stored in registers B and D, respectively, design a binary-to-BCD converter.

  (a) Illustrate the algorithm starting with the binary number 8'hcf, showing the details of registers B and D at each step.
  (b) Draw the block diagram of the binary-to-BCD converter.
  (c) Derive an ASM chart of the controller unit.
  (d) Write a Verilog HDL module to describe the binary-to-BCD converter.
  (e) Write a test bench to verify the functionality of the resulting module.

11-23 A 4×16 multiplier. This problem involves the design of an unsigned multiplier that performs the multiplication of a 4-bit number with a 16-bit number to give a 20-bit product. To speed up the operation, a 4×4 unsigned array multiplier is used so that the multiplication of 4 bits rather than only 1 bit is performed at each step. Supposing that a 20-bit accumulator, an 8-bit adder, the 4×4 array multiplier, and a controller are used to construct the system, design this 4×16 multiplier.

  (a) Draw the block diagram and label each component properly to indicate its functionality and size.
  (b) Derive the ASM chart for the controller unit.
  (c) Write a Verilog HDL module to describe the 4×16 multiplier.
  (d) Write a test bench to verify the functionality of the resulting module.

11-24 A 16 × 16 signed multiplier. This problem considers the design of a signed multiplier that multiplies two 16-bit signed numbers to give a 32-bit product. To use as little hardware as possible, a multicycle unsigned algorithm for multiplying two unsigned numbers based on the shift-and-add approach is used. Negative numbers are represented in 2's-complement form. The signed multiplier proceeds the multiplication of both input numbers as in the following steps: First, the multiplier and multiplicand are converted into unsigned numbers if they are negative. Next, the multiplication is performed on the unsigned numbers. Finally, the product is converted into 2's-complement form if necessary.

(a) Draw the block diagram and label each component appropriately to indicate its functionality and size.
(b) Derive an ASM chart of the controller unit.
(c) Write a Verilog HDL module to describe the 16 × 16 signed multiplier.
(d) Write a test bench to verify the functionality of the resulting module.

11-25 An automated traffic light controller. This problem involves the design of traffic light controllers at the intersection of streets A and B. There are two sensors at the intersection, denoted by S_A and S_B, respectively. When a vehicle approaches the intersection along street A, sensor S_A is triggered and outputs a pulse to the controller. The same situation is applied to sensor S_B. Street A is the main street and has a green light until a vehicle from street B approaches it. Then the traffic lights change. The traffic light of street A changes to a red light after 12 seconds, including a 5-second blinking yellow light. The street B now has a green light, which lasts for 30 seconds, and then changes to a blinking yellow light for 5 seconds before changes to a red light. When there is a vehicle on street B but none on A approaches the intersection, street B remains at a green light until there is a vehicle on street A approaching the intersection. At the time that the traffic light at street B changes to a red light, the street A gets a green light again. The green light at street A at least sustains for 50 seconds and changes only when a vehicle approaches the intersection along street B.

(a) Draw the block diagram and label each component appropriately to indicate its functionality and size.
(b) Derive an ASM chart of the controller unit.
(c) Write a Verilog HDL module to describe the traffic light controller.
(d) Write a test bench to verify the functionality of the resulting module.

11-26 An integer square-root circuit. In this problem, we are concerned with the design of a square-root circuit that finds the integer square root of an input number with a repeated subtraction algorithm. In this algorithm, the input number is subtracted by consecutive odd numbers starting from 1 until the result becomes zero or negative, namely, less than zero. The number of subtraction performed before the result becomes zero or negative is the integer square root of the input number. Assume that the input number is an n-bit binary number.

(a) Illustrate the square-root algorithm with the number 78, showing the details of each step.
(b) Draw the block diagram and label each component appropriately to indicate its functionality and size.
(c) Derive an ASM chart of the controller unit.
(d) Write a Verilog HDL module to describe the square-root circuit.
(e) Write a test bench to verify the functionality of the resulting module.

12
Synthesis

THE success of logic synthesis has dramatically cut the design time and pushed the HDLs into the forefront of digital designs. Accompanying with this success, a large number of consumer's products, such as MP3 players and DVD players, are produced and change people's daily life. For this reason, we begin to introduce in this chapter the general ASIC- and FPGA-based synthesis flows, including the general synthesis flow, RTL synthesis flow, and physical synthesis flow. In addition, the principle of timing-driven placement is also introduced.

In order to make effective use of a synthesis tool, we need to provide the synthesis tool the design environment and timing constraints along with the RTL code of a design as well as the needed technology libraries. The design environment provides the process parameters, I/O-port attributes, and statistical wire-load models for the synthesis tool to synthesize a design. The timing constraints specify the clock-related parameters, input and output delays, and timing exceptions.

The principles of logic synthesis and the general architecture of synthesis tools are dealt with in depth. Synthesis can be generally cast into logic synthesis and high-level synthesis. The former transforms an RTL representation into gate-level netlists while the latter transforms a high-level representation, such as a C-language algorithm, into an RTL representation. Logic synthesis can be generally further partitioned into two steps: technology-independent synthesis and technology-dependent synthesis. The former restructures the logic function to satisfy the target criteria. The latter takes into account design constraints and maps the structured logic function to the devices in the target library.

The successful synthesis of a design strongly depends on the proper partitioning of the design along with a good HDL coding style. To this end, we explore some examples of language structure synthesis and provide guidelines about how to write a good Verilog HDL code acceptable by most logic synthesis tools and capable of achieving the best compilation times and synthesis results. These guidelines include clock signals, reset signals, the partition of a design, and synthesis for power optimization.

Finally, a case study is given to present an overview of the Xilinx CPLD/FPGA design flow and reveal the principles of XST. Related design constraints of XST are also explored to some extent along with examples.

12.1 Synthesis Flow of ASICs and FPGA-Based Systems

We often follow a synthesis flow when designing an ASIC or an FPGA-based system. A *synthesis flow* is a set of procedures that allows designers to proceed from the specification of a desired system to the final chip or FPGA implementation in an efficient and error-free way. In this section, we describe the general synthesis flow of designing ASICs and FPGA-based systems.

12.1.1 General Synthesis Flow

The general synthesis flow of ASIC and FPGA-based design is shown in Figure 12.1. It can be partitioned into two major parts: the *front end* and the *back end*. The front end is target-independent and includes three steps, starting from the product requirement, the behavioral/RTL description, and ending with RTL synthesis, and generates a gate-level netlist. Remember that RTL modeling uses RTL components and is a mixed style combining the structural, dataflow, and behavioral styles with the constraint that the resulting description must be acceptable by synthesis tools. The back end is target-dependent and mainly comprises physical synthesis, which accepts the structural description of the gate-level netlist and generates a physical description. In other words, RTL synthesis is at the heart of the front end while physical synthesis is the essential part of the back end.

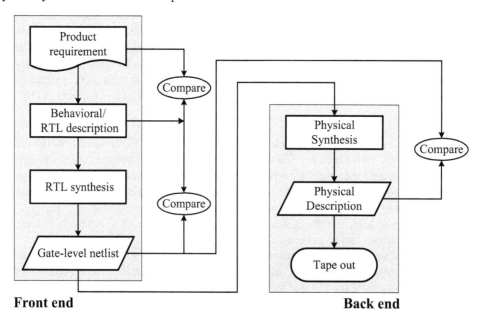

Figure 12.1: The general synthesis flow of ASIC- and FPGA-based design.

12.1.1.1 RTL Synthesis Flow The general RTL synthesis flow is shown in Figure 12.2. The RTL synthesis flow begins with the product requirement, which is converted into a *design specification*. The design specification is then described at the RTL in behavioral style using Verilog HDL, SystemVerilog, or VHDL. The results are then verified by a set of test benches written using an HDL or HDLs. This verifying process is called *RTL functional verification*.

After the function of an RTL design has been verified correctly, a synthesis process, referred to as *RTL synthesis* or *logic synthesis,* is performed by a logic synthesizer to convert the RTL module into a gate-level netlist. The essential function of logic synthesizers is to convert an RTL description into a network of generic gates and registers, and then optimize the generating logic to improve speed, area, or both. In addition, finite-state machine decomposition, datapath optimization, and power optimization may also be performed at this stage. In general, a logic synthesizer accepts three inputs, an *RTL code*, one or more *technology libraries*, and *design environment and constraints*, and generates a gate-level netlist.

Section 12.1 Synthesis Flow of ASICs and FPGA-Based Systems

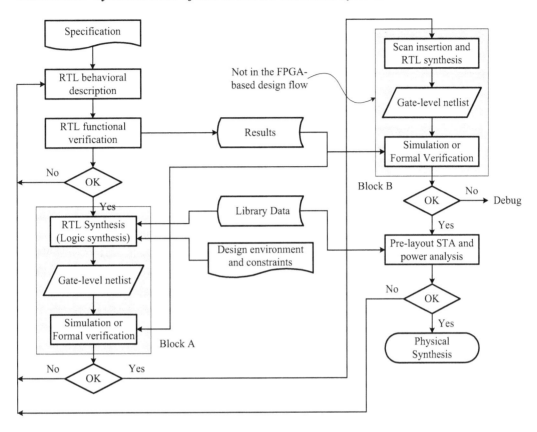

Figure 12.2: The general flow of RTL synthesis.

After a gate-level netlist is generated, it is necessary to rerun the test benches used in the stage of RTL functional verification to see if they produce exactly the identical output for both behavioral and structural descriptions or to perform an RTL versus gate equivalence checking process to ensure the logical equivalence between the two descriptions.

The next three steps often used in ASIC design (namely, the cell-based design) but not in FPGA-based designs are *scan-chain logic insertion*, resynthesis, and verification, as shown in the shaded block B. In fact, this block may be combined with block A into one. The step of scan-chain (or test logic) insertion is to insert or modify logic and registers in order to aid the manufacturing test. Automatic test pattern generation (ATPG) and the built-in self-test (BIST) are usually used in most modern ASIC designs. The details of these topics will be addressed in Chapter 16.

The final stage of the RTL synthesis flow is pre-layout static timing analysis (STA) and power dissipation analysis. Static timing analysis checks the temporal requirements of the design. Through detailed STA, many timing problems can be found and fixed and system performance might also be optimized. More details of STA will be described in Section 13.5. Power analysis estimates the power dissipation of the design. The power dissipation of a circuit depends on the activity factors [15] of the gates in the circuit. Power analysis can be performed for a particular set of test vectors by running the simulator and evaluating the total capacitance switched at each clock transition of each node.

12.1.1.2 Physical Synthesis Flow The second part of the synthesis flow of an ASIC- or FPGA-based system is physical synthesis. Irrespective of the ASIC- or FPGA-based system, physical synthesis can be further subdivided into two major stages: *placement* and *routing*, as

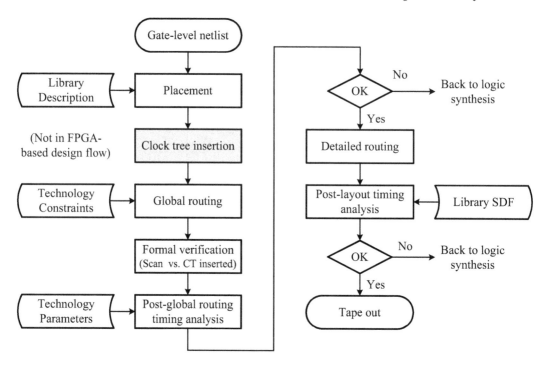

Figure 12.3: The general flow of physical synthesis.

shown in Figure 12.3. For this reason, physical synthesis is usually called the *place and route* (PAR) in most CAD tools.

In the placement stage, logic cells (standard cells or building blocks) are placed at the fixed positions of a chip in order to minimize the total area and interconnect lengths. In other words, the placement stage defines the locations of logic cells (modules) on a chip and sets aside spaces for the interconnect to each logic cell (module). This stage is generally a combination of three substeps: *partitioning, floorplanning,* and *placement.* Partitioning divides the circuit into parts such that the sizes of components (modules) are within prescribed ranges and the number of connections between two components is minimized. Floorplanning determines the appropriate (relative) location of each module in a rectangular chip area. Placement finds the best position of each module on the chip such that the total chip area is minimized or the total length of wires is minimized. Of course, not all CAD tools have their placement divided into the above three substeps. Some CAD tools simply combine these three substeps into one big step known as *placement.*

After placement, a clock tree is inserted in the design. In this step, a clock tree is generated and routed along with the required buffers to provide the required current driving capability or to balance propagation delays among different paths. A clock tree is often placed before the completion of main logic placement and routing to minimize the clock skew of the design. This step is not necessary in the FPGA-based design flow in which a clock distribution network is already fixed on the chip.

The next big stage is known as *routing*, which is used to complete the connections of signal nets among the cell modules placed by the placement step. This stage is further partitioned into two substages: *global routing* and *detailed routing*. Global routing decomposes a large routing problem into many smaller and manageable subproblems (i.e., detailed routing) by finding a rough path for each net in order to reduce the chip size, to shorten the total length of wires, and to evenly distribute the congestion of the design over the routing area. Detailed routing carries out the actual connections of signal nets among modules.

Section 12.1 Synthesis Flow of ASICs and FPGA-Based Systems

After global and detailed routing, post-global-routing timing analysis and post-layout timing analysis are separately performed. Both of them rerun static timing analysis with actual routing loads associated with gates to see if timing constraints are still valid.

The final tape-out stage has different meanings for cell-based (cell library or standard cells) and gate-array-based synthesis as well as CPLD/FPGA-based synthesis. For cell-based and gate-array-based synthesis, the tape-out stage generates a set of photomasks so that the resulting design can be "programmed" (manufactured) in an IC (Integrated Circuit) foundry. For CPLD/FPGA-based synthesis, the tape-out stage generates a programming file in order to program the device. Of course, after fabricating a device or programming a CPLD or an FPGA device, we often test the device in a real-world environment to see if it indeed works as expected.

■ Review Questions

12-1 What is a synthesis flow?
12-2 Which steps are included in the front end of the general synthesis flow?
12-3 Why is the front end often called logic synthesis?
12-4 Which steps are included in the back end of the general synthesis flow?
12-5 Why is the back end often called physical synthesis?
12-6 What are the two major stages of physical synthesis?

12.1.2 Timing-Driven Placement

A problem with the PAR and timing analysis described previously is that timing information can only be obtained after the layout has been done. As a result, in order to satisfy the post-layout timing requirements, it may go back to the step of logic synthesis and starts there again, as illustrated in Figure 12.3. Even worse, this process may need to be repeated several times. One approach to solving this is by incorporating timing analysis into the placement stage so that the *critical path* can be placed into the layout with a high priority. This results in an approach called the *timing-driven placement*. In this section, we will introduce a simple approach of timing-driven placement based on the concept of *slack time*.

Before defining slack time, we need to define *arrival time* and *required time*, as shown in Figure 12.4. Assume that d_i, for $0 \leq i \leq m$, are net delays and D is the delay of the gate containing net j. The arrival time of net j is defined as

$$t_A^j = \max_{i=1}^{m}(t_A^i + d_i) + D$$

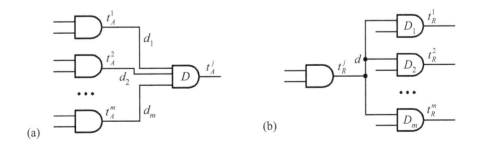

Figure 12.4: The definition of (a) arrival and (b) required times.

Assume that D_i, for $0 \leq i \leq m$, are gate propagation delays and d is the delay of the net that connects the gate to its fanout gates. Here, the term *fanout gates* refers to those gates connected to the output of a gate. The required time of net j is defined as

$$t_R^j = \max_{i=1}^{m}(t_R^i - D_i) - d$$

The arrival time of a signal at the output of a gate is known only when the arrival time of a signal at every input of that gate has been computed.

Assume that t_A and t_R are the arrival and required times of a signal at a given net, respectively. The slack time of the signal at the given net is defined as the difference between the required time and arrival time of the signal and can be expressed as follows:

$$t_{slack} = t_R - t_A$$

The features of slack time are as follows: First, the paths with the smallest slack time are called *timing critical paths* or simply, *critical paths*. Second, the negative value of slack time means that the associated paths are the critical paths. Thus, slack time can be used to analyze the critical paths of a design and is the method widely used in timing analysis tools.

The concept of slack time can also be used in constructing timing-driven placement algorithms. A simple version, called the *zero-slack algorithm*, is illustrated next to give the reader a flavor of how timing-driven placement works.

In order to describe the zero-slack-time algorithm [20], suppose that the arrival time of a signal at each primary input and the required time of a signal at each primary output are known. Then, the zero-slack-time algorithm can be described as follows.

■ Algorithm 12-1: Zero-slack-time algorithm

Begin

1. repeat

    1.1 Compute all slack times of all nets of of the logic circuit in question;

    1.2 Find the minimum positive slack time;

    1.3 Find a path with all slack times equal to the minimum slack time;

    1.4 Distribute the slack times into nets as the net propagation delay along the path;

  until (there exists no positive slack time);

End

In other words, the zero-slack-time algorithm aims to compute and distribute the slack time of a signal evenly in the nets along the path from a primary input to a primary output, provided that the arrival times of all primary inputs and the required times of all primary outputs are known in advance. That is, it transforms timing constraints into the net-length constraints. As an illustration to explore the detailed operations of this algorithm, consider the following example.

■ Example 12.1: An example of the zero-slack-time algorithm.

This example illustrates the operations of the zero-slack-time algorithm. As shown in Figure 12.5(a), the arrival times of all primary inputs and the required times of all primary outputs are known in advance. In addition, the propagation delays of all gates are shown in the figure. In the first iteration, the propagation delay of each net is assumed to be zero.

By definition, the arrival time of a signal at each node is computed from a primary input to a primary output and the required time of the signal is computed backward from a primary output to a primary input. Once both arrival and required times are found, the slack time of a signal at each node can then be determined. The results are shown in Figure 12.5(b). The

Section 12.1 Synthesis Flow of ASICs and FPGA-Based Systems

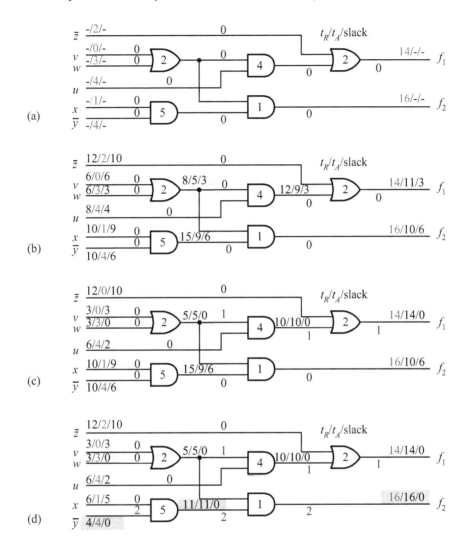

Figure 12.5: An illustration of the zero-slack-time algorithm: (a) original status; (b) after step 1 of the first iteration; (c) after the first iteration; (d) after the second iteration.

minimum positive slack time is 3. This slack time is evenly distributed in each net from the primary output backward to the primary input, as shown in Figure 12.5(c), along with the recomputed arrival, required, and slack times of the signal at each node.

The same process is applied to the remaining portion of the figure. The final result is shown in Figure 12.5(d). This result is used as a reference to place the gates, as depicted in Figure 12.6. The net with a greater slack time may use a longer wire and the net with a smaller slack time needs to use a shorter wire.

■ Review Questions

12-7 Define the arrival time of a signal at a net.
12-8 Define the required time of a signal at a net.
12-9 Define the slack time of a signal at a net.

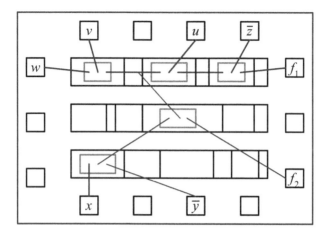

Figure 12.6: The final placement result of the logic circuit shown in Figure 12.5.

12-10 What are the features of slack time?
12-11 What does a critical path mean?
12-12 How could slack time be used in timing-driven placement?

12.2 RTL Synthesis Flow

The design environment, design constraints [2, 22], RTL codes, and a technology library are required for a design to be synthesized. The *design environment* provides synthesis tools with the operating conditions and wire-load models (cell-based design only) that a design should be operated. Design constraints are the instructions given by the designer to control what the synthesis tool can or cannot do with a design or how the synthesis tool should behave. In this section, we first introduce the general design environment and design constraints. Then, we briefly describe the general architecture of logic synthesizers and the principles of logic synthesis.

12.2.1 Design Environment

After a design has been partitioned, coded, and verified, the next step is to input the design environment and design constraints of the design along with the RTL code and a technology library to the synthesis tool, as shown in Figure 12.7, so that the design can be synthesized into a gate-level netlist.

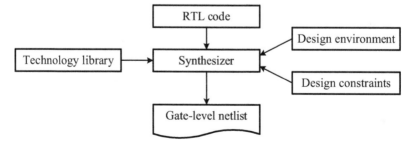

Figure 12.7: The general design environment of synthesis tools.

Section 12.2 RTL Synthesis Flow

The design environment[1] is one that a design is supposed to operate. The general design environment of digital systems includes *specifying operating conditions* and *modeling wire loading*.

12.2.1.1 Specifying Operating Conditions The operating conditions deal with the variations in process, voltage, and temperature (called *PVT variations*) that a design is expected to encounter. These variations are taken into consideration with the operating condition specifications in the technology library. Operating conditions can be the worst case, the typical case, or the best case. The cell and wire (i.e., interconnect) delays are scaled in terms of these conditions accordingly.

A typical technology library usually includes the worst-case, the typical-case, and the best-case libraries. The worst-case condition is used during the pre-layout synthesis stage to optimize the design for the maximum setup time whereas the best-case condition is commonly used to fix the hold-time violations. More details about this can be referred to Lin [15]. The typical case is mostly ignored because it is already covered by the analyses of the worst and best cases.

12.2.1.2 Modeling Wire Loading The *wire-load model* provides a wire-load model for processing the pre-layout static timing analysis of a cell-based design. Recall that after placement, only the propagation delays of logic cells are known. To proceed the pre-layout static timing analysis (STA), a wire-load model must be provided so that the timing analysis tool can use it to calculate the wire (interconnect) delays. Wire load models are used to estimate the effects of interconnect nets on capacitance, resistance, and area before realistic data can be obtained from the actual layout. These models are statistical models in nature and estimate the net delays as a function of capacitive loading. Through the detailed pre-layout STA, it proves useful to find the timing violations of a design at the early stage of the design and fix the bugs accordingly.

■ Example 12.2: Design environment.

As described, the design environment includes the setting of operating conditions and the selection of a wire load model. Examples of the setting of a design environment are as follows:

```
set_operating_condition -lib my_library_25C TYPICAL
set_wire_load_model -lib my_library_25C -name "suggested_10K"
set_wire_load_mode top
```

The above examples are the basic constraints that should be set on each design. However, before moving on to the optimization step, we often put together all commands as a *script file*, which is a text file containing a sequence of synthesizer's commands and variables.

12.2.2 Design Constraints

Design constraints are the instructions given by the designer to control what the synthesis tool can or cannot do with a design or how the synthesis tool should behave. Usually this information can be derived from the various design specifications (e.g., timing specification) of a design.

There are basically two types of design constraints: *design rule constraints* and *optimization constraints*. Generally, a synthesis tool tries to meet both design rule constraints and optimization constraints but design rule constraints always have precedence over optimization

---

[1] The design environment and constraints introduced in this section are based on the features of Synopsys Design Compiler and applicable to cell-based/gate-array-based design, referring to [22] for more details. Synopsys design constraints have become the industry standard and are widely used in the Altera Quartus II system and Xilinx Vivado IDE.

constraints. This means that synthesis tools may violate optimization constraints if necessary to avoid violating design rule constraints.

12.2.2.1 Design Rule Constraints Design rule constraints are implicit constraints defined by ASIC vendors in their technology libraries. These rules cannot be discarded or overridden. By specifying the technology library that a synthesis tool should use, all design rules can also be specified in that library. Design rule constraints include the following major parts: *maximum transition time, maximum fanout, maximum* (and *minimum*) *capacitance*, and *cell degradation*.

Maximum transition time means the longest time allowed for a driving source of a net to change its logic value. This generally includes both rise and fall times. The transition time of a driving source is approximated by the product of the drive resistance and the capacitance load of the driving source.

Maximum fanout defines the maximum fanout that a driving source can drive. Here, the fanout means that the maximum number of logic gate inputs that an identical gate can drive. Refer to Lin [15] for more details.

Maximum (and minimum) capacitance defines the maximum (and minimum) total load capacitance that an output source can drive. The total capacitance consists of the load capacitance of the output source and the interconnect capacitance associated with the output source.

Cell degradation means that the propagation delay of a cell is proportional to the transition time of its input signal. To take into account this effect, some technology libraries contain cell degradation tables to list the maximum capacitance that can be driven by a cell as a function of the transition times at the inputs of the cell.

12.2.2.2 Optimization Constraints Optimization constraints are explicit ones set by the designer. They describe the design goals, including maximum area and timing constraints, which are set for the design by the designer, and instruct the synthesis tool how to synthesize the design. The most common timing constraints include the following ones.

System clock definition. Clock constraints are the most important ones in an ASIC- or FPGA-based design. The clock signal is the signal to synchronize the operation of the system and defines timing requirements for all paths in the design. Most of the other timing constraints are related to the clock signal. Generally, a clock signal is specified by a number of factors, including the *clock period, clock transition time, clock latency,* and *clock uncertainty*. Each of these is described below briefly.

- Clock period defines the time duration at which the clock signal repeats itself.
- Clock transition time defines both the rise and fall times of the clock signal.
- Clock latency specifies the time it takes the clock signal to propagate from the clock source to the clock capture points.
- Clock uncertainty may be used to model both clock skew and clock jitter. Clock skew is the variations of the clock propagation delays in the different branches of unequal length of the clock tree. Clock jitter means the variations in the clock period. That is, the clock period may shrink or expand on a cycle-by-cycle basis.

Input delay and output delay. Remember that the minimum period of the system clock signal is determined by the longest delay of the combinational logic between two flip-flops in the system (see Section 9.1.2). At the interface between two modules, two constraints, *input delay* and *output delay*, are used to constrain external path delays relative to a reference edge, usually the positive edge, of the clock signal, as illustrated in Figure 12.8. The input delay specifies how much time is used by external logic before a signal arrives at the input of the system. As shown in Figure 12.8, this external delay is subtracted from the clock period and the margin specifies how much time is left for internal logic. This margin is used to check the setup-time requirement of the flip-flop for input signals. The setup time of a register is the

Section 12.2 RTL Synthesis Flow

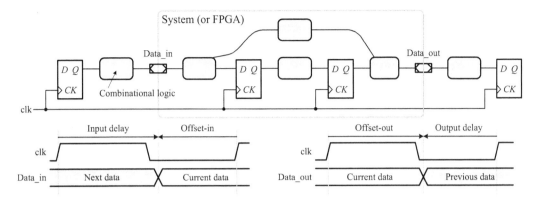

Figure 12.8: Input delay and output delay and their relationships to offset-in and offset-out: (a) logic circuit; (b) input delay; (c) output delay.

amount of time that its input data must be stable before the active clock transition is coming. The output delay specifies how much time is needed by the external block before the signal can be captured at the path endpoint, as shown in Figure 12.8. It is used to check the hold-time requirement of the flip-flop for output signals. The hold time of a register is the amount of time that its input data must remain stable after the active clock transition. More details about the timing of flip-flops can be referred to in Section 9.1.2.

Both input delay and output delay constrains take part in the determination of the minimum clock period of the design. Note that some synthesis tools such as Xilinx synthesis technology (XST) (used in the ISE) use both *offset-in* and *offset-out* constraints in place of input delay and output delay constraints. The relationships between the input delay and offset-in and between the output delay and offset-out are also illustrated in Figure 12.8. Note the sum of both the offset-in and the input delay is the clock period; similarly, the sum of the offset-out and the output delay is also the clock period.

Minimum and maximum path delays. Minimum and maximum path delays allow us to constrain individual paths and to set specific timing constraints on these paths. The minimum and maximum path delay specifications can only be applied to combinational logic blocks. That is, the path from an input port (pad) to an output port (pad) without passing through any register, as shown in Figure 12.8. The minimum path delay specification defines the required minimum path delay in terms of time units for a particular path. The maximum path delay specification defines the required maximum path delay in terms of time units for a particular path.

Input transitions and output load capacitance. These two constraints can be used to constrain the input slew rates and the output capacitance on the input and output ports (pads or pins) of the underlying design, respectively.

False paths. A *false path* is identified by STA tools as a timing failing path but it does not actually propagate signals. For instance, a path that is not activated by any combination of inputs is a false path. To further illustrate this, consider the logic circuit depicted in Figure 12.9. The path indicated by a dark line is the longest path among the four possible paths with a delay of 20 ns. Nevertheless, it is never activated because both multiplexers cannot select their inputs 0 to their output end at the same time. Such a timing path is referred to as a false path. In other words, a false path is the path that static timing analysis tools identify to be timing-failure, but the designer knows that it is not actually failure because it is never the path being to propagate the desired signals. False-path setting is used to instruct the synthesis tool to ignore a particular (false) path or a group of (false) paths for timing optimization. It is worth noting that false paths in a logic circuit may be removed by synthesis tools during logic optimization

(Problems 12-18 and 12-19).

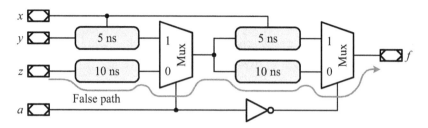

Figure 12.9: An example of false paths.

Multicycle paths. A *multicycle path* means that the signal on such a path needs more than one clock cycle to propagate from the start point to the endpoint of the path, as illustrated in Figure 12.10. The multicycle path constrain is used to inform the synthesis tool about the fact of the number of clock cycles required by a particular path to reach its endpoint. Most synthesis tools always assume that all paths take a single cycle by default. Nevertheless, this may not be true in a realistic design. Whenever this situation occurs, we have to tell the synthesis tools how to treat this as a timing exception. Refer to in Section 13.5.3 for more details about timing exceptions.

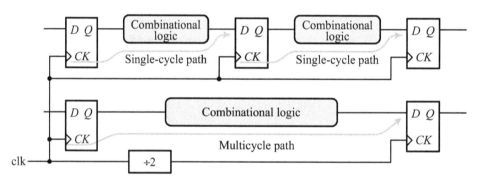

Figure 12.10: An example of single-cycle and multicycle paths.

■ Example 12.3: Example of design environment and constraints.

A typical script file in *tool command language* (TCL) of a design environment and design constraints is as follows. The more details of related TCL commands should be referred to the appropriate reference manuals of synthesis tools [1, 22, 26].

```
/* Set design environment */
set_operating_condition -lib my_library_25C TYPICAL
set_wire_load_model -lib my_library -name "suggested_10K"
set_wire_load_mode top

/* set parameters for clock signal */
create_clock clk -name clk -period 10 -waveform {0 5} {clk}
set_clock_uncertainty 0.5 clk
set_clock_latency 0.5 clk
set_clock_transition 0.5 clk
set_dont_touch_network clk
```

Section 12.2 RTL Synthesis Flow

```
/* set parameters for input ports, x and y, and output port z */
set_driving_cell -library my_library_25C
                 -lib_cell HDDFFPB1 -pin Q [get_ports x]
set_driving_cell -library my_library_25C
                 -lib_cell HDINVD1 -pin Z [get_ports y]
set_load [load_of my_library_25C/HDDFFPB1/D] [get_ports z]

/* set input and output delays */
set_output_delay 3.75 -clock clk {r_wb}
set_input_delay 7 -clock clk all_inputs() - {clk, rst_n, ready}
set_output_delay 5 -clock clk all_outputs()-{r_wb}

/* set max and min path delays  */
set_max_delay -from data_a -to out_x 2.25
set_min_delay -from data_b -to out_y 0

/* Asynchronous input signals */
set_false_path -from rst_n
set_false_path -from ready

/* Don't change reset signal */
set_dont_touch rst_n
set_ideal_network rst_n
```

■ Review Questions

12-13 What are usually specified in the design environment?
12-14 What are usually specified in timing constraints?
12-15 What are the major parts of design constraints?
12-16 What do optimization constrains mean?
12-17 What is the meaning of false paths?
12-18 What is the meaning of multicycle paths?

12.2.3 Architectures of Logic Synthesizers

As mentioned previously, a synthesis tool (also called a *silicon compiler* or a *logic synthesizer*) usually accepts an RTL code, a technology library, a design environment, and a set of timing constraints, and generates a gate-level netlist.

The general architecture of synthesis tools is depicted in Figure 12.11. It consists of two major parts: the *front end* and the *back end*. The front end is further partitioned into two steps: *parsing* and *elaboration*; the back end contains three steps: *analysis/translation*, *logic synthesis* (*logic optimization*), and *netlist generation*.

12.2.3.1 The Front End At the parsing step, the parser is responsible for checking the syntax of the source code and creating internal components to be used by the next step. At the elaboration step, the elaborator constructs a representation of the input circuit by connecting the internal components, unrolling loops, expanding generate-loops, setting up parameters passing for tasks and functions, and so on. The end result from the elaboration step is a complete description of the input circuit, which can then be input to the back end for generating the final gate-level netlist.

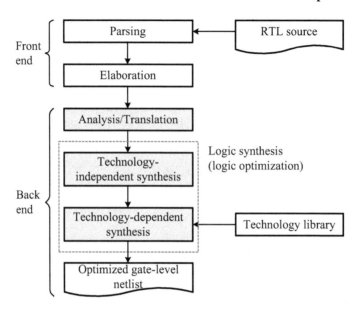

Figure 12.11: The general architecture of logic synthesis tools.

12.2.3.2 The Back End The analysis/translation step of the back end extracts the logic functions from a language structure and prepares for technology-independent logic synthesis. Its major operations include managing the design hierarchy, extracting finite-state machines (FSMs), exploring resource sharing, and so on. The logic synthesis (often called *logic optimization*) is the heart of synthesis tools. It creates a new gate network which computes the functions specified by a set of Boolean functions, one per primary output. Synthesizing a gate network needs to balance a number of different concerns, which can be categorized into two classes: *functional metrics*, such as fanins, fanouts, and others, as well as *non-functional metrics*, such as area, power, and delay.

Logic synthesis is usually partitioned into two steps: *technology-independent* synthesis and *technology-dependent* synthesis. The former restructures the logic function of the design to satisfy the target criteria while the latter takes into account design constraints and maps the structured logic function of the design to the devices in the target library. The process of translating a logic network from its technology-independent form into a technology-dependent form is called *library binding*. These two steps will be dealt with in more details later in this chapter.

After library binding, a gate-level netlist is generated and optimized. The gate-level netlist is the description of the original design and is represented in terms of the cells from the technology-dependent library. It is closer to the realistic hardware than the original design at the RTL.

■ Review Questions

12-19 What are the functions of the parser of logic synthesizers?
12-20 What are the functions of the elaborator of logic synthesizers?
12-21 What does library binding mean?
12-22 Describe the function of technology-independent synthesis.
12-23 Describe the function of technology-dependent synthesis.

12.2.4 Optimization of Logic Synthesis

The objective of optimization[2] is to minimize the area of a design but maintain the target timing requirements through synthesis and optimization. During optimization, the HDL code of the design in question is assumed to be frozen. The factors that can be traded off are among area, delay, and constraints. In general, the area of a design increases considerably with tightening constraints whereas the propagation delay of logic of a design decreases with tightening constrains. The propagation delay of logic of a design increases with loosened constrains.

As mentioned, hierarchy is an effective way to design a large system and synthesis tools by default maintain the original hierarchy of the design in question. Nevertheless, this would result in a design that is often not optimized because the designer often creates unnecessary hierarchy. To overcome this, a technique known as *removing hierarchy* is widely used to allow synthesis tools to optimize the design across the hierarchy, namely, crossing the module boundaries. Refer to Section 12.6.3 for more about the discussion of hierarchy.

12.2.4.1 Architectural Optimization Architectural optimization is carried out on the HDL code of the design at hand. This optimization generally includes the following tasks: *arithmetic optimization, resource sharing, sharing common subexpressions, reordering operators,* and *selecting default implementations.*

Arithmetic optimization uses the rules of algebra to improve the implementation of the design. That is, the synthesis tool may rearrange the operations in arithmetic expressions according to the design constraints to minimize the area or timing of the design.

Resource sharing tries to reduce the amount of hardware by sharing hardware resources with multiple operators in an HDL description. For instance, a single adder component may be shared with multiple addition operators in an HDL code. Without resource sharing, each operator in the HDL code will result as a separate hardware component in the final gate-level netlist.

Extracting sharing common subexpressions is a well-known synthesis optimization technique that reduces hardware by avoiding the repetition of the same code. For example,

$s1 = x + y + u$
$s2 = x + y + v$
$s3 = x + y + w$

where the common subexpression $x + y$ can be extracted and defined as a new term, say, t and realized by an adder. After this, the above computations for $s1$, $s2$, and $s3$ can then be expressed in terms of t as follows:

$t = x + y$
$s1 = t + u$
$s2 = t + v$
$s3 = t + w$

As a result, two adders can be saved without increasing the propagation delays significantly.

A reordering operator is one that changes the computational order of operators in order to optimize speed. As an illustration, consider the following computation

$t = w + x + y + z$

which is usually proceeded from left to right. If all inputs have the same delay, then synthesis tools will use a binary tree to compute t; that is, t can be equivalently expressed as

$t = (w + x) + (y + z)$

---

[2] The optimization principles are based on the features of Synopsys Design Compiler. Refer to [22] for more details.

Nevertheless, if the input w has a large delay, then the computation of t can be reordered to optimize speed as follows:

$$t = w + (x + (y + z))$$

It should be noted that the operators cannot be reordered if the original expression is overridden by the use of parentheses in Verilog HDL. Moreover, it should bear in mind that *HDL coding can force a specific topology to be synthesized*. Based on this, the above sharing common subexpressions and reordering operators can be done by the designer in their HDL codes explicitly.

Selecting a default implementation means that the implementation of a specific resource is left to the synthesizer. For example, the technology library (such a library is often called an *IP library*) contains three implementations (ripple, carry-lookahead, and Brent-Kung adders) for the addition (+) operator. As selecting default implementation is specified, the synthesizer considers all available implementations and makes its selection in accordance with the specified design constraints. At this point, the design is represented by a generic technology-independent netlist.

12.2.4.2 Logic-level Optimization Logic-level optimization is independent of technology library and is carried out on Boolean equations (namely, generic technology-independent netlists). It consists of two processes: *structuring* and *flattening*. Structuring is constraint-based and defaulted by most synthesis tools.

Structuring is used to extract the shared logic from a design that contains regular structured logic. That is, it evaluates the design equations denoted by the generic technology-independent netlist and tries to factor out common subexpressions in these equations by using Boolean algebra. The subexpressions that have been identified and factored out can then be shared between the equations. The result has an impact on the total delay of logic.

The structuring technique can be performed on the basis of either timing (default) optimization or Boolean optimization. Boolean optimization is often used to reduce the area but may increase the propagation delay of the resulting logic. Therefore, it is more suitable for non-critical timing circuits such as random logic and finite-state machine structures.

The flattening technique is used to reduce logic with intermediate variables and parentheses to a pure two-level sum-of-product (SOP) representation. As a consequence, the resulting logic has fewer logic levels between the input and output and hence produces faster logic at the expense of area increase. Flattening is carried out independently of constraints and recommended for those designs containing unstructured or random logic.

12.2.4.3 Technology-Mapping Optimization Technology-mapping optimization carries out on the technology-independent netlist and maps it to the library cells to produce a technology-dependent gate-level netlist. Technology-mapping optimization includes the following processes: *technology mapping, delay optimization, design rule fixing*, and *area optimization*.

The technology-mapping process binds the cells from a technology-independent netlist to the cells in a specified technology library.

The delay optimization process fixes the timing violations caused by the technology-mapping process. However, it does not mend design rule violations or meet area constraints.

The design rule fixing process fixes the design rule violations in the design through the insertion of buffers or resizes existing cells by the synthesizer. Note that the design rule fixing is allowed to break timing constraints.

Area optimization is the last step that the synthesizer performs on the design under consideration. During this process, only the optimization that does not break design rules or timing constraints is allowed. It should be noted that the area optimization performed or not performed by the synthesizer depends on the design constraints. As a consequence, *setting realistic constraints is one of the most important tasks in synthesizing a design*.

12.2.4.4 Clock Networks
The design of clock networks is one of the most important design issues in an ASIC design. It is also one of the hardest operations to be performed. A *big-buffer approach* is only for use in feature sizes of 0.35 μm and above. For deep submicron (DSM) technologies, with feature sizes equal to and below 0.25 μm, *clock tree synthesis* (CTS) is a dominant method because it has minimal clock latency and clock skew as compared with the big-buffer approach. The details of clock networks can be referred to Lin [15].

A final comment is that most synthesis tools try to optimize the design for timing by default. Designs that are non-timing critical but area intensive should be optimized for area.

■ Review Questions

12-24 Explain why removing hierarchy can help optimize a design.
12-25 What does optimization mean?
12-26 What are the major types of architectural optimization?
12-27 What does flattening mean?
12-28 What does structuring mean?
12-29 What is logic-level optimization?
12-30 What is technology-mapping optimization?

12.3 Technology-Independent Synthesis

Technology-independent synthesis can be divided into two-level and multilevel logic synthesis. The two-level logic synthesis includes *node minimization* [19], also called *simplification*, to reduce the complexity of a given Boolean network by using a Boolean minimization technique on its nodes. The complexity is measured by the number of literals. The multilevel logic synthesis contains *restructuring operations* to modify the structure of the Boolean network by introducing new nodes, eliminating others, and by adding and removing edges as well. In this section, we begin with the introduction of two-level logic synthesis and then deal with the principles of multilevel synthesis.

12.3.1 Two-Level Logic Synthesis

Before proceeding, we begin with the definition of several related terms often used in logic synthesis and then briefly discuss two well-known two-level logic synthesis approaches: Quine-McCluskey and Espresso algorithms.

12.3.1.1 Basic Terminology As known, a logic or Boolean variable x takes on one of two values 0 and 1. The complement of the variable x is denoted by $\bar{x}$. Both x and $\bar{x}$ are referred to as *literals*. A Boolean function $f : \{0,1\}^n \to \{0,1\}$ is a binary function of logic variables. It is convenient to represent the n-dimensional Boolean space by an n-dimensional hypercube of 2^n nodes. In general, the set of nodes on an n-dimensional hypercube can be divided into the following three sets. The set of nodes of the hypercube with a function value 1 is referred to as the *on-set*, the set of nodes with a function value 0 is referred to as the *off-set*, and the set of nodes with unspecified function values is referred to as the *don't-care-set* or *dc-set*.

A *product term* is a product of literals. A *cube* of a logic function f is a product term whose on-set does not have vertices in the off-set of f. A $minterm$ is a cube in which all variables are assigned a value 0 or 1. In other words, a minterm is a single on-set node of an n-dimensional hypercube. A product term p is denoted by an *implicant* of f if for every input combination such that $p = 1$ then $f = 1$ too. Hence, a cube is an implicant of f. A *prime implicant* is an implicant of f such that if any variable is removed from p, then the resulting product term does not imply f again. A *cover* is an expression which contains all minterms.

Note that a cover is not necessary minimal. It can be shown that the minimal sum (minimal cover) is the sum of prime implicants.

A Boolean network [19] is a standard technology-independent model for a logic circuit. A Boolean network $B = (V, E)$ is a directed acyclic graph, where each of nodes $v \in V$ represents a primary input, primary output, or function, and each of edges $e \in E$ denotes the relationship between nodes. Each node is referred to as a variable name. An example of Boolean networks is shown in Figure 12.12. There is an edge from node j to node i if the function represented by node i depends explicitly on the function represented by j. Node j is said to be a fanin of node i and node i is said to be a fanout of node j. There are two sets of special nodes: input and output nodes. Input nodes with no incoming edges represent primary inputs and output nodes with no outgoing edges represent primary outputs.

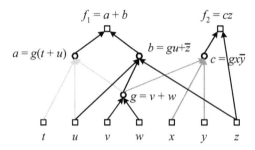

Figure 12.12: An example of Boolean networks.

12.3.1.2 Quine-McCluskey Algorithm Now that we have defined terms related to logic synthesis, we could describe the two-level logic optimization methods. The most well-known two-level logic optimization method is the Quine-McCluskey algorithm, which covers two major steps:

1. Generates all prime implicants.
2. Finds the minimum-cost cover of all minterms from prime implicants.

A major disadvantage of this method is its high time and space complexities. The detailed operations of the Quine-McCluskey algorithm can be found in Lin [14] or other textbooks related to digital logic.

12.3.1.3 Espresso Algorithm Another method widely used in two-level logic optimization is the Espresso [3] algorithm. The Espresso algorithm starts with a given cover and repeatedly reduces it in size using a reduce-expand-irredundant loop. This heuristic algorithm has three basic steps:

1. *Reduce*: The reduce step reduces the size of cubes in the cover; thereby, the number of literals in the cover may be increased temporarily.
2. *Expand*: The expand step makes each cube as large as possible without covering a node in the off-set.
3. *Irredundant*: The irredundant step throws out smaller (i.e., redundant) cubes so that there are no redundant cubes left.

The rationale behind the Espresso algorithm is to throw out the smaller cubes covered by larger cubes. The following example illustrates how the Espresso algorithm works.

Section 12.3 Technology-Independent Synthesis

■ Example 12.4: The operations of the Espresso algorithm.

As mentioned above, the basic steps of the Espresso algorithm are reduce, expand, and irredundant. Suppose that the initial cover is $f(x,y,z) = \bar{x}\bar{y} + \bar{x}\bar{z} + xy + xz$. As depicted in Figure 12.13(a), the initial hypercube contains six on-set nodes. After the reduce step, the minterm $\bar{x}y\bar{z}$ becomes an implicant, as shown in Figure 12.13(b). Now, there are four cubes in total. The next step is to expand the implicant $\bar{x}y\bar{z}$ with other minterms to form a new prime implicant, which results in Figure 12.13(c). Finally, the irredundant step removes redundant prime implicants, leaving behind an irredundant cover, $f(x,y,z) = \bar{x}\bar{y} + y\bar{z} + xz$, as shown in Figure 12.13(d).

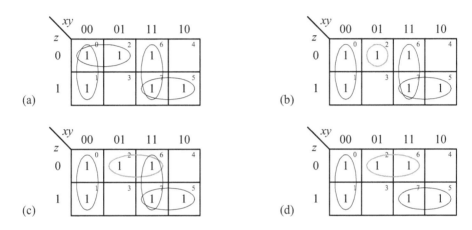

Figure 12.13: An example illustrating the operations of the Espresso algorithm: (a) start; (b) reduce; (c) expand; (d) irredundant.

The input to the Espresso algorithm is typically an encoded truth table. Hence, we use the above example to illustrate how the Espresso algorithm can work with truth tables.

■ Example 12.5: The Espresso algorithm with truth tables.

Figure 12.14 illustrates the set of transformations of the above example. Initially, the switching function contains six minterms as given in Figure 12.14(a). After the expand and irredundant steps, the initial covering consists of four prime implicants $\bar{x}\bar{y}$, $\bar{x}\bar{z}$, xz, and xy, as shown in Figure 12.14(b). Then, the prime implicant $\bar{x}\bar{z}$ is reduced into $\bar{x}y\bar{z}$, as shown in Figure 12.14(c). After expanding with the minterm $xy\bar{z}$, the new prime implicant $y\bar{z}$ is resulted, as depicted in Figure 12.14(d). Finally, the irredundant step removes the prime implicant xy, leaving behind an irredundant cover $\bar{x}\bar{y}$, $y\bar{z}$, and xz, as shown in Figure 12.14(e).

■ Review Questions

12-31 What are the cube, minterm, implicant, and prime implicant?
12-32 What is the minimum cover?
12-33 What is a Boolean network?
12-34 Define the on-set, off-set, and don't-care-set (dc-set).
12-35 What is the difference between a literal and a variable?
12-36 Describe the basic operations of the Espresso algorithm.

x y z	f
0 0 0	1
0 0 1	1
0 1 0	1
1 0 1	1
1 1 0	1
1 1 1	1

(a)

$\Rightarrow$

x y z	f
0 0 -	1
0 - 0	1
1 - 1	1
1 1 -	1

Expand and irredundant
(b)

$\Rightarrow$ Reduce

x y z	f
0 0 -	1
0 1 0	1
1 - 1	1
1 1 -	1

(c)

$\Rightarrow$ Expand

x y z	f
0 0 -	1
- 1 0	1
1 - 1	1
1 1 -	1

(d)

$\Rightarrow$ Irredundant

x y z	f
0 0 -	1
- 1 0	1
1 - 1	1

(e)

Figure 12.14: An example illustrating the operations of the Espresso algorithm using encoded truth tables.

12.3.2 Multilevel Logic Synthesis

After understanding the two-level logic synthesis, we are now in a position to introduce the principles of multilevel logic synthesis. To proceed this, we first address the motivation of multilevel logic synthesis, then deal with general restructuring operations, and finally introduce the kernel approach for synthesizing a multilevel logic circuit from two-level logic functions.

12.3.2.1 Motivation The motivation of multilevel logic synthesis is that there are many functions that are too expensive, in terms of area and propagation delays, to implement in two-level logic. As an example, consider the following two switching functions

$$f(w,x,y,z) = wx + xy + xz$$
$$g(w,x,y,t) = w\bar{x} + \bar{x}y + \bar{x}t \qquad (12.1)$$

which contain 12 literals and need eight gates, including two three-input gates and six two-input gates. However, they require much less hardware when using three-level logic structure by factoring both functions and extracting the common factor $w + y$.

$$\begin{aligned} f(w,x,y,z) &= wx + xy + xz \\ &= x(w+y) + xz \\ &= x \cdot k + xz \\ g(w,x,y,t) &= w\bar{x} + \bar{x}y + \bar{x}t \\ &= \bar{x}(w+y) + \bar{x}t \\ &= \bar{x} \cdot k + \bar{x}t \\ k(w,y) &= w + y \end{aligned}$$

The result has 10 literals and needs only seven two-input gates.

12.3.2.2 Restructuring Operations The rationale behind the multilevel logic synthesis is on the basis of restructuring operations [19], which include *decomposition, extraction, factorization, substitution,* and *collapsing*. In what follows, we briefly describe each of these.

Decomposition. Decomposition is a process that expresses a given logic function in terms of a number of new functions. Decomposition replaces a divisor of a function by a new literal. For example,

$$f = wyz + xyz + uv$$

Let $g = yz$ and $h = w + x$, and then function f can be represented as a function of both g and h and is as follows

$$f = gh + uv$$

where the common factor yz is extracted from the first two terms of function f and defined as a new function g, and another new function h is also defined as well. The function f is then represented as the product of functions g and h plus the original term uv of function f. Decomposition is an essential step in logic optimization for FPGAs.

Extraction. Extraction replaces a common divisor of two functions by a new literal. Extraction is closely related to decomposition but operates on a number of given functions. With extraction, the given functions are expressed in terms of newly created intermediate functions and variables. For example,

$$f = xyz + uw$$
$$g = xyz + uv$$

The product term xyz in both f and g functions can be extracted and denoted by a new function, say h. Functions f and g can then be represented by using the newly created function h as follows:

$$h = xyz$$
$$f = h + uw$$
$$g = h + uv$$

Factorization. Factorization is usually based on Boolean division to extract common factors (new nodes) and then used to restructure the network. Factorization generally takes three steps. It first generates all potential common factors and estimates how many literals can be saved for each when these common factors are substituted into the network, then it chooses which factors to substitute into the network, and finally it restructures the network by adding the new factors and modifying the other functions to use these factors. For example,

$$f = wyz + xyz + uv$$

can be factored as

$$f = yz(w + x) + uv$$

For the situation that propagation delay is the major consideration, the factorization of a subnetwork is dedicated to reduce the number of function nodes through which delay-critical signals must pass.

Substitution. Substitution, also called *resubstitution*, replaces a divisor of a function with an existing literal and then resubstitutes the literal into the original function. For example, let $g = yz$ then

$$f = wyz + xyz + uv$$

can be rewritten as follows:

$$f = gw + gx + uv$$

Collapsing. Collapsing, also called *elimination* or *flattening*, is the inverse operation of substitution. It removes a literal from a function by replacing it with its corresponding function. For example,

$$g = yz$$
$$f = g(w + x) + uv$$

By substituting function g into f, we obtain

$$f = wyz + xyz + uv$$

12.3.2.3 The Kernel Approach The general operations involved in multilevel synthesis are to minimize two-level logic functions, find common subexpressions, substitute one expression into another, and factor each function. In what follows, we introduce a systematic approach to synthesize a multilevel logic circuit from two-level logic functions. This approach is often referred to as the *kernel approach* [3, 4]. Before going to an in-depth description of the approach, we need to define some terminology first.

The *divisors* of f are defined as the set

$$D(f) = \{g | f/g \neq \phi\}$$

The *primary divisors* of an expression f are defined as the set

$$P(f) = \{f/c | c \text{ is a cube}\}$$

For example, if

$$f = wxy + wxzt$$

then

$$f/w = xy + xzt$$

is a primary divisor. Every divisor of f is contained in a primary divisor. If g divides f, then $g \subseteq p \in P(f)$. We say g divides f evenly if the remainder is ϕ. g is said to be *cube-free* if the only cube dividing g evenly is 1. In other words, an expression is cube-free if no cube divides the expression evenly. For example, $xy + z$ is cube-free while $xy + xz$ and xyz are not cube-free. Note that a cube-free expression must have more than one cube.

The *kernels* of an expression f are the cube-free primary divisors of f. The kernels of f are defined as

$$K(f) = \{k | k \in P(f) \text{ and } k \text{ is cube-free}\}$$

For example, consider the logic function $f = wxy + wxzt$, then $f/w = xy + xzt$ is a primary divisor but is not cube-free because x is a factor of $f/w = x(y + zt)$. A cube c used to obtain the kernel $k = f/c$ is called a *cokernel* of k. For example, $f/wx = y + zt$ is a kernel and wx is the cokernel. The cokernel of a kernel is not unique in general.

■ **Example 12.6: Kernels and cokernels.**

Consider the following switching function:

$$f(w,x,y,z) = xz + yz + wxy$$

It contains 7 literals. To find the cokernels and kernels, we proceed the following division operations

$$f/w = xy$$
$$f/x = z + wy$$
$$f/y = z + wx$$
$$f/z = x + y$$

where w is not a cokernel because kernel xy is not cube-free, which has x or y as a factor. Consequently, the cokernel set is $\{x,y,z\}$ and the kernel set is $\{z + wy, z + wx, x + y\}$.

The cokernels of a switching function can also be found by using a *cokernel table*, in which all nontrivial product terms are listed one by one in both the column header and row

Section 12.3 *Technology-Independent Synthesis*

	twy	txy	uwy	uxy	vwy	vxy
twy	*					
txy	ty	*				
uwy	wy	y	*			
uxy	y	xy	uy	*		
vwy	wy	y	wy	y	*	
vxy	y	xy	y	xy	vy	*

	ty	uy	vy	wy	xy
ty	*				
uy	y	*			
vy	y	y	*		
wy	y	y	y	*	
xy	y	y	y	y	*

Figure 12.15: The cokernel tables used in a multilevel logic synthesis example.

header of the table, as shown in Figure 12.15. Each entry in the table is the intersection of the corresponding product term in the column header and the one in the row header. Because of the symmetrical property in nature, only the lower half of the table is needed to construct. All entries in the cokernel table except the ϕ term are cokernels. If all entries in the cokernel table are a single literal or ϕ, then we have done the job; otherwise, the entries not in single-literal form are collected and another cokernel table is constructed based on these entries to find the cokernels associated with the new cokernel table again. This process is repeated until all entries in the cokernel table are left with ϕ or literals. The union of all cokernels found in all cokernel tables is the cokernel set of the switching function in question.

In the kernel approach, we often find the cokernel set first and then obtain the kernel set by dividing the switching function by its corresponding cokernels, one by one. To see this and to illustrate how the cokernel set can be found with cokernel tables and how the kernel approach can be used to decompose a two-level switching function into a multilevel one, consider the following example.

■ Example 12.7: An example of multilevel logic synthesis.

Consider the following switching function:

$$f(t,u,v,w,x,y,z) = twy + txy + uwy + uxy + vwy + vxy + z$$

It contains 19 literals. To find the cokernels, we construct the cokernel tables shown in Figure 12.15. We obtain the cokernel set $\{ty, uy, vy, wy, xy, y\}$ from the first cokernel table. Because some entries contain two literals, another cokernel table is needed to construct based on the set $\{ty, uy, vy, wy, xy\}$, from which the cokernel y is found. Combining the above two cokernel sets, the resulting cokernel set is $\{ty, uy, vy, wy, xy, y\}$. The kernel corresponding to each cokernel can be calculated as follows:

$$f/ty = w + x = K_1$$
$$f/uy = w + x = K_1$$
$$f/vy = w + x = K_1$$
$$f/wy = t + u + v = K_2$$
$$f/xy = t + u + v = K_2$$
$$f/y = tw + tx + uw + ux + vw + vx$$
$$= t(w+x) + u(w+x) + v(w+x) = (w+x)(t+u+v)$$
$$= K_3 = K_1 K_2$$

The function f can then be reduced to

$$f = K_3 y + z = (w+x)(t+u+v)y + z$$

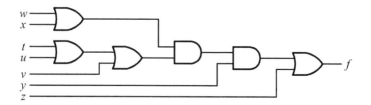

Figure 12.16: An example of multilevel logic synthesis.

It contains only 7 literals. The resulting logic diagram is shown in Figure 12.16.

The kernel approach can also be used to decompose multiple-output switching functions. The rationale behind this is as follows. The objective of multilevel logic synthesis is to find the common divisors of two (or more) functions f and g. Two switching functions f and g have a non-trivial common divisor d ($d \neq$ cube) if and only if there exist kernels $k_f \in K(f)$ and $k_g \in K(g)$ such that $k_f \cap k_g$ is non-trivial, i.e., not a cube. Hence, we can use kernels of f and g to extract common divisors. An illustration is explored in the following example.

■ Example 12.8: Multiple-output multilevel logic synthesis.

Consider the following three switching functions:

$$f_1(t, u, v, w) = tv + tw + uv + uw$$
$$f_2(v, w, x, y) = vx\bar{y} + wx\bar{y}$$
$$f_3(u, v, w, x, y, z) = uv + uw + \bar{z}$$

There are 19 literals. The cokernel set is found to be

$$C_{f_1} = \{t, u, v, w\}$$
$$C_{f_2} = \{x\bar{y}\}$$
$$C_{f_3} = \{u\}$$

The kernel set is obtained by dividing each switching function by its corresponding cokernels, one by one. The resulting kernel sets for the three switching functions are as follows:

$$K_{f_1} = \{v + w, t + u\}$$
$$K_{f_2} = \{v + w\}$$
$$K_{f_3} = \{v + w\}$$

From the above three kernel sets, it is clear that the common divisor is $v + w$. The switching functions can then be expressed as functions of the common divisor and the results are as follows

$$f_1(t, u, v, w) = g(t + u)$$
$$f_2(v, w, x, y) = gx\bar{y}$$
$$f_3(u, v, w, x, y, z) = gu + \bar{z}$$

where $g = v + w$. There are only 11 literals. The resulting logic diagram is shown in Figure 12.17.

Section 12.3 *Technology-Independent Synthesis*

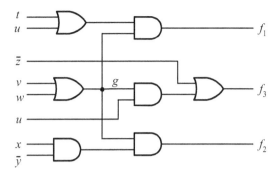

Figure 12.17: An example of multiple-output multilevel logic synthesis.

12.3.2.4 Don't Cares Don't cares [3] are very important in minimizing two-level and multi-level logic. There are three types of don't cares that are widely used to simplify logic functions. These are *output don't cares*, *observability don't cares*, and *satisfiability don't cares*.

- An *output don't care* occurs when an output function is not specified for some input values. For example, in a microprocessor design, certain instruction codes may not be used and therefore will never occur as a valid input.
- An *observability don't care* occurs when an intermediate variable value does not affect the primary outputs of the Boolean network. As an illustration, consider the case that has two logic blocks in which the first computes an arithmetic function and the second implements an enable signal which controls whether the arithmetic result is latched at the outputs or not. Clearly, the output of the arithmetic function is not observable under the conditions which disable the register.
- *Satisfiability don't cares* mean that intermediate variable values are inconsistent with its function inputs. In other words, satisfiability don't cares represent the combinations of variables of the Boolean network that can never occur because of the structure of the network itself.

To see how satisfiability don't cares can be used to optimize logic during the course of network optimization, consider the following example.

■ Example 12.9: An example of the satisfiability don't cares.

An example of satisfiability don't cares is as follows. If node i of the network carries out the Boolean function $f(a,b)$, where $a = x \cdot y$, $b = (x+y) \cdot z$, and x, y, z are the primary inputs of the Boolean network, then in some intermediate node with a Boolean function of variables a, x, and y, the combination $a = 0$, $x = 1$, and $y = 1$ will never occur. Generally, since $a = x \cdot y$, $a \neq x \cdot y$ will never occur. This can be expressed as $a \cdot \overline{x \cdot y} + \bar{a}(x \cdot y)$, which is a satisfiability don't care because there exists no input combinations of x and y to satisfy it and hence it can be removed from the network. The same argument can be applied to the node with a Boolean function of variables b, x, y, and z and obtain the satisfiability don't care, $b \cdot \overline{(x+y) \cdot z} + \bar{b}[(x+y) \cdot z]$.

The don't cares have been proved to be very effective for a wide variety of cases and are very often the only Boolean operation performed during the course of network optimization.

■ Review Questions

12-37 What is the rationale behind multilevel logic synthesis?
12-38 Describe the operation of decomposition, extraction, and factorization.
12-39 Describe the operation of substitution and collapsing.
12-40 Describe the meaning of the term cube-free.
12-41 Define the divisor and the prime divisor.
12-42 What is a satisfiability don't care?

12.4 Technology-Dependent Synthesis

The technology-dependent synthesis (or simply *technology mapping* or *library binding*) is the optimization problem of finding a minimum-cost cover of the technology-independent logic network by choosing from the collection of primitive logic elements in the target library. During technology-dependent synthesis, the area and/or timing (i.e., propagation delays) of a design is optimized. In this section, we first introduce a widely used technology mapping approach, referred to as *network covering* and then deal with three simple approaches for LUT-based FPGA architectures, including the two-step approach, the FlowMap method, and Shannon's expansion approach. The two-step approach and FlowMap method are based on the multilevel logic circuit while Shannon's expansion approach is based on the two-level logic circuit.

12.4.1 Network Covering

Network covering [12] is a means by which each subnetwork of the technology-independent logic network is replaced with cells from the underlying cell library such that the whole network is covered and the desired objective is met.

The network covering generally follows the following four steps:

1. Decomposing a network into base functions
2. Partitioning a network into subject graphs
3. Obtaining matches
4. Obtaining a network cover.

12.4.1.1 Base Functions The first step of the network covering approach is to decompose the technology-independent logic network into a graph such that each node can be covered by a set of *base functions*. The graph is called a *subject graph*. The set of base functions must be functionally complete so that it can implement any logic network. A simple set of base functions is {NOT, NAND2}. To implement the subject graph with the targeted technology, a technology library, referred to as the *cell library*, is created. A technology library typically comprises hundreds of cells, including gates and sequential elements like latches and flip-flops. Each cell in the cell library is represented as a graph, called the *pattern graph*, in which each node is also derived from the set of base functions. Typically, all base functions are also implemented as cells, thereby guaranteeing that the cell library can realize any switching function. For the ease of optimization, both area and delay costs are associated with each pattern graph. An example of a cell library based on {NOT, NAND2} is given in Figure 12.18.

As an illustration of how to map a technology-independent logic network onto basic functions, consider the following example.

Section 12.4 Technology-Dependent Synthesis

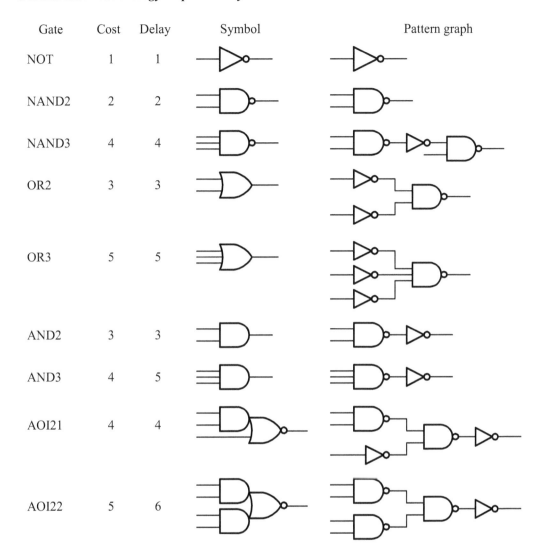

Figure 12.18: Some typical pattern graphs and their related parameters based on the set of {NOT, NAND2}.

■ Example 12.10: An example of technology mapping.

Figure 12.19 shows a simple technology mapping example. The technology-independent logic network is first converted into a logic network realized by the base functions. The resulting logic network is shown in Figure 12.19(b). Here, the set of base functions is assumed to be {NOT, NAND2}. Next, the resulting logic network is mapped to cells in the target cell library. Depending on the mapping approach used, different results may be obtained. For instance, Figure 12.19(c) is the result obtained by trivially mapping each base function to its corresponding cell. The area cost is 10. However, if we use the three-input NAND gate (NAND3), a logic network with area cost of 9 is resulted, as exhibited in Figure 12.19(d).

12.4.1.2 Subject Graph After a technology-independent logic network is decomposed into a network composed of base functions, the next step is to partition the resulting logic network into subnetworks such that each subnetwork forms a subject graph. The subject graph is one

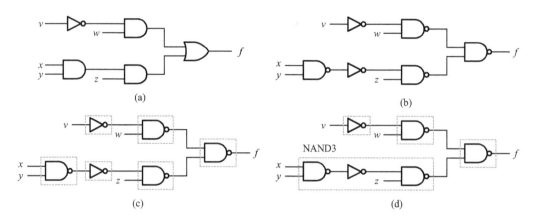

Figure 12.19: A technology mapping example: (a) technology-independent network; (b) NAND implementation; (c) technology mapping with area cost 10; (d) technology mapping with area cost 9.

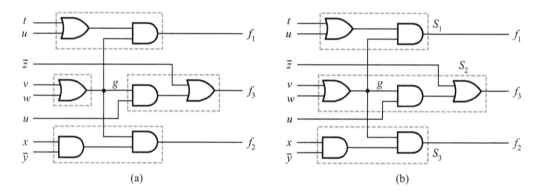

Figure 12.20: An example of (a) legal and (b) illegal subject graphs.

that is acyclic and does not have internal fanouts. The tree matching and network cover are then separately applied to each subject graph and the resulting logic networks are combined together to realize the original logic network. As illustrations of subject graphs, refer to Figure 12.20 and consider the following example.

■ Example 12.11: An example of subject graphs.

Figure 12.20 illustrates the legal and illegal subject graphs. By definition, a subject graph needs to be acyclic and has no internal fanouts. As a consequence, Figure 12.20(a) is a legal subject graph but Figure 12.20(b) is not because node g in the subject graph S_2 has an internal fanout of 3. This means that node g cannot be connected to subject graphs S_1 and S_3. Notice that in this example we assume that the base functions are {NOT, AND2, OR2}.

12.4.1.3 Tree Matching In tree matching, all possible ways that pattern graphs can match each node of the subject graph are searched and listed as a table. Because base functions are also implemented as cells, it is guaranteed that at least one way can be found.

Section 12.4 Technology-Dependent Synthesis

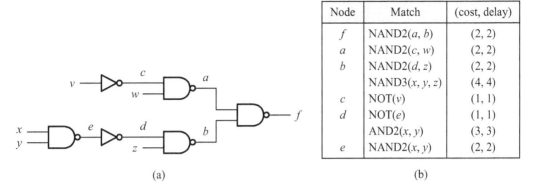

Figure 12.21: An example of tree matching: (a) subject graph and (b) tree matches.

■ Example 12.12: An example of tree matching.

Figure 12.21 illustrates the operation of tree matching. In tree matching, each node of the subject graph is replaced with a pattern graph that is functionally equivalent. The tree matching can be proceeded by starting from either a primary input or primary output of the subject graph. Assume that we begin with the output node f of the subject graph in Figure 12.21(a). The results are given in Figure 12.21(b). Node b may match either NAND2 or NAND3 and node d may match NOT or AND2. Both area and delay costs are also indicated in Figure 12.21(b).

12.4.1.4 Network Cover
The goal of technology mapping is to minimize the area cost or delay cost of the logic circuit in question, or to minimize area (delay) under delay (area) constraints. For this objective, the *dynamic programming method* is often used in the network cover step to choose a network cover optimally in terms of area cost or delay cost. Here, the network cover means a collection of pattern graphs such that every node of the subject graph is covered by at least one pattern graph. To further illustrate the operation of technology mapping, consider the subject graph shown in Figure 12.22(a), which is an implementation of Figure 12.16 with base functions of {NOT, NAND2}. The tree matches are shown in Figure 12.22(b).

■ Example 12.13: An example of area optimization.

Given in Figure 12.22(a) is the subject graph with the matched pattern graphs from the assumed cell library depicted in Figure 12.18. To find the optimal area matching, we need to search all possible options of each node with a minimum accumulated area cost until reach the node under consideration. To achieve this, we search from the primary inputs toward the primary output. The result is listed as a table shown in Table 12.1. At node l, there are two options, NAND2(m,n) and OR2(t,u), with area costs of 4 and 3, respectively. Hence, the area cost 3 will be used in the later calculation. In general, the optimal area cost is the sum of the area of the matched pattern graph plus the optimal area costs of the inputs of the matched pattern graph. For instance, at node j, the accumulated area cost from primary inputs t and u is equal to $3+1=4$ if it matches NOT(l) gate and is 5 if it matches AND2(m,n) gate. Based on this idea, the final optimal area cost is found to be 16. The cells used are indicated with shaded boxes.

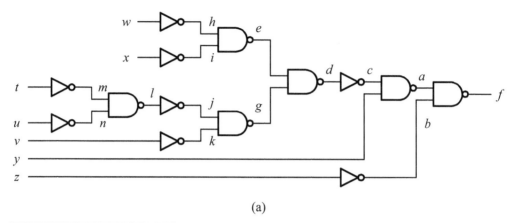

Node	Match	(cost, delay)	Node	Match	(cost, delay)
f	NAND2(a, b)	(2, 2)	g	NAND2(j, k)	(2, 2)
a	NAND2(c, y)	(2, 2)		OR2(l, v)	(3, 3)
	NAND3(e, g, y)	(4, 4)	h	NOT(w)	(1, 1)
b	NOT(z)	(1, 1)	i	NOT(x)	(1, 1)
c	NOT(d)	(1, 1)	j	NOT(l)	(1, 1)
	AND2(e, g)	(3, 3)		AND2(m, n)	(3, 3)
	AOI22(h, i, j, k)	(5, 6)	k	NOT(v)	(1, 1)
d	NAND2(e, g)	(2, 2)	l	NAND2(m, n)	(2, 2)
e	NAND2(h, i)	(2, 2)		OR2(t, u)	(3, 3)
	OR2(w, x)	(3, 3)	m	NOT(t)	(1, 1)
			n	NOT(u)	(1, 1)

(b)

Figure 12.22: An example of tree matching for using in the network cover step: (a) subject graph and (b) tree matches.

■ **Example 12.14: An example of timing optimization.**

Generally, the delay at the output of a matched pattern graph is the sum of the delay of the matched pattern graph itself plus the maximum delay of the inputs of the matched pattern graph. Based on this rule, the optimal delay cover of the subject graph given in Figure 12.22(a) can be found from the primary inputs toward the primary output. For example, at node d the delay is the sum of NAND2(e,g) and the maximum delay of its inputs, namely, $2 + \max\{3,6\} = 8$. The resulting optimal delay cover of the subject graph given in Figure 12.22(a) is shown in Table 12.2. The cells used are indicated with shaded boxes. In this example, it happens to be the same as that of the area optimal cover. Generally, they may be different.

12.4.2 A Two-Step Approach

The two-step approach proposed in [19, 24] is one proposed to carry out the technology-dependent synthesis for FPGAs. This two-step approach works as follows. In the first step, the underlying technology-independent network is decomposed into nodes with no nodes more

Section 12.4 Technology-Dependent Synthesis

Table 12.1: The network cover with an optimal area cost.

Node	Match	Area cost	Node	Match	Area cost
m	NOT(t)	1	g	NAND2(j,k)	7
n	NOT(u)	1		OR2(l,v)	6
l	NAND2(m,n)	4	d	NAND2(e,g)	11
	OR2(t,u)	3	c	NOT(d)	12
h	NOT(w)	1		AND2(e,g)	12
i	NOT(x)	1		AOI22(h,i,j,k)	12
j	NOT(l)	4	b	NOT(z)	1
	AND2(m,n)	5	a	NAND2(c,y)	14
k	NOT(v)	1		NAND3(e,g,y)	13
e	NAND2(h,i)	4	f	NAND2(a,b)	16
	OR2(w,x)	3			

Table 12.2: The network cover with an optimal delay cost.

Node	Match	Delay cost	Node	Match	Delay cost
m	NOT(t)	1	g	NAND2(j,k)	6
n	NOT(u)	1		OR2(l,v)	6
l	NAND2(m,n)	3	d	NAND2(e,g)	8
	OR2(t,u)	3	c	NOT(d)	9
h	NOT(w)	1		AND2(e,g)	9
i	NOT(x)	1		AOI22(h,i,j,k)	10
j	NOT(l)	4	b	NOT(z)	1
	AND2(m,n)	4	a	NAND2(c,y)	12
k	NOT(v)	1		NAND3(e,g,y)	10
e	NAND2(h,i)	3	f	NAND2(a,b)	12
	OR2(w,x)	3			

than k inputs, where k is determined by how many number of inputs each LUT can have. Using this idea, the number of nodes is reduced. In the second step, the number of nodes is further reduced by combining some of them, taking into account the special features of LUTs. The following example illustrates the idea of the two-step approach.

■ **Example 12.15: An example of the two-step approach.**

Suppose that each LUT has 4 inputs as in most of widely used FPGA architectures. By using the two-step approach, the gate network in Figure 12.17 can be easily fitted into four 4-input LUTs, indicated with dashed boxes as shown in Figure 12.23(a). However, if we pay more attention to it, we can find that it only requires three 4-input LUTs, as illustrated in Figure 12.23(b). In general, the mapping result depends on how much the effort is used to search the netlist. Generally, the more effort, the better result. Nonetheless, it is a time-consuming process and may take much more computation time.

In practice, most synthesis tools by default try to optimize the design for timing. Nonetheless, the synthesis tools may also optimize the design for area instead of timing if they are instructed to do this. When this is the case, as stated above the synthesis tools would yield the better results if more time is allowed to search the netlist.

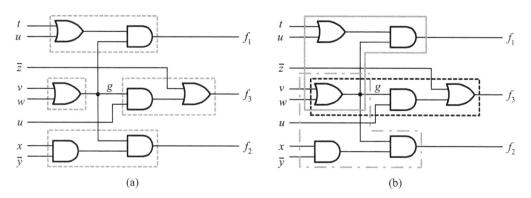

(a)   (b)

Figure 12.23: An example illustrating the operations of the two-step approach: (a) mapping I requires four 4-input LUTs and (b) mapping II only needs three 4-input LUTs.

12.4.3 The FlowMap Method

The FlowMap [6, 24] method is a delay-optimized technology-mapping algorithm for LUT-based FPGA architectures. The essential idea of the FlowMap method is to break the network into LUT-sized blocks by using an algorithm to find a minimum-height k-feasible cut in the network. It uses network flow algorithms to optimally find the minimum-height cut and uses heuristics to maximize the amount of logic fitted into each cut to reduce the needed number of LUTs. An illustration of the FlowMap method is given in the following example.

■ **Example 12.16: An example of the FlowMap method.**

The FlowMap method can be revealed by the logic circuit depicted in Figure 12.24. Assume that 4-input LUTs are available for implementing the gate network. The method proceeds from the primary output f backward to the primary inputs. It calculates the cut at each gate level until a cut can fit into an LUT optimally. Then, it removes the portion and processes the rest of the gate network repeatedly until it reaches the primary inputs. As a result, three LUTs are required to implement the gate network.

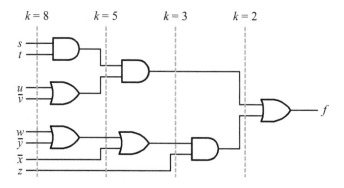

Figure 12.24: An example illustrating the operations of the FlowMap method.

It is instructive to apply the FlowMap method to the logic circuit shown in Figure 12.17. By applying the FlowMap method to the output functions one by one, it can be easily to find that each function only needs one 4-input LUT and thus three 4-input LUTs are enough.

Section 12.4 *Technology-Dependent Synthesis*

12.4.4 Shannon's Expansion Approach

Shannon's expansion theorem [14, 15] can also be used to carry out technology-dependent synthesis for FPGAs. Shannon's expansion (or decomposition) theorem is stated as follows:

■ **Theorem 12-1: Shannon's Expansion Theorem**

Let $f(x_{n-1},\cdots,x_{i+1},x_i,x_{i-1},\cdots,x_0)$ be an n-variable switching function. Then f can be decomposed with respect to variable x_i, where $0 \leq i < n$, as follows:

$$f(x_{n-1},\cdots,x_{i+1},x_i,x_{i-1},\cdots,x_0) = x_i \cdot f(x_{n-1},\cdots,x_{i+1},1,x_{i-1},\cdots,x_0)+$$
$$\bar{x}_i \cdot f(x_{n-1},\cdots,x_{i+1},0,x_{i-1},\cdots,x_0) \quad (12.2)$$

The proof of this theorem is quite simple and hence we omit it here. Nevertheless, we give an example to reveal the validity of this theorem. For convenience, the subfunctions $f(x_{n-1},\cdots,0,\cdots,x_1,x_0)$ and $f(x_{n-1},\cdots,1,\cdots,x_1,x_0)$ are referred to as the *residues* of f with respect to $\bar{x}_i$ and x_i, respectively.

■ **Example 12.17: Shannon's expansion theorem.**

Using the following switching function shows the validity of Shannon's expansion (or decomposition) theorem:

$$f(x,y,z) = xy + yz + xz$$

Solution: Supposing that we decompose the above switching function with respect to switching variable x, the switching function can then be expressed as

$$f(x,y,z) = \bar{x} \cdot f(0,y,z) + x \cdot f(1,y,z)$$
$$= \bar{x}(yz) + x(y+z)$$
$$= \bar{x}yz + xy + xz$$

which can be easily proved to be equivalent to the original switching function.

To use Shannon's expansion theorem to synthesize a two-level switching function with a k-input LUT-based FPGA device, the switching function is repeatedly decomposed into subfunctions with respect to switching variables selected in an arbitrary way or according to some criterion until all subfunctions are only left with k switching variables, where k is the allowed number of inputs of LUTs associated with the underlying FPGA device. This idea is illustrated by the following example.

■ **Example 12.18: Synthesis with Shannon's expansion theorem.**

Supposing that 4-input LUTs are available, synthesize the following switching function by using Shannon's expansion theorem.

$$f(t,u,v,w,x,y,z) = twy + txy + uwy + uxy + vwy + vxy + z$$

Solution: Because the switching function contains seven variables, it is needed to decompose with respect to three selected variables so that all residues are only a function of four variables. Assuming that the switching function is decomposed with respect to variables, t, u, and v, the residues of the switching function can then be found by substituting the value of each combination of these variables into the switching function and are found to be

$$f(0,0,0,w,x,y,z) = z = h$$

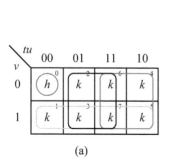
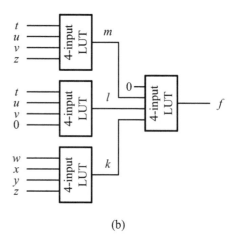

Figure 12.25: A synthesis example using Shannon's expansion theorem: (a) Karnaugh map and (b) realized with 4-input LUTs.

$$f(0,0,1,w,x,y,z) = wy + xy + z = k$$
$$f(0,1,0,w,x,y,z) = wy + xy + z = k$$
$$f(0,1,1,w,x,y,z) = wy + xy + z = k$$
$$f(1,0,0,w,x,y,z) = wy + xy + z = k$$
$$f(1,0,1,w,x,y,z) = wy + xy + z = k$$
$$f(1,1,0,w,x,y,z) = wy + xy + z = k$$
$$f(1,1,1,w,x,y,z) = wy + xy + z = k$$

As a result, the switching function can be represented as follows with the help of the Karnaugh map shown in Figure 12.25(a):

$$\begin{aligned} f(t,u,v,w,x,y,z) &= \bar{t}\bar{u}\bar{v}h + (t+u+v)k \\ &= \bar{t}\bar{u}\bar{v}z + (t+u+v)(wy + xy + z) \\ &= m + l \cdot k \end{aligned}$$

Hence, it can be realized with four 4-input LUTs, as shown in Figure 12.25(b).

Another example gives the realization of a 9-bit even-parity generator, which can be expressed as follows:

$$ep = x[0] \oplus x[1] \oplus x[2] \oplus x[3] \oplus x[4] \oplus x[5] \oplus x[6] \oplus x[7] \oplus x[8]$$

Even though this switching function can still be realized by using Shannon's expansion theorem to decompose into several subfunctions and then fit into LUTs, it can be simply realized by using three 4-input LUTs. More about the use of Shannon's expansion theorem to realize switching functions with function generators such as multiplexers can be referred to Lin [14, 15].

■ Review Questions

12-43 Describe the functions of technology-dependent mapping.
12-44 Describe the operations of the FlowMap method.

Table 12.3: The synthesizable operators of Verilog HDL by most synthesis tools.

Arithmetic	Bitwise	Reduction	Relational
+: add	~: negation (not)	&: and	>: greater than
-: subtract	&: and	|: or	<: less than
*: multiply	|: or	~&: nand	>=: greater than or equal to
/: divide	^: xor	~|: nor	<=: less than or equal to
%: modulus	^~, ~^: xnor	^: xor	
**: power (exponent)		^~, ~^: xnor	
Logical equality	**Logical**	**Shift**	**Miscellaneous**
==: equality	&&: logical and	<<: logical left shift	{ , }: concatenation
!=: inequality	||: logical or	>>: logical right shift	{const{expr}}: replication
Case equality	!: logical negation (not)	<<<: arithmetic left shift	? : : conditional
===: equality		>>>: arithmetic right shift	
!==: inequality			

12-45 Describe the operations of the two-step approach.
12-46 Apply the two-step approach to the circuit of Figure 12.24.
12-47 Apply the FlowMap method to the circuit of Figure 12.17.
12-48 Prove the correctness of Shannon's expansion theorem.

12.5 Language Structure Synthesis

From the viewpoint of users, a synthesis tool at least performs the following critical tasks: to detect and eliminate redundant logic and combinational feedback loops, to explore don't-care conditions, to detect unused states and collapse equivalent states, to make state assignments, and to synthesize optimum multilevel logic subject to design constraints.

In this section, we address the operations of synthesis tools from the viewpoint of Verilog HDL programmers. In other words, we describe how synthesis tools deal with the language structures, including assignment statements, **if-else** statements, **case** statements, loop statements, **always** blocks, and memory synthesis approaches.

12.5.1 General Considerations of Language Synthesis

The most straightforward language structures in Verilog HDL are assignment statements, including **assign** continuous assignments and procedural assignments. As introduced previously, an **assign** continuous assignment is basically an expression, which in turn comprises operands and operators. Except for the divide, exponent, and modulus operators, all other operators in Verilog HDL are synthesizable, as illustrated in Table 12.3. Most synthesis tools limit the second operand of divide and modulus operators and the first operand of the exponent operator to an integer power of 2. The operands can be **wire** and **tri** nets, **reg** variables, parameters, and functions. Note that the initial values associated with nets or registers may be ignored or unacceptable by synthesis tools.

Module instances, gate primitive instances, functions, and tasks are synthesizable with that the delay timing constructs are ignored. Procedural statements, including **always**, **if-else**, **case**, **casex**, and **casez**, are all synthesizable but the synthesis of the **initial** block is not supported. Procedural blocks, including the **begin-end** and named blocks, and the **disable** statement are also synthesizable. Nevertheless, the **disable** statement is generally not suggested to use in a synthesizable module.

Loop statements, including **for**, **while**, and **forever**, are synthesizable except that **while** and **forever** statements must contain event timing control @(**posedge** clk) or @(**negedge** clk).

12.5.2 Synthesis of Selection Statements

Recall that **if-else** and **case** statements are used to perform two-way and multiway selections, respectively. An **if-else** statement is usually synthesized into a 2-to-1 multiplexer and a nested **if-else** statement is synthesized into a priority-encoded, cascaded combination of multiplexers.

12.5.2.1 if-else Statements For many applications, one might inadvertently use only the **if** part without the **else** part of an **if-else** statement, thereby leading to an incomplete **if-else** statement. In this case, the synthesized results of combinational and sequential logic will be different. In a combinational circuit, a logic circuit accompanied with a latch will be inferred whereas in a sequential circuit, a correct result is obtained.

To get more insight into this, consider the following statement:

```
always @(enable or data)
   if (enable) y = data;  // infer a latch
```

Because the y value is only updated whenever the control signal enable is true, the synthesis tools assume that the y value is unchanged otherwise by default. As a result, a latch is needed and inferred to keep the y value unchanged as the enable signal is false. The synthesized result is shown in Figure 12.26. Of course, whenever a latch is indeed needed, this statement can be used to create one appropriately. Nonetheless, as we design a digital system with the HDL-based synthesis flow, it is good practice to avoid the use of latches for the reason of making the resulting system more testable.

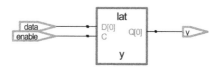

Figure 12.26: The synthesized result of an incomplete **if-else** statement.

The inference of a latch in an *incomplete* **if-else** statement describing a combinational circuit can be avoided by either of the following two ways:

- Complete the **if-else** statement by adding the **else** part. For example,

  ```
  always @(enable or data)
     if (enable) y = data;
     else y = 1'b0; // to avoid the inferred latch
  ```

- Initialize the output of the combinational circuit before set its value conditionally. For example,

  ```
  always @(enable or data)
     y = 1'b0;         // initialize y
     if (enable) y = data;
  ```

The use of the **if-else** statement in sequential logic is quite different from that in combinational logic. To illustrate this, consider the following statement:

```
always @(posedge clk)
   if (enable) y <= data;
   else y <= y;   // a redundant expression
```

Section 12.5 *Language Structure Synthesis*

A warning message of "removing a redundant expression..." will be obtained when the **else** part of the above **if-else** statement is specified because for sequential logic, the value of a variable is unchanged by default unless it is changed otherwise explicitly. As a consequence, the synthesis tools will remove the redundant expression in the optimization step.

12.5.2.2 case Statements Remember that the **case** statement is often used to model multiplexers and can be viewed as a concise representation of the nested **if-else** statement. A **case** statement will be inferred as a multiplexer and has the same consideration as an **if-else** statement in the use of a complete or an incomplete specified format. An illustration of a latch inferred due to an incomplete **case** statement is explored in the following example.

■ Example 12.19: Latch inference—an incomplete case statement.

This example intends to model a 3-to-1 multiplexer; hence, three case items are naturally used to select and route the input data to the output of the multiplexer. However, it leaves a combination (i.e., 2'b11) of select signals unspecified, thereby leading to a latch to be inferred. To avoid this, a **default** case item may be used as the last one to completely specify the **case** statement.

```
// the example of an incompletely specified case
// which will be inferred a latch by synthesis tools
module latch_infer_case(
       input   [1:0] select,
       input   [2:0] data,
       output reg y);

// the body of the 3-to-1 MUX
always @(select or data)
   case (select)
       2'b00: y = data[select];
       2'b01: y = data[select];
       2'b10: y = data[select];
   endcase
endmodule
```

The synthesized result of the above module with an incomplete **case** statement is shown in Figure 12.27. It is instructive to synthesize this module with and without the **default** case item, respectively, and see the difference between them.

■ Coding Style

1. *It should avoid the use of any latches in a design for the reason of testability.*
2. *Complete **if-else** and **case** statements should be used when describing combinational circuits to avoid an inferred latch.*
3. *It is good practice to assign outputs for all input conditions to avoid inferred latches.*

■ Review Questions

12-49 Explain why the exponent operator is generally unsynthesizable.
12-50 Explain why the modulus operator is in general unsynthesizable.
12-51 Explain the meaning of an incomplete **if-else** statement.

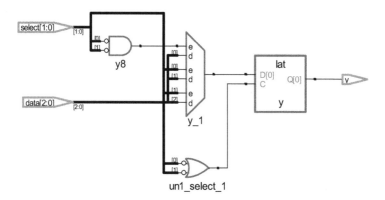

Figure 12.27: The synthesized result of an incomplete **case** statement.

12-52 Explain the meaning of an incomplete **case** statement.
12-53 Explain why an incomplete combinational **if-else** statement will be inferred a latch.

12.5.3 Delay Values

Synthesis tools ignore the delays appearing in RTL descriptions because the ultimate delays of the logic circuits will be determined by the realistic delays of gates used to implement the gate-level netlist. However, the delays are often used to mimic the propagation and transport delays of logic gates and wires or the period of a clock signal during simulation.

An illustration of demonstrating how synthesis tools ignore the delays and generate a meaningless logic circuit is explored in the following example.

■ Example 12.20: An unsynthesizable four-phase clock generator.

This example intends to model a four-phase clock generator. Phase 0 starts at time unit 0 from the positive edge of the clock signal. Each phase lasts for 5 time units. Although this module may generate the desired timing during simulation, it cannot be synthesized into a realistic logic circuit because all delays are ignored by synthesis tools. The synthesized result is only the last statement; namely, it assigns 4'b0001 to the phase_out output after logic optimization.

```
// a four-phase clock example---an incorrect hardware
module four_phase_clock_wrong(
       input clk,
       output reg [3:0] phase_out);   // phase output

// the body of the four-phase clock
always @(posedge clk) begin
   phase_out <= #0  4'b0001;
   phase_out <= #5  4'b0010;
   phase_out <= #10 4'b0100;
   phase_out <= #15 4'b1000;
   phase_out <= #20 4'b0001;
end
endmodule
```

Section 12.5 *Language Structure Synthesis*   547

Owing to ignoring the delays associated with statements by synthesis tools, in the case needing timing control, event timing control rather than delay timing control should be used to yield a synthesizable description. An illustration of this idea is explored as follows.

■ **Example 12.21: A four-phase clock generator—a correct version.**

This example uses a ring-counter-like structure to model a four-phase clock generator. At each positive edge of the clock signal clk, the output signal phase_out changes its value one time; namely, the value 1 is shifted to the next higher significant bit. The output signal phase_out returns to its original status after four clock cycles through the **default** case item. As a result, it generates the desired four-phase clock signals.

```
// a four-phase clock example--- a synthesizable version
module four_phase_clock_correct(
        input clk,
        output reg [3:0] phase_out);   // phase output

// the body of the four-phase clock
always @(posedge clk)
   case (phase_out)
      4'b0000: phase_out <= 4'b0001;
      4'b0001: phase_out <= 4'b0010;
      4'b0010: phase_out <= 4'b0100;
      4'b0100: phase_out <= 4'b1000;
      default: phase_out <= 4'b0000;
   endcase
endmodule
```

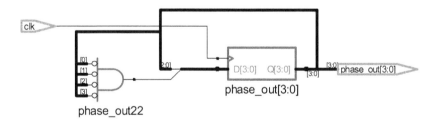

Figure 12.28: The synthesized result of the four-phase clock generator.

The synthesized result of the above module is revealed in Figure 12.28; it is indeed a four-phase clock generator based on the ring-counter structure. A comment on the above two examples is as follows. The second example of the four-phase clock generator does not exactly realize the requirements of the first example. Why? Try to explain it. If it is indeed needed to meet the requirement closely, some other ways, such as to use gate delays rather than simply to use event timing control, have to be used.

12.5.4 Synthesis of Positive and Negative Signals

For a sequential circuit, it is often required that a positive- or negative-edge clock signal be used as an event to trigger a set of operations to be performed in order. In Verilog HDL, there are no constraints on the use of **posedge** and **negedge** in the same **always** block. The combination of two or more edge-triggered signals with different polarities in the same sensitivity list of an **always** block is also allowed by synthesis tools. However, mixing edge-triggered with level-sensitive signals in the same sensitivity list of an **always** block is generally prohibited by synthesis tools. The following example gives some more insight into this.

■ Example 12.22: Mixing edge-triggered with level-sensitive signals.

Although the use of both edge-triggered and level-sensitive signals in the same sensitivity list of an **always** block is so convenient and legal in Verilog HDL as well as can be simulated in behavioral style, it is considered as a bad coding style because this type of statement is often unacceptable by synthesis tools. To explore this, consider the following module, where the **always** block uses an edge-triggered signal clk and a level-sensitive signal reset. As we synthesize this module, an error message is generally obtained from the synthesis tool: "Error: Can't mix posedge/negedge use with plain signal references."

```
// mixing edge-triggered with level-sensitive signals
module DFF_bad(
        input clk, reset, d,
        output reg q);

// the body of the D flip-flop
always @(posedge clk or reset)
   if (reset) q <= 1'b0;
   else       q <= d;
endmodule
```

Although mixing an edge-triggered signal with a level-sensitive signal is generally not accepted by a synthesis tool, the combination of **posedge** and **negedge** signals is commonly accepted by synthesis tools. Many examples of this have been used to model D flip-flops with asynchronous or synchronous reset. See Section 9.1.1 for more details.

■ Example 12.23: The combination of posedge and negedge signals.

In this example, we use a positive-edge (**posedge**) event of the clock signal clk to trigger the sampling operation of the D flip-flop and a negative-edge (**negedge**) event of the reset signal reset_n to clear the flip-flop. This is a way commonly used to model a D flip-flop with asynchronous reset, as we have described in Section 9.1.1.

```
// mixing posedge with negedge signals
module DFF_good(
        input clk, reset_n, d,
        output reg q);

// the body of the D flip-flop
always @(posedge clk or negedge reset_n)
   if (!reset_n) q <= 1'b0;
   else          q <= d;
endmodule
```

The reader is strongly recommended to review Section 9.1.1 for the details of how to model a D flip-flop with asynchronous or synchronous reset.

12.5.5 Synthesis of Loop Statements

Recall that in Section 4.5, we introduced four iteration (loop) statements: **for**, **while**, **repeat**, and **forever**. The loop statements, **for** and **while** as well as **forever**, are generally synthesizable but the **repeat** loop statement is generally unsynthesizable.

To synthesize a **for** loop statement, the elaborator of the synthesizer unrolls the **for** loop and then the synthesizer proceeds with analysis/translation and logic optimization. Based on this, *the number of times that a* **for** *loop to be executed must be known at the elaboration time.* An illustration of this idea is explored in the following example.

■ **Example 12.24: An example illustrating the loop statements.**

In this example, we are concerned with an n-bit adder, which adds two n-bit operands and produces an $(n+1)$-bit sum. The default value of n is 4. During the elaboration phase, the **always** block is expanded into the following four statements

```
{co, sum[0]} = x[0] + y[0] + c_in;
{co, sum[1]} = x[1] + y[1] + co;
{co, sum[2]} = x[2] + y[2] + co;
{c_out, sum[3]} = x[3] + y[3] + co;
```

where each statement corresponds to a full adder. The four full adders are then cascaded together as a 4-bit ripple-carry adder. The synthesized result is shown in Figure 12.29.

```
// an N-bit adder using a for loop
module nbit_adder_for #(parameter N = 4)(// set the default size
        input    [N-1:0] x, y,  // declare as an N-bit array
        input    c_in,
        output reg [N-1:0] sum, // declare as an N-bit array
        output reg c_out);
reg co;
integer i;
// specify an N-bit adder using a for loop
always @(x or y or c_in) begin
   co = c_in;
   for (i = 0; i < N; i = i + 1)
      {co, sum[i]} = x[i] + y[i] + co;
   c_out = co;
end
endmodule
```

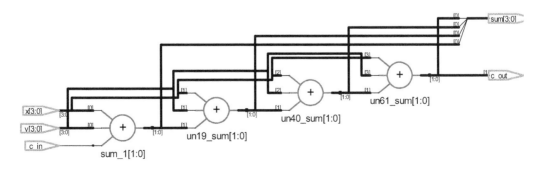

Figure 12.29: The synthesized result of a 4-bit adder using a **for** loop statement.

Another illustration of using a **for** loop statement is considered as follows. Assume that we would like to accumulate a number data_b into the sum total under the control of the data_a input. The number of 1 bits of the input number data_a controls the number of times that the number data_b is added to the sum total.

∎ Example 12.25: Incorrect single-cycle multiple-iteration module.

Assume that we want to compute the sum total at each positive edge of the clock signal. The sum total is initially set to 0 by the reset signal, reset_n. Then, at each incoming positive edge of the clock signal, clk, a **for** loop statement is used to detect the bit value of the input data data_a from the LSB to the MSB in sequence. If the bit is 1, then data_b is added to the sum total; otherwise, nothing needs to do. However, it does not work as expected. Actually, it adds data_b to the sum total at every positive edge of the clock signal whenever not all bits of data_a are zeros.

```verilog
// a multiple iteration example---this is an
// incorrect version. please try to correct it!
module multiple_iteration_example_a
       #(parameter M = 4,
         parameter N = 8)(
         input   clk, reset_n,
         input   [M-1:0] data_a,
         input   [N-1:0] data_b,
         output  reg [N-1:0] total);
integer i;
// what does the following statement do?
always @(posedge clk or negedge reset_n) begin
   if (!reset_n) total <= 0;
   else for (i = 0; i < M; i = i + 1)
          if (data_a[i] == 1) total <= total + data_b;
end
endmodule
```

The synthesized result of the multiple_iteration_example_a module is shown in Figure 12.30. This is certainly not the result as what we expect. Try to explain why the synthesis tool produces such a result. Moreover, what would happen if we replace all nonblocking assignments with blocking assignments within the **always** block? Try to synthesize it on your own system to confirm whatever you obtain. Of course, we can write a module to correctly carry out the desired operations. However, we would like to leave this for the reader as an exercise (Problem 12-20).

12.5.6 Memory and Register Files

Memories, registers, and register files often play their important roles in designing a digital system. At present, static random-access memories (SRAMs) are the most common memory modules in those digital system designs based on the synthesis flow. Memory modules have a wide variety of types, including RAMs, register files, first-in first-out (FIFO) buffers, and dual-port RAMs.

A desired memory module can usually be constructed in many ways. One way is to use flip-flops or latches to construct the desired memory module. This approach is independent of any synthesis software and the type of target system, cell-based or FPGA-based. It is easy to use but is inefficient in terms of area (or PLBs in FPGAs) because a flip-flop may generally take up much more area than a 6-transistor static RAM cell, usually 10 to 20 times. Another

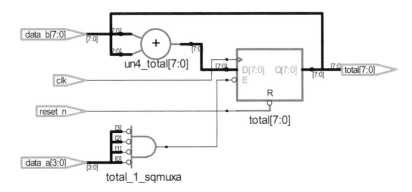

Figure 12.30: The synthesized result of the `multiple_iteration_example_a` module.

way is by using standard memory components supplied by cell-library and FPGA vendors. Of course, the availability of these standard components totally depends on the cell-library and FPGA device types. The third way is to use the RAM compiler supplied by most cell-library vendors to generate a desired memory module. This may be the most area-efficient approach because the size of the memory module can be generally customized to fit into the actual requirements.

Another type of memory module often used in digital system design is the register file. A register file is usually generated by using a synthesis directive approach or a hand instantiated RAM. More details of this can be referred to in Section 9.2.2.

In summary, the flip-flop- or latch-based memory module or the register is only applied to the case in which the required memory capacity is small. For the case where a large capacity of the memory module or register is required, it is better to use the RAM compiler or standard memory components supplied by vendors in order to save the hardware cost.

■ Review Questions

12-54 Explain why synthesis tools ignore the delay timing control in RTL descriptions.

12-55 Explain the operations of loop unrolling.

12-56 What approaches can be used to create a memory module?

12-57 Explain why the `multiple_iteration_example_a` module is synthesized into the result depicted in Figure 12.30.

12.6 Coding Guidelines

Successful synthesis strongly depends on the proper partitioning of the design along with a good HDL coding style. A good coding style not only results in the reduction of the chip area (namely, hardware cost, the number of LUTs) and aids in the top-level timing but also produces faster logic (namely, performance improvement). A frequent obstacle to writing HDL code is the software mind-set. *You should think hardware* when designing a hardware module. In this section, we introduce some coding guidelines to help the reader write a code that achieves the best synthesis result and reduction in compile time [2, 10].

12.6.1 Guidelines for Clock Signals

Clock signals are at the heart of any digital system. For a digital system to work properly, the clock signals must be applied to the system very carefully. To see this, a number of issues related to clock signals are to be discussed. These are using a single global clock signal, avoiding the use of gated clock signals, avoiding mixing positive- with negative-edge-triggered flip-flops, and avoiding the use of internally generated clock signals.

12.6.1.1 Using a Single Global Clock Signal In designing a digital system, it should use a single global clock signal as the preferred one and use positive-edge-triggered flip-flops as the only memory components, as shown in Figure 12.31(a). For ASIC designs, it is necessary to avoid instantiating clock buffers in the RTL code manually because clock buffers are normally inserted after synthesis as part of physical synthesis, as we have mentioned earlier. See Figure 12.3 for more details.

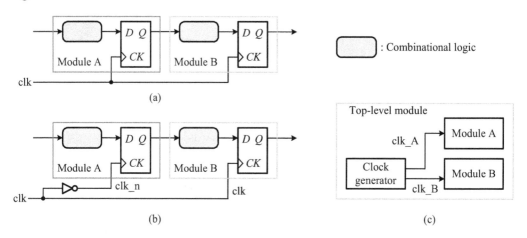

Figure 12.31: The clocking schemes for general digital systems: (a) an ideal clock scheme; (b) an example of mixing positive- with negative-edge-triggered flip-flops; (c) using a separate clock module at the top level.

12.6.1.2 Avoiding the Use of Gated Clock Signals We should avoid coding gated clock signals in an RTL code for the following reasons: First, clock-gating circuits tend to be timing-dependent and technology-specific. Second, gated clock signals may cause clock skews of local clock signals which in turn may cause hold-time violations. Third, gated clock signals limit the testability because the logic clocked by a gated clock signal cannot be made as part of a scan chain. A multiplexer may be used to bypass the gated clock signal in the test mode if the gated clock signal is required for some reasons, such as to optimize power dissipation. See Section 12.6.4 for more details.

12.6.1.3 Avoiding Mixing Positive- with Negative-Edge-Triggered Flip-Flops Although the combination of both types of positive- and negative-edge-triggered flip-flops in a design is so convenient, nowadays it is considered as a bad coding style since it needs to tackle stringent timing requirements. However, if this is indeed needed due to performance reasons, the following two problems must be addressed with caution: First, the duty cycle of the clock signal now becomes a critical issue in the timing analysis. Second, most scan-based testing methodologies require separate handling of positive- and negative-edge-triggered flip-flops. A good example of using both types of positive- and negative-edge-triggered flip-flops is depicted in Figure 12.31(b).

Section 12.6 *Coding Guidelines*

12.6.1.4 Avoiding the Use of Internally Generated Clock Signals The final concern about the issues of clock signals is to avoid using internally generated clock signals in the design as much as possible. Internally generated clock signals at least have the following two drawbacks: First, they make it more difficult to constrain the design for synthesis. Second, they limit the testability of the design because the logic driven by the internal clock signal cannot be made as part of a scan chain much like the case of gated clock signals. When an internally generated clock, reset, or gated clock signal is required, it is good practice to keep the clock and/or reset generation circuit as a separate module at the top level of the design, as shown in Figure 12.31(c).

■ Review Questions

12-58 Explain why we should avoid instantiating clock buffers manually in ASIC designs.

12-59 Explain why we should avoid using gated clock signals in an RTL code.

12-60 Explain why we should avoid mixing positive- with negative-edge-triggered flip-flops in a design.

12-61 Explain why we should avoid using internally generated clock signals in a design.

12.6.2 Guidelines for Reset Signals

The reset signal plays an important role in any digital system because it initializes the system to its known initial state by clearing all flip-flops. From the viewpoint of designers, the basic design issues related to reset signals are asynchronous versus synchronous, internal or external power-up, and hard versus soft. More about reset signals can be referred to in Section 15.2.3.

12.6.2.1 Asynchronous versus Synchronous Reset Asynchronous and synchronous reset signal generations have their own features. As mentioned, asynchronous reset means that it occurs randomly and is independent of the clock signal. Hence, an asynchronous reset signal must be modeled by putting it within the sensitivity list of the **always** block being used to describe a *D* flip-flop or a register so that it can trigger the **always** block as it occurs. That is,

```
// an asynchronous reset
always @(posedge clk or posedge reset)
   if (reset) ...
   else ...
```

Asynchronous reset has the following advantages:

- An asynchronous reset signal has priority over any other signals and the datapaths of the underlying system are always clear of reset signals.
- An asynchronous reset signal does not require a free-running clock signal and does not affect the data timing of flip-flops.
- No coercion of the synthesis tool is needed for correct synthesis.

The drawbacks of asynchronous reset are

- A global asynchronous reset signal is a special signal like the clock signal, requiring a tree of buffers. Routing and buffering of a reset tree is almost as critical as a clock tree.
- The capacitive load of a reset line is very large since the reset line goes to every flip-flop, possibly hundreds of thousands.
- Asynchronous reset signals make both static timing analysis (or cycle-based simulation) and automatic scan-chain (test structure) insertion more difficult.
- Reset deassertion to all flip-flops must occur in less than one clock cycle and an enough amount of time must be reserved for the reset recovery time.

- The reset line is vulnerable to glitches at any time.

For typical CPLD/FPGA devices, a set of global reset signals is generally provided to asynchronously reset all flip-flops on the devices. For a cell-based design, even though all flip-flops have asynchronous reset signals, it is still harder to implement the system reset circuit because a global reset signal is a special signal like the clock signal, requiring a tree of buffers to be inserted at the place-and-route (namely, the physical synthesis) stage.

Synchronous reset means that it occurs in a predictable way, namely, under the control of the clock signal. For this reason, to model a synchronous reset signal of a D flip-flop or a register, it needs to put the reset signal outside the sensitivity list of the **always** block being used to describe the D flip-flop or register so that the reset signal is merely checked after a specified clock edge occurs. That is,

```
// a synchronous reset
always @(posedge clk)
   if (reset) ...
   else ...
```

Some advantages of using synchronous reset are as follows:

- A synchronous reset signal is easy to be synthesized because it is just another synchronous signal to the input.
- With synchronous reset, the entire system is completely synchronous.
- With synchronous reset, the flip-flops or registers of the underlying system are less complex and thus are smaller in area.
- Synchronous reset inherently has the capability of noise filtering for the reset line.

The shortcomings of synchronous reset are

- It requires a free-running clock signal, in particular, at power-up, for the reset to occur.
- All cell-based or FPGA-based flip-flops do not support this type of reset mechanism.
- The complexity of the reset circuit is moved from flip-flops to the combinational logic and hence the combinational logic grows and may offset the benefit.
- The reset signal may take the fastest path to flip-flops or registers, even needed to be stretched so that it is wide enough to be detected at the clock edge of interest.
- The reset buffer tree may have to be pipelined so as to keep the reset of all circuits occurring within the same clock cycle.

Refer to Section 9.1.1 for more details and more complete examples about the reset signals.

12.6.2.2 Avoid Internally Generated Conditional Reset Signals The reset signal may be generated internally or externally when the power supply of the system is turned on. However, it is necessary to avoid internally generated conditional reset signals if possible. When a conditional reset signal is required, it is necessary to create a separate signal for the reset signal and to isolate the conditional reset logic in a separate logic block in order to improve the synthesis result and make the code more readable. For instance, in the following **always** block, the wr signal is a conditional reset signal used to clear the timer register whenever it is asserted.

```
always @(posedge clk or negedge reset_n or posedge wr)
   if (!reset_n || wr) timer <= 0;
   else if (!wr) timer <= timer + 1;
```

A better coding style is to combine the system reset signal !reset_n with the conditional reset signal wr into one separate logic block, say, timer_reset, and then apply it to the timer register. An illustration is given as follows:

Section 12.6 *Coding Guidelines*

```
assign timer_reset = !reset_n || wr;
always @(posedge clk or posedge timer_reset)
   if (timer_reset) timer <= 0;
   else if (!wr) timer <= timer + 1;
```

More detailed examples about this topic may be found in Section 15.4. Also refer to in Section 9.1.1 for more details about flip-flop modeling.

■ Review Questions

12-62 Compare the features of asynchronous and synchronous reset signals.

12-63 How would you code an asynchronous reset signal? Explain it.

12-64 How would you code a synchronous reset signal? Explain it.

12-65 Explain why we need avoid internally generated conditional reset signals in a design.

12.6.3 Partitioning for Synthesis

Because of the increasing complexity of digital systems, it is necessary to partition a design into many smaller modules so that they can be handled easily. In addition, the purposes of partition are to obtain a faster compile time and a better synthesis result as well as the capability of using simpler synthesis strategies to meet timing requirements. In this section, we give some guidelines for carrying out partition.

12.6.3.1 Registering All Outputs In order to make the output drive strengths and input delays predictable, all outputs are needed to register, as shown in Figure 12.32. In this case, there are no combinational logic being placed between the register and the output port (pad). Refer to Section 12.2.2 for related issues, input and output delays.

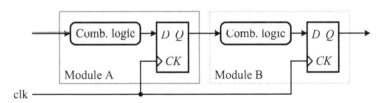

Figure 12.32: Registering all outputs in a digital system.

12.6.3.2 Keeping Related Logic within the Same Module When partitioning a design into smaller modules, one needs to keep in mind that it should put as much related combinational logic into the same module as possible. For example, as shown in Figure 12.33(a), the combinational logic A and B are displaced in two separate modules. It is better to combine them into one module, as shown in Figure 12.33(b). Here, the resulting two modules are also output registered.

12.6.3.3 Separating Structural Logic from Random Logic Another guideline for partitioning a design is to separate structural logic from random logic. When partitioning a design, we must bear in mind the following features at all times: to limit a reasonable block size, to partition the top-level module, and to separate I/O pads of boundary scan from core logic. In addition, it is necessary to remember that it should not use glue logic at the top-level module. Moreover, it should avoid using asynchronous logic in a design.

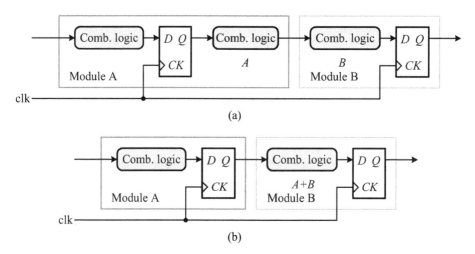

Figure 12.33: Keeping all related combinational circuits together: (a) a bad style and (b) a good style.

12.6.3.4 Synthesis Tools Tend to Maintain the Original Hierarchy One important feature of synthesis tools is that they tend to maintain the original hierarchy of a design and only optimize the codes within the same module (see Section 12.2.4). This feature has two implications at least: First, the codes in a design with different design goals should be separated into different modules so that they can be separately optimized. For example, for a module with the critical path logic, speed optimization is applied to optimize the logic. For the module with noncritical path logic, area optimization is applied to optimize the logic.

Second, in order for synthesis tools to consider resource sharing, all relevant resources are needed to put in the same module. For example, the two adders shown in Figure 12.34(a) cannot be shared because they are in separate modules. However, the two adders shown in Figure 12.34(b) can be shared because they are in the same module.

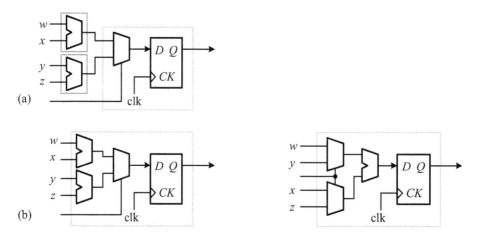

Figure 12.34: Partition for resource sharing: (a) resources in different modules cannot be shared and (b) resources in the same module can be shared.

■ Review Questions

12-66 Explain why we usually register all outputs in a design.

12-67 Explain why we keep related logic within the same module.

12-68 Describe why we separate structural logic from random logic in a design.

12-69 What are implications that synthesis tools tend to maintain the original hierarchy of a design?

12.6.4 Synthesis for Power Optimization

The power dissipation [15] of a CMOS logic circuit mainly depends on the load capacitance, voltage swing (usually the power-supply voltage), and operating frequency. It can be expressed as follows

$$P = C_L V_{DD}^2 f \tag{12.3}$$

where C_L is the load capacitance, V_{DD} is the power-supply voltage, and f is the operating frequency. It should be noted that here we intend to omit the static power dissipation, composed of various leakage currents.

To take a closer look at the ingredients of power dissipation, the above equation can be further refined as follows

$$P_d = V_{DD}^2 f \sum_n \alpha_n C_n \tag{12.4}$$

$$= \alpha V_{DD}^2 f C \quad (\text{if } n = 1) \tag{12.5}$$

where α_n denotes the node switching activities, C_n represents the node capacitance, and n represents the number of nodes. From this equation, it can be seen that in order to reduce power dissipation, we may address the following issues:

- Reduction of voltage swing
- Reduction of switching activity
- Reduction of switched capacitance

It is worth noting that the V_{DD} factor within the dissipation equation is indeed the output voltage swing, ΔV_{swing}, because the equation is derived from the charging and discharging the output capacitance of an inverter, where the output voltage swing is equal to V_{DD}.

In this section, we address the last two issues; namely, we explore how to reduce unwanted (or unnecessary) transitions and hence remove a substantial amount of unnecessary power dissipation in a design or an implementation at the RTL.

12.6.4.1 Reduction of Unwanted Transitions In designing a digital system, the reduction of unwanted transitions may significantly reduce power dissipation. The general coding guidelines in writing a Verilog HDL code in this respect are as follows.

Finite-state machine encoding. Recall that the Gray encoding scheme merely makes a single-bit change while transitioning from one state to another and hence consumes less power than the binary encoding scheme (see Section 11.1.1). Therefore, this encoding scheme is often used in FSMs as the power consumption is the major concern of the underlying system.

Avoidance of unwanted signal transitions. Unwanted signal transitions are often referred to as *glitches* and may be caused by a variety of factors, including the *imbalance of signal paths* and *unnecessary signal transitions* in logic circuits [15].

The unbalanced propagation delays of signal paths in a logic circuit may be alleviated by either restructuring the logic circuit or adding buffers to the shorter-delay signal paths. To get more insight into this, consider the 4-bit parity generator shown in Figure 12.35(a), which is implemented with a linear structure and results in the imbalance of signal paths. To

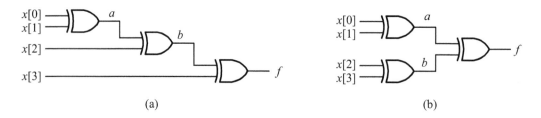

Figure 12.35: Balancing signal paths by logic restructuring: (a) linear structure and (b) binary-tree structure.

balance signal paths and hence remove unwanted glitches, the binary-tree structure shown in Figure 12.35(b) may be used instead. Unfortunately, *logic restructuring* cannot be applicable to all logic circuits with unbalanced signal paths. As this is the case, buffer insertion may be used instead. With buffer insertion, the signal paths are balanced by adding buffers to increase the propagation delays of the shorter-delay signal paths so that they can be compatible with others, thereby leading to balanced signal paths.

The removal of unnecessary signal transitions in logic circuits can be illustrated by the circuit in Figure 12.36(a). The arithmetic unit is only enabled when the input 0 of the multiplexer is selected. Hence, by gating the inputs to the arithmetic unit appropriately as shown in Figure 12.36(b), the unnecessary signal transitions can be significantly reduced.

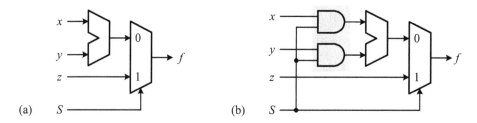

Figure 12.36: The removal of unnecessary signal transitions in logic circuits: (a) original circuit and (b) gated-input circuit.

Control over counters. In many applications, a counter does not need to run at all times but instead only needs to run for a specific period controlled by the start and stop signals on demand. In these cases, the counter may unnecessarily keep running at all times if both start and stop conditions are not dealt with carefully.

As an illustration, consider the monostable circuit shown in Figure 12.37. Suppose that an output pulse with a width of 4'b1011 clock cycles is generated each time the `start` signal is asserted. To this end, the binary counter can be coded in at least two ways in accordance with the logic circuit exhibited in Figure 12.37. One is simply to let the counter run freely and the other is to make the counter only run once. More details about the monostable circuits can be referred to in Section 9.6.2. In what follows, we give two examples to illustrate these two ways in depth.

■ Example 12.26: Control over counters.

In this example, the counter repeatedly counts from 4'b0000 toward 4'b1011 and generates a cnt_out pulse each time the count reaches 4'b1011. However, only the first output pulse right after the `start` signal is indeed needed to clear the output of the JK flip-flop. As a consequence, this encoding style consumes unnecessary power.

Section 12.6 Coding Guidelines

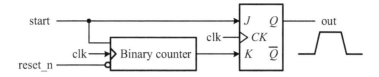

Figure 12.37: An illustration of control over counters.

```
// an illustration of unnecessary transitions
module counter_control(
      input  clk, reset_n, start,
      output cnt_out,
      output reg [3:0] qout);

// the body of the 4-bit binary counter
assign cnt_out = (qout == 4'b1011);
always @(posedge clk or negedge reset_n)
   if   (!reset_n)  qout <= 4'b0;
   else if ((qout == 4'b1011) || start)
            qout <= 4'b0000;
         else qout <= qout + 1'b1;
endmodule
```

■ Example 12.27: Control over counters.

In this example, the counter starts to count from 4'b0000 toward 4'b1011 as the start signal is activated. It stays at 4'b1011 and yields an output pulse thereafter. The unnecessary transitions are removed and hence it consumes less power than the previous module.

```
// an illustration of removing unnecessary transitions
module counter_control(
      input  clk, reset_n, start,
      output cnt_out,
      output reg [3:0] qout);
// the body of the 4-bit binary counter
assign cnt_out = (qout == 4'b1011);
always @(posedge clk or negedge reset_n)
   if   (!reset_n) qout <= 4'b0000;
   else if (start) qout <= 4'b0000;
         else if (qout < 4'b1011)
                 qout <= qout + 1'b1;
endmodule
```

Minimizing data transitions on bus. In many applications, it is often to leave the data bus continuously to changing values due to no default state being assigned to the data bus. This might not affect the functionality of the system if there exist some signals to indicate that the data on the data bus are valid. However, the unnecessary data bus transitions indeed consume power. To see this, consider the following example.

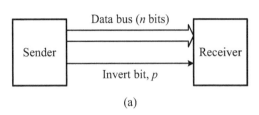

Data	d	p	Data$_{enc}$
1001_0011	-	0	1001_0011
0101_0111	3	0	0101_0111
1001_1101	4	0	1001_1101
0100_0011	6	1	1011_1100

(a)   (b)

Figure 12.38: Bus-invert coding: (a) a conceptual block diagram; (b) an example.

■ **Example 12.28: Minimize data bus transitions.**

The following code assumes that the data bus should be returned to its default state after the data_valid signal is deasserted. If this is not indeed the case, the data bus may hold its previous value after the data_valid signal is deasserted by removing the last **else** statement so as to reduce the bus transitions and hence power dissipation.

```
always @(posedge clock or negedge reset_n)
   if (!reset_n) data_bus <= {N{1'b0}};
   else if (data_valid) data_bus <= output_data;
      [else data_bus <= {N{1'b0}};]
```

Using Gray encoding for addressing memory modules. The single-transition feature between two adjacent codewords of the Gray code also proves useful to reduce power dissipation significantly in the access of memory modules due to the less number of transitions needed to carry out by the address counter.

Using bus-invert coding for I/Os or long data paths. Bus-invert coding is a technique that inverts the bits to be transmitted on the bus if the *Hamming distance* (d) between the current data and the previous one is more that $n/2$ (where n is the bus width). Here, the Hamming distance between two data words means the number of bit positions that the corresponding bits differ. Based on this, the number of transitions on the bus can be minimized. Of course, in using this technique a control bit must come along with the data to notify the receiver whether the data is inverted or not. A conceptual block diagram of bus-invert coding is depicted and illustrated in Figure 12.38.

Pipeline structures versus systolic arrays. Remember that pipelining registers the inputs at regular intervals and hence shortens the overall net lengths and minimizes glitches, thereby leading to the reduction of power consumption. Systolic architectures are indeed a type of iterative logic and hence have high modularity and help reduce long interconnect delays. Depending on the requirements of latency and hardware, one of these two techniques may be used for a particular system.

Removal of redundant logic. Any logic circuit always consumes a definite amount of power due to inherent resistance and capacitance associated with it. Thus, the removal of redundant logic proves useful in the reduction of power dissipation. To accomplish this, the RTL coding should be carried out very carefully so that there are no unwanted or redundant logic circuits. An illustration is given in Figures 12.34(a) and (b), where the redundant logic in Figure 12.34(a) cannot be removed by synthesis tools while that in Figure 12.34(b) can.

12.6.4.2 Low-Power Synthesis Certain constraints and coding styles which reduce area utilization or improve logic optimization should be followed in designing a digital system. To this end, modern logic synthesis tools also pay their attention to power optimization. The two most widely used techniques are *clock gating* and *power gating*. More details can be referred to [15].

Clock gating. In practical digital systems, not all modules or circuits are active at all times. Some are active now and the others next. For those modules or circuits that are inactive for a period of time, their clock signals can be temporarily inhibited to reduce power dissipation. Such a technique is known as clock gating and is proved to be an effective way to reduce the dynamic power dissipation of a digital system.

Power gating. One drawback of the clock-gating technique is that it can be only applied to reduce dynamic power dissipation. It cannot help remove the static power dissipation of a logic circuit. To reach this, another technique known as power gating may be applied to shut down a logic circuit when the result of the logic circuit is not needed. The rationale behind the power-gating technique is to add a power switch to control the power-supply voltage applied to the logic circuit. If the power switch is turned on, the power-supply voltage is applied to the logic circuit; otherwise, the power-supply voltage is cut off from the logic circuit. Hence, the logic circuit consumes no power when it is inactive.

■ Review Questions

12-70 What are the factors affecting the power dissipation of a CMOS logic circuit?
12-71 How would you balance the signal paths?
12-72 What is the meaning of bus-invert coding?
12-73 Explain the meaning of clock gating.
12-74 Explain the meaning of power gating.

12.7 Summary

In this chapter, we first introduced the general ASIC- and FPGA-based synthesis flows, including the general synthesis flow, RTL synthesis flow, and physical synthesis flow. In addition, the principle of timing-driven placement was also presented. Then, we dealt with the principles of logic synthesis and described how to write a good Verilog HDL code acceptable by most logic synthesis tools. Besides, we gave some comments about the constructs that can and cannot be synthesized. Finally, we gave a case study to present an overview of the Xilinx CPLD/FPGA design flow and the principles of XST.

Synthesis can be generally cast into logic synthesis and high-level synthesis. The former transforms an RTL representation into a gate-level netlist while the latter transforms a high-level representation into an RTL representation. Logic synthesis is the most common approach in designing general digital systems whereas high-level synthesis is only successfully applied to specific domains in which intensive computation is required, such as in digital signal processing (DSP) and multimedia applications.

To properly synthesize a design, we need to provide a synthesis tool the design environment and design constraints together with the RTL code and technology libraries. The design environment includes the operating conditions and statistical wire-load models. The design constraints include design rule constraints and optimization constraints. The former includes maximum transition time, maximum fanout, maximum (and minimum) capacitance, and cell degradation; the latter covers the system clock definition, input and output delays, maximum and minimum path delays, input transitions and output load capacitance, and timing exceptions.

The general architecture of synthesis tools can be partitioned into two parts: the front end and the back end. The front end consists of two steps, parsing and elaboration, while the back end contains three steps, analysis/translation, logic synthesis (logic optimization), and netlist generation. From the viewpoint of users, a synthesis tool at least performs the following critical tasks: to detect and eliminate redundant logic and combinational feedback loops, to explore don't-care conditions, to detect unused states and collapse equivalent states, to make state assignments, and to synthesize optimum multilevel logic subject to constraints.

Some guidelines about how to write a good Verilog HDL code that can be acceptable by most logic synthesis tools and can achieve the best compilation times and synthesis results were given. These guidelines also include clock signals, reset signals, the partition of a design, and synthesis for power optimization.

Finally, a case study was given to present an overview of the Xilinx CPLD/FPGA design flow and reveal the principles of XST. Related design constraints of XST were also explored to some extent along with examples.

References

1. Altera, *Quartus II Handbook Volume 3: Verification*, June, 2014.
2. H. Bhatnagar, *Advanced ASIC Chip Synthesis: Using Synopses Design Compiler and Prime time*, Boston: Kluwer Academic Publishers, 1999.
3. R. K. Brayton, C. McMullen, G. D. Hachtel, and A. Sangiovanni-Vincentelli, *Logic Minimization Algorithms for VLSI Synthesis*, Kluwer Academic Publisher, Norwell, MA, 1984.
4. R. K. Brayton, R. Rudell, A. Sangiovanni-Vinventell, and A. R. Wang, "MIS: a multiple-level logic optimization system," *IEEE Transactions on Computer-Aided Design*, Vol. 6, No. 6, pp. 1062–1081, November 1987.
5. R. K. Brayton, G. D. Hachtel, and A. L. Sangiovanni-Vinventell, "Multilevel Logic Synthesis," *Proceedings of the IEEE*, Vol. 78, No. 2, pp. 264–300, February 1990.
6. J. Cong and Y. Ding, "FlowMap: an optimum technology mapping algorithm for delay optimization in lookup-table based FPGA designs," *IEEE Transactions on Computer-Aided Design of Integrated Circuits and Systems*, Vol. 13, No. 1, pp. 1–12, January 1994.
7. S. H. Gerez, *Algorithms for VLSI Design Automation*, New-York: John Wiley & Sons, 1999.
8. IEEE 1364-2001 Standard, *IEEE Standard Verilog Hardware Description Language*, 2001.
9. IEEE 1364-2005 Standard, *IEEE Standard for Verilog Hardware Description Language*, 2006.
10. M. Keating and P. Bricaud, *Reuse Methodology Manual: For System-on-a-Chip Designs*, Boston: Kluwer Academic Publishers, 2002.
11. K. Keutzer, "DAGON: technology binding and local optimization by DAG matching," *Proceedings of the 24th ACM/IEEE Design Automation Conference*, pp. 341–347, June 1987.
12. Z. Kohavi and N. K. Jha, *Switching and Finite Automata Theory*, 3rd ed., New York: Cambridge University Press, 2010.
13. E. Lehman, Y. Watanabe, J. Grodstein, and H. Harkness, "Logic decomposition during technology mapping," *IEEE Transactions on Computer-Aided Design of Integrated Circuits and Systems*, Vol. 16, No. 8, pp. 813–834, August 1997.
14. M. B. Lin, *Digital System Design: Principles, Practices, and Applications*, 4th ed., Chuan Hwa Book Ltd. (Taipei, Taiwan), 2010.

15. M. B. Lin, *Introduction to VLSI Systems: A Logic, Circuit, and System Perspective*, CRC Press, 2012.
16. F. Mailhot and G. De Micheli, "Algorithms for technology mapping based on binary decision diagrams and on Boolean operations," *IEEE Transactions on Computer-Aided Design of Integrated Circuits and Systems*, Vol. 12, No. 5, pp. 599–620, May 1993.
17. A. Mishchenko, S. Chatterjee, and R. K. Brayton, "Improvements to technology mapping for LUT-Based FPGAs," *IEEE Transactions on Computer-Aided Design of Integrated Circuits and Systems*, Vol. 26, No. 2, pp. 240–253, February 2007.
18. S. Palnitkar, *Verilog HDL: A Guide to Digital Design and Synthesis*, 2nd ed., SunSoft Press, 2003.
19. A. Sangiovanni-Vincentelli, A. El Gamal, and J. Rose, "Synthesis methods for field programmable gate arrays," *Proceedings of the IEEE*, Vol. 81. No. 7. pp. 1057–1083, July 1993.
20. M. Sarrafzadeh and C. K. Wong, *An Introduction to VLSI Physical Design*, New-York: The McGraw-Hill Companies, Inc., 1996.
21. M. R. Stan and W. P. Burleson, "Bus-invert coding for low-power I/O," *IEEE Transactions on Very Large Scale Integration (VLSI) Systems*, Vol. 3, No. 1, pp. 49–58, March 1995.
22. Synopsys, *Design Compiler User Guide*, June 2010.
23. L. T. Wang, Y. W. Chang, and K. T. Cheng, *Electronic Design Automation*, New York: Morgan Kaufmann, 2009.
24. W. Wolf, *FPGA-Based System Design*, Upper Saddle River, New Jersey: Prentice-Hall, 2004.
25. Xilinx, *XST User Guide for Virtex-6, Spartan-6, and 7 Series Devices*, October, 2011.
26. Xilinx, *Vivado Design Suite User Guide–Using Constraints*, September, 2012.

Problems

12-1 Suppose that switching functions f, g, and h are as follows:

$$f(v,w,x,y,z) = vxy + wxy + z$$
$$g(v,w,x,y,z) = v + wx$$
$$h(v,w,x,y,z) = v + w$$

Find the quotient functions of f/g and f/h.

12-2 Find the kernel and cokernel sets of the following switching expression:

$$f(s,t,u,v,w,x,y,z) = tsu + tsv + wz + xz + yz + xy$$

12-3 Find the kernel and cokernel sets of each of the following switching expressions:

(a) $f(t,u,v,w,x,y,z) = twy + txy + uwy + uxy + vwy + vxy + z$
(b) $g(t,u,v,w,x,y,z) = tx + ty + uvx + uwx + uvy + uwy + z$

12-4 Use the kernel approach to implement each of the following multiple-output switching functions with a multilevel logic circuit and calculate the number of literals of each multiple-output switching function.

(a) $f_1(v,w,x,y,z) = vx + vy + vz$
$f_2(v,w,x,y,z) = wx + wy + wz$

(b) $f_1(u,v,w,x,y,z) = uy + uz + vy + vz$
$f_2(u,v,w,x,y,z) = uy + uz + wy + wz$
$f_3(u,v,w,x,y,z) = vy + vz + xy + xz$

12-5 Use the kernel approach to implement each of the following multiple-output switching functions with a multilevel logic circuit and calculate the number of literals of each multiple-output switching function.

(a) $f_1(v,w,x,y,z) = v\bar{w} + \bar{v}w$
$f_2(v,w,x,y,z) = wz + \bar{v}z + \bar{v}xy + wxy$

(b) $f_1(u,v,w,x,y,z) = v + w$
$f_2(v,w,x,y,z) = vx + vy + wx + wy + z$

12-6 Use the kernel approach to implement the following multiple-output switching function with a multilevel logic circuit and calculate the number of literals of the multiple-output switching function.

$$f_1(t,u,v,w,x,y,z) = tuvwz + tuvxz + tuvyz$$
$$f_2(t,u,v,w,x,y,z) = tuvwz + tuwxz + tuwyz$$

12-7 Use the two-step approach described in Section 12.4.2 to map Problem 12-4(a) into 4-input LUTs.

(a) How many LUTs are required for the original switching expressions?
(b) How many LUTs are required for the results after being implemented by using the kernel approach?

12-8 Use the two-step approach described in Section 12.4.2 to map Problem 12-4(b) into 4-input LUTs.

(a) How many LUTs are required for the original switching expressions?
(b) How many LUTs are required for the results after being implemented by using the kernel approach?

12-9 Use the two-step approach described in Section 12.4.2 to map Problem 12-5(a) into 4-input LUTs.

(a) How many LUTs are required for the original switching expressions?
(b) How many LUTs are required for the results after being implemented by using the kernel approach?

12-10 Use the two-step approach described in Section 12.4.2 to map Problem 12-5(b) into 4-input LUTs.

(a) How many LUTs are required for the original switching expressions?
(b) How many LUTs are required for the results after being implemented by using the kernel approach?

12-11 Use the two-step approach described in Section 12.4.2 to map Problem 12-6 into 4-input LUTs.

(a) How many LUTs are required for the original switching expressions?
(b) How many LUTs are required for the results after being implemented by using the kernel approach?

12-12 Using Shannon's expansion approach, synthesize the following switching function with 4-input LUTs:

$$f(t, u, v, w, x, y, z) = tuvwz + tuvxz + tuvyz$$

12-13 Using Shannon's expansion approach, synthesize the following switching function with 4-input LUTs:

$$f(t, u, v, w, x, y, z) = tuvwz + tuwxz + tuwyz$$

12-14 Using Shannon's expansion approach, synthesize the following switching function with 4-input LUTs:

$$f(v, w, x, y, z) = vx + vy + wx + wy + z$$

12-15 If we want to use four inverters cascaded together in some applications, write a Verilog HDL module to describe it. Synthesize your design with an available FPGA device and see what happens.

12-16 A simple frequency doubler is a circuit that uses an XOR gate to extract the edge information from an input clock signal, such as the one shown in Figure 12.39.

Figure 12.39: A simple frequency doubler.

  (a) Describe the circuit shown in Figure 12.39 in Verilog HDL.
  (b) Synthesize your module with an available FPGA device and see what happens.

12-17 Synthesize the following two arithmetic expressions separately and see what happens.

$$f(w, x, y, z) = w + x + y + z;$$
$$g(w, x, y, z) = (w + x) + (y + z);$$

12-18 Considering the Figure 12.40, answer each of the following questions.
  (a) Show the false path of the logic circuit.
  (b) Write a Verilog HDL module to describe the logic circuit.
  (c) Synthesize the module and check the synthesized result to see whether the false path still exists or not.
  (d) Explain why the synthesized result is.

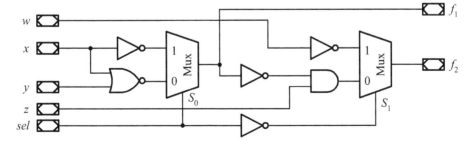

Figure 12.40: Another example of false paths.

12-19 Considering the Figure 12.40 and assuming that both 2-to-1 multiplexers are individually selected by S_0 and S_1 instead of the single signal *sel*, answer each of the following questions.
  (a) In what situation, does the logic circuit have a false path? What is it?
  (b) Write a Verilog HDL module to describe the logic circuit.
  (c) Synthesize the module and check the synthesized result to see whether the false path still exists or not.
  (d) Explain why the synthesized result is.

12-20 For every cycle, we want to compute the following for-loop statement and then store the result at the next positive edge of the clock signal, clk.

```
for (i = 0; i < m; i = i + 1)
   if (data_a[i] == 1) then total = total + data_b;
```

  (a) Explain the reason why the multiple_iteration_example_a module described in Section 12.5.5 cannot correctly compute the desired results. Of course, you may synthesize it by using a synthesis tool and examine the synthesized result carefully. Perhaps this may help you explore why it cannot work appropriately.
  (b) Of course, if we change the nonblocking assignment operators into blocking ones, the results will be correct. Please explain why?
  (c) Assume that we want to insist with the coding style set in this book, namely, still using a nonblocking assignment. Rewrite or modify the code so that it works properly.

12-21 Considering the following module written by Professor xyz, answer each of the following short questions:

```
module xyz(
      input   [1:0] a,
      input   in,
      output reg [3:0] y);
integer i;
//
always @(*)
   for (i = 0; i < 4; i = i + 1) begin
      if (a == i) y[i] = in; else y[i] = 1'b0; end
endmodule
```

  (a) What is the function of above module? Explain it.
  (b) Can we replace the **always** block with the following one:

```
always @(*)
   for (i = 0; i < 4; i = i + 1) y[i] = in;
```

  Explain it regardless of whatever answer you give.

12-22 In this problem, we are concerned with two simple AND-based logic circuits shown in Figure 12.41.
  (a) Draw the timing diagram of each circuit to show the relationship among variables, $x, y,$ and f.
  (b) Describe each logic circuit shown in Figure 12.41 in Verilog HDL in behavioral style.
  (c) Synthesize your modules with an available FPGA device and see what happens.

Figure 12.41: Two simple AND-based logic circuits.

12-23 In this problem, we are concerned with two simple NAND-based logic circuits shown in Figure 12.42.

Figure 12.42: Two simple NAND-based logic circuits.

(a) Draw the timing diagram of each circuit to show the relationship among variables, x, y, and f.

(b) Describe each logic circuit shown in Figure 12.42 in Verilog HDL in behavioral style.

(c) Synthesize your modules with an available FPGA device and see what happens.

12-24 In this problem, we are concerned with two simple XOR-based logic circuits shown in Figure 12.43.

Figure 12.43: Two simple XOR-based logic circuits.

(a) Draw the timing diagram of each circuit to show the relationship among variables, x, y, and f.

(b) Describe each logic circuit shown in Figure 12.43 in Verilog HDL in behavioral style.

(c) Synthesize your modules with an available FPGA device and see what happens.

12-25 In this problem, we are concerned with two simple clock-gating logic circuits shown in Figure 12.44.

(a) Draw the timing diagram of each circuit to show the relationship among variables, Clk, E_{clk} or $\overline{E_{clk}}$, and Gated Clk.

(b) Describe each logic circuit shown in Figure 12.44 in Verilog HDL in behavioral style.

(c) Synthesize your modules with an available FPGA device and see what happens.

12-26 In this problem, we are concerned with two *integrated clock-gating* (ICG) logic circuits shown in Figure 12.45.

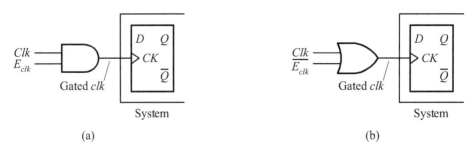

Figure 12.44: Two simple clock-gating logic circuits: (a) a bad circuit; (b) a better circuit.

(a) Draw the timing diagram of each circuit to show the relationship among variables, Clk, E_{clk}, and Gated Clk.
(b) Describe each logic circuit shown in Figure 12.45 in Verilog HDL in behavioral style.
(c) Synthesize your modules with an available FPGA device and see what happens.

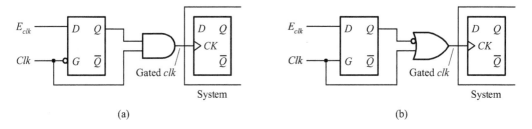

Figure 12.45: Two integrated clock-gating (ICG) logic circuits built on (a) an active-low gated D latch and (b) an active-high gated D latch.

13
Verification

VERIFICATION is a necessary process to ensure that a design can meet its specifications. Because of the inherent features of timing relationships between various components of the hardware module in question, the verification process needs to ensure that the design works correctly in both function and timing. As a consequence, the verification of a design can be divided into two types: functional verification and timing verification. Functional verification only considers whether the logic function of the design meets the specifications and can be done either by simulation or by formal proof. Timing verification considers whether the design meets the timing constraints and can be performed through dynamic timing simulation (DTS) or static timing analysis (STA).

In order to use dynamic timing simulation to verify the timing constraints of a design, two essential components are needed. The first is a test bench which generates and controls stimuli, monitors and stores responses, and checks the simulation results. The second is a file, known as a standard delay format (SDF) file, which contains the timing information of all cells in the design and can be produced by place-and-route (PAR) tools. During simulation, a value change dump (VCD) file may also be produced to provide valuable information about value changes on selected variables in the design. The purpose of VCD files is to provide information for debug tools in order to help remove the function bugs and timing violations of the design.

In this chapter, we deal with the above-mentioned issues in more detail and give a comprehensive example based on the FPGA synthesis flow to illustrate how to enter, synthesize, implement, and configure the underlying FPGA device of a design. Along the synthesis flow, static timing analyses are also given and explained. In addition, design verification through dynamic timing simulation, incorporating into the delays of logic elements and interconnect, is introduced.

13.1 Functional Verification

A design is the process that transforms a set of specifications into a fesible implementation. Verification is the reverse process of the design; that is, it begins with an implementation and confirms whether the implementation meets the specifications. The goal of verification is to ensure a design 100% correct in both of its functionality and timing. On average, design teams usually spend about 50–70% of their time to verify their designs. As a consequence, verification is a very important process to justify whether a design is successful.

Functional verification can be done by simulation or formal proof (called formal verification). In simulation-based functional verification, the design is placed under a test bench, input stimuli are applied to the design, and the outputs from the design are compared with the reference outputs. Formal verification formally proves that a protocol, an assertion, a property, or a design rule holds for all possible cases in the design. In other words, simulation-based functional verification is based on the analysis of the simulation results and code coverage of a design whereas formal verification is built on the basis of mathematical algorithms.

13.1.1 Models of Design Units

Before addressing simulation-based verification, we need to explore the models of the *design under test* (DUT), or called the *design under verification* (DUV). From the viewpoint of verification, a design under test can be considered as one of the following models based on whether the details of its internal functionality can be known from outside or not.

- *Black box model*: In this model, only the external interface (namely, the input and output behavior of the design) of the DUT are known. The internal signals and construct of the DUT are unknown (namely, black). Most simulation-based verification tools begin with this model.
- *White box model*: In this model, both the external interface and internal structure of the DUT are known. Most formal verification environments use this model.
- *Gray box model*: This model is a combination of both black box and white box models. In this model, some of the internal signals in addition to the external interface of the DUT are known. Most simulation-based verification environments use this model.

The major advantage of the black box model is that any structural changes inside the design (namely, DUT) would have little impact on the verification code because the function of a design is independent of implementation. The disadvantage is that it lacks the control and observation points of its internal structure.

In contrast, the white box model has the capability of direct measurement of the DUT. Thus, it is more flexible and powerful to verify the design. It can directly flag a bug in the source code instead of indirectly capturing the symptoms in the black box environment. The disadvantage of the white box model is that the verification code depends on the internal structure of the DUT. If any signal name of the internal constructs changes, the checker component must also change.

13.1.1.1 Assertion-Based Verification Assertion checking is a direct application of the white box model. The main purpose of assertion checking is to improve the *observability* of a design. Observability is the ease that incorrect behavior of a design can be identified. In general, a systematic application of assertions may catch about 30% design bugs on large industrial projects.

Assertions are statements about a designer's intended behavior. They are usually used in a source description to describe the intended behavior of a group of statements. Some good examples are that an FSM state should always use one-hot encoding, and the full and empty flags of a FIFO should never be asserted at the same time.

Assertions can be classified into the following two types:

- *Static assertions*: A static assertion is an atomic and simple check for the absence of events. These events do not relate to any other events.
- *Temporal assertions*: Several events occur in sequence and many events have to occur before the final asserted event can be checked. That is, they have timing relationships.

■ Review Questions

13-1 Distinguish between design and verification.
13-2 Describe the meaning of the black box model.

Section 13.1 *Functional Verification*

13-3 Describe the meaning of the white box model.
13-4 Describe the meaning of the gray box model.
13-5 What are the essential components of simulation-based verification?
13-6 What are the two types of assertions?

13.1.2 Simulation-Based Verification

Simulation-based verification is the most widely used approach for verifying whether a design is functionally correct. The generic verification flow based on simulation is shown in Figure 13.1. Before a good design specification is obtained, the system architect needs to survey many possible ways to achieve the same purpose through simulating the architecture models of the design.

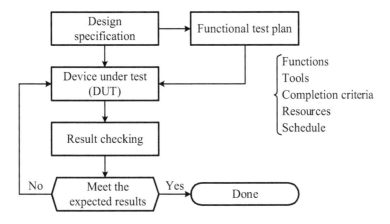

Figure 13.1: A simulation-based functional verification flow.

When the design specification is ready, a *functional test plan* is needed to create. The functional test plan must at least include the following features: functions to be verified, tools required to verify it, completion criteria, and resource and schedule details. The functions to be verified are derived by extracting the functionality from architecture specifications, prioritizing functionality, and creating the test cases. The functional test plan is the framework of functional verification of the design.

According to the framework of functional verification, test signals can be applied to the DUT and the results can be stored and analyzed. If both result checking and code coverage analysis meet the expected results, then we have done it; otherwise, the DUT is needed to modify until it may meet the expected results.

13.1.2.1 Hierarchy of Functional Verification In general, the hierarchy of functional verification of an ASIC chip can be divided into many levels as follows:

- *Designer level* (or block level): In this level, verification is usually done by the designer using Verilog HDL, SystemVerilog, or VHDL for both design and verification.
- *Unit level*: A unit usually contains multiple designer blocks and has a more stable interface of it. Hence, in this level, verification is usually done by the design team using randomized stimuli and autonomous checking.
- *Core level*: A core usually contains multiple units and has a completely functional specification and a stable interface. It is a special reusable unit. In this level, verification employs a well-defined process coupled with a well-documented specification, including functions and interface, and possibly coverage items.

- *Chip level*: A chip usually contains multiple cores and has very well-defined interface boundaries. In this level, verification is to ensure that all units are properly connected and the design adheres to the interface protocols of all units.

Functional verification usually combines directed tests with random tests to achieve the maximum percentage of correctness. Directed tests are used to test a specific behavior of the design whereas random tests are used to simulate corner cases which may be missed by designers.

The set of verification tests of a design includes the following ones:

- *Compliance tests* are used to verify that the design complies with the specifications.
- *Corner-case tests* try to find the corner cases that are most likely to break the design.
- *Random tests* are essentially a complement to compliance and corner case tests.
- *Real-code tests* are used to uncover the errors that may have arisen from misunderstanding the specification by hardware designers.
- *Regression tests* are those tests generated for previous versions of a design. Using regression tests helps ensure that old bugs do not reappear and helps uncover new bugs.
- *Property check* is used to check certain properties of a design, such as a particular FIFO should never be read when it is empty.

13.1.2.2 Tools for Writing Test Benches The essential component of simulation-based verification is the test bench. A test bench can be written by using Verilog HDL or other verification-assisted tools. As mentioned previously, Verilog HDL is not only a modeling language but also a verification language. Therefore, in a design description based on Verilog HDL, the test bench is usually written in Verilog HDL too.

However, with the advent of the system-on-chip (SoC) era, a design may exceed million gates. The test bench built on Verilog HDL becomes less effective due to the following reasons: First, because of the decreasing controllability and observability (to be defined in Section 16.1.2) of the design, it becomes harder and more time-consuming to write a test bench and more difficult to verify the correct behavior of the design. Second, the maintenance of the test benches becomes difficult because their sizes increase dramatically with the increasing complexity of chip function and size.

Because of above difficulties, many verification-assisted tools have been introduced recently. These are often called *high-level verification languages* (HVLs). Among these, the most popular one is SystemC. HVLs are programming languages combining the object oriented approach of C++ with the parallelism and timing constructs in HDLs. They help in the automatic generation of test stimuli and provide an integrated environment for functional verification, including input drivers, output drivers, coverage, and so on.

SystemC is initiated by Open SystemC Initiative (OSCI). It is a large C++ library that supports high-level hardware design, modeling, simulation, and verification. Although SystemC originally targeted mostly the simulation and specification of designs in C++ as its main goals, the synthesis of a subset of SystemC to RTL code is available today.

13.1.2.3 An HDVL Language Recently, a unified hardware description and verification language (HDVL) standard known as *SystemVerilog* has been established as the IEEE 1800-2005 Standard. As depicted in Figure 13.2, SystemVerilog is a superset of Verilog HDL and provides a set of extensions to improve the capability of hardware description and enhance the capability of verification of Verilog HDL; namely, it unifies the flow of design and verification. On the design hand, through the use of advanced design constructs provided by SystemVerilog, designers can model their designs in a more readable and concise way. On the verification hand, through coverage-driven, constrained-random test benches and assertions, designers may locate bugs easily from SystemVerilog reports without tracing back through the output waveforms. The details of SystemVerilog are beyond the scope of this book and can be found in LRM [7].

Section 13.1 *Functional Verification* 573

Figure 13.2: The relationship between Verilog HDL and SystemVerilog.

13.1.3 Formal Verification

Formal verification uses mathematical techniques to prove assertions or properties of the design without the need of technological considerations, such as timing and physical effects. It proves a design property by exploring all possible ways to manipulate the design. Formal verification has the advantage that it can prove the correctness of a design without doing any simulation.

At present, one of the most popular applications of formal verification is *equivalence checking*, as shown in Figure 13.3. Equivalence checking is to validate an RTL code against another RTL code, a gate-level netlist against an RTL code, or the comparison between a gate-level netlist and a physical description. RTL to RTL verification is used to verify the new RTL code against the old function-correct RTL code. This often occurs when new features are needed to add to a function-correct RTL code. Formal verification can then be used to verify whether the old function remains valid.

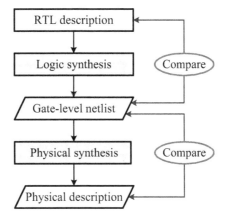

Figure 13.3: The operations of equivalence checking.

After a gate-level netlist created by a logic synthesizer and a physical description generated by a physical synthesis (place and route) tool, it is necessary to check whether these implementations meet the functionality of the original RTL description. The straightforward approach is to rerun the test bench and apply all test stimuli used for RTL verification again with the gate-level netlist and the physical description.

An alternative method is by using formal verification to prove that their functionalities are equivalent. Equivalence checking ensures that the gate-level netlist or the physical description has the same functionality as the RTL description that was simulated. In other words, equivalence checking verifies whether the logic synthesis tool accurately synthesizes the logic function described in RTL code and the physical synthesis (place and route) tool properly transforms the gate-level netlists into a physical layout. In order to work properly, equivalence checkers build a logical model of both the RTL and gate-level representations of the design and prove that they are functionally equivalent in a mathematical way.

■ Review Questions

13-7 Describe the meaning of formal verification.
13-8 Describe the functions of equivalence check.
13-9 Explain why high-level verification languages have become increasingly popular recently.
13-10 Describe the meaning of regression tests.

13.2 Simulation

Up to now, simulation is still an efficient and dominant way for verifying both the functionality and timing of a design. Therefore, in this section, we deal with the simulation of a design. We begin with a description of the generic types of simulations often encountered in designs, the simulation approaches in verifying digital systems, and the architectures of simulators. Then, we introduce the principles of event-driven simulation and cycle-based simulation.

13.2.1 Types of Simulation and Simulators

In general, the simulation of a digital system design can be performed in a variety of levels as described as follows:

- *Behavioral simulation*: Behavioral simulation models large pieces of a system as black boxes with inputs and outputs accessible externally.
- *Functional simulation*: Functional simulation ignores the timing of gates and assumes that each gate has a unit delay.
- *Gate-level (logic) simulation*: This type of simulation uses logic gates or logic cells as basic black boxes, where each of them may contain delay information.
- *Switch-level simulation*: In this level of simulation, transistors are considered as switches; namely, they can only be turned on or off but may associate with resistances.
- *Circuit-level (transistor-level) simulation*: In this level of simulation, transistors are considered as a set of nonlinear equations, which describe their nonlinear voltage and current characteristics.

For digital system designs, using simulation to verify whether a design can meet its specification is usually carried out at the behavioral, functional, and gate level. In this context, a design can be simulated in one of the following three ways:

- *Software simulation*
- *Hardware acceleration*
- *Hardware emulation*

13.2.1.1 Software Simulation Software simulation is typically used to run designs described in Verilog HDL, SystemVerilog, or VHDL. It runs a software simulator on a personal computer (PC), a workstation, or a server. The verification environment based on software simulation is shown in Figure 13.4. At the heart of software simulation is a simulation engine, which is a program that runs on a PC, a workstation, or a server. To simulate an RTL (source) code, the RTL code must be first transformed into models recognizable by the simulation engine. Generally speaking, software simulation is suitable for small designs and may be the most widely used approach for verifying a design. However, software simulation is a

Section 13.2 *Simulation*

time-consuming process; as a design exceeds a few million gates, it starts to consume large amounts of time and memory space and becomes a bottleneck in the verification process. In such a case, the other two approaches are preferred.

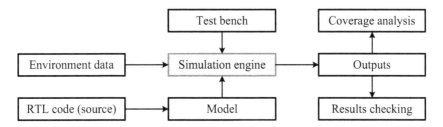

Figure 13.4: The simulation-based verification environment.

13.2.1.2 Hardware Acceleration To speed up the simulation of a design, an approach based on a mixed system of hardware and software, referred to as *hardware acceleration*, is often used. In this method, the design is divided into two parts: *synthesizable* and *unsynthesizable*. The synthesizable part is mapped onto a reconfigurable hardware system, called a *hardware accelerator*, which usually consists of CPLD, or FPGA devices, as shown in Figure 13.5. The unsynthesizable part runs on a software simulator, which may be a Verilog HDL simulator or an HVL simulator. The interaction between the simulator and the hardware accelerator produces the final results, which can then be checked and analyzed. The advantages of hardware acceleration are as follows: First, it can speed up simulation by two to three orders of magnitude in comparison with software-based simulation. Second, it can run sequences of random transactions during functional verification.

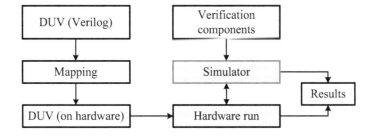

Figure 13.5: The environment of hardware acceleration.

13.2.1.3 Hardware Emulation The third widely used approach to verifying a design is known as *hardware emulation*. At present, hardware emulation systems are often built-on reconfigurable CPLD or FPGA devices. Sometimes one or more programmable system chips (PSCs) are used. As introduced before, a PSC is also a CPLD- or FPGA-based device. The general block diagram of hardware emulation is shown in Figure 13.6. The operation of hardware emulation is as follows. The design to be simulated is first mapped into the hardware platform. The application software then run on the software platform which in turn runs on the hardware platform to produce the required results. Hardware emulation is a dominant approach used to verify the design in a real-world environment with realistic system software running on the system. It proves useful to verify a design by hardware emulation before the design is transformed to an ASIC.

A hardware emulation system can be used either as the prototype of a product or as an emulation engine for verifying a design. When it is used as an emulation engine, the application

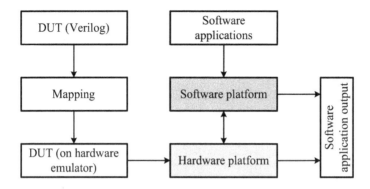

Figure 13.6: The environment of hardware emulation.

software runs exactly as it would do on the real chip in a real circuit. As a result, the integration of software and hardware can start before the actual hardware is available. Hardware emulation systems can run at a speed of hundreds of megahertz, even more faster, depending on the speed of the underlying hardware devices.

13.2.2 Architecture of HDL Simulators

The general architecture of HDL simulators is shown in Figure 13.7. An HDL simulator usually consists of a compiler and a simulation engine. The compiler is further composed of two parts: the front end and the back end. The front end consists of two phases: *parsing* and *elaboration*. In the parsing phase, the parser is responsible for checking the syntax of the source code and creates internal components to be used by the next phase. In the elaboration phase, the *elaborator* constructs a representation of the input circuit by connecting the internal components, unrolling loops, expanding generate loops, setting up parameters passing for tasks and functions, and so on. The end results from an elaborator is a complete description of the input circuit, which can then be input to the back end for generating the final code.

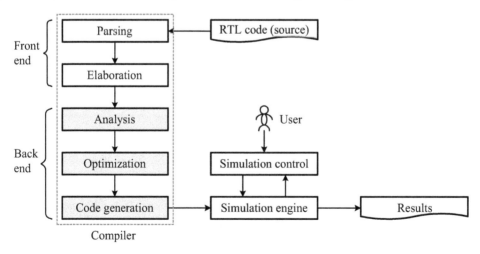

Figure 13.7: The general architecture of HDL simulators.

The back end determines the type of simulator and is mainly composed of three phases: *analysis*, *optimization*, and *code generation*. The detailed operations of the analysis phase depend on the type of simulation engine used. For example, in the cycle-based simulator, it carries out the levelization and the clock domain analysis of the design. The code generation phase generates the code to be used in a simulation engine. According to the types of simulation engines used, three classes of generated codes may be created: the *interpreted code*, *compiled code*, and *native code*.

13.2.2.1 Types of Simulators According to the types of simulation engines used and the code generated, the simulators can be classified into three types: interpreted, compiled code, and native code simulators. Each of these is described briefly as follows:

- *Interpreted simulators*: An interpreted simulator reads in the design description in Verilog HDL, creates data structures in memory, and runs the simulation interpretively. The most common example is the Cadence Verilog-XL simulator.
- *Compiled code simulators*: A compiled code simulator reads in the design description in Verilog HDL and converts it into an equivalent C code. The C code is then compiled by a standard C compiler to obtain an executable binary code. The binary code is executed to carry out the desired simulation. An example of this type of simulator is the Synopsys VCS simulator.
- *Native code simulators*: A native code simulator reads in the design description in Verilog HDL and converts it directly into a binary code for a specific machine platform. For example, the Cadence Verilog-NC simulator belongs to this type of simulator.

13.2.2.2 Another Classification of Simulators Verilog HDL simulators can also be categorized into two types based on the way that they are triggered. These two types are event-driven simulators and cycle-based simulators. Each of these is described in brief as follows:

- *Event-driven simulators*: As the name implies, an event simulator processes elements in the design only whenever one or more events occur; namely, any net or variable within elements changes its values. Event-driven simulators are best suitable for implementing general delay models and detecting hazards. They are also suitable for circuits with low activity, such as power-aware circuits with the clock-gating technique.
- *Cycle-based simulators*: A cycle based simulator processes all elements in the design on a cycle-by-cycle basis, irrespective of changes in signals. In other words, cycle-based simulators are triggered by clock signals—that can be considered as a special, regular event. The essential operations of such simulators are that they collapse combinational logic into equations. Cycle-based simulators are useful for synchronous designs where operations happen only at active clock edges. However, timing information between two clock edges is lost. As a consequence, they cannot detect any hazard occurs in a circuit. In order to make up this flaw, most commercial cycle-based simulators are often integrated with an event-driven simulator.

In the rest of this section, we will describe each of these two types of simulators in more detail.

■ Review Questions

13-11 Distinguish between design and verification.
13-12 Distinguish between behavioral simulation and functional simulation.
13-13 Describe the features of software simulation.
13-14 Describe the features of hardware acceleration.
13-15 Describe the general architecture of HDL simulators.
13-16 What are the three types of HDL simulators?

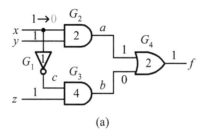

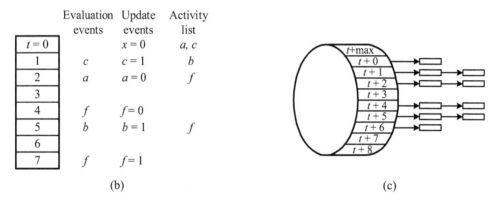

Figure 13.8: An illustration of an event-driven simulator: (a) a circuit example; (b) scheduled events and the activity list; (c) a timing wheel.

13.2.3 Event-Driven Simulation

The basic concept of event-driven simulation is that the evaluation of a gate or a block of code is performed only when an event occurs. Recall that an event is a change in value of a net or a variable. To keep track of the events and to ensure that they are processed in a correct order, all events are kept on a circular event queue, ordered by the *simulation time*. The simulation time is maintained by the simulator to model the actual time that would be taken by the circuit being simulated. The general structure of an event-driven simulator is shown in Figure 13.8, which is based on a data structure, known as a *timing wheel*. The timing wheel has a set of time slots, with each pointing to a linear event queue which stores events occurred at that time.

The timing wheel is indeed a two-dimensional queue consisting of a circular time queue and many linear event queues. The circular time queue is used to schedule events that occur at different times whereas the linear event queue is employed to schedule events that occur at the same time. The operation of putting an event on the queue is said to *schedule* the event. Events can occur at different times. Events that occur at the same time are processed in an arbitrary order.

Before describing the detailed operations of an event-driven simulator, the following two terms are needed to define first: the *update event* and the *evaluation event*. An update event means a change in value of a net, variable, or named event in the circuit being simulated. When a net, variable, or named event changes its value, all processes dependent on it must be evaluated. Here, a process means an **initial** or **always** block. The evaluation of a process is called an evaluation event.

The operations of an event-driven simulator can be outlined as follows:

Section 13.2 Simulation

■ Algorithm 13-1: Event-driven simulation I

> **while** (there are events)
>> **if** (there are no events at the current simulation time)
>>> advance simulation time;
>>
>> **for** (each event at the current simulation time)
>>> **if** (the event is update) { // an update event
>>>> update the variables or nets dependent on it;
>>>> schedule evaluation events; }
>>>
>>> **else** { // an evaluation event
>>>> evaluate the processes;
>>>> schedule update events; }
>
> **endwhile**

The simulation proceeds along time slots. At each time slot, the events in the event queue are processed one at a time until the queue is empty. If an event is an update event, then it updates all variables or nets dependent on it and schedules evaluation events. If an event is an evaluation event, then it evaluates all related processes and schedules update events.

■ Example 13.1: The operations of event-driven simulation.

Consider the circuit shown in Figure 13.8(a), the event that x changes its value from 1 to 0, is an update event. When this happens, the evaluation events a and c are scheduled at simulation times 2 and 1, respectively, due to the propagation delays of their individual gates. Since there exists no other event in the current simulation time, the simulation time is advanced to the next time slot. The two evaluation events, a and c, then trigger update events a and c, which in turn schedules two evaluation events b and f at simulation times 5 and 4, respectively. After the evaluation events b and f are evaluated, two new update events b and f are triggered and a new evaluation event f is scheduled at simulation time 7. This evaluation event is evaluated and the update event f is executed. The above process repeats until it reaches the end of simulation time, set by the **$finish** or **$stop** system task. The details are shown in Figure 13.8(b).

In the above discussion, we assume that all events in the same time slot are processed in an arbitrary order. In practice, they may be processed in some predefined order according to the features of events. In Verilog HDL, events at the same simulation time are classified into the following five types in accordance with the order of processing:

1. Active events
2. Inactive events
3. Nonblocking assignment update events
4. Monitor
5. Future events events

An *active event* is one that occurs at the current simulation time and can be processed in any order. The processing of all active events is defined as a *simulation cycle*. Events that occur at the current simulation time, but will be processed after all active events have been processed, are called *inactive events*. An example of such an event is an explicit zero-delay assignment (Section 4.3.1), such as "#0 c = b + 1;," which occurs at the current simulation time but will be processed after all active events at the current simulation time are processed. Recall that nonblocking assignment events are processed in three steps: read, evaluation and schedule, and assignment (Section 4.2.4). The first two steps are active events and the last step is executed only after both active and inactive events have been processed. The monitor events generated by system tasks, such as **$monitor** or **$strobe**, are processed as the last events so as to capture the stable values of variables at the current simulation time. The more detailed operations of a Verilog HDL event-driven simulator [4] can be described as follows.

■ Algorithm 13-1: Event-driven simulation II

    while (there are events)
        if (there are no active events) {
            if (there are inactive events)
                activate all inactive events;
            else if (there are nonblocking assignment update events)
                activate all nonblocking assignment update events;
            else if (there are monitor events)
                activate all monitor events;
            else // future events
                advance the simulation time and activate all inactive events; }
        for (each active event)
            if (event is update) // an update event
                update the modified objects and schedule evaluation events;
            else // an evaluation event
                evaluate the process and schedule update events;
    endwhile

13.2.4 Cycle-Based Simulation

Although event-driven simulators are universal in the sense that they can be applied to a broad variety of applications, their simulation efficiency is quite low. A specialized technique widely used to improve the simulation efficiency is *cycle-based simulation*. Cycle-based simulation is a technique that does not simulate detailed circuit timing, but instead computes the steady state responses of a circuit at each clock cycle.

The advantage of cycle-based simulation is that it can achieve a speedup factor of 100 or above as compared to event-driven simulation. This speedup factor is achieved by the fact that it uses simpler algorithms and performs the total optimization toward a synchronous hardware design style. However, because of this, cycle-based simulation puts severe constraints on the DUT. These constraints are as follows: First, it ignores delay control and hence cannot detect the glitch behavior of signals between clock cycles. Second, it limits sequential constructs to only synchronous ones. Third, it does not allow most test bench features.

The principles of cycle-based simulation can be best illustrated by way of an example. To see this, consider the logic circuit depicted in Figure 13.8(a) again. As a cycle-based simulator runs simulation on the circuit, it only evaluates the net values of a, c, b, and f at each clock cycle. In other words, the combinational logic is evaluated at each clock cycle boundary and each gate is evaluated only once. In contrast, in the case of an event-driven simulator, the net values of a, c, b, and f are evaluated not only at the clock cycle boundary, but also at the time whenever any event of the logic circuit occurs.

In order to evaluate a stable value, all gates within a combinational circuit must be evaluated in an appropriate order. To illustrate this, consider the logic circuit shown in Figure 13.8(a) again. The proper evaluation order of nets a, b, c, and f are as follows: net c is first evaluated, then nets a and b or nets b and a follow, and finally net f is evaluated. In other words, it is necessary to arrange the gates into levels. This operation is referred to as *levelization*. One method

Section 13.2 Simulation

Table 13.1: The detailed operations of the logic levelization example.

Step	Nets							Queue (Q)
	x	y	z	a	b	c	f	
0	0	0	0					G_2, G_1, G_3
1	0	0	0	1				G_1, G_3, G_4
2	0	0	0	1		1		G_3, G_4
3	0	0	0	1	2			G_4
5	0	0	0	1	2	1	3	

to arrive at this is by using a *topological sort* to transform a directed acyclic graph (DAG) into a linearly ordered list. Another feature of the topological sort is that it can also detect the cycle in a DAG. That is, it can also detect the feedback loop in a combinational circuit.

Another widely used logic levelization algorithm [13] is based on the breadth-first search of graphs and can be described as follows:

■ Algorithm 13-1: Logic levelization

Begin

  Assign level 0 to all primary inputs;
  enqueue (Q, all fanout gates of primary inputs);
  while (queue (Q) is not empty) **do**
    g = dequeue (Q);
    if (all driving gates of g are levelized)
      l = max{driving gate levels};
      Assign $l + 1$ to g;
      enqueue (Q, all fanout gates of g);
    else enqueue (Q, g);
End // end of the algorithm

■ Example 13.2: An example of logic levelization

To illustrate how the logic levelization algorithm works, let us consider Figure 13.8(a) again. To begin with, all primary inputs are set to level 0 and their fanout gates, G_2, G_1, and G_3 are enqueued in an arbitrary order, as shown in Table 13.1. Next, gate G_2 is dequeued and net a is assigned to 1. The fanout gate, G_4, of net a is enqueued. After gate G_1 is dequeued, net c is assigned to 1. The other two steps dequeue gates G_3 and G_4 and assign levels 2 and 3 to nets b and c, respectively. The output of the logic levelization algorithm for this instance is: $G_2 G_1 G_3 G_4$. The other possible result is: $G_1 G_2 G_3 G_4$.

When multiple clock signals are used in a logic circuit, not all gates are required to evaluate at each clock cycle. Only those gates associated with the same clock signal are evaluated at the clock transition. Consequently, we have to determine the part of the logic circuit that requires to be evaluated at each clock cycle. This process is called the *clock domain analysis*. The set of gates belonging to the same clock signal is called the *clock domain*.

In summary, for cycle-based simulation, the back end of the simulator analyzes the clock domain, topologically sorts all gates of the underlying logic circuit into a linearly ordered list, and generates code accordingly for the execution by the simulator.

■ Review Questions

13-17 What is an event-driven simulator?
13-18 Describe the operations of a general event-driven simulator.
13-19 What is a cycle-based simulator?
13-20 What are the features of cycle-based simulators?
13-21 What is the meaning of logic levelization?
13-22 What is the meaning of clock domains?

13.3 Test Bench Design

Recall that a test bench comprises three major components: generating and controlling stimuli, monitoring and storing responses, and checking the results. In this section, we discuss how to generate stimuli and check results, how to generate clock signals, and how to generate reset signals. In addition, verification coverage is introduced briefly.

13.3.1 Design of Test Benches

The basic design principles of test benches are that a test bench should generate stimuli and check responses in terms of test cases, a test bench should employ reusable verification components whenever possible, rather than coding from scratch each time, and the response checking must be automatic (i.e., a self-checking test bench). Generally speaking, a *test case* is a set of statements used to exercise a set of stimuli for checking a specific function of the device under test (DUT). A test bench is usually composed of many test cases.

13.3.1.1 Types of Test Benches The two most common types of test benches are *non-self-checking* and *self-checking*. The non-self-checking test benches are often used to verify the basic functionality of the DUT at an early stage of the development cycle. The self-checking test benches are those that place the knowledge of the DUT function into the test bench environment in a way such that they can automate the tedious result checking process. The self-checking test benches are used to verify the final functionality of the DUT.

In practice, the three widely used types of verification models based on test benches are *golden vectors*, the *reference model*, and the *transaction-based model*, as shown in Figure 13.9.

- *Golden vectors*: In an environment of using golden vectors, some known output vectors are stored somewhere for comparing with the outputs from the DUT, as shown in Figure 13.9(a), and the comparison result is indicated as pass or fail.
- *Reference model*: In the reference model, a module or device called the *reference model* is used to accept the same stimuli as the DUT. The outputs from the reference model are then compared with those from the DUT to determine whether the outputs of the DUT are correct or not, as illustrated in Figure 13.9(b).
- *Transaction-based model*: In the transaction-based model, the DUT is supposed to have identifiable transactions in which commands and data act on and are forwarded to appropriate output signals. A *scoreboard* is used to keep track commands and data driven on the inputs of the DUT. In this model, the test bench generates stimuli and checks responses in terms of transactions. The known transaction results are then compared with the outputs from the DUT, as shown in Figure 13.9(c). Examples that must use such a checking are network protocol devices, including Ethernet, USB, and IIC, that route and forward data packets of data.

Section 13.3 *Test Bench Design*

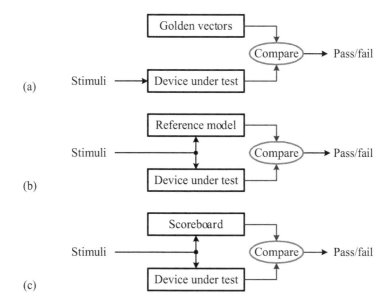

Figure 13.9: Verification models based on test benches: (a) golden vectors; (b) reference model; (c) transaction-based model.

13.3.1.2 Stimulus Generation and Response Checking To verify a design with simulation, a set of stimuli is needed to apply to the design. Stimuli can be generated either in a deterministic manner or in a random way, and they can be generated on the fly during simulation or prior to simulation. As a result, two basic choices of stimulus generation are as follows:

- *Deterministic* generation versus *random* stimulus generation.
- *Pregenerated test* case generation versus *on-the-fly test* case generation.

The result checking can be done during simulation or at the end of a test case, which are called *on-the-fly checking* and *end-of-test checking*, respectively.

- *On-the-fly checking*: In this method, the result checking is done throughout the life of a test case. On-the-fly checking is suitable for transaction-based operations. It has an advantage of that it requires less memory for simulation because it need not store the entire simulation outputs and runs faster than the end-of-test checking.
- *End-of-test checking*: In end-of-test checking, the result checking is done at the end of a test case. End-of-test checking is often used when the checking components require to check the state of the test bench after the test is complete. It generally needs more memory for simulation and runs slower than the on-the-fly checking.

The data values and data protocols may be generally analyzed by either one of the following approaches: the waveform viewer and the log file.

- A *waveform viewer* is used to view the dump files. It is suitable for debugging a small size code.
- *Log files* contain traces of the simulation run. They provide valuable information for debug tools to debug a large design.

13.3.1.3 Test Bench Examples To illustrate how the above-mentioned features of test benches can be applied to verify a practice design, we address in what follows a number of test bench examples to illustrate the possible ways of choosing test signals in the test sets: the exhaustive test, the random test, and golden vectors.

■ Example 13.3: Test bench example 1: Exhaustive test.

In this example, deterministic test stimuli (signals) are generated on the fly during simulation. Of course, generating stimuli exhaustively can only be applied to a small design with a small number of inputs. In addition, the end-of-test checking is used.

```
// test bench design example 1: an exhaustive test
`timescale 1ns / 100ps
module nbit_adder_for_tb;
parameter N = 4;
reg  [N-1:0] x, y;
reg  c_in;
wire [N-1:0] sum;
wire c_out;
// Unit Under Test port map
   nbit_adder_for UUT (
       .x(x), .y(y), .c_in(c_in), .sum(sum), .c_out(c_out));
reg [2*N-1:0] i;
initial
   for (i = 0; i <= 2**(2*N)-1; i = i + 1) begin
       x[N-1:0] = i[2*N-1:N]; y[N-1:0] = i[N-1:0];
       c_in =1'b0;
       #20;
   end
initial
   #1280 $finish;
initial
   $monitor($realtime,"ns %h %h %h %h",x,y,c_in,{c_out, sum});
endmodule
```

An illustration of using random stimulus generation on the fly during simulation is explored in the following example.

■ Example 13.4: Test bench example 2: Random test.

In this example, the test stimuli (signals) are generated randomly on the fly during simulation. In addition, the on-the-fly checking is used; that is, it is self-checking.

```
// test bench design example 2: a random test
`timescale 1ns / 100ps
module nbit_adder_for_tb1;
parameter N = 4;
reg  [N-1:0] x, y;
reg  c_in;
wire [N-1:0] sum;
wire c_out;
// Unit Under Test port map
   nbit_adder_for UUT (
       .x(x), .y(y), .c_in(c_in), .sum(sum), .c_out(c_out));
integer i;
```

Section 13.3 *Test Bench Design*

```verilog
reg [N:0] test_sum;
initial
   for (i = 0; i <= 2*N ; i = i + 1) begin
      x = $random % 2**N;
      y = $random % 2**N;
      c_in =1'b0;
      test_sum = x + y;
      #15;
      if (test_sum != {c_out, sum})
         $display("Error iteration %h", i);
      #5;
   end
initial
   #200 $finish;
initial
   $monitor($realtime,"ns %h %h %h %h",x,y,c_in,{c_out,sum});
endmodule
```

An illustration of combining golden vectors with the on-the-fly checking is given in the following example. Other combinations are also possible in practical applications.

■ Example 13.5: Test bench example 3: Golden vectors.

In this example, the golden vectors are read in through using the **$readmemh** system task and stored in memory. The stimuli are applied to the DUT one by one and the outputs from the DUT are then compared with the golden vectors on the fly during simulation. If any mismatch is detected, then a message is displayed on the standard ouput.

```verilog
// test bench design example 3: using golden vectors
`timescale 1ns/100ps
module nbit_adder_for_tb2;
//Internal signals declarations:
parameter N = 4;
parameter M = 8;
reg   [N-1:0] x, y;
reg   c_in;
wire [N-1:0] sum;
wire c_out;
// Unit Under Test port map
   nbit_adder_for UUT (
        .x(x), .y(y), .c_in(c_in), .sum(sum), .c_out(c_out));
integer i;
reg [N-1:0] x_array [M-1:0];
reg [N-1:0] y_array [M-1:0];
reg [N:0] expected_sum_array [M-1:0];
initial begin   // reading verification vector files
   $readmemh("inputx.txt", x_array);
   $readmemh("inputy.txt", y_array);
   $readmemh("sum.txt", expected_sum_array);
end
initial
   for (i = 0; i <= M - 1 ; i = i + 1) begin
```

```
        x = x_array[i];
        y = y_array[i];
        c_in =1'b0;
        #15;
        if (expected_sum_array[i] != {c_out, sum})
            $display("Error iteration %h", i);
        #5;
    end
initial
    #200 $finish;
initial
    $monitor($realtime,"ns %h %h %h %h",x,y,c_in,{c_out,sum});
endmodule
```

The contents of files: inputx.txt, inputy.txt, and sum.txt, are as follows:

inputx.txt	inputy.txt	sum.txt
4	1	05
9	3	0c
d	d	1a
5	2	07
1	d	0e
6	d	13
d	c	19
9	6	0f

■ **Review Questions**

13-23 What are the major components of a test bench?
13-24 What is a self-checking test bench?
13-25 What do golden vectors mean?
13-26 What are the two types of result checking?

13.3.2 Clock Signal Generation

The clock signal is the essential part of any digital system. The operations of a digital system are triggered by a clock signal in a cycle-by-cycle manner. Hence, to make the system work properly, the clock signal must be generated delicately. Besides, many types of clock signals are often required in specific digital systems. Some of them are as follows:

- General clock signals
- Aligned derived clock signals
- Clock multipliers
- Asynchronous clock signals
- Retriggerable monostable signals

To simulate these systems, the above types of clock signals must be generated in the test benches. Hence, in this section we address each of these in depth.

Section 13.3 Test Bench Design

13.3.2.1 General Clock Signals Several ways can be used to generate a (general) clock signal for simulation. Among these, the simplest one is to use an **initial** block to set the initial value of a clock signal clk and an **always** block to toggle the clock signal in a specified time interval. For instance,

```
reg  clk;
initial clk <= 1'b0;
always #10 clk <= ~clk;
```

Another commonly used scheme is to assign 1 and 0 to the clock signal explicitly. As an illustration, consider the following statements.

```
always begin
   #5 clk <= 1'b0;
   #5 clk <= 1'b1;
end
```

The third approach is to use a **forever** statement inside an **initial** block. To see this, consider the following example.

```
initial begin
   clk <= 1'b0;
   forever #10 clk <= ~clk;
end
```

where the **forever** statement functions exactly as an **always** block; however, it must be put inside an **initial** or **always** block. It is worth noting that the clock signal clk must have an initial value; otherwise, the simulator will not generate a clock signal and instead will generate an unknown value at all times.

In most practical applications, it is better to define the period of a clock signal as a parameter. In these cases, we need to pay some attention to the timescale, which may affect the timing of the clock edge. Usually, two errors may be raised when division is used to derive the duty cycle of the clock signal from a defined parameter. These two errors are called the *truncation error* and *rounding error*, respectively. The truncation error may occur as integer division is used. As an example, consider the following program segment:

```
`timescale 1ns/1ns
reg  clk;
parameter clk_period = 25;
// derive the clock signal
always begin
   #(clk_period/2) clk <= 1'b0;
   #(clk_period/2) clk <= 1'b1;
end
```

Because of the effect of time precision and the truncation error of integer division, the high-level half cycle of the clock signal is 12 time units.

The rounding error may occur as real division is used. To see this, consider the following example:

```
`timescale 1ns/1ns
reg  clk;
parameter clk_period = 25;
// derive the clock signal
always begin
   #(clk_period/2.0) clk <= 1'b0;
```

Figure 13.10: The proper way to generate two derived clock signals.

```
    #(clk_period/2.0) clk <= 1'b1;
end
```

The high-level half cycle of the clock signal is 13 time units due to the effect of time precision and the rounding error of real division.

In order to avoid the truncation or rounding error, it needs to use proper time precision in the **'timescale** compiler directive. As an example, consider the following statements

```
'timescale 1ns/100ps
reg clk;
parameter clk_period = 25;

always begin
    #(clk_period/2) clk <= 1'b0;
    #(clk_period/2) clk <= 1'b1;
end
```

Although the time precision in the above program segment is set to 100 ps, a truncation error due to integer division still occurs. Hence, the high-level half cycle of the clock signal is still 12 time units. However, if real division is used instead, the resulting high-level half cycle of the clock signal becomes 12.5 time units and no rounding error would occur.

13.3.2.2 Aligned Derived Clock Signals Because the purpose of a test bench is to mimic the real-world environment where the DUT will work, it often needs to generate two aligned clock signals that hardware actually does. An improper approach may arise a delta delay (not a real delay but only for ordering events) between the derived clock signals. For example,

```
always begin
    if (clk == 1'b1) clk2 <= ~clk2;
end
```

where the clock clk2 toggles whenever the clock signal clk is 1. As a result, there is some delay between clk and clk2; such an infinitesimal, virtual delay is called a *delta delay*.

To avoid the delta delay, it needs to mimic the operations of a hardware device used to generate the aligned derived clock signals, as depicted in Figure 13.10. From this figure, we can see that both two clock signals clk1 and clk2 are derived from a common clock signal clk. Based on this idea, a proper approach for generating a set of derived clock signals is as follows:

```
// both clk1 and clk2 are derived from clk
always begin
    clk1 <= clk;
    if (clk == 1'b1) clk2 <= ~clk2;
end
```

Explain why there is no delta delay when two clock signals are derived in this way.

13.3.2.3 Clock Multipliers The clock multipliers in real-world circuits are implemented using internal or external phase-locked loops (PLLs). A clock signal can be derived from

Section 13.3 Test Bench Design

another via a frequency divider or a frequency multiplier. To model such a circuit, the combination of both **repeat** and **forever** statements is often used. As an illustration, consider the following example, which generates a clock signal clk1 with the one-fourth frequency of the clock signal clk4.

```
initial begin
   clk1 <= 1'b0;
   clk4 <= 1'b0;
   forever begin
      repeat(4) #(period_clk4/2) clk4 <= ~clk4;
      clk1 <= ~clk1;
   end
end
```

Note that generating clock multipliers by division does not need to know the frequency of the reference clock signal. To generate a clock signal with a frequency n times the reference clock signal, the general approach is to use the **repeat** statement to repeatedly generate $2n$ transitions for each clock cycle. For example,

```
always @(posedge clk)
   repeat(2N) clkN <= #(period_clk/2N) ~clkN;
```

Another approach is to use a **forever** statement. An illustration is given in the following program segment.

```
initial
   forever clkN <= #(period_clk/2N) ~clkN;
```

13.3.2.4 Asynchronous (Unrelated) Clock Signals The word "asynchronous" means randomness. In principle, there is no way that can accurately model unrelated clock signals. Nevertheless, in practice, unrelated clock signals can be modeled as a separate **initial** or **always** block. As an illustration, consider the following example.

```
initial begin
   clk100 <= 1'b0;
   #2;
   forever #5 clk100 <= ~clk100;
end

initial begin
   clk33 <= 1'b0;
   #5;
   forever #15 clk33 <= ~clk33;
end
```

In order to emphasize the random effect, an amount of jitter may be added to one of the clock signals to mimic the nondeterministic effect of the phase. Some amount of jitter may be generated by using the random number generator in Verilog HDL, namely, the **$random** system function.

13.3.2.5 Retriggerable Monostable Signals Remember that a monostable circuit generates a high-level pulse with a specified width whenever it is triggered. During simulation, a few simple statements are sufficient to model such a retriggerable monostable circuit. As an illustration, consider the following monostable example which outputs a high-level pulse with a width of 100 time units when it is triggered. The input trigger signal retrig restarts the monostable time period. If retrig continually occurs within 100 time units, then the output qout will remain at 1; namely, it is retriggerable.

```
initial qout = 0;
always begin: monostable
   #100 qout = 0;
end
always @(posedge retrig) begin
   disable monostable;
   qout <= 1;
end
```

The reader is encouraged to modify the above retriggerable monostable example into a non-retriggerable one. It should be noted that the above retriggerable monostable example cannot be synthesized into a real retriggerable monostable circuit. Refer to in Section 9.6.2 for the details of synthesizable monostable circuits.

■ Review Questions

13-27 Give at least two approaches to generate a specific clock signal.
13-28 How would you generate two aligned clock signals?
13-29 Describe how to generate two asynchronous clock signals.
13-30 Describe how to generate a clock signal, clk5, with five times the frequency of the original clock signal, clk.

13.3.3 Reset Signal Generation

The hardware reset signal is the first signal that is applied to a system to set the system to a known state. Consequently, it is of importance to create the reset signal to be applied to the DUT in the test bench to mimic the function of the hardware reset signal during simulation. For the generation of such a reset signal, the following related issues are needed to pay more attention: the *race problem*, *maintainability*, and *reusability*.

13.3.3.1 Race Problems The race problem is often created unintentionally between two or more synchronized signals, such as the clock and reset signals. To see this, consider the following example. Suppose that a clock signal clk with a period of 10 time units and a 50% duty cycle is to be generated by an **always** block using blocking assignments. The reset signal reset is also generated by using blocking assignments within an **initial** block and intends to last for 40 time units, say, from simulation time 20 to 60 time units. Depending on the execution order of the **always** and **initial** blocks, a race problem between the clock and reset signals, clk and reset, may occur and cause system hazards.

```
always begin
   #5 clk = 1'b0;
   #5 clk = 1'b1;
end
initial begin // has the race problem
   reset = 1'b0;
   #20 reset = 1'b1;
   #40 reset = 1'b0;
end
```

As described previously, the race problem between the clock and reset signals can be solved by replacing the blocking assignments with nonblocking ones, as stated as follows:

```
always begin
   #5 clk <= 1'b0;
```

```verilog
      #5 clk <= 1'b1;
   end
   initial begin  // no race problem
      reset <= 1'b0;
      #20 reset <= 1'b1;
      #40 reset <= 1'b0;
   end
```

It is worth comparing the above two examples very carefully. In particular, the effect of the race problem caused by the first example and how it is solved by the second example. Note that it is not good practice to make the reset signal aligned with the clock signal. Usually, the reset signal should be deasserted a definite time before the active clock edge in order to prevent the underlying register or flip-flop from being subject to metastability (Section 15.2.3).

13.3.3.2 Maintainability It is often desirable to change the clock period during the developing stage of a design. As this is the case, the reset signal generated by using absolute values of the simulation time might not properly reset the device under test (DUT). As a result, a functional error may arise, especially, at the startup time during simulation. This is the case of introducing functional error due to the lack of maintainability of the reset signal. In order to avoid such a problem, the reset signal must be written in such a way that it is a function of the clock signal rather than an absolute value of the simulation time. For example, the reset signal must last for two clock periods instead of 40 time units.

As an illustration of how to write a reset signal with maintainability, consider the following example.

■ Example 13.6: A reset signal with maintainability.

In this example, the `reset` signal is set to 1 for two clock cycles, starting at the negative edge of `clk`. Consequently, it is independent of the period of the clock signal `clk`. Of course, because of the asynchronous feature of the reset signal, it is often desirable to add some phase jitter to the reset signal `reset`. This can be done by using a random number generator, i.e., the **$random** system function, to generate a random number as the value of delay control. The details are left for the reader as an exercise.

```verilog
always begin
   #(clk_period/2) clk <= 1'b0;
   #(clk_period/2) clk <= 1'b1;
end
// the reset signal lasts for two clock cycles
initial begin
   reset <= 1'b0;
   wait (clk !== 1'bx); // wait until the clock is active
   // set reset to 1 for two clock cycles
   // starting at the falling edge of clk
   repeat (3) @(negedge clk) reset <= 1'b1;
   reset <= 1'b0;
end
```

13.3.3.3 Reusability In verifying a design, a reset signal along with a process may need to be applied many times during simulation. To reuse such a reset process, it is convenient to encapsulate the generation of the reset signal with a task. This idea is manifested in the following example.

Example 13.7: A reset signal generated by using a task.

This example is simply to rewrite the **initial** block used to generate the reset signal in the preceding example as a task. The hardware_reset task can then be invoked to generate a reset signal in any place within a test bench whenever it is needed.

```
always begin
   #(clk_period/2) clk <= 1'b0;
   #(clk_period/2) clk <= 1'b1;
end
// define a synchronous reset process as a task
task hardware_reset;
begin
   reset <= 1'b0;
   wait (clk !== 1'bx);
   // set reset to 1 for two clock cycles starting
   // at the falling edge of clk
   repeat (3) @(negedge clk) reset <= 1'b1;
   reset <= 1'b0;
end
endtask
```

Review Questions

13-31 Explain why the race problem may occur between the reset and clock signals.
13-32 Why is the maintainability so important in generating a reset signal in a test bench?
13-33 Why is the reusability so important in generating a reset signal in a test bench?
13-34 What is the purpose of the hardware reset signal in a digital system?

13.3.4 Verification Coverage

Verification coverage is a measurement of state space that the simulation-based verification process has touched. The percentage of verification coverage only means what fraction of coverage points is included in the simulation; it does not mean the percentage of functional correctness in the design. As a matter of fact, the quality of verification coverage strongly depends on how well the test bench is. Generally, verification coverage includes two major types: *structural coverage* and *functional coverage*. Structural coverage denotes the representation of the design to be covered whereas functional coverage means the semantics of the design implementation to be covered.

13.3.4.1 Structural Coverage Structural (code) coverage deals with the structure of the Verilog HDL code and indicates which key parts of that structure have been exercised. Structural coverage mainly includes the following types: *statement coverage, branch coverage, condition coverage, expression coverage, finite-state machine coverage,* and *toggle coverage*.

Statement coverage. Statement coverage gives statistics about executable statements that are executed in simulation. In other words, it gives the count of how many times statements were executed. The rationale behind the importance of statement coverage is based on the assumption that unexercised code potentially bears bugs. The statement coverage is defined as

$$\text{Statement coverage } (\%) = \frac{\text{Number of statements executed}}{\text{Total number of executable statements}} \times 100 \qquad (13.1)$$

It should be noted that statement coverage is useful but not a complete metric.

Table 13.2: The results of code coverage analysis of the Booth algorithm.

Module	Statement count	Statement hits	Statement miss	Statement %
Three-step Booth	35	35	0	100%
Controller	21	21	0	100%
Datapath	5	5	0	100%

Branch coverage. Branch coverage verifies that all branch sub-conditions have triggered the conditional branch and gives a count of how many times each condition occurred. In other words, branch coverage counts the execution of each **if-else** conditional statement and each **case** statement and indicates when a true or false condition has not executed. The branch coverage is defined as

$$\text{Branch coverage } (\%) = \frac{\text{Number of branches taken}}{\text{Total number of possible branches}} \times 100 \qquad (13.2)$$

Condition coverage. Condition coverage is an extension to branch coverage and analyzes the decision made in **if, else-if, else** and ternary (?:) statements. Condition coverage verifies all logical operators (i.e., !, ||, and &&) and gives a count of how many times each condition occurred. It can be used to detect errors, such as the use of a wrong logical operator or the incorrect placement of brackets. For example, the logical or (||) is used instead of logical and (&&), and the !(x||y) is used instead of !(x)||(y).

Expression coverage. Expression coverage is similar to condition coverage and analyzes the expressions on the right-hand side of assignment statements. In other words, expression coverage measures various ways of an expression that paths through the code are executed during simulation.

Toggle coverage. Toggle coverage measures how many times the signals, including nets and variables, have changed their logic values during simulation. It can identify the signals which fail to be initialized or toggled by a test case.

In summary, an analysis of code coverage is only to let you know if you have done your job. A 100% code coverage is by no means an indication that the job is over or the design can perfectly work correctly. As an illustration of code coverage analysis, consider Table 13.2, which shows the results of code coverage analysis of the Booth algorithm from a typical simulation tool. Here, the coverage of all three statements is 100% due to the simplicity of the design; in other words, all statements in the design are touched by the verification test.

13.3.4.2 More Structural Coverage
Sometimes, the following two types of structural coverage are also considered: *path coverage* and *trigger coverage*.

Path coverage. Path coverage measures all possible ways that can execute a sequence of statements during simulation. It is a refinement of branch coverage. Owing to infeasible paths, 100% path coverage generally cannot be achieved. An infeasible path means the one consisting of branches that are never taken.

Trigger coverage. Trigger coverage counts how many times an **always** block that was activated by each signal in the sensitivity list changes its value during simulation. In other words, it simply measures the number of exercised variables in the sensitivity list of an **always** block. For instance, in the following **always** block

    always @(a or b or c or d)...

if only variables a and d change their values during simulation, then trigger coverage is only 50%.

13.3.4.3 Finite-State Machine Coverage
Finite-state machine (FSM) coverage usually contains two types of coverage: *state coverage* and *transition (arc) coverage*. The state cov-

erage measures which of the states have been discovered and transition coverage measures which of the state transitions have been visited.

13.3.4.4 Functional Coverage Functional coverage ensures that all possible legal values of input stimuli are exercised in all possible combinations at all possible times. Functional coverage lets you know whether you have done the job.

Item coverage. Item coverage records the individual scalar values, such as the packet length, instruction opcode, interrupt level, and so on.

Cross coverage. Cross coverage measures the presence or occurrence of combinations of values.

Transition coverage. Transition coverage measures the presence or occurrence of sequences of values.

■ Coding Style

1. *All response checking should be done automatically, rather than have the designer view waveforms and determine whether they are correct.*
2. *The time unit set in the '**timescale** compiler directive must match the real-world propagation delay of gate-level circuitry.*
3. *The reset signal must be set properly. Especially, the time interval of the reset signal must be large enough; otherwise, the initial operation of gate-level simulation may not work properly.*

■ Review Questions

13-35 What does 100% code coverage mean?
13-36 What does structural coverage mean?
13-37 What does functional coverage mean?
13-38 Explain the meaning of statement coverage.
13-39 Explain the meaning of toggle coverage.

13.4 Dynamic Timing Analysis

So far, the most widely used approach to verifying both the functionality and timing of a design is still accomplished by simulation. In the previous sections, we have dealt with functional verification through the use of simulation and of formal proof. In this section, we address the timing verification used to verify whether the timing of a design can meet its timing specification or not.

13.4.1 Basic Concepts of Timing Analysis

Roughly speaking, the goal of timing analysis is to estimate when the output of a given circuit becomes stable. As illustrated in Figure 13.11, it is necessary to make the output of the combinational logic, i.e., the data input D of the second flip-flop, stable at the time $t = T$ for the correct functionality. However, how could we ensure this? To achieve this, at least two approaches may be used to carry out timing analysis: *dynamic* and *static*. *Dynamic timing analysis* (DTA) is done by carrying out timing analysis through the simulation of the design in a cycle-by-cycle manner whereas *static timing analysis* (STA) is by performing timing analysis on the basis of signal paths. In other words, the timing analysis with simulation is called dynamic timing analysis and without simulation is referred to as static timing analysis.

Section 13.4 Dynamic Timing Analysis

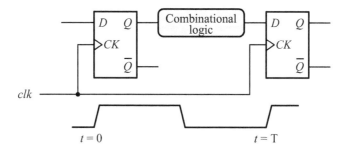

Figure 13.11: The meaning and requirement of timing analysis.

In practical applications, timing analysis has two objectives: *timing verification* and *timing optimization*. Timing verification verifies whether a design meets the given timing constraints, such as the cycle-time constraint, whereas timing optimization tries to identify the critical part, such as *critical paths*, of a design for further optimization. A critical path is a timing-critical signal path that limits the performance of the entire design. Probably, many critical paths may exist in a design at the same time.

13.4.1.1 Delay Back-Annotation To verify the timing relationships between various components of a design, a file in *standard delay format* (SDF) containing timing information of all cells in the design is used to provide timing information for simulating the gate-level netlist or performing the timing analysis of signal paths without doing simulation.

The process by which timing information from the SDF file updates specify path delays, **specparam** values, timing constraint values, and interconnect delays is known as *delay back-annotation*. An illustration of dynamic timing simulation using delay back-annotation is shown in Figure 13.12. Both pre-layout (after placement is completed) and post-layout (after routing is completed) delays are extracted from their corresponding outputs, as depicted in this figure. These delays are stored in individual SDF files and annotated backward to the pre-layout and post-layout gate-level netlists, respectively, so as to take into account the actual cell (gate) delays of the design. Detailed examples of how to use an SDF file in dynamic timing simulation can be found in Section 13.7.2.

13.4.2 Standard Delay Format Files

The standard delay format (SDF) file [6] contains timing information of all cells in a design. The timing information includes values for specify path delays, **specparam** values, timing check constraints, and interconnect delays. The timing information in an SDF file usually comes from the ASIC delay calculation tool that takes into account connectivity, technology, and layout geometry information of underlying cells.

The basic timing information of an SDF file is composed of the following ones:

- *IOPATH delay*: The IOPATH delay specifies the cell delay, which is computed based on the transition of the input signal and the output wire loading.
- *INTERCONNECT delay*: The INTERCONNECT delay is a point-to-point, path-based delay, including the RC delay between the driving gate and the driven gate.
- *Timing checks*: Timing checks contain values that determine the required setup time and hold time of each sequential cell. These values are based on the characterized values in the technology library.

The SDF file is used to provide timing information in dynamic timing analysis for simulating the gate-level netlist. The SDF file can be generated in two phases:

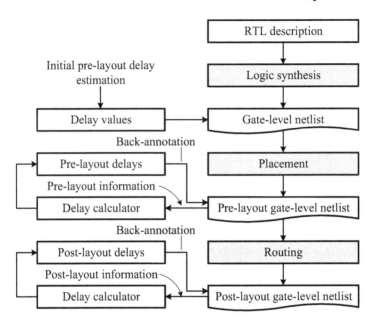

Figure 13.12: The back-annotation of pre-layout and post-layout delays.

- *Pre-layout SDF file*: In cell-library-based design, the pre-layout SDF file is generated by using a proper wire-load model but does not include clock trees in general. Hence, the delays contained in the pre-layout SDF file only include the logic-cell delays and the estimated interconnect delays. In FPGA-based design, the pre-layout SDF file merely includes the logic-cell delays without having the interconnect delays. An example is the *_map.sdf file (contains gate delays only) in the ISE synthesis flow.
- *Post-layout SDF file*: The post-layout SDF file contains delays that are based on the actual layout, including logic-cell delays, interconnect delays, and clock tree delays. For example, the *_timesim.sdf file (contains both gate and interconnect delays) in the ISE synthesis flow.

13.4.3 Delay Back-Annotation

As described above, during delay back-annotation, timing values are read from an SDF file into a specified region of the design to update specify path delays, **specparam** values, timing constraint values, and interconnect delays. This operation can be accomplished by using the **$sdf_annotate** system task. The **$sdf_annotate** system task has the general form

```
$sdf_annotate("sdf_file" [, [module_instance][,
    ["config_file"][, ["log_file" ][, ["mtm_spec"]
    [,[ "scale_factors"][, ["scale_type"]]]]]]]);
```

where sdf_file is the SDF file generated from delay calculators; module_instance is the name of the design module. The remaining arguments are optional.

The config_file parameter is the configuration file and is used to provide detailed control over annotation. log_file is the log file used to record results from each individual annotation of timing data during SDF annotation. mtm_spec specifies which of the min/typ/max triples is used to annotate. It can be any of MAXIMUM, MINIMUM, TOOL_CONTROL (default), and TYPICAL, and overrides any MTM_SPEC in the configuration file. scale_factors specifies the scale factors to be used and the default values are 1.0:1.0:1.0. scale_type specifies how the scale factors should be applied to the min/typ/max triples. It can be any of FROM_MAXIMUM,

Section 13.4 *Dynamic Timing Analysis* 597

FROM_MINIMUM, FROM_MTM (default), and FROM_TYPICAL. scale_factors and scale_type arguments override the SCALE_FACTORS and SCALE_TYPE in the configuration file, respectively.

As an example, to annotate the pre-layout delay information back to the pre-layout gate-level netlist, the following statement is included in the test bench.

$sdf_annotate ("four_bit_adder_map.sdf", four_bit_adder);

where the first argument, four_bit_adder_map, denotes the file name of the pre-layout SDF file of the design and the second argument, four_bit_adder, is the file name of the design to be annotated. To annotate the post-layout delay information, the following statement is included in the test bench.

$sdf_annotate ("four_bit_adder_timesim.sdf", four_bit_adder);

where four_bit_adder_timesim is the name of the post-layout SDF file of the design. The above *_map.sdf and *_timesim.sdf files are generated automatically by place and route (PAR) tools.

The detailed pre-layout timing simulation and post-layout timing simulation are to be discussed in more detail in Section 13.7.2.

13.4.4 Details of Standard Delay Format Files

The SDF is a file format used to convey timing information to the simulator. The SDF is based on the IEEE 1497 Standard specification [6], which describes an ASCII file format that contains the propagation delays of cells, interconnect delays, and timing constraints.

An SDF file usually contains two parts: a header and cell descriptions. As illustrations of typical SDF files, consider the following three examples. The first one describes the header part of an SDF file. The second one reveals how a combinational logic cell is described in the SDF file. The third one presents the details of how a sequential logic cell is described in a typical SDF file.

■ Example 13.8: A typical SDF file.

An SDF file contains a header and one or many cell descriptions. The header begins with DELAYFILE and usually ends with TIMESCALE. Except for SDFVERSION all other fields are optional. SDFVERSION specifies the version of the SDF format used. The next two fields, DESIGN and DATE, describe the name of the design and the date of the file created, respectively. The VENDOR, PROGRAM, and VERSION fields describe the originator, the name, and the version of the program used to generate the file, respectively. The DIVIDER field is used to separate the elements of the hierarchical path to each cell. It has two optional values "." and "/". It defaults to "." if the separator is omitted. The VOLTAGE, PROCESS, and TEMPERATURE fields separately specify the voltage, process, and temperature conditions. The TIMESCALE field specifies the units for all values used in the SDF file. The default value is 1 ns. In this example, it is redefined to 1 ps.

The rest of an SDF file is a list of cells. The cells may be listed in any order. A cell description includes a cell type, an instance name, and delay specifications. It also includes a timing specification if the cell is a sequential one.

```
(DELAYFILE
  (SDFVERSION "3.0")
  (DESIGN "CRC_MSB")
  (DATE "Sat Dec 06 17:44:48 2014")
  (VENDOR "Xilinx")
  (PROGRAM "Xilinx SDF Writer")
```

```
(VERSION "M.63c")
(DIVIDER /)
(VOLTAGE 1.14)
(TEMPERATURE 85)
(TIMESCALE 1 ps)
  ...
(CELL (CELLTYPE "X_LUT4")
  ...
(CELL (CELLTYPE "X_FF")
  ...
```

The following example is an instance of a combinational logic cell.

■ Example 13.9: A typical combinational cell description.

A cell description begins with CELL and is followed by CELLTYPE and INSTANCE. The rest of a cell description is timing specifications. The CELLTYPE field is the name of the component model appearing in the Verilog HDL netlist. The INSTANCE field indentifies the particular instance of the cell, including the hierarchical cell path. The DELAY part specifies ABSOLUTE, PORT, and IOPATH delays. The IOPATH field specifies the delay of an input-output path. It is followed by the names of input and output ports and an ordered list of delay specifications. In this example, two delays are used and each delay uses the min:typ:max delay specifier.

```
(CELL (CELLTYPE "X_LUT4")
  (INSTANCE Mxor_d_Result1)
    (DELAY
      (ABSOLUTE
        (PORT ADR1 (672:672:840))
        (PORT ADR2 (345:345:432))
        (IOPATH ADR0 O (415:519:519)(415:519:519))
        (IOPATH ADR1 O (415:519:519)(415:519:519))
        (IOPATH ADR2 O (415:519:519)(415:519:519))
        (IOPATH ADR3 O (415:519:519)(415:519:519))
      )
    )
)
```

In general, DELAY has four types [6]: PATHPULSE, PATHPULSEPERCENT, ABSOLUTE, and INCREMENT. The keywords PATHPULSE and PATHPULSEPERCENT are used to specify how pulses will propagate across paths in this cell. The keywords ABSOLUTE and INCREMENT mean that the delay data in the SDF file will replace and add to the existing delays in the design during annotation, respectively.

The following example is an instance of a sequential logic cell.

■ Example 13.10: A typical sequential cell description.

In a typical sequential cell description, a timing specification is included to specify the timing constraints. It begins with the keyword TIMINGCHECK. The timing specification is usually used to check the following timing constraints: the setup time, hold time, clock period, recovery time, and pulse width. Details can be found in Section 7.3.3.

```
(CELL (CELLTYPE "X_FF")
  (INSTANCE qout_0)
```

```
      (DELAY
        (ABSOLUTE
          (IOPATH CLK O (500:626:626)(500:626:626))
          (IOPATH SET O ( 0 )( 0 ))
          (IOPATH RST O (1039:1299:1299)(1039:1299:1299))
        )
      )
      (TIMINGCHECK
        (SETUPHOLD (posedge I) (posedge CLK) (56:70:70)(264:331:331))
        (SETUPHOLD (negedge I) (posedge CLK) (56:70:70)(264:331:331))
        (PERIOD (posedge CLK) (1092:1366:1366))
        (RECOVERY (negedge RST) (posedge CLK) (614:768:768))
        (WIDTH (posedge RST) (600:751:751))
      )
    )
)
```

■ Review Questions

13-40 What can dynamic timing simulation do?
13-41 Explain the meaning of delay back-annotation.
13-42 Distinguish the differences between the DTA and STA.
13-43 Define the term: the critical path of a design.
13-44 What does an SDF file contain?

13.5 Static Timing Analysis

As mentioned, timing verification is often performed through dynamic timing simulation. However, dynamic timing simulation has the following drawbacks: First, it poses a bottleneck for large, complex designs because simulation requires a lot of time. Second, it relies on the quality and coverage of the test bench being used for verification. Third, it is very difficult to figure out the critical paths of a design through the results from dynamic timing simulation. Hence, in this section, we introduce an alternative approach to verifying the timing specification of a logic circuit without performing simulation. This approach is called *static timing analysis* (STA).

13.5.1 Fundamentals of Static Timing Analysis

Static timing analysis (STA) is an alternative approach to dynamic timing analysis (DTA). It is widely used to determine if a logic circuit meets timing constraints without having to simulate the design in a cycle-by-cycle manner. In other words, STA analyzes a design in a static way. It computes the delay for each (signal) path of the design. As a result, the critical paths can be easily pointed out. Unlike DTA, STA is independent of input values.

To perform STA, the following two basic assumptions are made:

- There are no combinational feedback loops.
- All register feedback paths are broken by the clock boundary.

Based on these two assumptions, the delay of each path can then be easily calculated. All path delays are checked to see whether timing constraints have been met. However, it is necessary to note that the comprehensive sets of test benches are still needed to verify the functionality of the design because STA can only verify the timing specification of the design. As discussed previously, a formal verification technique, referred to as *equivalence checking*, may also be used as an alternative way to verify the functionality of the gate-level netlist against the RTL source code of the design.

In STA, a design is broken into a number of sets of signal paths, where each path has a start point and an endpoint, as shown in Figure 13.13. The start point may be either an input port or the clock input of a register, and the endpoint may be either an output port or the data input of a register. Based on the definitions of start points and endpoints, there are four types of paths that can be combined from the start points and endpoints. As illustrated in Figure 13.13, these paths are described as follows:

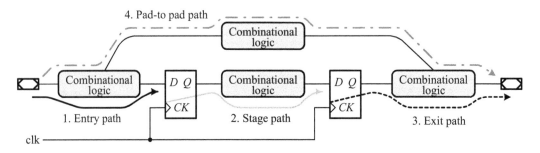

Figure 13.13: Four analysis types of static timing analysis.

- An *entry path* (an input-to-D path) starts at an input port (pad) and ends at the data input of a register, such as a flip-flop or latch.
- A *stage path* (a register-to-register path or clock-to-D path) starts at the clock input of a register and ends at the data input of another register.
- An *exit path* (a clock-to-output path) starts at the clock input of a register and ends at an output port.
- A *pad-to-pad path* (a port-to-port path) starts at an input port (pad) and ends at an output port (pad).

The above paths are defined over a single clock domain. When a circuit has multiple independent clock domains, the paths are grouped in terms of the clock signals controlling their endpoints, as shown in Figure 13.14; namely, the paths associated with the same clock signal are grouped together. Two new groups of paths are defined: the *path group* and *default path group*. A path group means that a set of paths associates with the same clock domain and the default path group comprises all paths not associated with any clock domain.

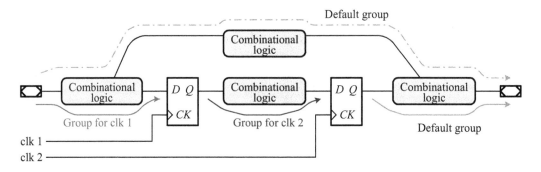

Figure 13.14: Clock domains and their associated groups.

13.5.2 Timing Specifications

Several timing-related constraints are defined when STA is used to verify the timing specifications of a design. The most commonly used constraints can be divided into two classes: clock-related constraints and port (or pad)-related constraints. The clock-related constraints contain the clock period, clock transition time, clock latency, and clock uncertainty. The port (or pad)-related constraints are the input delays (offset-in), output delays (offset-out), input-output (pad-to-pad) delays, minimum and maximum path delays, input transition, and output load capacitance. More details can be referred to in Section 12.2.2.

13.5.2.1 Timing Analysis After a timing specification is given, static timing analysis can be proceeded to find the critical paths. A *critical path* is the signal path of the longest propagation delay in a design. More formally, a critical path can be defined in the context of the slack time as follows. A critical path is a combinational logic path that has a negative or the smallest slack time, where the slack time is defined as

$$\text{slack time} = \text{required time} - \text{arrival time}$$
$$= \text{requirement} - \text{datapath (in ISE)}$$

Critical paths limit system performance. Critical paths not only tell us the system cycle time but also point out which part of the combinational logic should be modified or restructured to improve the performance of the system under consideration.

13.5.3 Timing Exceptions

Recall that timing analysis tools usually treat each path in the design as a single-cycle path by default and perform STA accordingly. However, in most designs, there may exist some paths that exhibit timing exceptions. The two common types of timing exceptions are *false paths*, *multicycle paths*, and *max/min path delays*. More details can be referred to in Section 12.2.2.

A false path is identified as a timing failing path that does not actually propagate a signal, as shown in Figure 12.9. A multicycle path means that its input data must take more than one clock cycle to reach the destination. An illustration is shown in Figure 12.10. As an illustration of this, consider the following example.

■ Example 13.11: A multicycle path example.

In this example, variable qout_a is updated every clock cycle, variable qout_b is updated every two clock cycles, and variable qout_c is updated every three clock cycles. The synthesized result is depicted in Figure 13.15.

```verilog
// a multiple cycle example
module multiple_cycle_example #(parameter N = 32)(
       input  clk, reset, load,
       input  [N-1:0] data_a, data_b, data_c,
       output reg [N-1:0] qout_a,
       output reg [2*N-1:0] qout_b,
       output reg [2*N-1:0] qout_c);
reg [N-1:0] data_qa, data_qb, data_qc;
reg [1:0] count_b;
wire clk2, clk3;
// sample input data
always @(posedge clk or posedge reset)
   if (reset) begin data_qa <= 0; data_qb <= 0; data_qc <= 0; end
   else if (load) begin
```

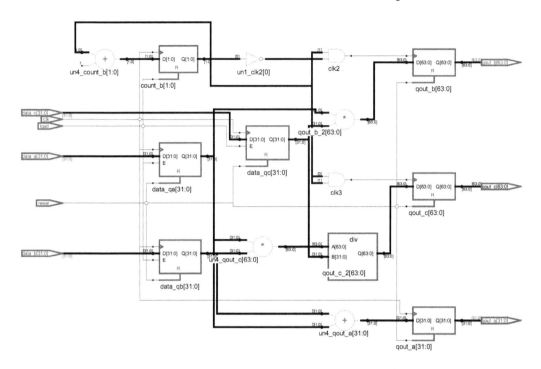

Figure 13.15: The synthesized result of the multicycle path example.

```
      data_qa <= data_a; data_qb <= data_b;
      data_qc <= data_c; end
// trivial one-cycle logic
always @(posedge clk or posedge reset)
   if (reset) qout_a <= 0;
   else qout_a <= data_qa + data_qb;
// two-cycle logic
assign clk2 = (count_b == 2'b10);
always @(posedge clk or posedge load) begin
   if (load) count_b <= 0;
   else count_b <= count_b + 1;
end
always @(posedge clk2 or posedge reset) begin
   if (reset) qout_b <= 0;
   else qout_b <= data_qa * data_qc;
end
// triple-cycle logic
assign clk3 = (count_b == 2'b11);
always @(posedge clk3 or posedge reset) begin
   if (reset) qout_c <= 0;
   else qout_c <= data_qa * data_qb / data_qc;
end
endmodule
```

Because STA tools usually treat all paths in a design to be one cycle by default and perform STA accordingly, it is required to tell them which paths are multicycle paths in order to avoid getting results falsely, namely, obtaining a timing violation. For example, as shown

Section 13.5 Static Timing Analysis

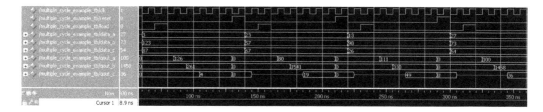

Figure 13.16: The simulation results of the multicycle path example.

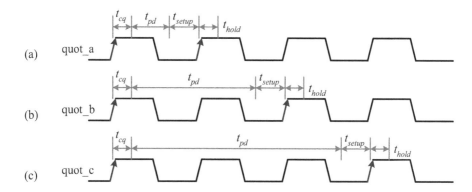

Figure 13.17: Multicycle path timing exceptions: (a) a single-cycle timing relationship; (b) a two-cycle timing relationship; (c) a three-cycle timing relationship.

in Figure 13.15, both paths from inputs data_b, and data_c to their destinations qout_b and qout_c, respectively, will be treated as a single-cycle path beginning from input data_a to its destination qout_a as no timing exception is set. However, as can be seen from Figures 13.16 and 13.17, the path from input data_b to qout_b is a two-cycle path and the path from input data_c to qout_c is a three-cycle path. They have more loose timing constraints than the path from input data_a to qout_a.

■ Coding Style

1. *It should avoid multicycle paths and other timing exceptions in the design.*
2. *If timing exceptions must be used in a design, it should be sure to use start points and endpoints which are guaranteed to exist and be valid at the chip level.*

■ Review Questions

13-45 What are the limitations of dynamic timing analysis?
13-46 What are the four types of paths of STA?
13-47 What do the port-related constraints include?
13-48 What do the clock-related constraints include?
13-49 Explain the meanings of false paths.
13-50 Explain the meanings of multicycle paths.

13.6 Value Change Dump (VCD) Files

A value change dump (VCD) file is a text file that contains information about value changes on selected variables in a design. The major purpose of VCD files is to provide information for other post-processing tools, such as debug tools. Two types of VCD files are as follows:

- *Four-state VCD file*: It contains selected variables with value changes in {0, 1, x, z} without any strength information.
- *Extended VCD file*: It contains selected ports with value changes in all states and strength information.

13.6.1 Four-State VCD Files

Verilog HDL provides the following system tasks to dump the desired value changes in a design.

13.6.1.1 Specifying Dump File The **$dumpfile** system task specifies the name of the VCD file. It has the form:

```
$dumpfile ("file_name");
```

13.6.1.2 Specifying Variables to be Dumped The **$dumpvars** system task is used to select which variables to be dumped into the specified file. It can be used with or without arguments and has the forms:

```
$dumpvars;
$dumpvars (levels[, modules_or_variables]);
```

When no arguments are specified, the **$dumpvars** system task dumps all variables to the VCD file by default. When arguments are specified, the **$dumpvars** system task dumps variables in the specified module and in all modules specified by the number of levels below. The `levels` argument specifies how many levels of the hierarchy below each specified module instance to be dumped to the VCD file. A value 0 of the `levels` argument means all variables in the specified module and in all module instances below the specified module. The rest of the arguments can specify entire modules or individual variables within a module. The following example addresses some uses of the **$dumpvars** system task:

```
// all levels below adder_nbit_tb
$dumpvars (0, adder_nbit_tb);
// one level below module adder_nbit_tb
$dumpvars (1, adder_nbit_tb);
$dumpvars(0, UUT.sum, UUT.c_in); // all levels below
```

13.6.1.3 Stopping and Resuming Dump The dump operations can be paused and resumed later by using the **$dumpoff** and **$dumpon** system tasks, respectively. Namely, these two system tasks provide a mechanism to control the simulation period. The **$dumpoff** system task stops the dump whereas the **$dumpon** system task resumes the dump. They have the forms:

```
$dumpoff;
$dumpon;
```

When the **$dumpoff** system task is executed, a checkpoint is made by dumping an x value to every selected variable. When the **$dumpon** system task is executed later, each variable is dumped with its value at that time. No value changes are dumped to the specified file between the **$dumpoff** and **$dumpon** system tasks.

Section 13.6 Value Change Dump (VCD) Files

13.6.1.4 Checkpoint Generation The **$dumpall** system task creates a checkpoint, by dumping the values of all selected variables at that time regardless of whether the port values have changed. It has the form:

$dumpall;

13.6.1.5 Setting Maximum Size of Dump File The maximum size of the specified VCD file can be set by using the **$dumplimit** system task. It has the form:

$dumplimit(file_size);

The unit of file_size is bytes. The dumping stops and a comment is inserted in the file when the size of the VCD file reaches this number of bytes.

13.6.1.6 Empty File Buffer The **$dumpflush** system task empties the VCD file buffer of the operating system to the VCD file so that no value changes are lost. It has the form:

$dumpflush;

The following example describes how the above system tasks can be used to create a VCD file in practice.

■ Example 13.12: A VCD file Example.

In this example, we suppose that the adder_nbit module is instantiated as the unit under test. A VCD file named "adder_nbit.vcd" is opened by the **$dumpfile** system task. The file size is limited to 4k bytes by the **$dumplimit** system task. The selected variables to be dumped are those input and output ports of the adder_nbit module. The value changes observed are the first and the last 20 time units of the simulation time. During the duration between them, the dump operations are turned off and on by the **$dumpoff** and **$dumpon** system tasks, respectively.

```
`timescale 1 ns / 1ns
// an example illustrates the VCD file
module adder_nbit_vcd_tb;
// internal signals declarations
reg [3:0] x;
reg [3:0] y;
reg c_in;
wire [3:0] sum;
wire c_out;
// Unit Under Test port map
   adder_nbit UUT (
      .x(x), .y(y), .c_in(c_in), .sum(sum), .c_out(c_out));

reg [7:0] i;
initial begin
   $dumpfile("adder_nbit.vcd");
   $dumplimit(4096);
   $dumpvars(0, UUT.sum, UUT.c_out, UUT.x, UUT.y, UUT.c_in);
end
initial
   for (i = 0; i <= 255; i = i + 1) begin
      x[3:0] = i[7:4]; y[3:0] = i[3:0]; c_in =1'b0;
      #5 ;
```

```
      end
initial begin
   #20    $dumpoff;
   #1000  $dumpon;
end
initial #1040 $dumpflush;
initial
   #1280 $stop;
initial
   $monitor($realtime,"ps %h %h %h %h", x, y, c_in, {c_out, sum});
endmodule
```

13.6.2 VCD File Format

A VCD file consists of three parts: *header information*, *node information*, and *value changes*. The header information gives the date, the version number of the simulator, and the timescale used. The node information contains definitions of the scope and type of a variable being dumped. The part of value changes comprises their values of those variables with values being changed at each simulation time. The contents of the VCD file obtained from the preceding example are shown in Figure 13.18.

Because VCD files are usually processed by debug tools but not analyzed manually, we do not discuss it furthermore. The detailed file format can be found in the LRM [4].

13.6.3 Extended VCD Files

The system tasks used to create and dump extended VCD files are roughly the same as four-state VCD files with the following exceptions:

- Each system task must also provide a specified file name except the **$dumpportsflush** system task, which means all files when no argument is specified. This also means that many extended VCD files may coexist at the same time during simulation.
- The variables that can be dumped are only those ports of the specified modules.
- The values dumped not only include all states {0, 1, x, z} but also contain strength information.

13.6.3.1 Specifying Dump file and Ports The **$dumpports** system task specifies the name of the extended VCD file and the ports to be dumped. It has the form

```
$dumpports (module_names, "file_name");
```

where `module_names` are one or more module identifiers, separated by a comma.

13.6.3.2 Stopping and Resuming Dump The dump operations can be paused and resumed later by using the **$dumpportsoff** and **$dumpportson** system tasks, respectively. That is, they provide a mechanism to control the simulation period. The **$dumpportsoff** system task stops the dump whereas the **$dumpportson** system task resumes the dump. They have the forms:

```
$dumpportsoff("file_name");
$dumpportson("file_name");
```

When the **$dumpportsoff** system task is executed, a checkpoint is made by dumping an x value to every port. When the **$dumpportson** system task is executed later, each port is dumped with its value at that time. No value changes are dumped to the specified file between the **$dumpportsoff** and **$dumpportson** system tasks.

Section 13.6 *Value Change Dump (VCD) Files*

$date	0(	$end
Sat Dec 06 1 8:05:37 2014	0'	#1020
$end	0&	$dumpon
$version	0-	1$
ModelSim Version 6.5	0,	1#
$end	0+	1"
$timescale	0*	0!
1ns	0.	1%
$end	$end	0)
$scope module adder_nbit_tb $end	#5	0(
$scope module UUT $end	1-	1'
$var wire 1 ! sum [3] $end	1$	1&
$var wire 1 " sum [2] $end	#10	1-
$var wire 1 # sum [1] $end	0-	1,
$var wire 1 $ sum [0] $end	1,	0+
$var wire 1 % c_out $end	0$	1*
$var wire 1 & x [3] $end	1#	0.
$var wire 1 ' x [2] $end	#15	$end
$var wire 1 (x [1] $end	1-	0-
$var wire 1) x [0] $end	1$	0,
$var wire 1 * y [3] $end	#20	1+
$var wire 1 + y [2] $end	$dumpoff	0$
$var wire 1 , y [1] $end	x$	0#
$var wire 1 - y [0] $end	x#	0"
$var wire 1 . c_in $end	x"	1!
$upscope $end	x!	#1025
$upscope $end	x%	1-
$enddefinitions $end	x)	1$
#0	x(	#1030
$dumpvars	x'	0-
0$	x&	1,
0#	x-	0$
0"	x,	1#
0!	x+	#1035
0%	x*	1-
0)	x.	1$
(Continued next column)	(Continued next column)	

Figure 13.18: A VCD file example.

13.6.3.3 Checkpoint Generation The **$dumpportsall** system task creates a checkpoint, by dumping the values of all ports at that time regardless of whether the port values have changed. It has the form:

    $dumpportsall(`"file_name"`);

13.6.3.4 Setting Maximum Size Dump Files The maximum size of the specified extended VCD file can be set by using the **$dumpportslimit** system task. It has the form:

    $dumpportslimit(`file_size, "file_name"`);

The unit of `file_size` is bytes. The dumping stops and a comment is inserted in the file when the size of the extended VCD file reaches this number of bytes.

13.6.3.5 Empty File Buffer The **$dumpportsflush** system task empties the extended VCD file buffer of the operating system to the extended VCD file so that no value changes are lost. It has the form:

$dumpportsflush("file_name");

When no argument is specified, the extended VCD buffers for all files opened by the **$dumpports** system task are flushed.

The following example describes how the above system tasks can be used to create an extended VCD file in practice.

■ Example 13.13: An extended VCD file Example.

In this example, we suppose the adder_nbit module is instantiated as the unit under test. An extended VCD file named "adder_nbit_evcd.vcd" is opened by the **$dumpports** system task, which also specifies the ports of the module to be dumped. The file size is limited to 4k bytes by the **$dumpportslimit** system task. The value changes observed are the first and the last 20 time units of the simulation time. During the duration between them, the dump operations are turned off and on by the **$dumpportsoff** and **$dumpportson** system tasks, respectively.

```
`timescale 1 ns / 1 ns
// an example illustrates the extended VCD file
module adder_nbit_evcd_tb;
// internal signals declarations
reg [3:0] x;
reg [3:0] y;
reg c_in;
wire [3:0] sum;
wire c_out;
// Unit Under Test port map
   adder_nbit UUT (
      .sum(sum),.c_out(c_out),.x(x),.y(y),.c_in(c_in));

reg [7:0] i;

initial begin
   $dumpports(adder_nbit, "adder_nbit_evcd.vcd");
   $dumpportslimit(4096, "adder_nbit_evcd.vcd");
end

initial
   for (i = 0; i <= 255; i = i + 1) begin
      x[3:0] = i[7:4]; y[3:0] = i[3:0]; c_in =1'b0;
      #5 ;
   end

initial begin
   #20    $dumpportsoff("adder_nbit_evcd.vcd");
   #1000  $dumpportson("adder_nbit_evcd.vcd");
end

initial #1040 $dumpportsflush("adder_nbit_evcd.vcd");
```

```
initial
   #1280 $stop;
initial
   $monitor($realtime,"ps %h %h %h %h", x, y, c_in, {c_out, sum});
endmodule
```

It is instructive to run this program and observe the differences between the extended VCD file and the VCD file.

■ Review Questions

13-51 Describe the function of a value change dump (VCD) file.
13-52 What are the contents of a VCD file?
13-53 What are differences between the four-state and extended VCD files?
13-54 Describe the operations of the **$dumpfile** system task.
13-55 Describe the operations of the **$dumpports** system task.

13.7 Case Study — FPGA Design and Verification Flow

In this section, we describe the CPLD/FPGA-based synthesis flow and verification through simulation. First, we summarize the basic operations of the Xilinx's Integrated Software Environment (ISE), which includes design entry, synthesis, implementation, and configure CPLD/FPGA. Along the synthesis flow, static timing analyses are also given and explained. Next, we consider how to verify the design through dynamic timing simulation, incorporating the delays of logic elements and interconnect. Finally, we give a general RTL-based verification flow.

13.7.1 ISE Design Flow

When using a CPLD or an FPGA device to implement a design, we follow a synthesis flow that optimizes the logic to fit into the logic elements, places logic elements in the CPLD logic structures or FPGA fabrics, routes the wires to connect the logic elements, and generates a programming file. To see this, consider the Xilinx's ISE synthesis flow [14] shown in Figure 13.19, which can be divided into four major stages:

1. *Design entry*: Two design entry methods, including HDL (such as Verilog HDL and VHDL) and schematic drawings, are allowed to use.
2. *Synthesis*: This stage translates the design in HDL and/or schematic files into a file in industry standard format, known as the *electronic design interchange format* (EDIF) file (*.edf) (in SynplifyPro or others), or a *native generic code* (NGC) file (in XST). The output of this stage is a gate-level netlist, consisting of cells, pins, ports, and nets. Cells are design objects, including instances of user modules, instances of library elements such as LUTs, FFs, RAMS, DSP cells, generic technology representations of hardware functions, and black boxes. Pins are connection points on cells while ports are the top-level ports of a design. Nets make connections between pins and between pins and ports.
3. *Implementation*: In FPGAs, it includes three main steps: translate, map, and place and route (PAR). In CPLDs, it only has a CPLD fitter step.
4. *Configure CPLD/FPGA*: It generates and downloads a BIT stream file into a specified CPLD/FPGA device so that the CPLD/FPGA device can function as desired.

In what follows, we describe each of these steps in more detail.

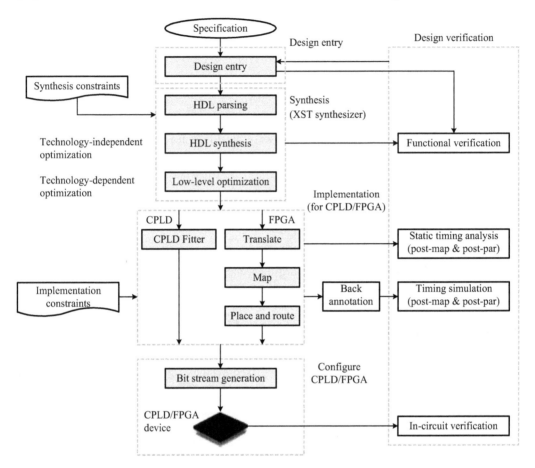

Figure 13.19: The block diagram of the ISE design flow.

13.7.1.1 Design Entry The most commonly used design entry approach is based on HDL, including Verilog HDL and VHDL, although schematic drawings can also be used as a design entry. After a design is entered the synthesis flow, the functional verification of the design is usually followed through simulation or formal proof, as described previously.

13.7.1.2 Synthesis The function of the synthesis stage is to extract logic from the HDL model and translate the design in HDL and/or schematic files into an EDIF file (*.edf) (in SynplifyPro or others) or an NGC file (in XST). The synthesis stage can be divided into three main steps: HDL parsing, HDL synthesis, and low-level optimization. In the HDL parsing step, possible syntactic errors are identified. The functions of the HDL synthesis step are to identify macros, extract finite-state machines, and identify resources that can be shared. The operations of the low-level optimization step include macro implementation, timing optimization, LUT mapping, and register replication.

Macro identification (or inference) consists of two major steps: macro recognition (in the HDL synthesis step) and macro implementation (in the low-level optimization step). In macro identification, Xilinx Synthesis Technology (XST) tries to recognize as many macros as possible. In macro implementation, XST makes technology dependent choices, which are based on macro types and sizes in accordance with the criteria that improve the performance and reduce the area of the design.

At present, most synthesis tools, including XST and SynplifyPro, are able to recognize finite-state machines from the source description regardless of which modeling style is used. In

Section 13.7 Case Study — FPGA Design and Verification Flow

addition, these synthesis tools are capable of optimizing the finite-state machines on the basis of state assignment codes and the flip-flop type. The most commonly used state assignment codes include the one-hot code for speed optimization, the binary code for area optimization, the Gray code for hazards and glitches minimization, and other user-defined codes, such as the Johnson code. Refer to in Section 11.1.1 for more details.

The output from synthesis is an EDIF file (*.edf) (in SynplifyPro or others) or an NGC file (in XST). A portion of an EDIF file is shown as follows:

```
...
(cell OBUF (cellType GENERIC)
   (view PRIM (viewType NETLIST)
     (interface
       (port O (direction OUTPUT))
       (port I (direction INPUT))
     )
   )
)
(cell LUT3 (cellType GENERIC)
   (view PRIM (viewType NETLIST)
     (interface
       (port I0 (direction INPUT))
       (port I1 (direction INPUT))
       (port I2 (direction INPUT))
       (port O (direction OUTPUT))
     )
   )
)
```

13.7.1.3 Implementation
The implementation stage includes three main steps: translate (NGDBuild), map, and place and route (PAR).

Translate. The translate step prepares for the map step, including managing the design hierarchy; namely, it merges multiple design files into a single gate-level netlist. The translate step is carried out by the NGDBUILD program and converts an NGC netlist (an output of the XST synthesizer) into a *native generic database* (NGD) netlist. The NGC netlist is built around the UNISIM component library, designed for behavioral simulation, whereas the NGD netlist is built on the SIMPRIM library. The produced NGD netlist contains some initial, approximate information about the propagation delays of logic components.

Map. The map step groups logical symbols from the gate-level NGD netlist into physical components. More precisely, it optimizes and fits the SIMPRIM primitives from an NGD netlist into specific device resources: LUTs, flip-flops, BRAMs, I/O blocks (IOBs), and others. The map step is performed by the MAP program and outputs a *native circuit description* (NCD) netlist. The NCD netlist contains precise information about the propagation delays of logic components, but no information about interconnect delays because at this time, the interconnect between physical components has not been set yet. Note that for Virtex-5 devices the map step also does placement. For other devices, placement is done by PAR. However, routing is done by PAR for all devices.

The map step finishes the logic synthesis operations. A resource used report is obtained after the map step finishes. The resource used by the 4×4 `booth_array_multiplier` example is as follows:

```
Design Summary
--------------
Number of errors:      0
```

```
Number of warnings:      0
Logic Utilization:
  Number of 4 input LUTs:        31 out of 3,072 1%
Logic Distribution:
  Number of occupied Slices: 17 out of 1,536 1%
  Number of Slices containing only related logic:
      17 out of 17  100%
  Number of Slices containing unrelated logic:
      0 out of 17    0%
*See NOTES below for an explanation of the effects
 of unrelated logic
Total Number 4 input LUTs:      31 out of 3,072 1%

  Number of bonded IOBs:         16 out of    200 8%

Total equivalent gate count for design:    186
Additional JTAG gate count for IOBs:   768
Peak Memory Usage:   137 MB
```

From this report, we can see the feasibility of design. In the case when the LUTs used are more than expected, the design is needed to modify or even restructure.

In addition, the map step also outputs a pre-layout static timing analysis about the design. A portion of the post-map STA of the 4×4 booth_array_multiplier is shown as follows:

Data Sheet report:
────────────────────

All values displayed in nanoseconds (ns)

Pad to Pad

Source Pad	Destination Pad	Delay
x[0]	product[0]	4.943
...		
x[0]	product[6]	7.625
x[0]	product[7]	7.625
x[1]	product[1]	4.943
...		
x[1]	product[6]	7.625
x[1]	product[7]	7.625
x[2]	product[2]	4.943
...		
x[2]	product[6]	6.731
x[2]	product[7]	6.731
x[3]	product[3]	4.943
...		
x[3]	product[6]	5.837
x[3]	product[7]	5.837
...		
y[0]	product[6]	7.625
y[0]	product[7]	7.625
y[1]	product[1]	4.943
...		

Section 13.7 Case Study — FPGA Design and Verification Flow

y[1]	|product[6]	| 7.625|
y[1]	|product[7]	| 7.625|
y[2]	|product[2]	| 5.837|
	...	
y[2]	|product[6]	| 7.625|
y[2]	|product[7]	| 7.625|
y[3]	|product[3]	| 6.284|
	...	
y[3]	|product[6]	| 7.178|
y[3]	|product[7]	| 7.178|

From this report, we can see that one critical path is from x[1] to product[6], which needs 7.625 ns. Because in the map step the design is only mapped onto CLBs without placing into a real FPGA device, the delay value obtained is only the cumulative delay of CLBs.

Place and route (PAR). Place and route is the most important and time-consuming step of the implementation. In this step, logic elements are placed onto the device and connected according to the required function. Place and route is performed by the PAR program and outputs a netlist in NCD format. It places synthesized logic components into FPGA fabrics of the underlying FPGA device, connects them, and extracts timing data into reports. It comprises placement and routing. Placement chooses which fabrics will hold each of logic elements while routing chooses which wire segments will be used to make necessary connections between the fabrics mapped to logic elements. PAR considers timing constraints (via the UCF) set up by the FPGA designer.

After the PAR step, the implementation tools extract timing data and give a post-layout static timing analysis about the design. For example, what follows is a portion of the post-route (also called post-PAR) STA report of the 4×4 booth_array_multiplier.

```
Data Sheet report:
-------------------

All values displayed in nanoseconds (ns)

Pad to Pad
```

Source Pad	Destination Pad	Delay
x[0]	|product[0]	| 7.620|
	...	
x[0]	|product[6]	| 12.866|
x[0]	|product[7]	| 12.385|
x[1]	|product[1]	| 7.181|
	...	
x[1]	|product[6]	| 13.102|
x[1]	|product[7]	| 12.621|
x[2]	|product[2]	| 7.563|
	...	
x[2]	|product[6]	| 11.470|
x[2]	|product[7]	| 10.989|
x[3]	|product[3]	| 6.848|
	...	
x[3]	|product[6]	| 9.851|
x[3]	|product[7]	| 9.370|
	...	
y[0]	|product[6]	| 12.643|

y[0]	|product[7]	| 12.162|
y[1]	|product[1]	| 7.330|
	...	
y[1]	|product[6]	| 12.551|
y[1]	|product[7]	| 12.070|
y[2]	|product[2]	| 8.670|
	...	
y[2]	|product[6]	| 12.298|
y[2]	|product[7]	| 11.817|
y[3]	|product[3]	| 9.459|
	...	
y[3]	|product[6]	| 12.251|
y[3]	|product[7]	| 11.770|

From this report, we can see that the critical path is from x[1] to product[6], which needs about 13.102 ns. This delay is the actual delay of the path, which includes both CLBs and interconnect (wire) delays. The interconnect delays along each path can be easily estimated by subtracting the delay obtained from the post-map output from that obtained from the post-PAR output. For example, the interconnect delay of the path from x[1] to product[6] is equal to $13.102 - 7.625 = 5.477$ ns.

13.7.1.4 Configure FPGA This stage generates a programming file (.bit), called a BIT file, which can be used to configure the specified CPLD/FPGA device so that it would work as expected. The BIT file can be directly downloaded to the CPLD/FPGA device, or converted into a PROM (Flash memory) file which stores the programming information.

13.7.1.5 Timing Constraints Even though HDLs, including Verilog HDL and VHDL, capture the function of the logic, some mechanism is still needed to specify non-functional requirements: timing, area, and power dissipation. These non-functional requirements are often carried to the synthesis tools and implementation tools in a file, known as the *XST constraint file* (XCF) and the *user constraint file* (UCF), respectively.

For FPGA devices, a user constraint file includes timing constraints, area constraints, and I/O pin assignments. The timing constrains improve design performance by placing logic elements closer together so that shorter routing wires may be used. Timing constraints cover pad-to-pad constraints and period constraints as well as offset-in and offset-out constraints. The pad-to-pad constraints are essential for completely specifying a design. The period constraint improves the path delays between synchronous elements. The offset-in constraint improves path delays from input ports to synchronous elements and the offset-out constraint improves path delays from synchronous elements to output ports.

The area constraint is achieved by carefully placing logic blocks on the underlying FPGA device. Placement can also affect the system performance (timing) and I/O pin assignments of the final system.

I/O pin assignments not only assign I/O blocks (IOBs) to I/O ports of the design but also allow one to specify the signal direction (input, output, or bidirectional), slew rates, drive strengths, and I/O standard (voltage levels) of I/O pins. I/O pin assignments are essential to the design of a printed-circuit board (PCB) on which the FPGA device will be used and I/O ports must be assigned to fixed I/O pins of the FGPA device. An example can be referred to in Section 10.5.3.

■ Review Questions

13-56 What is the basic synthesis flow of the Xilinx's ISE?
13-57 Describe the function of the synthesis step of the synthesis stage.

Section 13.7 Case Study — FPGA Design and Verification Flow

13-58 Describe the function of the optimization step of the synthesis stage.
13-59 What are the three major steps of the implementation stage?
13-60 Describe the operations of the place-and-route step.

13.7.2 Dynamic Timing Simulation

As described previously, dynamic timing simulation is still an effective and widely used approach for verifying a design. In this section, we introduce how to perform a variety of dynamic simulations along the synthesis flow. As shown in Figure 13.20, there are four types of simulations along the synthesis flow. Except for the behavioral (also called functional) simulation used to verify the functional correctness of a design at the beginning of the synthesis flow, the remaining three types of simulations are at the gate level and are based on the outputs of various steps of the implementation stage of the ISE. As a result, different but functionally equivalent source files are used in different steps. Nevertheless, the same test bench is used in all steps, from the behavioral level to the post-route (also called post-PAR) gate level.

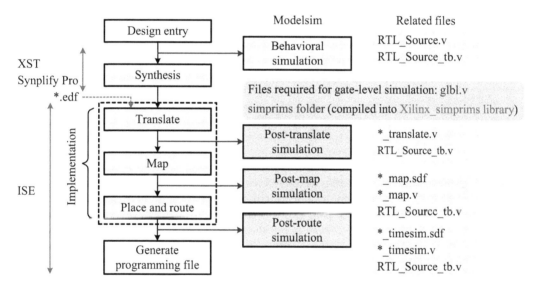

Figure 13.20: The ISE-based simulation flow and its related files.

There are three types of timing simulations at the gate level. Before the timing simulation can be done, one file and one folder must be prepared: *glbl.v* and *simprims folder*. The glbl.v file is usually compiled into the work library and the simprims folder is compiled into a library, say, Xilinx_simprims.

13.7.2.1 Post-Translate Simulation The translate step prepares the design for the map step. It expands the design hierarchy and reads timing constraints. The translate step generates a post-translate simulation model, which is built from SIMPRIM primitives that more closely relate to the structure of the CLBs of the FPGA device under consideration. A snapshot of the result from the translate step is as follows:

```
X_LUT2 #(.INIT ( 4'hD ))
  \Madd_AUX_13_addsub0001_xor<0>111  (
    .ADR0(x_3_IBUF_56),
    .ADR1(x_2_IBUF_55),
    .O(N2) );
X_LUT3 #(.INIT ( 8'h18 ))
```

```
    AUX_14_and00001 (
        .ADR0(y_1_IBUF_62),
        .ADR1(x_2_IBUF_55),
        .ADR2(x_3_IBUF_56),
        .O(AUX_14_and0000) );
...
X_BUF x_0_IBUF (
        .I(x[0]),
        .O(x_0_IBUF_53) );
```

Of course, this is different from the original source description. However, they are functional equivalent. At this point, it is worth noting that if you find a discrepancy between your source description and the post-translate model, it may be because some constructs in your source description do not exactly translate in the way that you thought they should do.

The above X_LUT2, X_LUT3, and X_BUF cells can be observed from the Xilinx simprims folder. The X_BUF cell is listed as follows:

```
module X_BUF (O, I);
parameter LOC = "UNPLACED";
output O;
input I;
buf (O, I);
specify
    (I => O) = (0:0:0, 0:0:0);
specparam PATHPULSE$ = 0;
endspecify
endmodule
```

The timing information in each cell is described with a specify block in which template module path delays are defined with a default value of 0 to indicate the rise and fall times in mintypmax format. In addition, the minimum path width is also defined with the **specparam** parameter. Refer to Section 7.3.2 for more details about the specify block. The delays specified in the specify block will be annotated by the realistic cell delays described in the sdf file generated in the map step (*_map.sdf) and in the place and route step (*_timesim.sdf).

In order to perform a post-translate simulation (a gate-level simulation), the following files are required. In what follows, the Booth array multiplier, booth_array_multiplier, is used as the running example.

- **Post-synthesis source file**: booth_array_multiplier.vm (SynplifyPro) or booth_array_multiplier_translate.v (Xinlix ISE)
- **Library files**:
    1. glbl.v ($Xilinx/verilog/src/) and
    2. Xilinx_simprims (simprims folder ($Xilinx/verilog/src/simprims) compiled into the Xilinx_simprims library)
- **Test bench file**: booth_array_multiplier_TB.v

The simulation results of the 4×4 booth_array_multiplier are shown in Figure 13.21. The translate step does not take into account the delays of cells; the post-translate simulation is only used for verifying whether the function is still equivalent to that of the design entry.

13.7.2.2 Post-Map Simulation The second type of gate-level simulation is the post-map simulation, which incorporates into the delays of CLBs but not of interconnect. To do this, a post-map simulation model is generated at the map step. It is also built from SIMPRIM primitives and more closely relates to the structure of the CLBs of the FPGA device under

Section 13.7 Case Study — FPGA Design and Verification Flow

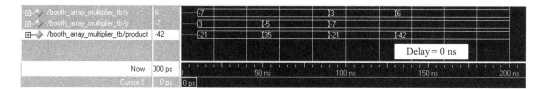

Figure 13.21: The results of a post-translate simulation.

consideration. Like the post-translate simulation model, this synthesized model is functionally equivalent to our original design but one more step closely matches the structure of the CLBs of the underlying FPGA device. The SDF file back-annotates the delays of CLBs into the simulation model. A snapshot of the result from the map step is as follows:

```
X_LUT4 #(.INIT ( 16'hDDDD ))
  \Madd_AUX_13_addsub0001_xor<0>111   (
     .ADR0(x_3_IBUF_230),
     .ADR1(x_2_IBUF_229),
     .ADR2(VCC),
     .ADR3(VCC),
     .O(N2_pack_1) );
X_LUT4 #(.INIT ( 16'h1818 ))
  AUX_14_and00001 (
     .ADR0(y_1_IBUF_232),
     .ADR1(x_2_IBUF_229),
     .ADR2(x_3_IBUF_230),
     .ADR3(VCC),
     .O(AUX_14_and0000) );
...
X_BUF \x<0>/IFF/IMUX   (
     .I(\x<0>/INBUF ),
     .O(x_0_IBUF_227) );
```

where X_LUT2 and X_LUT3 used to implement XOR gate and AND gate in the translate step are individually mapped to an X_LUT4. It is instructive to note that at this step, cells are only mapped into FPGA fabrics without assigning their exact locations.

In order to perform the post-map simulation, two files are required besides the library files: a post-map simulation model (*_map.v) and a standard delay format (*_map.sdf). These two files are generated automatically by the implementation tool at the map step.

- **Post_map Source file**: booth_array_multiplier_map.v
- **Library files**:
    1. glbl.v ($Xilinx/verilog/src/) and
    2. Xilinx_simprims (simprims folder ($Xilinx/verilog/src/simprims) compiled into the Xilinx_simprims library)
- **SDF file**: booth_array_multiplier_map.sdf
- **Test bench file**: booth_array_multiplier_TB.v

The simulation results of the 4 × 4 booth_array_multiplier are shown in Figure 13.22. Although the map step does include the delay of CLBs, it is still very difficult to find the longest delay of logic elements because a lot of data are needed to search manually. Consequently, the post-map simulation is only used for verifying whether the function is still equivalent to that of the design entry.

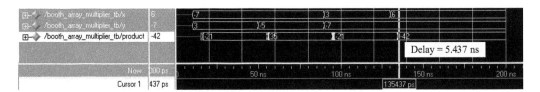

Figure 13.22: The results of a post-map simulation.

13.7.2.3 Post-Route Simulation The third and final type of gate-level simulation is the post-route (i.e., post-PAR) simulation, which incorporates into both the delays of CLBs and interconnect. To do this, a post-route simulation model is generated at the place-and-route (PAR) step. It is built from SIMPRIM primitives and exactly matches the structure of the CLBs of the FPGA device under consideration. Like the post-map simulation model, this synthesized model is functionally equivalent to our original design but most closely matches the structure of the CLBs of the underlying FPGA device. The SDF file back-annotates the delays of the CLBs and interconnect into the simulation model. A snapshot of the result from the PAR step is as follows:

```
X_LUT4 #(.INIT ( 16'hCCFF ),
        .LOC ( "SLICE_X10Y44" ))
  \Madd_AUX_13_addsub0001_xor<0>111 (
     .ADR0(VCC),
     .ADR1(x_2_IBUF_229),
     .ADR2(VCC),
     .ADR3(x_3_IBUF_230),
     .O(N2_pack_1)
  );
X_LUT4 #(.INIT ( 16'h2424 ),
        .LOC ( "SLICE_X11Y44" ))
  AUX_14_and00001 (
     .ADR0(y_1_IBUF_232),
     .ADR1(x_3_IBUF_230),
     .ADR2(x_2_IBUF_229),
     .ADR3(VCC),
     .O(AUX_14_and0000)
  );
...
X_BUF #(.LOC ( "PAD25" ))
  x_0_IBUF (
     .I(x[0]),
     .O(\x<0>/INBUF )
  );
```

where each cell, such as X_LUT4 and X_BUF, is exactly mapped into a physical LUT4 or an I/O pad on the underlying FPGA device.

In order to perform this simulation, two files are required in addition to the library files: a post-route simulation model (*_timesim.v) and a standard delay format (*_timesim.sdf). These two files are generated automatically by the implementation tool at the PAR step.

- **PAR Source file**: booth_array_multiplier_timesim.v
- **Library files**:
    1. glbl.v ($Xilinx/verilog/src/) and

Section 13.7 Case Study — FPGA Design and Verification Flow

2. Xilinx_simprims (simprims folder ($Xilinx/verilog/src/simprims) compiled into the Xilinx_simprims library)

- **SDF file**: booth_array_multiplier_timesim.sdf
- **Test bench file**: booth_array_multiplier_TB.v

The simulation results of the 4×4 booth_array_multiplier are shown in Figure 13.23. Although the PAR step does take into account both the delays of CLBs and interconnect, like the post-map simulation it is still very difficult to find the longest delay of logic elements and its associated interconnect, namely, the critical path, because a lot of data are needed to search manually. Consequently, the post-route simulation is only used for verifying whether the function is equivalent to that of the design entry. As for the critical path, it usually resorts to static timing analysis (STA).

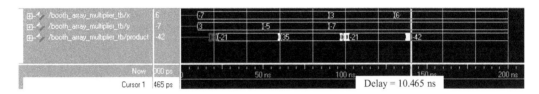

Figure 13.23: The results of a post-route simulation.

After the PAR step, we can also examine the power dissipation of the design. The power report from the implementation tool of the ISE is shown as follows:

On–Chip Power Summary

On–Chip	Power (mW)	Used	Available	Utilization (%)
Clocks	0.00	0	—	—
Logic	0.00	38	3840	1
Signals	0.00	43	—	—
IOs	0.00	16	141	11
Quiescent	41.02			
Total	41.02			

13.7.3 An RTL-Based Synthesis and Verification Flow

In what follows, we first summarize the generic considerations of FPGA-based system design and then introduce an RTL-based verification flow suitable for using in ASIC design.

13.7.3.1 Generic Considerations of FPGA-Based Design Informally, the four fundamental issues to build a generic FPGA-based system after a design has been described in Verilog HDL are as follows:

1. *Synthesis:* This step is to synthesize all the RTL source codes using one of industry standard tools so as to create a set of netlists, consisting of gates, RAMs, LUTs, and so on, that implementation tools can read and recognize.

2. *Implementation* (also called *place and route*): A place-and-route (PAR) tool accepts the synthesized netlist and carries out the floorplanning, placement, and routing to generate a gate-level netlist and a BIT-stream (programming) file. The BIT-stream file can be used to configure an FPGA device so that the FPGA device can function as desired.

3. *Constraint file:* A constraint file describing I/O-pad assignments and timing constraints is also needed in synthesizing and implementing a design into a realistic FPGA device. The I/O-pad assignments map input/output ports of the top-level module at the RTL to the physical I/O pins of the FPGA device used. The proper drive strengths and voltage levels of I/O pads are also specified in the file. The timing constraints cover at least clock-related timing constraints, input- and output-delay constraints, and timing exceptions.

4. *Static timing analysis:* STA is the most important step to analyze the final netlist against the constraints specified in the constraint file. In this step, timing violations, especially setup-time and hold-time violations, are looked for and recognized. Any significant timing violations are needed to analyze carefully and correct accordingly.

13.7.3.2 An RTL-Based Verification Flow As we have discussed in the previous sections, before an RTL-based design is said to be complete, many verifications are still needed to carry out. The general RTL-based verification flow includes three major steps: simulation, static timing analysis, and prototyping. We list each of these in detail as follows:

1. Simulation
    (a). Functional (behavioral) simulation
    (b). Code coverage analysis
    (c). Assertion (property) checking
    (d). Gate-level simulation
    (e). Dynamic timing simulation (gate-level simulation + SDF back annotation)
2. Static timing analysis
    (a). Critical paths
    (b). Timing violations
3. Prototyping
    (a). FPGA prototyping
    (b). Cell-based prototyping

■ **Review Questions**

13-61 What are the differences between the post-map and post-route SDF files?

13-62 Describe the differences between the post-map simulation and post-route simulation in terms of timing information.

13.8 Summary

Verification is a necessary process that ensures a design can meet its specifications. Because of the inherent features of the timing relationship between various components of the hardware module in question, the verification process needs to assure that the design works correctly in both function and timing. As a consequence, the verification of a design can be divided into two types: functional verification and timing verification. Functional verification merely takes into account whether the logic function of the design meets the specifications and timing verification considers whether the design meets the timing constraints.

Functional verification can be done either by simulation or by formal proof. In simulation-based functional verification, the design is placed under a test bench, input stimuli are applied to the design, and the outputs from the design are compared with the reference outputs. Formal verification proves mathematically that protocols, assertions, properties, or design rules hold for all possible cases in the design.

Simulation-based verification is based on the analysis of simulation results and code coverage. The quality of this type of verification considerably relies on the test bench, which generates and controls stimuli, monitors and stores responses, and checks the simulation results. The two most common types of test benches are non-self-checking and self-checking. The non-self-checking test benches are often used to verify the basic functionality of device under test (DUT) in an early stage of the development cycle. The self-checking test benches are used to verify the final functionality of the DUT.

Formal verification can be further cast into equivalence checking and property verification. Equivalence checking determines whether two implementations are equivalent and property verification determines whether a property exists in a design. The major drawback of formal verification is that it can only be applied to a design of limited size.

Timing verification is often carried out through simulation, known as dynamic timing simulation, which can be used to verify both the functionality and timing of a design. However, dynamic timing simulation has the following shortcomings: First, it poses a bottleneck for large complex designs because simulation requires a lot of time. Second, it relies on the quality and coverage of the test bench used for verification. Third, it is very difficult to figure out a critical path through using the results from dynamic timing simulation. Hence, an alternative way, called static timing analysis (STA), is widely used to determine if a circuit meets timing constraints without having to simulate the design in a cycle-by-cycle manner. In other words, static timing analysis analyzes a design in a static way and computes the delay for each path of the design. Therefore, the critical paths can be easily pointed out.

In order to use dynamic timing simulation to verify the timing constraints of a design, besides test benches, an SDF file is also needed. The SDF file contains timing information of all cells in the design and can be created by PAR tools automatically. In addition, when simulation is used to verify the function and timing of a design, a value change dump (VCD) file may also be produced during simulation in order to provide valuable information about value changes on selected variables in the design for debug tools to help remove the function bugs and timing violations of the design.

A comprehensive example based-on the FPGA synthesis flow was given to illustrate how to enter, synthesize, implement, and configure the underlying FPGA device of a design. Along the synthesis flow, static timing analyses were also given and explained. In addition, design verification through dynamic timing simulation, incorporating the delays of CLBs and interconnect, was introduced.

References

1. J. Bergeron, *Writing Testbenches: Functional Verification of HDL Models,* 2nd ed., Boston: Kluwer Academic Publishers, 2003.
2. J. Bhasker, *A Verilog HDL Primer,* 3rd ed., Star Galaxy Publishing, 2005.
3. H. Bhatnagar, *Advanced ASCI Chip Synthesis: Using Synopses Design Compiler and Prime time,* Boston: Kluwer Academic Publishers, 1999.
4. IEEE 1364-2001 Standard, *IEEE Standard Verilog Hardware Description Language,* 2001.
5. IEEE 1364-2005 Standard, *IEEE Standard for Verilog Hardware Description Language,* 2006.
6. IEEE 1497-2001 Standard, *IEEE Standard for Standard Delay Format (SDF) for the Electronic Design Process,* 2001.

7. IEEE 1800-2005 Standard, *IEEE Standard for SystemVerilog — Unified Hardware Design, Specification, and Verification Language,* 2005.

8. M. Keating and P. Bricaud, *Reuse Methodology Manual: For System-on-a-Chip Designs,* Boston: Kluwer Academic Publishers, 2002.

9. W. K. Lam, *Hardware Design Verification: Simulation and Formal Method-Based Approaches,* Upper Saddle River, New Jersey: Prentice-Hall, 2005.

10. R. Munden, *ASIC and FPGA Verification: A Guide to Component Modeling,* Morgan Kaufmann, 2005.

11. S. Palnitkar, *Verilog HDL: A Guide to Digital Design and Synthesis,* 2nd ed., SunSoft Press, 2003.

12. B. Wile, J. C. Goss, and W. Roesner, *Comprehensive Functional Verification: The Complete Industry Cycle,* Morgan Kaufmann, 2005.

13. L. T. Wang, Y. W. Chang, and K. T. Cheng, *Electronic Design Automation,* New York: Morgan Kaufmann, 2009.

14. Xilinx, *XST User Guide for Virtex-6, Spartan-6, and 7 Series Devices,* October, 2011.

Problems

13-1 Use the logic circuit shown in Figure 13.24 as an example to show the operations of the event-driven algorithm. Draw a table like the one shown in Figure 13.8(b) to indicate all scheduled events, assuming that all NAND gates have the same propagation delay of 5 time units and input x changes from 1 to 0 while input y remains at 1.

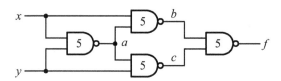

Figure 13.24: The implementation of an XOR gate with four 2-input NAND gates.

13-2 Use the logic circuit shown in Figure 13.24 as an example to show the operations of the logic levelization algorithm. Draw a table like the one shown in Table 13.1.

13-3 Replace the parameter definitions in the n-bit CRC generator example in Section 9.5.2 with the following two values:

    parameter N = 16;
    parameter [N:0]tap = 17'b11000000000000101;

(a) Describe how to verify whether the CRC-16 circuit works correctly.

(b) Write a test bench to generate the stimuli, monitor the responses, and check the results at the end of the test.

13-4 Considering Problem 13-3 again, write a self-checking test bench to read the stimuli from a file, monitor the responses, and check the results on the fly.

13-5 Considering Problem 13-3 again, write a self-checking test bench to randomly generate the stimuli, monitor the responses, and check the results on the fly.

13-6 Instantiate the Booth array multiplier discussed in Section 11.3.4 with the following statement:

booth_array_multiplier #(8, 8) booth_array
                (.x(x), .y(y), .product(product));

  (a) Describe how to verify whether the booth_array_multiplier module works correctly.
  (b) Write a test bench to generate the stimuli, monitor the responses, and check the results at the end of the test.

13-7 Considering Problem 13-6 again, write a self-checking test bench to read the stimuli from files, monitor the responses, and check the results on the fly.

13-8 Considering Problem 13-6 again, write a self-checking test bench to randomly generate the stimuli, monitor the responses, and check the results on the fly.

13-9 Considering Problem 13-6 again, properly set the options of the simulator to activate the code coverage analysis of the booth_array_multiplier module and then observe the results.

13-10 Write a Verilog HDL module to describe an 8-bit adder with two inputs x and y and one output sum and then write a self-checking test bench with each of the following specified functions to verify the 8-bit adder.
  (a) Use the **$random** system function to generate two random numbers as the inputs and then use the **$monitor** system task to display the result on the standard output.
  (b) Use the **$fgets** and **$sscanf** system functions to read two numbers from the standard input and use the **$display** and **$monitor** system tasks to display the result on the standard output.

13-11 Write a Verilog HDL module to describe a two-digit BCD adder with two inputs x and y and one output sum and then write a self-checking test bench with each of the following specified functions to verify the two-digit BCD adder.
  (a) Use the **$random** system function to generate two random numbers as the inputs and then use the **$monitor** system task to display the result on the standard output.
  (b) Use the **$fgets** and **$sscanf** system functions to read two numbers from the standard input and use the **$display** and **$monitor** system tasks to display the result on the standard output.

13-12 In order to mimic the jitter effects of two asynchronous clock sources in practical applications, a random number is often used to generate a random phase of the clock signals. Modify the following two **initial** blocks so that their clock signals contain random phases. Assume that the random phase of each clock signal is within 5% of the clock period.

```
initial begin
    clk100 <= 1'b0;
    forever #5 clk100 <= ~clk100;
end
// the other clock source
initial begin
    clk33 <= 1'b0;
    forever #15 clk33 <= ~clk33;
end
```

13-13 Considering the tutorial example discussed in Section 1.4.3, modify the test bench to dump out variables within a module you chose using the **$dumpvars** system task.

  (a) Dump out variables under the scope of the current module.

  (b) Dump out variables at all levels under the scope of the current module.

13-14 Considering the parameterizable multiplexer discussed in Section 8.3.1, instantiate it with the size specified as follows. Then synthesize it and observe the number of LUTs used as well as the critical path delay. Finally, perform a post-PAR simulation.

  (a) An 8-to-1 multiplexer

  (b) A 16-to-1 multiplexer

  (c) A 32-to-1 multiplexer

13-15 Considering the parameterizable magnitude comparator discussed in Section 8.5.1, instantiate it with the size specified as follows. Then synthesize it and observe the number of LUTs used as well as the critical path delay. Finally, perform a post-PAR simulation.

  (a) An 8-bit magnitude comparator

  (b) A 16-bit magnitude comparator

  (c) A 32-bit magnitude comparator

13-16 Considering the Booth algorithm described in Section 11.2.2, instantiate it with the size specified as follows. Then synthesize it and observe the number of LUTs used as well as the minimum clock period. Finally, perform a post-PAR simulation.

  (a) An 8×8 multiplier

  (b) A 16×16 multiplier

  (c) A 32×32 multiplier

13-17 Considering the Booth array multiplier discussed in Section 11.3.4, instantiate it with the size specified as follows. Then synthesize it and observe the number of LUTs used as well as the critical path delay. Finally, perform a post-PAR simulation.

  (a) An 8×8 multiplier

  (b) A 16×16 multiplier

  (c) A 32×32 multiplier

13-18 Within a test bench, write a program segment to generate two asynchronous clock signals. One generates a clock signal with the frequency of 25 MHz and the other of 50 MHz.

13-19 Within a test bench, write a program segment to generate two synchronous clock signals. One generates a clock signal with the frequency of 20 MHz and the other of 100 MHz.

13-20 Within a test bench, write a program segment to realize a retriggerable monostable. The out output generates a pulse of 50 time units each time it is triggered.

13-21 Within a test bench, write a program segment to realize a non-retriggerable monostable. The out output generates a pulse of 20 time units when it is triggered each time.

14
Arithmetic Modules

ARITHMETIC operations are critical in most digital systems. Hence, we will examine in this chapter many frequently used arithmetic modules, including addition, multiplication, division, arithmetic-and-logic units (ALUs), shifters, and two digital-signal processing (DSP) filters. By way of the introduction of these arithmetic operations and their algorithms, we also reemphasize the concept that a hardware algorithm can usually be realized by using either a multicycle or single-cycle structure.

The bottleneck of a conventional n-bit ripple-carry adder is on the generation of the carry needed in each bit stage. To overcome this, many schemes have been proposed. Among these, carry-lookahead (CLA) adders, block CLA adders, and two parallel-prefix adders, Kogge-Stone adders and Brent-Kung adders, are to be explored.

The shift-and-add (subtract) technique is widely applied to develop both multiplication and division algorithms, which may process unsigned and signed numbers. To explore how this technique can be used to build multipliers and dividers, two array multipliers, including an unsigned array multiplier and the modified Baugh-Wooley signed array multiplier as well as two basic division algorithms, restoring division and non-restoring division, are considered in depth in this chapter.

The ALU is often the major component for the datapath of many applications, in particular, for central processing unit (CPU) design. An ALU comprises a number of arithmetic and logic operations. In some applications, shifts, including both logical and arithmetic, are also included as an important part of their ALU functions. To equip such needs, a multiplexer-based single-cycle structure, known as a barrel shifter, is considered in detail. Finally, two basic DSP techniques, finite-impulse response (FIR) and infinite-impulse response (IIR) filters, are addressed briefly.

14.1 Addition and Subtraction

The importance of addition is not only manifested itself but also exhibited on the construction of other related operations, such as multiplication and division. In modern digital systems, subtraction is usually accomplished by 2's-complement arithmetic, which is also based on addition. Hence, the term *addition* is often used to mean both addition and subtraction. In this section, we deal with carry-lookahead (CLA) adders and parallel-prefix adders.

14.1.1 Carry-Lookahead Adders

Recall that the performance bottleneck of an n-bit ripple-carry adder is on the generation of the carry needed in each stage. To explore the carry generation problem, consider the ith full

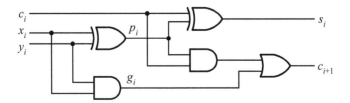

Figure 14.1: The logic diagram of the ith full adder in an n-bit adder.

adder of an n-bit adder, as shown in Figure 14.1, which is the one implemented in terms of two half adders that we have introduced in Section 1.3.2.

Before proceeding, let us make the following observations. At the ith stage of an n-bit adder, the carry-out, c_{i+1}, is generated if both inputs, x_i and y_i are 1, regardless of what value of the carry-in, c_i, is. The carry-in, c_i, is propagated to the output if either input, x_i or y_i, is 1. Consequently, we can define two new functions: *carry generate* (g_i) and *carry propagate* (p_i) in terms of inputs x_i and y_i as follows:

$$g_i = x_i \cdot y_i$$
$$p_i = x_i \oplus y_i \tag{14.1}$$

Based on these two functions, the sum and carry-out of the ith-stage full adder can be separately represented as a function of both g_i and p_i and are equal to

$$s_i = p_i \oplus c_i$$
$$c_{i+1} = g_i + p_i \cdot c_i \tag{14.2}$$

As a result, the carry-in of the $(i+1)$th-stage full adder can be generated by using the recursive equation of the c_{i+1}. For example, the first four carry-in signals are as follows:

$$c_1 = g_0 + p_0 \cdot c_0 \tag{14.3}$$

$$c_2 = g_1 + p_1 \cdot c_1$$
$$= g_1 + p_1(g_0 + p_0 \cdot c_0) = g_1 + p_1 \cdot g_0 + p_1 \cdot p_0 \cdot c_0 \tag{14.4}$$

$$c_3 = g_2 + p_2 \cdot c_2$$
$$= g_2 + p_2(g_1 + p_1 \cdot g_0 + p_1 \cdot p_0 \cdot c_0)$$
$$= g_2 + p_2 \cdot g_1 + p_2 \cdot p_1 \cdot g_0 + p_2 \cdot p_1 \cdot p_0 \cdot c_0 \tag{14.5}$$

$$c_4 = g_3 + p_3 \cdot c_3$$
$$= g_3 + p_3 \cdot (g_2 + p_2 \cdot g_1 + p_2 \cdot p_1 \cdot g_0 + p_2 \cdot p_1 \cdot p_0 \cdot c_0)$$
$$= g_3 + p_3 \cdot g_2 + p_3 \cdot p_2 \cdot g_1 + p_3 \cdot p_2 \cdot p_1 \cdot g_0 + p_3 \cdot p_2 \cdot p_1 \cdot p_0 \cdot c_0 \tag{14.6}$$

The resulting carry-in signals are only functions of both inputs x and y, and the primary input carry-in (c_0). These carries can be realized by a two-level logic circuit, referred to as a *carry-lookahead (CLA) generator*, as shown in Figure 14.2. By using this CLA generator, a 4-bit adder may be expressed as a function of the carry propagate signals p_i and carry signals c_i, as indicated by Equation (14.2). The resulting circuit is shown in Figure 14.3.

To model the above 4-bit carry-lookahead adder, **assign** continuous assignments and generate-loop statements may be combined to use. Two illustrations are given in the following two examples: One uses **assign** continuous assignments and the other uses generate-loop statements. As usually, the use of generate-loop statements implies that a parameterizable module is resulted.

Section 14.1 Addition and Subtraction

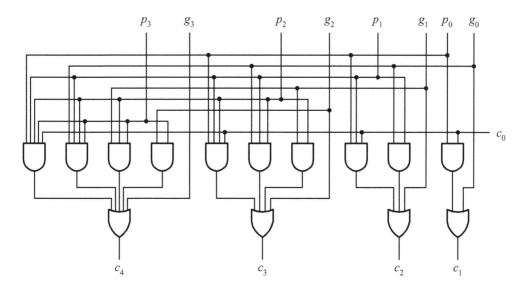

Figure 14.2: The logic diagram of a 4-bit carry-lookahead generator.

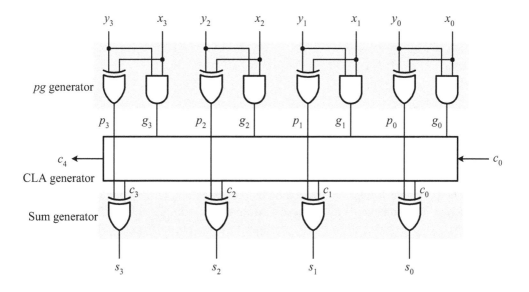

Figure 14.3: The logic diagram of a 4-bit CLA adder.

■ Example 14.1: A 4-bit CLA adder.

In this example, a 4-bit carry-lookahead adder is modeled by using **assign** continuous assignments to describe the logic diagrams shown in Figures 14.2 and 14.3 directly.

```
// a 4-bit CLA adder using assign statements
module cla_adder_4bits(
       input  [3:0] x, y,
       input  cin,
       output [3:0] sum,
       output cout);
// internal wires
```

```verilog
wire p0,g0, p1,g1, p2,g2, p3,g3;
wire c4, c3, c2, c1;
// compute the p for each stage
assign p0 = x[0] ^ y[0],
       p1 = x[1] ^ y[1],
       p2 = x[2] ^ y[2],
       p3 = x[3] ^ y[3];
// compute the g for each stage
assign g0 = x[0] & y[0],
       g1 = x[1] & y[1],
       g2 = x[2] & y[2],
       g3 = x[3] & y[3];
// compute the carry for each stage
// note that c_in is equivalent c0 in the arithmetic equation
// for carry-lookhead computation
assign c1 = g0 | (p0 & cin),
       c2 = g1 | (p1 & g0) | (p1 & p0 & cin),
       c3 = g2 | (p2 & g1) | (p2 & p1 & g0) |
            (p2 & p1 & p0 & cin),
       c4 = g3 | (p3 & g2) | (p3 & p2 & g1) |
            (p3 & p2 & p1 & g0) | (p3 & p2 & p1 & p0 & cin);
// compute the sum
assign sum[0] = p0 ^ cin,
       sum[1] = p1 ^ c1,
       sum[2] = p2 ^ c2,
       sum[3] = p3 ^ c3;
// assign the carry-out
assign cout = c4;
endmodule
```

The use of **assign** continuous assignments to model the CLA adder limits the resulting module to be parameterizable. To overcome this drawback and write a parameterizable CLA adder, generate-loop statements may be used. As an illustration, consider the following example.

■ Example 14.2: A parameterizable CLA adder.

In this example, we rewrite the operations of each **assign** continuous assignment in the preceding example with a generate-loop statement. By properly setting the parameter N, a CLA adder with an arbitrary bits can be obtained.

```verilog
// an N-bit CLA adder using generate loops
module cla_adder_generate #(parameter N=4)(// set the default size
        input   [N-1:0] x, y,
        input   cin,
        output  [N-1:0] sum,
        output  cout);
// internal wires
wire [N-1:0] p, g;
wire [N:0] c;
// assign carry-in
assign c[0] = cin;
```

Section 14.1 *Addition and Subtraction*

```
genvar i;
// compute carry generate and carry propagate functions
generate for (i = 0; i < N; i = i + 1) begin: pq_cla
   assign p[i] = x[i] ^ y[i];
   assign g[i] = x[i] & y[i];
end endgenerate
// compute the carry for each stage
generate for (i = 1; i < N+1; i = i + 1) begin: carry_cla
   assign c[i] = g[i-1] | (p[i-1] & c[i-1]);
end  endgenerate
// compute the sum
generate for (i = 0; i < N; i = i + 1) begin: sum_cla
   assign sum[i] = p[i] ^ c[i];
end endgenerate
// assign the final carry-out
assign cout = c[N];
endmodule
```

14.1.1.1 Block CLA Generators Even though a wide CLA adder can be simply generated by setting the parameter N of the preceding example, the hardware cost of the resulting CLA adder increases considerably. To reduce hardware cost, the concept of hierarchical structure may be used instead. To see this, let us reconsider the 4-bit CLA generator and adder again. The carry-in c_4 can be rewritten as follows

$$c_4 = g_3 + p_3 g_2 + p_3 p_2 g_1 + p_3 p_2 p_1 g_0 + p_3 p_2 p_1 p_0 c_0$$
$$= g_{[3:0]} + c_0 p_{[3:0]} \tag{14.7}$$

where $g_{[3:0]}$ and $p_{[3:0]}$ are called *group-carry generate* and *group-carry propagate*, respectively, and are defined as

$$g_{[3:0]} = g_3 + p_3 g_2 + p_3 p_2 g_1 + p_3 p_2 p_1 g_0$$
$$p_{[3:0]} = p_3 p_2 p_1 p_0 \tag{14.8}$$

More generally, both group-carry generate and group-carry propagate can be redefined as

$$g_{[i+3:i]} = g_{[i+3]} + p_{[i+3]} g_{[i+2]} + p_{[i+3]} p_{[i+2]} g_{[i+1]} + p_{[i+3]} p_{[i+2]} p_{[i+1]} g_{[i]}$$
$$p_{[i+3:i]} = p_{[i+3]} p_{[i+2]} p_{[i+1]} p_{[i]} \tag{14.9}$$

Based on this definition, the group carries can then be expressed in the same way as the carries generated by the 4-bit CLA generator described previously. As illustrations, carries, c_8, c_{12}, and c_{16}, can be expressed as follows

$$c_8 = g_{[7:4]} + g_{[3:0]} p_{[7:4]} + c_0 p_{[3:0]} p_{[7:4]}$$
$$c_{12} = g_{[11:8]} + g_{[7:4]} p_{[11:8]} + g_{[3:0]} p_{[7:4]} p_{[11:8]} + c_0 p_{[3:0]} p_{[7:4]} p_{[11:8]}$$
$$c_{16} = g_{[15:12]} + g_{[11:8]} p_{[15:12]} + g_{[7:4]} p_{[11:8]} p_{[15:12]} +$$
$$g_{[3:0]} p_{[7:4]} p_{[11:8]} p_{[15:12]} + c_0 p_{[3:0]} p_{[7:4]} p_{[11:8]} p_{[15:12]} \tag{14.10}$$

which can then be implemented by the 4-bit CLA generator with a minor modification. The resulting CLA generator is called a *block CLA generator* [3] and is exhibited in Figure 14.4. Through a proper combination of block CLA generators, a wide adder can be implemented.

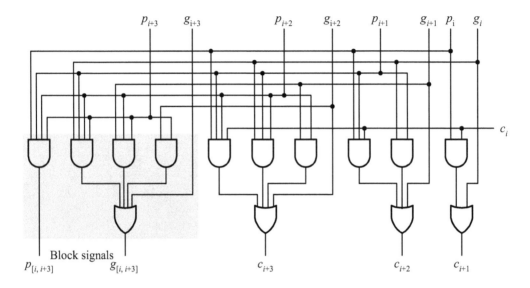

Figure 14.4: The logic circuit of a block CLA generator.

■ Example 14.3: A 16-bit CLA adder.

A 16-bit CLA adder constructed with block CLA generators is shown in Figure 14.5. Each 4-bit group generates the group-carry generate and group-carry propagate signals, and these signals are then combined together by a second-level 4-bit block CLA generator to yield the input carries, c_4, c_8, and c_{12}, for all first-level groups to compute their sums. By repeatedly applying this method, wider adders can be readily built by using more levels of block CLA generators. For instance, a 64-bit CLA adder can be formed by four 16-bit CLA adders along with an extra block CLA generator. The details of this adder is left for the reader as an exercise.

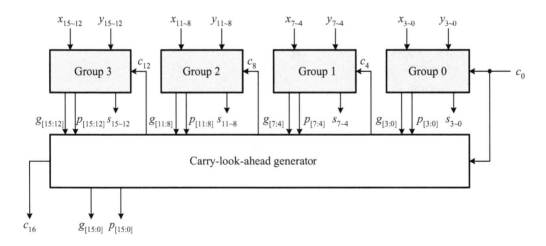

Figure 14.5: The logic block diagram of a 16-bit CLA adder.

14.1.2 Parallel-Prefix Adders

Although in principle the CLA generator can be implemented in two-level logic, in practice it cannot work as fast as we expected. The reason is that in practice the propagation delay of a gate is not independent of the number of its inputs (namely, fanin); instead, it is a strong function of its fanin. As a result, bounded-fanin rather than unbounded-fanin gates are preferred when we design wide-width adders. In this section, we consider an alternative method that is also frequently used to design wide adders. This method is known as the *parallel-prefix approach* [6], which is described in detail by way of two examples in the rest of this section.

14.1.2.1 Principles of Parallel-Prefix Adders Before proceeding, we first define the prefix sum of a sequence of n elements as follows. Consider a sequence of n elements $\{x_{n-1}, \cdots, x_1, x_0\}$ with an associative binary operator, denoted by $\odot$. The *prefix sums* of this sequence are the n partial sums defined by

$$s_{[i,0]} = x_i \odot \cdots \odot x_1 \odot x_0 \tag{14.11}$$

where $0 \leq i < n$.

Let $g_{[i,k]}$ and $p_{[i,k]}$ denote the carry generation and carry propagation functions from bits k to i, respectively. Both $g_{[i,k]}$ and $p_{[i,k]}$ can then be defined recursively as follows

$$g_{[i,k]} = g_{[i,j+1]} + p_{[i,j+1]} g_{[j,k]}$$
$$p_{[i,k]} = p_{[i,j+1]} p_{[j,k]} \tag{14.12}$$

where $0 \leq i < n$, $k \leq j < i$, $0 \leq k < n$, $g_{[i,i]} = x_{[i]} \cdot y_{[i]}$, and $p_{[i,i]} = x_{[i]} \oplus y_{[i]}$. Based on this definition, the carry-in of the $(i+1)$th-bit adder $c_{i+1} = g_i + p_i \cdot c_i$ can be written as $g_{[i,0]} = g_{[i,j+1]} + p_{[i,j+1]} \cdot g_{[j,0]}$.

In order to parallelize the operations, let us group $g_{[i,j]}$ and $p_{[i,j]}$ together as a group and denote it as $w_{[i,j]} = (g_{[i,j]}, p_{[i,j]})$. It is easy to show that the binary operator $\odot$ in the following equation

$$\begin{aligned} w_{[i,k]} &= w_{[i,j+1]} \odot w_{[j,k]} \\ &= (g_{[i,j+1]}, p_{[i,j+1]}) \odot (g_{[j,k]}, p_{[j,k]}) \\ &= (g_{[i,j+1]} + p_{[i,j+1]} g_{[j,k]}, p_{[i,j+1]} p_{[j,k]}) \end{aligned} \tag{14.13}$$

is an associative binary operator, where $k \leq j < n$. The logic diagram used to implement the binary operator $\odot$ is shown in Figure 14.6.

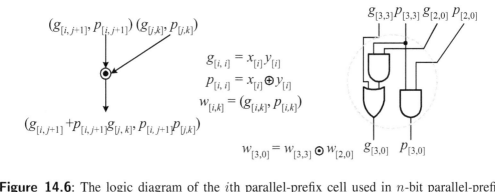

Figure 14.6: The logic diagram of the ith parallel-prefix cell used in n-bit parallel-prefix adders.

■ Example 14.4: The CLA generator as prefix sums.

Let $w_{[i,i]} = (g_{[i,i]}, p_{[i,i]})$, for all $0 \leq i \leq 3$. By the definition of the binary operator $\odot$, the $w_{[i,0]}$, where $0 \leq i \leq 3$, can be calculated in a variety of ways. Some of them are shown below:

$$w_{[1,0]} = w_{[1,1]} \odot w_{[0,0]} = (g_{[1,1]}, p_{[1,1]}) \odot (g_{[0,0]}, p_{[0,0]})$$
$$= (g_{[1,1]} + p_{[1,1]}g_{[0,0]}, p_{[1,1]}p_{[0,0]})$$
$$w_{[3,2]} = w_{[3,3]} \odot w_{[2,2]} = (g_{[3,3]}, p_{[3,3]}) \odot (g_{[2,2]}, p_{[2,2]})$$
$$= (g_{[3,3]} + p_{[3,3]}g_{[2,2]}, p_{[3,3]}p_{[2,2]})$$
$$w_{[2,0]} = w_{[2,2]} \odot w_{[1,0]}$$
$$= (g_{[2,2]}, p_{[2,2]}) \odot (g_{[1,1]} + p_{[1,1]}g_{[0,0]}, p_{[1,1]}p_{[0,0]})$$
$$= (g_{[2,2]} + p_{[2,2]}g_{[1,1]} + p_{[2,2]}p_{[1,1]}g_{[0,0]}, p_{[2,2]}p_{[1,1]}p_{[0,0]})$$
$$w_{[3,0]} = w_{[3,2]} \odot w_{[1,0]}$$
$$= (g_{[3,3]} + p_{[3,3]}g_{[2,2]}, p_{[3,3]}p_{[2,2]}) \odot (g_{[1,1]} + p_{[1,1]}g_{[0,0]}, p_{[1,1]}p_{[0,0]})$$
$$= (g_{[3,3]} + p_{[3,3]}g_{[2,2]} + p_{[3,3]}p_{[2,2]}g_{[1,1]} +$$
$$p_{[3,3]}p_{[2,2]}p_{[1,1]}g_{[0,0]}, p_{[3,3]}p_{[2,2]}p_{[1,1]}p_{[0,0]})$$

As mentioned, $g_{[i,0]}$ ($c_0 = 0$) corresponds to the carry-in c_{i+1} of the $(i+1)$th-bit adder. The reader may compare the above result with the carries derived from the recursive equation c_{i+1}. Note that $g_{[i,i]} = x_i \cdot y_i$ and $p_{[i,i]} = x_i \oplus y_i$. Also, this structure cannot have the primary carry-in, cin.

There are many possible parallel-prefix networks on which fast adders can be based. In what follows, we introduce the two most common networks [8]: the *Kogge-Stone network* and the *Brent-Kung network*. More about parallel-prefix networks can be referred to Lin [6].

14.1.2.2 Kogge-Stone Parallel-Prefix Adders A 16-input Kogge-Stone parallel-prefix network is shown in Figure 14.7, which has a 4-level propagation delay and 49 parallel-prefix cells. In general, an n-input Kogge-Stone parallel-prefix network has a propagation delay of $\log_2 n$ levels and needs $n\log_2 n - n + 1$ cells.

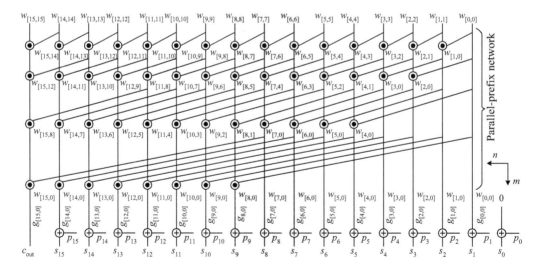

Figure 14.7: The logic diagram of the 16-input Kogge-Stone parallel-prefix network.

Section 14.1 *Addition and Subtraction*

An example showing how to model an n-bit fast adder by using the Kogge-Stone parallel-prefix network is explored in the following example.

■ Example 14.5: The Kogge-Stone parallel-prefix adder.

From the parallel-prefix network shown in Figure 14.7, we know that there are $m = \log_2 n$ levels for an n-input network. Level 0 is the input and level m is the output. Level j starts from the cell with an index of 2^{j-1} and each cell within it combines itself with the 2^{j-1}th previous cell's output. After completing the last-stage computing, the sum can be simply derived using the following two equations:

$sum[0] = p[0][0];$
$sum[i] = g[i-1][m] \oplus p[i][0];$ for $0 < i < n$

```
// an N-bit Kogge-Stone adder using generate loops
module Kogge_Stone_adder
        #(parameter   N = 16, // define the default size
          parameter   M = 4)( // M = log2 N
          input     [N-1:0] x, y,
          output    [N-1:0] sum,
          output    cout);
// internal wires
wire    p[N-1:0][M:0];
wire    g[N-1:0][M:0];

// generate variables
genvar i, j;
// compute carry generate and carry propagate functions
generate for (i = 0; i < N; i = i + 1) begin: pg_KoggeStone
   assign p[i][0] = x[i] ^ y[i];
   assign g[i][0] = x[i] & y[i];
end endgenerate
// compute the carry for each stage
generate for (j = 1; j <= M; j = j + 1) begin: carry_prefix
   for (i = 0 ; i < N; i = i + 1) begin: parallel_prefix
      if (i < 2**(j-1)) begin
         assign p[i][j] = p[i][j-1];
         assign g[i][j] = g[i][j-1]; end
      else begin
         assign p[i][j] = p[i][j-1] & p[i-2**(j-1)][j-1];
         assign g[i][j] = g[i][j-1] | (p[i][j-1] &
                          g[i-2**(j-1)][j-1]); end
   end
end   endgenerate
// compute the sum and the final-stage carry-out
generate for (i = 1; i < N; i = i + 1) begin: sum_KoggeStone
   assign sum[i] = g[i-1][M] ^ p[i][0];
end endgenerate
assign sum[0] = p[0][0];
assign cout = g[N-1][M];
endmodule
```

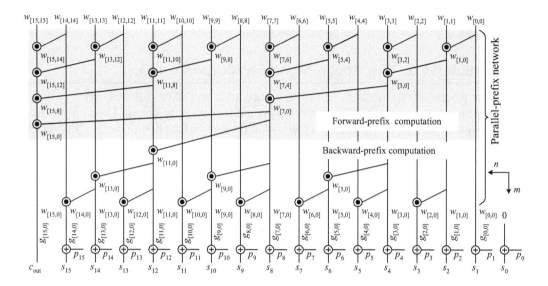

Figure 14.8: The logic diagram of the 16-input Brent-Kung parallel-prefix network.

14.1.2.3 Brent-Kung Parallel-Prefix Adders A 16-input Brent-Kung parallel-prefix network is shown in Figure 14.8, which has a 6-level propagation delay and 26 parallel-prefix cells. Generally, an n-input Brent-Kung parallel-prefix network has a propagation delay of $2\log_2 n - 2$ levels and a cost of $2n - 2 - \log_2 n$ parallel-prefix cells.

An illustration showing how to model an n-bit fast adder by using the Brent-Kung parallel-prefix network is explored in the following example.

■ Example 14.6: The Brent-Kung parallel-prefix adder.

Although from Figure 14.8 there are only $2m - 2 = 2\log_2 n - 2$ levels for an n-input network, we use $2m - 1$ levels, where m and $m - 1$ levels are used for forward- and backward-prefix computations, respectively. Level 0 is the input and level m is the output. In the forward-prefix computation, the cell with index $2^j - 1$ at level j is combined with its 2^{j-1}th previous cell's output; otherwise, it keeps the value from its previous level. In the backward-prefix computation, the cells with indices that have $m - j$ lower bits equal to $2^{m-j-1} - 1$ at level j are combined with their $(i - 2^{m-j-1})$th previous cell if the value of $i - 2^{m-j-1}$ is greater than 0; otherwise, they keep the values from their previous level.

```
// an N-bit Brent-Kung adder using generate loops
module Brent_Kung_adder
       #(parameter  N = 16, // define the default size
         parameter  M = 4)( // M = log2 N
         input   [N-1:0] x, y,
         output  [N-1:0] sum,
         output  cout);
// internal wires
wire    p[N-1:0][2*M-1:0];
wire    g[N-1:0][2*M-1:0];
// generate variables
genvar i, j;
// compute carry generate and carry propagate functions
generate for (i = 0; i < N; i = i + 1)begin:pg_BrentKung
```

```
      assign p[i][0] = x[i] ^ y[i];
      assign g[i][0] = x[i] & y[i];
end endgenerate
// compute the carry for each stage
generate for (j = 1; j <= M; j = j + 1)begin:forward_prefix
   for (i = 0; i < N; i = i + 1) begin: prefix_a
      if ((i & {j{1'b1}}) != (2**j - 1)) begin
         assign p[i][j] = p[i][j-1]; // copy all ps and gs
         assign g[i][j] = g[i][j-1]; end
      else begin
         assign p[i][j] = p[i][j-1] & p[i-2**(j-1)][j-1];
         assign g[i][j] = g[i][j-1] | (p[i][j-1] &
               g[i-2**(j-1)][j-1]); end
   end
end endgenerate
generate for (j = 1; j < M; j = j + 1)begin:backward_prefix
   for (i = N - 1; i >= 0; i = i - 1) begin: prefix_b
      if (((i & {(M-j){1'b1}}) == (2**(M-j-1)-1)) &&
         (i-2**(M-j-1)) > 0) begin
         assign p[i][M+j] = p[i][M+j-1] &
                     p[i-2**(M-j-1)][M+j-1];
         assign g[i][M+j] = g[i][M+j-1]|(p[i][M+j-1] &
                     g[i-2**(M-j-1)][M+j-1]); end
      else begin
         assign p[i][M+j] = p[i][M+j-1];// copy all ps and gs
         assign g[i][M+j] = g[i][M+j-1]; end
   end
end endgenerate
// compute the sum and the final-stage carry-out
generate for (i = 1; i < N; i = i + 1) begin: sum_BrentKung
   assign sum[i] = g[i-1][2*M-1] ^ p[i][0];
end endgenerate
assign sum[0] = p[0][0];
assign cout = g[N-1][2*M-1];
endmodule
```

■ Review Questions

14-1 Explain the meanings of carry generate and carry propagate functions.
14-2 What are the propagation delay and hardware cost of an n-bit CLA adder.
14-3 What are the propagation delay and hardware cost of an n-bit Kogge-Stone adder.
14-4 What are the propagation delay and hardware cost of an n-bit Brent-Kung adder.

14.2 Multiplication

The basic operations of the multiplication algorithms considered in this chapter are based on the shift-and-add (subtract) technique and may process unsigned and signed numbers. A multiplication algorithm may be usually realized by using either multicycle or single-cycle structure. The multicycle structure is sequential in nature. The single-cycle structure naturally evolves into array structure, such as unsigned array multipliers and modified Baugh-Wooley signed array multipliers. These two types of array multipliers are to be discussed in great depth in this section.

14.2.1 Unsigned Multiplication

Basically, the principle of multiplication in hardware is exactly like the "pencil-and-paper" method that we use daily. This method is often called the *shift-and-add approach*. In this section, we first present the approach and then derive an array multiplier based on it. In addition, two variants, including an $(\frac{n}{k})$-cycle structure and a pipeline structure, of the basic array multiplier are addressed in detail.

14.2.1.1 Shift-and-Add Multiplication The basic operation of shift-and-add multiplication is based on the following observation. The rule for the multiplication of a multiple-bit multiplicand and a 1-bit multiplier is as follows:

- The partial product is the same as the multiplicand if the multiplier is 1 and is 0 otherwise.

As the multiplier is also a multiple-bit number, the above rule is applied to each individual bit of the multiplier, and then all partial products are added together accordingly to obtain the final product. The detailed operations are summarized as the following algorithm.

■ Algorithm 14-1: Shift-and-add multiplication

Input: An m-bit multiplicand and an n-bit multiplier.
Output: The $(m+n)$-bit product.
Begin

1. Load the multiplicand and multiplier into registers M and Q, respectively; clear register A and set loop count CNT equal to n.
2. **repeat**

    2.1 if $(Q[0] == 1)$ **then** $A = A + M$;

    2.2 Shift register pair $A:Q$ right one bit;

    2.3 $CNT = CNT - 1$;

    until $(CNT == 0)$;

End

A multicycle hardware structure, i.e., a sequential implementation, that directly realizes the shift-and-add multiplication algorithm is shown in Figure 14.9. The detailed ASM chart and Verilog HDL module are left for the reader as exercises. Another consequence of the above shift-and-add algorithm is that it can be readily adapted to perform the bit-serial multiplication in which the multiplier is input in bit-serial form. Since this multiplier is intuitively simple, its implementation details are also left for the reader as an exercise.

14.2.1.2 Array Multipliers As introduced earlier, multicycle structure can also be implemented by using iterative logic structure. To illustrate this, consider the 4×4 multiplication example shown in Figure 14.10, which can be thought of as the case that we expand the shift-and-add multiplication in the temporal dimension and arrange all partial products in accordance with their proper positions without doing any summation. Each partial product is simply generated by ANDing an appropriate bit of the multiplier with the multiplicand. For an $m \times n$ multiplier, all partial products can be yielded with $m \times n$ two-input AND gates at the same time.

Section 14.2 Multiplication

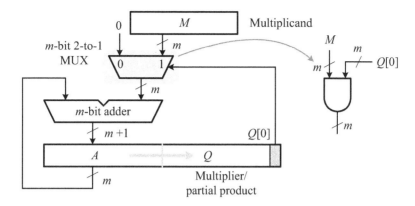

Figure 14.9: The sequential implementation of the shift-and-add multiplication algorithm.

$$
\begin{array}{r}
x_3 \; x_2 \; x_1 \; x_0 = X \text{ (multiplicand)} \\
\times \; y_3 \; y_2 \; y_1 \; y_0 = Y \text{ (multiplier)} \\
\hline
x_3y_0 \; x_2y_0 \; x_1y_0 \; x_0y_0 \\
x_3y_1 \; x_2y_1 \; x_1y_1 \; x_0y_1 \\
x_3y_2 \; x_2y_2 \; x_1y_2 \; x_0y_2 \\
+ \; x_3y_3 \; x_2y_3 \; x_1y_3 \; x_0y_3 \\
\hline
P_6 \; P_5 \; P_4 \; P_3 \; P_2 \; P_1 \; P_0 \quad \text{Product}
\end{array}
$$

Partial product

Figure 14.10: An example illustrating the idea of array multipliers.

After all partial products are generated, the next step is to add them together to obtain the final product. Depending on how to add up partial products, a number of algorithms are resulted. In what follows, we only consider two basic types of multipliers: *RCA array multipliers* and *CSA array multipliers*. Both types are based on the straightforward method and use full adders but with a slightly different concept. RCA array multipliers use *ripple-carry adders* (RCAs, also called *carry-ripple adders*) and CSA array multipliers use *carry-save adders* (CSAs). RCAs (along with CLA adders) are often called *carry-propagate adders* (CPAs) because in these adders the carry-out from the lower-order bit is added to its next higher-order bit in contrast to carry-save adders in which the carry-out from each bit is left as another output just like the sum without being added into its next higher-order bit.

RCA array multipliers. To illustrate the structure of RCA array multipliers, consider the 4×4 RCA array multiplier shown in Figure 14.11. In the array, each cell corresponds to an item in Figure 14.10, and consists of an AND gate that generates a partial product and a full adder to add the partial product into the running sum. All cells are arranged in the same way as shown in Figure 14.10. In practical applications, we often need a parameterizable array multiplier so that we may generate an array multiplier with the desired size on demand. As an illustration of how to describe an $m \times n$ array multiplier using generate-loop statements, consider the following example.

■ Example 14.7: An m×n RCA array multiplier.

Since the multiplication structure is a 2-D array, we need a nested generate-loop structure. To memorize internal temporary net values, two 2-D arrays, `sum` and `carry`, are declared.

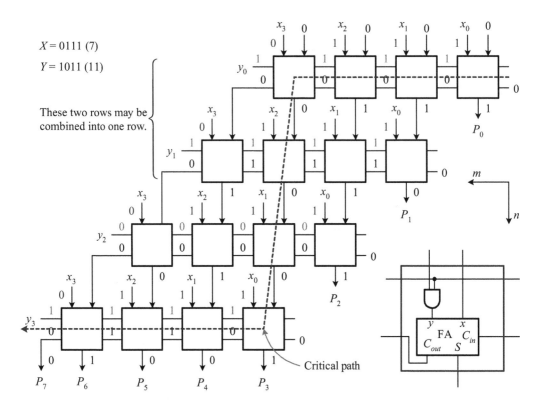

Figure 14.11: The logic diagram of a 4 × 4 RCA array multiplier.

Because each cell of the array multiplier only consists of a full adder and an AND gate, it is simply described by an **assign** continuous assignment. Like iterative logic, this array multiplier has two types of cells: boundary cells and internal cells. The boundary cells include all cells of the first row as well as each LSB cell and each MSB cell of the other rows. After carefully considering these cells, the remaining cells can be easily described.

```
// an M-by-N unsigned array multiplier using generate loops
module unsigned_array_multiplier_generate
       #(parameter M = 4,
         parameter N = 4)(
         input   [M-1:0] x,
         input   [N-1:0] y,
         output  [M+N-1:0] product);
// internal wires
wire sum   [M-1:0][N-1:0]; // declare internal nets
wire carry [M-1:0][N-1:0];
genvar i, j;
generate for (i = 0; i < N; i = i + 1) begin: multiplier
   if (i == 0)  // the first row
      for (j = 0; j < M; j = j + 1) begin: first_row
         assign sum[j][i] = x[j] & y[i], carry[j][i]=0; end
   else          // the remaining rows
      for (j = 0; j < M; j = j + 1) begin: rest_rows
         if (j == 0) // the LSB of each row
```

Section 14.2 *Multiplication*

```
                assign {carry[j][i],sum[j][i]} =
                    sum[j+1][i−1] + (x[j]&y[i]);
            else if (j == M − 1) // the MSB of each row
                assign {carry[j][i],sum[j][i]}=
                    (x[j]&y[i]) + carry[M−1][i−1] +
                    carry[j−1][i];
            else      // the other bits of each row
                assign {carry[j][i],sum[j][i]}=
                    (x[j]&y[i])+sum[j+1][i−1]+carry[j−1][i];
        end
end endgenerate
// generate product bits
generate for (i = 0; i < N ; i = i + 1) begin: lower_product
    assign product[i] = sum[0][i];
end endgenerate
generate for (i = 1; i < M ; i = i + 1) begin: higher_product
    assign product[N−1+i] = sum[i][N−1];
end endgenerate
    assign product[M+N−1] = carry[M−1][N−1];
endmodule
```

One disadvantage of RCA array multipliers is the long propagation delay associated with it. To see this, consider the critical path depicted by the dashed line of the 4×4 RCA array multiplier shown in Figure 14.11. The propagation delay is $[2(4-1)+4]t_{FA} = 10 t_{FA}$. In general, for an $m \times n$ RCA array multiplier, the propagation delay of the critical path is $[2(m-1)+n]t_{FA}$ when the first two rows are not combined and is $[2(m-1)+(n-1)]t_{FA}$ when the first two rows are combined, where t_{FA} is the propagation delay of a full adder. The hardware cost of an $m \times n$ array multiplier is as follows. It needs $m \times n$ AND gates, and $m \times n$ full adders when the first two rows are not combined and $m \times (n-1)$ full adders when the first two rows are combined.

CSA array multipliers. Another type of widely used array multiplier uses carry-save adders (CSAs) to replace the RCAs used in the RCA array multipliers. The resulting structure is known as the *CSA array multiplier*. A CSA is simply an n-bit full adder placed in parallel. It is similar to a RCA but simply leaves the carry-out of each cell to the output rather than inputs to the next higher significant bit cell. As a result, an n-bit CSA accepts three n-bit inputs and generates two n-bit outputs: carry and the sum. To combine these two n-bit outputs into an $(n+1)$-bit result, a CPA (RCA or CLA adder) is needed to add up both the n-bit carry-out and the sum. As an illustration of CSA array multipliers, consider the 4×4 CSA unsigned array multiplier shown in Figure 14.12. Like RCA array multipliers, we often need a parameterizable $m \times n$ CSA array multiplier in practical applications. Since it has the similar structure as the parameterizable $m \times n$ RCA array multiplier, we would like to leave its description in Verilog HDL for the reader as an exercise.

The critical path of an $m \times n$ CSA array multiplier is not too hard to be figured out. As an illustration of this, consider the 4×4 CSA array multiplier shown in Figure 14.12. The critical path of this array multiplier is marked by the dashed line depicted in this figure. The propagation delay of this critical path is $(4+4)t_{FA} = 8 t_{FA}$. In general, the critical path of an $m \times n$ CSA array multiplier is composed of n or $(n-1)$ CSA cells, depending on whether the first two rows are combined, and an m-bit CPA (assuming that a RCA is used), namely, $(m+n)$ or $(m+n-1)$ full adders in total. That is, the critical path has the propagation delay of $(m+n)t_{FA}$, when the first two rows are not combined and of $[m+(n-1)]t_{FA}$, when the first two rows are combined. Compared to the $m \times n$ RCA array multiplier, the $m \times n$ CSA

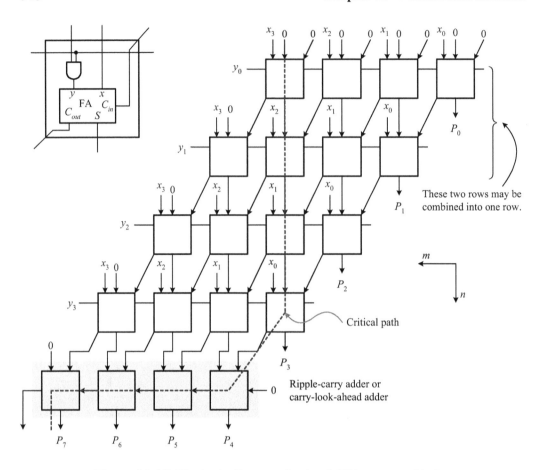

Figure 14.12: The logic diagram of a 4 × 4 CSA array multiplier.

array multiplier takes less propagation delay by a factor of $(m-2)$ at the cost of m full adders, namely, an m-bit RCA.

■ **Review Questions**

14-5 Explain the operations of Figure 14.9.
14-6 What are the features of RCAs and CLA adders?
14-7 Why are both RCAs and CLA adders called CPAs?
14-8 Define an n-bit CSA.
14-9 What are the differences between RCAs and CSAs?

14.2.1.3 Variants of Array Multipliers Now that we have considered both the multicycle (sequential) and single-cycle (array) structures of the multiplication algorithm based on the shift-and-add approach, in what follows we first explore a compromised approach known as an $\left(\frac{n}{k}\right)$-*cycle structure* and then describe how to pipeline the single-cycle structure by using the RCA array multiplier as an example.

$\left(\frac{n}{k}\right)$-**cycle multiplication structure.** The general $\left(\frac{n}{k}\right)$-cycle structure of $m \times n$ multiplication is shown in Figure 14.13. Compared to the circuit shown in Figure 14.9, an $m \times k$ array multiplier is used here instead of the trivial $m \times 1$ array multiplier, which is actually an m-bit adder with the multiplicand input ANDed by the proper bit of the multiplier. The operation of

Section 14.2 Multiplication

Figure 14.13 is similar to that of Figure 14.9 except that now it processes k bits rather than a single bit each time. Hence, the accumulator (A) concatenated with the multiplier register (Q) is shifted right k-bit positions after each time the partial product is loaded into the accumulator. The multiplication is finished after $(\frac{n}{k})$ steps, with each step needing two clock cycles, one for the computation of the $m \times k$ partial product and loading into the accumulator and the other for the k-bit right shift. In fact, these two steps can be combined into one step and hence only one clock cycle is needed. It is worth noting that the $(\frac{n}{k})$-cycle structure is reduced to the multicycle structure depicted in Figure 14.9 when k is set to 1 and reduced to the array structure when k is set to n. This means a trade-off existing between the hardware cost and performance of the multiplier.

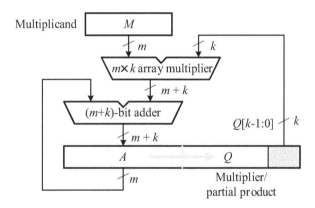

Figure 14.13: An $(\frac{n}{k})$-cycle structure for an $m \times n$ unsigned multiplier.

The hardware required for this $(\frac{n}{k})$-cycle structure can be calculated as follows. In addition to three registers, an m-bit multiplicand register, an $(m+k)$-bit accumulator, and an n-bit multiplier/partial product register, there is an $m \times k$ array multiplier which requires $m \times k$ full adders and AND gates. The total running time of this structure is $2(\frac{n}{k})t_{clk}$, without considering various housekeeping operations, such as clearing the accumulator as well as loading the multiplicand and the multiplier into their individual registers, where t_{clk} is determined by the propagation delay of the $m \times k$ array multiplier as well as the t_{cq} and t_{setup} of the registers used. Hence, it needs $2(\frac{n}{k})\{[2(m-1)+k]t_{FA} + t_{cq} + t_{setup}\}$. The constant factor 2 counts both the computation of the $m \times k$ partial product as well as the right-shift k-bit positions of the multiplier register and the accumulator, as it can be seen from Figure 14.13. Certainly, this factor can be eliminated through a more careful design.

Pipeline multiplication structure. Another variant of the array multiplier structure is the pipeline multiplication structure. The general pipeline structure for $m \times n$ unsigned multipliers is shown in Figure 14.14. Basically, this pipeline structure can be regarded either as the expansion of the $(\frac{n}{k})$-cycle multiplier structure in temporal dimension or as the partition of an $m \times n$ unsigned array multiplier into $(\frac{n}{k})$ stages of $m \times k$ unsigned array multipliers with pipeline registers being inserted after each stage.

As in the $(\frac{n}{k})$-cycle structure, the value of k in the pipeline structure may affect both the pipeline clock period, T_i, and the pipeline latency. Hence, its value should be chosen very carefully in accordance with the actual requirements. The pipeline clock period is determined by the propagation delay of the $m \times k$ unsigned array multiplier and the t_{cq} and t_{setup} of the registers used. That is, T_i is equal to $[2(m-1)+k]t_{FA} + t_{cq} + t_{setup}$. The hardware required in the pipeline multiplier structure is calculated as follows. In addition to the $m \times n$ full adders and AND gates. Three types of registers are required. They are the $(\frac{n}{k} - 1)$ m-bit multiplicand register, the $\frac{1}{2}(\frac{n}{k} - 1)$ n-bit multiplier register, and the partial product register of

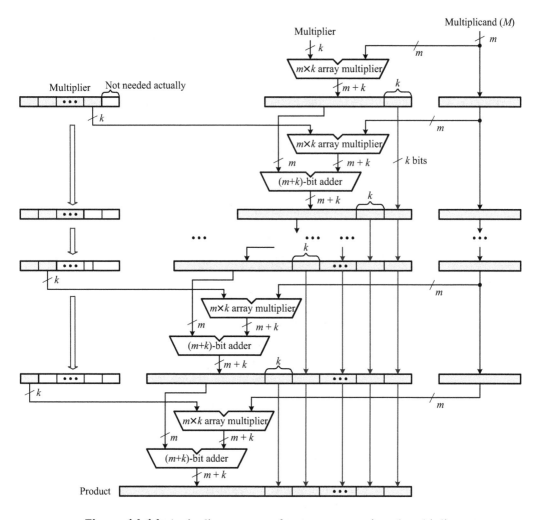

Figure 14.14: A pipeline structure for an $m \times n$ unsigned multiplier.

$\left(\frac{n}{k}\right)(m+k) + \frac{1}{2}\left(\frac{n}{k}-1\right) \times n$ bits. The total number of register bits required is then equal to $\left(\frac{n}{k}\right)(m+k) + \left(\frac{n}{k}-1\right)(m+n)$ bits.

■ Review Questions

14-10 Design a shift register with the capability of shifting its content right k-bit positions in one clock cycle, where k is a constant.

14-11 Explain the operation of the $\left(\frac{n}{k}\right)$-cycle structure of the $m \times n$ unsigned multiplier shown in Figure 14.13.

14-12 Show that the total number of register bits required in the pipeline structure of an $m \times n$ unsigned multiplier is $\left(\frac{n}{k}\right)(m+k) + \left(\frac{n}{k}-1\right)(m+n)$ bits.

14.2.2 Signed Multiplication

Like unsigned multiplication, signed multiplication is also widely used in digital systems. There are many algorithms that can be used to implement signed 2's-complement multiplication. In this book, we only consider the two most common approaches: the *Booth algorithm*

Section 14.2 *Multiplication*

$$
\begin{array}{rrrrrrrrr}
 & & & & x_3 & x_2 & x_1 & x_0 & = X \text{ (multiplicand)} \\
 & & & \times & y_3 & y_2 & y_1 & y_0 & = Y \text{ (multiplier)} \\
\hline
 & & & 1 & \overline{y_3x_0} & x_2y_0 & x_1y_0 & x_0y_0 & \\
 & & & & \overline{y_3x_1} & x_2y_1 & x_1y_1 & x_0y_1 & \\
 & & & & \overline{y_3x_2} & x_2y_2 & x_1y_2 & x_0y_2 & \\
+ & & 1 & y_3x_3 & \overline{y_2x_3} & \overline{y_1x_3} & \overline{y_0x_3} & & \\
\hline
 & P_7 & P_6 & P_5 & P_4 & P_3 & P_2 & P_1 & P_0 \quad \text{Product}
\end{array}
$$

Figure 14.15: The operation of a 4×4 modified Baugh-Wooley array multiplier.

and the *modified Baugh-Wooley algorithm*. The Booth algorithm has been dealt with in detail in Sections 11.2.2 and 11.3.4. Hence, in what follows, we only focus on the modified Baugh-Wooley algorithm.

14.2.2.1 A Modified Baugh-Wooley Algorithm In order to illustrate the operation of the modified Baugh-Wooley algorithm, let the multiplicand and the multiplier be $X = (x_{m-1} \cdots x_1 x_0)$ and $Y = (y_{n-1} \cdots y_1 y_0)$, respectively. Recall that the most significant bit (MSB) of a number has a negative weight in the 2's-complement representation. Based on this property, the two numbers X and Y can be represented as follows

$$X = -x_{m-1}2^{m-1} + \sum_{i=0}^{m-2} x_i 2^i,$$

$$Y = -y_{n-1}2^{n-1} + \sum_{j=0}^{n-2} y_j 2^j,$$

and the product of X and Y can then be expressed as

$$
\begin{aligned}
P = XY &= \left(-x_{m-1}2^{m-1} + \sum_{i=0}^{m-2} x_i 2^i\right)\left(-y_{n-1}2^{n-1} + \sum_{j=0}^{n-2} y_j 2^j\right) \\
&= \sum_{i=0}^{m-2}\sum_{j=0}^{n-2} x_i y_j 2^{i+j} + x_{m-1}y_{n-1}2^{m+n-2} - \\
&\quad \left(\sum_{i=0}^{m-2} x_i y_{n-1} 2^{i+n-1} + \sum_{j=0}^{n-2} x_{m-1} y_j 2^{j+m-1}\right)
\end{aligned}
\tag{14.14}
$$

An illustration of the modified Baugh-Wooley algorithm using a 4×4 example is shown in Figure 14.15. An array implementation of the modified Baugh-Wooley algorithm is depicted in Figure 14.16. It is easy to see that two kinds of cells are required in the array. Some cells use NAND gates instead of AND gates to generate partial products. By comparing both Figures 14.16 and 14.11, we can see that the unsigned array and the modified Baugh-Wooley array are so similar that a single array can be possibly used for both purposes if XOR gates are used to conditionally complement some terms, which are dependent on the operation mode.

In what follows, we only give an example to show how to design a parameterizable 2's-complement array multiplier based on the modified Baugh-Wooley algorithm. The problem

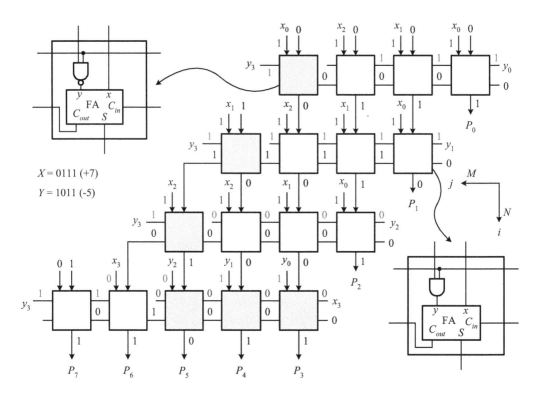

Figure 14.16: The block diagram of a 4 × 4 modified Baugh-Wooley array multiplier.

of using a single array for performing both unsigned and signed multiplication is left for the reader as an exercise.

■ Example 14.8: An m×n modified Baugh-Wooley multiplier.

Like the unsigned array multiplier, the modified Baugh-Wooley array multiplier also needs $m \times n$ cells. As shown in Figure 14.16, the structure of the modified Baugh-Wooley array multiplier is much similar to that of the unsigned array multiplier except that it requires two types of cells: NAND and AND. The shaded cells use NAND gates rather than AND gates to produce partial products. By properly coding these cells as special cases, a parameterizable module is obtained.

```
// an M-by-N 2's-complement array multiplier
module modified_Baugh_Wooley
        #(parameter M = 4, N = 4)(
           input   [M-1:0] x,
           input   [N-1:0] y,
           output  [M+N-1:0] product);
// internal wires
wire sum   [M-1:0][N-1:0]; // declare internal nets
wire carry [M-1:0][N-1:0];
genvar i, j;
generate for (i = 0; i < N; i = i + 1) begin: multiplier
   if (i == 0)  // describe the first row
      for (j = 0; j < M; j = j + 1) begin: first_row
         if (j == M - 1)
```

Section 14.2 *Multiplication*

```
              assign sum[j][i]=!(x[i]&y[N-1]),carry[j][i]=1;
          else
              assign sum[j][i]=x[j]&y[i],carry[j][i]=0; end
       else if (i == N - 1) // describe the last rows
          for (j = 0; j < M; j = j + 1) begin: last_row
              if (j == M - 1)
                 assign {carry[j][i],sum[j][i]}=(x[M-1]&y[N-1])+
                         carry[j][i-1] + carry[j-1][i];
              else if (j == 0)
                 assign {carry[j][i],sum[j][i]} = !(x[M-1]& y[j]) +
                         sum[j+1][i-1];
              else
                 assign {carry[j][i],sum[j][i]} = !(x[M-1]& y[j]) +
                         sum[j+1][i-1] + carry[j-1][i]; end
       else                         // the other rows
          for (j = 0; j < M; j = j + 1) begin: rest_rows
              if (j == 0) // the LSB of each row
                 assign {carry[j][i],sum[j][i]} = (x[j]& y[i]) +
                         sum[j+1][i-1];
              else if (j == M - 1) // the MSB of each row
                 assign {carry[j][i],sum[j][i]} = !(x[i]& y[N-1])+
                         carry[M-1][i-1]+ carry[j-1][i];
              else // the other bits of each row
                 assign {carry[j][i],sum[j][i]}= (x[j]& y[i])+
                         sum[j+1][i-1] + carry[j-1][i]; end
end endgenerate
// generate product bits
generate for (i = 0; i < N ; i = i + 1) begin: product_lower
    assign product[i] = sum[0][i];
end endgenerate
generate for (i = 1; i < M ; i = i + 1) begin: product_upper
    assign product[N-1+i] = sum[i][N-1];
end endgenerate
    assign product[M+N-1] = carry[M-1][N-1] + 1'b1;
endmodule
```

Unlike the unsigned array multiplier and the Booth array multiplier being able to accommodate the situation that $m \neq n$, the above-discussed modified Baugh-Wooley array multiplier can only work correctly when $m = n$. However, if some proper modifications on the array shown in Figure 14.16 were made, the modified Baugh-Wooley algorithm could also work well for any m and n [8]. Please examine the example given in Figure 14.15 very carefully and explore how the inverted partial products are generated, especially, in the case of $m \neq n$. The detailed operations when $m \neq n$, along with its Verilog HDL module, are left as exercises for the reader.

14.2.2.2 Mixing Unsigned with Signed Multipliers In many applications, such as microprocessors, providing both signed and unsigned multipliers on the same system is often necessary. Depending on how these multipliers are used, many different schemes are possible. If both types of multipliers are only used exclusively, we may then design a multiplier capable of carrying out both signed and unsigned multiplication under the control of a mode selection signal. One possible candidate for such a scheme is the aforementioned scheme, a combination of the unsigned array multiplier and the Baugh-Wooley array multiplier. Another possible scheme is to use an $[(m+1) \times (n+1)]$-bit Booth multiplier to serve as the desired $(m \times n)$-bit unsigned and signed multiplier. Using this scheme, the unsigned multiplication is carried out simply by setting the highest bit to 0 and the signed multiplication is performed by extending the sign bit to the highest bit. Then, the Booth algorithm is performed and the lowest $(m+n)$-bit result is the desired product.

■ **Review Questions**

14-13 Compare the unsigned array multiplier with the modified Baugh-Wooley array multiplier in terms of the cost and propagation delay.

14-14 How many inverted partial product terms are needed in an $m \times m$ modified Baugh-Wooley array multiplier?

14-15 How many AND gates are required in an $m \times n$ unsigned array multiplier?

14-16 How many full adders are needed in an $m \times n$ unsigned array multiplier?

14.3 Division

The essential operations of multiplication are a sequence of additions. In contrast, the essential operations of division are a sequence of subtractions. Based on this idea, we introduce in this section two basic division algorithms known as the *restoring division algorithm* and the *non-restoring division algorithm*, respectively. For simplicity, we only consider the case of unsigned input numbers. Like multiplication algorithms, a division algorithm may also be realized by using either multicycle or single-cycle structure.

14.3.1 Restoring Division Algorithm

When division is performed in the binary system, we generally compare the magnitude of both the dividend and the divisor. The quotient bit is set to 1 and the divisor is subtracted from the dividend if the dividend is greater than or equal to the divisor; otherwise, the quotient bit is cleared to 0 and the next bit is proceeded. In digital systems, the comparison is usually carried out by subtraction to simplify the circuit design.

Based on the above idea, the basic design techniques of division circuits can be classified into two types: the restoring division method and the non-restoring division method. In the restoring division method, the quotient bit is set to 1 whenever the result after the divisor is subtracted from the dividend is greater than or equal to 0; otherwise, the quotient bit is cleared to 0 and the divisor is added back to the dividend to restore the original dividend before proceeding the next bit. This process is repeated until the required number of times is reached. The detailed operations of restoring division can be summarized as the following algorithm.

■ **Algorithm 14-1: Unsigned restoring division**

Input: An n-bit dividend and an m-bit divisor.
Output: The quotient and remainder.
Begin

1. Load divisor and dividend into registers M and D, respectively; clear partial-remainder register R and set loop count CNT equal to $n-1$.
2. Shift register pair $R:D$ left one bit.
3. **repeat**

Section 14.3 *Division*

    3.1 Compute $R = R - M$;

    3.2 if $(R < 0)$ **begin** $D[0] = 0$; $R = R + M$; **end**
        else $D[0] = 1$;

    3.3 Shift register pair $R : D$ left one bit;

    3.4 $CNT = CNT - 1$;

  until $(CNT == 0)$

End

14.3.2 Non-Restoring Division Algorithm

In the restoring division method, the divisor (M) is needed to add back to the dividend (D) in order to restore the dividend to its initial value before the subtraction is performed again for the next bit whenever $D - M < 0$. As a result, it requires $\frac{3}{2}n$ m-bit additions and subtractions on average to complete an n-bit division operation.

In the non-restoring division method, the divisor is not added back to the dividend if $D - M < 0$. Instead, the result, $D - M$, is shifted left one bit (corresponding to shifting M right one bit) and then added to the divisor, M. The result is the same as that the divisor (M) is added back to the dividend (D) and then performs the next subtraction. To see this, let $X = D - M$ and assume that $X < 0$. In the restoring division method, the divisor is added back to X and then the result is shifted left one bit and subtracted by M, namely, $2(X + M) - M = 2X + M$, which is equivalent to shifting X left one bit and then being added with the divisor, M. Hence, both approaches yield the same result. However, the non-restoring division method needs only n m-bit additions and subtractions. The detailed operations of the non-restoring division method are summarized as the following algorithm.

■ Algorithm 14-1: Unsigned non-restoring division

Input: An n-bit dividend and an m-bit divisor.
Output: The quotient and remainder.
Begin

1. Load divisor and dividend into registers M and D, respectively;
    clear partial-remainder register R and set loop count CNT equal to $n - 1$.
2. Shift register pair $R : D$ left one bit.
3. Compute $R = R - M$;
4. **repeat**

    4.1 if $(R < 0)$ **begin** $D[0] = 0$; shift register pair $R : D$ left one bit; $R = R + M$; **end**
        else begin $D[0] = 1$; shift register pair $R : D$ left one bit; $R = R - M$; **end**

    4.2 $CNT = CNT - 1$;

    until $(CNT == 0)$

5. **if** $(R < 0)$ **begin** $D[0] = 0$; $R = R + M$; **end**
    else $D[0] = 1$;

End

A numerical example to illustrate the operations of the non-restoring division algorithm is shown in Figure 14.17. The reader is encouraged to trace the operations of this example in accordance with the detailed steps described in the algorithm.

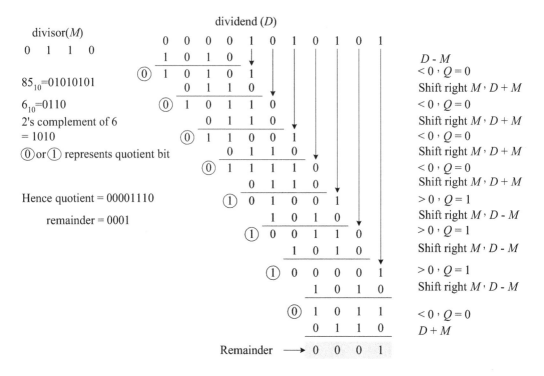

Figure 14.17: A numerical example of illustrating the operations of unsigned non-restoring division.

Like other arithmetic operations dealt with in this chapter, the non-restoring division algorithm can also be implemented by using either multicycle or single-cycle structure. A basic multicycle structure for the non-restoring division algorithm is depicted in Figure 14.18. It consists of an m-bit 2's-complement adder, two m-bit registers, including M for storing the divisor, R for the partial remainder, and one n-bit register D for the dividend and quotient register. The partial remainder R and register D also form an $(m+n)$-bit left-shift register. At each round, the quotient bit is inserted into $D[0]$. After n cycles, the division operation is completed. The detailed ASM chart and its related Verilog HDL module can be easily derived from the non-restoring division algorithm and the logic diagram depicted in Figure 14.18. Consequently, they are left for the reader as exercises.

14.3.3 Non-Restoring Array Dividers

As mentioned previously, an array (single-cycle) structure can be obtained by expanding its multicycle structure in the temporal dimension. Based on this idea, we obtain a 4-by-4 unsigned array divider, as shown in Figure 14.19. The basic limitations of this array divider are as follows: First, it does not check the divide-by-0 case. Second, the MSB of the divisor must be set to 0. Third, the bit length of the dividend must not be less than that of the divisor.

An illustration shows how to describe an $m \times n$ unsigned non-restoring array divider based on the algorithm described above using generate-loop statements is explored in the following example.

■ **Example 14.9: A non-restoring array divider.**

As shown in Figure 14.19, only one boundary condition and the remainder correction step are required to specially consider. In the first row, one operand is the divisor and the other operand

Section 14.3 *Division*

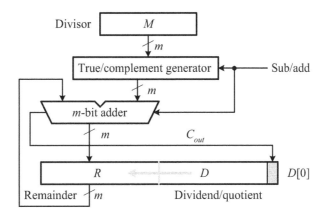

Figure 14.18: The logic diagram of a sequential implementation of unsigned non-restoring division.

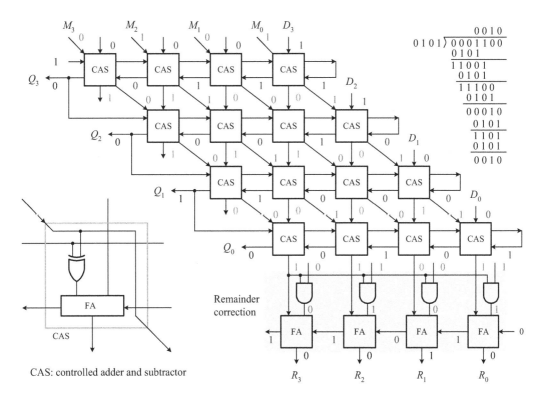

Figure 14.19: The logic diagram of a 4×4 unsigned non-restoring array divider.

has all bits set to 0s except the LSB, which is the MSB of the dividend. The carry-in must be set to 1 so that subtraction can be performed in the first row. Starting from the second row to the last row, one operand is the divisor and the other operand is the sum from its preceding row except that the LSB is a dividend bit. The carry-in is the carry-out of the MSB of its preceding row. The remainder correction step is simply to add 0 or the divisor to the remainder depending on whether the sign bit of the remainder is 0 or 1.

```
// an M-by-N unsigned nonrestoring array divider
module unsigned_array_divider
```

```verilog
   #(parameter M = 4,  // the default divisor width
     parameter N = 4)( // the default dividend width
     input  [M-1:0] x, // divisor
     input  [N-1:0] y, // dividend
     output [N-1:0] quotient,
     output [M-1:0] remainder);
// the body of the M-by-N unsigned array divider
wire  [M-1:0] rem_carry;
// internal wires
wire sum   [M-1:0][N-1:0]; // declare internal nets
wire carry [M-1:0][N-1:0];
genvar i, j;
generate for (i = N - 1; i >= 0; i = i - 1) begin: divider
   if (i == N - 1)   // the first row
      for (j = 0; j < M; j = j + 1) begin: first_row
         if (j == 0)
            assign {carry[j][i],sum[j][i]}=y[i]+!x[j]+1;
         else
            assign {carry[j][i],sum[j][i]}=!x[j]+carry[j-1][i];
      end
   else              // the other rows
      for (j = 0; j < M; j = j + 1) begin: rest_rows
         if (j == 0) // the LSB of each row
            assign {carry[j][i],sum[j][i]} =  y[i] +
                   (x[j]^carry[M-1][i+1]) + carry[M-1][i+1];
         else        // the other bits of each row
            assign {carry[j][i],sum[j][i]} = sum[j-1][i+1]+
                   (x[j]^carry[M-1][i+1]) + carry[j-1][i];
      end
end endgenerate
// generate the quotient
generate for (i = 0; i < N ; i = i + 1) begin: quotient
   assign quotient[i] = carry[M-1][i];
end endgenerate
// generate and adjust the final remainder
generate for (j = 0; j < M ; j = j + 1) begin: R_adjust
   if (j == 0)
      assign {rem_carry[j],remainder[j]} = sum[j][0] +
             (sum[M-1][0] & x[j]);
   else
      assign {rem_carry[j],remainder[j]} = sum[j][0] +
             (sum[M-1][0] & x[j]) + rem_carry[j-1];
end endgenerate
endmodule
```

In the above discussion, we only consider the cases that both restoring and non-restoring division algorithms merely process unsigned input numbers. As the non-restoring division algorithm needs to process signed input numbers, it needs some minor modifications. The interested reader can be referred to Parhami [8] for details.

■ Review Questions

14-17 Describe the basic operation of the restoring division algorithm.

14-18 Explain the rationale behind the non-restoring division algorithm.

14-19 Describe the basic operation of the non-restoring division algorithm.

14-20 In what case is the remainder needed to correct in the non-restoring division algorithm?

14.4 Arithmetic-and-Logic Units

An arithmetic-and-logic unit (ALU) is often the major component for the datapath of many applications, especially, for central processing unit (CPU) design. An ALU contains two portions: an *arithmetic unit* and a *logic unit*. The arithmetic unit sometimes also includes multiplication, even division, in addition to the two basic operations: addition and subtraction; the logic unit simply comprises three basic Boolean operators: AND, OR, and NOT. In some more general applications, a shifter is often an important part of their ALUs too. Therefore, we first consider shifts in what follows.

14.4.1 Shifts

In this section, we first introduce various types of shifts briefly and then consider the implementations of shifts concisely. More about shifts can be referred to [5, 6].

14.4.1.1 Types of Shifts A shift is to move an input left or right a specified number of bit positions with zeros or the sign bit filled in the vacant bits. Depending on whether the vacant bits are filled with zeros or the sign bit, shifts can be cast into *logical shifts* and *arithmetic shifts*.

Logical shifts. In logical shifts, the vacant bits are filled with zeros. Depending on the shift direction, logical shifts can be divided into the following two types:

- *Logical left shift:* The input is shifted left a specified number of bit positions and all vacant bits are filled with zeros.
- *Logical right shift:* The input is shifted right a specified number of bit positions and all vacant bits are filled with zeros.

Arithmetic shifts. The basic features of an arithmetic shift are that its shifted result is the input number divided by 2 when a right shift is executed and is the input number multiplied by 2 when a left shift is carried out. In order to maintain these features, the vacant bits of the shifted result are filled with zeros when the left shift is carried out and filled with the sign bit when the right shift is executed. More formally, arithmetic shifts can be divided into the following two types:

- *Arithmetic left shift:* The input is shifted left a specified number of bit positions and all vacant bits are filled with zeros. Indeed, this is exactly the same as the logical left shift.
- *Arithmetic right shift:* The input is shifted right a specified number of bit positions and all vacant bits are filled with the sign bit.

14.4.1.2 Barrel Shifters The device used to perform the shift, arithmetic or logical, described above is known as a *shifter*. In general, an n-bit shifter is a device that is capable of shifting its input at most n bits. Like other arithmetic operations, an n-bit shifter can also be implemented with either a multicycle or a single-cycle structure.

The basic multicycle structure is a universal shift register, such as the one described in Section 9.3.2. It loads the input to be shifted and then performs the desired number of shifts at the cost of the equal number of clock cycles. Detailed operations of the universal shift register can be reviewed in the related section again.

Like other arithmetic operations discussed in this book, the multicycle structure is essentially a sequential circuit. By contrast, the single-cycle structure is simply a combinational circuit in which any number of bit positions can be shifted in one cycle. To see how such a combinational circuit can be used to realize a shifter, in what follows, we introduce a combinational circuit, known as a *barrel shifter*, to implement a shifter with an arbitrary number of shifts.

A barrel shifter is a device that can shift its input the number of bit positions specified by another input. It consists of an input $I = \langle I_{n-1}, I_{n-2}, \cdots, I_1, I_0 \rangle$, an output $O = \langle O_{n-1}, O_{n-2}, \cdots, O_1, O_0 \rangle$, and a quantity of shifts $S = \langle S_{m-1}, S_{m-2}, \cdots, S_1, S_0 \rangle$. For a logical-left shifter, the relationship between the output and input can be described as follows

$$O_i = \begin{cases} 0 & \text{for all } i < s \\ I_{i-s} & \text{for all } i \geq s \end{cases} \quad (14.15)$$

where $i = 0, 1, \cdots, n-1$, and $s = 0, 1, \cdots, m-1$.

As an example, consider the 8-bit logical-left barrel shifter shown in Figure 14.20. It consists of three multiplexer stages, with each stage containing eight 2-to-1 multiplexers. Each multiplexer stage is controlled by $s[i]$ and shifts its input a number of $s[i] \times 2^i$ bit positions. In other words, each stage shifts its input a number of 0 or 2^i bit positions depending on whether the value of $s[i]$ is 0 or 1.

As usual, a parameterizable module is often required in practical applications. To achieve this, we give an example to show how to describe a parameterizable logical-left barrel shifter using generate-loop statements.

■ Example 14.10: A parameterizable logical-left barrel shifter.

As described above, in a barrel shifter, each stage shifts its input a number of 0 or 2^i bit positions depending on whether the value of $s[i]$ is 0 or 1. Besides, the first stage accepts the data input and the last stage outputs the shifted result to the data output. By combining the above two observations, the resulting module can be obtained.

```
// a logical-left barrel shifter
module left_logical_barrel_shifter
        #(parameter N = 8,    // the default data width
          parameter S = 3)(   // S = log2 N
          input   [N-1:0] din,      // data input
          input   [S-1:0] shifts,   // the number of shifts
          output  [N-1:0] dout);    // data output
// internal wires
wire    [N-1:0] temp_data [S-1:0];
genvar i, j;
generate for (i = 0; i < S ; i = i + 1) begin: shifter
   if (i == 0)   // the first column
      for (j = 0; j < N; j = j + 1) begin: first_column
         if (j < 2**i)
            assign temp_data[i][j]=(shifts[i]) ? 1'b0 : din[j];
         else
            assign temp_data[i][j]=
                  (shifts[i]) ? din[j-2**i] : din[j];
      end
   else
```

Section 14.4 *Arithmetic-and-Logic Units*

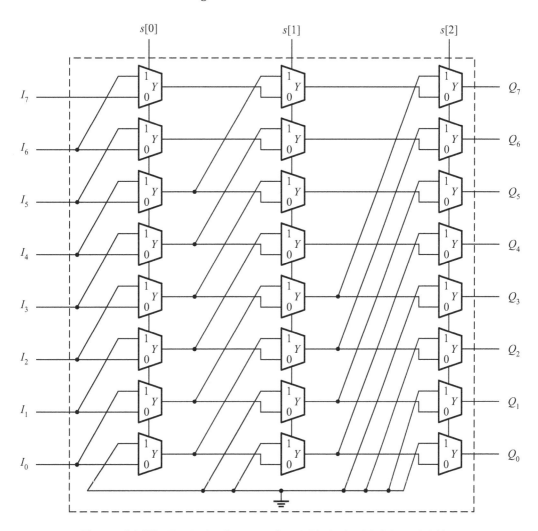

Figure 14.20: The logic diagram of an 8-bit logical-left barrel shifter.

```
        for (j = 0; j < N; j = j + 1) begin: rest_columns
          if (j < 2**i)
             assign temp_data[i][j] =
                    (shifts[i]) ? 1'b0 : temp_data[i-1][j];
          else
             assign temp_data[i][j] =
                    (shifts[i]) ? temp_data[i-1][j-2**i]:
                                  temp_data[i-1][j];
        end
end endgenerate
// generate data output
assign dout = temp_data[S-1];
endmodule
```

14.4.2 ALUs

An ALU comprises two major function groups: arithmetic and logic. Sometimes, a shifter is also incorporated into an ALU to facilitate both logical and arithmetic shifts. In this section, we address two simple ALUs. One is constructed in a straightforward manner and the other uses the logic function of a full-adder to implement the required logic operations.

14.4.2.1 A Simple ALU The logic diagram of a simple ALU is shown in Figure 14.21, which is composed of an ALU, a shifter, and a flag register. The ALU can carry out both the arithmetic and logic operations selected by the ALU_mode input. The logic operations can be simply implemented by using the three basic gates: AND, OR, and NOT. Because of their intuitive simplicity, we will not further consider their details here. The shifter is a barrel shifter capable of performing logical and arithmetic shifts. Both types of shifts may be in both directions: left and right. These options are selected by the Shifter_mode input and the number of shifts is specified by the Shift_number input. In some applications, the shifter is placed on one of the ALU's inputs, A or B.

The flag register includes the four standard flags: *negative* (N), *zero* (Z), *overflow* (V), and *carry* (C). It indicates the status of the ALU after an arithmetic or a logic operation is carried out. In some applications, if the status of a shifted result is needed to indicate, these flag circuits should be applied to the output of the shifter instead of the output of the ALU.

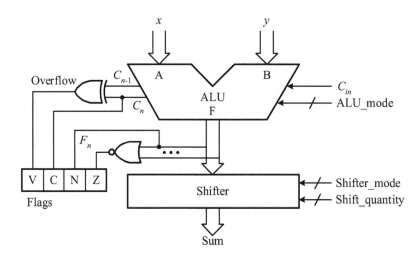

Figure 14.21: The logic diagram of a simple ALU.

The flags N, Z, V, and C are defined as follows:

- *Negative* (N) flag: It is set to 1 when the ALU result is negative; namely, the MSB bit is 1. Otherwise, it is cleared.
- *Zero* (Z) flag: It is set to 1 when the ALU result is zero; namely, all bits are zero. Otherwise, it is cleared.
- *Overflow* (V) flag: It is set to 1 when the sign bit of the ALU result is changed; otherwise, it is cleared. In other words, an overflow occurs whenever the number of carry-ins to the sign bit (MSB) does not match the number of carry-outs from the sign bit.
- *Carry* (C) flag: It is set to 1 when the carry-out from the MSB of the ALU result is 1. Otherwise, it is cleared.

As an illustration, consider a simple ALU with two arithmetic operations, addition and subtraction, and three logical operations, AND, OR, and XOR. The details of this ALU is described in the following example.

Section 14.4 *Arithmetic-and-Logic Units*

■ Example 14.11: A simple ALU.

The following module describes a simple ALU, which simply performs two arithmetic operations, addition, subtraction, as well as three logic operations, AND, OR, and XOR. Besides, four standard flags, N, Z, V, and C, are provided and output to indicate the status of the ALU results.

```
// a simple-alu example
module simple_alu #(parameter M = 4)(// the default word width
    input [M-1:0] x, y,
    input [2:0] alu_op,
    output reg [M-1:0] sum,
    output reg c_out,
    output [3:0] flags);

// the body of the ALU
localparam ADD = 3'b000, SUB = 3'b001, AND = 3'b010,
           OR = 3'b011, XOR = 3'b100;
reg  c_tmp;
wire N, Z, V, C;    // N, Z, V, and C flags
// the definition of flags
assign N = sum[M-1],       // negative flag
       Z = ~(|sum),        // zero flag
       V = c_tmp ^ c_out,  // overflow flag
       C = c_out;          // carry flag
assign flags = {N, Z, V, C};
// the operations of the arithmetic-and-logic unit
always @(*)
  case (alu_op)
    ADD: begin
        {c_tmp, sum[M-2:0]} = x[M-2:0] + y[M-2:0];
        {c_out, sum[M-1]} = x[M-1] + y[M-1] + c_tmp; end
    SUB: begin
        {c_tmp, sum[M-2:0]} = x[M-2:0] - y[M-2:0];
        {c_out, sum[M-1]} = x[M-1] - y[M-1] - c_tmp; end
    AND: begin
        {c_tmp, sum[M-2:0]} = x[M-2:0] & y[M-2:0];
        {c_out, sum[M-1]} = x[M-1] & y[M-1]; end
    OR: begin
        {c_tmp, sum[M-2:0]} = x[M-2:0] | y[M-2:0];
        {c_out, sum[M-1]}   = x[M-1] | y[M-1]; end
    XOR: begin
        {c_tmp, sum[M-2:0]} = x[M-2:0] ^ y[M-2:0];
        {c_out, sum[M-1]} = x[M-1] ^ y[M-1]; end
    default: begin {c_out, sum} = 0; c_tmp = 0; end
  endcase
endmodule
```

14.4.2.2 An FA-Based ALU Another ALU example relies on the function of a full adder (FA) to implement the three basic logic operations. To see this, recall that the sum and carry switching functions of a full adder can be expressed as follows

$$S = (x \oplus y) \oplus C_{in}$$

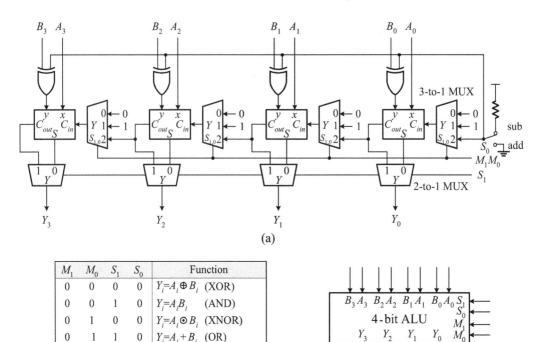

Figure 14.22: The (a) logic diagram, (b) function table, and (c) logic symbol of a FA-based ALU.

$$C_{out} = x \cdot y + (x \oplus y) \cdot C_{in} \tag{14.16}$$

where S implies both XOR and XNOR operations and C_{out} implies the OR and AND operations. As a consequence, all three logic operations can be realized through appropriately controlling the value of carry-in C_{in}.

- If $C_{in} = 0$,
$$\begin{aligned} S &= (x \oplus y) & \text{(XOR)} \\ C_{out} &= x \cdot y & \text{(AND)} \end{aligned} \tag{14.17}$$

- If $C_{in} = 1$,
$$\begin{aligned} S &= (x \odot y) & \text{(XNOR)} \\ C_{out} &= x + y & \text{(OR)} \end{aligned} \tag{14.18}$$

As a result, by properly controlling the value of C_{in} and routing S and C_{out} to the output, the desired logic operations can be readily obtained. A complete logic diagram of a 4-bit FA-based ALU is depicted in Figure 14.22(a). Figures 14.22(b) and (c) give its function table and logic symbol, respectively. It is instructive to verify the function table shown in Figure 14.22(b) by checking the details of the logic circuit depicted in Figure 14.22(a).

As always, generate-loop statements are used to describe iterative logic. To reemphasize this, we use a generate-loop statement to describe the FA-based ALU shown in Figure 14.22 in the following example. Even though the resulting module is very simple, the module can indeed describe an n-bit ALU, where n is an arbitrary positive integer.

■ Example 14.12: An n-bit FA-based ALU.

As shown in Figure 14.22, each bit of the ALU is composed of a 3-to-1 multiplexer, an XOR gate, a 2-to-1 multiplexer, and a full adder. These components are all combinational circuits in nature and hence easily described by using **assign** continuous assignments. The resulting module is as follows:

```
// an N-bit ALU using a generate-loop statement
module ALU_generate #(parameter N = 4)(// the default size
       input   [N-1:0] A, B,
       input   S1, S0, M1, M0,  // mode selection
       output  Cout,
       output  [N-1:0] Y);
// internal wires
wire    [N-1:0] cin, c, s;
// using a generate block to generate the entire ALU
genvar i;
generate for (i = 0; i < N; i = i + 1) begin: ALU
   if (i == 0)  // specify the LSB
      assign cin[i] = (~M1 & ~M0 & 1'b0) |
                      (~M1 &  M0 & 1'b1) |
                      ( M1 & ~M0 & S0);
   else          // specify the other bits
      assign cin[i] = (~M1 & ~M0 & 1'b0) |
                      (~M1 &  M0 & 1'b1) |
                      ( M1 & ~M0 & c[i-1]);
   assign {c[i], s[i]} = A[i] + (B[i] ^ S0) + cin[i];
   assign Y[i] =  (~S1 & s[i]) | (S1 & c[i]);
end endgenerate
assign Cout = c[N-1];  // assign the final carry-out
endmodule
```

The functions of a practical ALU are often much more complicated than what we have dealt with in this section. However, their operations are quite similar to those introduced in this section.

■ Review Questions

14-21 Describe the functions of a typical ALU.
14-22 Describe how the FA-based ALU performs logical operations.
14-23 What is the operation of barrel shifters?
14-24 What is the function of the flag register associated with an ALU?
14-25 How would you detect whether an overflow occurs in an ALU operation?
14-26 Show that $C_{out} = x \cdot y + (x \oplus y)C_{in} = x + y$ as $C_{in} = 1$.

14.5 Digital-Signal Processing Modules

With the mature development of *digital-signal processing* (DSP) techniques and algorithms over the past decades, DSP has become a ubiquitous technique in designing modern digital systems, including image processing, instrumentation, control, speech, audio, military, and telecommunications, and biomedical applications. Hence, we introduce in this section the

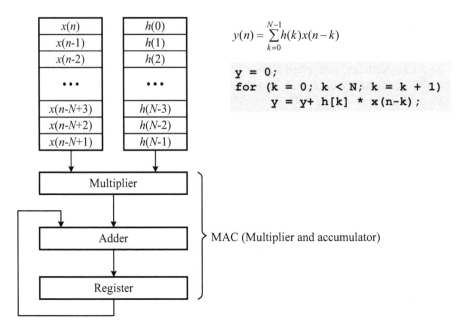

Figure 14.23: The general datapath structure of a DSP processor for implementing an Nth-order FIR filter.

most basic DSP techniques which are widely used in many practical systems. These basic techniques include *finite-impulse response* (FIR) and *infinite-impulse response* (IIR) filters.

14.5.1 Finite-Impulse Response Filters

In an FIR filter, the output is a weighted sum of the present and past samples of its inputs. The output of an FIR can be generally described by the following equation

$$y(n) = \sum_{k=0}^{N-1} h(k)x(n-k) \qquad (14.19)$$

where $h(k)$ represent the tap (weighting) coefficients and $x(n-k)$ are input samples. Tap coefficients determine the features of an FIR. The details of how to determine the tap coefficients $h(k)$ from a given specification are beyond the scope of this book. However, the interested reader can be referred to related DSP textbooks such as Oppenheim and Schafer [7] and Schiling and Harris [10].

Figure 14.23 shows the general datapath structure for implementing FIR filters. It comprises two memory modules, a tap coefficient memory and an input sample memory, a multiplier, an adder, and a register. The symbol z^{-1} denotes a unit-time delay and hence corresponds to a single-stage D flip-flops. It is easy to derive an ASM chart of the controller for this datapath and use Verilog HDL to model it. However, they are left for the reader as exercises.

The single-cycle implementation of the above general datapath structure of FIR filters is shown in Figure 14.24. Here, three types of basic components, a register, a multiplier, and an adder, are repeated in use at each stage. Because of its shape, the structure is often called a *transversal filter*. The multiplier and adder associated with each tap is often combined into one basic unit, known as a *multiplier and accumulator* (MAC), which is usually constructed from a special hardware design technique in order to speed up its operations.

In what follows, we give a simple example of a constant-coefficient FIR filter to illustrate how to describe an FIR filter in practice.

Section 14.5 Digital-Signal Processing Modules

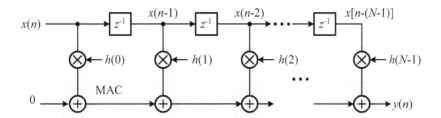

Figure 14.24: The general structure of an Nth-order FIR transversal filter.

■ Example 14.13: An Nth-order constant-coefficient FIR filter.

In this example, we assume that tap coefficients are constant, which is often the case in practice. Hence, they are defined by using a **localparam** statement. The evaluation of the FIR equation is then performed by a simple **assign** continuous assignment. Besides, it is necessary to sample the new inputs at each clock cycle and shift the old samples one step forward. These two operations are completed by using an **always** block. Note that the input to register x[1] is data_in.

```
// an Nth order W-bit generic FIR filter with constant
// coefficients
module FIR #(parameter N = 4,    // define the order
             parameter W = 8)(   // define the word size
             input clock, reset,
             input  [W-1:0] data_in,
             output [W-1:0] data_out);
reg  [W-1:0] x[N:1]; // samples
// define coefficients for the FIR filter
localparam h0 = 8'd51, h1 = 8'd23, h2 = 8'd50, h3 = 8'd26,
           h4 = 8'd34;
integer i;
// the body of the FIR filter
assign
   data_out = h0 * data_in + h1 * x[1] +
              h2 * x[2] + h3 * x[3] + h4 * x[4];
// update input samples
always @(posedge clock or posedge reset)
   if (reset) for (i = 1; i <= N; i = i + 1)
      x[i] <= 0;
   else begin
      x[1] <= data_in;
      for (i = 2; i <= N; i = i + 1) x[i] <= x[i-1];
   end
endmodule
```

It is worth noting that the **for** loop in the **else** part of the above module cannot be replaced with a single assignment: x[N:2] <= x[N-1:1]. Why? Try to explain it.

14.5.2 Infinite-Impulse Response Filters

Another widely used digital filter structure is the IIR filter. In IIR filters, the output at a given time is a function of their inputs and their previously computed outputs. The output of an IIR

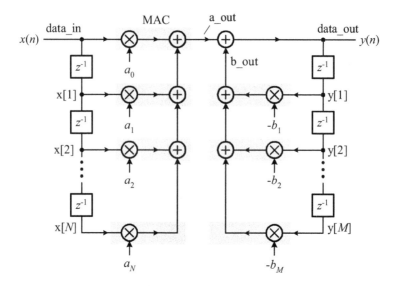

Figure 14.25: The general structure of the direct-form realization of an IIR filter.

filter can be represented as a weighted sum of inputs according to the recursive difference equation shown below.

$$y(n) = \sum_{k=0}^{N-1} a(k)x(n-k) - \sum_{k=0}^{M} b(k)y(n-k) \qquad (14.20)$$

For simplicity, we assume that $N = M$. A direct-form realization of an IIR filter is shown in Figure 14.25. Like the traversal FIR filter, an Nth-order IIR filter in direct form is also composed of three types of basic components: registers, multipliers, and adders. The multiplier and adder associated with each tap is also often combined into a single MAC unit. Consequently, regardless of whether the structure is for FIR or IIR, only two basic types of components, registers and MACs, are required.

Next, we give a simple example of a constant-coefficient IIR filter to illustrate how to describe an IIR filter in practice.

■ Example 14.14: An Nth-order constant-coefficient IIR filter.

Like the preceding example, the tap coefficients are assumed to be constant, which is often the case in practice. Hence, they are defined by using two individual **localparam** statements. The evaluation of the IIR equation is then performed by two simple **assign** continuous assignments and uses a third **assign** continuous assignment to combine them together. Besides, it is necessary to sample the new inputs at each clock cycle and shift the old samples one step forward. These two operations are completed by using an **always** block. Note that in a direct-form IIR filter, the input to register x[1] is data_in and the input to register y[1] is data_out, namely, a_out - b_out.

```
// an Nth order generic IIR filter with constant
// coefficients (assume that N = M)
module IIR #(parameter N = 4,    // define the order
             parameter W = 16)(  // define the word size
            input clock, reset,
            input [W-1:0] data_in,
```

```verilog
              output [W-1:0] data_out);
wire [2*W-1:0] a_out, b_out;
reg  [W-1:0] x[N:1]; // input samples
reg  [W-1:0] y[N:1]; // output samples
integer i;

// define coefficients for the IIR filter
localparam a0 = 16'd51, a1 = 16'd78, a2 = 16'd12, a3 = 16'd84,
           a4 = 16'd25;
localparam b1 = 16'd23, b2 = 16'd58, b3 = 16'd26, b4 = 16'd45;

// compute the data_out from a_out and b_out
assign a_out = a0 * data_in + a1 * x[1] +
               a2 * x[2] + a3 * x[3] + a4 * x[4];
assign b_out = b1 * y[1] + b2 * y[2] + b3 * y[3] + b4 * y[4];
assign data_out = a_out - b_out;

// update input samples
always @(posedge clock or posedge reset)
   if (reset) for (i = 1; i <= N; i = i + 1) begin
      x[i] <= 0;
      y[i] <= 0; end
   else begin
      x[1] <= data_in;
      y[1] <= data_out;
      for (i = 2; i <= N; i = i + 1) x[i] <= x[i-1];
      for (i = 2; i <= N; i = i + 1) y[i] <= y[i-1];
   end
endmodule
```

■ Review Questions

14-27 What are the basic arithmetic operations of an FIR filter?

14-28 What are the basic arithmetic operations of an IIR filter?

14-29 How many registers are required in an Nth-order FIR traversal filter?

14-30 How many multipliers are required in an Nth-order FIR traversal filter?

14-31 How many registers are required in an Nth-order direct-form IIR filter?

14-32 How many multipliers are required in an Nth-order direct-form IIR filter?

14.6 Summary

In this chapter, we have examined many frequently used arithmetic modules, including addition, multiplication, division, ALUs, shifters, and two digital-signal processing (DSP) filters. By way of the introduction of these arithmetic operations and their algorithms, we also emphasized again the concept that a hardware algorithm can usually be realized by using a multicycle or a single-cycle structure.

The bottleneck of a conventional n-bit ripple-carry adder is on the generation of the carry needed in each stage. To overcome this, many schemes have been proposed. Among these, the carry-lookahead (CLA) generators and its associated adders are first considered. Then, two parallel-prefix carry generators and their associated adders, the Kogge-Stone adder and the Brent-Kung adder, were discussed.

The essential operations of multiplication are a sequence of additions. In this chapter, we considered the shift-and-add multiplication algorithm, which is essentially the "pencil-and-paper" algorithm that we use daily. The single-cycle structure of this algorithm evolves naturally into array structure, including an unsigned array multiplier and a modified Baugh-Wooley signed array multiplier. These two multipliers have the similar structure and can be combined into a single one with a mode selection to choose whether the desired operation is unsigned or signed. Moreover, an $\left(\frac{n}{k}\right)$-cycle structure and a pipeline structure of the $m \times n$ unsigned multiplication are explored in great detail. Similarly, the essential operations of division are a sequence of subtractions. Based on this idea, two basic division algorithms, known as the restoring division algorithm and the non-restoring division algorithm, were introduced.

An arithmetic-and-logic unit (ALU) is often the major component for the datapath of many applications, especially, for central processing unit (CPU) design. An ALU contains two portions: arithmetic-and-logic units. The arithmetic unit sometimes also includes multiplication, even division, in addition to the two basic operations: addition and subtraction; the logic unit simply comprises three basic Boolean operators: AND, OR, and NOT. In some more general applications, a shifter is also included as an important part of their ALU functions.

A shift is to shift an input, left or right, a specified number of bit positions with zeros or the sign bit filled in the vacant bits. Depending on whether the vacant bits are filled with zeros or the sign bit, shifts can be cast into logical shifts and arithmetic shifts. Multicycle and single-cycle structure can be employed to implement these shifts. The former is sequential in nature and based on universal shift registers while the latter is combinational naturally and based on multiplexer-based networks, known as barrel shifters.

With the mature development of digital-signal processing (DSP) techniques and algorithms over the past decades, DSP has become a ubiquitous technique in designing modern digital systems, ranging from image processing, instrumentation/control, speech/audio, military, and telecommunications, to biomedical applications. For this reason, the two most basic DSP techniques frequently used in many practical systems were introduced. These two basic techniques are finite-impulse response (FIR) and infinite-impulse response (IIR) filters.

References

1. C. Baugh and B. Wooley, "A two's complement parallel array multiplication algorithm," *IEEE Trans. on Computers*, Vol. 22, No. 12, pp. 1045–1047, December 1973.

2. A. Booth, "A signed binary multiplication technique," *Quarterly Journal of Mechanics and Applied Mathematics*, Vol. IV, Part 2, pp. 236–240, June 1951.

3. K. Hwang, *Computer Arithmetic Principles, Architecture, and Design*, New York: John Wiley & Sons, 1979.

4. E. C. Ifeachor and B. W. Jervis, *Digital Signal Processing: A Practical Approach*, Reading, Massachusetts: Addison-Wesley, 1993.

5. M. B. Lin, *Digital System Design: Principles, Practices, and Applications*, 4th ed., Chuan Hwa Book Ltd. (Taipei, Taiwan), 2010.

6. M. B. Lin, *Introduction to VLSI Systems: A Logic, Circuit, and System Perspective*, CRC Press, 2012.

7. A. V. Oppenheim and R. W. Schafer, *Discrete-Time Signal Processing*, 2nd ed., Upper Saddle River, New Jersey: Prentice-Hall, 1999.

8. B. Parhami, *Computer Arithmetic: Algorithms and Hardware Designs*, New-York: Oxford University Press, 2000.

9. B. Parhami, *Computer Architecture: From Microprocessors to Supercomputers*, New-York: Oxford University Press, 2005.

10. R. J. Schilling and S. L. Harris, *Fundamentals of Digital Signal Processing Using Matlab*, Thomson Publisher, 2005.

Problems

14-1 Suppose that 4-bit block CLA generators are used. Referring to Figure 14.5, answer each of the following questions.

  (a) Draw the logic block diagram of a 64-bit CLA adder. Label each component appropriately to indicate its size and functionality.

  (b) Write a Verilog HDL module to describe the above 64-bit CLA adder and write a test bench to verify its functionality.

  (c) Write a parameterizable module to describe an n-bit multiple-level CLA adder using 4-bit block CLA generators.

  (d) Write a test bench to verify the functionality of the parameterizable module.

14-2 This problem compares the hardware cost and propagation delays of both n-bit single-level and n-bit multiple-level CLA adders, where n is 8, 16, 32, and 64.

  (a) Referring to Figure 14.5, draw the logic block diagrams of the desired n-bit multiple-level CLA adders. Label each component appropriately to indicate its size and functionality.

  (b) Write one parameterizable or more non-parameterizable Verilog HDL modules to describe the above n-bit multiple-level CLA adders and write test benches to verify their functionality.

  (c) Synthesize the above n-bit multiple-level CLA adders. Compare the hardware cost in terms of the number of LUTs of these n-bit multiple-level CLA adders with that of the n-bit single-level CLA adders.

  (d) Perform the post-PAR simulation of both n-bit multiple-level and n-bit single-level adders and compare their propagation delays.

14-3 Considering the three fast adders discussed in Section 14.1, compare the performance of these three fast adders in terms of the number of LUTs and propagation delays with different word widths, ranging from 4 to 64 spaced with 2^k, where k is 3, 4, or 5.

14-4 Considering the logic diagram of the 16-input hybrid Brent-Kung/Kogge-Stone parallel-prefix network shown in Figure 14.26, answer each of the following questions.

  (a) Write a parameterizable Verilog HDL module to describe an n-bit adder based on this hybrid parallel-prefix network.

  (b) Write a test bench to verify the functionality of the parameterizable module.

14-5 Consider the conversion problem from a Gray code to a binary code. As we have described previously, the rule of converting a Gray code into its equivalent binary code is that if the number of 1s of the input Gray code counting from the MSB to the current position i is odd, then the binary bit is 1; otherwise, the binary bit is 0.

  (a) Show that this Gray code to binary code conversion problem can be cast into a prefix-sum problem.

  (b) Draw a parallel-prefix network to show how to compute each binary bit from the Gray code in parallel.

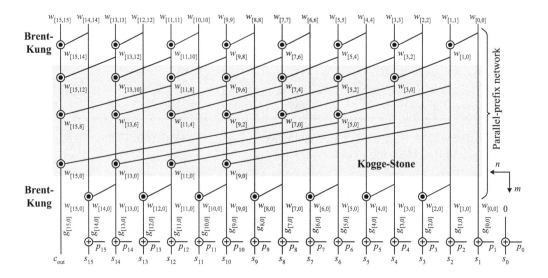

Figure 14.26: The logic diagram of a 16-input hybrid Brent-Kung/Kogge-Stone parallel-prefix network.

- (c) Write a parameterizable module to describe the conversion operations using the parallel-prefix approach.
- (d) Write a test bench to verify the functionality of the parameterizable module.

14-6 Considering the shift-and-add multiplier described in Section 14.2.1, answer each of the following questions.
- (a) Derive an ASM chart from the algorithm.
- (b) Write a parameterizable Verilog HDL module to describe the ASM chart obtained.
- (c) Write a test bench to verify the functionality of the module.

14-7 Considering the 4-bit bit-serial multiplier shown in Figure 14.27, answer each of the following questions.

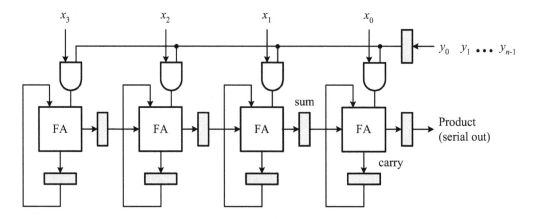

Figure 14.27: The logic diagram of a 4-bit bit-serial multiplier.

- (a) Derive an ASM chart from the observation of the figure.

Problems **665**

(b) Write a parameterizable Verilog HDL module to describe the ASM chart obtained.

(c) Write a test bench to verify the functionality of the module.

14-8 Write a parameterizable module to describe the unsigned CSA array multiplier shown in Figure 14.12. You also need to write a test bench to verify the functionality of the module.

14-9 Considering the $(\frac{n}{k})$-cycle structure for an $m \times n$ multiplier described in Figure 14.13, answer each of the following questions.

(a) Derive an ASM chart by referring to the algorithm given in Section 14.2.1.

(b) Write a parameterizable Verilog HDL module to describe the ASM chart obtained. The parameters used are m, k, and n.

(c) Write a test bench to verify the functionality of the module.

14-10 Write a parameterizable module to describe the pipeline structure of the $m \times n$ unsigned multiplier shown in Figure 14.14. The parameters used are m, k, and n. You also need to write a test bench to verify the functionality of the module.

14-11 Considering an $m \times n$ unsigned CSA array multiplier and supposing that an m-bit RCA is used in the last stage to add up the sum and carry in order to yield the final product, answer each of the following questions.

(a) Partition the array into l stages with the same or almost the same propagation delay. Denote l as a function of m and n.

(b) Insert pipeline registers and draw the logic diagram of the resulting pipeline structure.

(c) Write a parameterizable Verilog HDL module to describe the above pipeline structure.

(d) Write a test bench to verify the functionality of the module.

14-12 Referring to Figure 14.13, design an $(\frac{n}{k})$-cycle structure for the $m \times n$ Booth multiplier described in Section 11.2.2.

(a) Derive an ASM chart by referring to the Booth array structure given in Section 11.3.4.

(b) Write a parameterizable Verilog HDL module to describe the ASM chart obtained. The parameters used are m, k, and n.

(c) Write a test bench to verify the functionality of the module.

14-13 Referring to Figure 14.14, pipeline the structure of the $m \times n$ Booth array multiplier described in Section 11.3.4 and write a parameterizable module to describe the resulting pipeline structure. The parameters used are m, k, and n. You also need to write a test bench to verify the functionality of the module.

14-14 Write a parameterizable $m \times n$ array multiplier module that is capable of performing both unsigned and signed multiplication controlled by a `mode` input and only using a single array. Write a test bench to verify the functionality of the module.

14-15 Considering the modified Baugh-Wooley module discussed in Section 14.2.2, answer each of the following questions.

(a) Give a numerical example to explore the operations of a modified Baugh-Wooley algorithm with the case of $m \neq n$.

(b) Modify the modified Baugh-Wooley module so that it can accommodate the situation of $m \neq n$.

(c) Write a test bench to verify the functionality of the module.

14-16 Considering the unsigned non-restoring division algorithm described in Section 14.3.2, answer each of the following questions.

(a) Derive an ASM chart from the algorithm given in Section 14.3.2.

(b) Write a parameterizable Verilog HDL module to describe the above ASM chart.

(c) Write a test bench to verify the functionality of the module.

14-17 Referring to Figure 14.13, design an $(\frac{n}{k})$-cycle structure for the non-restoring division algorithm discussed in Section 14.3.2.

(a) Derive an ASM chart by referring to the non-restoring array divider described in Figure 14.19.

(b) Write a parameterizable Verilog HDL module to describe the ASM chart obtained. The parameters used are m, k, and n.

(c) Write a test bench to verify the functionality of the module.

14-18 Referring to Figure 14.14, pipeline the $m \div n$ structure of the non-restoring array divider exemplified in Figure 14.19 and write a parameterizable module to describe the resulting pipeline structure. The parameters used are m, k, and n. You also need to write a test bench to verify the functionality of the module.

14-19 In a logical-right barrel shifter, the relationship between the output and the input can be described as follows:

$$O_i = \begin{cases} 0 & \text{for all } i > (n-1) - s \\ I_{i+s} & \text{otherwise} \end{cases} \quad (14.21)$$

where $i = 0, 1, \cdots, n-1$, and $s = 0, 1, \cdots, m-1$.

Design a barrel shifter that can perform the logical-right shift of its input.

(a) Draw the block diagram of the logical-right barrel shifter. Label each component appropriately to indicate its functionality and size.

(b) Write a parameterizable Verilog HDL module to describe it.

(c) Write a test bench to verify its functionality.

14-20 In a arithmetic-right barrel shifter, the relationship between the output and the input can be described as follows:

$$O_i = \begin{cases} I_{n-1} & \text{for all } i > (n-1) - s \\ I_{i+s} & \text{otherwise} \end{cases} \quad (14.22)$$

where $i = 0, 1, \cdots, n-1$, and $s = 0, 1, \cdots, m-1$.

Design a barrel shifter that can perform the arithmetic-right shift of its input.

(a) Draw the block diagram of the arithmetic-right barrel shifter. Label each component appropriately to indicate its functionality and size.

(b) Write a parameterizable Verilog HDL module to describe it.

(c) Write a test bench to verify its functionality.

14-21 This problem involves the design of a barrel shifter that can perform both logical left and right shifts.
- **(a)** Draw the block diagram of the barrel shifter. Label each component appropriately to indicate its functionality and size.
- **(b)** Write a parameterizable Verilog HDL module to describe it.
- **(c)** Write a test bench to verify its functionality.

14-22 This problem concerns the design of a barrel shifter that can perform both arithmetic and logical shifts, including both left and right directions.
- **(a)** Draw the block diagram of the barrel shifter. Label each component appropriately to indicate its functionality and size.
- **(b)** Write a parameterizable Verilog HDL module to describe it.
- **(c)** Write a test bench to verify its functionality.

14-23 Suppose that a datapath consisting of a register file, an ALU, and a shifter is required. The ALU has the four arithmetic functions, +, -, *, and /, as well as the three basic logic operations, AND, OR, and NOT. The register file has eight 8-bit registers. The shifter can perform arithmetic and logical shifts, including left and right shifts.
- **(a)** Draw the block diagram of the datapath. Label each component appropriately to indicate its functionality and size.
- **(b)** Model this datapath in behavioral style.
- **(c)** Write a test bench to verify its functionality.

14-24 Considering the general datapath structure for the FIR filters shown in Figure 14.23, answer each of the following questions.
- **(a)** Using the datapath and controller approach, derive an ASM chart of the controller.
- **(b)** Write Verilog HDL modules to describe the controller and the datapath, respectively.
- **(c)** Write a test bench to verify the functionality of the FIR filter module.

14-25 Observing the direct form of the IIR filter shown in Figure 14.25, answer each of the following questions.
- **(a)** Design a general datapath structure (modify the one depicted in Figure 14.23) suitable for implementing IIR filters.
- **(b)** Using the datapath and controller approach, derive an ASM chart of the controller.
- **(c)** Write Verilog HDL modules to describe the controller and the datapath, respectively.
- **(d)** Write a test bench to verify the functionality of the IIR filter module.

14-26 In floating-point multiplication, the exponents of the two operands are added and their significands are multiplied; that is,

$$(\pm 2^{e1} s1) \times (\pm 2^{e2} s2) = \pm 2^{e1+e2}(s1 \times s2) \tag{14.23}$$

Supposing that both inputs and the result are represented as the single-precision format of the IEEE-754 Standard, as shown in Figure 14.28, design a floating-point multiplication circuit.
- **(a)** Draw the block diagram of the floating-point multiplication circuit. Label each component appropriately to indicate its functionality and size.

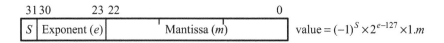

Figure 14.28: The single-precision format of the IEEE 754 floating-point standard.

   (b) Model this datapath in behavioral style.
   (c) Write a test bench to verify its functionality.

14-27 In floating-point addition, the exponents of the two operands are equalized and their significands are added; namely,

$$(\pm 2^{e1} s1) + (\pm 2^{e1}(s2/2^{e1-e2})) = \pm 2^{e1}(s1 \pm s2/2^{e1-e2})) \qquad (14.24)$$

Assume that $e1 > e2$. Supposing that both inputs and the result are represented as the single-precision format of the IEEE-754 Standard, as shown in Figure 14.28, design a floating-point addition circuit.
   (a) Draw the block diagram of the floating-point addition circuit. Label each component appropriately to indicate its functionality and size.
   (b) Model this datapath in behavioral style.
   (c) Write a test bench to verify its functionality.

14-28 In floating-point division, the exponents of the two operands are subtracted and their significands are divided; that is,

$$(\pm 2^{e1} s1)/(\pm 2^{e2} s2) = \pm 2^{e1-e2}(s1/s2) \qquad (14.25)$$

Supposing that both inputs and the result are represented as the single-precision format of the IEEE-754 Standard, as shown in Figure 14.28, design a floating-point division circuit.
   (a) Draw the block diagram of the floating-point division circuit. Label each component appropriately to indicate its functionality and size.
   (b) Model this datapath in behavioral style.
   (c) Write a test bench to verify its functionality.

15

Design Examples

W E begin this chapter with an introduction of the bus structure and its related issues, such as bus arbitration and data transfer modes. The bus structure can be either a tristate bus or a multiplexer bus. In order to schedule the use of a shared bus, bus arbitration logic is required for the bus system. The two most widely used bus schedule schemes are daisy-chain arbitration and radial arbitration. The essential of bus arbitration is the rule to choose a device from many ones that are requesting the use of a bus at the same time. To this end, both fixed priority and round-robin priority schemes can be used. Data transfer on a bus can be proceeded in a synchronous or an asynchronous manner and in a bundle of bits (namely, parallel) or a single bit at a time (namely, serial).

After studying bus structures and data transfer, we consider a real-world example that illustrates the design of a small microcontroller (μC) system, which is the most complex design example presented in this book. This system includes a 16-bit central processing unit (CPU), a general-purpose input and output (GPIO), a timer, and a universal asynchronous receiver and transmitter (UART), which are connected by a system bus composed of an address bus, a data bus, and a control bus. The GPIO is an 8-bit parallel input/output port that can be used as a single 8-bit port or as eight 1-bit ports. A timer is a device that can be used to count events. The basic operations of timers and how to design such timers are discussed in depth. A UART is a device that supports the serial communication between two devices in an asynchronous manner. An example is given to illustrate how to design such a device. The 16-bit CPU is the heart of the system, which provides 27 instructions and 7 addressing modes. A detailed description of it is given in this chapter.

15.1 Bus

A bus is a set of wires used to transport information between two or more devices in a digital system. In practice, multiple devices are usually required to connect on a bus in order to transfer messages to their destinations, thereby leading to a *bus structure*. However, this may give rise to the so-called *multiple-driver problem*, meaning that multiple circuits output their signals on the bus at the same time. To solve this problem, a technique known as *multiplexing* is often used. Through the use of multiplexing, only one device is allowed to use the bus at a time for sending messages to its specified destination. The multiplexing technique mainly consists of two related issues: the *bus structure* and the *bus arbitration* scheme. The bus structure is the physical organization of the bus and the bus arbitration scheme is a mechanism used to schedule the use of the bus, i.e., only allowing one device to send messages at a time. In this section, we deal with these two issues in detail.

15.1.1 Bus Structure

In practice, a bus structure can be organized by using either tristate buffers or multiplexers. The bus structure is called a *tristate bus* when it is constructed with tristate buffers and a *multiplexer bus* when it is built with multiplexers. The tristate bus is often called the *bus* for short. In this section, we address these two bus structures in greater detail.

15.1.1.1 Tristate Bus A typical tristate bus used in digital systems is illustrated in Figure 15.1, where n modules are connected to a shared bus. Through a bidirectional interface, each module is connected to the bus and enabled to drive a signal T onto the bus when the transmit enable signal TE is asserted and to sample a signal from the bus onto an internal signal R when the receive enable signal RE is asserted. For instance, when module 1 wants to send a message a to module 2, module 1 asserts its transmit enable TE to drive its transmit signal T onto the bus. During the same cycle, module 2 asserts its receive enable signal RE to sample the signal from the bus onto its internal received signal, R.

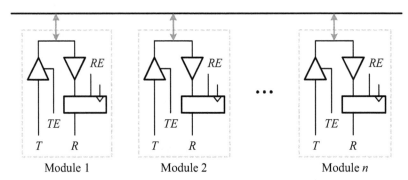

Figure 15.1: The block diagram of a typical tristate bus structure.

In what follows, we first give an example to explain how to model an n-bit tristate buffer and then use another example to describe the construction of a tristate bus.

■ Example 15.1: A tristate buffer.

Suppose that an n-bit tristate buffer is to be modeled. When the enable input is asserted, the buffer is enabled and places its input data on the output. Otherwise, the buffer output is in a high-impedance state.

```
// an N-bit tristate buffer
module tristate_buffer #(parameter N=2)(// the bus width
       input    enable,
       input    [N-1:0] data,
       output   [N-1:0] qout);

// the body of the tristate bus
assign qout = enable ? data : {N{1'bz}};
endmodule
```

The synthesized result is depicted in Figure 15.2. It is instructive to try to synthesize it and check the result. The following example further explores how to use the tristate buffers to construct a tristate bus.

Section 15.1 Bus

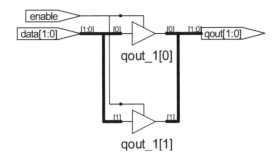

Figure 15.2: The block diagram of a tristate buffer.

■ Example 15.2: A tristate bus with four drivers.

Suppose that there are four n-bit source modules to be connected on the same bus, namely, a tristate bus in which four n-bit tristate buffers are connected together. Each tristate buffer has its own enable input to control its operation. As the enable input is asserted, the tristate buffer is enabled and places its input data onto the bus. Otherwise, the tristate buffer places its output in a high-impedance state. To avoid the data conflict on the bus, at any time only one tristate buffer at most can be enabled; namely, all tristate buffers are at their high-impedance states or only one of them is enabled and places its input data onto the bus. Based on this discussion, the resulting module is as follows. In the module, qout is needed to declare as a **tri** net because it is driven by multiple drivers.

```
// a tristate bus with four drivers
module tristate_bus #(parameter N = 2)(// the bus width
     input   [3:0] enable,
     input   [N-1:0] data0, data1, data2, data3,
     output tri [N-1:0] qout);

// the body of the tristate bus
assign qout = enable[0] ? data0 : {N{1'bz}},
       qout = enable[1] ? data1 : {N{1'bz}},
       qout = enable[2] ? data2 : {N{1'bz}},
       qout = enable[3] ? data3 : {N{1'bz}};
endmodule
```

It is instructive to synthesize the above module and check the results. All tristate buffers driving the same bus must be turned on exclusively; otherwise, the data on the bus will be in an erroneous state due to the conflict caused by multiple driving sources. In addition, to model a net with multiple drivers, the net has to be declared as a **tri** net data type (Section 2.1.2).

Using a proper combination of tristate buffers, a bidirectional bus system can be readily constructed. For example, a bidirectional bus system can be obtained by connecting two tristate buffers in a back-to-back manner. An illustration is exploited in the following example.

■ Example 15.3: A bidirectional bus.

A bidirectional bus is indeed a combination of two tristate buffers in a way such that the input of one buffer is connected to the output of the other and the enable input of each buffer is used to control the data transfer direction. Note that, in this example, we intend to leave the enable inputs of both tristate buffers as two individual enable inputs, denoted by send and receive. The synthesized result of the module is depicted in Figure 15.3.

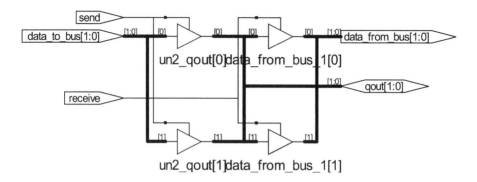

Figure 15.3: The block diagram of a bidirectional bus.

```
// a bidirectional bus example
module bidirectional_bus #(parameter N = 2)(// the bus width
    input   send, receive,
    input   [N-1:0] data_to_bus,
    output wire [N-1:0] data_from_bus,
    inout  wire [N-1:0] qout ); // the bidirectional bus

// the body of the bidirectional bus
assign data_from_bus = receive ? qout : {N{1'bz}};
assign qout = send ? data_to_bus : {N{1'bz}};
endmodule
```

The tristate bus has the same objective as a multiplexer—to select one data source from multiple ones and then route the data to its destination. When a tristate bus is used, each module on the bus only needs a bidirectional interface and a proper bus arbiter to schedule the sequence of bus activities. However, in some applications using a tristate bus may not be a good choice, especially, when the capacitive loading of the driver within the bidirectional interface of the active module is large. For example, in the typical tristate bus structure shown in Figure 15.1, each transmit buffer needs to drive an amount of $n \times (C_{bout} + C_{bin} + C_{wire})$ capacitive load, where C_{bout} is the output capacitance of the tristate output buffer, C_{bin} is the input capacitance of the input buffer, and C_{wire} is the parasitic capacitance of the bus. This amount of capacitive load may not be tolerable in some applications [7].

15.1.1.2 Multiplexer Bus An alternative approach to avoiding the large amount of capacitive load associated with the tristate bus is to use a multiplexer to enforce the multiplexing, as illustrated in Figure 15.4. The output signals T_i of n modules are routed to their destination through an individual multiplexer tree. It can be shown that the propagation delay of a multiplexer tree is much less than that of the bus structure as the number of modules attached to it is large enough [6]. Consequently, for many practical systems, such as ARM Cortex processors, multiplexer trees are often used instead of tristate buses to obtain better performance.

■ Coding Style

1. *It should only allow one output driver on a tristate bus to be activated at a time.*
2. *It should use the **tri** net data type to model a net with multiple drivers.*

Section 15.1 Bus

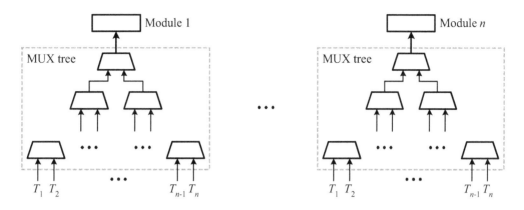

Figure 15.4: The typical multiplexer bus structure.

15.1.2 Bus Arbitration

Because a bus is a shared and an exclusively used resource among modules, at any time only one module may send messages onto the bus. To make a bus work properly, the bus activities are needed to sequence, namely, to be scheduled appropriately as multiple transmitters drive the bus at the same time. Generally speaking, bus activities may be sequenced internally or externally [4]. As bus activities are sequenced internally, each module generates its own TE and RE control signals according to a bus protocol. As bus activities are sequenced externally, a central controller is needed to manage the TE and RE control signals of all modules appearing on the bus. Both strategies are often used in digital systems. For instance, the internal bus of a microprocessor is usually sequenced externally whereas the bus between processor and its peripherals, such as memory modules, a *direct memory access* (DMA) controller, or a *universal serial bus* (USB) controller, is often sequenced internally.

From the above discussion, there must exist some mechanisms to arbitrate the use of the bus when multiple transmitters initiate their data transfers at the same time. The operation that chooses one transmitter from multiple ones attempting to transmit data on the bus is called the *bus arbitration*. The device used to perform the function of bus arbitration is known as a *bus arbiter*.

Currently, the two most widely used bus arbitration schemes are *daisy-chain arbitration* and *radial arbitration*, as shown in Figure 15.5. In what follows, we describe each of these concisely.

15.1.2.1 Daisy-Chain Arbitration
When daisy-chain arbitration is used, as shown in Figure 15.5(a), each module has an arbitration logic circuit called *daisy-chain logic*, consisting of two inputs and two outputs. The inputs are *carry-in* (c_in) and *request* (Req); the outputs are *carry-out* (c_out) and *grant* (Grant). The carry-in input indicates whether the preceding stage has granted the bus. As it is high, the preceding stage has not granted the bus; otherwise, the preceding stage has granted the bus. When both the carry-in and request inputs of the module are high, the grant signal is set to a high value to indicate that the bus is granted by the module. At the same time, the carry-out output is set to a low value to inhibit the succeeding stage to grant the bus.

One widely used implementation of the daisy-chain logic is shown in Figure 15.6(a), which is an iterative logic circuit and consists of two basic gates. The logic expressions of grant g_i and carry-out c_{i+1} are as follows:

$$g_i = r_i \cdot c_i$$
$$c_{i+1} = \bar{r}_i \cdot c_i \qquad (15.1)$$

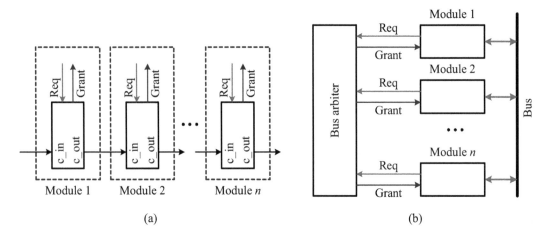

Figure 15.5: The block diagrams of two bus arbitration schemes: (a) daisy-chain arbitration and (b) radial arbitration.

An example of daisy-chain logic using this iterative logic cell is shown in Figure 15.6(b). It is a 4-request daisy-chain bus arbiter. Request r_0 has the highest priority while request r_3 has the lowest priority. When a device with request r_i requests the use of the bus, it will inhibit its succeeding devices to request the use of the bus with request r_{i+1} by clearing carry-out c_{i+1}. For example, when device 0 requests the bus, it inhibits the requests of other devices by clearing carry-out c_1 to 0.

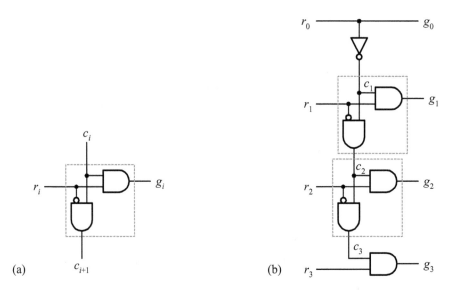

Figure 15.6: The logic diagrams of (a) a single-bit cell and (b) a 4-request daisy-chain bus arbiter.

The daisy-chain arbitration has the following two features: First, the priority is fixed with the highest priority associated with the first module. Second, the bus arbitration time is determined by the modules cascaded in the daisy-chain logic. This can be easily seen from Figure 15.5(a). The second feature may limit the system performance when a large number of modules are connected onto the bus.

Section 15.1 Bus

15.1.2.2 Radial Arbitration The second bus arbitration scheme is radial arbitration, also known as *independent-requested line arbitration*, which uses separate request and grant lines for each module. As shown in Figure 15.5(b), each module has a set of request and grant lines. The request lines of all modules sharing the bus are connected to a bus arbiter through which at most one grant line can be activated at a time. The structure of a bus arbiter with fixed-priority is simply composed of a priority encoder and a decoder. The priority encoder determines which module is granted for using the bus and the decoder generates the grant signal for the module. This fixed-priority logic is quite easily to be implemented and therefore is left for the reader as an exercise.

Instead of using fixed priority, an alternative approach is to use *variable priority*. As an example of this, consider the logic diagrams shown in Figure 15.7, where a little modification of the iterative logic cell in Figure 15.6(a) is made by adding an OR gate along with the priority input, p_i. Each priority input, p_i, determines the priority of the device associated with it.

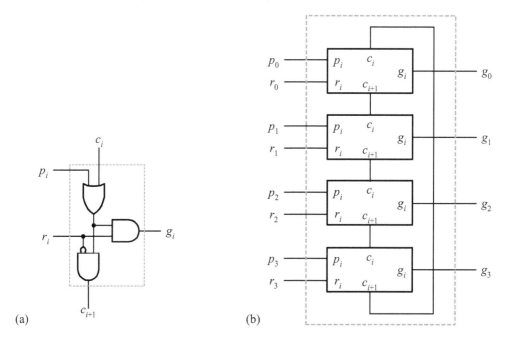

Figure 15.7: The logic diagrams of (a) a single-bit cell and (b) a 4-request variable-priority arbiter.

The means to generate priority p_i determines the priority of devices. For example, the priority of devices is to shift one position if a shift register is used to generate the priority p_i. The priority is random if it is generated randomly by a random number generator. The random number can be readily generated by using the PR-sequence generators described in Section 9.5.1.

Although the above two possible variable-priority schemes can determine the priority dynamically, they are still not good enough for practical use because they could not provide strong fairness. To see this, consider the case that two adjacent devices with requests r_i and r_{i+1} repeatedly request the bus. Request r_{i+1} wins the arbitration only when p_{i+1} is true and request r_i wins the arbitration for the other possible priority inputs. Thus, the request r_i wins the arbitration $n-1$ times often than request r_{i+1}.

To overcome the above unfairness, in most digital systems a priority scheme known as *round-robin priority* is often used. In this scheme, the device just being served is made the lowest priority whereas its succeeding device is made the highest priority. Based on this rule,

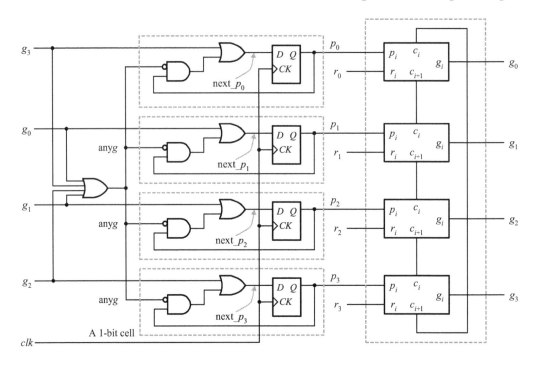

Figure 15.8: The logic diagram of a 4-request round-robin bus arbiter.

an iterative logic cell can be constructed, as shown on the left of Figure 15.8. The priority p_i is set to 1 if its preceding device has been granted its request on the use of the bus. The logic function of the next-p_i is

$$\text{next-}p_i = \overline{anyg} \cdot p_i + g_{(i-1) \bmod n} \tag{15.2}$$

An example of a 4-request round-robin priority bus arbiter is shown in Figure 15.8. If no grant was issued, the $anyg$ signal is 0, thereby causing the priority to remain unchanged. If a grant was issued at the current cycle, one of g_i must be 1. Hence, $anyg$ will be 1 and cause p_{i+1} to be 1 at the next cycle. As a result, the request next to the one receiving the grant has the highest priority and the request being served has the lowest priority.

■ Review Questions

15-1 Explain what a bus and a bus structure are.
15-2 What are the two basic bus structures?
15-3 Compare the differences between the tristate bus and multiplexer bus structures.
15-4 Explain the operation of the 4-request round-robin bus arbiter in Figure 15.8.
15-5 What are the two most widely used bus arbitration schemes?
15-6 What is the meaning of round-robin priority?

15.2 Data Transfer

When the data transfer between two devices is synchronized by a common clock signal, it is called a *synchronous mode*; otherwise, it is called an *asynchronous mode*. The actual data transfer can be proceeded in parallel or in series. Parallel data transfer means that a set of data

Section 15.2 Data Transfer

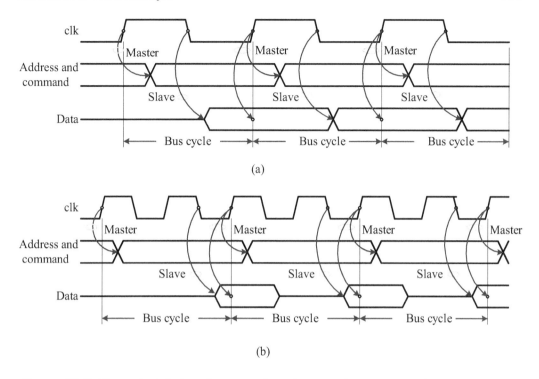

Figure 15.9: The typical timing diagrams of (a) single-cycle and (b) multicycle bus cycles.

bits is transferred at the same time while serial data transfer means that the data is transferred in a bit-by-bit manner. In this section, we deal with all these types of data transfer.

15.2.1 Synchronous Data Transfer

As we have discussed, a bus is a set of wires and has the goal to transfer messages from one source to one or more destinations. In general, a bus works in units of cycles, messages, and transactions. A message is a logical unit of information transferred between a source and one or more destinations. A transaction consists of a sequence of messages that are strongly related. It is initiated by one message and consists of a chain of related messages generated in response to the initiating message.

As mentioned before, a bus must be scheduled before it can transfer messages. Once a bus is sequenced, the transfer of messages onto it can then be proceeded in a synchronous or an asynchronous way. Regardless of which way is used, there must be a signal or means to indicate when the data on the bus are valid or stable. In synchronous data transfer, the data is transferred under the control of the clock signal directly, whereas in asynchronous data transfer, the data is transferred under the control of either a *strobe* or a *handshake*.

15.2.1.1 Parallel Data Transfer In the synchronous data transfer mode, each transfer is in synchronism with a clock signal, as shown in Figure 15.9. In other words, the receiver samples and latches the data at the specified edge, positive or negative, of the clock signal. In a synchronous bus system, the device generating addresses and commands to the bus is called a *bus master* and the device receiving addresses and commands from the bus is called a *bus slave*.

Synchronous bus transfers can be further divided into two types: the single-cycle and multicycle bus cycles. The single-cycle bus cycle, as shown in Figure 15.9(a), needs only one clock cycle to complete a data transfer. At the positive edge of the clock signal, the bus master sends

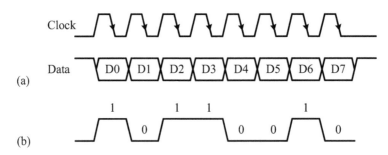

Figure 15.10: The data format of synchronous serial data transfer with: (a) explicitly clocking and (b) implicitly clocking (NRZ code).

the address and command, and latches the data read on the previous cycle; at the negative edge of the clock signal, the bus slave sends the data.

Figure 15.9(b) shows an example of multicycle bus cycles, which requires two clock cycles to complete a data transfer. At the positive edge of the first clock signal, the bus master sends the address and command, and latches the data read on the previous cycle; at the negative edge of the second clock signal, the bus slave sends the data. The number of clock cycles required is usually 2 or 3. Sometimes much more clock cycles are required in a multicycle bus cycle. The actual number of clock cycles needed is determined by the operating speeds of the devices on the bus.

15.2.1.2 Serial Data Transfer Like parallel data transfer, serial data transfer can also be proceeded in either an asynchronous or a synchronous way. In synchronous serial data transfer, the clock signal is sent along with data implicitly or explicitly. Sending the clock signal with data explicitly as a separate signal is an intuitive solution for synchronous serial data transfer and is often referred to as an *explicitly clocking scheme*. A simple example is shown in Figure 15.10(a), where the negative edges of the clock signal are used to capture the data. The major drawback of an explicitly clocking scheme is that at high signaling speeds, where the wire delays become significant, the different delays of both data and clock wires may cause the incorrect data to be sampled at the destination. Thereby, the explicitly clocking scheme is usually limited its use in low-speed serial data transfer, especially for those cases below several Mbps.

To overcome the above difficulty and to improve data rates, a widely used approach is that the clock signal is encoded into the data being sent at the transmitter and then extracted at the receiver before the data are sampled. Such an approach is known as an *implicitly clocking scheme*. A widely used example is the *non-return-to-zero* (NRZ) code, as illustrated in Figure 15.10(b). A NRZ code is a binary code in which a logic 1 is denoted by a positive voltage and a logic 1 is represented by a negative voltage. The implicitly clocking scheme is an approach widely used in high-speed serial data transfer, especially for those applications adopting modern high-speed I/O buses.

15.2.1.3 Serdes Interface The term "SerDes" means the *serializer/deserializer* interface that is a pair of functional blocks commonly used in high-speed communications over a single/differential line in order to minimize the number of I/O pins and interconnects. The pair of functional blocks convert data between serial data and parallel interfaces in each direction. As illustrated in Figure 15.11, the basic SerDes function is made up of two functional blocks: the parallel-in, serial-out (PISO) shift register (i.e., parallel-to-serial converter) and the serial-in, parallel-out (SIPO) shift register (i.e., serial-to-parallel converter). The input and output shift registers may be replaced with a multiplexer and a demultiplexer, respectively, if the input data is already latched and the output data will be latched somewhere immediately.

Section 15.2 Data Transfer

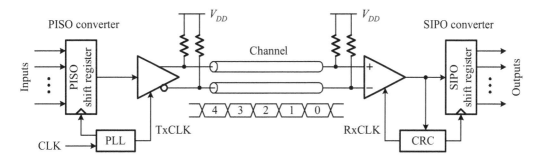

Figure 15.11: A conceptual illustration of the SerDes interface.

At the input end, the input parallel data are converted into the serial output via the PISO shift register and transmit to the output end along with the clock signal generated by a *phase-locked loop* (PLL) through the channel via LVDS signals. At the output end, the clock embedded in the data stream is first recovered with a *clock-recovery circuit* (CRC) and then used to extract the bit data from the data stream. After this, the bit data are output in parallel to the destination module or system through the SIPO shift register.

■ Review Questions

15-7 Describe the operations of synchronous bus transfer.
15-8 Describe the meaning of explicitly clocking schemes.
15-9 Describe the meaning of implicitly clocking schemes.
15-10 What is the major drawback of explicitly clocking schemes?
15-11 Describe the two kinds of synchronous serial data transfer.

15.2.2 Asynchronous Data Transfer

In asynchronous data transfer, each data transfer occurs at random; that is, it cannot be predicted in advance. The data transfer may be controlled by using either a strobe or handshake. It is worth noting that both the strobe and the handshake are used extensively on numerous occasions that require the transfer of data between two asynchronous (namely, independent) devices. In this section, we consider these two schemes in detail.

15.2.2.1 Strobe In the strobe scheme, only one control signal, known as a *strobe* signal, is used. When there are data to be transferred between two devices, a strobe signal is enabled by either the source or destination device, depending on the actual application.

Source-initiated transfer. The detailed operations of source-initiated strobe-con-trolled data transfer are depicted in Figure 15.12(a). The source device places the data onto the data bus and then asserts the strobe signal to notify the destination device that the data is available. The destination device then samples and stores the data onto its internal register at the negative edge of the strobe signal. The strobe signal is deasserted by the source device and the data transfer cycle is completed. The data transfer from the CPU to a memory location is such an example, where the write control signal is the strobe signal.

Destination-initiated transfer. The detailed operations of destination-initiated strobe-controlled data transfer are depicted in Figure 15.12(b). The destination device asserts the strobe signal to request data from the source device. Once the source device has received this strobe signal, it places the data onto the data bus for a duration long enough for the destination device to read. Usually, the destination device samples and stores the data onto its internal register at

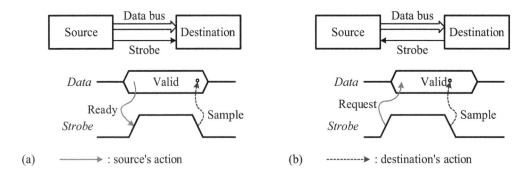

Figure 15.12: Timing diagrams of (a) source-initiated and (b) destination-initiated strobe-controlled data transfer.

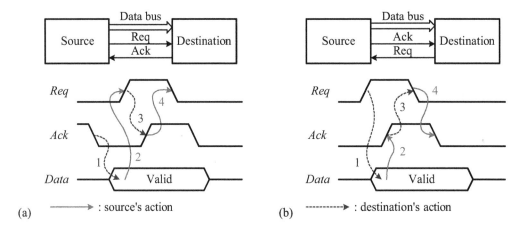

Figure 15.13: Timing diagrams of (a) source-initiated and (b) destination-initiated handshaking data transfer.

the negative edge of the strobe signal. The strobe signal is deasserted by the destination device and the data transfer cycle is completed. The data transfer from a memory location to the CPU is such an example, where the read control signal is the strobe signal.

In summary, the strobe scheme presumes that the requested device is always ready for transferring data once it receives the request. Unfortunately, this is not always the case. In many applications, the requested device is not ready when it is being requested to send data. Therefore, another control signal must be used for the requested device to indicate whether it is ready. The resulting scheme is known as the handshake, which is a two-way control scheme. In contrast, the strobe scheme is a one-way control scheme.

15.2.2.2 Handshake Conceptually, a *handshake* is a technique that provides a two-way control scheme for asynchronous data transfer. In this kind of data transfer, each data transfer is sequenced by the edges of two control signals, request (*req* or valid) and acknowledge (*ack*), as shown in Figure 15.13. In handshaking data transfer, four events are proceeded in a cyclic order. These events are *ready (request)*, *data valid*, *data acceptance*, and *acknowledge*.

Source-initiated transfer. Like the strobe scheme, an asynchronous data transfer controlled by the handshake can be initiated by either a source device or a destination device. A conceptual illustration of source-initiated handshaking data transfer is shown in Figure 15.13(a). The sequence of events is as follows:

Section 15.2 Data Transfer

1. *Ready*: The destination device deasserts the acknowledge signal and is ready to accept the next data.
2. *Data valid*: The source device places the data onto the data bus and asserts the request (req, or valid) signal to notify the destination device that the data on the data bus are valid.
3. *Data acceptance*: The destination device accepts (latches) the data from the data bus and asserts the acknowledge signal.
4. *Acknowledge*: The source device invalidates data on the data bus and deasserts the request (valid) signal.

Because of the inherent back and forth operations of the scenario, the asynchronous data transfer described above is referred to as the *handshaking protocol* or *handshaking data transfer*.

Destination-initiated transfer. A conceptual illustration of destination-initiated handshaking data transfer is shown in Figure 15.13(b). The sequence of events is as follows:

1. *Request*: The destination device asserts the request signal to request data from the source device.
2. *Data valid*: The source device places the data on the data bus and asserts the acknowledge (ack, or valid) signal to notify the destination device that the data are valid now.
3. *Data acceptance*: The destination device accepts (latches) the data from the data bus and deasserts the request signal.
4. *Acknowledge*: The source device invalidates data on the data bus and deasserts the acknowledge (ack, or valid) signal to notify the destination device that it has removed the data from the data bus.

Two-phase handshake. The above handshaking scheme is referred to as the *four-phase handshake* because it needs four events to complete an active cycle. In fact, a handshake can be accomplished by merely using two events. This kind of handshake is known as the *two-phase handshake*.

In the two-phase handshake, only two phases of operations can be distinguished, namely, the active cycle of the source device and the active cycle of the destination device. As shown in Figure 15.14(a), once the source device has recognized the acknowledge signal from the destination device, it places valid data onto the data bus and then notifies the destination device by generating a transition on the request signal. The destination device makes a transition on the acknowledge signal to recognize this event and latch into the data on the data bus. Consequently, the request event terminates the active cycle of the source device and the acknowledge event ends the active cycle of the destination device. Figure 15.14(b) exhibits the timing diagram of the four-phase handshaking scheme in two cycles for comparison.

The important features of the two-phase handshake are of being simple and fast. However, it needs the detection of transitions, which may not be applicable to some applications, such as modern CMOS technologies that typically tend to be sensitive to levels or transitions on only one direction. As a result, the (four-phase) handshaking data transfer depicted in Figures 15.13 and 15.14(b) is more widely used.

Failures of handshakes. A major problem of handshaking data transfer, regardless of whether the four-phase or two-phase handshake is used, is that the data transfer will not be finished if one device is faulty. Fortunately, such an error can be detected by using a *timeout mechanism* to give rise an alarm if the destination device does not respond within a specified time interval. In practice, the timeout mechanism may be realized by using a timer that is restarted each time the source/destination device begins a data transfer. A broad variety of timers will be introduced in detail later in this chapter.

15.2.2.3 Asynchronous Serial Transfer In asynchronous serial data transfer, the clock signal is not sent with data. Instead, the receiver generates its own local clock signal which

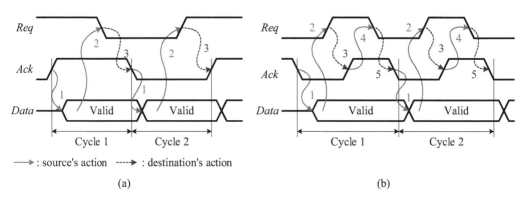

Figure 15.14: Timing diagrams of (a) two-phase and (b) four-phase handshaking data transfer.

is then used to capture the data being received. To make the asynchronous serial data transfer between two devices more reliable, the data are transmitted one byte at a time and are encapsulated in the standard data format shown in Figure 15.15. Each byte is packed as a frame consisting of a start bit, a number of data bits, an optional parity bit, and one or more stop bits. To guarantee at least one transition for each byte (namely, each frame), the start bit is set to a logic 0 whereas the stop bit is set to a logic 1. Whenever there are no data to be sent, the transmitter continuously sends 1s in order to maintain a continuous communication channel. The receiver monitors the channel continuously until the start bit is detected. Then, it starts to receive a new data frame.

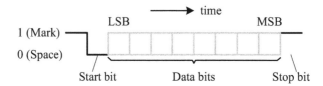

Figure 15.15: The data format of asynchronous serial data transfer.

Because in asynchronous serial data transfer, the data being received (called RxD) and the sampling pulse (known as RxC) are not synchronized, the sampling point might occur in any position of the bit time of the receiving data. In order to sample the data being received as at or nearby the center of the bit time as possible, the sampling clock frequency (RxC) at the receiver is often set to n times the transmission clock frequency, TxC, where n may be 1, 4, 16, or 64. Based on this idea, the receiver samples the start bit after $n/2$ counts of the RxC cycles once the negative edge of the RxD is detected (i.e., the beginning of the start bit). Then, it samples each bit every n RxC cycles thereafter.

The need that a sampling clock frequency is often higher than the transmission clock frequency can be illustrated by Figure 15.16. Figure 15.16(a) shows the case where the sampling clock frequency is identical to the transmission clock frequency. In this case, only one sampling pulse is available for each bit time. At the worst case, the sampling pulse might occur at the edge of the bit time so that the resulting sampled value may be easily erroneous. One way to alleviate this situation is to increase the sampling clock frequency. To see this, consider Figure 15.16(b), where each bit time contains four RxC pulses. The maximum deviation of the sampling pulse is only 25% of the bit time in this case. Generally, the situation is much better as a higher sampling clock frequency is used.

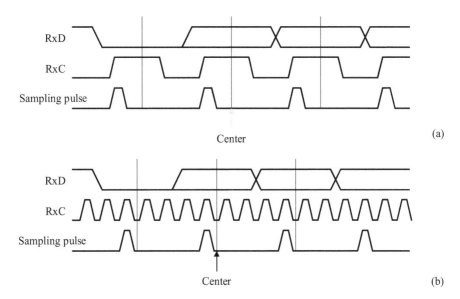

Figure 15.16: The reception of asynchronous serial data transfer at two different sampling frequencies: (a) RxC = TxC and (b) RxC = 4×TxC.

■ **Review Questions**

15-12 Describe the operations of asynchronous data transfer.
15-13 What are the four events associated with handshaking data transfer?
15-14 Describe the operations of two-phase handshaking data transfer.
15-15 Describe the operations of four-phase handshaking data transfer.
15-16 Distinguish between synchronous and asynchronous serial data transfer.

15.2.3 Multiple-Clock Domains

In many practical complex digital systems, *multiple-clock domain* (MCD) (also called *crossing clock domains*) design [1, 3] is ubiquitous and inevitable. In multiple-clock domain designs, we need to ensure that signals/data crossing clock domains are properly synchronized. Here, the word "synchronized" means that asynchronous signals/data are sampled in such a way that they have transitions synchronized to a local or sample clock signal. To achieve this, the following techniques are generally combined into use: synchronizers (Section 9.1.3), handshakes, and FIFOs. In what follows, we address these in the context of reset signals, detection of signals, passing control/data signals, and datapath synchronization.

15.2.3.1 Reset Signals Generally, the entire digital system with multiple-clock domains may be reset by a single reset signal. Like other signals crossing clock domains, the reset signal is also subject to metastability and must be protected by synchronizers. Since by definition all flip-flops (registers) are reset to their initial values and the reset signal will be activated for an enough number of cycles to allow any metastability to settle out, it is no need to synchronize on the activation edge of the reset signal. Nevertheless, as a reset signal is deactivated, it must be properly synchronized so that flip-flops do not become metastable when they come out of the reset state. To this end, a separate synchronizer for the reset signal as it enters each clock domain in the digital system should be used.

Recall that reset signals can be synchronous or asynchronous, depending on the chip, board, or system requirements. As reset signals are deasserted asynchronously or synchronously, two timing constraints, *reset recovery time* and *reset removal time*, need to be considered [1]. These two timing constraints may be checked by using timing checks provided by Verilog HDL, referring to Section 7.3.3 for details.

Reset recovery time ($t_{recovery}$) refers to the minimum time required between the deassertion of the reset signal and the arrival of the active clock edge. It is similar to the setup-time requirement in flip-flops. Basically one should not release the reset signal during this time period. Violating the reset recovery time may cause metastability on flip-flop outputs. *Reset removal time* ($t_{removal}$) is the minimum time required between the active clock edge and the deassertion of the reset signal that follows immediately. This is similar to the hold-time requirement in flip-flops. One should not release the reset signal in this time period. Reset removal-time violations may occur if there are slight differences in propagation delay in either (or both) the reset signal or (and) the clock signal, which may cause some flip-flops to exit the reset state before the others. To ensure that the reset signal does not have these two timing-related problems, ASIC designers often spend a lot of time to ensure their reset logic works correctly by meticulously running simulations on both RTL and gate-level netlists.

Synchronizing an asynchronous reset. The approach to synchronizing an asynchronous reset signal is conceptually illustrated in Figure 15.17 by using a two-stage synchronizer. As mentioned, since a single-stage synchronizer may cause its output to be metastable for a period of time, two-stage synchronizers will be used in the rest of this section. From Figure 15.17, the reset signal (*reset_n*) from the outside of the system is connected to the *CLR* inputs of D flip-flops in the synchronizer. The D-input of the first-stage D flip-flop is tied to V_{DD}. As the *reset_n* signal is deasserted asynchronously, the *master_reset_n* signal is removed synchronously with the synchronizer with a guaranteed full clock cycle recovery-time and removal-time requirement within the system. Since the *master_reset_n* signal needs to drive a large number of flip-flops, a reset buffer tree, a structure like a clock tree, is often used to distribute the *master_reset_n* signal within the system.

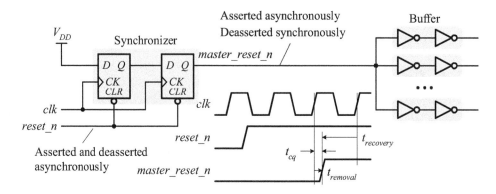

Figure 15.17: An example of synchronizing an asynchronous reset signal.

Synchronizing an asynchronous reset in MCDs. Once a global asynchronous reset (*reset_n*) signal is synchronized with a system reference clock (*clk*) signal, the synchronized reset (*master_reset_n*) signal can be used in a system with multiple-clock domains, as shown in Figure 15.18. Within each clock domain, the globally synchronized reset (*master_reset_n*) signal is synchronized with its local clock signal again by using a synchronizer clocked by its local clock signal.

At the top level, the global *reset_n* signal is asynchronously asserted and synchronously deasserted. When the reset (*reset_n*) signal is asserted, all synchronizers are reset asynchronously. The synchronously deasserted (*reset_n*) signal propagates through the synchronizers in modules A and B, where they are reset down to the user logic via *clk_a_reset_n* and *clk_b_reset_n*,

Section 15.2 Data Transfer

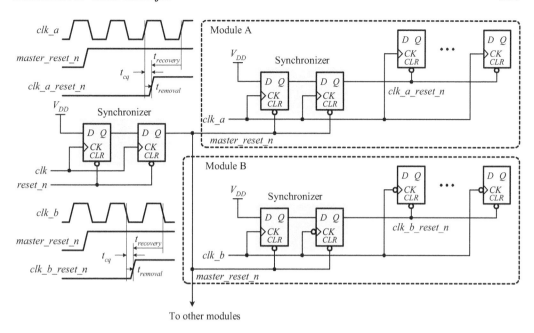

Figure 15.18: An example of synchronizing asynchronous resets in multiple-clock domains.

respectively. When the *reset_n* signal is deasserted, both D flip-flops in the synchronizer still hold their outputs low until the next active clock edge of *clk* is arrived. When the active *clk* edge arrives close to the time that *reset_n* is removed from both the D flip-flops of the synchronizer, there is no metastability risk on the second-stage D flip-flop since the D-input value of the first-stage D flip-flop is already at a stable value [7] and thereby, even unfortunately entering the metastable state, its output will quickly recover to a logic 1 unambiguously. Because the synchronizer is essentially a two-stage shift register, after two *clk* clock cycles, the synchronizer outputs a logic 1. The reset (*clk_a_reset_n*) signal of Module A is also metastability safe for the same reason as the top-level reset (*reset_n*) circuit, regardless of when *clk_a* arrives relative to the reference clock signal, *clk*.

As both clock edges are needed within a system or module due to design requirements, a half clock cycle recovery-time and removal-time requirement is often difficult to meet in high-speed clock domains. One way to solve this is to split the second-stage D flip-flop of the synchronizer by the clock edge [1], as depicted in Module B of Figure 15.18, so as to allow the reset (*clk_b_reset_n*) signal of Module B to have a full clock cycle to meet the recovery-time and removal-time requirement.

Sequencing resets in a MCD design. As many modules in a MCD design need to be reset in sequence, the scheme shown in Figure 15.19 may be employed to ensure that data are not accessed from a module until that module completely comes out of reset. In this scheme, the modules within the *clk_a* domain are reset first, followed by the modules within the *clk_b* domain, and so on.

15.2.3.2 Detection of Signals
As a signal passing from a clock domain to another, we often need to ensure that the signal is properly sampled and captured. To illustrate this, in what follows we cast the discussion into two cases: the *level-detection method* and the *edge-detection method*.

The level-detection method. While passing a signal from a slow clock domain to a fast clock domain, the control signal usually has a pulse width much wider than one clock cycle of the fast clock domain. In this case, the control signal will be sampled many times and hence

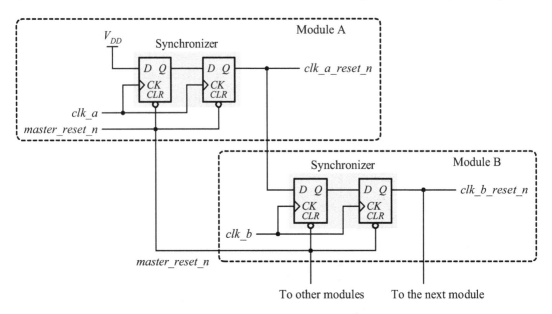

Figure 15.19: An example of the proper sequence of resets between MCD modules.

be regarded as many control events, which may cause some undesired effects, such as to make a counter count up many times. To avoid this, a single-cycle generator needs to be used.

On the other hand, while passing a signal from a fast clock domain to a slow clock domain, the control signal usually has a pulse width much narrower than one clock cycle of the slow clock domain. In this case, the control signal may not be detected and sampled by the synchronizer within the slow clock domain. One way to solve this is to stretch the control signal within the fast clock domain before it is sent out. The rule of thumb is to ensure that the input control signal must be stable for at least three clock edges of the clock signal in the slow clock domain. In other words, the control signal must be wider than 1.5 times the clock period within the receiving clock domain.

An illustration of the level-detection method is shown in Figure 15.20, where the single-cycle generator consists of a synchronizer and a single-cycle pulser. Here, we assume that the control signal y must be wider than 1.5 times the clock period clk_b within the receiving clock domain. The synchronizer within Module B captures the input control signal y reliably while the single-cycle pulser generates a single-cycle output at the second positive edges of clk_b after the input control signal y becomes inactive.

■ Example 15.4: Detection of a wide control signal.

In this example, we assume that clk_a has a period 1.5 times larger than clk_b at least. As a consequence, the control signal y is wider than 1.5 times the clock period of clk_b at least. In the mcd_signal_wide_a module, an **always** block describes the D flip-flop. In the mcd_-signal_wide_b module, an **always** block describes the three-stage shift register and an **assign** continuous assignment describes the NOT gate and the AND gate, which combine with the third D flip-flop to form the single-cycle pulser.

```
// an example of detecting a wide signal in MCD
// module a
module mcd_signal_wide_a(
    input  clk, reset_n, x,
    output reg y);    // the output signal
```

Section 15.2 Data Transfer

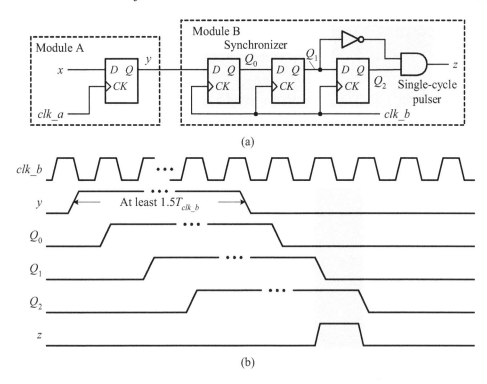

Figure 15.20: An illustration of the level-detection method: (a) logic circuit; (b) timing diagram.

```
// the body of module a
always @(posedge clk or negedge reset_n)
   if (!reset_n) y <= 1'b0;
   else          y <= x;
endmodule

// module b
module mcd_signal_wide_b(
      input   clk, reset_n, y,
      output  z);  // the output signal
// the body of module b
reg [0:2] qout;   // an alias of Q[0:2]
// a synchronized pulse generator
always @(posedge clk or negedge reset_n)
   if (!reset_n) qout <= 0;
   else          qout <= {y, qout[0:1]};
assign z = ~qout[1] & qout[2];
endmodule
```

The edge-detection method. In the level-detection method, the input control signal to be synchronously sampled must be wide enough so that it may be detected and captured reliably by the receiving module. In other words, the transmitting module must know the clock period of the receiving module and stretches the signal to be sent. Unfortunately, this is generally not an easy task. To relax this restriction, a *pulse stretcher* may be added in front of the syn-

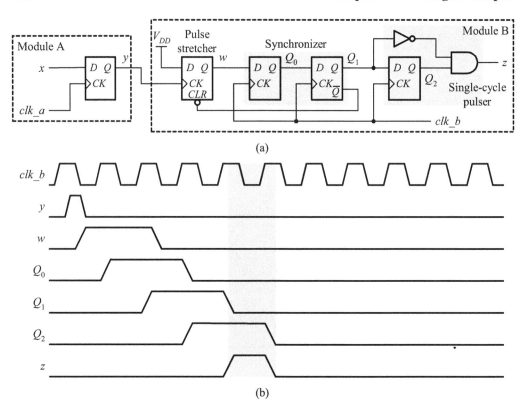

Figure 15.21: An illustration of the edge-detection method: (a) logic circuit; (b) timing diagram.

chronizer to capture the edge transition of the input control signal without concerning the pule width. The resulting scheme is referred to as the *edge-detection method*.

As an illustration, consider the logic circuit depicted in Figure 15.21. A pulse stretcher is added in front of the synchronizer. The pulse stretcher consists of a D flip-flop with its D input being connected to V_{DD} and its *CLR* being connected to the complemented output of the synchronizer. It captures the positive edge of control signal y and generates a pulse with a width of two clock cycles. The synchronizer and the single-cycle pulser then capture and convert the two-cycle pulse into a single-cycle control signal z. This method has one advantage that it does not need to know the clock periods of both sending and receiving clock domains in advance; thereby it may operate with much wider range of clock frequencies. Of course, it may also be applied to the case that the input control signal is wider than the clock period of the receiving module.

■ Example 15.5: Detection of a narrow control signal.

In this example, we assume that *clk_a* has a period smaller than *clk_b*. Hence, the control signal y is narrower than the clock period of *clk_b*. In the mcd_signal_narrow_a module, an **always** block describes the D flip-flop. In the mcd_signal_narrow_b module, the first **assign** continuous assignment and the first **always** block describe the pulse stretcher. The second **always** block describes the three-stage shift register and the second **assign** continuous assignment describes the NOT gate and the AND gate, which combine with the third D flip-flop to form the single-cycle pulser.

```
// an example of detecting a narrow signal in MCD
```

Section 15.2 Data Transfer

```verilog
// module a
module mcd_signal_narrow_a(
      input   clk, reset_n, x,
      output reg y);    // the output signal
// the body of module a
always @(posedge clk or negedge reset_n)
   if (!reset_n) y <= 1'b0;
   else          y <= x;
endmodule

// module b
module mcd_signal_narrow_b(
      input   clk, reset_n, y,
      output z);    // the output signal
// the body of module b
reg [0:2] qout;    // an alias of Q[0:2]
reg w;
wire clear_n;
// a pulse stretcher
assign clear_n = reset_n & ~qout[1];
always @(posedge y or negedge clear_n)
   if (!clear_n) w <= 0;
   else          w <= 1'b1;
// a synchronized a single-cycle pulser
always @(posedge clk or negedge reset_n)
   if (!reset_n) qout <= 0;
   else          qout <= {w, qout[0:1]};
assign z = ~qout[1] & qout[2];
endmodule
```

15.2.3.3 Passing Control/Data Signals While synchronizing a signal from a clock domain to another, we need to ensure that the signal is properly to be captured in the second clock domain. To guarantee this, two general approaches are often used: the *strobe* (i.e., an open-loop control scheme) and the *handshake* (i.e., a closed-loop control scheme). The former does not involve the feedback acknowledgment but the latter does. Despite which approach is used, only the *control signals* need to be synchronized with synchronizers.

Strobe. As mentioned, the strobe is a one-way control method and hence it is an open-loop control scheme. To illustrate this, consider the conceptual block diagram depicted in Figure 15.22, where a *synchronized signal sampler* in Module B is utilized to sample the control signal E_a from module A and generate a synchronized control signal E_b to load the control signals from Module A into a register, thereby resulting in the synchronized control signals in Module B. The synchronized signal sampler can be the logic circuit designed with the level-detection method or the edge-detection method, depending on which is feasible and cost-effective.

■ **Example 15.6: An example of open-loop control.**

An example of the strobe approach is illustrated in Figure 15.23, assuming that *clk_a* is much slower than *clk_b*. A synchronized signal sampler in Module B is utilized to sample the control signal E_a from module A and generate a synchronized control signal *CE* to load the data from Module A into a register, thereby resulting in the synchronized D_{out} in Module B. In this example, a double-edge detector is used instead of the single-cycle pulser described previously.

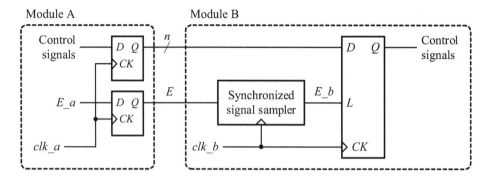

Figure 15.22: A conceptual block diagram of strobe control.

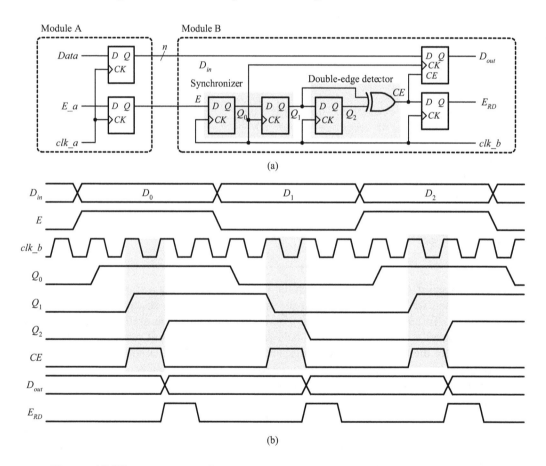

Figure 15.23: An example of strobe control: (a) logic circuit; (b) timing diagram.

The double-edge detector detects each edge of the E signal and generates a single-cycle pulse output. From Figure 15.23(b), the double-edge detector outputs a single-cycle pulse at the second positive edges after detecting the transition of the E signal. At the following positive edge of clk_b, the synchronized data output D_{out} are obtained in Module B.

Handshake. Another approach to passing control/data signals from one module to another is by using the handshake. As mentioned, the handshake is a two-way control method which

Section 15.2 Data Transfer

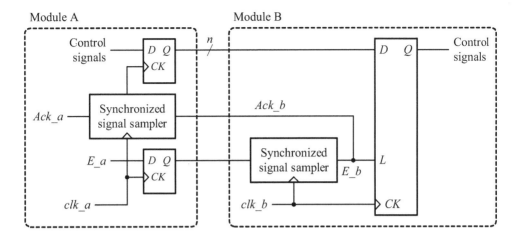

Figure 15.24: A conceptual block diagram of handshaking control.

can ensure that the destination clock domain properly captures the control/data signals. As shown in Figure 15.24, after capturing the signal coming from Module A, the captured signal E_b is sent back to Module A as an acknowledgment (ack_b) using another synchronized signal sampler. If the ack_b signal does not come back within a certain period of time (or the specific number of clock cycles), the signal E_a is resent to Module B.

15.2.3.4 Datapath Synchronization As multiple data bits, such as in the datapath, are involved, using synchronizers proves impractical since in this case, skew in the data bits may lead to data being captured incorrectly. As a consequence, in practice the datapath synchronization between clock domains often uses handshakes or FIFOs

Handshake. Whenever data are passed between clock domains, a handshake may be used to ensure proper synchronization. Nonetheless, as mentioned the disadvantage of using the handshake is that the more the control signals are used for the handshake, the longer the latency is to pass data between clock domains. For this reason, a simple two-phase handshaking sequence, depicted in Figure 15.14(a), is utilized in most practical cases and is proved sufficient.

To illustrate how the handshaking control works, consider the one shown Figure 15.25. In module A (i.e., the source clock domain), both the data and the control signal (Req_a / Req) are sent to module B (i.e., the destination clock domain). After synchronizing the control signa and hence generated Req_b, module B captures the data into a register. To properly synchronize, the control signal (Req) sent out from module A must be greater than one clock cycle of module B. The synchronized control signal (Req_b) is then sent back to module A as an acknowledgement signal (Ack_b). As the Ack_a signal is generated, module A may change the value being driven on the data bus. The control signal (Req_a) from module A may be deasserted as the Ack_a signal is received from module B. This ensures that control signal has been captured correctly in module B.

■ Example 15.7: A handshaking control example.

In this example, a two-phase handshake is used to transfer data from Module A to Module B. Both handshaking control signals, Req and Ack, have the active width much wider than the individual clock period of clk_a and clk_b. In the Module A, the first **always** block describes the data register. The second **always** block describes the T flip-flop, which is employed to generate the Req output signal. The third **always** block models the synchronized double-edge detector that detects the transition edges of the Ack signal from Module B and generates a single-cycle pulse (*Ack_a*) on each edge of the *Ack* signal. Similarly, in the Module B, the first **always** block describes the data register. The second **always** block describes the *T* flip-flop, which is employed to generate the *Ack* output signal. The third **always** block models the synchronized double-edge detector that detects the transition edges of the *Req* signal from Module A and generates a single-cycle pulse (*Req_b*) on each edge of the *Req* signal.

```
// an example of handshaking control
// module a
module data_synchronization_handshaking_a
        #(parameter n = 8)( // the default size
          input clk, reset_n, Ack_in,
          output reg Req_out,   // the Req generator
          output reg [n-1:0] Data_out);  // the output data
// the body of module a
reg [0:2] qout;
wire Ack_a, Req_a, DT;
// the data register
always @(posedge clk or negedge reset_n)
   if (!reset_n)   Data_out <= 0;
   else   if (Ack_a) Data_out <= {$random % 2^n};
// the T flip-flop
assign DT = Req_out ^ Req_a;
assign Req_a = Ack_a;
always @(posedge clk or negedge reset_n)
   if (!reset_n) Req_out <= 1'b0;
   else          Req_out <= DT;
// a synchronized double-edge detector
always @(posedge clk or negedge reset_n)
   if (!reset_n) qout <= 0;
   else          qout <= {Ack_in, qout[0:1]};
assign Ack_a = qout[1] ^ qout[2];
endmodule

// module b
module data_synchronization_handshaking_b
        #(parameter n = 8)(    // the default size
```

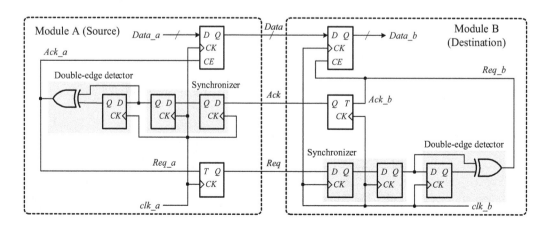

Figure 15.25: Datapath synchronization using a two-phase handshaking control scheme.

Section 15.2 Data Transfer

```verilog
            input  clk, reset_n, Req_in,
            input  [n-1:0] Data_in,
            output reg Ack_out); // the acknowledgement
// the body of module b
reg [0:2] qout;
reg [n-1:0] Data_b;
wire Ack_b, Req_b, DT;
// the data register
always @(posedge clk or negedge reset_n)
   if (!reset_n)     Data_b <= 0;
   else if (Req_b) Data_b <= Data_in;
//   the T flip-flop
assign DT = Ack_out ^ Ack_b;
assign Ack_b = Req_b;
always @(posedge clk or negedge reset_n)
   if (!reset_n) Ack_out <= 1'b1;
   else          Ack_out <= DT;
// a synchronized double-edge detector
always @(posedge clk or negedge reset_n)
   if (!reset_n) qout <= 0;
   else          qout <= {Req_in, qout[0:1]};
assign Req_b = qout[1] ^ qout[2];
endmodule
```

In summary, as data are passed crossing clock domains, the handshake can ensure proper synchronization. Nevertheless, the drawback of using the handshake is that the more the control signals are used for the handshake, the longer the latency is to pass data between clock domains. Fortunately, in most practical cases a simple two-phase handshake is sufficient. In those cases that a low latency is needed, a FIFO can be employed.

FIFO. Another way to pass data between clock domains is to use a FIFO. A FIFO or queue is an array of memory that is commonly used in hardware or software as an *elastic buffer* to coordinate different data access rates between two modules. A FIFO may be synchronous or asynchronous. In an asynchronous FIFO, there is no clock signal. In a synchronous FIFO, there may be one or two clock signals since some FIFOs have separate clocks for read and write operations. Asynchronous FIFOs are not commonly used nowadays because synchronous FIFOs have improved interface timing.

Generally, a FIFO can be realized in many different ways. One simple way is to use a register file or a synchronous memory with two rotating pointers to indicate the read and write addresses, referred to as the *read pointer* (rd_ptr) and the *write pointer* (wr_ptr), respectively, as shown in Figure 15.26. The item pointed by the write pointer is often referred to as the *head* while by the read pointer as the *tail*. At beginning, both the read and write pointers of the FIFO point to the same location. In a FIFO with n items, after writing the $(n-1)$th item, the write pointer points to the 0th item again.

Except for the register file or the synchronous memory and two binary counters used for write and read pointers, the write enable signal (wr_en) is also needed to detect and reshape into a single-cycle pulse, wr_en_one_clock, in synchronism with the clk signal so as to enable the write pointer and the write operation of the FIFO properly. Similarly, the read enable signal (rd_en) is also needed to detect and reshape into a single-cycle pulse, rd_en_one_clock, in synchronism with the clk signal so that it can properly enable the read pointer. To realize the *full* (fifo_full) and *empty* (fifo_empty) flags, a FIFO counter (fifo_counter) is utilized to memorize the number of items keeping in the FIFO up to now. This FIFO counter counts up

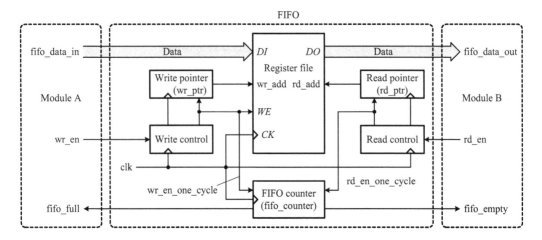

Figure 15.26: Datapath synchronization using a synchronous FIFO.

by one enabled by the wr_en_one_clock signal and counts down by one enabled by the rd_en_one_clock signal. As long as its value is zero, the empty (fifo_empty) flag is set to one and its value is n, the full (fifo_full) flag is set to one.

As an illustration of how to describe the synchronous FIFO depicted in Figure 15.26, consider the following example.

■ Example 15.8: A synchronous FIFO example.

This module consists of seven **always** blocks. The first **always** block updates both empty (fifo_empty) and full (fifo_full) flags by detecting the value of the FIFO counter (fifo_counter). The second and third **always** blocks individually convert the write enable (wr_en) and read enable (rd_en) signals into single-cycle pulses, (wr_en_one_clock) and (rd_en_one_clock). The fourth **always** block updates the FIFO counter (fifo_counter). The next two **always** blocks read and write the FIFO, respectively. The final **always** block updates the read and write pointer, rd_ptr and wr_ptr.

```verilog
// an example of a synchronous FIFO
module fifo_synchronous
        #(parameter FIFO_SIZE = 16,
          parameter FIFO_WIDTH = 8,
          parameter PTR_SIZE = 4)(
          input clk, reset, wr_en, rd_en,
          input [FIFO_WIDTH-1:0] fifo_data_in,
          output reg [FIFO_WIDTH-1:0] fifo_data_out,
          output reg fifo_empty, fifo_full,
          output reg [PTR_SIZE:0] fifo_counter);
// internal variable declarations
reg [PTR_SIZE-1:0] rd_ptr, wr_ptr;
reg [FIFO_WIDTH-1:0] fifo_mem[FIFO_SIZE-1:0];
reg wr_state, rd_state;   // one clock cycle pulse
reg wr_en_one_clock, rd_en_one_clock;
localparam S0 = 1'b0, S1 = 1'b1;

// the body of the synchronous FIFO
// update empty and full flags
```

Section 15.2 Data Transfer

```verilog
always @(fifo_counter) begin
   fifo_empty = (fifo_counter == 0);
   fifo_full = (fifo_counter == FIFO_SIZE);
end
// convert wr_en into one clock cycle signal wr_en_one_clock
always @(posedge clk or posedge reset)
   if (reset) begin wr_en_one_clock <= 1'b0;
               wr_state <= S0; end
   else case (wr_state)
      S0: if (wr_en)begin wr_state <= S1;
               wr_en_one_clock <= 1'b1;end
          else wr_state <= S0;
      S1: begin wr_en_one_clock <= 1'b0;
               if (wr_en) wr_state <= S1;
               else wr_state <= S0; end
   endcase
// convert rd_en into one clock cycle rd_en_one_clock
always @(posedge clk or posedge reset)
   if (reset) begin rd_en_one_clock <= 1'b0;
                    rd_state <= S0; end
   else case (rd_state)
      S0: if (rd_en)begin rd_state <= S1;
               rd_en_one_clock <= 1'b1;end
          else rd_state <= S0;
      S1: begin rd_en_one_clock <= 1'b0;
               if (rd_en) rd_state <= S1;
               else rd_state <= S0; end
   endcase
// update the fifo counter
always @(posedge clk or posedge reset) begin
   if (reset) fifo_counter <= 0;
   else if (wr_en_one_clock && !fifo_full)
      fifo_counter <= fifo_counter + 1;
   else if (rd_en_one_clock && !fifo_empty)
      fifo_counter <= fifo_counter - 1;
end
// read the FIFO
always @( posedge clk or posedge reset)
   if (reset) fifo_data_out <= 0;
   else if (rd_en_one_clock && !fifo_empty)
         fifo_data_out <= fifo_mem[rd_ptr];
// write the FIFO
always @(posedge clk)
   if (wr_en_one_clock && !fifo_full)
      fifo_mem[wr_ptr] <= fifo_data_in;
// update write and read pointers
always @(posedge clk or posedge reset)
   if (reset) begin
      wr_ptr <= 0;
      rd_ptr <= 0; end
   else begin
```

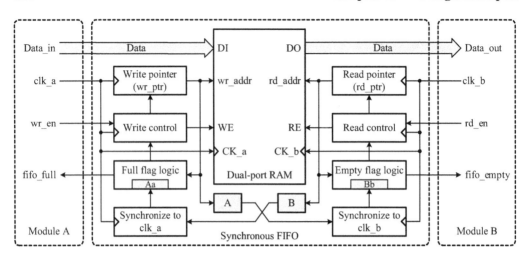

Figure 15.27: Datapath synchronization using a synchronous FIFO with a synchronous dual-port memory (Details of blocks A and B as well as Aa and Bb are left for the reader as an exercise, Problem 15-9).

```
      if (!fifo_full && wr_en_one_clock)  wr_ptr <= wr_ptr + 1;
      if (!fifo_empty && rd_en_one_clock) rd_ptr <= rd_ptr + 1;
   end
endmodule
```

Another way to implement a synchronous FIFO is to use a dual-port memory for storage. As depicted in Figure 15.27, in such a FIFO, one port of the dual-port memory is clocked by the sender (producer or source; here Module A) and the other by the receiver (consumer or destination; here Module B). Except for the dual-port memory, one write (wr_ptr) pointer and one read (rd_ptr) pointer along with a piece of necessary control logic are also needed. Module A writes data into the FIFO using the write enable (wr_en) and clock (clk_a) signals while module B reads data out from the FIFO using the read enable (rd_en) and clock (clk_b) signals. Two status flags, the full flag (fifo_full) and the empty flag (fifo_empty) are utilized to indicate the status of the FIFO. Some designers also use two more flags to indicate when the FIFO is almost full and almost empty. The design of this FIFO is much more complex than the one shown in Figure 15.26 and we would like to leave it for the aggressive, interested reader (Problem 15-9).

■ Review Questions

15-17 What are the multiple-clock domains?

15-18 Explain the rationale behind the use of synchronizers together with reset signals.

15-19 What are the two methods that can be used to reliably detect asynchronous signals in MCD systems?

15-20 What are the two basic control schemes for passing control/data signals in MCD systems?

15.3 General-Purpose Input and Output

In any general μC system, both input and output ports are commonly used in most applications. Hence, a device that can configure its interface as an input, an output, or a bidirectional port

Section 15.3 General-Purpose Input and Output

is often a necessary module in such a system. This device is known as a *general-purpose input and output* (GPIO) port (or module). In practice, a GPIO port often contains 8 to 32 I/O pins with each being able to be programmed as an input or output port on demand. To illustrate how to design such a GPIO port, in this section we detail the design of an 8-bit GPIO module.

15.3.1 Basic Principles

As mentioned previously, a *port* is a physical place where a signal can be passed through between two modules. A port is called a *parallel port* when it can process a bundle of signals as a single unit at the same time. A port is called a *serial port* when it can merely process signals in a serial way, that is, one bit at a time. A port that only allows the signal to be input from the outside world of the system is called an *input port*; a port that only allows the signal to flow out of the system is referred to as an *output port*. A port that can be used as an input or an output port, but not both at the same time, is known as a *bidirectional port*.

15.3.1.1 Architecture of GPIOs
A GPIO port is a device that can be used as an input or output port on demand. It generally has eight bidirectional input/output (I/O) pins and two 8-bit registers: a *data direction register* (DDR) and a *port* (PORT) *register*. The DDR controls the data direction of the bidirectional I/O pins; that is, it determines which I/O pins are inputs and which are outputs. The PORT register holds the data written to the port. It may or may not be read back, depending on how the GPIO port is designed.

Generally speaking, in using a general-purpose device, the two common user's perspectives of the device are of importance: the hardware interface, also called the *hardware model*, and the software interface, also referred to as the *programming model*. For example, consider the 8-bit GPIO port shown in Figure 15.28. The hardware model of the GPIO port is shown in Figure 15.28(a). It has two groups of interface signals. One group is the interface signals to the system, such as the CPU, and the other is the port pins for use with I/O devices. The interface with the system includes an 8-bit *data bus*, an *address signal* (addr) for selecting the PORT register or DDR, a *chip select* (cs) input to enable the chip, a *read and write* (r_w) control input, a *clock input* (clk), and a *reset* (reset_n) control signal. The function table of the GPIO port is shown in Figure 15.28(c). The programming model of a device is the user's view of the device as the device is used. Based on this, the programming model of the GPIO port consists of two registers, the data direction register (DDR) and the port (PORT) register, as shown in Figure 15.28(b).

15.3.1.2 Design Issues
In order to design the GPIO port described above, the following issues are often taken into consideration:

- *Readback capability of the PORT register:* The port register usually holds the output data written into port pins. It may or may not be allowed to read back, depending on how the GPIO port is designed.
- *Group or individual bit control:* Several options can be made to control the programmability of the data direction of port pins. One extreme is to allow the data direction of entire-port pins to be controlled by a single-bit DDR; the other extreme is to use one DDR bit for each port pin, thus allowing the data direction of each port pin to be set individually. Of course, a compromise is to group port pins into several disjoint sets and then allow each set to be programmed into input or output as a unit. The cardinality of the set is usually set to 2 or 4.
- *Selection of the value of the DDR:* Although the value of the DDR used to control the data direction of a GPIO pin is arbitrary, one way is to use 0 for output and 1 for input. The other way uses the reverse values. The former is usually found in devices from Intel and the latter from Motorola (now is Freescale).

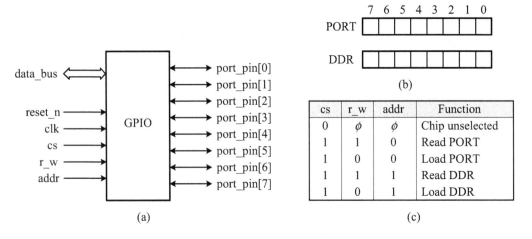

Figure 15.28: The (a) hardware model, (b) programming model, and (c) function table of an 8-bit GPIO port.

- *Handshaking control:* The handshake may also be provided with the GPIO port. These control signals can be provided by two specific I/O pins, which may reside in another GPIO port, or use two dedicated I/O pins, in addition to the GPIO pins.
- *Readback capability of the DDR:* The data written into the DDR may or may not be allowed to read back, depending on how the GPIO port is designed.
- *Input latch:* The input data may also be latched. In some applications, it proves much useful to latch the input data. Hence, the latch capability of input data can be considered as a useful feature of the GPIO port.
- *Input/output pull-up capability:* For a bidirectional I/O port, it usually needs to pull up the input/output pin of the GPIO port to avoid the floating condition as the bidirectional I/O port is used as an output port.
- *Current-driving capability:* What is the amount of current that an input/output pin can sink and source? The I/O pins are usually used to drive external LEDs, or other devices, such as relays. As the current-driving capability is insufficient, some extra circuits are needed to compensate for this and hence increases the cost of the application systems.

One example of GPIO ports is shown in Figure 15.29(a), which is the port-1 structure of the MCS-51 system. It only contains a port register without the DDR. When a logic 1 is written into the port register, the nMOS transistor is turned off and hence the port_pin[i] is pulled up to a high-level voltage (usually, V_{DD}). When a logic 0 is written into the port register, the nMOS transistor is turned on and hence the port_pin[i] is pulled down to the ground level. The port register and the port_pin[i] can be read by enabling RD_port and RD_pin control signals, respectively.

Another example is shown in Figure 15.29(b), which is found in the pin structure of port RB2:RB0 in the PIC 18Fxx2 system. From this figure, it is clear that both the port register and the DDR are used and both registers not only can be written into but can also be read back under the control of appropriate control signals. In addition, the output also has an active pull-up transistor controlled by $\overline{\text{RBPU}}$ when the port is configured as an input. Moreover, both TTL input and Schmitt-trigger buffers are used for driving two different internal signals from the same external pin. The input signal is also latched when the port pin is read.

Section 15.3 General-Purpose Input and Output

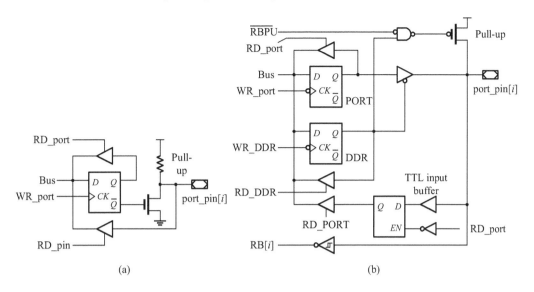

Figure 15.29: The logic diagram of the ith-bit of two GPIO examples: (a) MCS-51 Port 1 and (b) PIC 18Fxx2 Port RB2:RB0.

15.3.2 A Design Example

In this section, we consider a simple GPIO port which is widely used in most microcontroller systems, ranging from 8 bits to 32 bits. In this GPIO, suppose that the data direction of each port pin is separately controlled by an individual DDR bit, as shown in Figure 15.30. The port_pin[i] is configured as an output when the DDR is 1 and as an input when the DDR is 0. The DDR is cleared when the reset_n is activated so as to configure the port_pin[i] as an input for the reason of failure safe. The port register does not allow the data to be read back whereas the DDR does. The shaded multiplexer should be removed and the output of the active-low tristate buffer is connected directly to the data_bus[i] if the DDR does not allow the data to be read back.

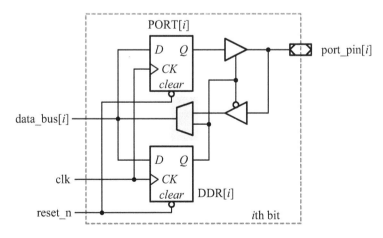

Figure 15.30: The logic diagram of the ith-bit of the parallel port example.

■ Example 15.9: A GPIO module.

In this example, an 8-bit GPIO port with its ith-bit cell depicted in Figure 15.30 is designed. The hardware model and programming model are shown in Figures 15.28(a) and (b), respectively. The module first generates four internal control signals with an **assign** continuous assignment. Then, it uses a generate-loop statement to produce an interface logic circuit and employs an **always** block to clear or update the port register and DDR.

```
// an example of the parallel port
module parallel_port(
      input reset_n, clk, r_w, cs, addr0,
      inout  [7:0] data_bus,
      output [7:0] port_pin);
// internal control signals
reg  [7:0] DDR, PORT;// directional and output registers
wire read_DDR, write_DDR, read_PORT, write_PORT;

// generate internal control signals
assign read_DDR   = ((cs == 1) && (addr0 == 1) && (r_w == 1)),
       write_DDR  = ((cs == 1) && (addr0 == 1) && (r_w == 0)),
       read_PORT  = ((cs == 1) && (addr0 == 0) && (r_w == 1)),
       write_PORT = ((cs == 1) && (addr0 == 0) && (r_w == 0));

// generate the interface logic circuit
genvar i;
generate for (i = 0; i < 8; i = i + 1) begin: interface
   assign port_pin[i] = DDR[i] ? PORT[i] : 1'bz ;
   assign data_bus[i] = read_DDR ? DDR[i] :
                        (read_PORT ? port_pin[i] : 1'bz);
// if not allow to read DDR, use the following statement
// assign data_bus[i] = read_PORT ? port_pin[i] : 1'bz;
end endgenerate

// clear or update the port register and DDR
always @(posedge clk or negedge reset_n)
   if (!reset_n) begin
      DDR <= 8'b0; PORT <= 8'b0; end
   else begin
      if (write_DDR)  DDR  <= data_bus;
      if (write_PORT) PORT <= data_bus;
   end
endmodule
```

■ Review Questions

15-21 Describe the hardware model of a typical GPIO port.
15-22 What is the programming model of a typical GPIO port?
15-23 Describe the operations of the port structure shown in Figure 15.29(a).
15-24 Describe the operations of the port structure shown in Figure 15.29(b).
15-25 Describe the operations of the port structure shown in Figure 15.30.

15.4 Timers

Timers are essential modules in any μC system because they provide at least the following important applications: time-delay creation, event counting, time measurement, period measurement, pulse-width measurement, time-of-day tracking, waveform generation, and periodic interrupt generation. In this section, we deal with the basic operations of timers and the issues of how to design such timers.

15.4.1 Basic Timer Operations

Recall that a counter is a device used to count events. As a counter operates at a clock signal with a known or fixed frequency, it is called a *timer*. Consequently, both timers and counters have exactly the same operations. In practice, the timers used in most μC systems are counters with programmable operation modes. The most basic operation modes of timers are as follows: *terminal count* (binary/BCD event counter), *rate generation, (digital) monostable* (or called one-shot), and *square-wave generation*.

15.4.1.1 Architecture of Timers A typical timer like the 82C54 is a 16-bit presettable counter and has three terminals: *clk, gate,* and *out*. In addition, the timer has a data bus and a set of control signals, which enable the access to the internal registers of the timer, as shown in Figure 15.31. The hardware model includes the CPU interface and I/O interface, as depicted in Figure 15.31(a). The CPU interface consists of data_bus, reset_n, *chip select* (cs), *read and write control* (r_w), and *register select* (A1 and A0). The I/O interface includes a clock input, clk, an enable input, gate, and an output, out. The programming model consists of three register: a mode register, a latch register, and a timer register. These three registers are accessed through the data bus and specified by the register select inputs, A1 and A0. Figure 15.31(c) is the function table of the timer.

The timer consists of two 16-bit registers: a *latch* register and a *timer* register. The latch register stores the initial value to be loaded into the timer register for counting down. The initial value may be reloaded into the timer register as many times as needed or only once depending on the operating mode of the timer. The timer register performs the actual counting operations. The initial value, called the *count*, is written into the latch register and then written into the timer register in the next clock cycle automatically. In what follows, we illustrate the detailed operations of each basic operating mode of the timer in sequence.

15.4.1.2 Terminal Count The terminal-count mode is used to count a specific number of clock cycles. After the write operation is complete, the count in the latch register is loaded into the timer register on the next clock pulse at the clock (clk) input and the count is decremented by 1 for each clock pulse that follows. When the count reaches 0, the terminal-count signal, i.e., a 0-to-1 transition, occurs at the out terminal. The gate input is an enable input. When the gate input is 1, the timer is enabled; otherwise, it is disabled. Figure 15.32(a) displays a waveform to illustrate how the timer operates.

The logic circuit used to implement the above operations is shown in Figure 15.32(b). The timer_load control signal is generated by an FSM described by the ASM chart shown in Figure 15.32(c). An illustration of how to model a timer operated in the terminal-count mode is explored in the following example.

■ Example 15.10: A timer operated in the terminal-count mode.

This module consists of four **always** blocks. The first **always** block describes the latch register, which is loaded from the data bus under the control of wr. This wr signal also activates the timer_load control signal for one clock cycle. The second **always** block describes the timer register, which is a binary-down counter with parallel load controlled by timer_load

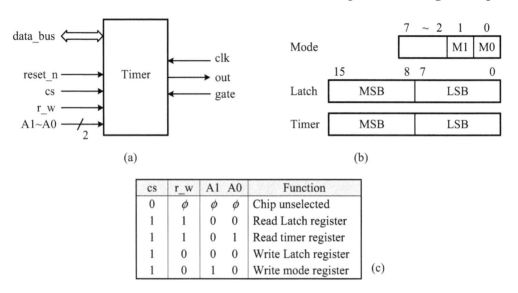

Figure 15.31: The (a) hardware model, (b) programming model, and (c) function table of a typical timer.

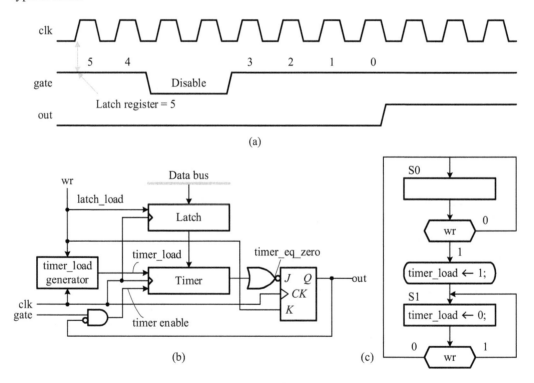

Figure 15.32: The illustration of a timer operated in the terminal-count mode: (a) an example; (b) block diagram; (c) ASM chart for generating the one-cycle timer_load pulse.

and enabled by `timer_enable`. The third **always** block generates the `out` signal. The fourth **always** block yields a one-cycle `timer_load` signal triggered by the `wr` input signal.

```
// an example of timer with the terminal-count mode
module timer_tc #(parameter N = 4)(  // the default size
```

Section 15.4 Timers

```verilog
           input clk, reset_n, gate, wr,
           input [N-1:0] data_bus,
           output reg out);
// local declarations
reg   [N-1:0] timer, latch;
wire  timer_enable, timer_eq_zero;
reg   timer_load;
reg   state;
localparam S0 = 1'b0, S1 = 1'b1;

// describe the latch register
always @(posedge clk or negedge reset_n)
   if (!reset_n) latch <= 0;
   else if (wr)  latch <= data_bus;
// describe the timer register
always @(posedge clk or negedge reset_n)
   if (!reset_n)         timer <= 0;
   else if (timer_load)  timer <= latch;
         else if (timer_enable) timer <= timer - 1;
assign timer_enable = gate & ~out;

// generate the out signal using a JK flip-flop
assign timer_eq_zero = (timer == 0);
always @(negedge clk or negedge reset_n)
  if (!reset_n) out <= 1'b0;
  else out <= (timer_eq_zero & ~out) | (~wr & out);

// generate the one-cycle timer_load control signal
always @(posedge clk or negedge reset_n)
   if (!reset_n) begin timer_load <= 1'b0; state <= S0; end
   else case (state)
      S0: if (wr) begin state <= S1; timer_load <= 1'b1; end
           else state <= S0;
      S1: begin timer_load <= 1'b0;
           if (wr) state <= S1; else state <= S0; end
   endcase
endmodule
```

15.4.1.3 Rate Generation In this mode, the out terminal outputs a one-cycle pulse per N clock cycles, as shown in Figure 15.33. In other words, it works exactly as a modulo-N counter. The gate input should be connected to a logic 1 to enable the timer. This mode can be easily implemented by modifying the timer with the terminal-count mode in a way such that the timer register is reloaded from the latch register whenever the terminal count is reached. Because of its intuitive simplicity, the implementation details of this mode are left for the reader as an exercise.

15.4.1.4 Retriggerable Monostable (One-Shot) In this mode, the out terminal outputs a high-level pulse for a duration equal to the number of clock pulses defined by the number (N) preloaded in the timer register. The gate input is used as the trigger signal. As the gate (trigger) input transitions from 0 to 1, the out terminal switches from 0 to 1 on the next pulse and remains in that logic value until the timeout of the timer occurs. This mode is generally retriggerable. An illustration of this operation mode is shown in Figure 15.34(a).

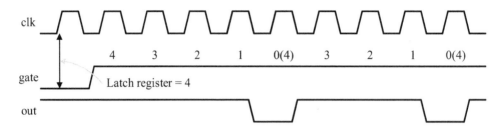

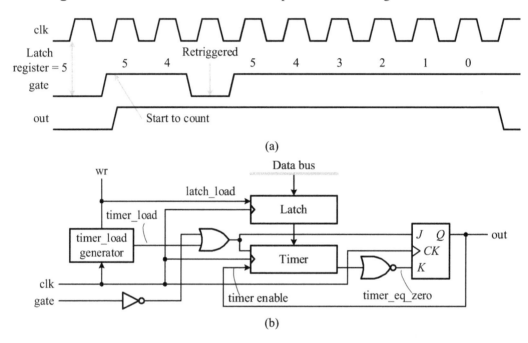

Figure 15.34: The illustration of a timer operated in the retriggerable monostable (one-shot) mode: (a) an example and (b) block diagram.

Figure 15.34(b) is the block diagram of a timer operated in the one-shot mode. The desired duration of the output pulse is stored in the latch register. This duration is loaded into the timer register under the control of the `timer_load` signal each time the trigger input, `gate`, transitions from 0 to 1. At the same time, the `out` terminal is set to a high-level voltage, which in turn enables the timer to count down. Once the timer counts down to 0, the JK flip-flop is cleared, thereby resetting the `out` terminal to 0 and disabling the timer. It is instructive to compare this monostable circuit with the one described in Section 9.6.2.

■ Example 15.11: A timer operated in the monostable mode.

As described above, the timer register is reloaded from the latch register under the control of the `timer_load` signal each time the `gate` trigger input is switched from 0 to 1, which is generated by the last **always** block. The first two **always** blocks describe the latch and timer registers, respectively. The third **always** block describes the output JK flip-flop, which is set to 1 when the timer is loaded a new value and reset to 0 when the timer counts down to 0. The fourth **always** block yields a one-cycle `timer_load` signal triggered by the `wr` input signal.

```
// an example of timer with the retriggerable one-shot mode
```

Section 15.4 Timers

```verilog
module timer_one_shot #(parameter N = 4)( // the default size
       input clk, reset_n, gate, wr,
       input [N-1:0] data_bus,
       output reg out);
// local declarations
wire timer_eq_zero;
reg  [N-1:0] timer, latch;
reg  timer_load, state;
localparam S0 = 1'b0, S1 = 1'b1;

// describe the latch register
always @(posedge clk or negedge reset_n)
   if (!reset_n) latch <= 0;
   else if (wr)  latch <= data_bus;
// describe the timer register
always @(posedge clk or negedge reset_n)
   if (!reset_n) timer <= 0;
   else if (timer_load || ~gate) timer <= latch;
        else if (out) timer <= timer - 1;

// generate the output pulses using a JK flip-flop
always @(posedge clk or negedge reset_n)
   if (!reset_n) out <= 1'b0;
   else out <= ((timer_load || ~gate) & ~out) |
              (~timer_eq_zero & out);
assign timer_eq_zero = (timer == 0);

// generate the one-cycle timer_load control signal
always @(posedge clk or negedge reset_n)
   if (!reset_n) begin timer_load <= 1'b0; state <= S0; end
   else case (state)
      S0: if (wr) begin state <= S1; timer_load <= 1'b1; end
          else state <= S0;
      S1: begin timer_load <= 1'b0;
          if (wr) state <= S1; else state <= S0; end
   endcase
endmodule
```

15.4.1.5 Square-Wave Generation In the square-wave mode, the *out* terminal produces a square wave with a 50% duty cycle whenever the timer register is loaded with an even number. If an odd number (N) instead of an even number is loaded into the timer register, the *out* terminal will be low for $N/2$ clock pulses and high for $(N+1)/2$ clock pulses. An illustration is given in Figure 15.35(a).

The block diagram of one possible implementation of the square-wave mode is depicted in Figure 15.35(b). The basic idea behind the operation of this mode is on the value being loaded into the timer register each time the timer counts down to 1. According to the above description, the output is low for $N/2$ cycles and high for $N/2$ or $(N+1)/2$ cycles, depending on whether the N is an even or odd number. In other words, the value in the latch register is shifted right one-bit position and then the result is loaded into the timer register when the *out* output is low or added with the LSB of the latch register before it is loaded into the timer register when the *out* output is high. This is done by the block named shift plus LSB in the

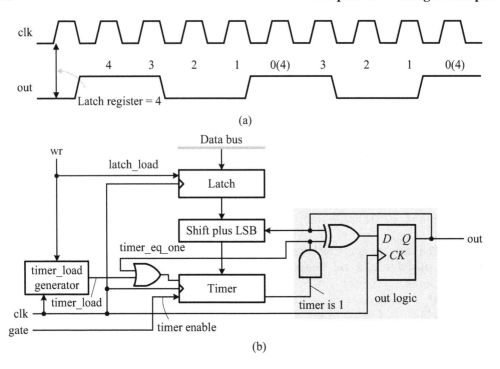

Figure 15.35: The illustration of a timer operated in the square-wave mode: (a) an example and (b) block diagram.

figure.

An illustration showing how to model a timer operated in the square-wave mode is explored in the following example.

■ Example 15.12: A timer operated in the square-wave mode.

The out output logic consists of a T flip-flop triggered by the terminal count with 1 of the timer register. This output pulse is also used to determine whether the value loaded into the timer is $N/2$ or $N/2+$ LSB. The T flip-flop is implemented using a D flip-flop with an XOR gate. The out logic is accomplished by the third **always** block and the function of the shift and plus LSB block is combined into the second **always** block directly.

```
// an example of timer with the square-wave mode
module timer_square_wave #(parameter N = 4)(// the default size
       input clk, reset_n, gate, wr,
       input [N−1:0] data_bus,
       output reg out);
// local declarations
reg   [N−1:0] timer, latch;
wire  timer_eq_one;
reg   timer_load;
reg   state;
localparam S0 = 1'b0, S1 = 1'b1;

// describe the latch register
always @(posedge clk or negedge reset_n)
```

Section 15.4 Timers

```verilog
      if (!reset_n) latch <= 0;
      else if (wr)  latch <= data_bus;
// describe the timer register
always @(posedge clk or negedge reset_n)
   if (!reset_n) timer <= 0;
   else if (timer_load || timer_eq_one) begin
         if (out == 0) timer <= (latch >> 1);
         else timer <= (latch >> 1) + latch[0]; end
      else if (gate) timer <= timer - 1;
assign timer_eq_one = (timer == 1);

// generate square-wave output pulses
always @(posedge clk or negedge reset_n)
   if (!reset_n) out <= 0;
   else if (wr)  out <= 0;
   else out <= (timer_eq_one) ^ out;

// generate the one-cycle timer_load control signal
always @(posedge clk or negedge reset_n)
   if (!reset_n) begin timer_load <= 1'b0; state <= S0; end
   else case (state)
      S0: if (wr) begin state <= S1; timer_load <= 1'b1; end
          else state <= S0;
      S1: begin timer_load <= 1'b0;
             if (wr) state <= S1; else state <= S0; end
   endcase
endmodule
```

15.4.2 Advanced Timer Operations

In addition to the four basic operation modes described previously, modern timers are often equipped with the following three additional modes: the *input capture mode*, *output compare mode*, and *pulse-width modulation* mode. Next, we discuss these modes in detail.

15.4.2.1 Input-Capture Mode In the input-capture mode, the timer loads the contents of the timer register into a capture register whenever the predefined event occurs. The event is often represented as the positive or negative edge of an external signal. By capturing the timer value, many common measurements can then be made, including period, pulse width, duty cycle, timing reference, and event arrival time.

It is quite easy to implement the input-capture mode. A possible way is shown in Figure 15.36, where a free-running timer and a capture register coupled with a few logic gates are used. The n-bit timer is a modulo-2^n binary-up counter, which is cleared when the reset signal is asserted or each time the write signal, wr, is activated. Once the write signal, wr, returns to its inactive state, the counter starts to count up. The contents of the timer register are captured into the capture register whenever the positive edge of the gate trigger input is detected. The captured value can then be read by asserting the read signal, rd.

■ **Example 15.13: A timer operated in the input-capture mode.**

This example only contains two **always** blocks. The first one describes the timer register and the second one describes the capture register. The timer register is also reset each time the wr signal is activated in addition to being reset by the reset_n signal. As a consequence, we

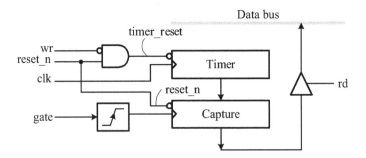

Figure 15.36: The block diagram of a timer operated in the input-capture mode.

may use the wr signal to control the operation of the timer register. Note that in this example, we also demonstrate how to read the captured value back to the data bus through a tristate buffer. Since we assume that the data bus is a shared bus for both read and write operations, the data_bus must be declared with the keyword **inout** as a bidirectional bus. The reader is encouraged to write a test bench to verify this module, in particular, to pay attention to how to drive the data_bus by noting that how to declare the data type of the data_bus within the test bench and how to drive the data_bus in an **initial** or **always** block.

```
// an example of timer with the capture mode
module timer_capture #(parameter N = 4)(// the default size
      input clk, reset_n, gate, wr, rd,
      inout [N−1:0] data_bus);
// local declarations
reg  [N−1:0] timer, capture;
wire timer_reset;

// read the capture register
assign data_bus = rd ? capture : {N{1'bz}};

// describe the timer register
assign timer_reset = !reset_n || wr;
always @(posedge clk or posedge timer_reset)
   if (timer_reset) timer <= 0;
   else if (!wr) timer <= timer + 1;

// capture the timer value into the capture register
always @(posedge gate or negedge reset_n)
   if (!reset_n) capture <= 0;
   else capture <= timer;
endmodule
```

15.4.2.2 Output-Compare Mode In the output-compare mode, the timer compares the contents of the timer register with the values stored in the latch register, as illustrated in Figure 15.37. The out terminal is set to 1 for one clock cycle if both values are matched; otherwise, it is cleared to 0. The applications of the output-compare mode at least include the following ones: to generate a time delay, to trigger an action at some future time, and to generate a digital waveform.

A possible implementation of this mode is shown in Figure 15.37, which is composed of a latch register, a timer register, and a comparator. The latch register stores the values to be

Section 15.4 Timers

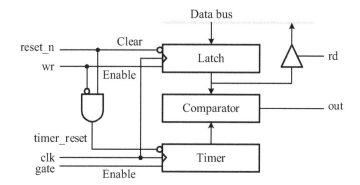

Figure 15.37: The block diagram of a timer operated in the output-compare mode.

compared with the contents of the timer register. The timer register, which is an n-bit binary-up counter enabled by the gate control input, starts to count from 0. The timer register is reset to 0 each time the reset signal is asserted or a new value is written into the latch register. The comparator compares the contents of both the latch register and the timer register. It generates a one-cycle high-level output pulse when they are matched; otherwise, it yields a low-level output. The following example illustrates how to model such a timer operated in the output-compare mode.

■ Example 15.14: A timer operated in the output-compare mode.

This example contains two **always** blocks to separately describe the latch register and the timer register. The compare operation is performed by an **assign** continuous assignment. Note that we use a tristate buffer to read the contents of the latch register back to the data bus. Thus, the data bus must be declared as a bidirectional port with the keyword **inout**. The reader is encouraged to write a test bench to verify this module.

```
// an example of timer with the compare mode
module timer_compare #(parameter N = 4)(// the default size
      input clk, reset_n, gate, wr, rd,
      inout [N-1:0] data_bus,
      output out);
// local declarations
reg  [N-1:0] timer, latch;
wire timer_reset;

// read the latch register
assign data_bus = rd ? latch : {N{1'bz}};

// describe the latch register which stores the compare value
always @(posedge clk or negedge reset_n)
   if (!reset_n) latch <= 0;
   else if (wr) latch <= data_bus;

// describe the timer register --- reset by reset_n and wr
// signals and enabled by the gate input
assign timer_reset = !reset_n || wr;
always @(posedge clk or posedge timer_reset)
```

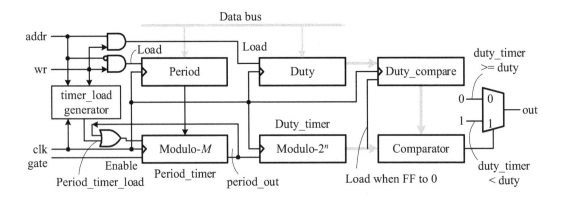

Figure 15.38: The block diagram of a timer operated in the PWM mode.

```
if (timer_reset) timer <= 0;
else if (gate) timer <= timer + 1;

// compare the timer against the value stored
// in the latch and output the results
assign out = (timer == latch);
endmodule
```

15.4.2.3 Pulse-Width Modulation (PWM) Mode In the PWM mode, the timer generates an output signal with the specified duty cycle and frequency. The PWM mode is employed in a broad variety of applications, ranging from measurement and communications to power control and conversion. PWM can be implemented with an analog circuit or a digital circuit. In what follows, we focus on the digital implementation.

In essence, a digital PWM circuit is a way of digitally encoding analog signals through the use of high-resolution counters. As it is implemented with timers, two timers are needed. One is for controlling the period and the other for controlling the *duty cycle*. The duty cycle of a signal is defined as follows:

$$\text{Duty cycle} = \frac{\text{Duration of high output}}{\text{Period}} \times 100\%. \tag{15.3}$$

One simple possible implementation of this mode is illustrated in Figure 15.38. Two timers are separately used to define the period and the duty cycle. The timer used to control the period is a module-M down counter, named the period_timer, which reloads the count value stored in the period register whenever it counts down to 0, in which a period_out signal is generated, or a new value is written into the period register.

The timer used to control the duty cycle is a module-2^n binary-up counter, which is denoted by the duty_timer. The contents of the duty_timer are compared with the duty values stored in the duty_compare register. The out terminal is set to high if the contents of the duty_timer are less than the duty value and set to low otherwise. In order to avoid the occurrence of output glitches, a duty register is used to accept a duty value from the data bus. The duty value is then loaded into the duty_compare register each time the output of the duty_timer changes from FF to 00. The out terminal yields a PWM pulse with a period equal to $(period + 1) \times 2^n$ clock cycles and a duty cycle of $(0 \text{ to } 2^n - 1)/2^n$ with a resolution of $(period + 1)$ clock cycles. The reader is encouraged to verify this.

The following example describes how to model the timer operated in the PWM mode depicted in Figure 15.38.

Section 15.4 Timers

■ Example 15.15: A timer operated in the PWM mode.

The first two **always** blocks describe the period and duty registers, respectively. The next **always** block models the duty_compare register, which is loaded when the duty_timer changes from FF to 00. The next two **always** blocks describe the period_timer and the duty_timer, respectively. The last **always** block yields a single-cycle control signal, period_-timer_load, for loading the period_timer. The output signal is generated by an **assign** continuous assignment.

```
// an example of timer being operated in pwm mode
module timer_pwm #(parameter N = 4)(// the default size
      input clk, reset_n, gate, wr, addr,
      input [N-1:0] data_bus,
      output out);
reg period_timer_load, state;
wire period_out, duty_out;
localparam S0 = 1'b0, S1 = 1'b1;
// local declarations
reg  [N-1:0] period_timer, period,
             duty_timer, duty, duty_compare;

// describe the period register which stores the period
always @(posedge clk or negedge reset_n)
   if (!reset_n) period <= 0;
   else if (wr && (addr == 0)) period <= data_bus;

// describe the duty register which stores the duty value
always @(posedge clk or negedge reset_n)
   if (!reset_n) duty <= 0;
   else if (wr && (addr == 1)) duty <= data_bus;
// describe the duty-compare register
always @(posedge clk or negedge reset_n)
   if (!reset_n) duty_compare <= 0;
   else if (duty_out) duty_compare <= duty;

// describe the period_timer, a presettable down counter
always @(posedge clk or negedge reset_n)
   if (!reset_n) period_timer <= 0;
   else if (period_timer_load || period_out)
          period_timer <= period;
        else if (gate) period_timer <= period_timer - 1;
assign period_out = ~|period_timer;

// describe the duty_timer, a modulo-2^N up counter
always @(posedge clk or negedge reset_n)
   if (!reset_n) duty_timer <= 0;
   else if (period_out) duty_timer <= duty_timer + 1;
assign duty_out = &duty_timer;

// generate the one-cycle period_timer_load control signal
always @(posedge clk or negedge reset_n)
   if (!reset_n) begin
```

```
                        period_timer_load <= 1'b0; state <= S0; end
   else case (state)
      S0: if (wr && (!addr)) begin state <= S1;
                 period_timer_load <= 1'b1; end
             else state <= S0;
      S1: begin period_timer_load <= 1'b0;
             if (wr && (!addr)) state <= S1;
             else state <= S0; end
   endcase

// generate the output signal
assign out = (duty_timer < duty_compare) ? 1'b1 : 1'b0;
endmodule
```

■ Review Questions

15-26 Explain the function of the `timer_load` generator in Figure 15.32.
15-27 Explain the rationale behind the timer operated in the one-shot mode.
15-28 Explain the rationale behind the timer operated in the square-wave mode.
15-29 Define the duty cycle of a signal.
15-30 Modify the `timer_capture` module so that it can capture the timer value at the negative edge of the external trigger signal.

15.5 Universal Asynchronous Receiver and Transmitter

Most μC systems have one or more serial data ports used to communicate with serial devices, such as μC emulators. The operation mode of serial data ports can be asynchronous or synchronous. In this section, we are concerned with the design of an asynchronous serial port.

15.5.1 UART

Currently, the *universal asynchronous receiver and transmitter* (UART) is still one of the most widely used devices for asynchronous serial data transmission and reception in many embedded microcomputer systems.

15.5.1.1 Architecture of UARTs The hardware and software models of a typical UART is shown in Figure 15.39. The hardware model includes the CPU interface and I/O interface, as depicted in Figure 15.39(a). The CPU interface consists of `data_bus`, `reset_n`, *chip select* (cs), *read and write control* (r_w), *clock* (clk), and *register select* (rs). The I/O interface includes two data lines, transmitter data line TxD and receiver data line RxD, as well as two clock signals, TxC and RxC. The transmitter data line, TxD, is employed to send data out of the device and the receiver data line, RxD, is used to receive data from the outside of the device. Both data lines, TxD and RxD, are controlled by two local clock signals: TxC and RxC, respectively. The frequency of the clock, RxC, is often much higher than that of TxC, by a factor of 4, 16, or 64.

The software model consists of four registers: the *receiver data register* (RDR), *transmitter data register* (TDR), *status register* (SR), and *control register* (CR), as shown in Figure 15.39(c). The RDR and the SR can only be read out while the TDR and the CR can only be written into. As a consequence, only one register select (rs) input is required. Both the SR and the CR are selected when rs is 0 and both the RDR and the TDR are selected when rs

Section 15.5 Universal Asynchronous Receiver and Transmitter

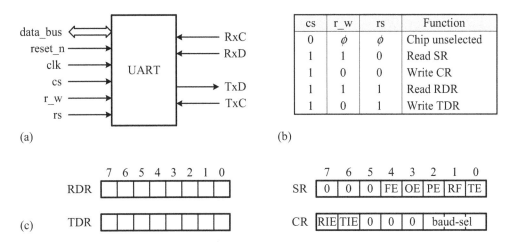

Figure 15.39: The (a) hardware model, (b) function table, and (c) programming model of a typical UART.

is 1. In addition, a chip select (cs) input is used to enable the chip; more precisely, it only controls the read and write of the four registers but does not affect the operations of TxD and RxD. The function table is shown in Figure 15.39(b). The details of each bit of both the SR and the CR will be described in their related subsections.

15.5.1.2 Design Issues of UARTs In order to design a useful UART device, the following design issues must be considered.

- *Baud rates*: The baud rate is the number of signaling events per unit time. If only one bit is sent per signaling interval, the baud rate is equal to bits per second (bps). For example, in a four-level voltage signaling scheme, the bit rate is twice the baud rate. In this book, we only consider the case of binary transmission, in which only one bit is sent per signaling interval. As a result, the baud rate is used interchangeably with the bit rate. For the most widely used devices nowadays, the baud rate ranges from 300 to 19200, or even more.

- *Sampling clock frequencies*: The sampling clock frequency determines the error tolerance of the mismatch between the transmitter and receiver clock signals. The most common sampling clock frequency is RxC = $n \times$ TxC, where n is often set to 1, 4, 16, or 64.

- *Stop bits*: The stop bits are mandatory and used along with the start bit to define a character (byte) frame, as shown in Figure 15.15. The number of stop bits may be one or two bits but one bit is the most common.

- *Parity check*: Parity check may be either even or odd. A parity is called *even parity* if the number of 1s of the information and the parity bit is even. Otherwise, the parity is called *odd parity*. For most UART devices, parity check is optional. In some UART devices, this bit can also be programmed as the ninth data bit.

In what follows, we deal with how to design such a serial data transmission and receiving device in detail. A typical UART device is composed of three major parts: a *transmitter*, a *receiver*, and a *baud-rate generator*. For simplicity, we will use a character frame that consists of one start bit, eight data bits, one even-parity bit, and one stop bit, as an example to illustrate how to design and implement such a device. In addition, the sampling clock frequency is assumed to be n (defaulted to 8) times the transmission clock frequency, i.e., RxC = $n \times$ TxC.

15.5.2 Transmitter

The essential component of the transmitter is a shift register which shifts right its contents out into TxD continuously with the LSB first and the MSB bit filled with 1. Figure 15.40 shows the basic components of a typical transmitter. The transmitter is composed of a TDR, a *transmitter shift data register* (TSDR), a *TDR empty flag* (TE), a *parity generator*, and a *transmitter control circuit*. The TDR accepts the data to be transmitted from the data bus. Once the TDR is filled, the contents of the TDR, along with its even-parity bit, are loaded into the TSDR at the next positive edge of the transmitter clock (TxC) signal. The contents of the TSDR are then continuously shifted out to TxD by the TxC clock signal.

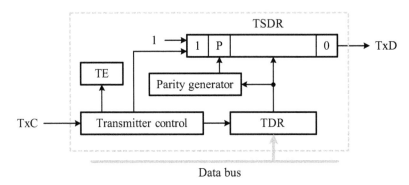

Figure 15.40: The block diagram of the transmitter of a typical UART.

The detailed operations of the transmitter are illustrated by the ASM chart shown in Figure 15.41. Figure 15.41(a) shows the operations of the CPU, which must detect the TE flag before it can write data into the TDR. Once the data is in the TDR, the data are transferred into the TSDR. Then, the TE flag is set and the transmitter starts to send the contents of the TSDR out through TxD in a bit-by-bit fashion. After the entire frame is sent, the transmitter goes back to the idle state and is ready for sending another byte. In the idle state, whenever there are no data to be sent, it continuously sends 1s.

The following example realizes the ASM chart shown in Figure 15.41(b). Here, the standard paradigm introduced in Section 11.2.2 is employed.

■ Example 15.16: The transmitter of the UART.

Basically, the module exactly follows the ASM chart depicted in Figure 15.41(b) except for the addition of the first two **always** blocks that are employed to load the TDR and update the TDR empty flag (TE), respectively.

```
// an example of a UART transmitter
module UART_transmitter
        // the frame size excluding the stop bit
        #(parameter M = 10)(
        input clk, reset_n, TxC, load_TDR,
        input [7:0] data_bus,
        output TxD,
        output reg TE);
// internal used registers
reg [7:0] TDR;
reg [M-1:0] TSDR;
reg [3:0] BitCnt;
```

Section 15.5 Universal Asynchronous Receiver and Transmitter

```
reg ps, ns, set_TE;
localparam idle = 1'b0, shift = 1'b1;

// load TDR when load_TDR is activated
always @(posedge clk or negedge reset_n)
   if (!reset_n) TDR <= 8'b0;
   else if (load_TDR) TDR <= data_bus;
// update TDR empty flag (TE)
always @(posedge clk or negedge reset_n)
   if (!reset_n) TE <= 1'b1;
   else TE <= (set_TE && !TE) || (!load_TDR && TE);

// load TSDR from TDR and perform data transmission
// step 1: initialize and update state registers
always @(posedge TxC or negedge reset_n)
   if (!reset_n) ps <= idle;
   else ps <= ns;
// step 2: compute the next state
always @(*)
   case (ps)
      idle: if (TE == 1) ns = idle;
            else ns = shift;
      shift: if (BitCnt < M - 1) ns = shift;
             else ns = idle;
   endcase
// step 3: execute RTL operations
always @(posedge TxC or negedge reset_n)
   if (!reset_n) begin TSDR <= {M{1'b1}};
```

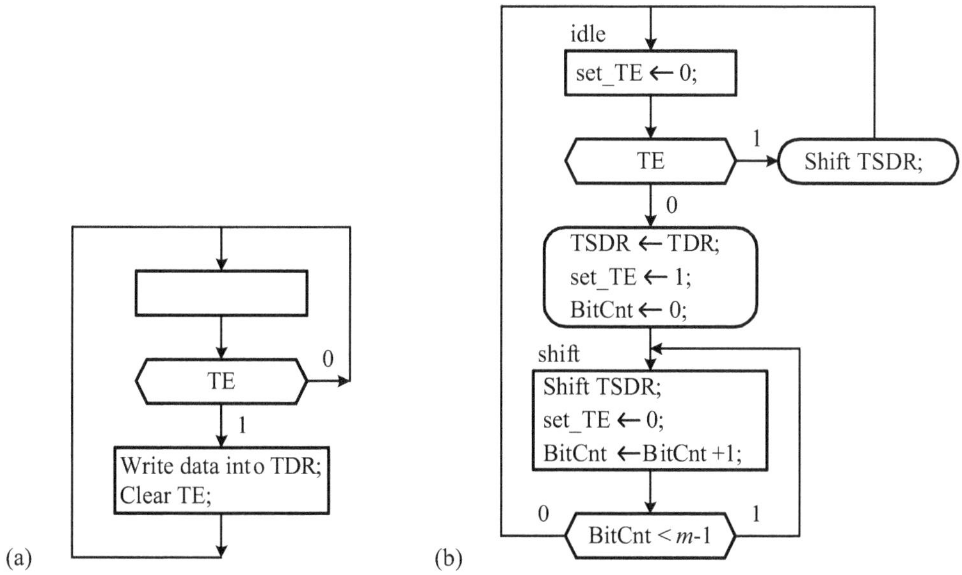

Figure 15.41: The ASM chart of the transmitter of a typical UART: (a) CPU operations and (b) TSDR operations.

```
        BitCnt <= 0; set_TE <= 1'b0; end
   else case (ps)
     idle: begin set_TE <= 1'b0;
             if (TE == 1) TSDR <= {1'b1, TSDR[M-1:1]};
             else begin
                TSDR <= {^TDR, TDR, 1'b0};
                set_TE <= 1'b1;
                BitCnt <= 0; end   end
     shift: begin
                TSDR <= {1'b1, TSDR[M-1:1]};
                set_TE <= 1'b0;
                BitCnt <= BitCnt + 1; end
   endcase
assign TxD = TSDR[0] & (ps == shift) | (ps == idle);
endmodule
```

15.5.3 Receiver

The essential component of the receiver is also a shift register that samples and shifts right the contents of RxD into a shift register with the LSB first once the negative edge of RxD has been detected. Figure 15.42 shows the basic components of a typical receiver. The receiver is composed of an RDR, a *receiver shift data register* (RSDR), an *RDR full flag* (RF) along with another three flags, and a *receiver control circuit*. Once the entire frame has been received, the data part is extracted from the RSDR and loaded into the RDR to be ready for reading.

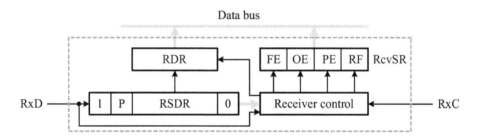

Figure 15.42: The block diagram of the receiver of a typical UART.

There are four flags associated with the receiver: *frame error* (FE), *overrun error* (OE), *parity error* (PE), and *RDR full* (RF).

- The FE flag indicates the fact that the frame is in error; that is, the stop bit is 0 rather than 1.
- The OE flag indicates that a new byte is received before the previous one is read; namely, the previous byte is overridden by the new byte.
- The PE flag indicates the parity of the byte received is incorrect.
- The RF flag is set to 1 whenever a new byte is extracted from the RSDR and loaded into the RDR.

The detailed operations of the receiver are illustrated by the ASM chart shown in Figure 15.43. Figure 15.43(a) shows the operations of the CPU, which must detect the RF flag before it can read data from the RDR. Each time the data byte is read, the read control signal also clears all four flags: FE, OE, PE, and RF.

Section 15.5 Universal Asynchronous Receiver and Transmitter

There are two counters required in the receiver module. One is employed to count the number of sampling clock cycles during each bit time, called the *bit-clock counter* (BckCnt), and the other counts the number of bits already received, called the *bit-number counter* (BitCnt). After reset, the BckCnt and BitCnt counters are set to $(n/2) - 1$ and 0, respectively, where n is the scale factor of the sampling clock frequency. The receiver starts from the idle state and enters the detected state once it has detected a negative edge transition of RxD. This event also triggers the BckCnt counter to decrease 1. The receiver will stay at this state until the BckCnt counter reaches 0. Then, the BckCnt counter is set to $n - 1$ and the receiver enters the shift state to start the reception of data. Once in the shift state, the receiver samples the RxD every n sampling clock cycles. This process is repeated until all data bits and the parity bit are received. Once the reception process is finished, the data is extracted from the RSDR and load into the RDR. At the same time, all four flags are updated. The RF flag is set to 1 to indicate a data byte is ready in the RDR for reading. The other flags are reflected the actual situations of the received data, that is, the frame error, overrun error, and parity error.

The following example realizes the ASM chart shown in Figure 15.43(b). Here, we also use the standard paradigm introduced in Section 11.2.2.

■ Example 15.17: The receiver of the UART.

Basically, this example exactly follows the ASM chart shown in Figure 15.43(b) except for the first two **always** blocks, which are used to update the status register, RcvSR, and generate a

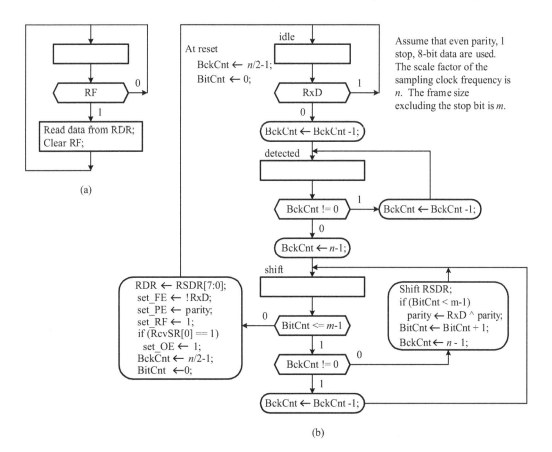

Figure 15.43: The ASM chart of the receiver of a typical UART: (a) CPU operations and (b) RSDR operations.

single-cycle signal, set_RF_1clk, from set_RF, respectively. The set_RF_1clk is employed to update the status register, RcvSR. This is required because the duration of set_RF is one RxC clock cycle, whose period is much larger than the system clock cycle. If we use set_RF directly as the load enable signal for the status register, RcvSR, then the read_RDR signal would not clear RcvSR correctly (Why?). Explain.

```verilog
// an example of a UART receiver -- RxC is 8 times TxC
// 1 stop, 8-bit data, and even parity are used
module UART_receiver
        // default frame size excluding the stop bit
        #(parameter M = 10,
          parameter N = 8)( // the default scale factor
           input  clk, reset_n, read_RDR, RxC, RxD,
           output reg [7:0] RDR,// the receiver data register
           output reg [3:0] RcvSR);
// internal used registers
localparam K = 3; // K = log2 N
localparam L = 4; // L = log2 M
reg [M-1:0] RSDR;   // the receiver shift data register
reg [L-1:0] BitCnt; // the number of bits received
reg [K-1:0] BckCnt; // the number of TxC clocks elapsed
reg [1:0] ps, ns;
wire RcvSR_reset;
reg parity, set_FE, set_OE, set_PE, set_RF, set_RF_1clk;
localparam idle = 2'b00, detected = 2'b01, shift = 2'b10;

// update the status register (RcvSR)
assign RcvSR_reset = !reset_n || read_RDR;
always @(posedge clk or posedge RcvSR_reset)
   if (RcvSR_reset) RcvSR <= 4'b0000;
   else if (set_RF_1clk)
           RcvSR <= {set_FE, set_OE, set_PE, set_RF};

// generate set_RF_1clk signal from set_RF
reg state;
localparam S0 = 1'b0, S1 = 1'b1;
always @(posedge clk or negedge reset_n)
   if (!reset_n) begin
         set_RF_1clk <= 1'b0; state <= S0; end
   else case (state)
      S0: if (set_RF) begin state <= S1;
              set_RF_1clk <= 1'b1; end
            else state <= S0;
      S1: begin set_RF_1clk <= 1'b0;
            if (set_RF) state <= S1; else state <= S0; end
   endcase

// receive data and load the RSDR into the RDR
// step 1: initialize and update the state register
always @(posedge RxC or negedge reset_n)
   if (!reset_n) ps <= idle;
   else ps <= ns;
```

```
// step 2: compute the next state
always @(*)
   case (ps)
      idle: if (RxD == 1) ns = idle;
            else ns = detected;
      detected: if (BckCnt != 0) ns = detected;
            else ns = shift;
      shift: if (BitCnt <= M−1) ns = shift;
            else ns = idle;
      default: ns = idle;
   endcase
// step 3: execute RTL operations
always @(posedge RxC or negedge reset_n)
   if (!reset_n) begin RSDR <= {M{1'b0}}; RDR <= 8'b00000000;
      BckCnt <= N/2−1; BitCnt <= 0;
      set_FE <= 1'b0; set_OE <= 1'b0;
      set_PE <= 1'b0; set_RF <= 1'b0; parity <= 0; end
   else case (ps)
      idle: begin
         set_FE <= 1'b0; set_OE <= 1'b0; set_PE <= 1'b0;
         set_RF <= 1'b0; parity <= 1'b0;
         if (RxD == 0) BckCnt <= BckCnt − 1; end
      detected: if (BckCnt != 0) BckCnt <= BckCnt − 1;
               else BckCnt <= N − 1;
      shift: begin
         if (BitCnt <= M − 1) begin
            if (BckCnt != 0) BckCnt <= BckCnt − 1;
            else begin // BckCnt = 0
               RSDR <= {RxD, RSDR[M−1:1]};
               if (BitCnt < M−1) parity <= RxD ^ parity;
               BitCnt <= BitCnt + 1; BckCnt <= N − 1; end
         end
         else begin
            RDR <= RSDR[7:0]; set_FE <= !RxD;
            set_PE <= parity; BckCnt <= N/2 − 1;
            set_RF <= 1'b1;
            if (RcvSR[0] == 1) set_OE <= 1'b1;
            BitCnt <= 0; end end
   endcase
endmodule
```

15.5.4 Baud-Rate Generator

The baud-rate generator plays an important role in the operations of any UART module because it provides clock sources, TxC and RxC, for the transmitter and receiver, respectively.

The two most common methods for designing baud-rate generators are *multiplexer-based* and *timer-based*, as shown in Figure 15.44. Of course, the essential components of both approaches are the same, that is, counters. Figure 15.44(a) shows the structure of a multiplexer-based baud-rate generator, which consists of a modulo-M prescalar counter, a baud-rate counter, and a multiplexer. The modulo-M prescalar counter scales down the system clock (clk) frequency so that its output is n times the maximum baud-rate allowed in the UART module,

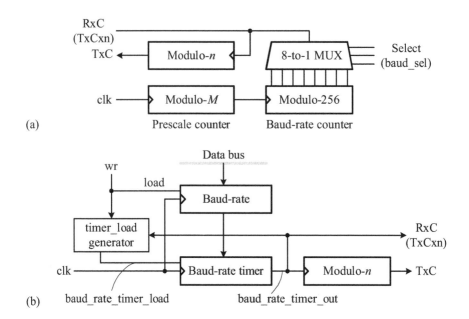

Figure 15.44: The block diagrams of (a) multiplexer-based and (b) timer-based baud-rate generators.

where n is the scale factor of the sampling clock (clk) frequency. The multiplexer aims to scale down this maximum baud-rate to many lower possible baud rates which can then be selected by the baud selection, baud_sel, set by the control register (CR).

Another widely used approach is by using a timer with a specific mode. This results in the so-called timer-based baud-rate generator, as shown in Figure 15.44(b). The timer-based baud-rate generator comprises a reloadable timer, called a *baud-rate timer*, and a modulo-n counter. The baud-rate timer is used to scale down the system clock frequency directly to the required baud rate times the scale factor (n) of the sampling clock frequency (RxC) with respect to the transmission clock frequency (TxC).

In what follows, we merely consider an implementation of the multiplexer-based baud-rate generator. The implementation of a timer-based baud-rate generator is left for the reader as an exercise.

■ Example 15.18: A multiplexer-based baud-rate generator.

The implementation of the multiplexer-based baud-rate generator is quite straightforward. Each **always** block corresponds to one functional block shown in Figure 15.44(a). Therefore, we will not further explain it here. Note that in this example, we assume that the sampling clock frequency is 8 times the transmission clock frequency; namely, n is set to 8.

```
// an example of the multiplexer-based baud-rate generator
module UART_baud_rate
       #(parameter BD_STEPS = 8)( // no. of baud-rate steps
         input clk, reset_n,
         input [2:0] baud_sel,
         output wire TxC,
         output reg TxCx8);
integer i;
```

Section 15.5 *Universal Asynchronous Receiver and Transmitter*

```
// internal used reg variables
wire prescale_out, TxCx8_clk;
reg  [2:0] prescale_counter;
reg  [BD_STEPS-1:0] baud_rate_counter;
reg  [2:0] mod8_counter;

// the prescale_counter, a presettable down counter
always @(posedge clk or negedge reset_n)
   if (!reset_n) prescale_counter <= 0;
   else prescale_counter <= prescale_counter + 1;
assign prescale_out = &prescale_counter;

// the a modulo-256 up counter
always @(posedge clk or negedge reset_n)
   if (!reset_n) baud_rate_counter <= 0;
   else if (prescale_out)
           baud_rate_counter <= baud_rate_counter + 1;

// the 8-to-1 mux for selecting the required baud rate
always @(baud_sel or baud_rate_counter) begin
   TxCx8 = 0;
   for (i = 0; i < BD_STEPS; i = i + 1)
       if (baud_sel == i) TxCx8 = baud_rate_counter[i]; end

assign TxCx8_clk = TxCx8;
// the modulo8_counter
always @(posedge TxCx8_clk or negedge reset_n)
   if (!reset_n) mod8_counter <= 0;
   else mod8_counter <= mod8_counter + 1;
assign TxC = mod8_counter[2];
endmodule
```

15.5.5 UART Top-Level Module

After considering the detailed operations of all three individual modules of a UART, the top-level module of the UART can then be readily constructed. As mentioned previously, there are four registers within the underlying UART. Except that the TDR and RDR are associated with the transmitter and the receiver, respectively, the status register, SR, is shared by both the transmitter and the receiver, and the control register, CR, is used to control all three modules. Hence, it is more convenient to put both SR and CR in the top-level module. In addition, an interrupt logic circuit is often associated with a useful UART module to make the CPU aware of events occurring in the UART, such as TDR empty or RDR full. This logic is also easier to be built in the top-level module. The resulting top-level module of the UART is exemplified as follows.

■ **Example 15.19: The top-level module of the UART.**

This top-level module, in addition to putting together the three individual modules discussed before, processes the interface logic, defines the control register and builds interrupt generation logic to form a practically useful UART module.

```
// a UART consists of a transmitter, a receiver,
```

```verilog
// and a baud-rate generator
module UART_top(
       input clk, reset_n, r_w, cs, rs, RxD,
       inout [7:0] data_bus,
       output TxD,
       output reg IRQ);
// internal used registers
wire [3:0] RcvSR;
wire RxC, TE, TxC;
wire read_UART, write_UART, load_TDR, read_SR, load_CR;
wire [7:0] RDR;
reg  [7:0] CR;

// instantiate a transmitter, a receiver, and a baud-rate generator
UART_transmitter my_transmitter
     (.clk(clk), .reset_n(reset_n), .TxC(TxC), .load_TDR(load_TDR),
      .data_bus(data_bus), .TxD(TxD), .TE(TE));
UART_receiver my_receiver
       (.clk(clk), .reset_n(reset_n), .read_RDR(read_RDR),
        .RxC(RxC), .RxD(RxD), .RDR(RDR), .RcvSR(RcvSR));
UART_baud_rate my_baud_rate
       (.clk(clk), .reset_n(reset_n), .baud_sel(CR[2:0]),
        .TxC(TxC), .TxCx8(RxC));

// generate interface control signals
assign read_UART  = (cs == 1) && (r_w == 1),
       write_UART = (cs == 1) && (r_w == 0);
assign read_RDR = read_UART && (rs == 1'b1),
       load_TDR = write_UART && (rs == 1'b1),
       read_SR  = read_UART && (rs == 1'b0),
       load_CR  = write_UART && (rs == 1'b0);
// read RDR or SR
assign data_bus = (read_RDR) ? RDR :
                  ((read_SR) ? {3'b000, RcvSR, TE}: 8'bz);
// define the control register
always @(posedge clk or negedge reset_n)
   if (!reset_n) CR <= 8'b0000_0000; // control register
   else if (load_CR) CR <= data_bus;
// irq generation logic
always @(posedge clk or negedge reset_n)
   if (!reset_n) IRQ <= 1'b0;
   else if ((((CR[7] == 1) && (RcvSR[0] == 1 || RcvSR[2] == 1)) ||
         (CR[6] == 1) && (TE == 1)) IRQ <= 1'b1;
        else IRQ <= 1'b0;
endmodule
```

■ **Review Questions**

15-31 Describe the operations of the transmitter of a typical UART.
15-32 Describe the operations of the receiver of a typical UART.
15-33 What is the distinction between the bit rate and the baud rate?

15-34 What are the two basic structures of baud-rate generators?

15.6 A Simple CPU Design

In this section, an *instruction set architecture* (ISA) from a simple commercial 16-bit CPU embedded in a TI MSP430 system [12] is used as an example to illuminate how to design a hardware architecture for realizing an ISA. In general, there are many implementation options, such as execution cycles and hardware resources, which can be taken into consideration for a specified ISA. Different considerations of implementation options would often result in different structures of the final hardware architecture.

15.6.1 Fundamentals of CPU

A basic μC system consists of a *center processing unit* (CPU), which is in turn composed of a datapath and a controller, a memory module, and a set of peripherals interconnected together with a system bus, as shown in Figure 15.45. The system bus includes a data bus, an address bus, and a control bus. Refer to Lin [8, 9] for more details about μC systems.

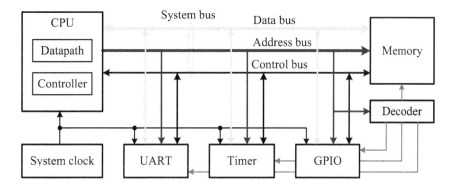

Figure 15.45: The block diagram of a typical small microcomputer system.

A CPU is a programmable digital module through which many different functions can be implemented by using appropriate subsets of instructions provided by the ISA of the CPU. For instance, multiplication can be carried out with a shift-and-add algorithm as we have described in the previous chapter by using a subset of instructions from the ISA.

The basic operations of a typical CPU are shown in Figure 15.46(a), in which two steps are repeated forever: *fetch an instruction* and *execute the instruction*. After the system is reset, the CPU starts to fetch an instruction from the memory of the μC system and executes the instruction. Then, the CPU fetches another instruction and repeats the above steps forever. More detailed operations of a CPU are depicted in Figure 15.46(b). Here, three steps are added to the basic flow shown in Figure 15.46(a): *decode the instruction, fetch a memory operand*, and *store the result*. After the CPU fetches an instruction, it decodes the instruction and determines whether a memory operand is required. If yes, the CPU fetches the memory operand before executing the instruction. After an instruction is executed, it may require to store the result back to the memory.

Now that we have realized the basic operations of a typical CPU, we are in a position to learn how to describe a CPU or how to use a CPU. As a CPU is described or used, the following issues are the most important to be taken into account: *the programming model, instruction formats, addressing modes*, and *the instruction set*.

724   Chapter 15   Design Examples

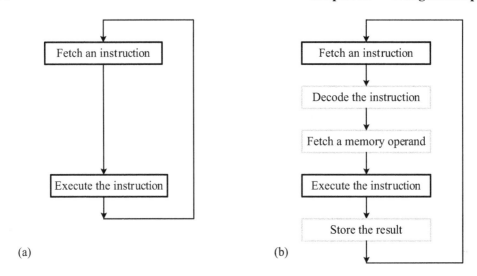

Figure 15.46: The (a) basic operations and (b) more detailed operations of a typical CPU.

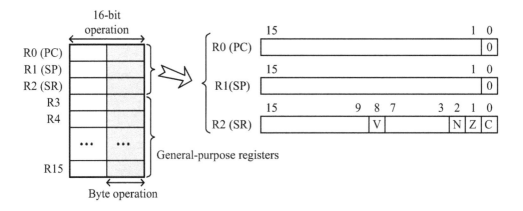

Figure 15.47: The programming model of the CPU under consideration.

15.6.1.1 Programming Model The programming model of a CPU is a set of registers that can be accessed by programmers through using the instruction set from the CPU. The programming model of the underlying 16-bit CPU is shown in Figure 15.47, which contains sixteen 16-bit registers. Among the registers, the first four registers, R0, R1, R2, and R3, have their own dedicated functions, while the other twelve registers, R4 to R15, are working registers for general purposes, including data registers, address pointers, and index values. Each of these sixteen registers can work as either a 16-bit or an 8-bit (lower byte) register.

The *program counter* (PC) (R0) always points to the next instruction to be executed. Each instruction uses an even number of bytes and the PC increments accordingly. In other words, the LSB of the PC is always set to 0.

The *stack pointer* (SP) (R1) is employed by the CPU to store the return addresses of subroutine calls and interrupts. In the underlying CPU, the SP is supposed to use a predecrement and postincrement scheme. In addition, the SP (R1) register, like other general-purpose registers, can be used with all instructions and addressing modes.

The *status register* (SR) (R2) is employed by the CPU to store the ALU flags: N, Z, V, and C (see Section 14.4.2 for their definitions), along with others. With the register addressing mode, register R2 may be used as a source or destination register; with other addressing modes,

Section 15.6 A Simple CPU Design

register R2 is utilized to generate constants, as shown in Table 15.1.

Table 15.1: Constant generation.

Register	As	Constant	Comments
R2	00	—	Register mode
R2	01	(0)	Absolute addressing mode
R2	10	0x0004	+4, bit processing
R2	11	0x0008	+8, bit processing
R3	00	0x0000	0, word processing
R3	01	0x0001	+1, bit processing
R3	10	0x0010	+2, bit processing
R3	11	0xFFFF	−1, word processing

Register R3, along with source-operand addressing modes, is employed to generate four useful constants, 0, +1, +2, and −1, as shown in Table 15.1. Consequently, register R3 cannot be used as a source register. It can only be used as a destination register. For software programming, it is common practice to leave the first four registers as special registers and only use the remaining twelve registers for general use.

15.6.1.2 Instruction Formats Any instruction is composed of two major parts: *opcode* and *operand*. The opcode field defines the operations of the instruction and the operand field specifies the operands to be operated by the instruction. There are three different instruction formats as shown in Figure 15.48. The first one is for double-operand instructions, which include data transfer, arithmetic operations, and logical operations. The second one is for single-operand instructions, which covers rotate and arithmetic right shift into carry, push onto stack, swap bytes, subroutine call, return from interrupt, and sign extension instructions. The third one is for jump instructions, which contain a 10-bit offset. The jump instructions include both conditional and unconditional types.

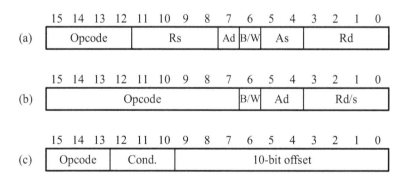

Figure 15.48: The (a) double-operand, (b) single-operand, and (c) jump instruction formats of the CPU under consideration.

15.6.1.3 Addressing Modes An addressing mode is the way that an operand is fetched. As listed in Table 15.2, the addressing modes of the underlying CPU include the following major types: *register*, *indexed*, *register indirect*, and *immediate*. The variants of indexed addressing include *symbolic addressing* and *absolute addressing*. *Autoincrement addressing* is a variant of register indirect addressing.

Register addressing (also known as *register direct addressing*) is probably the simplest method of referring to a data operand. In this addressing mode, the register contents are the operands.

Table 15.2: Addressing modes.

As/Ad	Addressing mode	Syntax	Comments
00/0	Register	Rn	Register contents are operands
01/1	Indexed	X(Rn)	(X+Rn) points to the operand
	Symbolic	ADDR	(X+PC) points to the operand
	Absolute	&ADDR	(X+SR) points to the operand
10/–	Register indirect	@Rn	Rn points to the operand
11/–	Autoincrement	@Rn+	Rn points to the operand and increments 1 or 2.
	Immediate	#N	@PC+

Indexed addressing uses a special index register. In this addressing mode, the address of the operand is formed by adding the contents of the index register with a base address. There are three variants of this addressing mode in terms of registers used.

- *Indexed addressing:* In this addressing mode, the address of the operand is at the memory location of X + Rn, where X is the next word and Rn is any register of the register file.
- *Symbolic addressing:* This is also called *PC-relative* addressing. In this mode, the PC (R0) is used as the index register. The address of the operand is at the memory location of X + PC, where X is the next word.
- *Absolute addressing:* In this addressing mode, the SR (R2) is used as the index register. The address of the operand is at the memory location of X + 0, where X is the next word.

Register indirect addressing is a way of referring to an operand in memory using a register. In this mode, the address of the operand in memory is the content of the specified register.

Autoincrement addressing (also known as *register indirect autoincrement addressing*) operates much like register indirect addressing except that the specified register is incremented afterward by 1 for .B instructions and 2 otherwise. One variant of this addressing mode is called *immediate addressing,* which is the case when the PC is used as the address register.

15.6.1.4 Instruction Set There are three major sets of instructions of the underlying CPU. These are double-operand, single-operand, and jump instruction sets.

The double-operand instruction set comprises twelve instructions, as shown in Table 15.3. They are employed to carry out data transfer, arithmetic, and logical operations. For arithmetic operations, only addition and subtraction are provided. Seven addressing modes may be applied to obtain the source operand and four (i.e., the first four addressing modes in Table 15.2) addressing modes can be used to specify the destination operand. All four types of operations, register to register, register to memory, memory to register, and memory to memory, are provided equally well.

The single-operand instruction set consists of seven instructions, as listed in Table 15.4. They are used to perform right rotate with carry (RRC) one-bit position, to perform arithmetic right shift into carry (RRA) one-bit position, to push an operand onto the stack (PUSH), to swap higher and lower bytes in a word (SWAP), to call a subroutine (CALL), to return from an interrupt (RETI), and to extend the sign bit of the lower byte into the higher byte (SXT). All instructions except the RETI instruction are allowed to use all addressing modes.

The jump instruction set includes eight instructions, as shown in Table 15.5. These instructions are utilized to test the status of a single ALU flag (C, Z, N) or a combination of two flags (N and V). In addition, an unconditional jump instruction is also provided. When the status of the flag being tested is valid, the value of the PC is added by a 10-bit offset. The unit of the 10-bit offset is in words; that is, the signed 10-bit offset is doubled before it is added to the PC. As a result, the possible jump range is from -511 to +512 words relative to the current PC value.

Section 15.6 A Simple CPU Design

Table 15.3: The double-operand instruction set.

Mnemonic	Operation	N	Z	V	C	Opcode
MOV(.B) src,dst	dst ← src	-	-	-	-	0x4xxx
ADD(.B) src,dst	dst ← src + dst	*	*	*	*	0x5xxx
ADDC(.B) src,dst	dst ← src + dst + C	*	*	*	*	0x6xxx
SUB(.B) src,dst	dst ← .not src + dst + 1	*	*	*	*	0x8xxx
SUBC(.B) src,dst	dst ← .not src + dst + C	*	*	*	*	0x7xxx
CMP(.B) src,dst	dst - src	*	*	*	*	0x9xxx
DADD(.B) src,dst	dst ← src + dst + C(decimal)	*	*	*	*	0xAxxx
BIT(.B) src,dst	dst .and src	*	*	0	*	0xBxxx
BIC(.B) src,dst	dst ← .not src .and dst	-	-	-	-	0xCxxx
BIS(.B) src,dst	dst ← src .or dst	-	-	-	-	0xDxxx
XOR(.B) src,dst	dst ← src .xor dst	*	*	*	*	0xExxx
AND(.B) src,dst	dst ← src .and dst	*	*	0	*	0xFxxx

Table 15.4: The single-operand instruction set.

Mnemonic	Operation	N	Z	V	C	Opcode
RRA(.B) dst	Arithmetic right shift, C ← LSB	*	*	0	*	0x110x
RRC(.B) dst	Right rotate through carry	*	*	*	*	0x100x
PUSH(.B) src	SP ← SP-2, @SP ←src	-	-	-	-	0x120x
SWAPB dst	Swap bytes in dst	-	-	-	-	0x108x
CALL dst	SP ← SP-2, @SP ← PC+2 PC ← dst	-	-	-	-	0x128x
RETI	SR ← TOS, SP ← SP+2 PC ← TOS, SP ← SP+2	*	*	*	*	0x130x
SXT dst	dst[15:8] ← dst[7]	*	*	0	*	0x118x

■ Review Questions

15-35 Describe the basic operations of a typical CPU.
15-36 Define the opcode, operand, and addressing mode.
15-37 In using a CPU, what are the important issues needed to consider?
15-38 What is the meaning of the programming mode of a CPU?
15-39 Explain the meaning of the PC-relative addressing mode.
15-40 Can immediate addressing be regarded as autoincrement addressing?

Table 15.5: The jump instruction set.

Mnemonic	Operation	N	Z	V	C	Opcode
JNE/JNZ label	Jump to label if the zero bit is reset	-	-	-	-	0x20xx
JNQ/JZ label	Jump to label if the zero bit is set	-	-	-	-	0x24xx
JC label	Jump to label if the carry bit is set	-	-	-	-	0x2Cxx
JNC label	Jump to label if the carry bit is reset	-	-	-	-	0x28xx
JN label	Jump to label if the negative bit is set	-	-	-	-	0x30xx
JGE label	Jump to label if (N .xor. V) = 0	-	-	-	-	0x34xx
JL label	Jump to label if (N .xor. V) = 1	-	-	-	-	0x38xx
JMP label	Jump to label unconditionally	-	-	-	-	0x3Cxx

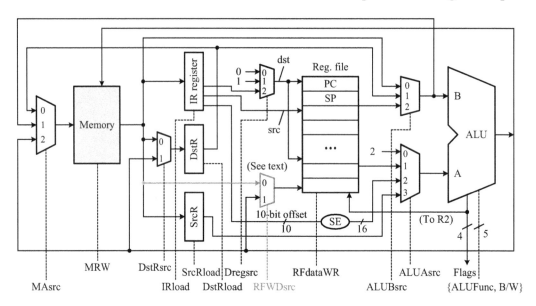

Figure 15.49: A possible datapath of the 16-bit ISA under consideration.

15.6.2 Datapath Design

The major components of the datapath of any CPU are a register file and an ALU along with some scratch registers and a few multiplexers used to route data to right places within the datapath. For a specified ISA, the datapath is usually not unique, which will be self-explanatory in the later part of this section. A possible datapath of the underlying ISA is shown in Figure 15.49. In this section, we will detail how to derive this datapath from the underlying ISA.

The datapath is a shared resource for the instruction set. Its function is to equip the required operations of each instruction in the ISA so that each instruction can be efficiently carried out. These operations are centered around some registers and an ALU. Therefore, the most straightforward way to obtain a datapath from an ISA is first to consider its core parts, the register file and the ALU, and then to explore the associated routing paths through the use of multiplexers. For this reason, the register file and the ALU generally play the most important roles in the datapath of any CPU. Their design must be considered very carefully in order to obtain the best performance of the CPU.

15.6.2.1 Register File From the double-operand instruction set, we know that to support the execution of the register-to-register operation in one cycle, the register file needs to have two read ports and one write port. Nonetheless, from the programming model depicted in Figure 15.47 and the constant generation shown in Table 15.1, both R2 and R3 registers are needed to consider specially. The R2 register contains the four ALU flags: N, Z, V, and C, which are needed to update each time the ALU performs an operation. Hence, the R2 register should be able to update in addition to the normal write operation through the write port. Furthermore, both R2 and R3 registers are used to generate constants specified by the combinations of the source addressing modes in the double-operand instruction set. In other words, we need to decode the source addressing modes and provide the proper constant outputs when a double-operand instruction specifies the R2 or R3 register as its source register.

There are many options that can be used to implement the above-mentioned register file. The most intuitive method is to employ flip-flops as building cells. Thus, each 16-bit register can be accessed independently. This method has one advantage that the problem of the R2 register can be easily solved. However, two 16-bit 16-to-1 multiplexers are required to provide

Section 15.6 A Simple CPU Design

Table 15.6: The ALU functions required for the underlying ISA.

Instruction	ALU function	Mode selection				B/W	Mnemonic
		m3	m2	m1	m0		
MOV(.B)	F ← A	0	0	0	0	-	Pass A
MOV(.B)	F ← B	0	0	0	1	-	Pass B
ADD(.B)	A+B	0	0	1	0	0/1	add (b/w)
ADDC(.B)	A+B+C	0	0	1	1	0/1	addc (b/w)
SUB(.B), CMP(.B)	A+ not B+1	0	1	0	0	0/1	sub (b/w)
SUBC(.B)	A+ not B+ not C	0	1	0	1	0/1	subc (b/w)
DADD(.B)	A+ 06H, A+60H on condition	0	1	1	0	0/1	dadd (b/w)
AND(.B), BIT(.B)	A and B	0	1	1	1	0/1	and (b/w)
BIC(.B)	not A and B	1	0	0	0	0/1	bic (b/w)
BIS(.B)	A or B	1	0	0	1	0/1	or (b/w)
XOR(.B)	A xor B	1	0	1	0	0/1	xor (b/w)
RRA(.B)	Arithmetic right shift into C	1	0	1	1	0/1	asrc (b/w)
RRC(.B)	Rotate right shift through C	1	1	0	0	0/1	rotate (b/w)
SWAP	Swap byte	1	1	0	1	0	swap

the two read ports. This may consume too much area in the cell-based or too many LUTs in the FPGA-based design and may be intolerable in some applications.

Another commonly used approach is to declare a multiple-port register file directly, like the one shown in Section 9.2.2, and then use an independent register for R2 to solve the update problem associated with it. This is a much better method because it consumes less area or needs fewer LUTs. Of course, there may exist many other possible implementation options that may be used to realize this register file.

15.6.2.2 ALU The second core part of the datapath of a CPU is the ALU module. The functions of an ALU used in a datapath can be obtained from the desired operations associated with the underlying ISA. By observing the instruction sets listed in Tables 15.3 to 15.5, the required ALU operations can be summarized and listed as in Table 15.6.

The simplest way to implement the ALU functions listed in this table is to utilize a **case** statement to describe the ALU functions in behavioral style. Another way is to modify either of the ALUs described in Section 14.4.2 to fit the required functions of Table 15.6. Because of the simplicity of these two implementations, their details are left for the reader as exercises.

The rest of the datapath besides the register file and the ALU are multiplexers and some scratch registers, as shown in Figure 15.49. These multiplexers and scratch registers are used to route the data to an appropriate place for carrying out the required operations needed by instructions. Two scratch registers, the SrcR and DstR, are used to provide the operations associated with memory operands. For the source memory operand, it is fetched and stored in the SrcR before the instruction is executed. For the destination memory operand, only the memory address of the operand is stored in the DstR and the actual operand is deferred to be fetched until the execution phase of the instruction. The memory address of the destination operand is formed by adding together a word from memory and the contents of a specified register. More details of the datapath will be explained in the subsection that follows immediately. The multiplexer with a lighter line is merely used for speeding up the execution of memory-to-register instructions. Here, we omit it for simplicity.

■ Review Questions

15-41 What is the major function of the datapath in a CPU?

15-42 Why do we say that the register file and ALU are the two core parts of a CPU?

15-43 Point out the possible path for implementing a memory-to-memory operation from the datapath shown in Figure 15.49.

15-44 Point out the possible path for implementing a memory-to-register operation from the datapath shown in Figure 15.49.

15-45 Point out the possible path for implementing a register-to-memory operation from the datapath shown in Figure 15.49.

15.6.3 Controller Design

The controller in a CPU aims to generate the control signals required for executing instructions by the datapath. The controller accepts the data from the instruction register (IR) as well as the ALU flags (N, Z, V, and C), and generates the control signals accordingly, as indicated in Figure 15.49.

The most straightforward way to design the controller of a CPU is to use the decoder-based approach, as shown in Figure 15.50, where an opcode decoder is used after the instruction fetch phase to decode (i.e., interpret) the instruction so that each instruction can then be decoded and executed accordingly. Although this approach is intuitively simple, the number of states required is rather large. As an illustration, in the running example, the total number of states will exceed seventy. The resulting state machine is not only hard to be handled but also costs too much hardware.

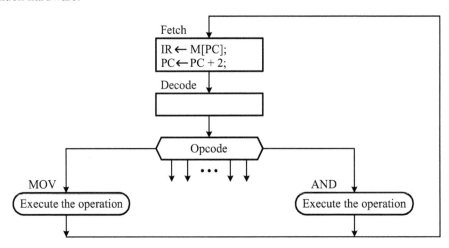

Figure 15.50: A possible controller structure of the CPU under consideration.

A widely used approach to overcoming the above difficulty is to group instructions in terms of addressing modes and the operations of instructions, as shown in Figure 15.51. Here, the instructions are first divided into three disjoint sets: double-operand, single-operand, and jump instructions. The instructions in the same set have very similar addressing modes and hence can be processed in a uniform way. For example, as shown in Figure 15.51, all double-operand instructions have the same addressing modes. Therefore, we first handle the addressing mode to prepare the desired operands for the execution phase of the instruction. The total number of states needed in this approach is the desired steps of the fetch phase plus the maximum number of execution steps, which is less than ten.

As mentioned before, each instruction before it can be executed has to be fetched from the memory first. The fetch step consists of the following two operations

$\quad$ IR $\longleftarrow$ M[PC];
$\quad$ PC $\longleftarrow$ PC + 2;

Section 15.6 A Simple CPU Design

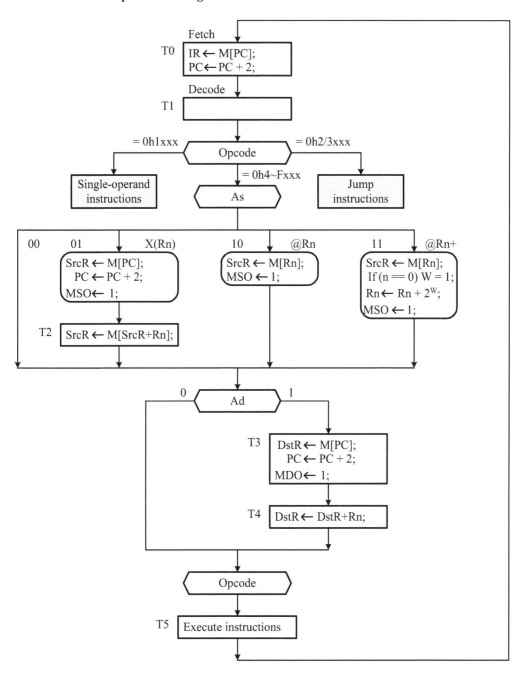

Figure 15.51: The ASM chart of the double-operand instruction set of the CPU under consideration.

which corresponds to the following control signals sent from the controller to the datapath: Dregsrc0, ALUBsrc2, MAsrc1, IRload, ALUAsrc0, {ALUFunc2,W} and RFdataWR.

15.6.3.1 Double-Operand Instructions The ASM chart of the double-operand instruction set is shown in Figures 15.51 and 15.52. Because two operands are needed in this instruction set, we first consider the source addressing mode and then deal with the destination addressing mode. As shown in Figure 15.51, for those instructions using a source operand from memory,

MOV (4xxx)   T5

```
case ({MSO,MDO})
00: Rd ← Rs;
01: M[DstR] ← Rs;
10: Rd ← SrcR;
11: M[DstR] ← SrcR;
```

DADD (Axxx)   T5

```
case ({MSO,MDO})
00: Rd ← Rd + Rs + C;
01: M[DstR] ← M[DstR] + Rs + C;
10: Rd ← Rd + SrcR + C;
11: M[DstR] ← M[DstR] + SrcR + C;
```

ADD (5xxx)   T5

```
case ({MSO,MDO})
00: Rd ← Rd + Rs;
01: M[DstR] ← M[DstR] + Rs;
10: Rd ← Rd + SrcR;
11: M[DstR] ← M[DstR] + SrcR;
```

BIT (Bxxx)   T5

```
case ({MSO,MDO})
00: Rd .and. Rs;
01: M[DstR] .and. Rs;
10: Rd .and. SrcR;
11: M[DstR] .and. SrcR;
```

ADDC (6xxx)   T5

```
case ({MSO,MDO})
00: Rd ← Rd + Rs + C;
01: M[DstR] ← M[DstR] + Rs + C;
10: Rd ← Rd + SrcR + C;
11: M[DstR] ← M[DstR] + SrcR + C;
```

BIC (Cxxx)   T5

```
case ({MSO,MDO})
00: Rd ← Rd .and. .not.Rs;
01: M[DstR] ← M[DstR] .and. .not.Rs;
10: Rd ← Rd .and. .not.SrcR;
11: M[DstR] ← M[DstR] .and. .not.SrcR;
```

SUBC (7xxx)   T5

```
case ({MSO,MDO})
00: Rd ← Rd + .not. Rs + C;
01: M[DstR] ← M[DstR]+ .not. Rs + C;
10: Rd ← Rd + .not. SrcR + C;
11: M[DstR] ← M[DstR] + .not. SrcR + C;
```

BIS (Dxxx)   T5

```
case ({MSO,MDO})
00: Rd ← Rd .or. Rs;
01: M[DstR] ← M[DstR] .or. Rs;
10: Rd ← Rd .or. SrcR;
11: M[DstR] ← M[DstR] .or. SrcR;
```

SUB (8xxx)   T5

```
case ({MSO,MDO})
00: Rd ← Rd + .not. Rs + 1;
01: M[DstR] ← M[DstR]+ .not. Rs + 1;
10: Rd ← Rd + .not. SrcR + 1;
11: M[DstR] ← M[DstR] + .not. SrcR + 1;
```

XOR (Exxx)   T5

```
case ({MSO,MDO})
00: Rd ← Rd .xor. Rs;
01: M[DstR] ← M[DstR] .xor. Rs;
10: Rd ← Rd .xor. SrcR;
11: M[DstR] ← M[DstR] .xor. SrcR;
```

CMP (9xxx)   T5

```
case ({MSO,MDO})
00: Rd + .not. Rs + 1;
01: M[DstR]+ .not. Rs + 1;
10: Rd + .not. SrcR + 1;
11: M[DstR] + .not. SrcR + 1;
```

AND (Fxxx)   T5

```
case ({MSO,MDO})
00: Rd ← Rd .and. Rs;
01: M[DstR] ← M[DstR] .and. Rs;
10: Rd ← Rd .and. SrcR;
11: M[DstR] ← M[DstR] .and. SrcR;
```

Figure 15.52: The detailed operations of the double-operand instruction set of the CPU under consideration.

we first fetch the operand from the specified memory location and store it in the scratch register SrcR to be ready for execution. In order to indicate this situation, a flag known as the memory source operand (MSO) is used.

The processing of the destination addressing mode is similar to that of the source addressing mode except that now we do not fetch the memory operand into the scratch register, DstR.

Section 15.6 *A Simple CPU Design*

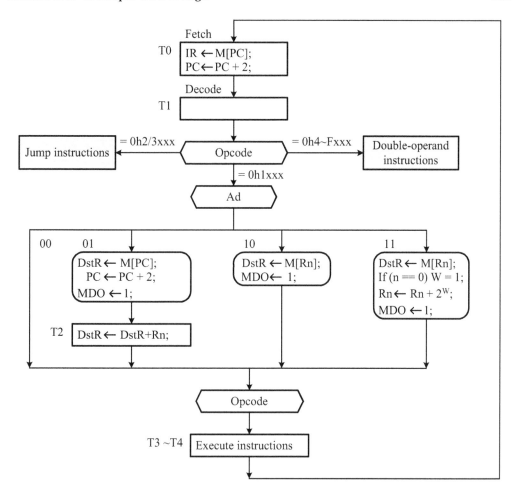

Figure 15.53: The ASM chart of the single-operand instruction set of the CPU under consideration.

Instead, we store the address of the destination operand in the DstR and set the memory destination operand flag MDO to indicate this situation.

The execution steps of the double-operand instruction set are illustrated in Figures 15.52. Each instruction has four cases controlled by the combinations of flags: MSO and MDO. As an illustration, consider the following instruction

```
ADD #5, &1000   ;M[1000] <- M[1000] + 5
```

which means an immediate data operand of 5 is added into the memory operand at location 1000. The As field of this instruction is 11 and the Ad field is 1. Thus, according to the ASM chart shown in Figure 15.51, at state T1 the scratch register, SrcR, stores 5 and at state T3 the DstR stores 1000. Both MSO and MDO flags are set to 1. At state T5, the data of 5 is added into the memory location at 1000, as shown in Figure 15.52.

15.6.3.2 Single-Operand Instructions The operations of the single-operand instruction set are described in the ASM chart shown in Figure 15.53. In these instructions, only source addressing modes are allowed. The operations of these addressing modes are exactly the same as those in the double-operand instructions depicted in Figure 15.51. The detailed operations of each instruction in this set are described in the ASM chart shown in Figure 15.54.

734 Chapter 15 Design Examples

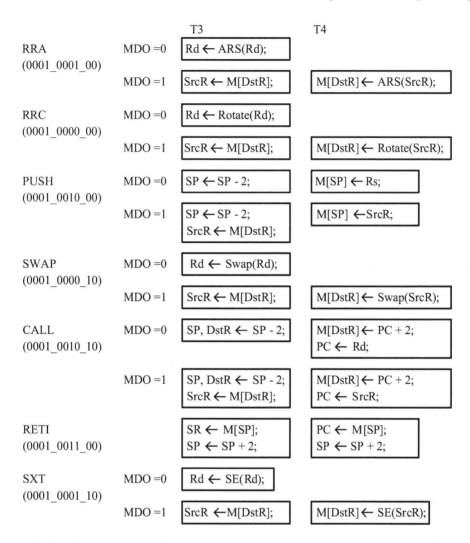

Figure 15.54: The detailed ASM chart of the single-operand instruction set of the CPU under consideration.

15.6.3.3 Jump Instructions The operations of the jump instruction set are described in the ASM chart shown in Figure 15.55. Instead of processing addressing modes, these instructions take care of the conditional flags. The resulting operations are quite simple—they only need one state—as shown in Figure 15.55.

■ Review Questions

15-46 Show how to execute the instruction MOV R4, &addr on the datapath shown in Figure 15.49.
15-47 Show how to execute the instruction ADD 3(R4), 8(R5) on the datapath shown in Figure 15.49.
15-48 Show how to execute the instruction RRA &addr on the datapath shown in Figure 15.49.
15-49 Show how to execute the instruction SWAP R8 on the datapath shown in Figure 15.49.
15-50 Show how to execute the instruction JGE 1200 on the datapath shown in Figure 15.49.

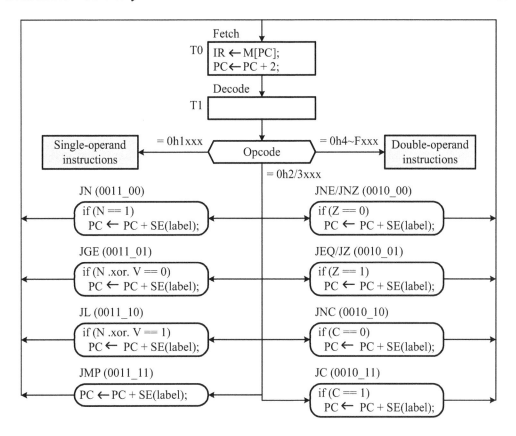

Figure 15.55: The ASM chart of the jump instruction set of the CPU under consideration.

15.7 Summary

The bus structure can be constructed based on tristate buffers or multiplexers. Since a bus is a shared and exclusively used resource, only one transmitter can place data onto the bus at any time. As a result, a bus arbitration logic circuit is required for any bus system to schedule the use of the bus. Daisy-chain arbitration and radial arbitration are the two most widely used bus schedule schemes. The former embeds a daisy-chain logic circuit in each module through which a priority can be determined in sequence. Each device in the daisy-chain logic has a fixed priority. In contrast, the latter uses individual request and grant lines for each module. The priority of each request line may be fixed or variable. The most widely used variable priority scheme is known as the round-robin priority, in which the device just being served is set the lowest priority whereas the device succeeding it is set the highest priority.

Data transfer on a bus can be proceeded in a synchronous or an asynchronous manner, and in a bundle of bits (namely, parallel) or a single bit at a time (namely, serial). When the data transfer between two devices is synchronized by a common clock signal, it is called the synchronous mode; otherwise, it is referred to as the asynchronous mode. When data are transferred in a bundle of bits at the same time, it is called parallel data transfer; when data are transferred in a bit-by-bit manner, it is known as serial data transfer.

In this chapter, we also used a real-world example to illustrate the design details of a small μC system. This system includes a GPIO port, a timer, a UART, and a system bus. The system bus includes an address bus, a data bus, and a control bus. The GPIO port is an 8-bit parallel input/output port which can be used as a single 8-bit port or as eight 1-bit ports. The basic operations of generic timers and how to design such timers are considered in detail. The UART is a device that supports the serial communication between two devices in an asynchronous manner. An example is given to illustrate how to design such a device.

The important issues of learning about and designing a CPU were dealt with in detail in this chapter too. These issues include the programming model, instruction formats, addressing modes, and the instruction set. To further illustrate how to design a CPU, a 16-bit CPU with 27 instructions and 7 addressing modes was considered. The associated 16-bit ISA was employed as an example to show how to design the related datapath and controller. For the datapath, we used a register file and an ALU as the core components and then illustrated how to derive the entire datapath from the operations of the ISA. For the controller, we showed how to design an efficient controller by using ASM charts.

References

1. Altera, "AN 545: Design Guidelines and Timing Closure Techniques for Hard Copy ASICs," July 2010, Altera Corporation.
2. J. Bhasker, *A Verilog HDL Primer,* 3rd ed., Star Galaxy Publishing, 2005.
3. C. E. Cummings, "Synthesis and scripting techniques for designing multi-asynchronous clock designs," SNUG San Jose 2001, Sunburst Design, Inc.
4. W. J. Dally and B. Towles, *Principles and Practices of Interconnection Networks,* San Francisco: Morgan Kaufmann Publishers, 2004.
5. J. P. Hayes, *Computer Architecture and Organization,* 3rd ed., New York: McGraw-Hill, 1997.
6. M. B. Lin, "On the design of fast large fan-in CMOS multiplexers," *IEEE Trans. on Computer-Aided Design of Integrated Circuits and Systems,* Vol. 19, No. 8, pp. 963-967, August, 2000.
7. M. B. Lin, *Introduction to VLSI Systems: A Logic, Circuit, and System Perspective,* CRC Press, 2012.
8. M. B. Lin, *Microprocessor Principles and Applications: x86/x64 Family Software, Hardware, Interfacing, and Systems,* 5th ed., Taipei, Taiwan: Chuan Hwa Book Ltd., 2012.
9. M. B. Lin and S. T. Lin, *Basic Principles and Applications of Microprocessors: MCS-51 Embedded Microcomputer System, Software, and Hardware,* 3rd ed., Chuan Hwa Book Ltd. (Taipei, Taiwan), 2013.
10. B. Parhami, *Computer Architecture: From Microprocessors to Supercomputers,* New-York: Oxford University Press, 2005.
11. C. H. Roth, Jr., *Digital Systems Design Using VHDL,* Boston: PWS Publishing Company, 1998.
12. Texas Instruments, *MSP430x4xx Family User's Guide,* 2003.

Problems

15-1 Suppose that a bus system uses a radial arbitration scheme and allows at most eight devices attached onto it at the same time. Use an 8-to-3 priority encoder and a 3-to-8 decoder to realize the bus arbitration logic.

    (a) Model your design using Verilog HDL.

    (b) Write a test bench to verify its functionality.

15-2 Considering the handshaking data transfer scheme shown in Figure 15.13, design a module to equip this kind of data transfer.
  (a) Model your design using Verilog HDL.
  (b) Write a test bench to verify its functionality.

15-3 Considering the case of the synchronous serial data transfer with an explicitly clocking scheme as shown in Figure 15.10(a), design a module to equip this kind of data transfer.
  (a) Model your design using Verilog HDL.
  (b) Write a test bench to verify its functionality.

15-4 Considering the block diagram shown in Figure 15.56, answer each of the following questions:
  (a) Explain why the grouped signals, x_a and y_a, in Module A cannot be correctly passed to Module B as grouped signals, x_b and y_b.
  (b) Give a correct solution for the above problem.

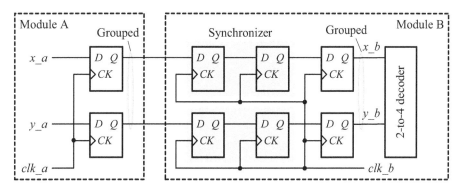

Figure 15.56: The block diagram for Problem 15-4.

15-5 Considering the block diagram shown in Figure 15.57, in which two synchronizers in module B are employed to sample the input signal x, answer each of the following questions:
  (a) What are the possible values of output signals, w and y? Can they obtain the same values for a given x value?
  (b) Give a correct solution for the above problem so that the output signals w and y can always have the same values for a given input x.

15-6 Considering the block diagram shown in Figure 15.58 and assuming that the strobe signal is one clock cycle, answer each of the following questions.
  (a) Design a scheme so that the destination device can capture the data into its internal register by the strobe signal from the source device.
  (b) Model your design using Verilog HDL.
  (c) Write a test bench to verify its functionality.

15-7 Considering the block diagram shown in Figure 15.59 and assuming that the strobe signal is one clock cycle, answer each of the following questions.
  (a) Design a scheme so that the source device can capture the data into its internal register by the strobe signal from the destination device.

738 Chapter 15 Design Examples

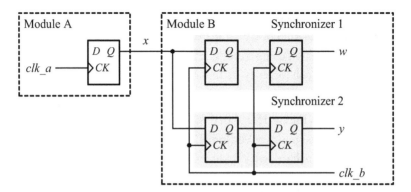

Figure 15.57: The block diagram for Problem 15-5.

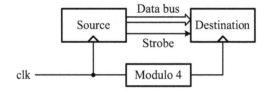

Figure 15.58: The block diagram for Problem 15-6.

- **(b)** Model your design using Verilog HDL.
- **(c)** Write a test bench to verify its functionality.

15-8 Referring to the synchronous FIFO with a register file shown in Figure 15.26, answer each of the following questions.
- **(a)** Design a parameterizable Verilog HDL module to describe a Gray-code counter.
- **(b)** Modify the module in Example 15.8 by replacing the read and write pointers with the Gray-code up counter.
- **(c)** Write a test bench to verify its functionality.

15-9 Considering the synchronous FIFO with a dual-port memory shown in Figure 15.27, answer each of the following questions.
- **(a)** Write a Verilog HDL module to describe this synchronous FIFO without the shaded logic blocks.
- **(b)** Write a test bench to verify its functionality.
- **(c)** Assume that each of blocks A and B is composed of a binary-to-Gray code converter and an output register; blocks Aa and Bb are Gray-to-binary code convert-

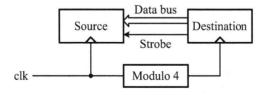

Figure 15.59: The block diagram for Problem 15-7.

Problems 739

ers. Design these logic circuits and explain why the resulting FIFO module will function correctly.

(d) Write a test bench to verify the functionality of the resulting module in (c).

15-10 Modify the GPIO port introduced in Section 15.3.2 by adding two extra control signals so that it can equip the handshaking capability.

(a) Model your design using Verilog HDL.
(b) Write a test bench to verify its functionality.

15-11 Combine the four basic timer operations described in Section 15.4.1 into one single timer module with the mode selection as shown in Figure 15.31.

(a) Model your design using Verilog HDL.
(b) Write a test bench to verify its functionality.

15-12 Design a timer module that performs the rate-generation function described in Figure 15.33.

(a) Draw the logic circuit of the timer at the RTL.
(b) Model it using Verilog HDL.
(c) Write a test bench to verify its functionality.

15-13 Suppose that a timer has two inputs, `clock` and `gate`, and one output, `out`, exactly the same as the one introduced in Figure 15.31, but only has two operation modes. It is a 16-bit binary-up counter in mode 0 and an 8-bit reloadable binary-up counter in mode 1.

(a) Draw the logic circuit of the timer at the RTL.
(b) Model it using Verilog HDL.
(c) Write a test bench to verify its functionality.

15-14 Design a real-time interrupt module that generates a periodic interrupt to remind the CPU to carry out some routine tasks, assuming that the interrupt interval may be programmable.

(a) Draw the logic circuit of the interrupt module at the RTL.
(b) Model it using Verilog HDL.
(c) Write a test bench to verify its functionality.

15-15 Considering the PWM timer shown in Figure 15.38 that operates at a frequency of 12 MHz, answer each of the following questions.

(a) What is the frequency range that the output signal, `out`, can be generated?
(b) What is the duty cycle that the output signal, `out`, can be generated?

15-16 Design a UART with the following features. There are 9 data bits without parity check; that is, the parity bit is used as the ninth data bit. The sampling clock frequency is 16 times the transmission clock frequency and the baud rate is fixed to 1/64 of the system clock frequency.

(a) Draw the logic circuit of the UART at the RTL.
(b) Model it using Verilog HDL.
(c) Write a test bench to verify its functionality.

15-17 Design a UART with the following features. There are 8 data bits with even-parity check. The sampling clock frequency is n times the transmission clock frequency, where n is the scale factor of the sampling clock frequency and can be set to 1, 4, 16, or 64. The baud rate can be set to any commonly used value.
  (a) Draw the logic circuit of the UART at the RTL.
  (b) Model it using Verilog HDL.
  (c) Write a test bench to verify its functionality.

15-18 The programmable interrupt controller (PIC) is a device or a module widely used in μC systems. Design a PIC module with the following features at least. The inputs are eight interrupt inputs from I/O devices, each of which may be level-sensitive or edge-triggered. The outputs are an interrupt (INT) to the CPU and an 8-bit data bus. The inputs are prioritized and numbered from 0 to 7 with an increasing priority order. Each input can be disabled if it is not allowed to generate an interrupt. The one-byte interrupt vector is output to the data bus as an acknowledge is received from the CPU.
  (a) Draw the block diagram of the programmable interrupt controller. Label each component properly to indicate its functionality and size.
  (b) Model it using Verilog HDL.
  (c) Write a test bench to verify its functionality.

15-19 Using D flip-flops as basic cells to construct the three-port register file required in the datapath shown in Figure 15.49, answer each of the following questions.
  (a) Draw the logic circuit of the resulting datapath at the RTL.
  (b) Model it using Verilog HDL.
  (c) Write a test bench to verify its functionality.

15-20 The register file may also be designed with a memory module instead of registers. Based on this idea, design the three-port register file required in the datapath shown in Figure 15.49. It is useful to take the R2 register as a special case and construct it with D flip-flops.
  (a) Draw the logic circuit of the resulting datapath at the RTL.
  (b) Model it using Verilog HDL.
  (c) Write a test bench to verify its functionality.

15-21 Considering the first ALU introduced in Section 14.4.2, answer each of the following questions.
  (a) Modify it to meet the functions required in Table 15.6. Draw the logic circuit of the result.
  (b) Model it using Verilog HDL.
  (c) Write a test bench to verify its functionality.

15-22 Considering the second ALU introduced in Section 14.4.2, answer each of the following questions.
  (a) Modify it to meet the functions required in Table 15.6. Draw the logic circuit of the result.
  (b) Model it using Verilog HDL.
  (c) Write a test bench to verify its functionality.

15-23 Considering the datapath shown in Figure 15.49, model it with Verilog HDL and write a test bench to verify its functionality.

Problems   741

15-24 Considering the case when the multiplexer with the lighter line shown in Figure 15.49 is used, analyze Tables 15.3 to 15.5 and show what kinds of instructions can be speeded up when they are executed.

15-25 Supposing that the CPU only supports the double-operand instructions as described in Figures 15.51 and 15.52, answer each of the following questions.
  (a) Design the controller and model it using Verilog HDL.
  (b) Write a test bench to verify its functionality.

15-26 Supposing that the CPU only supports the single-operand instructions as described in Figures 15.53 and 15.54, answer each of the following questions.
  (a) Design the controller and model it using Verilog HDL.
  (b) Write a test bench to verify its functionality.

15-27 Supposing that the CPU only supports the jump instructions as described in Figure 15.55, answer each of the following questions.
  (a) Design the controller and model it using Verilog HDL.
  (b) Write a test bench to verify its functionality.

15-28 The following **always** block intends to update a status register (SR). The SR is cleared if reset_n is asserted or it is read by the control signal, read_SR. The SR is updated with the new value of Flags each time the load control signal, set_Flag, is asserted. Suppose that the pulse width of the control signal, read_SR, is one system clock (clk) cycle when it is asserted and the pulse width of the control signal, set_Flag, is much wider than one system clock (clk) cycle.

```
assign SR_reset = !reset_n || read_SR;
always @(posedge clk or posedge SR_reset)
    if (SR_reset)      SR <= 4'b0000;
    else if (set_Flag) SR <= Flags;
```

  (a) Explain why the above **always** block cannot correctly clear the status register SR when the read_SR control signal is asserted.
  (b) Give a solution to correct the problem.

15-29 This problem involves the design of a circuit that converts an incoming binary stream into a non-return-to-zero-inverted (NRZI) code. The NRZI code maintains a constant voltage pulse for the duration of a bit time. The presence or absence of a signal transition at the beginning of the bit time denotes the binary data bit. A transition, from low to high or high to low, at the beginning of a bit time denotes a binary 1 for that bit time; no transition indicates a binary 0. Design a circuit to convert the incoming binary stream into its equivalent NRZI code.
  (a) Draw the ASM chart of the converter.
  (b) Write a Verilog HDL module to describe the converter.
  (c) Write a test bench to verify the module.

15-30 This problem concerns the design of a circuit that converts an incoming NRZI code into its equivalent binary data. The details of the NRZI coding scheme can be referred to the above problem. Design a circuit to convert the incoming NRZI code into its equivalent binary data.
  (a) Draw the ASM chart of the converter.
  (b) Write a Verilog HDL module to describe the converter.

(c) Write a test bench to verify the module.

15-31 This problem involves the design of a converter that converts an incoming binary stream to its equivalent Manchester code. In the Manchester code, there is a transition at the middle of each bit time. This transition serves as the clocking information and data: a low-to-high transition denotes a binary 1 and a high-to-low transition represents a binary 0. Design a circuit to convert an incoming binary stream into its equivalent Manchester code.
  (a) Draw the ASM chart of the converter.
  (b) Write a Verilog HDL module to describe the converter.
  (c) Write a test bench to verify the module.

15-32 This problem concerns the design of a converter that converts an incoming Manchester code to its equivalent binary stream. The details of the Manchester coding scheme can be referred to the above problem. Design a circuit to convert an incoming Manchester code into its equivalent binary stream.
  (a) Draw the ASM chart of the converter.
  (b) Write a Verilog HDL module to describe the converter.
  (c) Write a test bench to verify the module.

15-33 This problem considers the design of a converter that converts an incoming binary stream to a differential Manchester code. In the differential Manchester code, the transition at the middle of each bit time only serves as the clocking information; the transition, low-to-high or high-to-low, at the beginning of a bit time denotes the binary value. If there is a transition, the binary value at that bit time is 0; otherwise, the binary value is 1. Design a circuit to convert an incoming binary stream into the differential Manchester code.
  (a) Draw the ASM chart of the converter.
  (b) Write a Verilog HDL module to describe the converter.
  (c) Write a test bench to verify the module.

15-34 This problem involves the design of a converter that converts a differential Manchester code to its equivalent binary data. The details of the differential Manchester coding scheme can be referred to the above problem. Design a circuit to convert an incoming differential Manchester code into an equivalent binary stream.
  (a) Draw the ASM chart of the converter.
  (b) Write a Verilog HDL module to describe the converter.
  (c) Write a test bench to verify the module.

16

Design for Testability

TESTING is an essential step in any PCB-based, cell-based, or FPGA-based system, even in a simple digital logic circuit because the only way to ensure that a system or a circuit may properly function is through a delicate testing process. The goal of testing is to find any existing faults in a digital system or a logic circuit.

In order to test any faults in a digital system or a logic circuit, some fault models must be first defined. The most common fault models used in CMOS technology include stuck-at faults, bridge faults, and stuck-open faults. Based on a specific fault model, the system or circuit can then be tested by applying stimuli, observing the responses, and analyzing the results. The input stimulus special for use in a given digital system or a logic circuit is often called a test vector, which is a combination of input values. A collection of test vectors for a logic circuit is called the test set of the logic circuit. Excepting in the exhaustive test, a test set is usually a subset of all possible input combinations of a logic circuit. To illustrate how to generate test vectors, a method of test vector generation for combinational circuits based on path sensitization is described concisely.

Nowadays, the most effective way to test a sequential circuit is by adding some extra logic circuits to the circuit under test in order to increase both the controllability and observability of the sequential circuit. Such a design is known as the testable circuit design or design for testability. The widely used approaches include the ad hoc approach, scan-path method, and built-in self-test (BIST). The scan-path method is also extended to the system-level testing, such as SRAM, the core-based system, and the system-on-chip (SoC).

16.1 Fault Models

Any test must be based on some kind of fault model. Even though there are many fault models that have been proposed over the past decades, the most widely used fault model in logic circuits is the stuck-at fault model. Moreover, two additional fault models, the bridge fault and stuck-open fault, are also common in CMOS technology. In this section, we first introduce some useful terms along with their definitions [1, 3, 7]. Then, we deal with the fault detection and test vectors.

16.1.1 Fault Models

A *fault* is a physical defect in a circuit. The physical defect may be caused by process defects, material defects, age defects, or even package defects. When a fault manifests itself in the circuit, it is called a *failure*. When a fault manifests itself in the signals of a system, it is called an *error*. In other words, a defect means the unintended difference between the implemented hardware and its design. A fault is a representation of a defect at an abstract function level and an error is the wrong output signal produced by a defective circuit.

16.1.1.1 Stuck-at Faults The most common fault model used today is still the *stuck-at fault* model due to its simplicity and the fact that it can model many faults arising from physical defects, such as broken lines, opened diodes, shorted diodes, and short circuits between the power supply and ground. The stuck-at fault model can be *stuck-at-0*, *stuck-at-1*, or both. The "stuck-at" means that the net will adhere to a logic 0 or a logic 1 permanently. A stuck-at-0 fault is modeled by assigning a fixed logic value 0 to a signal line or net in the circuit. Similarly, a stuck-at-1 fault is modeled by assigning a fixed logic value 1 to a signal line or net in the circuit.

■ **Example 16.1: Stuck-at faults.**

Figure 16.1(a) shows a stuck-at-0 fault. Because the logic gate under consideration is an AND gate, its output is always 0 whenever one of its inputs is stuck-at-0, regardless of the other input value. Figure 16.1(b) shows a stuck-at-1 fault. Because the logic gate under consideration is an OR gate, its output is always 1 whenever one of its inputs is stuck-at-1, regardless of the other input value.

Figure 16.1: The (a) stuck-at-0 and (b) stuck-at-1 fault models.

When a circuit may have at most one net occurring a stuck-at fault, it is called a *single fault*; when a circuit may have many nets occurring stuck-at faults at the same time, it is known as *multiple faults*. For a circuit with n nets, it can have $3^n - 1$ possible stuck-at net combinations but at most $2n$ single stuck-at faults. Of course, a single fault is a special case of multiple faults. Besides, when all faults in a circuit are either stuck-at-0 or stuck-at-1, but cannot be both, the fault is named a *unidirectional fault*.

16.1.1.2 Bridge and Stuck-Open/Stuck-Closed Faults In CMOS circuits, there are two additional common fault models: *bridge faults* and *stuck-open faults*. A bridge fault is a short circuit between any two nets unintentionally, such as the two examples shown in Figure 16.2(a). In general, the effect of a bridge fault is completely determined by the underlying technique of the logic circuit. In CMOS technology, a bridge fault may be evolved into a stuck-at fault or a stuck-open fault, depending on where the bridge fault occurs.

A stuck-open fault [9] is a feature of CMOS circuits. It means that some net is broken during manufacturing or after a period of operation. A major difference between the stuck-at fault and stuck-open fault is that the circuit is still a combinational circuit when a stuck-at fault occurs but the circuit will be converted into a sequential circuit when a stuck-open fault occurs.

■ **Example 16.2: Bridge and stuck-open faults.**

Figure 16.2(a) shows two bridge faults. The pMOS block functions as $\overline{y+z}$ rather than its original function $\overline{x(y+z)}$ when the bridge fault S_1 occurred. The nMOS block functions as

Section 16.1 Fault Models

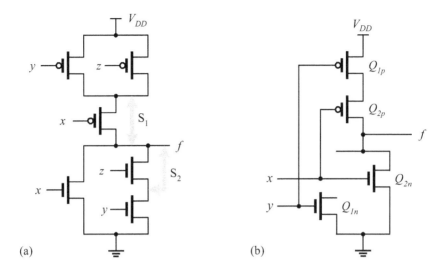

Figure 16.2: The (a) bridge and (b) stuck-open fault models.

$(x + y)$ rather than its original function $(x + yz)$ when the bridge fault S_2 occurred. Consequently, bridge faults often change the logic function of the circuit.

To see the effects of stuck-open faults, consider the two-input NOR gate in Figure 16.2(b). As the nMOS transistor Q_{1n} is at a stuck-open fault, the output f will change its function from $\overline{(x + y)}$ into a sequential logic function (Why? Try to explain it).

Sometimes a *stuck-closed* (*stuck-on*) fault is also considered in CMOS technology. A stuck-open fault is modeled as a switch being permanent in the open state whereas a stuck-closed (stuck-on) fault is modeled as a switch being permanent in the shorted state; that is, the switch cannot be turned off. The effect of a stuck-open fault is to produce a floating state at the output of the faulty logic circuit whereas the effect of a stuck-closed fault is to yield a conducting path between two nets or from the power-supply rail to ground.

16.1.1.3 Equivalent Faults Some stuck-at faults will cause the same output values regardless of what inputs are. Such indistinguishable stuck-at faults are referred to as *equivalent faults*. More precisely, two or more faults of a logic circuit are said to be equivalent if they transform the circuit in a way such that the two or more faulty circuits have the same output function. The size of a test set can be reduced considerably if we make use of the concept of equivalent faults. The process used to select one fault from each set of equivalent faults is known as *fault collapse*. The set of selected equivalent faults is called the set of *equivalent collapsed faults*. The metric of this is the *collapse ratio*, which is defined as

$$\text{Collapse ratio} = \frac{|\text{Set of equivalent collapsed faults}|}{|\text{Set of faults}|}$$

where $|x|$ denotes the cardinality of the x set.

■ Example 16.3: Examples of equivalent faults.

In Figure 16.1(a), the output of the AND gate is always 0 whenever one or both inputs a and b are stuck-at-0, or output f is stuck-at-0. These faults are indistinguishable and hence are equivalent faults. Similarly, in Figure 16.1(b), the output of the OR gate is always 1 whenever

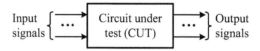

Figure 16.3: The basic model of a logic circuit under test.

one or both inputs a and b are stuck-at-1, or output f is stuck-at-1. Hence, they are equivalent faults.

■ Review Questions

16-1 Define stuck-at-0 and stuck-at-1 faults.
16-2 Define the bridge fault and stuck-open fault.
16-3 What is the meaning of a fault model?
16-4 How many single stuck-at faults can an m-input logic gate have?

16.1.2 Fault Detection

Before proceeding, we first define three related terms: a *test*, *fault detection*, and *fault location*. A test is a response generated by a procedure that applies proper stimuli to a specified fault-free circuit. Fault detection is a procedure which determines whether the circuit under test (CUT) is faulty by applying a test. The fault location means to locate the exact position, or nearby, of a fault. However, before a fault can be located, it is generally necessary to detect whether the fault has occurred. As a result, locating a fault often requires more resources and is much more difficult than detecting a fault.

As we have mentioned in Section 13.1.1, a logic circuit under test (CUT) can be considered as one of the following three models: a black box, a white box, and a gray box. In the rest of this chapter, we will assume that the gray box model is used throughout. In other words, the CUT is a black box with a known logic function, even the logic circuit, but all stimuli must be applied at the (primary) inputs and all responses must be detected at the (primary) outputs. The basic model of a logic circuit under test is shown in Figure 16.3.

To test whether a fault has occurred at a net, we need to apply an appropriate stimulus from the inputs to set the net with the fault to be detected to an opposite logic value and to propagate the net value to the outputs so that we can observe the net value and determine whether the fault has occurred. More precisely, the capability of fault detection is based on the following two features [7] of the circuit under test: *controllability* and *observability*.

- The *controllability* of a particular net in a logic circuit is a measure of the ease of setting the net to a logic 1 or a logic 0 from the primary inputs.
- The *observability* of a particular net in a logic circuit is the degree to which the net value can be observed at the primary outputs.

In a combinational circuit, if at least one test can be found to determine whether a specified fault has occurred, the fault is called a *detectable fault* or *testable fault*; otherwise, the fault is referred to as an *undetectable fault*. In other words, a detectable fault means that at least one input combination can be found to make the outputs of the faulty circuit different from those of the fault-free circuit; an undetectable fault means that we cannot find such an input combination. To evaluate the capability of detecting faults, a metric referred to as *fault coverage* is often used and is defined as follows

$$\text{Fault coverage} = \frac{\text{Number of detected faults}}{\text{Total number of faults}}$$

Section 16.1 *Fault Models*

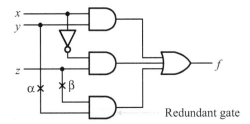

Figure 16.4: A redundant gate may cause some nets associated with it to be undetectable.

which means that what fraction of faults can be detected from all possible faults. Commercial CAD tools, such as *Sandia controllability and observability analysis program* (SCOAP), can be employed to analyze the fault coverage of a logic circuit. To achieve a world-class quality level, the fault coverage of a circuit has to be in excess of 98.5%. In principle, the condition that all stuck-at faults of a combinational circuit can be detected is the combinational circuit must be irredundant (or irreducible); namely, its logic function must be an irredundant switching expression [10].

■ **Example 16.4: Detectable and undetectable stuck-at faults.**

Figure 16.4 is a logic circuit with a redundant gate; that is, it is not an irredundant logic circuit. From the basic features of AND and OR gates, in order to pass the value set by a specified input of an AND gate to the output of the AND gate, the remaining inputs of the AND gate must be set to 1; to pass the value set by a specified input of an OR gate to the output of the OR gate, the remaining inputs of the OR gate must be set to 0. As a result, nets α and β can be set independently to 0 or 1. Both α_{s-a-0} and β_{s-a-0} are unobservable faults but α_{s-a-1} and β_{s-a-1} are observable faults (Why? Explain). Hence, α_{s-a-0} and β_{s-a-0} are undetectable faults but α_{s-a-1} and β_{s-a-1} are detectable faults.

From the above example, we can draw a conclusion. A fault of a CUT that is detectable or undetectable depends on the following two features of the fault: controllability and observability. As mentioned before, when a combinational circuit is irredundant, any physical defect occurring on any its net will cause the output value to deviate from its normal value. As a result, it can be detected at least by one input combination.

■ **Review Questions**

16-5 Define test, fault detection, and fault location.
16-6 Define detectable and undetectable faults.
16-7 Define controllability and observability.
16-8 How many single stuck-at faults do the Figure 16.4 have, once that the redundant AND gate has been removed? Can they all be detectable?

16.1.3 Test Vectors

In testing a logic circuit, it is necessary to find a simple set of input combinations that can be applied to the inputs of the logic circuit. This set of input combinations is called a *test set*. A combination of inputs used to test a specified fault is called a *test vector* (also called a *test pattern*) for the fault. A test set of a logic circuit is a union of test vectors for the circuit. If a test set can test all testable (detectable) faults of a logic circuit, the test set is referred to as a *complete test set* for the logic circuit.

In general, the truth table of a logic circuit is a complete test set of the logic circuit. However, there are 2^n possible input combinations of an n-input combinational circuit. Consequently, it is impossible or ineffective to use the entire truth table as a complete test set for a large n in practice. Fortunately, we generally do not need to do this. Through carefully examining the logic circuit, we often can find a complete test set with a size much smaller than the truth table. The following example illustrates this idea.

■ Example 16.5: An example of complete test sets.

Figure 16.5 shows a two-input NAND gate. As mentioned above, the simplest way to test this circuit is to apply the four combinations of inputs a and b one by one and then compare the results of the output with its truth table to determine whether it is faulty.

When the combination of inputs a and b is 01, the output is 1 if it is fault-free but is 0 if net α is stuck-at 1 or net γ is stuck-at 0. When the combination of inputs a and b is 10, the output is 1 if it is fault-free but is 0 if net β is stuck-at 1 or net γ is stuck-at 0. When the combination of inputs a and b is 11, the output is 0 if it is fault-free but is 1 if net α or net β is stuck-at 0 or net γ is stuck-at 1. Because the above three combinations have completely tested all possible single stuck-at faults, they constitute a complete test set. That is, the complete test set of the two-input NAND gate is {01, 10, 11}. It saves 25% test cost compared to an exhaustive test set with the whole truth table.

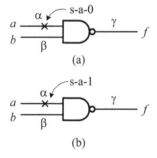

a	b	$f_{fault\text{-}free}$	f_{faulty}	Detectable faults
0	0	1	0	$\gamma/0$
0	1	1	0	$\alpha/1$, $\gamma/0$
1	0	1	0	$\beta/1$, $\gamma/0$
1	1	0	1	$\alpha/0$, $\beta/0$, $\gamma/1$

(c)

Figure 16.5: An example of complete test sets.

In summary, the technique of using all possible input combinations of a logic circuit to test the combinational circuit is known as an *exhaustive test*, belonging to the type of the complete test. However, it can usually find a complete test set with a size much smaller than the truth table if we examine the combinational circuit under test more carefully.

■ Review Questions

16-9 Define the test vector, test set, and complete test set.

16-10 Define the exhaustive test.

16-11 Why is the size of a complete test set usually smaller than that of the truth table of a combinational circuit?

16.2 Automatic Test Pattern Generation

As described previously, in order to test a combinational circuit, it is necessary to derive a test set with the minimal number of test vectors so as to save the test cost and speed up the test process. Over the past decades of development, a great number of test vector generation methods have been reported. Now, commercial automatic test vector generation tools (computer programs) have been invented, called *automatic test pattern generation* (ATPG). Although they are computation-intensive, they are able to achieve excellent code coverage. To illustrate how to generate a set of test vectors for a combinational circuit, in this section we only survey the *path sensitization* method and its derivation, a simplified D-algorithm. The interested reader can be referred to [1, 3, 7, 15] for more details. In industry and in the literature, the two terms, the *test vector* and *test pattern*, are often used interchangeably.

16.2.1 Path Sensitization

The essence of detecting a fault is to apply a proper combination of inputs (namely, a test vector) to set the net of interest with a value opposite to its fault value, propagate the fault effect to the primary outputs, and observe the output value to determine whether the fault at the net has occurred. The path sensitization approach [3, 13] is a direct application of this idea and has the following three steps:

1. *Fault sensitization:* This step sets the net to be tested (namely, the test point) to a value opposite to the fault value from the primary inputs. Fault sensitization is also known as *fault excitation*.
2. *Fault propagation:* This step selects one or more paths, starting from the test point (i.e., the specified net) to the primary outputs, in order to propagate the fault effect to the primary output(s). Fault propagation is also known as *path sensitization*.
3. *Line justification:* This step sets the primary inputs to justify the internal assignments previously made to sensitize a fault or propagate its effect. Line justification is also known as a *consistency operation*.

An illustration of how the path sensitization approach works is demonstrated in the following example.

■ **Example 16.6: The path sensitization approach.**

As mentioned above, the stuck-at-0 fault at net β of Figure 16.6 is tested by first setting the logic value of net β to 1, which in turn requires input x_1 or x_2 to be set to 1. Then, in order to propagate the logic value at net β to the output, a path from net β to the output needs to be established. This means that the primary input, x_3, has to be set to 0, which also sets the net γ to 0. As a result, the value of the primary input, x_4, is irrelevant. Therefore, the test vectors are the set: $\{(1, \phi, 0, \phi), (\phi, 1, 0, \phi)\}$.

A path is called a *sensitizable path* for a stuck-at-fault net if it can propagate the net value to the primary outputs with a consistency operation. Otherwise, the path is called an *unsensitizable path* for the stuck-at-fault net. The following example explores how the sensitizable and unsensitizable paths occur in an actual logic circuit.

■ **Example 16.7: The hazard of the path sensitization approach.**

When the stuck-at-0 fault at net h shown in Figure 16.7 is to be tested, both primary inputs, x_2 and x_3, have to be set to 0 in order to set the logic value at net h to 1. Two possible paths from net h could be selected as the sensitization paths to the primary output. If path 1 is used,

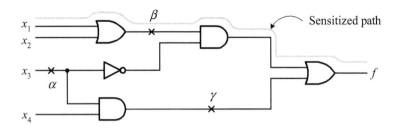

Figure 16.6: An example of path sensitization.

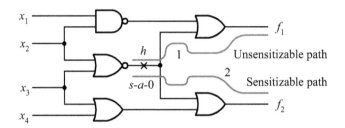

Figure 16.7: Unsensitizable path in the path sensitization approach.

both primary inputs, x_1 and x_2, are needed to set to 1 in order to propagate the logic value at net h to the primary output. As a result, we obtain an inconsistent setting of the primary input, x_2, and path 1 is an unsensitizable path for the stuck-at-0 fault at net h. If path 2 is used, both primary inputs, x_3 and x_4, are needed to set to 0 in order to propagate the logic value at net h to the primary output. This is a consistent operation. Therefore, the test vectors are the set: $\{(\phi, 0, 0, 0)\}$, which is independent of the primary input, x_1. Path 2 is a sensitizable path for the stuck-at-0 fault at net h.

In the path sensitization approach, as only one path is selected, the result is called the *one-dimensional* or *single-path sensitization* approach; as multiple paths are selected at the same time, the result is called the *multiple-dimensional* or *multiple-path sensitization* approach. The difficulty of the single-path sensitization approach and how it can be overcome by the use of the multiple-path sensitization approach is explored in the following example.

■ Example 16.8: Multiple-path sensitization.

As shown in Figure 16.8(a), an inconsistent setting of the primary input values, x_3 and $\bar{x}_3$, is obtained when testing the stuck-at-0 fault at the primary input, a. In this case, both primary inputs, x_3 and $\bar{x}_3$, must be set to 0. However, as illustrated in Figure 16.8(b), when all paths related to the primary input, x_1, are sensitized at the same time, both primary inputs, x_3 and $\bar{x}_3$, are separately set to 1 and 0, which is a consistent setting.

As a signal splits into several ones, which pass through different paths and then reconverge together later at the same gate, the signal is called a *reconvergent fanout signal*. For such a reconvergent fanout signal, as illustrated above, the single-path sensitization approach cannot be generally applied to detect the stuck-at faults along it due to the difficulty or impossibility of finding consistent input combinations. However, the multiple-path sensitization approach can be applied to sensitize all paths related to the signal at the same time to solve this difficulty.

Section 16.2 Automatic Test Pattern Generation

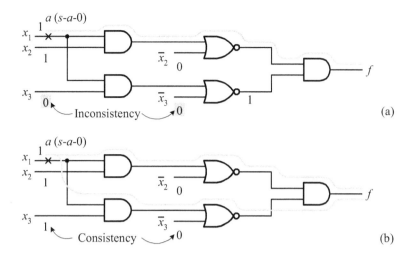

Figure 16.8: Comparison between (a) single and (b) multiple sensitized paths in the path sensitization approach.

This results in the well-known D-algorithm. For simplicity, a simplified version of the D-algorithm will be described in the rest of this section.

16.2.2 A Simplified D-Algorithm

The essential idea of the D-algorithm is based on the multiple-path sensitization approach. Nevertheless, its whole operations are quite complicated [3, 13]. Hence, in what follows, we merely present a simplified version of it to show how the multiple-path sensitization approach can be employed in practical applications. Before describing the simplified D-algorithm, it is necessary to define some basic operators of a logic gate: *primitive D-cube of fault* (or failure) (pdcf), *propagation of D-cube* (pdc), and *singular cover* (sc). At first glance, these are new operators. However, they are indeed the different views of the truth table of a specific logic gate.

The pdcf is used to model faults in a logic circuit and can model any stuck-at fault, bridge fault, or any change in the logic gate function. For example, the pdcf of the AND gate when its output is stuck-at-1 is "1 0 $\bar{D}$," where $\bar{D}$ represents an unknown value. Both ϕ and D symbols represent the value of 0 or 1. However, each ϕ symbol in a logic circuit can be set to an arbitrary value independently but all D symbols are needed to set to the same value. The pdcf of the seven most commonly used gates are shown in Figure 16.9.

The D-cube is a collapsed truth table that can be used to characterize an arbitrary logic gate or block. The pdc is used to propagate a D-cube through the gate under consideration. Recall that any basic gate can be considered as a controlled gate whose output receives the data from one of its inputs and is enabled by the other input(s) of the gate. For example, in order to propagate a D-cube through a two-input AND gate from one input of the AND gate, it is necessary to set the other input to 1, and the output of the AND gate is still D. If a two-input NAND gate is considered instead, the output is $\bar{D}$ because the input D is inverted. The pdc of the seven most commonly used gates are shown in Figure 16.10.

The singular cover (sc) of a logic gate is the input combinations required for a specific output value, including 0 and 1. It is indeed a compressed form of the truth table of the logic gate. For example, for an AND gate, the output is 0 whenever one of its two inputs is 0 and the output is 1 only when both of its inputs are 1. Hence, the singular cover of a two-input AND gate has three rows, as shown in Figure 16.11. In either of the first two rows, one input is 0 and

	x	z	Faults covered
	0	D	x/1, z/0
	1	$\bar{D}$	x/0, z/1

(a)

x	y	z	Faults covered
0	0	$\bar{D}$	z/1
0	1	$\bar{D}$	x/1, z/1
1	0	$\bar{D}$	y/1, z/1
1	1	D	x/0, y/0, z/0

(b)

x	y	z	Faults covered
0	0	$\bar{D}$	x/1, y/1, z/1
0	1	D	y/0, z/0
1	0	D	x/0, z/0
1	1	D	z/0

(c)

x	y	z	Faults covered
0	0	D	z/0
0	1	D	x/1, z/0
1	0	D	y/1, z/0
1	1	$\bar{D}$	x/0, y/0, z/1

(d)

x	y	z	Faults covered
0	0	D	x/1, y/1, z/0
0	1	$\bar{D}$	y/0, z/1
1	0	$\bar{D}$	x/0, z/1
1	1	$\bar{D}$	z/1

(e)

x	y	z	Faults covered
0	0	$\bar{D}$	x/1, y/1, z/1
0	1	D	x/1, y/0, z/0
1	0	D	x/0, y/1, z/0
1	1	$\bar{D}$	x/0, y/0, z/1

(f)

x	y	z	Faults covered
0	0	D	x/1, y/1, z/0
0	1	$\bar{D}$	x/1, y/0, z/1
1	0	$\bar{D}$	x/0, y/1, z/1
1	1	D	x/0, y/0, z/0

(g)

Figure 16.9: The primitive D-cube of fault (pdcf) of the seven most commonly used gates: (a) NOT; (b) AND; (c) OR; (d) NAND; (e) NOR; (f) XOR; (g) XNOR.

NOT		AND			OR			NAND			NOR			XOR			XNOR		
x	z	x	y	z	x	y	z	x	y	z	x	y	z	x	y	z	x	y	z
$\bar{D}$	D	$\bar{D}$	1	$\bar{D}$	$\bar{D}$	0	$\bar{D}$	$\bar{D}$	1	D	D	0	$\bar{D}$	D	0	$\bar{D}$	D	1	D
		1	$\bar{D}$	$\bar{D}$	0	$\bar{D}$	$\bar{D}$	1	$\bar{D}$	D	0	D	$\bar{D}$	0	D	$\bar{D}$	1	D	D
D	$\bar{D}$	D	1	D	D	0	D	D	1	$\bar{D}$	$\bar{D}$	0	D	$\bar{D}$	0	D	$\bar{D}$	1	$\bar{D}$
		1	D	D	0	D	D	1	D	$\bar{D}$	0	$\bar{D}$	D	0	$\bar{D}$	D	1	$\bar{D}$	$\bar{D}$

Figure 16.10: The propagation of D-cube (pdc) of the seven most commonly used gates.

the other input is ϕ (either 0 or 1), and in the third row both inputs are 1. The singular cover (sc) of the seven most commonly used gates are shown in Figure 16.11.

After defining the required operators, the basic operations of the simplified D-algorithm can be described as follows.

■ Algorithm 16-1: A simplified D-algorithm

1. *Fault sensitization*: Select a pdcf of the fault for which a test pattern is to be generated.
2. *D drive*: Use an appropriate pdc set to propagate the D signal to at least one primary outputs of the logic circuit.
3. *Consistency operations*: Use the sc of each logic gate to perform the consistency operation.

Section 16.2 Automatic Test Pattern Generation

NOT		AND			OR			NAND			NOR			XOR			XNOR		
x	z	x	y	z	x	y	z	x	y	z	x	y	z	x	y	z	x	y	z
1	0	0	ϕ	0	1	ϕ	1	0	ϕ	1	1	ϕ	0	0	0	0	0	0	1
0	1	ϕ	0	0	ϕ	1	1	ϕ	0	1	ϕ	1	0	0	1	1	0	1	0
		1	1	1	0	0	0	1	1	0	0	0	1	1	0	1	1	0	0
														1	1	0	1	1	1

Figure 16.11: The singular cover (sc) of the seven most commonly used gates.

An illustration of applying the simplified D-algorithm to find test vectors for a specified fault is explored in the following example.

■ Example 16.9: An example of the simplified D-algorithm.

As an illustration of the simplified D-algorithm, consider the logic circuit depicted in Figure 16.12. Suppose that the stuck-at-0 at net β in Figure 16.12 is to be tested. The first step of the simplified D-algorithm is fault sensitization. From Figure 16.9, we obtain the pdcf of the AND gate when one input stuck-at-0 is "1 1 D," which is shown in Figure 16.13 as the first row. The second step is D drive. In this step, we need to propagate the D cube of the AND gate to the output through an OR gate. Hence, the pdc of the OR gate is employed, which is "0 D D," as shown in the second row in Figure 16.13. The final step is consistency operations. Here, three gates are needed to find consistent assignments by using their singular covers. After applying the singular covers to the corresponding gates, the resulting set of test vectors for the stuck-at-0 at net β is $\{(1, \phi, 0, \phi), (\phi, 1, 0, \phi)\}$.

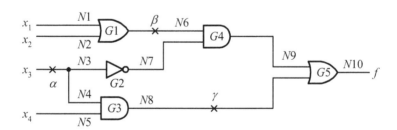

Figure 16.12: The logic circuit for illustrating the operation of the simplified D-algorithm.

In applying the simplified D-algorithm to find test vectors, it is necessary to find the test vectors for each fault, and then build a fault detection table to find the test set of the circuit under consideration. Refer to Lin [10] for more details about the simplified D-algorithm.

■ Review Questions

16-12 Describe the basic operations of the path-sensitization approach.

16-13 What is the major drawback of the single-path sensitization approach?

16-14 What are the basic differences between the single-path and multiple-path sensitization approaches?

16-15 Describe the basic operations of the simplified D-algorithm.

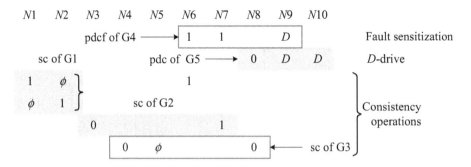

Figure 16.13: An illustration of the operations of the simplified D-algorithm.

16-16 What are the basic operators used in the simplified D-algorithm?

16.3 Testable Circuit Design

Recall that the controllability and observability of a net determine whether or not a fault at that net can be detected. To make a combinational circuit detectable, the combinational circuit must be irredundant. Nevertheless, this is not always the case and hence, not every fault of the combinational circuit is detectable. Moreover, it is very difficult to test a sequential circuit because the state diagram of the sequential circuit is not guaranteed to be a strongly connected graph; namely, it is generally hard, even impossible, to transfer the sequential circuit to a specified state at will.

For these reasons, the best way to equip the testability of a logic circuit is to add some extra logic circuits to or modify the original logic circuit in order to increase its controllability and observability. Such an approach by adding extra logic circuits to mitigate the test difficulty is known as the *testable circuit design* or *design for testability* (DFT).

In what follows, we briefly introduce the three most widely used DFT approaches: the *ad hoc approach*, the *scan-path method*, and the *built-in self-test* (BIST).

- The *ad hoc approach* uses some means to increase both the controllability and observability of the logic circuit under consideration.
- The *scan-path method* provides the controllability and observability of each register within the logic circuit by adding a 2-to-1 multiplexer to the input of each register so that all registers in the same *scan path* (also called the *scan chain*) may be configured into a shift register, thereby being able to be accessed externally.
- The *built-in self-test (BIST)* relies on some types of augmenting logic circuits that can perform operations to verify the correct operations of the logic circuit.

16.3.1 Ad hoc Approach

The basic principles behind the ad hoc approach are to increase both the controllability and observability of the logic circuit of interest by using some heuristic rules. The general guidelines are listed as follows:

- *Providing more control and test points:* Control points are used to set logic values of some selected signals while test points are used to observe the logic values of the selected signals.
- *Using multiplexers to increase the number of internal control and test points:* Multiplexers are used to apply external stimuli to the logic circuit and take out the responses from the logic circuit.

Section 16.3 Testable Circuit Design

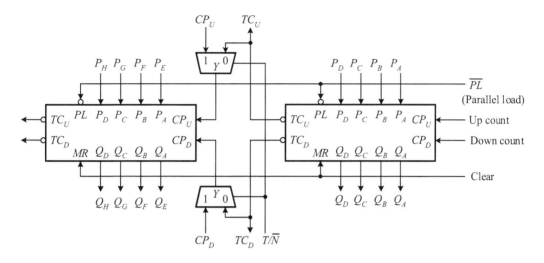

Figure 16.14: An example of the exhaustive test.

- *Breaking feedback paths:* For sequential circuits with feedback paths, multiplexers, AND gates, or other appropriate gates may be applied to break these feedback paths so as to transform sequential circuits into combinational circuits during testing.
- *Using state registers* to reduce the additional I/O pins required for testing signals.

In addition, the exhaustive test is feasible as the number of input signals of a logic circuit is not too large. For a logic circuit with a large number of input signals, the exhaustive test might still be feasible if the logic circuit can be partitioned into several smaller modules. These modules can then be separately tested exhaustively in a sequential or parallel fashion.

The following example demonstrates how it is possible to perform an exhaustive test by partitioning the logic circuit into two smaller modules and using multiplexers to increase the number of internal control and test points.

■ Example 16.10: An example of the exhaustive test method.

Figure 16.14 shows an 8-bit binary-up/down counter consisting of two 4-bit binary-up/down counter modules. Because of the symmetry operations of up and down counting, we only consider the up counting operation. By using the exhaustive test, we can continuously apply clock pulses to make it count from 0 to 255, which requires 256 clock cycles in total.

To reduce the test time, namely, the required number of clock cycles, two multiplexers are used to cut the connections between the two 4-bit binary-up/down counters, as depicted in Figure 16.14. As the source selection signal, $T/\overline{N}$ (Test/Normal), of the multiplexers is set to 1, the circuit is in the test mode. The two 4-bit binary counters are allowed to operate independently. They can be tested exhaustively in parallel and require 16 clock cycles. An additional clock cycle is required for testing the carry propagation. As the source selection signal, $T/\overline{N}$, of the multiplexers is set to 0, the circuit is in the normal mode. At this point, the two 4-bit binary-up/down counters are cascaded into an 8-bit binary up/down counter.

■ Review Questions

16-17 Describe the essence of testable circuit design.
16-18 What is the design for testability?

16-19 What are the three widely used DFT approaches?
16-20 Describe the basic rules of the ad hoc approach.
16-21 In what conditions, can the exhaustive test be applied to test a circuit?

16.3.2 Scan-Path Method

Recall that it is very difficult to test a sequential circuit [9, 10] without the aid of any extra logic circuit. The scan-path method is an approach to alleviate this difficulty by adding a 2-to-1 multiplexer at the data input of each D flip-flop to allow the access of the D flip-flop from an external logic circuit directly. As a result, both controllability and observability increase to the point where the states of the underlying sequential circuit can be completely controlled and observed externally.

In the scan-path method, the 2-to-1 multiplexer at the data input of each D flip-flop operates in either of two modes: *normal* and *test*. In the normal mode, the data input of each D flip-flop connects to the output of its associated combinational circuit so as to perform the normal operation of the sequential circuit. In the test mode, all D flip-flops are cascaded into a shift register (called a *scan chain* or *scan path*) through the use of those 2-to-1 multiplexers associated with the D flip-flops. As a consequence, all D flip-flops can be set to specific values or be read out to examine their values externally.

In general, by using the scan-path method, the test approach for a sequential circuit works as follows:

1. *Test the shift register (scan chain):* Set all D flip-flops into the test mode to form a shift register (scan chain). Shift a sequence of specific 0s and 1s into the shift register (scan chain) and then observe whether the shift-out sequence exactly matches the input sequence.
2. *Test the sequential circuit:* Use either of the following two methods to test the sequential circuit:

    (a). Test to see if each state transition of the state diagram is correct.

    (b). Test to see if any stuck-at fault exists in the combinational logic part.

As an illustration of the scan-path method, consider the following example, which is the controller of a sequential circuit by adding a 2-to-1 multiplexer to the data input of each D flip-flop.

■ **Example 16.11: An example of the scan-path method.**

Figure 16.15 shows a sequential circuit used as the controller [10] for some application. In the original controller, the 2-to-1 multiplexer at the data input of each D flip-flop does not exist. The addition of 2-to-1 multiplexers aims to form a scan chain. As the source selection signal $T/\overline{N}$ of all 2-to-1 multiplexers is set to 1, the sequential circuit is in the test mode. As indicated with the light line shown in the figure, all 2-to-1 multiplexers and D flip-flops form a 2-bit shift register with *ScanIn* as the input and *ScanOut* as the output. Therefore, any D flip-flop can be set to a specific value by a shift and the output of any D flip-flop can be shifted out as well. That is, each D flip-flop has the features of perfect controllability and observability. All D flip-flops are operated in their original function desired in the sequential circuit when the source selection signal, $T/\overline{N}$, of all 2-to-1 multiplexers is set to 0.

■ **Review Questions**

16-22 Describe the basic principles of the scan-path method.
16-23 In what conditions, can the scan-path method be applied to test a sequential circuit?

Section 16.3 Testable Circuit Design

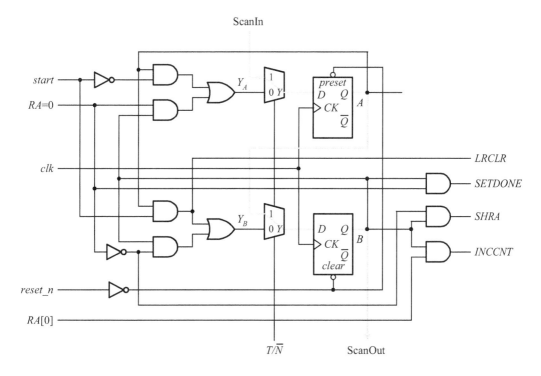

Figure 16.15: An example of a scan chain embedded in a controller.

16.3.3 BIST

The BIST principles can be best illustrated by using Figure 16.16, which shows the basic concepts of testing a digital system. Figure 16.16(a) shows the scheme using an *automated test equipment* (ATE) system to test a digital system. In a typical ATE system, the test vectors are stored in memory, then read out, and sent to the circuit under test (CUT) as stimuli by a μP. The response signals from the CUT are captured by the ATE system and stored in memory, and then compared with the expected results. If they are matched, the test is pass; otherwise, the test fails and hence the logic circuit is faulty.

Because of the increasing integration density of modern integrated circuits, the complexity of the circuit under test is much higher than before. In order to decrease the test cost, it is desirable to reduce the amount of time for using the ATE system or to incorporate the test vector (also referred to as the test pattern) generation circuits and the output response analyzer into the CUT. The approach to embedding these two types or other types of testing circuits into a logic circuit is referred to as the *built-in self-test* (BIST), as shown in Figure 16.16(b). The test vectors required for the test are generated automatically by a logic circuit and directly applied to the CUT. The response signals from the CUT are compressed into a signature by a response-compression circuit. The signature is then compared with the fault-free signature to determine whether the CUT is faulty. Since all of the above test vector generation and response-compression circuits are embedded into the CUT, they have to be simple enough so as not to increase too much cost of the CUT. The logic circuit used to generate the test vectors of the CUT is known as an *automatic test pattern generator* (ATPG) and the response-compression logic circuit is called a *signature generator*.

16.3.3.1 Random Test and ATPG For a complicated logic circuit, much time may be needed to generate the test set and it may not be feasible to incorporate a BIST logic circuit into the CUT due to too much hardware being required for storing test vectors. In such a case,

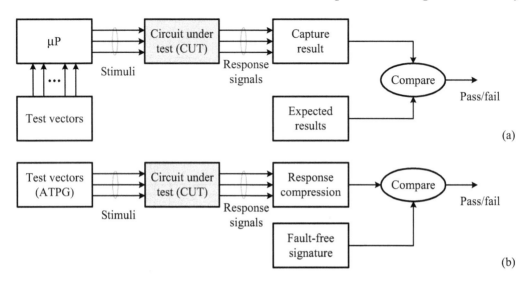

Figure 16.16: The (a) ATE system and (b) BIST principles of digital systems.

an alternative way, known as the *random test*, may be used instead. The random test only generates enough test vectors randomly.

In the random test, a maximum-length sequence generator, i.e., a PR-sequence generator (PRSG), is usually used to generate the stimuli. It is not an exhaustive test because it does not generate all input combinations and it does not guarantee that the logic circuit passing the test is completely fault-free. However, the detection rate of detectable faults will approach 100% when enough test vectors are applied.

The two implementations of a PR-sequence generator (PRSG), accompanied with sample primitive polynomials, have been discussed in Section 9.5.1. The interested reader may be referred to that section again for details. The output signals from a PRSG can be an output in one of two modes: *serial* and *parallel*. In the serial output mode, the output signal may be taken from any D flip-flop of the PRSG; in the parallel output mode, the output signals are taken from all or part of all D flip-flops of the PRSG.

16.3.3.2 Signature Generators/Analyzers As shown in Figure 16.16(b), in the BIST the response signals have to be compressed into a small amount of information, called a *signature*, and then compared with the expected result (i.e., fault-free signature) to determine whether the CUT is faulty. For now, the most widely used response-compression logic circuit is one using an n-stage LFSR, much like the CRC circuit introduced in Section 9.5.2.

Signature generators/analyzers can be classified into two types: *serial signature generators/analyzers* and *parallel signature generators/analyzers*. An n-stage *serial-input signature register* (SISR) is a PRSG with an additional XOR gate at the input for compacting an m-bit message M into the standard-form LFSR. An n-bit *multiple-input signature register* (MISR) is a PR-sequence generator with an extra XOR gate at the data input of each D flip-flop for compacting an m-bit message M into the standard-form LFSR. Examples of SISR and MISR circuits are shown in Figures 16.17(a) and (b), respectively. Except for the data inputs, the feedback functions of both circuits are the same as that of the PRSG circuits.

■ **Example 16.12: An application of signature analysis.**

An application of signature analysis is shown in Figure 16.18. The inputs, x, y, and z, are generated by a PRSG. The output response, f, of the logic circuit is sent to a 4-stage SISR for compressing into a 4-bit signature, as shown in Figure 16.18(a). Figure 16.18(b) lists the

Section 16.3 Testable Circuit Design

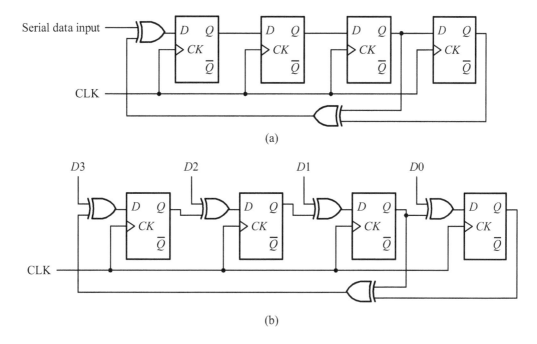

Figure 16.17: The (a) SISR and (b) MISR circuit examples of signature generators.

outputs and signatures of the fault-free circuit, stuck-at-0 faults at both nets α and β, stuck-at-1 fault at net α, and stuck-at-1 fault at net β under six input combinations. Consequently, whenever it is needed to test the logic circuit, the test vectors are first generated in sequence and applied to the logic circuit. Then, the signature of the output response is compared with that of the fault-free circuit to determine whether the logic circuit is faulty or not. If both signatures are matched, the logic circuit is fault-free; otherwise, the logic circuit is faulty.

The basic principle of signature analysis is to compress a long stream of information into a short one (i.e., a signature). Hence, it is possible to map many streams to the same signature. In general, if the length of the stream is m bits and the signature generator has n stages and hence the signature is n bits, then there are $2^{m-n} - 1$ erroneous streams that will yield the same signature. Since there are a total of $2^m - 1$ possible erroneous streams, the following theorem [1] follows immediately.

■ Theorem 16-1: Signature Analysis

For an m-bit input data stream, if all possible error patterns are equally likely, then the probability that an n-bit signature generator will not detect an error is

$$P(m) = \frac{2^{m-n} - 1}{2^m - 1} \tag{16.1}$$

which approaches 2^{-n}, for $m \gg n$.

Consequently, the fault detection capability of an n-stage signature generator will asymptotically approach to be perfect if n is large enough.

16.3.3.3 BILBO As mentioned above, both the PRSG and the signature generator need an n-stage shift register and an associated primitive polynomial. Therefore, both logic circuits

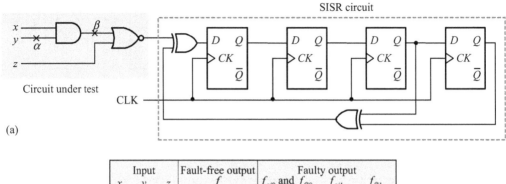

Figure 16.18: The (a) logic diagram and (b) a numeric example of a signature analysis.

can be combined together and use only one register. The resulting logic circuit is known as a *built-in logic block observer* (BILBO).

Figure 16.19 shows a typical BILBO logic circuit. The BILBO has the following four operation modes:

- In the *scan mode* ($M_1 M_0 = 00$), it is configured as a shift register.
- In the *multiple-input signature register (MISR) mode* ($M_1 M_0 = 01$), it supports the signature analysis.
- In the *clear mode* ($M_1 M_0 = 10$), it clears the contents of the register.
- In the *parallel-load mode* ($M_1 M_0 = 11$), it functions as a register with the parallel-load capability.

The function selection of the logic circuit shown in Figure 16.19(a) is summarized in Figure 16.19(b). A simple application is depicted in Figure 16.19(c). In the normal mode, the PRSG at the input of the combinational circuit is isolated and removed effectively by a multiplexer. The registers of the signature generator are used as the state registers and combined with the combinational circuit to form a sequential circuit. In the test mode, the inputs of the combinational circuit are the random test vectors generated by the PRSG and the responses from the combinational circuit are sent to the signature generator to perform the signature analysis.

16.3.3.4 Comments on DFT
In applying a scan-chain tool to insert scan-chain circuits into a design, it is often necessary to pay some attention to the following issues [2]:

- *Avoiding using the gated reset or preset control signal*: A flip-flop is unscannable if its reset is functionally gated in the design.
- *Avoiding using gated or generated clock signals*: A flip-flop is unscannable if its clock input is functionally gated in the design. When this is necessary, a multiplexer may be used to bypass the gated clock signal in the scan mode.
- *Using a single edge of the clock signal in a design*: The same clock edge (positive or negative) is used for the entire design when the design is in the scan mode.

Section 16.3 Testable Circuit Design

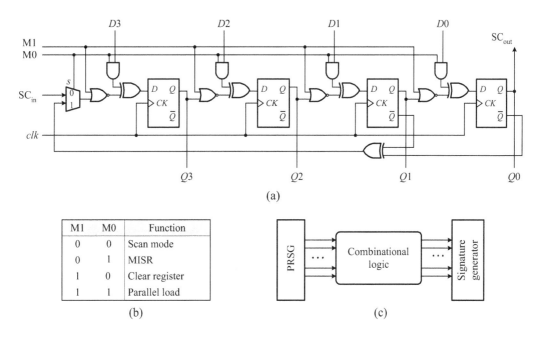

Figure 16.19: The (a) logic diagram, (b) function table, and (c) an application example of a BILBO logic circuit.

- *Needing multiple clock domains*: It is suggested to assign a separate scan chain for each clock domain since the combination of clock domains within a scan chain usually leads to timing problems. As multiple clock domains are needed, two solutions are possible. *Solution* 1 is to group together all flip-flops belonging to a common clock domain, and then connect them in series to form a single scan chain. *Solution* 2 is to use a clock multiplexer at the clock source, so that only one clock signal is appeared during the scan mode.

■ Review Questions

16-24 Describe the basic components of a BIST.
16-25 Describe the basic meaning of a random test.
16-26 Describe the basic principles of BILBOs.
16-27 What is the probability that an n-stage signature generator fails to detect a fault?

16.3.4 Boundary-Scan Standard—IEEE 1149.1 Standard

The test methods described previously are limited to a single chip or device. It is much more difficult to test the entire system on a printed-circuit board (PCB). To reduce the test cost of PCB systems, the *Joint Test Advisory Group* (JTAG) proposed a testable bus specification in 1988 and then it was defined as a standard known as the IEEE 1149.1 Standard by the IEEE [5, 15]. The IEEE 1149.1 Standard, also referred to as the *boundary-scan standard*, has become a standard for almost all integrated circuits that must follow.

The boundary-scan standard aims to provide a data transfer standard between the ATE system and the devices on a PCB, to provide a method of interconnecting devices on a PCB, and to provide a way of using the test bus standard or BIST hardware to find the faulty devices on a PCB. An example to reveal the boundary-scan architecture is given in Figure 16.20.

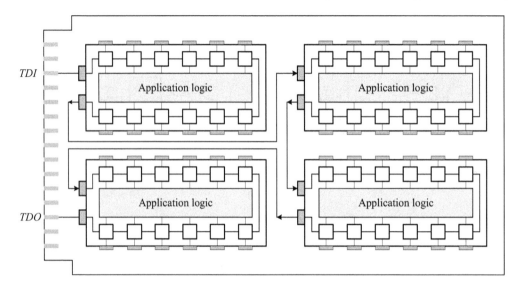

Figure 16.20: A boundary-scan architecture used to test the entire board-level module.

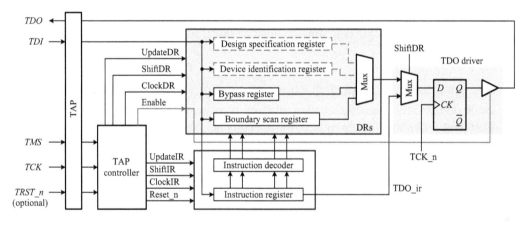

Figure 16.21: The basic structure of the boundary-scan standard.

Figure 16.20 shows a PCB-based system, consisting of four devices equipped with the boundary-scan standard. The boundary-scan cells of all four devices are cascaded into a single scan path (i.e., scan chain) by connecting the test data output (TDO) of a device to the test data input (TDI) of another device. The resulting system can perform the following test functions: the *interconnect test, normal operation data observation*, and *each device test*.

From the above discussion, it can be seen that the essential idea of the boundary-scan standard is to extend the scan-path method into the entire PCB system. To achieve this, each device has to provide a testable bus interface, as shown in Figure 16.21. The boundary-scan standard mainly include five components: a *test access port* (TAP), a set of *data registers*, a *TDO driver*, an *instruction register and decoder*, and a *TAP controller*.

The boundary-scan standard provides the following two major operation modes: the *non-invasive mode* and the *pin-permission mode*. In the non-invasive mode, the outside world is allowed to serially write in test data or instructions or serially read out test results in an asynchronous way. In the pin-permission mode, the boundary-scan standard takes over the control of IC input/output pins, thereby, disconnecting the system logic from the outside world. All ICs adhering to the boundary-scan standard must be designed to power up in the non-invasive mode. Moreover, the boundary-scan standard allows the delivery of BIST mode commands (e.g., RUNBIST) through JTAG hardware to the device under test.

16.3.4.1 Test Access Port (TAP) TAP defines four mandatory control signals: the *test clock input* (TCK), *test data input* (TDI), *test data output* (TDO), and *test mode select* (TMS). Their functions are as follows:

- TCK: The TCK signal is used to clock test data and results into and out of the device, respectively.
- TDI: The TDI is sampled at the positive edge of the TCK signal and used to input test data into the device.
- TDO: The TDO is a tristate signal and driven only when the TAP controller is shifting out the data from the device. It outputs test results and status out of the device at the negative edge of the TCK signal in order to avoid the race with the TDI.
- TMS: The TMS signal is sampled at the positive edge of the TCK signal and decoded by the TAP controller in order to control the test operations.

Besides, one optional control signal, *test reset* (TRST_n), is defined as follows:

- TRST_n (also TRST*): The TRST_n is an optional active-low asynchronous reset signal, used to reset the TAP controller externally.

In the normal mode, both TRST_n and TCK inputs are held low and the TMS input is held high to disable the boundary scan. To prevent race conditions, input data are sampled at the positive edge of the TCK signal whereas the output toggles at the negative edge of the TCK signal.

16.3.4.2 Data Registers Data registers (DRs) accept the test data from the TDI input. The boundary-scan standard defines two mandatory registers, a *bypass register* and a *boundary-scan register*, and one optional register, called the *device identification* (ID) register. For some special applications, an optional *design specification register* can also be used for special requirements. In what follows, we consider both the bypass register and the boundary-scan register in depth.

A 1-bit bypass register is used to bypass the boundary-scan cells in a device by way of connecting the TDO to the TDI through one D flip-flop. Its structure is shown in Figure 16.22(a). The use of the bypass register in the BYPASS instruction is shown in Figure 16.22(b). It is used in multichip modules (MCMs) or PCBs, where all devices have a boundary-scan chain connected serially but only one device is being tested.

The boundary-scan cell is indeed an extension of the scan-path register used in the scan-path method to test the interconnection among devices on a PCB-based system, to test external devices, and to sample the signals of application logic within the device. A possible boundary-scan cell is shown in Figure 16.23(a), which can be utilized as an input cell or an output cell.

When used as an input cell, the data input (Din) is connected to an input pad of the device and the data output (Qout) is connected to a normal input of the application logic circuit. When used as an output cell, the data input (Din) is connected to a normal output of the application logic circuit and the data output (Qout) is connected to an output pad of the device.

A boundary-scan cell can be generally operated in one of the following modes:

- In the *normal mode*, the mode selection M is set to 0. The data flow directly from the data input (Din) to the data output (Qout), as shown in Figure 16.23(a).
- In the *scan mode*, all boundary-scan cells in a device are cascaded into a shift register, as shown in Figure 16.23(a). The data path spans from the scan data input (ScanIn) to the scan data output (ScanOut) through the D flip-flop, QA.

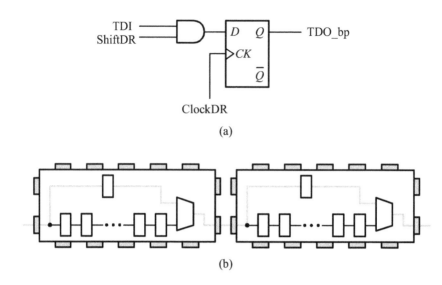

Figure 16.22: The (a) bypass register and (b) TDO driver of the boundary-scan standard.

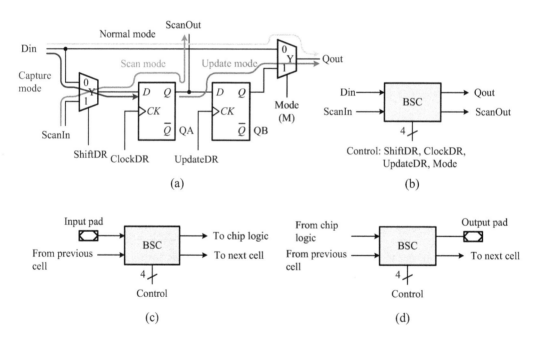

Figure 16.23: A possible boundary-scan cell: (a) circuit structure; (b) logic symbol; (c) input pad configuration; (d) output pad configuration.

- In the *capture mode*, the data input (Din) is sampled and stored in the D flip-flop QA, as shown in Figure 16.23(a). The data output (Qout) may be Din or from the output of the D flip-flop QB, depending on the value of mode selection M.
- In the *update mode*, the data stored in the D flip-flop QA may be transferred into the D flip-flop QB and the data output (Qout), as shown in Figure 16.23(a).

The boundary-scan cell can be represented by the logic symbol shown in Figure 16.23(b). Using this logic symbol, both input- and output-pad configurations can then be configured as

Section 16.3 Testable Circuit Design

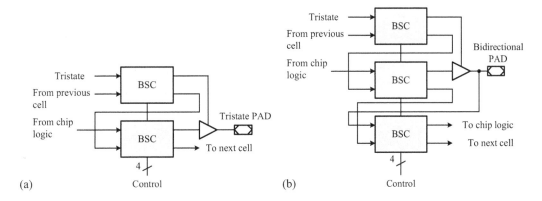

Figure 16.24: The boundary-scan cells used for (a) tristate and (b) bidirectional pads.

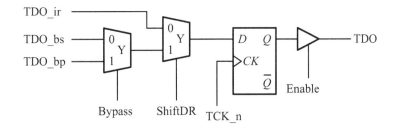

Figure 16.25: The TDO driver of the boundary-scan standard.

Figures 16.23(c) and (d), respectively. Each of these two simple pad configurations only needs one boundary-scan cell.

The boundary-scan cell shown in Figure 16.23(a) can also be used to configure the other two useful pads: tristate and bidirectional. As shown in Figure 16.24(a), each tristate pad needs two boundary-scan cells, one for the control signal and the other for the input of the tristate buffer. Each bidirectional pad needs three boundary-scan cells, the first for the control signal, the second for the output signal, and the third for the input signal of the tristate buffer, as shown in Figure 16.24(b).

16.3.4.3 TDO Driver The TDO driver is used to drive the TDI of another device in a cascaded chain. One possible structure of the TDO driver is depicted in Figure 16.25. Some other structures may not contain the negative-edge-triggered D flip-flop. The input data to the TDO register may come from one of the following sources: the instruction register (TDO_ir), boundary-scan register (TDO_bs), and bypass register (TDO_bp). To avoid the race problem, the TMS signal is sampled at the positive edge of the TCK signal whereas the TDO is sampled at the negative edge of the TCK signal. The following example explains how to model the data registers of the boundary-scan standard.

■ Example 16.13: A data register of the boundary-scan standard.

In this example, we assume that the application logic is an n-bit adder, which has two n-bit inputs, x and y, and one $(n+1)$-bit output, sum. The data registers include a boundary-scan register, a bypass register, and a TDO driver.

```
// a data register module
module data_reg
```

```verilog
    #(parameter N = 4)(// the default size of the N-bit adder
        input   TCK, TDI, TDO_IR, ClockDR, UpdateDR, ShiftDR,
        input   enable, test_mode, bypass,
        input   [N-1:0] din_a,         // din_a = input a
                        din_b,         // din_b = input b
        input   [N:0] fromCore_sum,    // sum = fromCore_sum
        output  TDO,
        output  [N:0] sum,
        output  [N-1:0] toCore_a, toCore_b);
// the body of the data register
reg [N-1:0] shift_reg_a, data_reg_a, shift_reg_b, data_reg_b;
reg [N:0] shift_reg_sum, data_reg_sum;
wire TDO_selected;
reg TDO_bypass, TDO_delayed;
// boundary-scan registers
always @(posedge ClockDR)
  {shift_reg_a, shift_reg_b, shift_reg_sum} <=
        ShiftDR ? {TDI, shift_reg_a, shift_reg_b,
                   shift_reg_sum[N:1]}: {din_a, din_b, fromCore_sum};
always @(posedge UpdateDR)
   {data_reg_a, data_reg_b, data_reg_sum} <=
        {shift_reg_a, shift_reg_b, shift_reg_sum};
assign {toCore_a, toCore_b} =
        test_mode ? {data_reg_a, data_reg_b} : {din_a, din_b};
assign sum = test_mode ? data_reg_sum : fromCore_sum;
// the bypass register
always @(posedge ClockDR)
   TDO_bypass <= TDI & ShiftDR;
// the TDO output driver
assign TDO_selected =
        ShiftDR ? (bypass ? TDO_bypass: shift_reg_sum[0]):
        TDO_IR;
always @(negedge TCK)
   TDO_delayed <= TDO_selected;
assign TDO = enable ? TDO_delayed: 1'bz;
endmodule
```

16.3.4.4 Instruction Register and Decoder The instruction register accepts and stores a test command, known as a *test instruction*, from the TDI at a time. The instruction is then decoded into control signals to select an appropriate data register or perform some specific operations. The structure of the instruction register is similar to that of the boundary-scan cell shown in Figure 16.23(a) with one exception that the output 2-to-1 multiplexer is removed, as shown in Figure 16.26. The instruction register has at least two bits in order to provide the following three mandatory instructions: BYPASS, EXTEST, and SAMPLE.

- BYPASS (the instruction code is all 1s) places the bypass register in the data register (DR) chain so that the path from TDI to TDO only involves one bit register.
- SAMPLE/PRELOAD places the boundary-scan registers in the DR chain. In the Capture-DR state, it copies the device's I/O values into DRs. New values are shifted into the DRs, but not driven onto the I/O pins.

Section 16.3 *Testable Circuit Design*

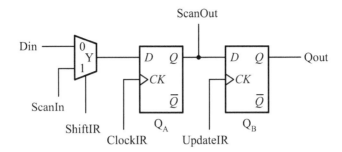

Figure 16.26: The instruction register of the boundary-scan standard.

- EXTEST (the instruction code is all 0s) places the boundary-scan registers in the DR chain and drives the values from the DRs onto the output pads. It allows to test the off-device circuitry.

The following instructions are optional but recommended:

- INTEST allows for the single-step testing of internal circuitry through the boundary-scan registers. It also drives the device core with signals from the DRs rather than from the input pads.
- RUNBIST is used to activate internal self-testing procedures within a device.
- CLAMP causes the state of all signals driven from the output pins of the device to be defined completely by the data held in the boundary-scan register.
- IDCODE is employed to connect the device ID register serially between the TDI and TDO for reading out in the Shift-DR state and for writing into the vendor identification code in the Capture-DR state. It is required when the device ID register is included in the devices. The layout of the device ID register is as follows: bit 0 (LSB) is always set to 1, bits 1 to 11 are manufacturer identifier, bits 12 to 27 are part numbers, and bits 28 to 31 are a version number.
- USERCODE is intended for programmable devices, such as FPGAs and Flash memories. The USERCODE instruction allows a user-programmable identification code to be loaded and shifted out for examination and thus allows the programmed function of the component to be determined. It is required when the device ID register is included in a user-programmable device. Like the IDCODE instruction, the USERCODE instruction connects the device ID register serially between the TDI and TDO for reading out in the Shift-DR state and for writing into the user-programmable identification code in the Capture-DR state.
- HIGHZ causes all output pins of the device to be placed in their high-impedance states.

As an illustration of the instruction register, consider the following example, which contains five instructions.

■ Example 16.14: An instruction register.

Because of only five instructions, the instruction register is defined to be 3 bits. In addition to the three mandatory instructions: BYPASS, EXTEST, and SAMPLE/PRELOAD, we add two other instructions: NOP and INTEST.

```
// an instruction register module
module instruction_reg (
        input   TDI, reset_n, ClockIR, UpdateIR, ShiftIR,
```

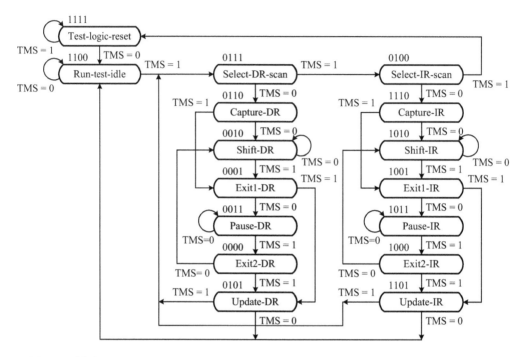

Figure 16.27: The state diagram of the TAP controller of the boundary-scan standard.

```
        output TDO_IR, test_mode, bypass);
// the body of the instruction register
reg [2:0] shift_reg, inst_reg;
// define instructions
localparam BYPASS         = 3'b111;
localparam EXTEST         = 3'b000;
localparam INTEST         = 3'b001;
localparam     NOP        = 3'b010;
localparam SAMPLE_PRELOAD = 3'b011;
// the body of the instruction module
always @(posedge ClockIR)
   shift_reg <= ShiftIR ? {TDI, shift_reg[2:1]}: NOP;
always @(posedge UpdateIR or negedge reset_n)
   if (!reset_n) inst_reg <= BYPASS;
   else inst_reg <= shift_reg;
assign TDO_IR    = shift_reg[0];
assign bypass    = (inst_reg == BYPASS);
assign test_mode = (inst_reg == INTEST);
endmodule
```

16.3.4.5 TAP Controller The TAP controller is a synchronous 16-state finite-state machine (FSM) with only one signal input, TMS, and one clock input, TCK. The details of the state diagram of the TAP controller is shown in Figure 16.27. The TMS signal is used to control the access of the data register (DR) and instruction register (IR). The data register is dispersed into each boundary-scan cell, as shown in Figure 16.23(a), and aims to set the test signals and store the test results. The instruction register stores the instruction being executed. In what follows, we describe each state in detail.

Section 16.3 *Testable Circuit Design*

Test-logic-reset. The TAP controller enters this state when the power supply is on or the optional TRST_n input is activated. In addition, no matter what the original state of the controller, it will enter the Test-logic-reset state as the TMS input is held high for at least five TCK cycles. When the test-logic is reset, the instruction register is set to the IDCODE instruction if the optional device identification register is provided and set to the BYPASS instruction, otherwise.

Run-test-idle. The TAP controller will pass through this state regardless of any operations. The controller will stay at this state as long as the TMS input remains 0.

Capture-DR. In this state, the data register selected by the test instruction will load the test data in parallel if any.

Shift-DR. In this state, new test data are shifted in from the TDI at the positive edge of the TCK signal and the old test data are shifted out from the TDO at the negative edge of the TCK signal. When the test data are needed to shift a number of bit positions, it only requires to maintain the TMS input at 0 and give the required cycles of the TCK signal.

Update-DR. In this state, the test data are latched into the data register at the negative edge of the TCK.

Capture-IR. In this state, the shift register contained in the instruction register loads a pattern of fixed logic values at the positive edge of the TCK signal. The current instruction and its selected test data register remain unchanged.

Shift-IR. In this state, a new test instruction is shifted in from the TDI at the positive edge of the TCK signal and the old test instruction is removed from the TDO at the negative edge of the TCK signal. When the test instruction is needed to shift a number of bit positions, it only needs to maintain the TMS input at 0 and give the required TCK cycles. In this state, the current instruction and its selected test data register remain unchanged.

Update-IR. In this state, the test instruction is latched into the instruction register at the negative edge of the TCK signal. The current instruction is updated to the new instruction.

Pause-DR and **Pause-IR.** In these states, the shifts can be temporarily suspended if the TMS input remains 0.

The remaining states: **Select-DR-scan, Select-IR-scan, Exit1-DR, Exit1-IR, Exit2-DR,** and **Exit2-IR**, are all temporary control states. They are branch points for changing the control flow of the TAP controller.

An illustration of a TAP controller is explored in the following example. Recall that the TAP controller is a 16-state finite-state machine. Hence, the following example indeed shows how to implement this finite-state machine.

■ Example 16.15: A TAP controller of boundary-scan standard.

All sixteen states are defined as local parameters with each associated a code that can be assigned arbitrarily. The details of the module are described in the state diagram of the TAP controller shown in Figure 16.27.

```
// a TAP controller --- IEEE 1149.1 standard
module tap_controller(
        input   TCK, TMS, TRST_n,
        output reg ShiftIR, ShiftDR,
        output  ClockIR, ClockDR, UpdateIR, UpdateDR,
        output reg reset_n, enable);
// the body of the TAP controller
reg [3:0] ps, ns;
// TAP controller states --- from IEEE 1149.1 standard
// actually, the following code assignment is arbitrary
localparam Test_logic_reset = 4'b1111;
```

```verilog
localparam Run_test_idle    = 4'b1100;
localparam Select_DR_scan   = 4'b0111;
localparam Capture_DR       = 4'b0110;
localparam Shift_DR         = 4'b0010;
localparam Exit1_DR         = 4'b0001;
localparam Pause_DR         = 4'b0011;
localparam Exit2_DR         = 4'b0000;
localparam Update_DR        = 4'b0101;
localparam Select_IR_scan   = 4'b0100;
localparam Capture_IR       = 4'b1110;
localparam Shift_IR         = 4'b1010;
localparam Exit1_IR         = 4'b1001;
localparam Pause_IR         = 4'b1011;
localparam Exit2_IR         = 4'b1000;
localparam Update_IR        = 4'b1101;
// update the next state
always @(posedge TCK or negedge TRST_n)
    if (!TRST_n) ps <= Test_logic_reset;
    else         ps <= ns;
// compute the next state
always @(ps or TMS) case(ps)
    Test_logic_reset: ns = (TMS)?Test_logic_reset:Run_test_idle;
    Run_test_idle   : ns = (TMS)?Select_DR_scan:Run_test_idle;
    Select_DR_scan  : ns = (TMS)?Select_IR_scan:Capture_DR;
    Capture_DR      : ns = (TMS)?Exit1_DR:Shift_DR;
    Shift_DR        : ns = (TMS)?Exit1_DR:Shift_DR;
    Exit1_DR        : ns = (TMS)?Update_DR:Pause_DR;
    Pause_DR        : ns = (TMS)?Exit2_DR:Pause_DR;
    Exit2_DR        : ns = (TMS)?Update_DR:Shift_DR;
    Update_DR       : ns = (TMS)?Select_DR_scan:Run_test_idle;
    Select_IR_scan  : ns = (TMS)?Test_logic_reset:Capture_IR;
    Capture_IR      : ns = (TMS)?Exit1_IR:Shift_IR;
    Shift_IR        : ns = (TMS)?Exit1_IR:Shift_IR;
    Exit1_IR        : ns = (TMS)?Update_IR:Pause_IR;
    Pause_IR        : ns = (TMS)?Exit2_IR:Pause_IR;
    Exit2_IR        : ns = (TMS)?Update_IR:Shift_IR;
    Update_IR       : ns = (TMS)?Select_DR_scan:Run_test_idle;
endcase
// clock the registers on the rising edge of TCK
// at the end of states
assign ClockIR = TCK|~((ps == Capture_IR)|(ps == Shift_IR));
assign ClockDR = TCK|~((ps == Capture_DR)|(ps == Shift_DR));
// update the registers on the falling edge of TCK
assign UpdateIR = ~TCK & (ps == Update_IR);
assign UpdateDR = ~TCK & (ps == Update_DR);
// change the control signals on the falling edge of TCK
always @(negedge TCK or negedge TRST_n)
    if (!TRST_n) begin ShiftIR <= 0; ShiftDR <= 0;
        reset_n <= 0; enable <= 0; end
    else begin
        ShiftIR <= (ps == Shift_IR);
```

Section 16.3 *Testable Circuit Design* 771

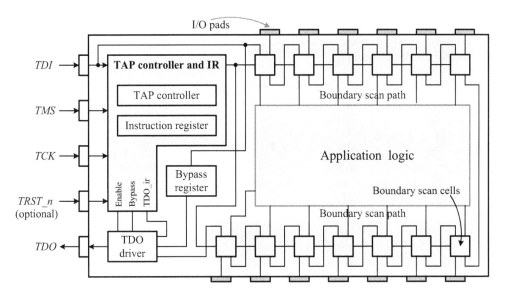

Figure 16.28: A complete boundary-scan example.

```
        ShiftDR <= (ps == Shift_DR);
        reset_n <= ~(ps == Test_logic_reset);
        enable  <= (ps == Shift_IR) | (ps == Shift_DR);
    end
endmodule
```

Once we have learned the structure of the boundary-scan standard and how it works, now we are ready to illustrate how the boundary-scan standard can be used to test an arbitrary logic circuit. To see this, consider the example depicted in Figure 16.28, which reveals a 4-bit adder as an application logic circuit being surrounded by a boundary-scan logic circuit. There are thirteen I/O terminals, including two 4-bit inputs, x and y, and one 5-bit output, sum. A parameterizable module is illustrated in the following example.

■ Example 16.16: A complete application example.

This is the top-level module of the complete example shown in Figure 16.28. In the top-level module, four modules: `nbit_adder`, `tap_controller`, `instruction_reg`, and `data_reg`, are instantiated. Moreover, the `nbit_adder` module is also shown in this example for illustration.

```
// an example to illustrate how the IEEE 1149.1 coworks
// with core logic
module ASIC_with_tap
        #(parameter N = 4)(// data width of core logic
           input   TCK, TMS, TDI, TRST_n,
           output  TDO,
           input   [N-1:0] a, b,// inputs of the N-bit adder
           output  [N:0] sum);  // output of the N-bit adder
// define port connection nets
wire [N-1:0] toCore_a, toCore_b;
wire [N:0] fromCore_sum;
wire UpdateIR, ShiftIR, ClockIR;
```

```verilog
wire UpdateDR, ShiftDR, ClockDR;
wire reset_n, enable, TDO_IR;
wire test_mode, bypass;

// an example of core logic instantiation
nbit_adder #(N) example_CoreLogic
        (toCore_a, toCore_b, fromCore_sum);
// the TAP controller
tap_controller tap_cntl
        (TCK, TMS, TRST_n, ShiftIR, ShiftDR, ClockIR,
         ClockDR, UpdateIR, UpdateDR, reset_n, enable);
// the instruction register
instruction_reg inst_reg
        (TDI, reset_n, ClockIR, UpdateIR, ShiftIR,
         TDO_IR, test_mode, bypass);
// test data registers
data_reg #(N) data_reg
        (TCK, TDI, TDO_IR, ClockDR, UpdateDR, ShiftDR,
         enable, test_mode, bypass, a, b, fromCore_sum,
         TDO, sum, toCore_a, toCore_b);
endmodule

// the example of the core logic module
module nbit_adder #(parameter N = 4)(
        input    [N-1:0] x, y,
        output   [N:0] sum);
// the function of an N-bit adder
assign sum = x + y;
endmodule
```

At present, the scan-path method is widely used in designing testable systems. Nevertheless, to make effective use of this method, the minimization of clock skews to avoid any hold-time violations in the scan path is extremely important. A practical approach is to order various scan paths in a design appropriately.

■ Coding Style

1. *It should order various scan paths in a design properly to minimize clock skew.*

■ Review Questions

16-28 What is the objective of the boundary-scan standard?
16-29 What are the ingredients of the boundary-scan standard?
16-30 What are the operation modes of boundary-scan cells?
16-31 What are the four mandatory test signals defined in the boundary-scan standard?
16-32 What instructions can the boundary-scan bus circuit perform?

Section 16.4 *System-Level Testing*

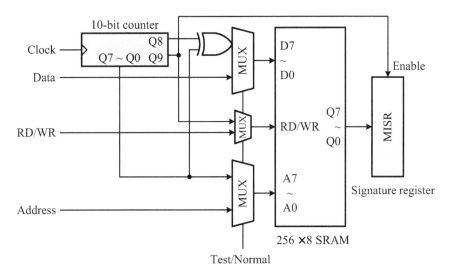

Figure 16.29: The SRAM BIST with an MISR.

16.4 System-Level Testing

Once we have understood both the scan-path method and the BIST technique, in this section, we further apply these two techniques to test system-level modules. To illustrate this, we are concerned with SRAM testing, the SRAM BIST, core-based testing, and SoC testing. Moreover, the IEEE 1500 Standard is introduced briefly.

16.4.1 SRAM BIST and March Test

As mentioned before, SRAM is an important component or module in most digital systems. In practice, SRAM modules with a variety of capacities are often embedded into a digital system. Therefore, they are needed to test properly; otherwise, they might become the faulty corner of the digital system which uses them.

At present, many schemes can be used to test a memory module, such as those reported in [3, 15]. Among these, testing a memory module at the system level with the BIST circuit is probably the most widely used approach. A simple example of using the BIST technique to test a memory module is shown in Figure 16.29. Here, a 10-bit counter generates the test data and an MISR is used to record the signature. Both the 10-bit counter and the MISR along with multiplexers comprise the BIST circuit. In the test mode, the 10-bit counter takes over the control of the SRAM module and generates all data required for testing the SRAM module, including both read and write modes. The test data are first written into the SRAM module and then read out and captured by the MISR to compute the signature. In the normal mode, all BIST hardware are effectively disabled or removed via the use of multiplexers.

Another famous algorithm widely used to test a memory module with the BIST is known as the *March algorithm* or *March test* [3, 14]. As shown in Figure 16.30, the required BIST hardware for the March algorithm includes the following ones:

- A BIST controller controls the test procedure.
- An address counter generates the required address during the testing procedure.
- A set of multiplexers (MUXs) is used to feed the memory module with the required data during the self-test from the controller.

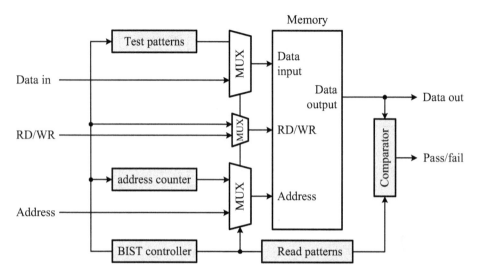

Figure 16.30: The BIST hardware architecture for testing SRAM based on the March test.

- A comparator checks the response from the memory module by comparing the data read from the memory module with the data patterns generated by the read pattern generator controlled by the BIST controller.
- A test pattern generator is used to generate the required test patterns for writing into the memory module.
- A read pattern generator is used to generate the required test patterns for comparing with the data read from the memory module.

Notations for the March+ test are as follows. Symbols **r0** and **r1** denote the read of a 0 and 1 from the specific memory location, respectively, while symbols **w0** and **w1** denote the write of a 0 and a 1 to the specific memory location, respectively. The arrows are used to the address related operations. Arrows ⇑ and ⇓ represent that the addressing orders are increasing and decreasing, respectively; arrow ⇕ denotes that the addressing order can be either increasing or decreasing.

The March test is applied to each cell in the memory module before proceeding to the next step, which means that if a specific pattern is applied to one cell, then it must be applied to all cells. For example, **M0:** ⇕(**w0**); **M1:** ⇑(**r0, w1**); **M2:** ⇓(**r1, w0**). The detailed operations of the above March test are shown in the following algorithm.

■ Algorithm 16-1: MATS+ March test

M0: /* March element ⇕ (w0) */

    for (i = 0; i $<= n$ - 1; i++) **write** 0 **to** A[i];

M1: /* March element ⇑ (r0, w1) */

    for (i = 0; i $<= n$ - 1; i++) **begin**
        read A[i]; /* expected value = 0 */
        write 1 **to** A[i];
    end

M2: /* March element ⇓ (r1, w0) */

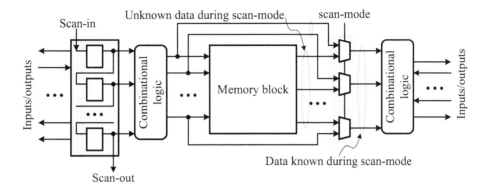

Figure 16.31: To bypass the BIST memory block for increasing the fault coverage of the design.

```
for (i = n - 1; i >= 0; i--) begin
    read A[i];   /* expected value = 1 */
    write 0 to A[i];
end
```

Even though the memory BIST circuitry may be used to test embedded memory blocks, during the scan mode the outputs of these memory blocks are still unknown. To solve this problem and increase the fault coverage of the design, a bypass logic circuit is often used, as shown in Figure 16.31. As the memory block is not bypassed, the fault coverage is low due to unknown data appearing at the outputs of the memory block. However, the fault coverage is much improved after adding the bypass logic circuit at the output of the memory block [2].

■ **Coding Style**

1. *It should bypass the BIST memory block for improving the fault coverage of a design.*

16.4.2 Core-Based Testing

As described previously, a core is an IP module of the vendor, which is a predesigned and verified functional block included on a chip. In general, the vendor provides tests for the core but the SoC designer must provide the boundary scan to a core embedded on the chip. It would become difficult, even impossible, to test the core and its surrounding logic if the core did not incorporate into the boundary-scan logic.

The most widely used approach to solving the above-mentioned difficulty is by surrounding the core with a test logic circuit, known as a *test wrapper* [3], as shown in Figure 16.32. Indeed, this is also an application of the concept and technique of the scan-path method. The test wrapper comprises a number of *wrapper elements* and a *wrapper test controller*.

The wrapper elements includes a cell for each I/O port of the core. For each input port of the core, the wrapper element has to provide three operation modes: the *normal mode*, *external test mode*, and *internal test mode*. The normal mode bypasses the test logic and allows the input port to work in its normal environment. The external test mode allows the test wrapper to observe the input port of the core for the interconnect test while the internal test mode allows the test wrapper to test the core function.

For each output port of the core, the wrapper elements also needs to provide three operation modes: the *normal mode*, *external test mode*, and *internal test mode*. The normal mode

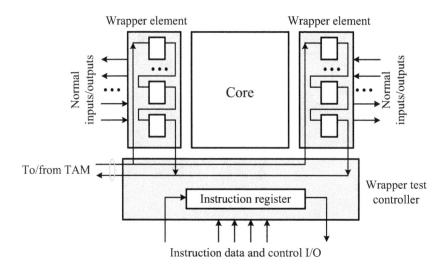

Figure 16.32: The general concept of core-based testing.

bypasses the test logic and allows the output port to work in its normal environment. The external test mode allows the test wrapper to observe the output port of the core for the interconnect test and the internal test mode allows the test wrapper to observe the core output for testing the core function.

The wrapper test controller consists of an instruction register along with some needed logic to accept instruction data from the outside of and input I/O data from and output I/O data to the *test access mechanism* (TAM).

16.4.3 SoC Testing

The test of an SoC module is an extension of the core-based test; that is, it uses the technique that is a combination of the test-wrapper concept and the boundary-scan approach. It also incorporates into the ATPG and signature analysis or other response checking techniques. The overall architecture was defined by the IEEE in 2005 as a standard known as the *IEEE 1500 Standard* [6, 15]. Here, we only briefly describe the basic concepts of the standard. The interested reader may be referred to the IEEE 1500-2005 Standard [6] for more details.

A system overview of the IEEE 1500 Standard is shown in Figure 16.33, where the system is composed of n cores, with each being wrapped with a 1500 wrapper. The 1500 standard provides both serial and parallel test modes. The serial test mode is equipped with the *wrapper serial port* (WSP), which is a set of I/O terminals of the wrapper for serial operations. It comprises the *wrapper serial input* (WSI), the *wrapper serial output* (WSO), and several *wrapper serial control* (WSC) signals. Each wrapper has a *wrapper instruction register* (WIR) used to store the instruction to be executed in the corresponding core.

The parallel test mode is achieved by incorporating a user-defined, parallel test access mechanism (TAM), as shown in Figure 16.33. Each core can have its own TAM-in and TAM-out ports, being composed of a number of data or control lines for parallel test operations. The test source includes counters, PRSGs, or test vectors stored in ROMs and the test sink contains the signature analysis or other response compressions.

■ Review Questions

16-33 How would you test a core that does not have the test access port?

Section 16.5 Summary

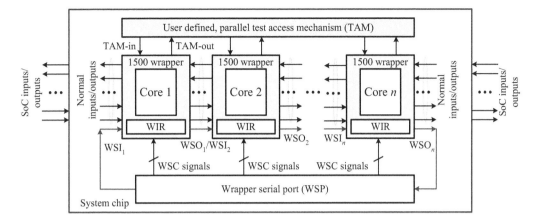

Figure 16.33: A system view of the IEEE 1500 Standard for testing an SoC.

16-34 Explain the operations of Figure 16.29.
16-35 What is the meaning of a test wrapper used in core-based testing?

16.5 Summary

Testing is an essential step in any PCB-based, cell-based, or FPGA-based system, even in a simple digital logic circuit because the only way to ensure that a system or a circuit can function correctly is through a careful testing process. Testing aims to find faults in a system or a logic circuit. A fault is a physical defect in a logic circuit. When a fault manifests itself in the logic circuit, it is called a failure. When a fault manifests itself in the signals of a system, it is called an error.

The test of any faults in a system or a circuit must be based on some predefined fault models. The most common fault models found in CMOS technology include stuck-at faults, bridge faults, and stuck-open faults. The "stuck-at" means that the net would adhere to a logic 0 or logic 1 permanently. The bridge fault means that a short circuit exists between any two lines (nets) unintentionally. A stuck-open fault is a feature of CMOS circuits and means that some net is broken during manufacturing or after a period of operation. To test a fault in a system or a logic circuit, we have to input stimuli, observe the responses, and analyze the results. The input stimulus special for use in a given system or a logic circuit is often called a test vector, which is an input combination. The collection of test vectors for testing detectable faults in a system or combinational circuit is called a test set. Except for the case of using the exhaustive test, the test set is a subset of all possible input combinations.

Both the controllability and observability of a net determine whether a fault at that net can be detected. To improve these two features of a logic circuit, the technique called the design for testability (DFT) or testable circuit design is developed. At present, DFT includes three types: the ad hoc approach, scan-path method, and built-in self-test (BIST). The ad hoc approach uses some means to increase both the test and observe points of the logic circuit. The scan-path method provides the capability of setting and observing the values of each register within the logic circuit. The built-in self-test (BIST) relies on augmenting a logic circuit to allow it to perform operations so as to prove the correctness of the logic circuit. The scan-path method is further developed into the boundary-scan standard (namely, the IEEE 1149.1 Standard).

The scan-path method also extends to system-level testing, including SRAM testing, the SRAM BIST, core-based testing, and SoC testing. A test wrapper is applied to test a core module. Both the test-wrapper concept and the boundary-scan approach are further combined to test SoC modules. In addition, the ATPG and signature analysis or other response checking techniques are also included in such testing. The overall architecture for testing an SoC module was integrated and defined by the IEEE in 2005 as a standard, known as the IEEE 1500 Standard.

References

1. M. Abramovici, M. A. Breuer, and A. D. Friedman, *Digital Systems Testing and Testable Design*, 2nd ed., IEEE Press, 1996.
2. H. Bhatnagar, *Advanced ASIC Chip Synthesis: Using Synopses Design Compiler and Prime time,* Boston: Kluwer Academic Publishers, 1999.
3. M. L. Bushnell and V. D. Agrawal, *Essentials of Electronic Testing for Digital, Memory & Mixed-Signal VLSI Circuits,* Boston: Kluwer Academic Publishers, 2000.
4. J. P. Hayes, *Introduction to Digital Logic Design,* Reading Massachusettes: Addison-Wesley, 1993.
5. IEEE Std 1149.1-2013 Standard, *IEEE Standard Test Access Port and Boundary-Scan Architecture,* New York: IEEE Press, 2013.
6. IEEE Std 1500-2005 Standard, *IEEE Standard for Embedded Core Test,* New York: IEEE Press, 2005.
7. B. W. Johnson, *Design and Analysis of Fault Tolerant Digital Systems,* Reading Massachusettes: Addison-Wesley, 1989.
8. Z. Kohavi and N. K. Jha , *Switching and Finite Automata Theory,* 3rd ed., New York: Cambridge University Press, 2010.
9. P. K. Lala, *Practical Digital Logic Design and Testing,* Upper Saddle River, New Jersey: Prentice-Hall, 1996.
10. M. B. Lin, *Digital System Design: Principles, Practices, and Applications*, 4th ed., Chuan Hwa Book Ltd. (Taipei, Taiwan), 2010.
11. E. J. McCluskey, *Logic Design Principles,* Englewood Cliffs, New Jersey: Prentice-Hall, 1986.
12. V. P. Nelson, H. T. Nagle, B. D. Carroll, and J. D. Irwin, *Digital Circuit Analysis & Design*, Upper Saddle River, New Jersey: Prentice-Hall, 1995.
13. J. P. Roth, "Diagnosis of automata failures: a calculus and a method," *IBM Journal of Research and Development*, Vol. 10, No. 4, pp. 278-291, July 1966.
14. D. S. Suk and S. M. Reddy, "A march test for functional faults in semiconductor random-access memories," *IEEE Trans. on Computers,* Vol. C-30, no. 12, pp. 982-985, 1981.
15. L. T. Wang, C. W. Wu, and X. Wen, *VLSI Test Principles and Architectures: Design for Testability,* New-York: Morgan Kaufmann Publishers, 2006.

Problems

16-1 Supposing that the stuck-at fault model is used, answer each of the following questions.
   (a) List all equivalent faults of a two-input NAND gate.
   (b) List all equivalent faults of a two-input NOR gate.

16-2 Considering the logic circuit shown in Figure 16.34, answer each of the following questions.

Problems

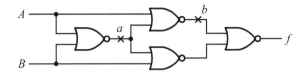

Figure 16.34: The logic circuit for Problem 16-2.

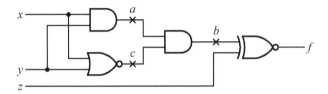

Figure 16.35: The logic circuit for Problem 16-3.

- (a) Using the single-path sensitization approach, find the test vector(s) for the stuck-at-0 fault at net a.
- (b) Using the multiple-path sensitization approach, find the test vector(s) for the stuck-at-0 fault at net a.
- (c) Using the path-sensitization approach, find the test vector(s) for the stuck-at-1 fault at net b.

16-3 Considering the logic circuit shown in Figure 16.35, answer each of the following questions.
- (a) Using the path sensitization approach, find the test vector(s) for the stuck-at-1 fault at net a.
- (b) Using the path sensitization approach, find the test vector(s) for the stuck-at-0 fault at net b.
- (c) Using the path-sensitization approach, find the test vector(s) for the stuck-at-0 fault at net c.

16-4 Considering the logic circuit shown in Figure 16.12, use the simplified D-algorithm to find the test vector(s) for each of the following specified stuck-at faults and nets.
- (a) The stuck-at-0 fault at net α
- (b) The stuck-at-1 fault at net α
- (c) The stuck-at-0 fault at net γ
- (d) The stuck-at-1 fault at net γ

16-5 This problem involves the use of an exhaustive test. Answer each of the following questions.
- (a) How many test vectors are required for testing an 8-bit ripple-carry adder?
- (b) Referring to the 8-bit counter example shown in Figure 16.14, introduced in Section 16.3.1, divide the 8-bit ripple-carry adder into two 4-bit ripple-carry adders and draw the resulting circuit.
- (c) How many test vectors are required for testing the above 8-bit ripple-carry adder constructed by cascading two 4-bit ripple-carry adders?

16-6 Letting $f(x) = 1 + x + x^5 + x^6 + x^8$, design a BILBO to function as follows:

$m_1 m_0 = 00$: ALFSR mode $m_1 m_0 = 10$: Parallel-load mode
$m_1 m_0 = 01$: Scan mode $m_1 m_0 = 11$: MISR mode

16-7 Letting $f(x) = 1 + x^3 + x^{10}$, design a BILBO to function as follows:

$m_1 m_0 = 00$: ALFSR mode $m_1 m_0 = 10$: Parallel-load mode
$m_1 m_0 = 01$: Scan mode $m_1 m_0 = 11$: MISR mode

16-8 Design a parameterizable BILBO to function as follows:

$m_1 m_0 = 00$: ALFSR mode $m_1 m_0 = 10$: Parallel-load mode
$m_1 m_0 = 01$: Scan mode $m_1 m_0 = 11$: MISR mode

16-9 Assume that a scan chain would like to be added into the $n \times n$ Booth multiplier introduced in Section 11.2.3.2. Modify the related Verilog HDL examples given in Section 16.3.4 and write a test bench to verify the functionality of the Booth multiplier through the scan-chain logic.

16-10 Assume that a scan chain would like to be added into the $m \times n$ unsigned array multiplier introduced in Section 14.2.1. Modify the related Verilog HDL examples given in Section 16.3.4 and write a test bench to verify the functionality of the unsigned array multiplier through the scan-chain logic.

Index

1-out-of-m code, 273, 280
D algorithm, 749, 751–754
SR-latch D latch, 311
μC system, 669, 723
μP/DSP approach, 371, 374

A keypad scanner and decoder, 414–424
    key code, 414
    keypad scanner, 414
    scan code, 414
Ad hoc approach, 743, 754–756
Adder, 4–25
    4-bit, 16, 22, 90
    4-bit (behavioral), 110
    4-bit 2's-complement, 90
    4-bit 2's-complement (behavioral), 112
    n-bit, 4
    n-bit 2's-complement, 210
    n-bit ripple-carry, 207–209, 211–213, 625
    BCD, 101, 102, 113, 131, 132
    Brent-Kung, 625, 634
    Brent-Kung network, 632, 634
    carry generate, 626
    carry propagate, 626
    carry-lookahead (CLA), 625–630
    carry-propagate (CPA), 637
    carry-ripple (RCA), 637
    carry-save (CSA), 637
    full adder, 15, 17, 18, 32
    group-carry generate, 629
    group-carry propagate, 629
    half adder, 15
    Kogge-Stone, 625, 632
    Kogge-Stone network, 632
    parallel-prefix, 625, 631–635
    parallel-prefix definition, 631
    prefix sums, 631
Addition, 625
Addressing modes, 725, 726
    absolute, 726
    autoincrement, 726
    immediate, 726
    indexed, 726
    PC-relative, 726
    register (register direct), 725
    register indirect, 726
    symbolic, 726
Algorithm, 446
    Booth, 446
Algorithmic state machine chart, *see* ASM chart
ALU, 625, 651–657, 724
    an example, 654
    carry (C) flag, 654
    FA-based, 655, 657
    flags, 724
    negative (N) flag, 654
    overflow (V) flag, 654
    zero (Z) flag, 654
Analysis/translation, 521
AND-OR logic structure, 374
Antifuse, 383
Application-specific integrated circuit (ASIC), 371, 374
Arithmetic shift, 651
Arithmetic-and-logic unit, *see* ALU
Array structure, 478
Arrival time, 513, 601
ASIC design, 377–382
ASM chart, 440–452
    ASM block, 442
    basic constructing rules, 442
    building block, 441
    conditional output box, 440
    decision box, 440
    invalid ASM blocks, 443
    Mealy machine, 443
    modeling style, 445–452
    Moore machine, 443
    multiple-bit signal, 441
    one-state register style, 445
    operation notation, 441
    register transfer, 442
    single-bit signal, 441
    state box, 440

two-state register style, 445
valid ASM blocks, 442, 443
ASMD, 453, 466
    definition, 466
    derived from ASM chart, 467
Assertion, 570
    static, 570
    temporal, 570
Assignment, 107
    blocking (=), 112
    blocking versus nonblocking, 114–119
    continuous, 71, 107
    nonblocking (three-step procedure), 114
    nonblocking (<=), 113
    procedural, 107, 111–119
    procedural continuous, 234–237
    procedural continuous (**assign-deassign**), 107
    procedural continuous (**force-release**), 107
Associativity, 73
Asynchronous mode, 676
Attribute, 11
Automated test equipment (ATE), 757
Automatic test pattern generation, 749–754
Automatic test pattern generation (ATPG), 511
Automatic test pattern generator (ATPG), 757
Autonomous linear feedback shift register (ALFSR), 349

Back end, 521
Back-end design, 6
Base format notation, 76
Base function, 534
BCD adders, 100–103
BCD code, 95
Begin-end block, 227
Big-endian form, 81
BILBO, 759–760
    a 4-bit example, 761
Binary encoding, 431
Binary-coded decimal (BCD) adder, 100
Binding, 215
BIST, 754, 757–761
    principle, 758
    SRAM, 773–775
Bistable device, 310
Block memory, 331
Blocking assignment, 111
Body, 193

Boolean function, 525
Boolean network, 526
Boolean variable, 525
Booth algorithm, 446–452, 642
    4×4 array multiplier, 479
    $m \times n$ array multiplier, 478
    a numerical example, 449
    array multiplier, 478–481
    ASM chart, 447, 448
    controller, 459
    datapath, 459
    definition, 446
    three-step paradigm example, 459–467
Boundary-scan standard, *see* IEEE 1149.1
Branch coverage, 592
Buffer, 31, 36–670
    tristate, 670
Buffer insertion, 558
Buffers, 39
Built-in logic block observer, *see* BILBO
Built-in self-test, *see* BIST
Built-in self-test (BIST), 511
Bus, 669, 670
    arbitration, 669, 673
    bidirectional, 671, 672
    definition, 669
    master, 677
    multiple driver, 669
    multiplexer, 669, 672, 673
    multiplexing, 669
    schedule scheme, 669
    shared, 669
    slave, 677
    structure, 669–672
    tristate, 669–671
Bus arbiter, 673
    4-request daisy-chain, 674
    4-request variable priority, 675
    fixed-priority, 675
    round-robin priority, 675
    variable priority, 675
Bus arbitration, 673–676
    daisy-chain, 669, 673
    independent-requested line, 675
    radial, 669, 673

Capacitive state, 59
Capacity, 387
Carry (C), 654
Cascadable magnitude comparator, 295
Case-sensitive language, 8
Cell library, 5, 534

Index

Cell-based approach, 374, 378
    standard cell library, 378
    standard cells, 378
Cell-based design, 6, 511, 513
Cell-based platform IP, 373
Center processing unit, *see* CPU
Charge decay, 61
Charge sharing problem, 60
Charge strength, 57
Charge-decay effect, 61
Charge-decay time, 61
Charge-sharing problem, 60
Circuit level, 372
Circuit under test (CUT), 757
Clock domain, 581, 600
Clock gating, 561
Clock network, 525
    big-buffer approach, 525
    clock tree, 512
    clock tree synthesis (CTS), 525
Clock signal generation, 586–590
    aligned derived, 586, 588
    asynchronous, 586, 589
    clock, 586, 587
    clock multiplier, 586, 588
Clock skew, 315
    negative, 319
    positive, 318
Clock timing, 318
    clock jitter, 318
    clock skew, 318
CMOS, 50–53
    inverter, 50, 51
    NAND, 51
    NOR, 52
Code, 357
    1-out-of-n, 357, 431
    binary, 431
    Gray, 431
    random, 431
Code conversion, 206
    Gray code to binary, 206, 207
Code converter, 273
Coding guidelines, 551–556
    clock signal, 552
    partition, 555
    reset signal, 553
Cokernel, 530
Cokernel table, 530
Collapse ratio, 745
Collapsing, 528

Combinational logic module, 273–308
Command register, 485
Comments, 9
    block comment, 9
    one-line comment, 9
Comparator, 295
Compiled code, 577
Compiler directive, 260–265
    `celldefine, 264
    `default_nettype, 265
    `define, 78, 196, 260
    `else, 261
    `elsif, 261
    `endcelldefine, 264
    `endif, 261
    `ifdef, 261
    `ifndef, 261
    `include, 261
    `line, 264
    `nounconnected_drive, 264
    `resetall, 265
    `timescale, 20, 263
    `unconnected_drive, 264
    `undef, 260
Complex PLD, *see* CPLD, *see* CPLD
Complex PLD (CPLD), 5
Complexity, 376
Concurrency, 4, 107
Condition coverage, 592
Conditioned event, 250
Configuration, 215–222
    cell statement, 216, 219
    default liblist statement, 219
    default statement, 216
    design statement, 216
    instance statement, 216, 220
    liblist clause, 216
    liblist statement, 216
    use clause, 216
    work library, 215
    binding, 215
    definition, 216
    hhierarchical, 221
    library, 215
    library declaration, 215
    library map file, 215
Constant, 79
    defined by compiler directive **`define**, 78
    defined by **localparam**, 78
    defined by **parameter**, 78

  integer, 9, 76
  real, 9, 77
  string, 9, 78
Contamination delay, 317
Control-point approach, 453
Controllability, 743, 746
Controlled gate, 35, 310
  AND/NAND, 35
  NAND, 310
  NOR, 310
  OR/NOR, 35
  XOR/XNOR, 36
Controller, 453
Core-based testing, 775
  general concept, 776
Cost, 376
  k point, 377
  non-recurring engineering (NRE), 376
  recurring, 376
Count trailing zeros, 137, 139, 140
Count zeros, 138, 140, 153, 159
Counter, 131, 343–349
  3-bit, ripple binary, 343
  3-bit, ripple binary, asynchronous reset, 344
  4-bit, Johnson, 359
  4-bit, ring, 358
  4-bit, ripple, 188
  4-bit, synchronous binary, 131
  n-bit, Johnson, 359
  n-bit, binary up/down, 346, 347
  n-bit, enable and synchronous reset, 346
  n-bit, ring, 358, 360
  n-bit, ripple, 212
  n-bit, ripple binary, 343, 345
  n-bit, up/down, cascade, 348
  Johnson, 359
  modulo r binary, 348
  Moebius, 359
  ring, 357
  ring, self-correcting, 358, 360
  standard ring, 357
  switched-tail, 359
  synchronous, 346
  twisted ring, 357
Counting loop, 139
Cover, 525
  minimal, 526
CPLD, 385, 400–403
  function block, 400

  I/O block, 385
  I/O block (IOB), 402
  input buffer, 402
  input/output block, 400
  interconnect, 385, 400
  macrocell, 385
  output driver, 402
  switch matrix, 402
CPU, 723–734
  addressing modes, 723, 725
  constant generation, 725
  controller, 730–734
  datapath, 728–730
  decode the instruction, 723
  execute instruction, 723
  fetch instruction, 723
  fetch memory operand, 723
  instruction formats, 723, 725
  instruction set, 723, 726
  operations, 724
  the programming model, 723, 724
CPU controller, 730
  double-operand instructions, 731
  double-operand instructions, ASM chart, 732
  jump instructions, 734
  jump instructions, ASM chart, 735
  single-operand instructions, 733
  single-operand instructions, ASM chart, 733, 734
  structure, 730
CPU datapath, 728
  ALU, 729
  register file, 728
CRC (clock-recovery circuit), 679
CRC code detector, 354
CRC code generator, 354
CRC code generator/detector, 354, 357
  4-bit, 355
  n-bit, 356
  CRC polynomial, 355
  general paradigm, 355, 356
CRC generator/detector, 354
  CRC algorithm, 354
  CRC generator, 354
Critical path, 513, 514, 595
Crossing clock domain, *see* Multiple-clock domain
Cube, 525
Cube-free, 530
Cycle-based simulation, 580–581

Index 785

clock domain, 581
clock domain analysis, 581
levelization, 580

Data event, 251
Data register, 485
Data transfer, 669, 676–683
    asynchronous, 669, 679–683
    handshake, 677, 680
    handshake, two-phase, 682
    multicycle, 677
    parallel, 677
    serial, 678
    serial, asynchronous, 681
    single-cycle, 677
    strobe, 677, 679, 680
    synchronous, 669, 677–679
Data type, 10, 79–80
    event, 125
    integer, 4, 10, 79, 80
    pulldown, 53
    pullup, 53
    realtime, 10, 79, 80
    real, 10, 79, 80
    reg, 10, 79
    supply0/1, 50
    time, 79, 80
    tri0/tri1, 40
    trireg, 59–62
    tri, 38, 671
    uwire, 41
    wand/triand, 39
    wire, 38
    wor/trior, 39
    net, 10
    variable, 10
Datapath, 453
Datapath and controller, 429, 453–467
    basic structure, 453
    control-point approach, 453
    controller, 453
    datapath structure, 453
    heuristic approach, 453, 481–498
    three-step paradigm, 453–467
Datapath optimization, 510
Decimal adder, 100
Decoder, 273–278
    2-to-4, active-high, 275
    2-to-4, active-low, 274, 275
    3-to-8, 277, 278
    m-to-n, parameterizable, 276
    BCD-to-seven-segment, 298, 300

block diagram, 274
decoding process, 273
definition, 273
expansion, 277
fully decoded, 274
parameterizable, 276
partially decoded, 274
Decoding process, 273
Decomposition, 528
Default working library, 215
Delay calculator, 3
Delay model, 237–240
    distributed delay, 237–238
    lumped delay, 237–239
    module path, 239
    path delay, 237, 239–240
Delay specification, 44
    delay specifier, 44
    gate delay, 44
    net delay, 45
    no delay, 44, 57
    one delay, 44, 57
    three delays, 44, 57
    two delays, 44, 57
Delay specifier, 44, 241
Delta delay, 588
DeMorgan's law, 394
Demultiplexer, 283, 289–294
    1-to-4, n-bit (case), 291
    1-to-4, n-bit (if-else), 289
    1-to-4, n-bit (with enable), 290
    1-to-4, (active-low), 289, 290
    1-to-8, 294
    1-to-m, parameterizable, 292
    block diagram, 289
    definition, 289
    demultiplexer tree, 293
    demultiplexing, 289
    expansion, 293
    parameterizable, 292
    versus decoder, 293
Design, 12, 569
    gate-array based, 6
    register transfer level(RTL), 71
    reuse, 5
Design constraints, 517–521
Design environment, 510, 516–525
    wire-load model, 517
Design flexibility, 376, 377
Design for testability, *see* DFT
Design under test (DUT), 570

Design under verification (DUV), 570
DFT, 754
    comments, 760
Digital-signal processing filters, *see* DSP filters
Dimension, 83
Direct memory access (DMA), 673
Direct-driven approach, 301
Directed acyclic graph (DAG), 581
Distributed delay model, 237
Distributed memory, 331, 407
Distribution function, 176
Dividers, 648–651
    non-restoring array, 648
    nonrestoring array, 649
Division, 625, 646–651
    a numerical example, non-restoring, 648
    non-restoring, 646–651
    restoring, 646–647
    sequential implementation, non-restoring, 649
Divisor, 530
    primary, 530
Don't cares, 533
Don't-care-set (dc-set), 525
Drive strength, 57
Driven state, 59
DSP filters, 657–661
    finite-impulse response (FIR), 625, 658–659
    infinite-impulse response (IIR), 625, 659–661
Duty cycle, 710
Dynamic hazard, 45
Dynamic programming method, 537
Dynamic timing analysis (DTA), 594–599
Dynamic timing simulation, 569, 615–619
    gate-level, 615
    post-map, 616, 618
    post-PAR, 618, 619
    post-translate, 615, 617

Elaboration, 521
Elaboration time, 205, 207, 213
Electronic design interchange format (EDIF), 609
Electrostatic discharge (ESD), 412
Embedded system, 375
Encoder, 273, 278–283
    4-to-2, 278, 279
    4-to-2 (case), 280
    4-to-2 (if-else), 279
    4-to-2, priority, 281
    4-to-2, priority (casex), 281
    4-to-2, priority (if-else), 281
    m-to-n, priority (for), 282
    m-to-n, priority (while), 283
    block diagram, 279
    definition, 278
    encoding process, 273
    parameterizable, 282
    priority, 280
Encoding process, 273
Equivalence checking, 573, 600
Erasable, 382
Erasable devices, 382
Erasable programmable ROM (EPROM), 387
Error, 744
Escaped identifier, 8
    \ (backslash), 8
    \b (blank space), 8
    \n (new line), 8
    \t (tab), 8
Event, 125, 701
Event list, 127
Event-driven simulation, 578
    active event, 579
    evaluation event, 578
    future event, 579
    inactive event, 579
    nonblocking assignment update, 579
    schedule, 578
    simulation time, 578
    simulator, 578
    timing wheel, 578
    update event, 578
Excess-3 code, 95
Exhaustive test, 743
Explicit finite-state machine (explicit FSM), 437, 438
Explicit FSM, 439
Explicit zero delay, 122
Explicitly clocking scheme, 678
Expression, 73
    delay, 74
    operands, 73
    operators, 73
Expression coverage, 592
Extraction, 528

Factorization, 528
Failure, 743
False path, 519

Index

Fanout gate, 513
Fault, 743
    detectable, 746
    detection, 746–747
    equivalent, 745
    location, 746
    multiple, 744
    single, 744
    testable, 746
    undetectable, 746
    unidirectional, 744
Fault collapse, 745
Fault coverage, 746
Fault model, 743
    bridge, 743, 744
    stuck-at, 743, 744
    stuck-at-0, 744
    stuck-at-1, 744
    stuck-closed, 745
    stuck-on, 745
    stuck-open, 743, 744
Field-programmable, 382
Field-programmable device, 371, 374
Field-programmable gate array, *see* FPGA
File descriptor (fd), 168
Finite-state machine, *see* FSM
Finite-state machine with datapath, *see* FSMD
Flash cell, 383
Flash memory, 387
Flip-flop, 309–315
    D, 312
    D, asynchronous reset, 313
    D, synchronous reset, 313
    basic timing, 316, 317
    bistable device, 310
    clear, 313
    clock-to-Q delay (t_{cq}), 316
    contamination of clock-to-Q delay, 317
    hold time, 315
    latch-based, 311
    master-slave, 312
    metastable state, 320
    minimum clock-to-Q delay, 317
    minimum period of clock, 316
    positive-edge-triggered, 312
    recovery time, 321
    reset, 313
    sampling window, 315
    setup time, 315
    timing parameters, definition, 316
    transparent property, 312

Floorplanning, 512
Fork-join block, 229
Four-phase handshake, 681
Four-valued logic, 9
FPGA, 371, 374, 385, 409
    carry chain, 406
    cascade chain, 406
    configurable logic block (CLB), 385
    distributed memory, 407
    fabrics, 385
    function block (CLB), 403
    horizontal routing channel, 385
    input/output block (IOB), 385, 403, 405
    interconnect, 385, 404
    logic element (LE), 385
    matrix type, 385
    PLB, 385
    programmable interconnect, 403
    programmable logic block (PLB), 375
    row type, 385
    switch matrix (SM), 404
    vertical routing channel, 385
Front end, 521
Front-end design, 6
FSM, 429–439
    0101 sequence detector, 433–437
    definition, 429
    explicit, 438, 439
    implicit, 437, 438
    Mealy machine, 430
    modeling style, 432
    Moore machine, 430
    one-state register style, 432
    state encoding options, 431
    two-state register style, 432
FSMD, 429, 453, 454
    block diagram, 453
    definition, 454
    Mealy machine, 454
    Moore machine, 454
Full-custom approach, 374, 377
Function, 157–163
    (static), 160
    automatic (recursive, dynamic), 160
    call, 158
    constant, 160
    definition, 157
    sharing, 162
    types of, 160
Function generator, 403, 542
Functional coverage, 592

Functional test plan, 571
Functional verification, 571
    chip level, 572
    core level, 571
    designer level (block level), 571
    unit level, 571
Fuses, 389

Gate arrays (GAs), 379
Gate delay, 41–45
    delay model, 41
    inertial, 41, 74
    propagation delay, 41
Gate level, 372
Gate primitive, 29–49
    and, 29
    buf, 31
    nand, 29
    nor, 29
    not, 31
    or, 29
    xnor, 29
    xor, 29
    and/or group, 30–35
    array instantiation), 35
    buf/not group, 36–39
    instantiation, 30
Gate-array-based approach, 374, 379
    gate array (GA), 379
    NAND gate, 381
    NOT gate, 381
    sea-of-gate (SoG), 379
Gate-array-based design, 513
General-purpose input and output, *see* GPIO
Generate region, 205–214
    genvar, 206
    generate-case, 213–214
    generate-if, 207–213
    generate-loop, 205
    generate-statements, 205
Generate-case, 205
Generate-if, 205
Generate-loop, 205
Generate-statements, 205
Glitches, 45, 256
Global routing area, 385
GPIO, 696–700
    an example, 699, 700
    basic principle, 697
    bidirectional port, 697
    data direction register (DDR), 697
    design issues, 697
    hardware model, 697
    input port, 697
    output port, 697
    port (PORT) register, 697
    programming model, 697
Gray encoding, 431
Ground, 50
Group-carry generate, 629
Group-carry propagate, 629

Hamming distance, 560
Handshake, 680
Handshaking, 126, 492, 496
    a simple example, 126
    source initiated, 128
    using **wait** statement, 129
Handshaking protocol, 681
Handshaking transfer, 681
Hardware description language, 1
Hardwired memory block, 331
Hazard, 45–49
    dynamic, 45, 47
    glitch, 45
    hazard-free, 47
    static, 45, 46
    static-0, 45
    static-1, 45
    timing, 45
    transient time, 45
    types of, 45
Hierarchical path name, 160, 162, 201, 204, 211, 222, 231, 232
Hierarchical structure, 372
Hierarchical system design, 372–373
    circuit level, 373
    design hierarchy, 372
    gate level, 373
    register-transfer level (RTL), 373
    system level, 372
High impedance, 36
High-level verification language, *see* HVL
High-speed transceiver, 408
Hold time, 250, 315
Hold-time failure, 318
HVL, 572
    SystemC, 572
    SystemVerilog, 572

I/O standard, 371, 409–412
    differential signal, 411
    Joint Electron Device Engineering Council (JEDEC), 409

Index **789**

  low-voltage CMOS (LVCMOS), 409
  low-voltage TTL (LVTTL), 409
  LVDS, 410
  single-ended signal, 410
I/O standards, 409
Identifier, 7
IEEE 1149.1, 761–772
  architecture, 762
  basic TAP structure, 762
  boundary-scan cell, 763, 764
  boundary-scan cell, operating modes, 763
  boundary-scan register, 763
  bypass register, 763
  data register, 762, 763
  design-specification register, 763
  device identification (ID) register, 763
  each device test, 762
  instruction register and decoder, 762, 766
  interconnect test, 762
  non-invasive mode, 762
  normal operation data observation, 762
  objectives, 761
  pin-permission mode, 762
  TAP controller, 762, 768
  TAP controller, state diagram, 768
  TDO driver, 762, 765
  test access port (TAP), 762, 763
  test clock input (TCK), 763
  test data input (TDI), 763
  test data output (TDO), 763
  test mode select (TMS), 763
  test reset (TRST), 763
IEEE 1500 standard, 776, 777
  test access mechanism (TAM), 776
  wrapper, 776
  wrapper instruction register (WIR), 776
  wrapper serial control (WSC), 776
  wrapper serial input (WSI), 776
  wrapper serial output (WSO), 776
  wrapper serial port (WSP), 776
Implementation, 5, 569
Implementation or (realization), 12
Implicant, 525
Implicit finite-state machine (implicit FSM), 437
Implicit net declaration, 72
Implicitly clocking scheme, 678
Indium tin oxide (ITO), 483
Inertial delay model, 41, 247

Input relationship, 295
Input voltage tolerance, 412
Instance, 14, 199
Instantiation, 14, 199
Instruction, 725
  opcode, 725
  operand, 725
Instruction set, 726
  double-operand, 726, 727
  jump, 726, 727
  single-operand, 726, 727
Instruction set architecture (ISA), 723
Integer, 76
Integer form, 76
  base format notation, 76
  simple decimal form, 76
Integration density, 376
Intellectual property (IP), 375
  hard, 375, 409
  hardwired, 375, 409
  soft, 375, 409
Interface, 193
Interpreted code, 577
Irredundant (or irreducible) logic circuit, 747
ISE, 609
  design flow, 610
  simulation flow, 615
ISE design flow, 609–615
ISE synthesis flow, 609–615
  configure FPGA, 609, 614
  design entry, 609, 610
  implementation, 609, 611
  synthesis, 609, 610
  timing constraints, 614
Iteration (loop) statement, 107, 137–143
  forever, 107, 142
  for, 107, 139–141
  repeat, 107, 141–142
  while, 107, 138–139
Iterative logic, 475
  0101 sequence detector, 475, 476
  one-dimensional, 475, 478
  two-dimensional, 478

Kernel, 530
Key code, 414
Keypad scanner, 414
Keyword, 8
  always, 17, 18, 107–111, 152, 205, 207, 213, 227–229, 233, 234
  assign-deassign, 107

assign/deassign, 227, 234
assign, 4, 5, 16, 38, 103, 107, 206
automatic, 160
begin-end, 108, 206
begin, 112, 227
casex, 107, 136
casez, 107, 136
case, 107, 132
config, 216
default, 213
defparam, 78, 196, 201
disable, 233
edge, 243, 250
endconfig, 216
endfunction, 157
endgenerate, 205
endmodule, 4, 5, 193
endprimitive, 181
endtable, 182
endtask, 151
end, 112, 227
force-release, 107
force/release, 38, 227, 235
forever, 142
fork-join, 108
fork, 112, 229
for, 139
function, 157
generate, 193, 205
genvar, 206
if-else, 107, 130, 207
include, 215
initial, 17, 107–111, 152, 186, 205, 207, 213, 227–229, 233, 234
inout, 13, 152, 157, 195
input, 13, 152, 157, 195
join, 112, 229
library, 215
localparam, 76, 78, 79, 88, 196
macromodule, 194
module, 4, 5, 193
negedge, 124, 243, 250
output, 13, 152, 157, 195
parameter, 76, 78, 88, 196, 201
posedge, 124, 243, 250
primitive, 181
repeat, 141
scalared, 82
signed, 38, 79, 195
table, 182
task, 151
vectored, 82
wait, 127, 157
while, 138
wire, 195

Language structure synthesis, 543–551
    negedge, 547
    posedge, 547
    an incomplete **case** statement, 545
    assignment, 543
    delay value, 546
    loop statement, 544, 549
    memory and register file, 550
    operators, 543
    procedural assignment, 543
    procedural block, 543
    selection statement, 544
Latch, 311
    D, 311
    SR, 311
    master, 312
    multiplexed, 311
    negative, 312
    positive, 312
    slave, 312
LCD, 481–498
    active, 482
    active-matrix, 482
    backlight, 482
    basic structure, active, 482, 483
    basic structure, reflective, 482
    cold cathode fluorescent lamp (CCFL), 482
    dynamic scattering device, 481
    field-effect device, 481
    LCD panel, 482
    material, 481
    reflective, 481, 482
    TFT panel, 482
    thin-film transistor (TFT), 482
    transmissive, 481
LCD module, 483–498
    4-digit driver, controller, 491
    4-digit driver, datapath, 490
    4-digit driver, system block diagram, 489
    busy flag (BF), 483
    character generator (CG) RAM, 483
    character generator (CG) ROM, 483
    commands, 485, 486
    controller, 491–498
    datapath, 488–491

display data (DD) RAM, 483
dot-matrix, 483
electrical interface, 484
Hitachi 44780, 483
initialization process, 484
pin assignments, 484
read and write timing, 484, 485
Library, 215
Library binding, 522, 534
Library declaration, 215
Light-emitting diodes (LEDs), 298
Linear feedback shift register (LFSR), 349
Linear multicycle structure, 470
Liquid-crystal display, *see* LCD
Literal, 525
Little-endian form, 81
Logic array personality, 393
Logic optimization, 521, 523
    flattening, 524
    removing hierarchy, 523
    structuring, 524
Logic restructuring, 558
Logic synthesis, 71, 521, 543
    analysis/translation, 522
    architecture, 521, 522
    elaboration, 521
    functional metric, 522
    library binding, 522
    logic optimization, 522
    logic synthesizer, 521
    node minimization, 525
    non-functional metric, 522
    parsing, 521
    restructuring operation, 525
    silicon compiler, 521
    technology-dependent, 522
    technology-independent, 522
Logical shift, 651
Loop unrolling, 206, 549
Low-power logic design, 557–561
    design issues, 557
Low-voltage differential signal
    see LVDS, 410
Lumped delay model, 237
LVDS, 410

Macro, 610
Macro expansion, 14
Macrocell, 385
Magnitude comparator, 294–297
    4-bit, 296
    n-bit, 295

block diagram, 8-bit, 297
block diagram, cascadable 4-bit, 296
cascadable, n-bit, 296
comparator, 295
definition, 295
input relationship, 295
output relationship, 295
March test (MATS+), 773, 774
Mask ROM, 387
Mask-programmable, 382
Mask-programmable gate array (MPGA), 375
Master-slave flip-flop, 311, 312
Max-delay failure, 316
MCD, *see* Multiple-clock domain
Mealy machine, 430, 454
Memory module, 453, 550
    ECC, 500
    SECMED, 501
Metastability, 591, 683
Metastable state, 316
MicroBlaze, 373, 408
Min-delay constraint, 318
Min-delay failure, 318
Minimum delay, 317
Minterm, 525
Model, 12
Model of design unit, 570
    black box, 570, 746
    gray box, 570, 746
    white box, 570, 746
Modified Baugh-Wooley algorithm, 642–646
    4×4 array, 644
    $m \times n$ array, 644
Modular design, 372
Module, 1, 7, 29, 193–205
    body, 193
    definition, 193
    hierarchical path name, 204
    instantiation, 199
    interface, 193
    interface (or ports), 7
    internal (or body), 7
    named association, 199
    parameter port, 197
    port connection rules, 199
    port declaration, 195
    port types, 195
    port-list declaration style, 193
    port-list style, 193

positional association, 199
Module modeling styles, 12–19
  behavioral, 17, 107–143
  behavioral or algorithmic, 13
  dataflow, 13, 16, 71–99
  hierarchically structural, 193–222
  mixed, 13, 18
  structural, 12, 14, 29–63
  structural (gate level), 29–41
  structural (switch level), 49–62
Module parameters, 196
Module path, 239
Modulus, 345
Moore machine, 430, 454
MSP430 system, 723
Multicycle path, 520
Multilevel logic synthesis, 528–534
  kernel approach, 530
Multiple channel descriptor (mcd), 168
Multiple cycles, 468
Multiple-clock domains, 683–696
  datapath synchronization, 691
  detection of signals, 685
  FIFO, 683, 693
  handshake, 683, 690, 691
  metastability, 683
  reset signal, 683
  single-stage synchronizer, 684
  strobe, 689
  synchronizer, 683
  the edge-detection method, 687
  the level-detection method, 685
  two-stage synchronizer, 684
Multiple-input signature register (MISR), 758
Multiple-output switching functions, 389
Multiplexed approach, 301
Multiplexed latch, 311
Multiplexer, 32, 283–288
  2-to-1, 39, 94
  2-to-1 (CMOS switch primitive), 55
  3-to-1, 134
  3-to-1 (a correct version), 134
  4-to-1, 32, 284, 285
  4-to-1 (behavioral), 109
  4-to-1 (case), 133
  4-to-1 (dataflow), 86
  4-to-1 (if-else), 130
  4-to-1 (nested case), 135
  4-to-1 (using conditional operator), 94
  4-to-1 (using equality operator), 97

  4-to-1, n-bit (case), 286
  4-to-1, n-bit (conditional operator), 284
  4-to-1, n-bit (if), 285
  5-to-1, 135
  8-to-1, 288
  m-to-1, parameterizable, 287
  block diagram, 284
  definition, 284
  demultiplexing, 283
  expansion, 287
  multiplexer tree, 288
  multiplexing, 283, 289
  parameterizable, 286
Multiplexer bus, 670
Multiplexers, 284
Multiplication, 625, 635–646
  sequential implementation, 636, 637
  shift-and-add, 636
  signed, 642–646
  unsigned, 636–640
Multiplier, 636
  $(\frac{n}{k})$-cycle structure, 640–641
  4×4 CSA array, 640
  4×4 RCA array, 638
  4×4 unsigned array, 637
  $m \times n$ unsigned array, 637
  array, 636, 637
  CSA array multiplier, 639
  pipeline structure, 641–642
Multiplier and accumulator (MAC), 658

Named association, 14, 199
Named block, 206, 231, 233
Named event, 124, 125
Native code, 577
Negative (N), 654
Nested block, 231
Net, 10
Net declaration assignment, 72
Net declaration delay, 74
Net delay, 45
Netlist, 3, 12, 149, 509–561, 573, 595, 609, 611, 613, 619
Netlist generation, 521
Network covering, 534
NIOS, 373
Node minimization, 525
  observability don't cares, 533
  output don't cares, 533
  satisfiability don't cares, 533
Nonblocking assignment, 111
Nonlinear multicycle structure, 470

Index 793

Notifier, 250
NRZ (non-return-to-zero) code, 678

Observability, 743, 746
Off-set, 525
On-set, 525
One-hot encoding, 431
One-time programmable (OTP), 382
Operand, 75–84
    array, 83
    array and memory, 83–84
    bit-select, 81–82
    constant, 76
    function call, 76
    part-select, 81–82
Operating conditions, 517
Operator, 84–99
    arithmetic, 87
    arithmetic shift, 98
    bit-wise, 84
    concatenation, 89
    conditional, 94
    equality, 96
    logical, 93
    logical shift, 98
    reduction, 91
    relational, 95
    shift, 98
Output relationship, 295
Output routing area, 385
Output voltage tolerance, 412
Overflow (V), 654

PAL, 391–392
    basic structure, 391
    features, 391
    PAL16R8, 393
    standard features, 392
PAL macro, 385
Parallel block, 112, 227–234
    definition, 229
Parallel port, 697
Parameter, 196
    module, 196
    named association, 202
    overridden by module instance parameter value assignment, 202
    overridden by **defparam**, 201
    overriding, 78, 201
    positional association, 202
    specify, 196
    types of, 196

Parameter port, 194, 197, 199
Parameterizable, 201, 276
Parameterizable module, 198, 201
Parity, 33, 713
    checker, 378
    even, 91, 190, 223, 500, 542, 713
    odd, 91, 191, 223, 713
Parity generator, 33, 714
    9-bit, 33
    even-parity, 91, 190, 223
    odd-parity, 91, 191, 223
    using reduction operator, 91
Parsing, 521
Partitioning, 512
Path coverage, 593
Path delay model, 237
Path sensitization, 749–754
    consistency operation, 749
    fault excitation, 749
    fault propagation, 749
    fault sensitization, 749
    line justification, 749
    multiple-dimensional, 750
    multiple-path, 750
    reconvergent fanout signal, 750
    sensitizable path, 749
    single-path, 750
    unsensitizable path, 749
Pattern graph, 534
PCB-based design, 373
Period, 250
Personality, 393
Personality memory, 395
Photomask, 513
Physical defect, 743
PicoBlaze, 408
Pipeline, 468, 472
    clock frequency, 472
    clock period, 641
    depth, 472
    latency, 472, 641
    register, 472
Pipeline register, 472
Pipeline structure, 318, 319
PLA, 389–390
    basic structure, 389, 390
    features, 389
    fuses, 389
PLA modeling, 392–400
    a PLA modeling example, array format, 398

a PLA modeling example, plane format, 399
array format, 395
array format, asynchronous, 396
array format, synchronous, 396
logic array personality, 393
NAND-NAND structure, 394
NOR-NOR structure, 394
personality memory, 395
personality, defined by file, 396
personality, defined by procedural assignments, 397
plane format, 395
product of sum (POS), 394
sum of product (SOP), 394
PLA symbolic diagram, 398
Place and route (PAR), 512
Placement, 511, 512
timing-driven, 513
Platform FPGA, 408
Platform-FPGA, 375
Platform-IP, 371, 374, 375
PLD, 383–385
basic structures, 384
erasable, 382
field-programmable, 382
mask-programmable, 382
one-time programmable (OTP), 382
programmable interconnect, 383, 384
programmer, 382
PLL (phase-locked loop), 679
Polynomial code, 350, 354
Port connection (association) rules, 14
named association, 14
positional association, 14
Port declaration, 13
inout, 13
input, 13
output, 13
port-list declaration style, 13, 109
port-list style, 13, 109
Port list, 197
Port-list declaration, 197
Positional association, 14, 199
Power analysis, 511
Power gating, 561
Power optimization, 510
Power supply, 50
PR-sequence, 349
PRBS, 349
Precedence of operators, 73

Primary divisor, 530
Prime implicant, 525
Primitive, 11
built-in, 11
combinational UDP, 182–185
gate, 11, 29
sequential UDP, 185–188
shorthand notation, 183
switch, 11, 29
UDP definition, 181
user-defined primitives (UDPs), 11, 29
Primitive polynomial, 350
Printed-circuit board (PCB), 375
Priority, 280, 669
fixed, 669
round-robin, 669
Priority encoder, 280
Procedural construct, 107–111
always block/construct, 107
always block/statement, 109
initial block/construct, 107
initial block/statement, 107
Process complexity, 376
Product of sum (POS), 394
Product term, 525
Program, 387
Programmable array logic, *see* PAL
Programmable array logic (PAL), 391
Programmable logic array, *see* PLA
Programmable logic array (PLA), 389
Programmable logic device, *see* PLD, *see* PLD
Programmable logic device (PLD), 5
Programmable ROM (PROM), 387
Programmable system chip (PSC), 373, 375
Programmer, 382
Programming, 5
Programming equipment, 382
Programming language interface (PLI), 3
Propagation delays, 41
Pseudo-nMOS NOR, 53
Pseudo-random sequence, *see* PR-sequence
Pulse-width limit control, 247
specparam PATHPULSE$, 247
error limit, 247
global, 247
non-path-specific **PATHPULSE$**, 247
path-specific **PATHPULSE$**, 247
rejection limit, 247
SDF annotation, 247
specify block, 247

Index

PVT variation, 317
PVT variations, 517

Quality, 25
    source code, 25

Race condition, 318
Race problem, 115
    blocking assignment, 115
    nonblocking assignment, 115
RAM, 332–338
    $\overline{CE}$ access time, 333
    $\overline{CE}$-controlled write, 333
    $\overline{OE}$ access time, 333
    $\overline{WE}$-controlled write, 333
    a RAM example, timing check, 336
    address access time, 332
    address hold time, 334
    address setup time, 334
    asynchronous, 328
    block diagram, 333
    data hold time, 335
    data setup time, 334
    pin assignment, 332, 334
    read cycle time, 333
    read-cycle timing, 332, 334
    synchronous, 328, 331
    write cycle time, 336
    write-cycle timing, 335
Random encoding, 431
Random number generator, 176
Random-access memory, *see* RAM
Read-only memory, *see* ROM
Real, 76
Real number, 9
    decimal notation, 9, 77
    scientific notation, 9, 77
Recovery time, 250, 321
Reduction of switched capacitance, 557
Reduction of switching activity, 557
Reduction of voltage swing, 557
Reference event, 251
Register, 328–330
    data, 328
    data, 4-bit, 328
    data, n-bit, 328
    data, n-bit, asynchronous reset, 329
    data, n-bit, synchronous load and asynchronous reset, 329
Register file, 328, 550
    n-word, single bidirectional port, 407
    n-word, three ports, 330

    n-word, two synchronous ports, 331
Register-transfer level, *see* RTL
Register-transfer level (RTL), 13, 26, 372, 439
Removal time, 250
Removing hierarchy, 523
Repeat event control, 142
Required time, 513, 601
Reset signal generation, 590–592
    maintainability, 590, 591
    race problem, 590
    reusability, 590, 591
Residue, 541
Restructuring operations, 528
    collapsing, 529
    decomposition, 528
    extraction, 529
    factorization, 529
    substitution, 529
Retriggerable monostable signal, 586, 589
ROM, 387–389
    basic structure, 387
    capacity, 387
    erasable programmable, 387
    flash memory, 387
    mask, 387
    program, 387
    programmable, 387
    size, 387
    word, 387
Rounding error, 587
Routing, 511
    detailed, 512
    global, 512
RTL, 439–481
    ASM chart, 440
    conditional statement, 440
    definition, 439
    DP+CU approach, 453
    features, 440
    multicycle structure, 470, 625
    pipeline structure, 471–475
    single-cycle structure, 468, 625
    unconditional statement, 440
RTL code, 510

Sales volume, 377
Sampling window, 315
Sandia controllability and observability analysis program (SCOAP), 747
Scalar, 10, 38, 79
Scan code, 414

Scan-chain logic insertion, 511
Scan-path method, 743, 754, 756
Scoreboard, 582
Script file, 517
SDF annotation, 172, 246, 247, 569, 595, 596
SDF file, 595
    delay back-annotation, 596
    format, 597–599
    INTERCONNECT delay, 595
    IOPATH delay, 595
    post-layout, 596
    pre-layout, 596
    timing check, 595
Sea-of-gates, 379
Selection statement, 107, 130–137
    casex, 136
    casez, 136
    case, 130, 132–137
    if-else, 130–132
Semi-custom approach, 374
Sensitivity list, 127
Sequence generator, 349–360, 758
    4-bit PR-sequence, 350, 351
    n-bit PR-sequence, 351
    n-bit PR-sequence, with self-start, 353
    n-bit PR-sequence, with start, 352
    block diagram, 350
    maximum-length sequence, 349
    modular format, 350
    PR-sequence, 349, 758
    standard format, 350
Sequential block, 112, 227–234
    definition, 228
Sequential logic module, 309–370
SerDes (serializer/deserializer), 678
Serial data transfer
    full-duplex, 410
    half-duplex, 410
    simplex, 410
Serial port, 697
Serial-input signature register (SISR), 758
Setup time, 250, 315
Setup-time failure, 316
Seven-segment display, 298
Seven-segment LED display, 298–304
    common-anode structure, 298
    common-cathode structure, 298
    direct-driven approach, 301
    display circuit, 299
    display code, 299, 300
    display codeword, 299
    multiplexed approach, 301
    multiplexed logic circuit, 302
    multiplexed system, 303
Shift, 651–653
    arithmetic, 651
    arithmetic left, 651
    arithmetic right, 651
    logical, 651
    logical left, 651
    logical right, 651
Shift register, 113, 338–343
    4-bit, 113, 338, 339
    4-bit, universal, 342
    n-bit, 338
    n-bit, parallel load, 340
    n-bit, universal, 341
    n-bit, universal (generate-loop), 341
    a correct version, 117
    an incorrect version, 116
    data format conversion, 338
    parallel in parallel out (PIPO), 340
    parallel in serial out (PISO), 340
    serial in parallel out (SIPO), 340
    serial in serial out (SISO), 340
Shift-and-add (subtract) technique, 625
Shift-and-add approach, 636
Shifter, 625, 651
    8-bit barrel, 652, 653
    barrel, 652
    definition of barrel, 652
    parameterizable logical left barrel, 652
Sign-bit extension, 98
Sign-extended, 85, 95, 96
Signal strength, 57–62
    large, 57
    medium, 57
    pull, 57
    small, 57
    strong, 57
    supply, 57
    weak, 57
    charge storage, 57
    driving, 57
    signal contention, 58
    types of, 57
Signature generator, 757, 758
    parallel, 758
    serial, 758
Simple decimal form, 76
Simulation, 19, 574–582

Index **797**

  behavioral, 574
  circuit-level (transistor-level), 574
  functional, 574
  gate-level (logic), 574
  hardware acceleration, 574, 575
  hardware emulation, 574, 575
  software, 574
  switch-level, 574
  types, 574
Simulator, 511, 576–577
  architecture, 576
  compiled code, 577
  cycle-based, 577
  elaborator, 576
  event-driven, 577
  interpreted, 577
  native code, 577
Single cycle, 468
Single-assignment rule, 440
Skew, 250
Slack time, 513, 601
SoC testing, 776
Specification, 569
Specify block, 237, 239–249
  ifnone statement, 244, 245
  if statement, 244
  specparam, 246
  *>, 241
  =>, 241
  min:typ:max, 245
  definition, 240
  delay specification, 245
  edge-sensitive path, 243
  full connection, 241, 242
  one delay, 246
  parallel connection, 241
  path declaration, 241
  simple path, 241
  six delays, 246
  state-dependent path, 243
  three delays, 246
  twelve delays, 246
  two delays, 246
Specify parameters, 196
Speed, 376
STA, 599–603
  basic assumptions, 599
  clock-to-D path, 600
  clock-to-output path, 600
  entry path, 600
  exit path, 600

  input-to-D path, 600
  pad-to-pad path, 600
  path group, 600
  port-to-port path, 600
  register-to-register path, 600
  stage path, 600
  types of paths, 600
Standard delay format (SDF), 595
Standard ICs, 371, 374, 375
State diagram, 429
State encoding, 431
  binary, 431
  Gray, 431
  one-hot, 431
  random, 431
State machine (SM) chart, 429, 440
State register, 430
State table, 182
State-dependent edge-sensitive path, 244
State-dependent path, 243
Statement coverage, 592
Static hazard, 45
Static RAM (SRAM), 383
Static timing analysis, *see* STA
Static timing analysis (STA), 511
Static-0 hazard, 45
Static-1 hazard, 45
Strength specification, 57
String, 76
Strobe, 679
Structural coverage, 592
Subject graph, 534
Substitution, 528
Subtraction, 625
Sum of product (SOP), 394, 401
Switch primitive, 29, 49–62
  cmos/rcmos, 54
  nmos/rnmos, 49
  pmos/rpmos, 49
  tran/rtran, 55
  tranif0/rtranif0, 55
  tranif1/rtranif1, 55
  bidirectional switch, 55
  CMOS switch, 54
  MOS switch, 49
Switch-debouncing circuit, 326–328
Synchronization failure, 321
Synchronizer, 321–326
  cascaded, 324
  cascaded, asynchronous reset, 324
  frequency-divided, 324

frequency-divided, asynchronous reset, 325
mean time between failure, (MTBF), 322
Synchronous mode, 676
Synthesis, 5, 12, 509
    gate-level netlist, 510, 511, 521, 546, 561
    high-level, 12, 509
    logic, 12, 510
    RTL, 510
Synthesis flow, 509–516
    ASIC, 509
    back end, 510
    design specification, 510
    FPGA-based system, 509
    front end, 510
    RTL, 510
    RTL functional verification, 510
System level, 372
System on a programmable chip (SoPC), 373, 375
System task and function, 163–181, 596
    $bitstoreal, 175, 200
    $clog2, 178
    $displayb, 164
    $displayh, 164
    $displayo, 164
    $display, 4, 20, 21, 58, 125, 161, 164
    $dist_, 176
    $dumpall, 605
    $dumpfile, 604
    $dumpflush, 605
    $dumplimit, 605
    $dumpoff, 604
    $dumpon, 604
    $dumpportsall, 607
    $dumpportsflush, 608
    $dumpportslimit, 607
    $dumpportsoff, 606
    $dumpportson, 606
    $dumpports, 606
    $dumpvars, 604
    $fclose, 168
    $fdisplay, 168
    $ferror, 173
    $fflush, 168
    $fgetc, 170
    $fgets, 171
    $finish, 20, 22, 142, 167
    $fmonitor, 168
    $fopen, 168
    $fread, 171
    $fscanf, 171
    $fseek, 172
    $fstrobe, 168
    $ftell, 172
    $fwrite, 168
    $itor, 175
    $monitorb, 165
    $monitorh, 165
    $monitoroff, 21, 165
    $monitoron, 21, 165
    $monitoro, 165
    $monitor, 20–22, 58, 125, 165
    $printtimescale, 165
    $q_add, 177
    $q_exam, 177
    $q_full, 177
    $q_initialize, 177
    $q_remove, 177
    $random, 176
    $readmemb, 171, 396
    $readmemh, 171, 396
    $realtime, 20–22, 166
    $realtobits, 175, 200
    $rewind, 172
    $rtoi, 175
    $sdf_annotate, 172, 173, 596
    $sformat, 173
    $signed, 89, 174
    $sscanf, 173
    $stime, 166
    $stop, 20, 22, 167
    $strobeb, 165
    $strobeh, 165
    $strobeo, 165
    $strobe, 58, 165
    $swriteb, 173
    $swriteh, 173
    $swriteo, 173
    $swrite, 173
    $test$plusargs, 178
    $timeformat, 165
    $time, 20, 21, 166
    $ungetc, 170
    $unsigned, 89, 174
    $value$plusargs, 179
    $writeb, 164
    $writeh, 164
    $writeo, 164
    $write, 164

Index

command line arguments, 178–181
continuous monitoring, 165
conversion, 174–176
display, 164–165
display and write, 164
display format, 164
file I/O, 167–173
file input, 170–173
file output, 168–170
math, 178
mathematical, 178
opening and closing files, 168
PLA modeling tasks, 395
probability distribution, 176–177
simulation control, 167
simulation time, 166
simulation-related, 164–167
stochastic analysis, 177
string formatting, 173–174
strobed monitoring, 165
timescale, 165
System-level testing, 773–777
System-on-chip (SoC), 375
System-on-chip FPGA (SoC FPGA), 375

Target-dependent, 5, 510
Target-independent, 5, 510
Task, 151–157, 233
(static), 155
automatic (reentrant, dynamic), 155
call (enable), 152
definition, 151
sharing, 162
types of, 155
Technology library, 510
Technology mapping, 534
FlowMap method, 540
Shannon's expansion theorem approach, 542
two-step approach, 538, 540
Test, 746
exhaustive, 748, 755, 758
random, 758
Test bench, 19, 572, 582–594
automated response checking, 583
basic principle, 582
design, 582–594
deterministic stimulus, 583
dynamic hazard, 49
end-of-test checking, 583
exhaustive test example, 584
golden vectors, 582

golden vectors example, 585
log file, 583
non-self-checking, 582
on-the-fly checking, 583
on-the-fly test, 583
pregenerated test, 583
random stimulus, 583
random test example, 584
reference model, 582
self-checking, 582, 584
structure, 19
test case, 582
transaction based, 582
transaction-based model, 582
waveform viewer, 583
Test pattern, 747
Test set, 743, 747
complete, 748
Test vector, 743, 747–748
Test-bench statements, 237
Testable circuit design, 743, 754–772
Time to market, 376
Timeout mechanism, 681
Timer, 343, 669, 701–712
advanced modes, 707–712
basic modes, 701–707
block diagram, 702
hardware model, 701
input capture, 707, 708
latch, 701
monostable, 701, 703, 704
one-shot, 701
output compare, 707–709
programming model, 701
pulse-width modulation (PWM), 707, 710, 711
rate generation, 701, 703, 704
square-wave generation, 701, 705, 706
terminal counter, 701, 702
timer, 701
Timing, 4, 107
Timing analysis, 5, 594, 595, 601
basic concepts, 594
critical path, 601
delay back-annotation, 595, 596
dynamic, 6, 594
slack time, 601
static, 6, 594
Timing check, 250–259, 595
$fullskew, 257, 258
$hold, 251

$nochange, 259
$period, 256
$recovery, 253
$recrem, 254
$removal, 253
$setuphold, 252
$setup, 251
$skew, 257, 258
$timeskew, 257, 258
$width, 256
    clock and signal, 250, 256–259
    conditioned event, 250
    event-based, 257
    hold time, 251
    negative, 254
    period, 256
    pulse width, 256
    recovery, 253
    recovery and removal, 254
    removal, 253
    setup and hold time, 252
    setup time, 251
    skew, 257
    timer-based, 257
    violation window, 254
    window-related, 250–254
Timing constraints, 510, 516–525
    duty cycle, 518
    input delay, 518
    maximum path delay, 519
    minimum path delay, 519
    multicycle path, 520
    offset-in, 519
    offset-out, 519
    output delay, 518
    period, 518
    skew, 518
    transition time, 518
Timing control, 107, 119–129
    delay, 107, 119–124, 228, 229
    edge-triggered, 124
    event, 107, 119, 124–129
    event or, 127
    implicit event list, 127
    intra-assignment delay control, 120, 229, 231
    intra-assignment timing control, 119
    level-sensitive, 124, 127
    named event, 125
    regular delay control, 119, 228, 229, 231

    regular timing control, 119, 143
    repeat event control, 142
Timing critical path, 514
Timing exceptions, 601–603
    false path, 519, 520, 565, 601
    max/min path delay, 601
    multicycle path, 601–603
Timing generator, 360–363
    binary counter with decoder, 361
    digital monostable, 362
    multiphase clock generator, 360
    non-retriggerable digital monostable, 363
    non-retriggerable monostable, 362
    retriggerable monostable, 362
Timing hazard, 45
Timing specifications, 601
Timing-driven placement, 513
Toggle coverage, 592
Top-level module, 204, 454, 463
Topological sort, 581
Transient time, 45
Transport delay, 41
Transport delay model, 42
Transversal filter, 658
Trigger coverage, 593
Tristate (three-state), 9, 36
Tristate buffers, 36–39
    active-high buffer (**bufif1**), 37
    active-high inverter (**notif1**), 37
    active-low buffer (**bufif0**), 37
    active-low inverter (**notif0**), 37
    instantiation, 37
Tristate bus, 670
Truncation error, 587
Turn-off time, 246
Two-level logic synthesis, 525–527
    Espresso algorithm, 526
    expansion, 526
    making irredundant, 526
    Quine-McCluskey algorithm, 526
    reduction, 526
Two-phase handshake, 681

UART, 712–723
    baud rate, 713
    baud-rate generator, 713, 719–721
    control register (CR), 712
    design issues, 713
    hardware model, 712
    logic diagram, 713
    parity check, 713

Index

receiver, 713, 716–719
receiver data register (RDR), 712
sampling clock frequency, 713
software model, 712
status register (SR), 712
stop bits, 713
top-level module, 721–723
transmitter, 713–716
transmitter data register (TDR), 712
UCF, 614
Uncommitted logic array (ULAs), 379
Uncommitted logic arrays (ULAs), 379
Universal asynchronous receiver and transmitter, *see* UART
Universal logic module, 403
Universal serial bus (USB), 673
Universal shift register, 340
Unstable circuit, 310
User-constrained file, 422
User-defined primitive (UDP), 29, 181–188

Value change dump file, *see* VCD file
Variable, 10
VCD file, 604–609
    an example, 605
    an example, extended, 608
    extended, 604, 606–609
    format, 606
    four-state, 604–606
Vector, 10, 38, 79
Verification, 569
    assertion checking, 570
    formal, 569, 573–574
    functional, 569–574
    observability, 570
    simulation, 594
    simulation-based, 571–572
    timing, 569
Verification coverage, 592–594
    branch, 593
    condition, 593
    cross, 594
    expression, 593
    finite-state machine (FSM), 593
    functional, 594
    item, 594
    path, 593
    statement, 592
    structural, 592
    toggle, 593
    transition, 594
    trigger, 593

Verification test, 572
    compliance, 572
    corner case, 572
    property check, 572
    random, 572
    real code, 572
    regression, 572
Violation window, 254
Voltage compliance, 412
Voltage tolerance, 371, 409, 412–413
    input, 412
    output, 413

White space, 8
Width, 250
Wire delay, 41
    inertial, 74
    transport, 41, 42
Wired logic, 39–41
    wired and, 39
Word, 387

XCF, 614
Xilinx's Integrated Software Environment, *see* ISE

Zero (Z), 654
Zero-delay control, 122
Zero-extended, 72, 85, 95, 96, 134, 135
Zero-slack algorithm, 514
    example, 514, 515
Zero/one detector, 92

Printed in the USA
CPSIA information can be obtained
at www.ICGtesting.com
LVHW080826141223
766408LV00005B/405